T0188190

Quantum Field Theory

Feynman Path Integrals and Diagrammatic Techniques in Condensed Matter

Lukong Cornelius Fai

CRC Press
Taylor & Francis Group
Boca Raton London New York

CRC Press is an imprint of the
Taylor & Francis Group, an **informa** business

CRC Press
Taylor & Francis Group
6000 Broken Sound Parkway NW, Suite 300
Boca Raton, FL 33487-2742

First issued in paperback 2021

© 2020 by Taylor & Francis Group, LLC
CRC Press is an imprint of Taylor & Francis Group, an Informa business

No claim to original U.S. Government works

ISBN 13: 978-0-367-77959-7 (pbk)
ISBN 13: 978-0-367-18574-9 (hbk)

This book contains information obtained from authentic and highly regarded sources. Reasonable efforts have been made to publish reliable data and information, but the author and publisher cannot assume responsibility for the validity of all materials or the consequences of their use. The authors and publishers have attempted to trace the copyright holders of all material reproduced in this publication and apologize to copyright holders if permission to publish in this form has not been obtained. If any copyright material has not been acknowledged, please write and let us know so we may rectify this in any future reprint.

Except as permitted under U.S. Copyright Law, no part of this book may be reprinted, reproduced, transmitted, or utilized in any form by any electronic, mechanical, or other means, now known or hereafter invented, including photocopying, microfilming, and recording, or in any information storage or retrieval system, without written permission from the publishers.

For permission to photocopy or use material electronically from this work, please access www.copyright.com (http://www.copyright.com/) or contact the Copyright Clearance Center, Inc. (CCC), 222 Rosewood Drive, Danvers, MA 01923, 978-750-8400. CCC is a not-for-profit organization that provides licenses and registration for a variety of users. For organizations that have been granted a photocopy license by the CCC, a separate system of payment has been arranged.

Trademark Notice: Product or corporate names may be trademarks or registered trademarks and are used only for identification and explanation without intent to infringe.

Library of Congress Cataloging-in-Publication Data

Names: Fai, Lukong Cornelius, author.
Title: Quantum field theory : Feynman path integrals and diagrammatic techniques in condensed matter / Lukong Cornelius Fai.
Description: Boca Raton, FL : CRC Press, Taylor & Francis Group, [2019] |
Includes bibliographical references and index.
Identifiers: LCCN 2019000921| ISBN 9780367185749 (hbk ; alk. paper) | ISBN 0367185741 (hbk ; alk. paper) | ISBN 9780429196942 (ebook) | ISBN 0429196946 (ebook)
Subjects: LCSH: Quantum field theory. | Feynman integrals. | Feynman diagrams. | Condensed matter.
Classification: LCC QC174.45 .F35 2019 | DDC 530.14/3–dc 3
LC record available at https://lccn.loc.gov/2019000921

Visit the Taylor & Francis Web site at
http://www.taylorandfrancis.com

and the CRC Press Web site at
http://www.crcpress.com

Contents

Preface

This book of quantum field theory (QFT) is the continuation of Chapter 13 (Functional Integration in Statistical Physics) of the book entitled *Statistical Thermodynamics: Understanding the Properties of Macroscopic Systems* published by CRC Press in 2002. QFT is a universal tool for the quantum mechanical description of processes permitting transitions among states that differ in their particle content and has applications ranging from condensed matter physics to elementary particle physics. As the quantum mechanics of an arbitrary number of particles, QFT provides an efficient tool describing quantum statistics of the particles. This implies antisymmetrization and symmetrization of the states of identical fermions or bosons, respectively, under interchange of pairs of identical particles. QFT facilitates the treatment of spontaneously symmetrical broken states, such as superfluids as well as critical phenomena with regard to phase transitions. This book uses the strength of Feynman functional and diagrammatic techniques as a presentation foundation that comfortably applies QFT to a broad range of domains in physics and shows the universality of the techniques for a broad range of phenomena. The powerful QFT functional techniques and the renormalization group techniques applicable to equilibrium as well as nonequilibrium field theory processes are extended to treat nonequilibrium states and subsequently transport phenomena. The Green's and correlation functions and the equations derived from them are used to solve real physical problems as well as to describe processes in real physical systems—in particular quantum fluid, electron gas, electron transport, optical response, superconductivity, and superfluidity.

This book should be of interest not only to condensed matter physicists but to other physicists as well because the techniques discussed apply to high-energy as well as soft condensed matter physics. The universality of the techniques is confirmed as a unifying tool in other domains of physics. This book is written for graduate students and researchers who are not necessarily specialists in QFT. It begins with elementary concepts and a review of quantum mechanics and builds the framework of QFT, which is now applied to current problems of utmost importance in condensed matter physics. In most cases, the problem sets represent an integral part of the book and provide a means of reinforcing the explanation of QFT in real situations. The material in this book is clear, and its illustrations emphasize the subject and aid the reader in an essential understanding of the important concepts. This book should be highly recommended for all theoretical physicists in QFT.

This book is the product of lecture notes given to graduate students at the Universities of Dschang and Bamenda, Cameroon. Chapter 1 studies symmetry requirements in quantum mechanics as well as bosonic and fermionic quantum fields operating on multiparticle state space (Fock space). The chapter examines creation and annihilation operators and applies the method of second quantization, a technique that underpins the formulation of quantum many-particle theories. It also treats, in a unified manner, systems of bosons (fermions) with a fixed or variable number of particles. Chapter 2 examines bosonic and fermionic coherent states. It also studies Grassmann algebra, Berezin integration, and Gaussian integrals as well as Wick theorem for multidimensional Grassmann integrals and the trace of a physical quantity. Chapter 3 studies the path integral approach for fermions and bosons considering the

Grassmann algebra. Apart from providing a global view of the entire system, this approach has proven to be an extremely useful tool for understanding and handling quantum mechanics, quantum field theory, and statistical mechanics. This chapter also examines the Green's functions that serves as tools for describing quantum dynamics of many-body systems. In addition, noninteracting particles are also studied, with their Green's function computed via path integrals as well as via generating functionals.

In Chapter 4, perturbation theory is comfortably constructed from the average value of a functional. This chapter studies the perturbation theory in many-particle systems based on Wick theorem, which is formulated in terms of Feynman functional integral and diagrammatic techniques that are very useful for providing an insight into the physical process that they represent. This chapter also examines the cornerstone of the functional technique, which is the concept of generating functionals sufficient to derive all propagators. Chapter 5 examines the (anti)symmetrized vertex, making it simpler and more convenient to formulate perturbation theory. Discussions are facilitated by introducing fully (anti)symmetrized vertices via a uniform and compact notation for the creation and annihilation fermion operators. Chapter 6 examines connected Green's functions with one-particle (1PI) and two-particle (2PI) irreducible vertices as well as the Dyson-Schwinger equations that are most conveniently studied via path integrals and that employ the approach of generating functions in the context of the path integral. The Luttinger-Ward functional and the 2PI vertices are used to set up approximations satisfying conservation laws as well as nonperturbative approaches. Chapter 7 examines the random phase approximation via the Feynman functional integral and diagrammatic technique: screened interactions and plasmons. Here, we study a model describing electrons in a metal that considers a system of electrons interacting with each other via the instantaneous Coulomb force (Jellium model).

Chapter 8 examines the theory of phase transitions and critical phenomena as well as Ginzburg-Landau phenomenology and the connection to statistical field theory. Chapter 9 examines weakly interacting Bose gas via quantum field theory. Application to Bogoliubov theory of the weakly interacting Bose gas and superfluidity is considered as well. This chapter also studies the path integral formalism for nonideal Bose gas considering the electron-electron interaction.

Chapter 10 studies superconductivity theory via the functional integral and diagrammatic approaches, where the statistical model is built on classical field configurations. The mean-field theory is also considered as well as its applications to Cooper instability and the BCS condensate. The vertex function for small momentum transfers is considered, that is, electron-electron interaction. Chapter 11 examines the path integral approach to the BCS theory where we study an accurate theory of interacting Fermi mixtures with spin imbalance. Chapter 12 discusses Green's functions' averages over impurities and, in particular, scattering potentials and disordered systems as well as disorder diagrams, perturbation series solution via T-matrix, and quenched and disorder averages. The diagrammatic cross technique is extended to superconductors considering the Nambu-Gorkov propagators.

Chapter 13 studies in detail the classical and quantum theory of magnetism and spin wave theory spin representations as well as spin liquids. The strongly interacting system and, in particular, the Kondo problem are studied in detailed where methods of quantum statistical field theory play a central role. Chapter 14 considers the nonequilibrium quantum field theory, where we study nonequilibrium Green's functions as well as Keldysh-Schwinger diagrammatic and 2PI effective action techniques relating to nonequilibrium dynamics.

I would like to acknowledge those who have helped at various stages of the elaboration and writing through discussions, criticism, and especially, encouragement and support. I single out Prof. Nicolas Dupuis (Directeur de Recherche at CNRS Laboratoire de Physique Théorique de la Matière Condensée, CNRS UMR, Université Pierre et Marie Curie Paris, France) for allowing me to use some of his pictures. I am very thankful to my wife, Dr. Mrs. Fai Patricia Bi, for all her support and encouragement, and to my four children (Fai Fanyuy Nyuydze, Fai Fondzeyuv Nyuytari, Fai Ntumfon Tiysiy, and Fai Jinyuy Nyuydzefon) for their understanding and moral support during the writing of this book. I acknowledge with gratitude the library support received from the Abdus Salam International Centre for Theoretical Physics (ICTP), Trieste, Italy.

About the Author

Lukong Cornelius Fai is professor of theoretical physics and founding head of the Condensed Matter and Nanomaterials as well as Mesoscopic and Multilayer Structures Laboratory at the Department of Physics, University of Dschang (UDs), Cameroon, where he has also served as chief of service for Research and chief of Division for Cooperation. He has been head of Department for Physics and later director of the Higher Teacher Training College of the University of Bamenda, Cameroon. He was senior associate at the Abdus Salam International Centre for Theoretical Physics (ICTP), Italy. He holds an MSc. in Physics and Mathematics (June 1991) and a Doctor of Science in Physics and Mathematics (February 1997) from the Department of Theoretical Physics, Faculty of Physics, Moldova State University. He is author of three textbooks and over a hundred and thirty scientific publications in the domain of Feynman functional integration, strongly correlated systems, and mesoscopic and nanophysics. He is a reviewer of several scientific journals. He has successfully supervised over 50 MSc. theses at UDs and twelve PhD theses at UDs, University of Yaoundé I, and the University of Angers-France. He is married and has four children.

1

Symmetry Requirements in QFT

Introduction

Quantum field theory (QFT) is a universal tool that has applications ranging from atomic, molecular, and particle physics to condensed matter and statistical physics as well as modern quantum chemistry. Recently, quantum field theory has also had an unexpected and profound impact on pure mathematics. Symmetries, which are at the heart of the universality shown by many physical systems, play a crucial role. Nowadays, quantum theory is the most complete microscopic method describing the physics of energy and matter. Considering the quantization of the electromagnetic field [1] and the representation of particles by quantized fields [2, 3] results in the development of quantum electrodynamics and quantum field theory. By convention, the original form of quantum mechanics is denoted by the first quantization, whereas quantum field theory is formulated in the language of second quantization that is an essential tool for the development of interacting many-body field theories.

The fundamental difference between classical and quantum mechanics relates to the concept of indistinguishability of identical particles. Each particle can be equipped with an identifying marker without influencing its behavior in classical mechanics. In addition, each particle follows its own continuous path in phase space. So, principally, each particle in a group of identical particles can be identified; but this is not the case in quantum mechanics. It is not possible to mark a particle without influencing its physical state. In addition, if a number of identical particles are brought to the same region in space, their wave functions will spread out rapidly and will overlap with one another. Eventually, it will be impossible to say which particle is where. One of the fundamental assumptions for n-particle systems, therefore, is that identical particles (i.e., particles characterized by the same quantum numbers such as mass, charge, and spin) are, in principle, indistinguishable.

1.1 Second Quantization

1.1.1 Fock Space

This chapter introduces and applies the method of second quantization, a technique that underpins the formulation of quantum many-particle theories. Second quantization formalism treats systems of bosons or fermions with a fixed or variable number of particles in a unified way. In this section, we review the key aspects of this formalism.

In quantum mechanics, the state of a system of n identical (indistinguishable) particles is described by a state vector belonging to a Hilbert space (a complete multiparticle system [Fock space]) as the direct sum

$$\mathrm{H} = \mathrm{H}_0 \oplus \mathrm{H}_1 \oplus \cdots = \oplus_{n=0}^{\infty} \mathrm{H}_n \tag{1}$$

The Hilbert space, H_n, corresponds to the n-particle states that are properly symmetrized (bosons) or antisymmetrized (fermions). It is a subspace of the direct tensor product

$$\bar{H}_n = H_1 \otimes H_1 \otimes \cdots H_1 \tag{2}$$

Also, H_0 corresponds to the vacuum state, $|0\rangle$, H_1 corresponds to the single-particle state, and so on.

Consider the orthogonal bases $\{|\alpha'\rangle\}$ and $\{|\alpha\rangle\}$ of H_1, where α' and α are discrete quantum numbers:

$$|\alpha\rangle = \sum_{\alpha'} |\alpha'\rangle\langle\alpha'|\alpha\rangle \tag{3}$$

Particles of the same species are completely indistinguishable in a quantum many-body system. Second quantization provides a general approach to many-body systems where the vector of state plays a minor role. Second quantization entails raising the Schrödinger vector of state to an operator that satisfies certain *canonical (anti)commutation* algebra. It is instructive to note that in first quantized physics, physical properties of a quantum particle such as density, kinetic energy, and potential energy can be expressed in terms of a one-particle vector of state. The essence of the second quantization is the elevation of each of these quantities to the status of an operator. This is done by replacing the one-particle vector of state with its corresponding field operator.

Knowing the orthonormal basis $\{|\alpha\rangle\}$ of H_1 allows us to obtain an orthonormal basis of $\bar{H}_n$ from the tensor product of the single-particle basis that is the n-particle state

$$|\alpha_1 \cdots \alpha_n) = |\alpha_1\rangle \otimes |\alpha_2\rangle \otimes \cdots \otimes |\alpha_n\rangle \tag{4}$$

where **the defined states utilize a curved bracket in the ket symbol**. The first ket on the right-hand side of this equation refers to particle 1, the second to particle 2, and so on. Then the **overlap of two vectors of the basis** is given as follows:

$$(\alpha_1' \cdots \alpha_n' | \alpha_1 \cdots \alpha_n) = \left(\langle\alpha_1'| \otimes \langle\alpha_2'| \otimes \cdots \otimes \langle\alpha_n'|\right)\left(|\alpha_1\rangle \otimes |\alpha_2\rangle \otimes \cdots \otimes |\alpha_n\rangle\right) = \langle\alpha_1'|\alpha_1\rangle \cdots \langle\alpha_n'|\alpha_n\rangle \tag{5}$$

From here, the **orthogonality relation** may be written as follows:

$$(\alpha_1' \cdots \alpha_n' | \alpha_1 \cdots \alpha_n) = \langle\alpha_1'|\alpha_1\rangle \cdots \langle\alpha_n'|\alpha_n\rangle = \delta_{\alpha_1'\alpha_1} \cdots \delta_{\alpha_n'\alpha_n} \tag{6}$$

where the Kronecker symbol is used to include the possibility of δ-function normalization for continuous quantum numbers.

The **completeness of the basis** is obtained from the tensor product of the completeness relation for the basis $\{|\alpha\rangle\}$ and yields the **closure relation**

$$\sum_{\alpha_1 \cdots \alpha_n} |\alpha_1 \cdots \alpha_n)(\alpha_1 \cdots \alpha_n| = \hat{1} \tag{7}$$

where $\hat{1}$ is the unit operator in $\bar{H}_n$. In the case of continuous quantum numbers, integration must be used in (7) instead of a summation; or a combination of both may be used in the case of mixed spectra.

Therefore, we see from the previous information that the Hilbert space describing the n-particle system is spanned by all the nth-rank tensors, such as in (4). We define the state in which the ith particle is localized at a point with radius vector $\vec{r}_i$ as:

$$\langle\vec{r}_1 \cdots \vec{r}_n| = \langle\vec{r}_1| \otimes \langle\vec{r}_2| \otimes \cdots \otimes \langle\vec{r}_n| \tag{8}$$

If we multiply the ket vector in equation (4) by the bra vector in (8), this permits us to express the n-particle **wave function in coordinate space:**

$$\psi_{\alpha_1 \cdots \alpha_n}(\vec{r}_1 \cdots \vec{r}_n) = (\vec{r}_1 \cdots \vec{r}_n | \alpha_1 \cdots \alpha_n) = \langle\vec{r}_1|\alpha_1\rangle\langle\vec{r}_2|\alpha_2\rangle \cdots \langle\vec{r}_n|\alpha_n\rangle = \varphi_{\alpha_1}(\vec{r}_1)\varphi_{\alpha_2}(\vec{r}_2) \cdots \varphi_{\alpha_n}(\vec{r}_n) \tag{9}$$

This is a squarable-integrable function and represents the probability amplitude for finding particles at the n positions $\vec{r}_1 \cdots \vec{r}_n$. It satisfies the following condition:

$$\int \left|\psi_{\alpha_1 \cdots \alpha_n}\left(\vec{r}_1 \cdots \vec{r}_n\right)\right|^2 d\vec{r}_1 \cdots d\vec{r}_n < +\infty \tag{10}$$

It is useful to note that a vector of state in quantum mechanics is a scalar product on Hilbert space of the corresponding state and eigenstates of the position operator, that is,

$$\varphi_\alpha(\vec{r}) = \langle \vec{r} | \alpha \rangle \tag{11}$$

Here, $\varphi_\alpha(\vec{r})$ is a single-particle wave function in the state $|\alpha\rangle$ that forms a complete set of orthonormal functions satisfying the following **orthonormal** and **completeness relations**:

$$\int d\vec{r} \varphi_\alpha^*(\vec{r}) \varphi_{\alpha'}(\vec{r}) = \delta_{\alpha\alpha'} \quad , \quad \sum_\alpha \varphi_\alpha^*(\vec{r}') \varphi_\alpha(\vec{r}) = \delta(\vec{r} - \vec{r}') \tag{12}$$

It is physically obvious that the space $\bar{\text{H}}_n$ is generated by linear combinations of products of single-particle wave functions as seen previously.

So far, in defining the Hilbert space $\bar{\text{H}}_n$, the symmetry properties of the wave function have not been taken into account. **We can define mathematically only those totally (anti)symmetric states observed in nature. This is in contrast to the multitude of pure and mixed symmetry states**. We find the basis of H_n by first (anti)symmetrizing the tensor product (4):

$$\left|\alpha_1 \cdots \alpha_n\right\} = \sqrt{n!}\hat{\text{P}}_\chi \left|\alpha_1 \cdots \alpha_n\right) = \frac{1}{\sqrt{n!}} \sum_{\text{P}} \chi^{\text{P}} \left|\alpha_{\text{P}(1)} \cdots \alpha_{\text{P}(n)}\right) \tag{13}$$

In these (anti)symmetrized states, we utilize the curly bracket in the ket symbol. Here, P runs through all permutations of n objects where $\chi = +1$ for **bosons** and $\chi = -1$ for **fermions**; the symbol χ^{P} equals unity for bosons and $(-1)^{\text{P}}$ for fermions, and $\hat{\text{P}}_\chi$ is the **symmetrization operator for bosons** and the **antisymmetrization operator for fermions**; $\frac{1}{\sqrt{n!}}$ is the normalization factor.

From (13), the **Pauli Exclusion Principle is automatically satisfied for antisymmetric states: Two fermions cannot occupy the same quantum state**. For example, let us say that two identical states $|\alpha_1\rangle = |\alpha_2\rangle = |\alpha\rangle$:

$$\left|\alpha_1, \alpha_2, \alpha_3, \cdots \alpha_n\right\} = \sqrt{n!}\hat{\text{P}}_\chi \left|\alpha_1, \alpha_2, \alpha_3, \cdots \alpha_n\right) = -\sqrt{n!}\hat{\text{P}}_\chi \left|\alpha_2, \alpha_1, \alpha_3, \cdots \alpha_n\right) = 0 \tag{14}$$

In this case, no acceptable many-fermion state exists.

We define the (anti)symmetrization operator $\hat{\text{P}}$ by its action on the many-body wave function $\psi(\vec{r}_1 \cdots \vec{r}_n)$ in (9) as

$$\psi_{\alpha_1 \cdots \alpha_n}\left(\vec{r}_1 \cdots \vec{r}_n\right) = \left(\vec{r}_1 \cdots \vec{r}_n \middle| \alpha_1 \cdots \alpha_n\right) \equiv \psi\left(\vec{r}_1 \cdots \vec{r}_n\right) \tag{15}$$

Considering (13), we have

$$\hat{\text{P}}_\chi \psi\left(\vec{r}_1 \cdots \vec{r}_n\right) = \frac{1}{n!} \sum_{\text{P}} \chi^{\text{P}} \psi\left(\vec{r}_{\text{P}(1)} \cdots \vec{r}_{\text{P}(n)}\right) \tag{16}$$

Then

$$\hat{\text{P}}_\chi^2 \psi\left(\vec{r}_1 \cdots \vec{r}_n\right) = \frac{1}{n!} \frac{1}{n!} \sum_{\text{PP}'} \chi^{\text{P}'} \chi^{\text{P}} \psi\left(\vec{r}_{\text{P}'\text{P}(1)} \cdots \vec{r}_{\text{P}'\text{P}(n)}\right) \tag{17}$$

Here, PP′ denotes the group composition of P′ and P. From $\chi^{P+P'} = \chi^{P'P}$, the summation over P and P′ can be swapped with the summation over M = P′P and P:

$$\hat{P}_\chi^2 \psi(\vec{r}_1 \cdots \vec{r}_n) = \frac{1}{n!}\sum_P \left(\frac{1}{n!}\sum_M \chi^M \psi\left(\vec{r}_{M(1)} \cdots \vec{r}_{M(n)}\right) \right) = \frac{1}{n!}\sum_P \hat{P}_\chi \psi(\vec{r}_1 \cdots \vec{r}_n) = \hat{P}_\chi \psi(\vec{r}_1 \cdots \vec{r}_n) \tag{18}$$

This equality holds for any wave function ψ, as well as for the operator itself, where the (anti)symmetrization operator is a projector.

It is easy to show that $\hat{P}_\chi^2 = \hat{P}_\chi$. For $\chi = +1$, the implication is that, for **bosons**, several particles can occupy the same one-particle state. This was shown empirically for the first time by the Indian physicist **Satyendra Nath Bose** (1894–1974) by proving the relation:

$$\hat{P}\psi(\vec{r}_1 \cdots \vec{r}_n) = \psi\left(\vec{r}_{P(1)} \cdots \vec{r}_{P(n)}\right) = \psi(\vec{r}_1 \cdots \vec{r}_n) \tag{19}$$

Examples of bosons are photons, pions, mesons, gluons, phonons, excitons, plasmons, magnons, cooper pair, and Helium-4 atoms. These are particles with integral spins $\{0,1,2,\cdots\}$. The wave function of n bosons is totally symmetric and satisfies the relation in (19).

From equation (19), we see that bosons are genuinely indistinguishable when enumerating the different possible states of the particles. For $\chi = -1$, the state (13) vanishes if two of α_i, s are identical. This implies that any two fermions cannot occupy the same particle state (**Pauli Exclusion Principle**). The **Pauli Exclusion Principle** was developed empirically by the German physicist **Wolfgang Pauli** (1900–1958) [4]. It is instructive to note that this principle follows directly from the symmetry requirements on vector states. The **Pauli Exclusion Principle** is a corollary to the principle of indistinguishability of particles. This principle poses a severe constraint on vector of states of many-fermion systems and limits the number of them that are physically admissible. **Fermions** take their name from the Italian physicist Enrico Fermi, who first studied the properties of fermion gases. Several important many-particle systems have fermions as their basic constituents. Examples of fermions are protons, electrons, muons, neutrinos, quarks, and helium-3 atoms. These are particles with half-integral spins $\left\{\frac{1}{2},\frac{3}{2},\cdots\right\}$. **Note that the statistics of composite particles are determined by the number of fermions. If the fermion number is odd, then the net result is a fermion. Otherwise, for energies sufficiently low compared to their binding energy, the net result is a boson.**

The states $\{\alpha_1 \cdots \alpha_n\}$ constitute a basis of H_n. So, the closure relation (7) in $\bar{H}_n$ becomes a closure relation in H_n:

$$\hat{1} = \sum_{\alpha_1 \cdots \alpha_n} P_\chi |\alpha_1 \cdots \alpha_n)(\alpha_1 \cdots \alpha_n| P_\chi = \frac{1}{n!}\sum_{\alpha_1 \cdots \alpha_n} |\alpha_1 \cdots \alpha_n\}\{\alpha_1 \cdots \alpha_n| \tag{20}$$

The overlap between two states constructed from the same basis $|\alpha\rangle$ is given by

$$\{\alpha_1' \cdots \alpha_n' | \alpha_1 \cdots \alpha_n\} = n!\left(\alpha_1' \cdots \alpha_n' | P_\chi^2 | \alpha_1 \cdots \alpha_n\right) = n!\left(\alpha_1' \cdots \alpha_n' | P_\chi | \alpha_1 \cdots \alpha_n\right) = \sum_P \chi^P \left(\alpha_1' \cdots \alpha_n' | \alpha_{P(1)} \cdots \alpha_{P(n)}\right) \tag{21}$$

From the orthogonality of the basis $|\alpha\rangle$, the only nonvanishing terms at the right-hand side of (21) are the permutations P contributing to the given sum:

$$\alpha' = \alpha_{P(1)} \quad , \quad \cdots \quad , \quad \alpha_n' = \alpha_{P(n)} \tag{22}$$

For **fermions**, all α_i,s should be different for each one-particle state $|\alpha\rangle$. So, there is only one such permutation P that transforms $\alpha_1 \cdots \alpha_n$ into $\alpha'_1 \cdots \alpha'$. The overlap (21) reduces to one term. If the states $|\alpha_i\rangle$ are normalized, then

$$\{\alpha'_1 \cdots \alpha'_n | \alpha_1 \cdots \alpha_n\} = (-1)^{\mathrm{P}} \tag{23}$$

For **bosons**, several particles can occupy the same one-particle state. So, any permutation that interchanges particles in the same state contributes to the sum in (21). The number of these permutations is $n_{\alpha_1}! \cdots n_{\alpha_n}!$, which transforms $\alpha_1 \cdots \alpha_n$ into $\alpha'_1 \cdots \alpha'_n$. Here, n_{α_i} is the number of bosons in the one-particle state $|\alpha_i\rangle$ where $\alpha_{\mathrm{P}(1)} \cdots \alpha_{\mathrm{P}(n)}$ are distinct with

$$\{\alpha'_1 \cdots \alpha'_n | \alpha_1 \cdots \alpha_n\} = n_{\alpha_{\mathrm{P}(1)}}! \cdots n_{\alpha_{\mathrm{P}(n)}}! \tag{24}$$

For both fermions and bosons, the sum of the occupation numbers that counts the total number of occupied states is equal to the number of particles:

$$n = \sum_{\alpha} n_\alpha \tag{25}$$

For bosons, these occupation numbers are a priori not restricted, whereas for fermions, they can take only the value 0 or 1.

If we use the convention that $0! = 1$, then formulae (23) and (24) yield the equivalent single expression:

$$\{\alpha'_1 \cdots \alpha'_n | \alpha_1 \cdots \alpha_n\} = \chi^{\mathrm{P}} \prod_{i=1}^{\mathrm{P}} n_{\alpha_i}! \tag{26}$$

So, the orthonormal basis for the Hilbert space H_n can be obtained by normalizing the states $|\alpha_1 \cdots \alpha_n\}$ with the help of (26):

$$|\alpha_1 \cdots \alpha_n\rangle = \frac{1}{\sqrt{\prod_{i=1}^{\mathrm{P}} n_{\alpha_i}!}} |\alpha_1 \cdots \alpha_n\} = \frac{1}{\sqrt{n! \prod_{i=1}^{\mathrm{P}} n_{\alpha_i}!}} \sum_{\mathrm{P}} \chi^{\mathrm{P}} |\alpha_{\mathrm{P}(1)} \cdots \alpha_{\mathrm{P}(n)}) \tag{27}$$

The prefactor $\frac{1}{\sqrt{n! \prod_{i=1}^{\mathrm{P}} n_{\alpha_i}!}}$ normalizes the many-body wave function. Here, n_{α_i} is the number of particles in the state α_i and, for fermions, considering the **Pauli Exclusion Principle**, $n_{\alpha_i} = 0,1$. The summation over $n!$ permutations P of $\{\alpha_1 \cdots \alpha_n\}$ is required by particle indistinguishability; the parity χ^{P} is the number of transpositions of two elements that brings permutations $(\mathrm{P}(1), \cdots, \mathrm{P}(n))$ back to ordered sequence $(1, \cdots, n)$. Note that the normalized (anti)symmetric state defined in (27) uses an angular bracket in the ket symbol in contrast to the states defined earlier in (4). Since orthonormality is used in the calculation of the normalization factor, then hereafter it is understood that whenever the symbol $|\alpha_1 \cdots \alpha_n\rangle$ is used, the basis $\{|\alpha_i\rangle\}$ is orthonormal.

The overlap of the tensor product $|\vec{r}_1 \cdots \vec{r}_n)$ and the (anti)symmetric state $|\alpha_1 \cdots \alpha_n\rangle$:

$$(\vec{r}_1 \cdots \vec{r}_n | \alpha_1 \cdots \alpha_n\rangle = \frac{1}{\sqrt{n! \prod_{i=1}^{\mathrm{P}} n_{\alpha_i}!}} \sum_{\mathrm{P}} \chi^{\mathrm{P}} \langle \vec{r}_1 | \alpha_{\mathrm{P}(1)}\rangle \cdots \langle \vec{r}_n | \alpha_{\mathrm{P}(n)}\rangle = \frac{1}{\sqrt{n! \prod_{i=1}^{\mathrm{P}} n_{\alpha_i}!}} \sum_{\mathrm{P}} \chi^{\mathrm{P}} \varphi_{\alpha_{\mathrm{P}(1)}}(\vec{r}_1) \cdots \varphi_{\alpha_{\mathrm{P}(n)}}(\vec{r}_n) \tag{28}$$

or

$$\left(\vec{r}_1\cdots\vec{r}_n\middle|\alpha_1\cdots\alpha_n\right)\equiv\frac{1}{\sqrt{n!\prod_{i=1}^{P}n_{\alpha_i}!}}\mathbf{S}\left(\left\langle\vec{r}_i\middle|\alpha_j\right\rangle\right)\equiv\frac{1}{\sqrt{n!\prod_{i=1}^{P}n_{\alpha_i}!}}\mathbf{S}\left(\mathbf{A}_{ij}\right) \tag{29}$$

Here, $\mathbf{S}(\mathbf{A}_{ij})$ is expressed in the following relation for fermions (bosons):

$$\left(\vec{r}_1\cdots\vec{r}_n\middle|\alpha_1\cdots\alpha_n\right)=\begin{cases}\frac{1}{\sqrt{n!}}\det\left[\mathbf{A}_{ij}\right]=\frac{1}{\sqrt{n!}}\sum_{P}(-1)^P\mathbf{A}_{1P(1)}\cdots\mathbf{A}_{nP(n)} & \text{, fermions}\\ \frac{1}{\sqrt{n!\prod_{i=1}^{P}n_{\alpha_i}!}}\operatorname{per}\left[\mathbf{A}_{ij}\right]=\frac{1}{\sqrt{n!\prod_{i=1}^{P}n_{\alpha_i}!}}\sum_{P}\mathbf{A}_{1P(1)}\cdots\mathbf{A}_{nP(n)} & \text{, bosons}\end{cases} \tag{30}$$

or

$$\left(\vec{r}_1\cdots\vec{r}_n\middle|\alpha_1\cdots\alpha_n\right)=\begin{cases}\frac{1}{\sqrt{n!}}\det\left[\varphi_{\alpha_i}\left(\vec{r}_j\right)\right] & \text{, fermions}\\ \frac{1}{\sqrt{n!\prod_{i=1}^{P}n_{\alpha_i}!}}\operatorname{per}\left[\varphi_{\alpha_i}\left(\vec{r}_j\right)\right] & \text{, bosons}\end{cases} \tag{31}$$

We therefore obtain a basis of **permanents for bosons (sign-less determinant)** and **Slater determinants for fermions** as seen earlier in equations (19) through (21). From (27) and the normalization in (20), we obtain the following closure relation:

$$\sum_{\alpha_1\cdots\alpha_n}\frac{\prod_{i=1}^{P}n_{\alpha_i}!}{n!}\left|\alpha_1\cdots\alpha_n\right\rangle\left\langle\alpha_1\cdots\alpha_n\right|=1 \tag{32}$$

We define an (anti)symmetrized many-particle state in the following coordinate representation via the states $|\vec{r},\sigma\rangle$ not normalized:

$$\left|\vec{r}_1,\sigma_1\cdots\vec{r}_n,\sigma_n\right\}=\frac{1}{\sqrt{n!}}\sum_{\{P\}}\chi^P\left|\vec{r}_{P(1)},\sigma_{P(1)}\cdots\vec{r}_{P(n)},\sigma_{P(n)}\right)=\frac{1}{V^{\frac{n}{2}}}\sum_{\vec{\kappa}_1,\cdots,\vec{\kappa}_n}\exp\left\{i\left(\vec{\kappa}_1\vec{r}+\cdots+\vec{\kappa}_n\vec{r}_n\right)\right\}\left|\vec{\kappa}_1,\sigma_1\cdots,\vec{\kappa}_n,\sigma_n\right\} \tag{33}$$

Here, σ denotes the spin index (as well as other discrete indices, if necessary) and $\vec{\kappa}$ the momentum.

1.1.2 Creation and Annihilation Operators

The formalism of second quantization treats systems of bosons (fermions) with a fixed or variable number of particles in a unified manner. In many physical processes, the particle number does change. Examples include electron-hole annihilations in metals or semiconductors, electron-phonon processes, and photon absorption or emission. To formulate statistical physics in terms of the grand canonical ensemble, we must deal with states having different numbers of particles. For Fock space to be a concept of interest, there must be operators connecting the different n-particle sectors; these are the creation $\hat{\psi}_\alpha^\dagger$ and annihilation $\hat{\psi}_\alpha$ operators (with each being the Hermitian adjoint of the other) that add a particle or remove a particle, respectively, in the one-particle state α thereby (anti)symmetrizing the resulting many-particle state:

$$\hat{\psi}_\alpha^\dagger|\alpha_1\cdots\alpha_n\} = |\alpha\alpha_1\cdots\alpha_n\} \quad , \quad \hat{\psi}_\alpha^\dagger|\alpha_1\cdots\alpha_n\rangle = \sqrt{n_\alpha+1}\,|\alpha\alpha_1\cdots\alpha_n\rangle \tag{34}$$

Here, n_α is the occupation number of the state $|\alpha\rangle$ in $|\alpha_1\cdots\alpha_n\rangle$. The annihilation operator $\hat{\psi}_\alpha$ is the adjoint of the creation operator $\hat{\psi}_\alpha^\dagger$. These operators provide a convenient representation of many-particle states and many-particle operators, generate the entire Hilbert space by their action on a single reference state, and provide a basis for the algebra of operators of the Hilbert space.

The operator $\hat{\psi}_\alpha^\dagger$ physically adds a particle in state $|\alpha\rangle$ to the state on which it operates and (anti)symmetrizes the new state. Because there can be at most one fermion in a given state, equation (34) takes the following form:

$$\hat{\psi}_\alpha^\dagger|\alpha_1\cdots\alpha_n\rangle = \begin{cases} |\alpha\alpha_1\cdots\alpha_n\rangle & , \ |\alpha\rangle \notin |\alpha_1\cdots\alpha_n\rangle \\ 0 & , \ |\alpha\rangle \in |\alpha_1\cdots\alpha_n\rangle \end{cases} \tag{35}$$

Writing $\hat{\psi}^\dagger(\vec{r})|0\rangle$ and $|\alpha\rangle = \xi_\alpha^\dagger|0\rangle$, where $|0\rangle$ is the **vacuum** state containing no particles at all and is distinguished from the zero of the Hilbert space, then we have the following relation between creation (annihilation) operators in the $\vec{r}$-basis and the α-basis:

$$\psi^\dagger(\vec{r}) = \sum_\alpha \varphi_\alpha^*(\vec{r})\xi_\alpha^\dagger \quad , \quad \psi(\vec{r}) = \sum_\alpha \varphi_\alpha(\vec{r})\xi_\alpha \tag{36}$$

The most relevant example is when $|\alpha\rangle = |\vec{\kappa},\sigma\rangle$, where $\vec{\kappa}$ and σ are the momentum and spin variables, respectively. It is important to note that like all other quantum variables, the quantum field in general is a strongly fluctuating degree of freedom. It only becomes sharp in certain special eigenstates. This function adds or subtracts particles to the system.

Since any basis vector $|\alpha_1\cdots\alpha_n\}$ or $|\alpha_1\cdots\alpha_n\rangle$ may be generated by the repeated action of a creation operator on the vacuum state

$$\hat{\psi}_\alpha^\dagger|0\rangle = |\alpha\rangle \tag{37}$$

then, generally,

$$|\alpha_1\cdots\alpha_n\} = \hat{\psi}_{\alpha_1}^\dagger\cdots\hat{\psi}_{\alpha_n}^\dagger|0\rangle \quad , \quad |\alpha_1\cdots\alpha_n\rangle = \frac{1}{\sqrt{\prod_{i=1}^{P} n_{\alpha_i}!}}|\alpha_1\cdots\alpha_n\} \tag{38}$$

We see that the creation operators generate the entire Fock space by repeated action on the vacuum state. From relation (34), we have

$$\{\alpha_1\cdots\alpha_n|\hat{\psi}_\alpha|\alpha_1'\cdots\alpha_m'\} = \left[\{\alpha_1'\cdots\alpha_m'|\hat{\psi}_\alpha^\dagger|\alpha_1\cdots\alpha_n\}\right]^* = \{\alpha\alpha_1\cdots\alpha_n|\alpha_1'\cdots\alpha_m'\} \tag{39}$$

This result can only be finite when $n = m-1$, so that $\hat{\psi}_\alpha$ removes a particle from the state on which it acts. In the case of a vacuum, this implies that $\hat{\psi}_\alpha|0\rangle = \langle 0|\hat{\psi}_\alpha^\dagger = 0$ for any state $|\alpha\rangle$. This is evidence that the vacuum is the kernel of the annihilation operators.

From the closure relation in the Fock space then

$$\sum_{n=0}^{\infty}\sum_{\alpha_1\cdots\alpha_n}\frac{1}{n!}\left|\alpha_1\cdots\alpha_n\right\}\left\{\alpha_1\cdots\alpha_n\right|=1 \tag{40}$$

Using (21), we have

$$\begin{aligned}\hat{\psi}_\alpha\left|\alpha_1'\cdots\alpha_m'\right\} &= \sum_{n-0}^{\infty}\sum_{\alpha_1\cdots\alpha_n}\frac{1}{n!}\left\{\alpha_1\cdots\alpha_n\right|\hat{\psi}_\alpha\left|\alpha_1'\cdots\alpha_m'\right\}\left|\alpha_1\cdots\alpha_n\right\} \\ &= \frac{1}{(m-1)!}\sum_{\alpha_1\cdots\alpha_{m-1}}\left\{\alpha\alpha_1\cdots\alpha_{m-1}\middle|\alpha_1'\cdots\alpha_m'\right\}\left|\alpha_1\cdots\alpha_{m-1}\right\}\end{aligned} \tag{41}$$

With the help of (6) and (21), we now have

$$\hat{\psi}_\alpha\left|\alpha_1'\cdots\alpha_m'\right\} = \frac{1}{(m-1)!}\sum_{\mathrm{P}}\chi^{\mathrm{P}}\left\langle\alpha\middle|\alpha_{\mathrm{P}(1)}'\right\rangle\left|\alpha_{\mathrm{P}(2)}'\cdots\alpha_{\mathrm{P}(m)}'\right\} \tag{42}$$

Because the permutation $\left(\mathrm{P}(2)\cdots\mathrm{P}(m)\right)\to\left(1,\cdots,\mathrm{P}(1)-1,\mathrm{P}(1)+1,\cdots,m\right)$ has the signature $\chi^{\mathrm{P}+\mathrm{P}(1)-1}$, this allows us to arrive at

$$\hat{\psi}_\alpha\left|\alpha_1'\cdots\alpha_m'\right\} = \frac{1}{(m-1)!}\sum_{\mathrm{P}}\chi^{\mathrm{P}(1)-1}\delta_{\alpha,\alpha_{\mathrm{P}(1)}'}\left|\alpha_1'\cdots\alpha_{\mathrm{P}(1)-1}',\alpha_{\mathrm{P}(1)+1}'\cdots\alpha_m'\right\} = \sum_{i=1}^{m}\chi^{i-1}\delta_{\alpha,\alpha_i'}\left|\alpha_1'\cdots\hat{\alpha}_i'\cdots\alpha_m'\right\} \tag{43}$$

Here, $\hat{\alpha}_i'$ indicates that α_i' is removed from the many-particle state $\left|\alpha_1'\cdots\alpha_m'\right\}$. For a similar result for the normalized state $\left|\alpha_1'\cdots\alpha_m'\right\rangle$ with occupation number n_α for the state α, we have

$$\hat{\psi}_\alpha\left|\alpha_1'\cdots\alpha_m'\right\rangle = \frac{1}{\sqrt{n_\alpha}}\sum_{i=1}^{m}\chi^{i-1}\delta_{\alpha,\alpha_i'}\left|\alpha_1'\cdots\hat{\alpha}_i'\cdots\alpha_m'\right\rangle \tag{44}$$

From here, we observe that the effect of $\hat{\psi}_\alpha$ acting on any state is to annihilate one particle in the state α from a given state. In the case of bosons, the general result is conveniently expressed in occupation number representation:

$$\left|n_{\alpha_1}\cdots n_{\alpha_\mathrm{P}}\right\rangle = \frac{1}{\sqrt{\prod_{i=1}^{\mathrm{P}}n_{\alpha_i}!}}\left(\hat{\psi}_{\alpha_1}^{\dagger}\right)^{n_{\alpha_1}}\cdots\left(\hat{\psi}_{\alpha_\mathrm{P}}^{\dagger}\right)^{n_{\alpha_\mathrm{P}}}\left|0\right\rangle \tag{45}$$

or

$$\left|n_{\alpha_1}\cdots n_{\alpha_\mathrm{P}}\right\rangle = \prod_{i=1}^{\mathrm{P}}\frac{1}{\sqrt{n_{\alpha_i}!}}\left(\hat{\psi}_{\alpha_i}^{\dagger}\right)^{n_{\alpha_i}}\left|0\right\rangle \tag{46}$$

Here, n_{α_i} is the occupation number of the one-particle state α_i. From (46), an arbitrary state in the Fock space can be obtained by acting on $\left|0\right\rangle$ with some polynomial of creation operators $\hat{\psi}_{\alpha_i}^{\dagger}$. This also follows that a fundamental property of creation (annihilation) operators is that they provide a basis for all operators in the Fock space. So, any operator can be expressed as a linear combination of the set of all products of the operators $\left\{\hat{\psi}_\alpha^{\dagger},\hat{\psi}_\alpha\right\}$.

The states (46) form an orthonormal basis for the complete multiparticle system and satisfy the following closure relation:

$$\sum_{\{\alpha_i\}} \left| n_{\alpha_1} \cdots n_{\alpha_i} \cdots \right\rangle \left\langle n_{\alpha_1} \cdots n_{\alpha_i} \cdots \right| = 1 \tag{47}$$

Let us now define the creation and annihilation operators $\hat{\psi}_\alpha^\dagger$ and $\hat{\psi}_\alpha$ for the αth type boson such that

$$\hat{\psi}_\alpha^\dagger \left| n_{\alpha_1} \cdots n_{\alpha_P} \right\rangle = \sqrt{n_\alpha + 1} \left| n_{\alpha_1} \cdots (n_\alpha + 1) \cdots n_{\alpha_P} \right\rangle \quad , \quad \hat{\psi}_\alpha \left| n_{\alpha_1} \cdots n_{\alpha_P} \right\rangle = \sum_{i=1}^{P} \delta_{\alpha,\alpha_i} \sqrt{n_\alpha} \left| n_{\alpha_1} \cdots (n_\alpha - 1) \cdots n_{\alpha_P} \right\rangle \tag{48}$$

Here, $\hat{\psi}_\alpha^\dagger$ increases, by one, the number of particles in the αth eigenstates; its adjoint $\hat{\psi}_\alpha$ reduces, by one, the number of particles. The operators (48) are therefore eigenstates of the operator $\hat{n}_\alpha = \hat{\psi}_\alpha^\dagger \hat{\psi}_\alpha$, which measures the number of particles in the one-body state $|\alpha\rangle$:

$$\hat{n}_\alpha \left| n_{\alpha_1} \cdots n_{\alpha_P} \right\rangle = \sum_{i=1}^{P} \delta_{\alpha,\alpha_i} n_\alpha \left| n_{\alpha_1} \cdots n_{\alpha_P} \right\rangle \tag{49}$$

1.1.3 (Anti)Commutation Relations

We now examine the (anti)commutation of creation and annihilation operators. The (anti)symmetry properties of the many-particle states impose (anti)commutation relations among the creation operators. We consider the case in which any two single-particle states $|\alpha\rangle$ and $|\alpha'\rangle$ belong to the orthonormal basis $\{|\alpha\rangle\}$ for any state $|\alpha_1 \cdots \alpha_n\}$, then

$$\hat{\psi}_\alpha^\dagger \hat{\psi}_{\alpha'}^\dagger \left|\alpha_1 \cdots \alpha_n\right\} = \left|\alpha\alpha'\alpha_1 \cdots \alpha_n\right\} = \chi \left|\alpha'\alpha\alpha_1 \cdots \alpha_n\right\} = \chi \hat{\psi}_{\alpha'}^\dagger \hat{\psi}_\alpha^\dagger \left|\alpha_1 \cdots \alpha_n\right\} \tag{50}$$

From here,

$$\left(\hat{\psi}_\alpha^\dagger \hat{\psi}_{\alpha'}^\dagger - \chi \hat{\psi}_{\alpha'}^\dagger \hat{\psi}_\alpha^\dagger\right) \left|\alpha_1 \cdots \alpha_n\right\} = 0 \tag{51}$$

So,

$$\hat{\psi}_\alpha^\dagger \hat{\psi}_{\alpha'}^\dagger - \chi \hat{\psi}_{\alpha'}^\dagger \hat{\psi}_\alpha^\dagger = 0 \tag{52}$$

and taking the adjoint of (52), then

$$\hat{\psi}_\alpha \hat{\psi}_{\alpha'} - \chi \hat{\psi}_{\alpha'} \hat{\psi}_\alpha = 0 \tag{53}$$

We may also write

$$\hat{\psi}_\alpha^\dagger \hat{\psi}_{\alpha'}^\dagger - \chi \hat{\psi}_{\alpha'}^\dagger \hat{\psi}_\alpha^\dagger = \left[\hat{\psi}_\alpha^\dagger, \hat{\psi}_{\alpha'}^\dagger\right]_{-\chi} = \hat{\psi}_\alpha \hat{\psi}_{\alpha'} - \chi \hat{\psi}_{\alpha'} \hat{\psi}_\alpha = \left[\hat{\psi}_\alpha, \hat{\psi}_{\alpha'}\right]_{-\chi} = 0 \tag{54}$$

In equations (52) and (53), for the case of bosons $\chi = +1$, we have the commutators

$$\hat{\psi}_\alpha^\dagger \hat{\psi}_{\alpha'}^\dagger - \hat{\psi}_{\alpha'}^\dagger \hat{\psi}_\alpha^\dagger = \hat{\psi}_\alpha \hat{\psi}_{\alpha'} - \hat{\psi}_{\alpha'} \hat{\psi}_\alpha = 0 \tag{55}$$

and for the case of fermions, we have the anticommutators

$$\hat{\psi}_{\alpha}^{\dagger}\hat{\psi}_{\alpha'}^{\dagger}+\hat{\psi}_{\alpha'}^{\dagger}\hat{\psi}_{\alpha}^{\dagger}=\hat{\psi}_{\alpha}\hat{\psi}_{\alpha'}+\hat{\psi}_{\alpha'}\hat{\psi}_{\alpha}=0 \tag{56}$$

Comparing

$$\hat{\psi}_{\alpha}\hat{\psi}_{\alpha'}^{\dagger}\left|\alpha_1\cdots\alpha_n\right\}=\hat{\psi}_{\alpha}\left|\alpha'\alpha_1\cdots\alpha_n\right\}=\delta_{\alpha\alpha'}\left|\alpha_1\cdots\alpha_n\right\}+\sum_{i=1}^{n}\chi^{i}\delta_{\alpha\alpha_i}\left|\alpha'\alpha_1\cdots\hat{\alpha}_i\cdots\alpha_n\right\} \tag{57}$$

and

$$\hat{\psi}_{\alpha'}^{\dagger}\hat{\psi}_{\alpha}\left|\alpha_1\cdots\alpha_n\right\}=\hat{\psi}_{\alpha'}^{\dagger}\sum_{i=1}^{n}\chi^{i-1}\delta_{\alpha\alpha_i}\left|\alpha_1\cdots\hat{\alpha}_i\cdots\alpha_n\right\}=\sum_{i=1}^{n}\chi^{i-1}\delta_{\alpha\alpha_i}\left|\alpha'\alpha_1\cdots\hat{\alpha}_i\cdots\alpha_n\right\} \tag{58}$$

then also

$$\hat{\psi}_{\alpha}\hat{\psi}_{\alpha'}^{\dagger}-\chi\hat{\psi}_{\alpha'}^{\dagger}\hat{\psi}_{\alpha}=\left[\hat{\psi}_{\alpha},\hat{\psi}_{\alpha'}^{\dagger}\right]_{\chi}=\delta_{\alpha\alpha'} \tag{59}$$

We consider relation (36) and calculate the following anti(commutation) relation

$$\left[\psi(\vec{r}'),\psi^{\dagger}(\vec{r})\right]_{\chi}=\sum_{\alpha\alpha'}\varphi_{\alpha}^{*}(\vec{r})\varphi_{\alpha'}(\vec{r}')\left\{\psi_{\alpha'}\psi_{\alpha}^{\dagger}-\chi\psi_{\alpha}^{\dagger}\psi_{\alpha'}\right\}=\sum_{\alpha\alpha'}\varphi_{\alpha}^{*}(\vec{r})\varphi_{\alpha'}(\vec{r}')\delta_{\alpha\alpha'}=\sum_{\alpha}\varphi_{\alpha}^{*}(\vec{r})\varphi_{\alpha}(\vec{r}') \tag{60}$$

So,

$$\psi(\vec{r}')\psi^{\dagger}(\vec{r})-\chi\psi^{\dagger}(\vec{r})\psi(\vec{r}')=\sum_{\alpha\alpha'}\varphi_{\alpha}^{*}(\vec{r})\varphi_{\alpha'}(\vec{r}')\delta_{\alpha\alpha'}=\sum_{\alpha}\varphi_{\alpha}^{*}(\vec{r})\varphi_{\alpha}(\vec{r}')=\delta(\vec{r}-\vec{r}') \tag{61}$$

This means that the operator function $\psi^{\dagger}(\vec{r})$ creates a particle with coordinate $\vec{r}$. It is instructive to note that if $\psi_{\alpha},\psi_{\alpha}^{\dagger}$ were not operators but the c-numbers, then $\psi(\vec{r})$ and $\psi^{\dagger}(\vec{r})$ in (36) would be considered as one-particle conjugated wave functions. Generally, to introduce wave functions for a particle, it is necessary to change from classical to quantum mechanics by performing a so-called first quantization. Therefore, in the transformation from wave functions to operator functions $\psi(\vec{r})$ and $\psi^{\dagger}(\vec{r})$, with the help of (61), we perform a so-called second quantization.

1.1.4 Change of Basis in Second Quantization

Different quantum operators are expressed most naturally in different representations, which makes basis changes an essential issue in quantum physics. Next, we characterize the Fock space bases introduced earlier to a full reformulation of many-body quantum mechanics and then introduce general transformation rules that will be exploited further in this book. To find out how changes from one ordered single-particle basis $\{|\alpha\rangle\}$ to another $\{|\tilde{\alpha}\rangle\}$ affect the operator algebra $\{\hat{\psi}_{\alpha}\}$:

$$|\tilde{\alpha}\rangle=\sum_{\alpha}|\alpha\rangle\langle\alpha|\tilde{\alpha}\rangle=\sum_{\alpha}\langle\tilde{\alpha}|\alpha\rangle^{*}|\alpha\rangle \tag{62}$$

For single-particle systems, we conveniently define creation operators $\hat{\psi}_{\tilde{\alpha}}^{\dagger}$ and $\hat{\psi}_{\alpha}^{\dagger}$, which correspond to the two basis sets $\{|\tilde{\alpha}\rangle\}$, to another $\{|\alpha\rangle\}$:

$$\hat{\psi}_{\tilde{\alpha}}^{\dagger}\left|\alpha_1\cdots\alpha_n\right\}=\left|\tilde{\alpha}\alpha_1\cdots\alpha_n\right\}=\sum_{\alpha}\langle\alpha|\tilde{\alpha}\rangle\left|\alpha\alpha_1\cdots\alpha_n\right\rangle=\sum_{\alpha}\langle\alpha|\tilde{\alpha}\rangle\hat{\psi}_{\alpha}^{\dagger}\left|\alpha_1\cdots\alpha_n\right\} \tag{63}$$

Therefore, the transformation rules for creation (annihilation) field operators

$$\hat{\psi}^{\dagger}_{\tilde{\alpha}} = \sum_{\alpha} \langle \alpha | \tilde{\alpha} \rangle \hat{\psi}^{\dagger}_{\alpha} \quad , \quad \hat{\psi}_{\tilde{\alpha}} = \sum_{\alpha} \langle \tilde{\alpha} | \alpha \rangle \hat{\psi}_{\alpha} \tag{64}$$

The general validity of equation (64) stems from applying the first quantization single-particle result in equation (62) to the n-particle first quantized basis states $|\alpha_1 \cdots \alpha_n\}$. This yields

$$\hat{\psi}^{\dagger}_{\tilde{\alpha}_1} \cdots \hat{\psi}^{\dagger}_{\tilde{\alpha}_n} |0\rangle = \left(\sum_{\alpha_1} \langle \alpha_1 | \tilde{\alpha}_1 \rangle \hat{\psi}^{\dagger}_{\alpha_1} \right) \cdots \left(\sum_{\alpha_n} \langle \alpha_n | \tilde{\alpha}_n \rangle \hat{\psi}^{\dagger}_{\alpha_n} \right) |0\rangle \tag{65}$$

The transformation rules (64) lead to two important results:

1. The basis transformation preserves the bosonic or fermionic particle statistics

$$\left[\hat{\psi}_{\tilde{\alpha}}, \hat{\psi}^{\dagger}_{\tilde{\alpha}'} \right]_{\chi} = \sum_{\alpha\alpha'} \langle \tilde{\alpha} | \alpha \rangle \langle \alpha' | \tilde{\alpha}' \rangle \left\{ \hat{\psi}_{\alpha} \hat{\psi}^{\dagger}_{\alpha'} - \chi \hat{\psi}^{\dagger}_{\alpha'} \hat{\psi}_{\alpha} \right\} = \sum_{\alpha\alpha'} \langle \tilde{\alpha} | \alpha \rangle \langle \alpha' | \tilde{\alpha}' \rangle \delta_{\alpha\alpha'} = \delta_{\tilde{\alpha}\tilde{\alpha}'} \tag{66}$$

2. The basis transformation leaves the total number of particles invariant

$$\sum_{\tilde{\alpha}} \hat{\psi}^{\dagger}_{\tilde{\alpha}} \hat{\psi}_{\tilde{\alpha}} = \sum_{\tilde{\alpha}\alpha\alpha'} \langle \alpha | \tilde{\alpha} \rangle \langle \tilde{\alpha} | \alpha' \rangle \hat{\psi}^{\dagger}_{\alpha} \hat{\psi}_{\alpha'} = \sum_{\alpha\alpha'} \langle \alpha | \alpha' \rangle \hat{\psi}^{\dagger}_{\alpha} \hat{\psi}_{\alpha'} = \sum_{\alpha\alpha'} \delta_{\alpha\alpha'} \hat{\psi}^{\dagger}_{\alpha} \hat{\psi}_{\alpha'} = \sum_{\alpha} \hat{\psi}^{\dagger}_{\alpha} \hat{\psi}_{\alpha} \tag{67}$$

When the new basis is orthonormal, the (anti)commutation relations are preserved and the following transformation is unitary:

$$\left\{ \hat{\psi}^{\dagger}_{\alpha}, \hat{\psi}_{\alpha} \right\} \to \left\{ \hat{\psi}^{\dagger}_{\tilde{\alpha}}, \hat{\psi}_{\tilde{\alpha}} \right\} \tag{68}$$

1.1.5 Quantum Field Operators

For many applications, the coordinate representation turns out to be suitable, which leads to the definition of quantum field operators. We define the **second-quantized field operators** at every point in space as follows

$$\begin{aligned} \hat{\psi}^{\dagger}_{\sigma}(\vec{r}) &= \sum_{\alpha} \langle \alpha | \vec{r}, \sigma \rangle \hat{\psi}^{\dagger}_{\alpha} = \frac{1}{\sqrt{V}} \sum_{\vec{\kappa}} \exp\{-i\vec{\kappa}\vec{r}\} \hat{\psi}^{\dagger}_{\sigma}(\vec{\kappa}) \quad , \quad \hat{\psi}_{\sigma}(\vec{r}) = \sum_{\alpha} \langle \vec{r}, \sigma | \alpha \rangle \hat{\psi}_{\alpha} \\ &= \frac{1}{\sqrt{V}} \sum_{\vec{\kappa}} \exp\{i\vec{\kappa}\vec{r}\} \hat{\psi}_{\sigma}(\vec{\kappa}) \end{aligned} \tag{69}$$

Here, the sum extends over all states α of the orthonormal basis. In (69), the last relations are achieved by selecting the momentum-representation basis $\{|\vec{\kappa}, \alpha\rangle\}$. The field operators satisfy the following (anti) commutation relations

$$\left[\hat{\psi}^{\dagger}_{\sigma}(\vec{r}), \hat{\psi}^{\dagger}_{\sigma'}(\vec{r}') \right]_{-\chi} = \left[\hat{\psi}_{\sigma}(\vec{r}), \hat{\psi}_{\sigma'}(\vec{r}') \right]_{-\chi} = 0 \quad , \quad \left[\hat{\psi}_{\sigma}(\vec{r}), \hat{\psi}^{\dagger}_{\sigma'}(\vec{r}') \right]_{-\chi} = \delta_{\sigma\sigma'} \delta(\vec{r} - \vec{r}') \tag{70}$$

In a way, the quantum field operators express the essence of the wave/particle duality in quantum physics. On the one hand, they are defined as fields. This implies some type of waves. However, on the other hand, they exhibit the commutator properties associated with particles.

Considering equations (34) and (43), we have

$$\hat{\psi}_{\sigma}^{\dagger}(\vec{r})\left|\vec{r}_1,\sigma_1,\cdots,\vec{r}_n,\sigma_n\right\} = \left|\vec{r},\sigma,\vec{r}_1,\sigma_1,\cdots,\vec{r}_n,\sigma_n\right\} \tag{71}$$

$$\hat{\psi}_{\sigma}(\vec{r})\left|\vec{r}_1,\sigma_1,\cdots,\vec{r}_n,\sigma_n\right\} = \sum_{i=1}^{n}\chi^{i-1}\delta(\vec{r}-\vec{r}_i)\left|\vec{r}_1,\sigma_1,\cdots,\widehat{\vec{r}_i,\sigma_i},\cdots\vec{r}_n,\sigma_n\right\} \tag{72}$$

where $\widehat{\vec{r}_i,\sigma_i}$ implies that $\vec{r}_i,\sigma_i$ is omitted. So, the field operator $\hat{\psi}_{\sigma}^{\dagger}(\vec{r})$ adds a spin-σ particle at point $\vec{r}$ and (anti)symmetrizes the resultant many-body state.

1.1.6 Operators in Second-Quantized Form

Second quantization provides a natural formalism for describing many-particle systems. In this section, we examine the case of a system of n interacting particles. We note that, in reality, particles do interact with one another. We present a general theory in which the particles not only interact with the external potential, say the operator $\hat{v}^{(1)}$, but also interact with each other via the potential, say $\hat{v}^{(2)}$. The state operators describing physical states should be (anti)symmetric under the exchange of two particles. This depends on the statistics of the particles and whether they are fermions or bosons. We note that any operator acting within the Fock space may be written in second quantization. When all operators are expressed in terms of the fundamental creation and annihilation operators, we consider the example of one-, two-, and n-body operators.

1.1.6.1 One-Body Operator

The operator $\hat{v}^{(1)}$ is a one-body operator that acts on each particle separately:

$$\hat{v}^{(1)}\left|\alpha_1\cdots\alpha_n\right) = \sum_{i=1}^{n}\hat{v}_i\left|\alpha_1\cdots\alpha_n\right) \quad , \quad \hat{v}^{(1)}\left|\alpha_1\cdots\alpha_n\right\} = \frac{1}{\sqrt{n!}}\sum_{\mathrm{P}}\chi^{\mathrm{P}}\sum_{i=1}^{n}\hat{v}_i\left|\alpha_{\mathrm{P}(1)}\cdots\alpha_{\mathrm{P}(n)}\right) \tag{73}$$

where $\hat{v}_i$ operates only on the ith particle. For example, say

$$\hat{v}_1\left|\alpha_1\cdots\alpha_n\right) = \left(\hat{v}\left|\alpha_1\right\rangle\right)\otimes\left|\alpha_2\right\rangle\otimes\cdots\otimes\left|\alpha_n\right\rangle \tag{74}$$

Suppose we first choose a basis where the operator $\hat{v}$ is diagonal:

$$\hat{v}\left|\alpha\right\rangle = \left\langle\alpha\right|\hat{v}\left|\alpha\right\rangle\left|\alpha\right\rangle \equiv v_{\alpha}\left|\alpha\right\rangle \tag{75}$$

So,

$$\hat{v}^{(1)}\left|\alpha_1\cdots\alpha_n\right\} = \frac{1}{\sqrt{n!}}\sum_{\mathrm{P}}\chi^{\mathrm{P}}\sum_{i=1}^{n}v_{\alpha_{\mathrm{P}(i)}}\left|\alpha_{\mathrm{P}(1)}\cdots\alpha_{\mathrm{P}(n)}\right) = \sum_{i=1}^{n}v_{\alpha_i}\left|\alpha_1\cdots\alpha_n\right\} = \sum_{\alpha}v_{\alpha}\hat{n}_{\alpha}\left|\alpha_1\cdots\alpha_n\right\} \tag{76}$$

Here, the sum extends over the complete set of one-body states α and the number operator $\hat{n}_{\alpha} = \hat{\psi}_{\alpha}^{\dagger}\hat{\psi}_{\alpha}$. From the aforementioned, we see that

$$\hat{v}^{(1)} = \sum_{\alpha}\left\langle\alpha\right|\hat{v}\left|\alpha\right\rangle\hat{n}_{\alpha} \tag{77}$$

To obtain the action of $\hat{v}^{(1)}$, we must sum over all states $\left|\alpha\right\rangle$, multiplying v_{α} by the number of particles in state $\left|\alpha\right\rangle$.

From equation (77), we arrive at the general expression for one-body operators that is valid in any complete basis in terms of the field operators:

$$\hat{v}^{(1)} = \sum_{\alpha\alpha'} \langle\alpha|\hat{v}|\alpha'\rangle \hat{\psi}^{\dagger}_{\alpha}\hat{\psi}_{\alpha'} \tag{78}$$

Formula (78) is precisely the expression of $\hat{v}^{(1)}$ in the **second quantization** that can be rewritten as follows

$$\hat{v}^{(1)} = \int d^d r d^d r' \sum_{\alpha\alpha'} \langle\vec{r},\alpha|\hat{v}|\vec{r}',\alpha'\rangle \hat{\psi}^{\dagger}_{\alpha}(\vec{r})\hat{\psi}_{\alpha'}(\vec{r}') \tag{79}$$

In order to obtain the action of $\hat{v}^{(1)}$, we have to sum over all states $|\alpha\rangle$ and multiplying v_α by the number of particles in state $|\alpha\rangle$. It is instructive to note that expressions (78) and (79) make no reference to the total number of particles actually present in the system.

We may compute another operator of basic importance—particle density at the point $\vec{r}$:

$$\hat{n}(\vec{r}) = \int d^d r_1 d^d r_2 \sum_{\alpha\alpha'} \langle\vec{r}_1,\alpha|\delta\left(\vec{r}-\hat{\vec{r}}\right)|\vec{r}_2,\alpha'\rangle \hat{\psi}^{\dagger}_{\alpha}(\vec{r}_1)\hat{\psi}_{\alpha'}(\vec{r}_2) = \sum_{\alpha} \hat{\psi}^{\dagger}_{\alpha}(\vec{r})\hat{\psi}_{\alpha}(\vec{r}) \tag{80}$$

The particle number operator:

$$\hat{N} = \int d^d r \hat{n}(\vec{r}) = \sum_{\alpha} \hat{\psi}^{\dagger}_{\alpha}\hat{\psi}_{\alpha} \tag{81}$$

This applies for any complete basis of the one-body states $|\alpha\rangle$.

Considering the one-body potential

$$\hat{v}^{(1)} = \sum_{i=1}^{n} v\left(\hat{\vec{r}}_i\right) \tag{82}$$

then

$$v^{(1)} = \int d^d r_1 d^d r_2 \sum_{\alpha\alpha'} \langle\vec{r}_1,\alpha|v\left(\hat{\vec{r}}\right)|\vec{r}_2,\alpha'\rangle \hat{\psi}^{\dagger}_{\alpha}(\vec{r}_1)\hat{\psi}_{\alpha'}(\vec{r}_2) = \int d^d r v(\vec{r}) \sum_{\alpha} \hat{\psi}^{\dagger}_{\alpha}(\vec{r})\hat{\psi}_{\alpha}(\vec{r}) = \int d^d r v(\vec{r})\hat{n}(\vec{r}) \tag{83}$$

Considering that $\hat{\mathrm{T}} = \sum_{i=1}^{n} \frac{\hat{p}_i^2}{2m}$ is the kinetic energy operator, then in the second-quantized form, one obtains the following:

$$\begin{aligned} \hat{\mathrm{T}} &= \sum_{\vec{\kappa}\vec{\kappa}'\alpha\alpha'} \langle\vec{\kappa},\alpha|\frac{\hat{p}^2}{2m}|\vec{\kappa}',\alpha'\rangle \hat{\psi}^{\dagger}_{\alpha}(\vec{\kappa})\hat{\psi}_{\alpha'}(\vec{\kappa}') = \sum_{\vec{\kappa}\alpha} \frac{\kappa^2}{2m} \hat{\psi}^{\dagger}_{\alpha}(\vec{\kappa})\hat{\psi}_{\alpha}(\vec{\kappa}) \\ &= \sum_{\vec{\kappa}\alpha} \in(\vec{\kappa})\hat{\psi}^{\dagger}_{\alpha}(\vec{\kappa})\hat{\psi}_{\alpha}(\vec{\kappa}) \quad , \quad \in(\vec{\kappa}) = \frac{\kappa^2}{2m} \end{aligned} \tag{84}$$

This expression is simple and intuitive because the underlying basis diagonalizes the kinetic energy.

1.1.6.2 Two-Body Operator

In this section, we introduce a two-body operator $\hat{v}^{(2)}$ acting on the state $|\alpha_1\cdots\alpha_n)$ of n particles as the sum of the $\hat{v}^{(2)}$ on all distinct pairs of particles:

$$\hat{v}^{(2)}|\alpha_1\cdots\alpha_n)=\sum_{\substack{i,j=1\\(i<j)}}^{n}\hat{v}_{ij}|\alpha_1\cdots\alpha_n) \quad , \quad \hat{v}^{(2)}|\alpha_1\cdots\alpha_n\}=\frac{1}{\sqrt{n!}}\sum_{\mathrm{P}}\chi^{\mathrm{P}}\sum_{\substack{i,j=1\\(i<j)}}^{n}\hat{v}_{ij}|\alpha_{\mathrm{P}(1)}\cdots\alpha_{\mathrm{P}(n)}) \tag{85}$$

Here, $\hat{v}_{ij}$ acts only on the i and j particles. The restriction $i<j$ results in a summation over distinct pairs. Considering the basis $\{|\alpha\rangle\}$ where $\hat{v}$ is diagonal, then

$$\hat{v}|\alpha\beta)=(\alpha\beta|\hat{v}|\alpha\beta)|\alpha\beta)\equiv v_{\alpha\beta}|\alpha\beta) \tag{86}$$

and

$$\begin{aligned}\hat{v}^{(2)}|\alpha_1\cdots\alpha_n\}&=\frac{1}{\sqrt{n!}}\sum_{\mathrm{P}}\chi^{\mathrm{P}}\sum_{\substack{i,j=1\\(i<j)}}^{n}\hat{v}_{\alpha_{\mathrm{P}(i)}\alpha_{\mathrm{P}(j)}}|\alpha_{\mathrm{P}(1)}\cdots\alpha_{\mathrm{P}(n)})\\&=\sum_{\substack{i,j=1\\(i<j)}}^{n}\hat{v}_{\alpha_i\alpha_j}|\alpha_1\cdots\alpha_n\}=\frac{1}{2}\sum_{i,j=1}^{n}\left(\hat{v}_{\alpha_i\alpha_j}-\delta_{ij}\hat{v}_{\alpha_i\alpha_i}\right)|\alpha_1\cdots\alpha_n\}\end{aligned} \tag{87}$$

or

$$\hat{v}^{(2)}|\alpha_1\cdots\alpha_n\}=\frac{1}{2}\sum_{\alpha\beta}v_{\alpha\beta}\left(\hat{n}_\alpha\hat{n}_\beta-\delta_{\alpha\beta}\hat{n}_\alpha\right)|\alpha_1\cdots\alpha_n\} \tag{88}$$

Here, the sum extends over all states α,β of the complete basis of the one-body states. From the (anti) commutation of the field operators, we find the operator $\hat{\wp}_{\alpha\beta}$ that counts the number of pairs in the states $|\alpha\rangle$ and $|\beta\rangle$. If $|\alpha\rangle$ and $|\beta\rangle$ are different, then the number of pairs is $n_\alpha n_\beta$; if $|\alpha\rangle=|\beta\rangle$, the number of pairs is $n_\alpha(n_\alpha-1)$. Hence, the operator counting pairs may be written

$$\hat{\wp}_{\alpha\beta}=\hat{n}_\alpha\hat{n}_\beta-\delta_{\alpha\beta}\hat{n}_\alpha=\psi_\alpha^\dagger\psi_\alpha\psi_\beta^\dagger\psi_\beta-\delta_{\alpha\beta}\psi_\alpha^\dagger\psi_\alpha=\psi_\alpha^\dagger\chi\psi_\beta^\dagger\psi_\alpha\psi_\beta=\psi_\alpha^\dagger\psi_\beta^\dagger\psi_\beta\psi_\alpha \tag{89}$$

then

$$\hat{v}^{(2)}=\frac{1}{2}\sum_{\alpha\beta}(\alpha\beta|\hat{v}|\alpha\beta)\psi_\alpha^\dagger\psi_\beta^\dagger\psi_\beta\psi_\alpha \tag{90}$$

Similar to the case of the one-body operator, the action of a two-body operator is obtained by summation over pairs of single-particle states $|\alpha\rangle$ and $|\beta\rangle$. Then, we multiply the matrix element $(\alpha\beta|\hat{v}|\alpha\beta)$ by the number of pairs of such particles present in the physical state.

Transforming from the diagonal representation to an arbitrary basis, we find the general expression of a two-body operator:

$$\hat{v}^{(2)}=\frac{1}{2}\sum_{\alpha_1\cdots\alpha_2'}(\alpha_1\alpha_2|\hat{v}|\alpha_1'\alpha_2')\psi_{\alpha_1}^\dagger\psi_{\alpha_2}^\dagger\psi_{\alpha_2'}\psi_{\alpha_1'} \tag{91}$$

or

$$\hat{v}^{(2)} = \frac{1}{2}\int d^d r_1 d^d r_2' \sum_{\alpha_1\cdots\alpha_2'} \left(\vec{r}_1,\alpha_1,\vec{r}_2,\alpha_2\middle|v\middle|\vec{r}_1',\alpha_1',\vec{r}_2',\alpha_2'\right)\hat{\psi}^\dagger_{\alpha_1}(\vec{r}_1)\hat{\psi}^\dagger_{\alpha_2}(\vec{r}_2)\hat{\psi}_{\alpha_2'}(\vec{r}_2')\hat{\psi}_{\alpha_1'}(\vec{r}_1') \tag{92}$$

Simplicity and convenience are among the great virtues of representing operators via creation and annihilation operators that concisely handle all the bookkeeping for Fermi and Bose statistics. Sometimes, it is convenient to use (92) where the matrix element is indeed (anti)symmetrized:

$$\hat{v}^{(2)} = \frac{1}{4}\sum_{\alpha_1\cdots\alpha_2'} \left\{\alpha_1\alpha_2\middle|\hat{v}\middle|\alpha_1'\alpha_2'\right\}\psi^\dagger_{\alpha_1}\psi^\dagger_{\alpha_2}\psi_{\alpha_2'}\psi_{\alpha_1'} \tag{93}$$

with

$$\left\{\alpha_1\alpha_2\middle|\hat{v}\middle|\alpha_1'\alpha_2'\right\} = \left(\alpha_1\alpha_2\middle|\hat{v}\middle|\alpha_1'\alpha_2'\right) + \chi\left(\alpha_1\alpha_2\middle|\hat{v}\middle|\alpha_2'\alpha_1'\right) \tag{94}$$

We examine the example of a two-body potential

$$\hat{v}^{(2)} = \sum_{(i<j)}^{n} \hat{v}\left(\hat{\vec{r}}_i - \hat{\vec{r}}_j\right) \tag{95}$$

such that

$$\hat{v}^{(2)} = \frac{1}{2}\int d^d r_1 d^d r_2' \sum_{\alpha_1\cdots\alpha_2'} \left(\vec{r}_1,\alpha_1,\vec{r}_2,\alpha_2\middle|v\left(\hat{\vec{r}} - \hat{\vec{r}}'\right)\middle|\vec{r}_1',\alpha_1',\vec{r}_2',\alpha_2'\right)\hat{\psi}^\dagger_{\alpha_1}(\vec{r}_1)\hat{\psi}^\dagger_{\alpha_2}(\vec{r}_2)\hat{\psi}_{\alpha_2'}(\vec{r}_2')\hat{\psi}_{\alpha_1'}(\vec{r}_1') \tag{96}$$

or

$$\hat{v}^{(2)} = \frac{1}{2}\int d^d r_1 d^d r_2 \sum_{\alpha_1\alpha_2} v(\vec{r}_1 - \vec{r}_2)\hat{\psi}^\dagger_{\alpha_1}(\vec{r}_1)\hat{\psi}^\dagger_{\alpha_2}(\vec{r}_2)\hat{\psi}_{\alpha_2}(\vec{r}_2)\hat{\psi}_{\alpha_1}(\vec{r}_1) \tag{97}$$

Here, we consider the fact that $\hat{v}(\vec{r}_1 - \vec{r}_2)$ is diagonal in the coordinate representation:

$$v\left(\hat{\vec{r}} - \hat{\vec{r}}'\right)\left|\vec{r}_1,\alpha_1,\vec{r}_2,\alpha_2\right) = v(\vec{r}_1 - \vec{r}_2)\left|\vec{r}_1,\alpha_1,\vec{r}_2,\alpha_2\right) \tag{98}$$

If we consider the Fourier space, then equation (97) yields the second-quantized form of the two-body potential in the momentum space:

$$\hat{v}^{(2)} = \frac{1}{2V}\sum_{\vec{\kappa},\vec{\kappa}',\vec{q},\alpha,\alpha'} v(\vec{q})\hat{\psi}^\dagger_{\alpha}(\vec{\kappa}+\vec{q})\hat{\psi}^\dagger_{\alpha'}(\vec{\kappa}'-\vec{q})\hat{\psi}_{\alpha'}(\vec{\kappa}')\hat{\psi}_{\alpha}(\vec{\kappa}) \tag{99}$$

Here, $v(\vec{q})$ is the Fourier transformed of the interaction potential $v(\vec{r})$. This may be thought of as one particle with initial momentum $\vec{\kappa}'$ interacting with another particle with initial momentum $\vec{\kappa}$ by exchanging a momentum $\vec{q}$; $v(\vec{q})$ is then the matrix element of such a process.

We may make a generalization for the case of an n-body operator in second-quantized form:

$$\begin{aligned}\hat{v}^{(n)} &= \frac{1}{n!}\sum_{\{\alpha_1,\alpha_1'\}} (\alpha_1\cdots\alpha_n|\hat{v}|\alpha_1'\cdots\alpha_n')\hat{\psi}^\dagger_{\alpha_1}\cdots\hat{\psi}^\dagger_{\alpha_n}\hat{\psi}_{\alpha_n'}\cdots\hat{\psi}_{\alpha_1'} \\ &= \frac{1}{(n!)^2}\sum_{\{\alpha_1,\alpha_1'\}} \{\alpha_1\cdots\alpha_n|\hat{v}|\alpha_1'\cdots\alpha_n'\}\hat{\psi}^\dagger_{\alpha_1}\cdots\hat{\psi}^\dagger_{\alpha_n}\hat{\psi}_{\alpha_n'}\cdots\hat{\psi}_{\alpha_1'}\end{aligned} \quad (100)$$

Next, it will be useful to examine the normal ordering for a many-particle operator: **An operator is normal-ordered if all the creation operators are to the left of all the annihilation operators.** As an example, the right-hand side of (89) is in normal order. Regardless, any expression not in normal order may be brought into normal order by a sequence of applications of (anti)commutation relations for creation and annihilation operators.

2

Coherent States

Introduction

The *minimum uncertainty wave packets* for the harmonic oscillator were published as *coherent states* by Schrödinger [5]. Recently, coherent states have played an important role in many branches of physics and, in particular, quantum field theory and quantum optics. We have seen earlier that second quantization is the natural formalism for studying many-particle systems. In this chapter, we show how to construct path integral representation using a closure relation based on the eigenstates of the creation or the annihilation operators. We show, in particular, how coherent states are used to obtain a path integral representation of the partition function as well as calculating directly the trace defining the partition function. Quantum field theory combines classical field theory with quantum mechanics and provides analytical tools to understand many-body and relativistic quantum systems. Recently, there have been many advances in controlled fabrication of phase coherent electron devices on the nanoscale and in the realization that ultracold atomic gases exhibit strong interaction and condensation phenomena in Fermi and Bose systems. These advances, along with many others, have resulted in new perspectives on quantum physics of many-particle systems.

This book aims to introduce the ideas and techniques of quantum field theory for the many-particle system. This begins with the introduction of path integrals that provide a description of quantum mechanical time evolution in terms of trajectories. Further, perturbation theory and Feynman diagrams, which provide powerful techniques to approximately evaluate path integrals of more complicated systems, are also introduced to further generalize path integral formalism to many-particle systems. Particular attention is paid to the treatment of fermionic many-particle systems because the corresponding path integral has to be formulated in terms of anticommuting (Grassmann) variables. This also aids us in examining the concept of supersymmetry. To begin, we look at bosonic coherent states and then generalize to the fermionic case.

2.1 Coherent States for Bosons

In the preceding chapter, we used permanents or Slater determinants as a natural basis for the Fock space

$$\mathrm{H} = \mathrm{H}_0 \oplus \mathrm{H}_1 \oplus \cdots = \oplus_{n=0}^{\infty} \mathrm{H}_n \tag{101}$$

Another useful basis of the Fock space is that of coherent states, which are an analog to the basis of position eigenstates in quantum mechanics. Though it is not an orthonormal basis, it spans the entire Fock space. Position states $|\vec{r}\rangle$ are defined as eigenstates of $\hat{\vec{r}}$, while coherent states are defined as eigenstates of the annihilation operators. To see why annihilation operators are selected rather than creation operators, we denote by $|\xi\rangle$ a general vector of the Fock space:

$$|\xi\rangle = \sum_{n=0}^{\infty} \sum_{\alpha_1 \cdots \alpha_n} \xi_{\alpha_1 \cdots \alpha_n} |\alpha_1 \cdots \alpha_n\rangle \tag{102}$$

So, $|\xi\rangle$ has a component with a minimum number of particles. Applying any **creation operator** $\hat{\psi}^\dagger$ to $|\xi\rangle$, the minimum number of particles in $|\xi\rangle$ is observed to be increased by one. Hence, the resulting state cannot be a multiple of the initial state, and a creation operator $\hat{\psi}^\dagger$ cannot have an eigenstate. Applying an **annihilation operator** $\hat{\psi}$ to $|\xi\rangle$ decreases the maximum number of particles in $|\xi\rangle$ by one. Since $|\xi\rangle$ may contain components with all particle numbers, nothing a priori prohibits $|\xi\rangle$ from having eigenstates. Our goal is to find eigenstates of the (non-Hermitian) Fock space operators $\hat{\psi}^\dagger$ and $\hat{\psi}$.

2.2 Coherent States and Overcompleteness

Coherent States

It is useful to note that an important property needed for setting up path integration is the completeness of the states. We now examine coherent states and define eigenstates of the operator $\hat{\psi}$. The physical meaning of the bosonic coherent states can be understood from a study of the system of harmonic oscillators described by the following Hamiltonian equation:

$$\hat{H} = \sum_{\alpha_k} \hat{H}_{\alpha_k} \quad , \quad \hat{H}_{\alpha_k} = \frac{\hat{P}_{\alpha_k}^2}{2m_k} + \frac{m_k \omega_{\alpha_k}^2 \hat{Q}_{\alpha_k}^2}{2} \tag{103}$$

with

$$\left[\hat{Q}_{\alpha_k}, \hat{P}_{\alpha_l}\right] = i\delta_{\alpha_k \alpha_l} \tag{104}$$

Here, m_k, ω_{α_k}, $\hat{Q}_{\alpha_k}$, and $\hat{P}_{\alpha_k}$ are the mass, frequency, position, and momentum operators, respectively, of the oscillator. The ladder operators

$$\hat{\psi}_{\alpha_k} = \sqrt{\frac{m\omega_{\alpha_k}}{2}}\left(\hat{Q}_{\alpha_k} + i\frac{\hat{P}_{\alpha_k}}{m\omega_{\alpha_k}}\right) \quad , \quad \hat{\psi}_{\alpha_k}^\dagger = \sqrt{\frac{m\omega_{\alpha_k}}{2}}\left(\hat{Q}_{\alpha_k} - i\frac{\hat{P}_{\alpha_k}}{m\omega_{\alpha_k}}\right) \quad , \quad \hat{P}_{\alpha_k} = -i\frac{\partial}{\partial Q_{\alpha_k}} \tag{105}$$

satisfy the canonical bosonic commutation relations

$$\left[\hat{\psi}_{\alpha_k}, \hat{\psi}_{\alpha_{k'}}\right] = \left[\hat{\psi}_{\alpha_k}^\dagger, \hat{\psi}_{\alpha_{k'}}^\dagger\right] = 0 \quad , \quad \left[\hat{\psi}_{\alpha_k}, \hat{\psi}_{\alpha_{k'}}^\dagger\right] = \delta_{\alpha_k \alpha_{k'}} \tag{106}$$

This permits us to rewrite the Hamiltonian equation as follows:

$$H = \sum_{\alpha_k} \omega_{\alpha_k}\left(\hat{\psi}_{\alpha_k}^\dagger \hat{\psi}_{\alpha_k} + \frac{1}{2}\right) \tag{107}$$

Singling out the ground state $|0\rangle$ then

$$\hat{\psi}_{\alpha_k}|0\rangle = 0 \quad , \quad \langle 0|\hat{P}_{\alpha_k}|0\rangle = \langle 0|\hat{Q}_{\alpha_k}|0\rangle = 0 \tag{108}$$

The eigenstates of the Hamiltonian equation (107) can be obtained from the tensor product of the states $|n_{\alpha_k}\rangle$:

$$|n_{\alpha_1}\cdots n_{\alpha_n}\rangle = |n_{\alpha_1}\rangle\otimes\cdots\otimes|n_{\alpha_n}\rangle = \frac{\left(\hat{\psi}^{\dagger}_{\alpha_1}\right)^{n_1}}{\sqrt{n_{\alpha_1}!}}\cdots\frac{\left(\hat{\psi}^{\dagger}_{\alpha_n}\right)^{n_{\alpha n}}}{\sqrt{n_{\alpha_n}!}}|0\rangle \quad , \quad \hat{\mathrm{H}}|n_{\alpha_1}\cdots n_{\alpha_n}\rangle = \sum_{\alpha_k}\omega_{\alpha_k}\left(\hat{n}_{\alpha_k}+\frac{1}{2}\right)|n_{\alpha_1}\cdots n_{\alpha_n}\rangle \tag{109}$$

It is instructive to note that one of the drawbacks of the given states is that they do not serve as eigenstates of either the position $\hat{\mathrm{Q}}$ or the momentum $\hat{\mathrm{P}}$ operator. Moreover, the commutation relation (104) prevents us from searching eigenstates for both operators. Notwithstanding, it is possible to define a **so-called coherent state** $|p_i,q_j\rangle$ having average position and momentum given by some classical value (p_i,q_j):

$$\langle p,q|\hat{\mathrm{P}}_{\alpha_i}|p,q\rangle = p_{\alpha_i} \quad , \quad \langle p,q|\hat{\mathrm{Q}}_{\alpha_i}|p,q\rangle = q_{\alpha_i} \tag{110}$$

We find such a state by considering

$$|p,q\rangle = \exp\left\{-\hat{\mathrm{A}}\right\}|\xi\rangle \tag{111}$$

and the identity

$$\exp\left\{\hat{\mathrm{A}}\right\}\hat{\mathrm{B}}\exp\left\{-\hat{\mathrm{A}}\right\} = \hat{\mathrm{B}}+\frac{1}{1!}\left[\hat{\mathrm{A}},\hat{\mathrm{B}}\right]+\frac{1}{2!}\left[\hat{\mathrm{A}},\left[\hat{\mathrm{A}},\hat{\mathrm{B}}\right]\right]+\cdots \tag{112}$$

that ends at the second term when $\left[\hat{\mathrm{A}},\hat{\mathrm{B}}\right]$ is the c-number. So (110) sets the condition

$$|\xi\rangle \equiv |0\rangle \tag{113}$$

and

$$\hat{\mathrm{A}} = -\sum_{\alpha_k} i\left(p_{\alpha_k}\hat{\mathrm{Q}}_{\alpha_k} - q_{\alpha_k}\hat{\mathrm{P}}_{\alpha_k}\right) \tag{114}$$

with

$$|p,q\rangle = \exp\left\{-\hat{\mathrm{A}}\right\}|0\rangle \tag{115}$$

Then, from (105) and

$$\xi_{\alpha_k} = \sqrt{\frac{m\omega}{2}}\left(q_{\alpha_k}+i\frac{p_{\alpha_k}}{m\omega}\right) \quad , \quad \xi^{\dagger}_{\alpha_k} = \sqrt{\frac{m\omega}{2}}\left(q_{\alpha_k}-i\frac{p_{\alpha_k}}{m\omega}\right) \tag{116}$$

we have

$$|p,q\rangle = \exp\left\{\sum_{\alpha_k}\left(\xi_{\alpha_k}\hat{\psi}^{\dagger}_{\alpha_k} - \xi^{*}_{\alpha_k}\hat{\psi}_{\alpha_k}\right)\right\}|0\rangle = \exp\left\{\hat{\psi}^{\dagger}\xi - \xi^{\dagger}\hat{\psi}\right\}|0\rangle \tag{117}$$

Here,

$$\hat{\psi} = \begin{bmatrix} \hat{\psi}_1 \\ \vdots \\ \hat{\psi}_{\alpha_P} \end{bmatrix}, \quad \xi = \begin{bmatrix} \xi_1 \\ \vdots \\ \xi_{\alpha_P} \end{bmatrix} \tag{118}$$

It can be easily verified that

$$\langle \xi | \left(\hat{P}_{\alpha_k} - p_{\alpha_k} \right)^2 | \xi \rangle = \langle \xi | \left(\hat{Q}_{\alpha_k} - q_{\alpha_k} \right)^2 | \xi \rangle = \frac{1}{2} \tag{119}$$

Therefore, $|\xi\rangle$ is as near as possible to a classical state. Considering (112) and (115), then we rewrite (117) as follows:

$$|\xi\rangle = \exp\{\hat{\psi}^{\dagger}\xi\}|0\rangle = \sum_{\alpha_1 \cdots \alpha_n} \frac{\left(\xi_{\alpha_1}\hat{\psi}^{\dagger}_{\alpha_1}\right)^{n_{\alpha 1}}}{n_{\alpha_1}!} \cdots \frac{\left(\xi_{\alpha_n}\hat{\psi}^{\dagger}_{\alpha_n}\right)^{n_{\alpha n}}}{n_{\alpha_n}!}|0\rangle \tag{120}$$

Since

$$|n_{\alpha_1} \cdots n_{\alpha_n}\rangle = \frac{\left(\hat{\psi}^{\dagger}_{\alpha_1}\right)^{n_1}}{\sqrt{n_{\alpha_1}!}} \cdots \frac{\left(\hat{\psi}^{\dagger}_{\alpha_n}\right)^{n_{\alpha n}}}{\sqrt{n_{\alpha_n}!}}|0\rangle \tag{121}$$

then the given coherent state is defined by the following eigenvector [6]:

$$|\xi\rangle = \sum_{\alpha_1 \cdots \alpha_n} \frac{\left(\xi_{\alpha_1}\right)^{n_{\alpha 1}}}{\sqrt{n_{\alpha_1}!}} \cdots \frac{\left(\xi_{\alpha_P}\right)^{n_{\alpha n}}}{\sqrt{n_{\alpha_n}!}}|n_{\alpha_1} \cdots n_{\alpha_n}\rangle \tag{122}$$

The eigenstate $|n_{\alpha_1} \cdots n_{\alpha_n}\rangle$, as seen earlier in equation (25), has a total number of particles, $n = \sum_{\alpha_k} n_{\alpha_k}$, while the Hilbert space (generically referred to as the Fock space) is written as the direct sum, such as in equation (1).

2.2.1 Overcompleteness of Coherent States

Assume we have constructed an eigenstate $|\xi\rangle$ of the annihilation operators $\hat{\psi}_{\alpha}$; then the eigenstates and eigenvalues of the bosonic operator can be obtained via the eigenvalue equation:

$$\hat{\psi}_{\alpha}|\xi\rangle = \hat{\psi}_{\alpha}\exp\{\hat{\psi}^{\dagger}\xi_{\alpha}\}|0\rangle = \hat{\psi}_{\alpha}\sum_{n=0}^{\infty}\frac{\xi_{\alpha}^{n}}{n!}\left(\hat{\psi}^{\dagger}\right)^{n}|0\rangle = \hat{\psi}_{\alpha}\sum_{n=0}^{\infty}\frac{\xi_{\alpha}^{n}}{\sqrt{n!}}|n\rangle = \sum_{n=0}^{\infty}\frac{\xi_{\alpha}^{n}}{\sqrt{n!}}\hat{\psi}_{\alpha}|n\rangle = \sum_{n=1}^{\infty}\frac{\xi_{\alpha}^{n}}{\sqrt{n!}}\sqrt{n}|n-1\rangle$$

$$= \sum_{n=1}^{\infty}\frac{\xi_{\alpha}^{n}}{\sqrt{(n-1)!}}|n-1\rangle \underset{n'=n-1}{=} \sum_{n'=0}^{\infty}\frac{\xi_{\alpha}^{n'+1}}{\sqrt{n'!}}|n'\rangle = \xi_{\alpha}\sum_{n=0}^{\infty}\frac{\xi_{\alpha}^{n}}{\sqrt{n!}}|n\rangle = \xi_{\alpha}\exp\{\hat{\psi}^{\dagger}\xi_{\alpha}\}|0\rangle = \xi_{\alpha}|\xi\rangle \tag{123}$$

Similarly, the adjoint

$$\langle \xi | \hat{\psi}^{\dagger}_{\alpha} = \langle \xi | \xi^{*}_{\alpha} \tag{124}$$

The action of a creation (annihilation) operator $\hat{\psi}_\alpha^\dagger\left(\hat{\psi}_\alpha\right)$ on a coherent state is obtained as

$$\hat{\psi}_\alpha^\dagger|\xi\rangle = \hat{\psi}_\alpha^\dagger \exp\left\{\sum_{\alpha'} \hat{\psi}_{\alpha'}^\dagger \xi_{\alpha'}\right\}|0\rangle = \frac{\partial}{\partial\xi_\alpha}|\xi\rangle \quad , \quad \langle\xi|\hat{\psi}_\alpha = \frac{\partial}{\partial\xi_\alpha^*}\langle\xi| \tag{125}$$

It is useful to show that the operators $\hat{\psi}_\alpha^\dagger$ and $\hat{\psi}_\alpha$ act in the coherent state representation in the same way as the operators $\hat{\vec{r}}$ and $\hat{\vec{p}}$ act in coordinate representation. From equations (123), (124), and (125), then

$$\langle\xi|\hat{\psi}_\alpha|\phi\rangle = \frac{\partial}{\partial\xi_\alpha^*}\phi\left(\xi^*\right) \quad , \quad \langle\xi|\hat{\psi}_\alpha^\dagger|\phi\rangle = \xi_\alpha^*\phi\left(\xi^*\right) \tag{126}$$

So, symbolically, we can write

$$\hat{\psi}_\alpha = \frac{\partial}{\partial\xi_\alpha^*} \quad , \quad \hat{\psi}_\alpha^\dagger = \xi_\alpha^* \tag{127}$$

This is consistent with the bosonic commutation rules:

$$\left[\xi_\alpha^*, \xi_{\alpha'}^*\right] = \left[\frac{\partial}{\partial\xi_\alpha^*}, \frac{\partial}{\partial\xi_\alpha^*}\right] = 0 \quad , \quad \left[\frac{\partial}{\partial\xi_\alpha^*}, \xi_{\alpha'}^*\right] = \delta_{\alpha\alpha'} \tag{128}$$

We observe that the behavior of $\hat{\psi}_\alpha$ an $\hat{\psi}_\alpha^\dagger$ in the coherent state representation is thus analogous to that of $\hat{\vec{r}}$ and $\hat{\vec{p}}$ in coordinate representation.

It is important to note that the coherent state with $\xi = 0$ is identical to the Fock vacuum $|0\rangle$. The eigenvalue ξ_α of the bosonic annihilation operator $\hat{\psi}_\alpha$ may be any real or complex number. What is unusual about this definition is that $\hat{\psi}_\alpha$ is not a Hermitian operator (and so is not observable in the usual sense). Nevertheless, the states $|\xi\rangle$ defined in such a way form **a complete set**—indeed **an overcomplete set**—and define a new representation, the **coherent state representation**. Introducing the **overcomplete base** of coherent states widens the concept of the **path integral formalism** in areas of many-particle systems [3, 7].

Because $|\xi\rangle$ is an eigenstate of the annihilation operators $\hat{\psi}_\alpha$, then the (anti)commutation relations

$$\left[\hat{\psi}_\alpha, \hat{\psi}_{\alpha'}\right]_{-\chi} = 0 \tag{129}$$

This implies that

$$\hat{\psi}_\alpha\hat{\psi}_{\alpha'}|\xi\rangle = \hat{\psi}_\alpha\xi_{\alpha'}|\xi\rangle = \xi_\alpha\xi_{\alpha'}|\xi\rangle = \chi\hat{\psi}_{\alpha'}\hat{\psi}_\alpha|\xi\rangle = \chi\hat{\psi}_{\alpha'}\xi_\alpha|\xi\rangle = \chi\xi_{\alpha'}\xi_\alpha|\xi\rangle \tag{130}$$

In the case of fermions, the main difficulty encountered is the lack of a classical limit for the eigenvalue ξ_α. This implies that c-numbers cannot reflect the anticommuting character of fermions. As a way of introducing anticommuting objects, we present anticommuting variables called **Grassmann numbers**, which follow the same concepts used in constructing coherent states for bosons.

2.2.2 Overlap of Two Coherent States

We examine the following inner product or overlap of two coherent states given by:

$$\langle\xi|\xi'\rangle = \sum_{n_{\alpha_1}\cdots n_{\alpha_P}} \sum_{n'_{\alpha_1}\cdots n'_{\alpha_P}} \frac{\left(\xi_{\alpha_1}^*\right)^{n_{\alpha_1}}}{\sqrt{n_{\alpha_1}!}} \cdots \frac{\left(\xi_{\alpha_P}^*\right)^{n_{\alpha_P}}}{\sqrt{n_{\alpha_P}!}} \frac{\left(\xi'_{\alpha_1}\right)^{n'_{\alpha_1}}}{\sqrt{n'_{\alpha_1}!}} \cdots \frac{\left(\xi'_{\alpha_P}\right)^{n'_{\alpha_P}}}{\sqrt{n'_{\alpha_P}!}} \langle n_{\alpha_1}\cdots n_{\alpha_P}|n'_{\alpha_1}\cdots n'_{\alpha_P}\rangle \tag{131}$$

Because the basis $|\alpha\rangle$ is orthonormal, the scalar product

$$\langle n_{\alpha_1}\cdots n_{\alpha_P}|n'_{\alpha_1}\cdots n'_{\alpha_P}\rangle=\delta_{n_{\alpha_1}n'_{\alpha_1}}\cdots\delta_{n_{\alpha_P}n'_{\alpha_P}} \tag{132}$$

and, so,

$$\langle\xi|\xi'\rangle=\exp\left\{\sum_{\alpha}\xi^*_\alpha\xi'_\alpha\right\} \tag{133}$$

It is instructive to note that the coherent states corresponding to two different values of ξ are not orthogonal states because they do not form a proper basis and are eigenvectors of a non-Hermitian operator. The overlap quickly falls off exponentially with the distance between the two points and gives a measure of the intrinsic uncertainty of the coherent state as probability amplitude in phase space.

From their definition, it is obvious that coherent states do not have a fixed number of particles but that the occupation numbers n_α in the coherent state $|\xi\rangle$ are Poisson distributed with mean values $|\xi_\alpha|^2$:

$$\left|\langle n_{\alpha_1}\cdots n_{\alpha_P}|\xi\rangle\right|^2=\prod_{\alpha}\frac{|\xi_\alpha|^{2n_\alpha}}{n_\alpha!} \tag{134}$$

So, the distribution of the total particle number has the average value

$$n=\langle\hat{n}\rangle=\frac{\langle\xi|\hat{n}|\xi\rangle}{\langle\xi|\xi\rangle}=\frac{\langle\xi|\hat{\psi}^\dagger_\alpha\hat{\psi}_\alpha|\xi\rangle}{\langle\xi|\xi\rangle}=\sum_{\alpha}\xi^*_\alpha\xi_\alpha \tag{135}$$

with the variance

$$\sigma^2=\left\langle\left(\hat{n}-\langle\hat{n}\rangle\right)^2\right\rangle=\sum_{\alpha}\xi^*_\alpha\xi_\alpha=\langle\hat{n}\rangle \tag{136}$$

In the thermodynamic limit, where $n\to\infty$, the relative width

$$\frac{\sigma}{n}=\frac{1}{\sqrt{n}}\to 0 \tag{137}$$

In this case, the particle number distribution becomes sharply peaked around n. This indicates that a product of Poisson distributions approaches a normal distribution.

2.2.3 Overcompleteness Condition

The most important property of coherent states is their **overcompleteness** in the Fock space, which implies that any vector of the Fock space can be expanded in terms of coherent states. To obtain a path integral representation, a closure relation for the coherent states is needed. We examine the closure relation (resolution identity) for the bosonic coherent states, which is defined as:

$$\int\prod_{\alpha}\frac{d\xi^*_\alpha d\xi_\alpha}{2\pi i}\exp\left\{-\sum_{\alpha}\xi^*_\alpha\xi_\alpha\right\}|\xi\rangle\langle\xi|=\hat{1} \tag{138}$$

In (138), $\hat{1}$ is the Fock space identity operator, and the measure is given by

$$\frac{d\xi^*_\alpha d\xi_\alpha}{2\pi i}=\frac{d(\mathrm{Re}\,\xi_\alpha)d(\mathrm{Im}\,\xi_\alpha)}{\pi} \tag{139}$$

As shown in (133), because the coherent states are in general not orthogonal, the set of coherent states is **overcomplete**, while formula (138) shows that the coherent states form a basis in the Fock space. Nonetheless the coherent states are very useful, particularly for deriving path integrals. In this book, they are important as an example of states where the creation and annihilation operators have nonvanishing expectation values. This factor will be essential in the discussion on Bose-Einstein condensates.

2.2.4 Closure Relation via Schur's Lemma

We prove the closure relation via **Schur's lemma**, which in this case states: **If an operator commutes with all creation and annihilation operators, then it is proportional to the unit operator in the Fock space.**

From equation (125), then

$$\left[\hat{\psi}_\alpha, |\xi\rangle\langle\xi|\right] = \left(\xi_\alpha - \frac{\partial}{\partial \xi_\alpha^*}\right)|\xi\rangle\langle\xi| \tag{140}$$

And, by evaluating the commutator (138) and performing integration by parts, we have

$$\left[\hat{\psi}_\alpha, \int \prod_{\alpha'} \frac{d\xi_{\alpha'}^* d\xi_{\alpha'}}{2\pi i} \exp\left\{-\sum_{\alpha'} \xi_{\alpha'}^* \xi_{\alpha'}\right\}|\xi\rangle\langle\xi|\right] = \int \prod_{\alpha'} \frac{d\xi_{\alpha'}^* d\xi_{\alpha'}}{2\pi i} \exp\left\{-\sum_{\alpha'} \xi_{\alpha'}^* \xi_{\alpha'}\right\}\left(\xi_\alpha - \frac{\partial}{\partial \xi_\alpha^*}\right)|\xi\rangle\langle\xi| = 0 \tag{141}$$

If we look at the adjoint of (141), we observe that the left-hand side of (138) commutes with all of the creation as well as the annihilation operators. Therefore, it must be proportional to the unit operator. We can calculate the proportionality factor by taking the expectation value of the left-hand side of (138) in the vacuum:

$$\int \prod_{\alpha} \frac{d\xi_\alpha^* d\xi_\alpha}{2\pi i} \exp\left\{-\sum_{\alpha} \xi_\alpha^* \xi_\alpha\right\}\langle 0|\xi\rangle\langle\xi|0\rangle = \int \prod_{\alpha} \frac{d\xi_\alpha^* d\xi_\alpha}{2\pi i} \exp\left\{-\sum_{\alpha} \xi_\alpha^* \xi_\alpha\right\}\langle 0|\xi\rangle\langle\xi|0\rangle = \hat{1} \tag{142}$$

Therefore, we prove the closure relation in (138).

For a single degree of freedom, we write ξ in polar form

$$\xi = |\xi|\exp\{i\varphi\} \quad , \quad \xi^* = |\xi|\exp\{-i\varphi\} \tag{143}$$

then this changes the variables from (ξ, ξ^*) to $(|\xi|, \varphi)$. Considering that the Jacobian determinant of this variable transformation is $2i|\xi|$, then the measure

$$\frac{d\xi d\xi^*}{2\pi i} = \frac{2i|\xi|\,d\xi d\varphi}{2\pi i} \tag{144}$$

Hence,

$$\int \frac{d\xi d\xi^*}{2\pi i} \exp\left\{-|\xi|^2\right\}|\xi\rangle\langle\xi| = \int_0^\infty \frac{2i|\xi|\,d\xi}{2\pi i} \sum_{n,m=0}^{\infty} \frac{|\xi|^{n+m}}{\sqrt{n!m!}} \exp\left\{-|\xi|^2\right\}|n\rangle\langle m| \int_0^{2\pi} \exp\{i(n-m)\varphi\}d\varphi \tag{145}$$

or

$$\int_0^\infty \frac{2i|\xi|d\xi}{2\pi i}\sum_{n,m=0}^{\infty}\frac{|\xi|^{n+m}}{\sqrt{n!m!}}\exp\left\{-|\xi|^2\right\}|n\rangle\langle m|2\pi\delta_{nm}=2\pi i\int_0^\infty \frac{2|\xi|d\xi}{2\pi i}\sum_{n=0}^{\infty}\frac{|\xi|^{2n}}{n!}\exp\left\{-|\xi|^2\right\}|n\rangle\langle n| \tag{146}$$

Changing variables again, $z=|\xi|^2$, and using the definition of the Gamma function,

$$\int_0^\infty dz\exp\{-z\}z^n=\Gamma(n+1)=n! \tag{147}$$

then

$$2\int_0^\infty |\xi|d\xi\sum_{n=0}^{\infty}\frac{|\xi|^{2n}}{n!}\exp\left\{-|\xi|^2\right\}|n\rangle\langle n|=\int_0^\infty dz\sum_{n=0}^{\infty}\frac{z^n}{n!}\exp\{-z\}|n\rangle\langle n|=\sum_{n=0}^{\infty}|n\rangle\langle n|=1 \tag{148}$$

For one degree of freedom and considering the position eigenstates $|q\rangle$ and the coherent states $|\xi\rangle$, then it makes sense to compute their inner product:

$$\langle q|\xi\rangle=\sum_{n=0}^{\infty}\frac{\xi^n}{\sqrt{n!}}\langle q|n\rangle \tag{149}$$

In this case, the wave function $\langle q|n\rangle$ of the nth excited state of the harmonic oscillator:

$$\langle q|n\rangle=\left(\frac{m\omega}{\pi\hbar}\right)^{\frac{1}{4}}\frac{\exp\left\{-\frac{m\omega q^2}{2\hbar}\right\}}{\sqrt{2^n n!}}\mathrm{H}_n\left(\left(\frac{m\omega}{\hbar}\right)^{\frac{1}{2}}q\right) \tag{150}$$

or

$$\langle q|n\rangle=\frac{\sqrt{\alpha}\exp\left\{-\frac{(\alpha q)^2}{2}\right\}}{\sqrt{2^n n!\sqrt{\pi}}}\mathrm{H}_n(\alpha q)\quad,\quad \alpha=\sqrt{\frac{m\omega}{\hbar}} \tag{151}$$

Here $\mathrm{H}_n(x)$ is the Hermite polynomial with argument x. The generating function for the Hermite polynomials $\mathrm{H}_n(x)$:

$$\exp\left\{2xy-y^2\right\}=\sum_{n=0}^{\infty}\mathrm{H}_n(x)\frac{y^n}{n!} \tag{152}$$

Therefore, from the aforementioned,

$$\langle q|\xi\rangle=\sum_{n=0}^{\infty}\frac{\xi^n}{\sqrt{n!}}\langle q|n\rangle=\sum_{n=0}^{\infty}\frac{\xi^n}{\sqrt{n!}}\frac{\sqrt{\alpha}\exp\left\{-\frac{(\alpha q)^2}{2}\right\}}{\sqrt{2^n n!\sqrt{\pi}}}\mathrm{H}_n(\alpha q) \tag{153}$$

or

$$\langle q|\xi\rangle = \sqrt{\frac{\alpha}{\sqrt{\pi}}}\exp\left\{-\frac{(\alpha q)^2}{2}\right\}\sum_{n=0}^{\infty}\frac{1}{n!}\left(\frac{\xi}{\sqrt{2}}\right)^n \mathrm{H}_n(\alpha q) = \sqrt{\frac{\alpha}{\sqrt{\pi}}}\exp\left\{-\frac{(\alpha q)^2}{2}\right\}\exp\left\{\sqrt{2}\alpha\xi q - \frac{\xi^*\xi}{2}\right\} \quad (154)$$

or

$$\langle q|\xi\rangle = \left(\frac{m\omega}{\pi\hbar}\right)^{\frac{1}{4}}\exp\left\{-\frac{\xi^*\xi}{2} - \frac{m\omega}{2\hbar}q^2 + \sqrt{\frac{2m\omega}{\hbar}}q\xi\right\} \quad (155)$$

2.2.5 Normal-Ordered Operators

In this section, we will examine one property of coherent states—the simple form of matrix elements of **normal-ordered** operators between coherent states. **An operator $\mathbf{A}(\hat{\psi}^\dagger_\alpha, \hat{\psi}_\alpha)$ is said to be normal-ordered when all creation operators stand to the left of the annihilation operators** $\hat{\psi}_\alpha$. The matrix element between coherent states of such an operator takes the form

$$\langle\xi|\mathbf{A}(\hat{\psi}^\dagger_\alpha, \hat{\psi}_\alpha)|\xi'\rangle = \exp\left\{\sum_\alpha \xi^*_\alpha \xi'_\alpha\right\}\mathbf{A}(\xi^*_\alpha, \xi'_\alpha) \quad (156)$$

One example is a two-body potential:

$$\langle\xi|\hat{v}^{(2)}|\xi'\rangle = \frac{1}{2}\sum_{\alpha\alpha'\beta\beta'}(\alpha\alpha'|v|\beta\beta')\langle\xi|\hat{\psi}^\dagger_\alpha\hat{\psi}^\dagger_{\alpha'}\hat{\psi}_{\beta'}\hat{\psi}_\beta|\xi'\rangle \quad (157)$$

or

$$\langle\xi|\hat{v}^{(2)}|\xi'\rangle = \frac{1}{2}\sum_{\alpha\alpha'\beta\beta'}(\alpha\alpha'|v|\beta\beta')\langle\xi|\xi^*_\alpha\xi^*_{\alpha'}\xi'_{\beta'}\xi'_\beta|\xi'\rangle\exp\left\{\sum_\alpha \xi^*_\alpha\xi'_\alpha\right\} \quad (158)$$

2.2.6 The Trace of an Operator

The overcompleteness relation can be used to represent a state of the extended Fock space in terms of coherent states. The completeness relation provides a useful expression for the trace of any operator **A**. Denoting $\{|n\rangle\}$ as a complete set of states, then

$$\begin{aligned}\mathrm{Tr}\,\mathbf{A} &= \sum_n \langle n|\mathbf{A}|n\rangle = \int\prod_\alpha \frac{d\xi^*_\alpha d\xi_\alpha}{2\pi i}\exp\left\{-\sum_\alpha \xi^*_\alpha\xi_\alpha\right\}\sum_n \langle n|\xi\rangle\langle\xi|\mathbf{A}|n\rangle = \\ &= \int\prod_\alpha \frac{d\xi^*_\alpha d\xi_\alpha}{2\pi i}\exp\left\{-\sum_\alpha \xi^*_\alpha\xi_\alpha\right\}\sum_n \langle\xi|\mathbf{A}|n\rangle\langle n|\xi\rangle = \int\prod_\alpha \frac{d\xi^*_\alpha d\xi_\alpha}{2\pi i}\exp\left\{-\sum_\alpha \xi^*_\alpha\xi_\alpha\right\}\langle\xi|\mathbf{A}|\xi\rangle\end{aligned} \quad (159)$$

or

$$\mathrm{Tr}\,\mathbf{A} = \int\prod_\alpha \frac{d\xi^*_\alpha d\xi_\alpha}{2\pi i}\exp\left\{-\sum_\alpha \xi^*_\alpha\xi_\alpha\right\}\langle\xi|\mathbf{A}|\xi\rangle \quad (160)$$

From quantum mechanics, the completeness of position eigenstates permits us to represent a state

$$|\hat{\psi}\rangle = \int d\vec{r}\, \hat{\psi}(\vec{r})|\vec{r}\rangle \tag{161}$$

Here, the coordinate representation of the state $|\hat{\psi}\rangle$ is

$$\hat{\psi}(\vec{r}) = \langle \vec{r}|\hat{\psi}\rangle \tag{162}$$

Similarly, equation (139) implies any state $|\hat{\psi}\rangle$ of Fock space can be represented:

$$|\hat{\psi}\rangle = \int \prod_{\alpha} \frac{d\xi_\alpha^* d\xi_\alpha}{2\pi i} \exp\left\{-\sum_\alpha \xi_\alpha^* \xi_\alpha\right\} |\xi\rangle\langle\xi|\hat{\psi}\rangle = \int \prod_{\alpha} \frac{d\xi_\alpha^* d\xi_\alpha}{2\pi i} \exp\left\{-\sum_\alpha \xi_\alpha^* \xi_\alpha\right\} \hat{\psi}(\xi^*)|\xi\rangle \tag{163}$$

where

$$\langle\xi|\hat{\psi}\rangle = \hat{\psi}(\xi^*) \tag{164}$$

by definition, is the coherent state representation of the state $|\hat{\psi}\rangle$, with ξ denoting the set $\{\xi_\alpha^*\}$. The coherent state representation for bosons often is referred to as the holomorphic representation. This is due to the fact that ψ is an analytic function of the variables ξ_α^*. Physically, the quantity $\hat{\psi}(\xi^*)$ simply is the wave function of the state $|\hat{\psi}\rangle$ in the coherent state representation. This implies the probability amplitude to find the system in the coherent state $|\xi\rangle$.

In the case of holomorphic functions $\hat{\psi}(\xi^*)$, the unit operator can be achieved via (138):

$$\langle\xi|\hat{\psi}\rangle = \int \prod_{\alpha} \frac{d\xi_\alpha'^* d\xi_\alpha'}{2\pi i} \exp\left\{-\sum_\alpha \xi_\alpha'^* \xi_\alpha'\right\} \langle\xi|\xi'\rangle\langle\xi'|\hat{\psi}\rangle \tag{165}$$

Therefore, it follows that

$$\hat{\psi}(\xi^*) = \int \prod_{\alpha} \frac{d\xi_\alpha'^* d\xi_\alpha'}{2\pi i} \exp\left\{-\sum_\alpha (\xi_\alpha'^* - \xi_\alpha^*)\xi_\alpha'\right\} \hat{\psi}(\xi'^*) \tag{166}$$

It is instructive to note that this simply is a general form in the complex plane for the familiar representation of a Dirac delta function:

$$\delta(x - x') = \int \frac{dk}{2\pi} \exp\{ik(x - x')\} \tag{167}$$

2.3 Grassmann Algebra and Fermions

2.3.1 Grassmann Algebra

The path integral approach is easy to employ for bosonic systems due to commuting functions instead of anticommuting operators [8, 9]. However, such an advantage is not obvious for fermionic systems because the integration variables are anticommuting. We will now discuss a fermionic system within the framework of the fermionic coherent state path integral. When dealing with fields instead of operators, we apply **Grassmann algebra**, which maintains the **Pauli Exclusion Principle**. Grassmann algebra allows us to elaborate all necessary calculation rules to derive the path integral and, subsequently, the Dyson equation for fermionic functionals that are accordingly functionals of Grassmann functions.

We consider the distribution law for Grassmann variables:

$$(\xi_1+\xi_2)\xi_3=\xi_1\xi_3+\xi_2\xi_3 \quad , \quad \xi_1(\xi_2+\xi_3)=\xi_1\xi_2+\xi_1\xi_3 \quad , \quad \lambda(\xi_1\xi_2)=(\lambda\xi_1)\xi_2=\xi_1(\lambda\xi_2) \tag{168}$$

as well as the anticommutative property

$$\{\xi_i,\xi_j\}=\xi_i\xi_j+\xi_j\xi_i=0 \tag{169}$$

This permits us to arrive at the square of any generator vanishing for any k:

$$\xi_k^2=0 \tag{170}$$

This is a particular important property of the anticommutation relation of equation 169.

We can construct a finite dimensional Grassmann algebra from n such elements, which are called generators $\{\xi_k\}$, $k=1,\cdots,n$. Then, from property (170), all elements of the given algebra can now be expressed via a linear combination of these generators:

$$\{1,\xi_{\lambda_1},\xi_{\lambda_1}\xi_{\lambda_2},\cdots,\xi_{\lambda_1},\xi_{\lambda_2}\cdots\cdots\xi_{\lambda_n}\} \tag{171}$$

Here, $0<\xi_k\leq n$, and we assume that the elements, by convention, are ordered by the indices $\lambda_1<\lambda_2<\cdots<\lambda_n$. From (170), there exists no element of the higher products containing more than one ξ_k. So now, any element of the n-dimensional Grassmann algebra can be expressed as a polynomial of first order in the generators:

$$f(\xi_1,\cdots,\xi_n)=f_0+\sum_{\alpha_1}f_{\alpha_1}\xi_1+\sum_{\alpha_1<\alpha_2}f_{\alpha_1\alpha_2}\xi_1\xi_2+\cdots+\sum_{\alpha_1<\alpha_2<\cdots<\alpha_n}f_{\alpha_1\alpha_2\cdots\alpha_n}\xi_{\alpha_1}\xi_{\alpha_2}\cdots\xi_{\alpha_n} \tag{172}$$

where the complex valued coefficients $f_k\in\mathbb{C}$ are complex numbers or complex functions. So, f is a function of the generators and a complex variable; therefore, we refer to objects shown in equation (172) as Grassmann functions.

In order to operate using Grassmann functions, it is necessary to define analog operations of differentiation and integration for Grassmann functions.

2.3.1.1 Differentiation over Grassmann Variables

We define differentiation over a Grassmann variable (generator) as:

$$\frac{d}{d\xi_j}\xi_{\lambda_1}\xi_{\lambda_2}\cdots\xi_{\lambda_n}=\delta_{j\lambda_1}\xi_{\lambda_2}\cdots\cdots\xi_{\lambda_n}-\delta_{j\lambda_2}\xi_{\lambda_1}\xi_{\lambda_3}\cdots\cdots\xi_{\lambda_n}+\cdots+(-1)^{n-1}\delta_{j\lambda_n}\xi_{\lambda_1}\xi_{\lambda_2}\cdots\xi_{\lambda_{n-1}} \tag{173}$$

The derivative over Grassmann numbers is precisely left-sided. In essence, we have to anticommute the variable to the left and apply the rules

$$\frac{d}{d\xi_i}1=0 \quad , \quad \frac{d}{d\xi_i}\xi_j=\delta_{ij} \tag{174}$$

This is a linear operation that is the same for ordinary numbers.

2.3.1.1.1 Grassmann Function Differentiation Rules

To examine differentiation rules for Grassmann functions, we consider some important properties of Grassmann generators. This can be done by first examining the following commutation of two Grassmann numbers with a single Grassmann number:

$$[\eta_1\eta_2,\xi]=\eta_1\eta_2\xi-\xi\eta_1\eta_2=-\eta_1\xi\eta_2-\xi\eta_1\eta_2=\xi\eta_1\eta_2-\xi\eta_1\eta_2=0 \tag{175}$$

From induction, it is obvious that any even number of Grassmann numbers commutes with a single Grassmann number:

$$[\xi_1\xi_2\cdots\xi_{2n},\eta]=0 \tag{176}$$

Similarly, we assume

$$[\eta_1\eta_2\cdots\eta_{2n},\xi]=0 \tag{177}$$

Then we examine the case:

$$\begin{aligned}[\eta_1\eta_2\cdots\eta_{2n+2},\xi]&=\eta_1\eta_2\cdots\eta_{2n}\eta_{2n+1}\eta_{2n+2}\xi-\xi\eta_1\eta_2\cdots\eta_{2n}\eta_{2n+1}\eta_{2n+2}=\\&=-\eta_1\eta_2\cdots\eta_{2n}\eta_{2n+1}\xi\eta_{2n+2}-\xi\eta_1\eta_2\cdots\eta_{2n}\eta_{2n+1}\eta_{2n+2}=\\&=\eta_1\eta_2\cdots\eta_{2n}\xi\eta_{2n+1}\eta_{2n+2}-\xi\eta_1\eta_2\cdots\eta_{2n}\eta_{2n+1}\eta_{2n+2}=(\eta_1\eta_2\cdots\eta_{2n}\xi-\xi\eta_1\eta_2\cdots\eta_{2n})\eta_{2n+1}\eta_{2n+2}\end{aligned} \tag{178}$$

With the swap in positions of η_{2n+2} and ξ in the first term, there is a sign change for that term and

$$\begin{aligned}[\eta_1\eta_2\cdots\eta_{2n+2},\xi]&=-\eta_1\eta_2\cdots\eta_{2n}\eta_{2n+1}\xi\eta_{2n+2}-\xi\eta_1\eta_2\cdots\eta_{2n}\eta_{2n+1}\eta_{2n+2}=\\&=\eta_1\eta_2\cdots\eta_{2n}\xi\eta_{2n+1}\eta_{2n+2}-\xi\eta_1\eta_2\cdots\eta_{2n}\eta_{2n+1}\eta_{2n+2}=[\eta_1\eta_2\cdots\eta_{2n},\xi]\eta_{2n+1}\eta_{2n+2}\end{aligned} \tag{179}$$

By induction from (177), then

$$[\eta_1\eta_2\cdots\eta_{2n+2},\xi]=0 \tag{180}$$

We also show that the commutation of an arbitrary number of even Grassmann numbers with any other number of Grassmann numbers:

$$[\eta_1\eta_2\cdots\eta_{2n},\xi_1\xi_2\cdots\xi_k]=0 \tag{181}$$

We examine the case:

$$[\eta_1\eta_2\cdots\eta_{2n},\xi_1\xi_2\cdots\xi_{k+1}]=\eta_1\eta_2\cdots\eta_{2n}\xi_1\xi_2\cdots\xi_k\xi_{k+1}-\xi_1\xi_2\cdots\xi_k\xi_{k+1}\eta_1\eta_2\cdots\eta_{2n} \tag{182}$$

Swapping the positions of ξ_k and ξ_{k+1} in the second term, then

$$[\eta_1\eta_2\cdots\eta_{2n},\xi_1\xi_2\cdots\xi_{k+1}]=\eta_1\eta_2\cdots\eta_{2n}\xi_1\xi_2\cdots\xi_k\xi_{k+1}-\xi_1\xi_2\cdots\xi_k\eta_1\eta_2\cdots\eta_{2n}\xi_{k+1} \tag{183}$$

or

$$[\eta_1\eta_2\cdots\eta_{2n},\xi_1\xi_2\cdots\xi_{k+1}]=[\eta_1\eta_2\cdots\eta_{2n},\xi_1\xi_2\cdots\xi_k]\xi_{k+1} \tag{184}$$

So, from induction and considering (181), then

$$[\eta_1\eta_2\cdots\eta_{2n},\xi_1\xi_2\cdots\xi_{k+1}]=0 \tag{185}$$

We show that two arbitrary odd numbers of Grassmann numbers anticommute:

$$\{\eta_1\eta_2\cdots\eta_{2n+1},\xi_1\xi_2\cdots\xi_{2k+1}\}=0 \tag{186}$$

One Grassmann number anticommutes with an odd number of Grassmann variables and we assume that

$$\{\eta_1\eta_2\cdots\eta_{2n+1},\xi\}=0 \tag{187}$$

Hence,

$$\{\eta_1\eta_2\cdots\eta_{2n+3},\xi\}=\eta_1\eta_2\cdots\eta_{2n+3}\xi+\xi\eta_1\eta_2\cdots\eta_{2n+3}=\eta_1\eta_2\cdots\eta_{2n+1}\eta_{2n+2}\eta_{2n+3}\xi+\xi\eta_1\eta_2\cdots\eta_{2n+1}\eta_{2n+2}\eta_{2n+3} \tag{188}$$

After swapping Grassmann variables in the first term of the last relation, then

$$\{\eta_1\eta_2\cdots\eta_{2n+3},\xi\}=\eta_1\eta_2\cdots\eta_{2n+1}\xi\eta_{2n+2}\eta_{2n+3}+\xi\eta_1\eta_2\cdots\eta_{2n+1}\eta_{2n+2}\eta_{2n+3}=\{\eta_1\eta_2\cdots\eta_{2n+1},\xi\}\eta_{2n+2}\eta_{2n+3} \tag{189}$$

From (187) and by induction, we arrive at

$$\{\eta_1\eta_2\cdots\eta_{2n+3},\xi\}=0 \tag{190}$$

Therefore, for an arbitrary $n>0$ and for any k by induction, we have

$$\{\eta_1\eta_2\cdots\eta_{2n+1},\xi_1\xi_2\cdots\xi_{2k+1}\}=0 \tag{191}$$

We now consider

$$\{\eta_1\eta_2\cdots\eta_{2n+1},\xi_1\xi_2\cdots\xi_{2k+3}\}=\eta_1\eta_2\cdots\eta_{2n+1}\xi_1\xi_2\cdots\xi_{2k+3}+\xi_1\xi_2\cdots\xi_{2k+3}\eta_1\eta_2\cdots\eta_{2n+1} \tag{192}$$

and insert the Grassmann variables ξ_{2k+1} and ξ_{2k+2} in the terms so that

$$\{\eta_1\eta_2\cdots\eta_{2n+1},\xi_1\xi_2\cdots\xi_{2k+3}\}=\eta_1\eta_2\cdots\eta_{2n+1}\xi_1\xi_2\cdots\xi_{2k+3}+\xi_1\xi_2\cdots\xi_{2k+1}\xi_{2k+2}\xi_{2k+3}\eta_1\eta_2\cdots\eta_{2n+1} \tag{193}$$

After swapping the Grassmann variables in the last term, we have

$$\{\eta_1\eta_2\cdots\eta_{2n+1},\xi_1\xi_2\cdots\xi_{2k+3}\}=\eta_1\eta_2\cdots\eta_{2n+1}\xi_1\xi_2\cdots\xi_{2k+3}+\xi_1\xi_2\cdots\xi_{2k+1}\eta_1\eta_2\cdots\eta_{2n+1}\xi_{2k+2}\xi_{2k+3} \tag{194}$$

This relation can also be rewritten as

$$\{\eta_1\eta_2\cdots\eta_{2n+1},\xi_1\xi_2\cdots\xi_{2k+3}\}=\{\eta_1\eta_2\cdots\eta_{2n+1},\xi_1\xi_2\cdots\xi_{2k+1}\}\xi_{2k+2}\xi_{2k+3} \tag{195}$$

So, by induction from (191), we have

$$\{\eta_1\eta_2\cdots\eta_{2n+1},\xi_1\xi_2\cdots\xi_{2k+3}\}=0 \tag{196}$$

The previous demonstrations allow us to summarize **some important properties of Grassmann generators**:

- **Two even numbers of Grassmann variables commute with each other:**

$$\left[\eta_1\eta_2\cdots\eta_{2n},\xi_1\xi_2\cdots\xi_{2k}\right]=0 \tag{197}$$

- **Even and odd numbers of Grassmann variables commute with each other:**

$$\left[\eta_1\eta_2\cdots\eta_{2n},\xi_1\xi_2\cdots\xi_{2k+1}\right]=0 \tag{198}$$

- **Two odd numbers of Grassmann variables anticommute with each other:**

$$\left\{\eta_1\eta_2\cdots\eta_{2n+1},\xi_1\xi_2\cdots\xi_{2k+1}\right\}=0 \tag{199}$$

Considering these properties and (172), then we see that two arbitrary Grassmann functions do not commute. Thus,

$$\left[f(\eta),g(\eta)\right]\neq 0 \tag{200}$$

Generally, this may lead to the definition of even and odd Grassmann functions.

We may also characterize an element of Grassmann algebra by introducing an automorphism, P, that acts as a parity operator

$$\mathrm{P}\left(\xi_{\lambda_1}\cdots\xi_{\lambda_n}\right)=(-1)^n\,\xi_{\lambda_1}\cdots\xi_{\lambda_n} \tag{201}$$

So, if f is an even function, then

$$\mathrm{P}(f)=f \tag{202}$$

and if it is an odd function, then

$$\mathrm{P}(f)=-f \tag{203}$$

Consider again the last relation in (174). In this case, the anticommutation relations (169) imply

$$\left\{\frac{\partial}{\partial\xi_i},\frac{\partial}{\partial\xi_j}\right\}_+=0 \tag{204}$$

So, the operator $\frac{\partial}{\partial\xi_i}$ is nilpotent, $\frac{\partial^2}{\partial\xi_i^2}=0$, thereby showing that the function $f(\xi_1,\cdots,\xi_n)$ is at most of order 1 in each variable. The last relation in (174) defines a left differentiation

$$\frac{\partial}{\partial\xi_i}\xi_j\xi_i=-\frac{\partial}{\partial\xi_i}\xi_i\xi_j=-\xi_j \quad , \quad (i\neq j) \tag{205}$$

When $\frac{\partial}{\partial\xi_i}$ acts on ξ_i, we shift first ξ_i to the left of the monomial. The **chain rule for differentiation** can then be written

$$\frac{\partial}{\partial\xi_i}f(g)=\frac{\partial g}{\partial\xi_i}\frac{\partial f}{\partial g} \tag{206}$$

However, contrary to ordinary variables, the order of the terms on the right-hand side is important.

Integration and differentiation for Grassmann numbers are identical:

$$\int d\xi_i f(\xi_1,\cdots,\xi_n) = \frac{\partial}{\partial \xi_i} f(\xi_1,\cdots,\xi_n) \tag{207}$$

This is a condition ensuring that two fundamental properties of ordinary integrals over functions vanishing at infinity are satisfied. The integral of an exact differential is zero:

$$\int d\xi_i \frac{\partial}{\partial \xi_i} f(\xi_1,\cdots,\xi_n) = 0 \tag{208}$$

The integral over ξ_i of $f(\xi_1,\cdots,\xi_n)$ is independent of ξ_i so that its derivative vanishes:

$$\frac{\partial}{\partial \xi_i} \int d\xi_i f(\xi_1,\cdots,\xi_n) = 0 \tag{209}$$

Properties in (208) as well as in (209) follow from (207) and the nilpotence of the differential operator $\frac{\partial}{\partial \xi_i}$. If we consider a Grassmann algebra with a single generator

$$f(\xi) = f_0 + f_1\xi \tag{210}$$

then

$$\frac{\partial}{\partial \xi} \int d\xi f = f_1 \tag{211}$$

2.3.1.2 Exponential Function of Grassmann Numbers

We examine the following exponential function of Grassmann variables:

$$\exp\left\{\sum_{\lambda_k} \xi_{\lambda_k}\xi_{\lambda_{k+1}}\cdots\xi_{\lambda_{2k}}\right\} = \prod_{\lambda_k} \exp\{\xi_{\lambda_k}\xi_{\lambda_{k+1}}\cdots\xi_{\lambda_{2k}}\} \equiv \exp\left\{\sum_{\lambda_k}^{2n}\prod_{k} \xi_{\lambda_k}\right\} = \prod_{\lambda_k}^{2n} \exp\left\{\prod_{k} \xi_{\lambda_k}\right\} \tag{212}$$

This permits us to write

$$\left[\exp\left\{\sum_{\lambda} \xi_{\lambda_1}\xi_{\lambda_2}\cdots\xi_{\lambda_{2k}}\right\},\eta_1\cdots\eta_n\right] = \prod_{\lambda_k}\left[\exp\{\xi_{\lambda_1}\xi_{\lambda_2}\cdots\xi_{\lambda_{2k}}\},\eta_1\cdots\eta_n\right] = \prod_{\lambda_k}\left[1-\xi_{\lambda_1}\xi_{\lambda_2}\cdots\xi_{\lambda_{2k}},\eta_1\cdots\eta_n\right] \tag{213}$$

or

$$\prod_{\lambda_k}\left[1-\xi_{\lambda_1}\xi_{\lambda_2}\cdots\xi_{\lambda_{2k}},\eta_1\cdots\eta_n\right] = \prod_{\lambda_k}\left(\left[1,\eta_1\cdots\eta_n\right]-\left[\xi_{\lambda_1}\xi_{\lambda_2}\cdots\xi_{\lambda_{2k}},\eta_1\cdots\eta_n\right]\right) \tag{214}$$

Because

$$\left[1,\eta_1\cdots\eta_n\right] = 0 \quad , \quad \left[\xi_{\lambda_1}\xi_{\lambda_2}\cdots\xi_{\lambda_{2k}},\eta_1\cdots\eta_n\right] = 0 \tag{215}$$

then

$$\left[\exp\left\{\sum_{\lambda}\xi_{\lambda_1}\xi_{\lambda_2}\cdots\xi_{\lambda_{2k}}\right\},\eta_1\cdots\eta_n\right]=0 \tag{216}$$

And this takes into consideration the relations

$$\left[\exp\left\{\sum_{\lambda}\xi_{\lambda}^{*}\xi_{\lambda}\right\},\eta\right]=0 \quad , \quad \left[\exp\left\{\sum_{\lambda}\xi_{\lambda}^{*}\xi_{\lambda}\right\},\eta_1\cdots\eta_n\right]=0 \tag{217}$$

Consider the general Hamiltonian determinant in normal order to be an operator of the form

$$\hat{\mathrm{H}}=\frac{1}{n!}\sum_{\lambda_1,\cdots\lambda_n\mu_1,\cdots\mu_n}\langle\lambda_1,\cdots\lambda_n|\mathrm{H}|\mu_1,\cdots\mu_n\rangle\xi_{\lambda_1}^{\dagger}\cdots\xi_{\lambda_n}^{\dagger}\xi_{\mu_n}\cdots\xi_{\mu_1} \tag{218}$$

Therefore, for any operator, there is always an even combination of creation and annihilation operators. Hence,

$$\left[\exp\left\{-i\varepsilon\mathrm{H}\left[\hat{\psi}^{\dagger},\hat{\psi}\right]\right\},\xi_k\right]=0 \tag{219}$$

Also, for any even Grassmann function $f(\xi)$ and any Grassmann function $g(\xi)$, it follows that

$$\left[\exp\left\{f(\xi)\right\},g(\xi)\right]=0 \tag{220}$$

2.3.1.3 Involution of Grassmann Numbers

We consider each even Grassmann algebra of $n=2l$ for which we introduce an involution operation. This is done by associating one generator ξ_k^{*} with each generator ξ_k and requesting that

$$(\xi_k)^{*}=\xi_k^{*} \quad , \quad (\xi_k^{*})^{*}=\xi_k \quad , \quad (\lambda\xi_k)^{*}=\lambda^{*}\xi_k^{*} \quad , \quad (\xi_{\lambda_1},\xi_{\lambda_2}\cdots\xi_{\lambda_n})^{*}=\xi_{\lambda_n}^{*}\xi_{\lambda_{n-1}}^{*}\cdots\xi_{\lambda_1}^{*} \tag{221}$$

Here, λ is complex valued index. Note that the generators ξ_k and ξ_k^{*} are completely independent. Hence, all rules previously derived are applicable. **Involuted Grassmann numbers as well as objects of the form $\xi_1+\xi_2$ are sometimes called complex Grassmann numbers.**

2.3.1.4 Bilinear Form of Operators

We simplify notation by considering the general element of Grassmann algebra with two generators $\{\xi^{*},\xi\}$, which are analytic functions of ξ^{*} and ξ with bilinear form:

$$f(\xi,\xi^{*})=f_0+f_1\xi^{*}+f_2\xi+f_{12}\xi^{*}\xi \tag{222}$$

Via the differentiation and integration rules, we find that

$$\frac{\partial f}{\partial\xi}=\int d\xi f=f_2-f_{12}\xi^{*} \quad , \quad \frac{\partial f}{\partial\xi^{*}}=\int d\xi^{*}f=f_1+f_{12}\xi \quad , \quad \frac{\partial^2 f}{\partial\xi^{*}\partial\xi}=\int d\xi^{*}d\xi f=-f_{12} \tag{223}$$

From here, we see that

$$\frac{\partial^2 f}{\partial\xi^*\partial\xi} = -\frac{\partial^2 f}{\partial\xi\partial\xi^*} = -f_{12} \tag{224}$$

and so

$$\left\{\frac{\partial}{\partial\xi^*}, \frac{\partial}{\partial\xi}\right\} = 0 \tag{225}$$

This again confirms the fact that the operator $\frac{\partial}{\partial\xi_i}$ is nilpotent, $\frac{\partial^2}{\partial\xi^2} = 0$, and that the function f is at most of order 1 in each variable. So, **for the two generators $\{\xi^*, \xi\}$ the algebra is generated by the four numbers $\{1, \xi, \xi^*, \xi^*\xi\}$. Further, the linear function in equation (210) will be the coherent state representation of a wave function, and the coherent state representation of an operator in Grassmann algebra will be a function of ξ^* and ξ that should have the form of equation (222).**

2.3.1.5 Berezin Integration

When explaining the definite integral, there is no equivalent for the familiar sum motivating the Riemann integral for ordinary variables. Therefore, integration over Grassmann variables can be defined as linear mapping with the fundamental property of ordinary integrals over functions vanishing at infinity. In this case, the integral of an exact differential form is zero. This constraint means the integral of 1 is zero because 1 is the derivative of ξ. The nonvanishing integral is only ξ^* because ξ is not a derivative. So, the Berezin definite integral over Grassmann numbers is defined as [10]:

$$\int d\xi 1 = 0 \quad , \quad \int d\xi\xi = 1 \tag{226}$$

In the case of a derivative, to apply the second equation in (226), we must first anticommute the variable ξ so as to bring it next to $d\xi$. This definition simply imitates Grassmann integration, which is equivalent to Grassmann differentiation. Because half of the generators ξ_i are defined arbitrarily to be conjugate variables but are otherwise equivalent to the generators ξ_i^*, we define integration over conjugate variables similarly:

$$\int d\xi^* 1 = 0 \quad , \quad \int d\xi^*\xi^* = 1 \tag{227}$$

This implies again that integration is equivalent to differentiation. The aforementioned definitions tailor integration to obey the usual rules of partial integration and in particular

$$\int d\xi \frac{\partial f(\xi)}{\partial\xi} = 0 \tag{228}$$

for any function $f(\xi) = f_0 + f_1\xi$. Note that condition (228) requires the satisfaction of the first equations in (226) and (227). Therefore, the last equations in (226) and (227) are solely for normalization purposes. Then the integral of the function $f(\xi)$:

$$\int d\xi f(\xi) = \int d\xi (f_0 + f_1\xi) = f_1 \tag{229}$$

From here, we see that for the Grassmann integral to have meaning, we have

$$\int d\xi f(\xi+\eta)=\int d\xi f(\xi) \tag{230}$$

where η is a Grassmann variable.

Similarly, we have

$$\int d\xi d\xi^*=0 \;,\; \int d\xi d\xi^*\xi=0 \;,\; \int d\xi d\xi^*\xi^*=0 \;,\; \int d\xi d\xi^*\xi^*\xi=1 \;,\; \int d\xi d\xi^*\xi\xi^*=-1 \tag{231}$$

2.3.1.6 Grassmann Delta Function

The Grassmann delta function can be defined by

$$\delta(\xi,\xi')=\int d\eta\exp\{-\eta(\xi-\xi')\}=\int d\eta\left(1-\eta(\xi-\xi')\right)=-(\xi-\xi') \tag{232}$$

Here, η is a Grassmann variable. We verify that this definition has the desired behavior by using the function $f(\xi)=f_0+f_1\xi$, and we have

$$\int d\xi'\delta(\xi,\xi')f(\xi')=-\int d\xi'(\xi-\xi')(f_0+f_1\xi')=f_0+f_1\xi=f(\xi) \tag{233}$$

2.3.1.7 Scalar Product of Grassmann Algebra

We define the scalar product in such a way that, on the one hand, it imitates the form of a scalar product with bosonic coherent states, and on the other hand, it allows functions of a Grassmann variable to have the structure of a Hilbert space.

$$\begin{aligned}\langle f|g\rangle&=\int d\xi d\xi^*\exp\{-\xi\xi^*\}f^*(\xi^*)g(\xi)=\int d\xi d\xi^*(1-\xi\xi^*)(f_0^*+f_1^*\xi^*)(g_0+g_1\xi)=\\&=-\int d\xi d\xi^*\xi\xi^*f_0^*g_0+\int d\xi d\xi^*\xi^*\xi f_1^*g_1=f_0^*g_0+f_1^*g_1\end{aligned} \tag{234}$$

The results presented for the two generators ξ^* and ξ can be applied to $2k$ generators $\xi_1\cdots\xi_{2k},\xi_1^*\cdots\xi_k^*$.

2.3.2 Fermions

Employing anticommuting Grassmann variables for calculating physical quantities for fermionic systems is a well-established technique [11]. This involves, in particular, the calculation of expectation values of quasifree (Gaussian) fermion states. This notion entails replacing the linear combinations of canonically anticommuting Fermi field operators with complex coefficients by linear combinations with coefficients that are **anticommuting Grassmann numbers** [11]. Consequently, these linear combinations achieve **canonical commutation relations**.

This section provides a mathematical background that imports the **Grassmann calculus** into the description of fermion systems. We set up a formalism that can be viewed as the fermionic analog of a quantum harmonic analysis on phase space. We present basic theorems that will aid in the understanding of **Grassmann algebra**.

Consider the anticommuting fermionic operators $\hat{\psi}^\dagger$ and $\hat{\psi}$ that create or annihilate, respectively, a fermion in state $|\xi\rangle$ and that serve as a **basis** for many-particle operators:

$$\hat{\psi}\hat{\psi}^\dagger + \hat{\psi}^\dagger\hat{\psi} = 1 \tag{235}$$

We find

$$\hat{n}^2 = \left(\hat{\psi}^\dagger\hat{\psi}\right)^2 = \hat{\psi}^\dagger\hat{\psi}\left(1 - \hat{\psi}\hat{\psi}^\dagger\right) \tag{236}$$

Then, from $\hat{\psi}^2 = 0$, we have

$$\hat{n}^2 = \left(\hat{\psi}^\dagger\hat{\psi}\right)^2 = \hat{\psi}^\dagger\hat{\psi}\left(1 - \hat{\psi}\hat{\psi}^\dagger\right) = \hat{\psi}^\dagger\hat{\psi} = \hat{n} \tag{237}$$

We see that $n^2 = n$. This implies that the **occupation number in each state can only be** $n = 0$ **or** $n = 1$.

Consider the eigenstates of the **fermionic** creation $\hat{\psi}^\dagger$ or annihilation operator $\hat{\psi}$:

$$\hat{\psi}^\dagger|0\rangle = |1\rangle \quad , \quad \hat{\psi}|1\rangle = |0\rangle \quad , \quad \hat{\psi}^\dagger = |1\rangle\langle 0| = \begin{pmatrix} 1 \\ 0 \end{pmatrix}\begin{pmatrix} 0 & 1 \end{pmatrix} = \begin{pmatrix} 0 & 1 \\ 0 & 0 \end{pmatrix} \quad , \quad \hat{\psi} = |0\rangle\langle 1| = \begin{pmatrix} 0 \\ 1 \end{pmatrix}\begin{pmatrix} 1 & 0 \end{pmatrix} = \begin{pmatrix} 0 & 0 \\ 1 & 0 \end{pmatrix} \tag{238}$$

where

$$|0\rangle = \begin{pmatrix} 0 \\ 1 \end{pmatrix} \quad , \quad |1\rangle = \begin{pmatrix} 1 \\ 0 \end{pmatrix} \tag{239}$$

From here, there should be a problem in dealing with the **anticommuting** behavior of the fermionic creation $\hat{\psi}^\dagger$ and annihilation $\hat{\psi}$ operators:

$$\hat{\psi}^\dagger\hat{\psi} = \begin{pmatrix} 0 & 1 \\ 0 & 0 \end{pmatrix}\begin{pmatrix} 0 & 0 \\ 1 & 0 \end{pmatrix} = \begin{pmatrix} 1 & 0 \\ 0 & 0 \end{pmatrix} \quad , \quad \hat{\psi}\hat{\psi}^\dagger = \begin{pmatrix} 0 & 0 \\ 1 & 0 \end{pmatrix}\begin{pmatrix} 0 & 1 \\ 0 & 0 \end{pmatrix} = \begin{pmatrix} 0 & 0 \\ 0 & 1 \end{pmatrix} \quad , \quad \hat{\psi}^\dagger\hat{\psi} + \hat{\psi}\hat{\psi}^\dagger = \begin{pmatrix} 1 & 0 \\ 0 & 1 \end{pmatrix} = \hat{1} \tag{240}$$

Therefore, we require Grassmann numbers to define fermionic coherent states where we resort to the **Grassmann numbers** ξ and ξ^* when considering the anticommuting eigenvalues of the operators $\hat{\psi}^\dagger$ and $\hat{\psi}$. The set $\{1, \xi, \xi^*, \xi^*\xi\}$, and their linear combinations with complex coefficients, form **Grassmann algebra** (this mimics the behavior of Fock space algebra). Therefore, the regular Fock space is a buildup of the direct sum of all n-dimensional Hilbert spaces (1) and has to be extended in order to contain Grassmann numbers. In this case, the extended Fock space is then formed by building the linear combination of the regular Fock space states and the Grassmann coefficients. Hence, the basis of Grassmann algebra is all distinct products, and we show that Grassmann numbers can be multiplied together and anticommute under multiplication from the following definitions:

$$\xi|0\rangle = |0\rangle\xi \quad , \quad \xi|1\rangle = -|1\rangle\xi \quad , \quad \xi^*\langle 0| = \langle 0|\xi^* \quad , \quad \langle 1|\xi^* = -\xi^*\langle 1| \tag{241}$$

$$\xi|0\rangle\langle 1| + |0\rangle\langle 1|\xi = 0 \quad , \quad \langle 0|\xi|0\rangle = -\langle 1|\xi|1\rangle \tag{242}$$

$$\{\xi, \hat{\psi}\} = \xi|0\rangle\langle 1| + |0\rangle\langle 1|\xi = |0\rangle\xi\left(-\langle 1|\xi\right) + |0\rangle\langle 1|\xi = 0 \quad , \quad \xi\langle 1| = -\langle 1|\xi \tag{243}$$

The presence of the sign change when there is a swap is a direct consequence of the anticommutation relations between Grassmann variables and Fock space operators as seen earlier. So, the Grassmann numbers can be multiplied together and anticommute under multiplication:

$$\{\xi,\xi\}=0 \quad , \quad \xi^2=0 \quad , \quad (\xi^*)^2=0 \tag{244}$$

This guarantees any analytic function will be linear in Grassmann algebra.

Grassmann numbers can be multiplied by complex numbers, and the multiplication by a complex number is distributive:

$$\alpha(\xi_1+\xi_2)=\alpha\xi_1+\alpha\xi_2 \tag{245}$$

Considering the fact that Grassmann numbers occur only inside time-ordered products, then it suffices to define the adjoint in such a way that it also anticommutes. So, the quantities

$$\xi=\xi_1+i\xi_2 \quad , \quad \xi^*=\xi_1-i\xi_2 \tag{246}$$

can be treated as independent Grassmann variables:

$$\{\xi,\xi^*\}=0 \tag{247}$$

Considering that the square and higher powers of a Grassmann number vanish, then the Taylor expansion of a wave function with Grassmann variables has only two terms. For example, say,

$$\exp\{\xi\}=1+\xi$$

2.4 Fermions and Coherent States

Considering the aforementioned, we see that fermionic systems such as electrons in a metal or ultracold fermionic atoms in a magnetic trap can be described by Grassmann variables. So, to construct coherent states for fermions, we must enlarge the fermion Fock space. It is useful to note that an important property needed for setting up path integration is the completeness of the states. We now examine coherent states and define them as eigenstates of the operator $\hat{\psi}$:

$$|\xi\rangle=\exp\{-\xi\hat{\psi}^\dagger\}|0\rangle=(1-\xi\hat{\psi}^\dagger)|0\rangle=|0\rangle-\xi|1\rangle \quad , \quad \langle\xi|=\langle 0|\exp\{-\xi^*\hat{\psi}\}=\langle 0|(1-\xi^*\hat{\psi})=\langle 0|-\langle 1|\xi^* \tag{248}$$

We apply $\hat{\psi}$ or $\hat{\psi}^\dagger$ on these states, and then we have

$$\hat{\psi}|\xi\rangle=\psi|0\rangle-\hat{\psi}\xi|1\rangle=\hat{\psi}|0\rangle+\xi\hat{\psi}|1\rangle=\xi(|0\rangle-\xi|1\rangle)=\xi|\xi\rangle \tag{249}$$

$$\hat{\psi}|\xi\rangle=\xi|\xi\rangle \tag{250}$$

$$\langle\xi|\hat{\psi}^\dagger=\langle\xi|\xi^* \tag{251}$$

Relations (250) and (251) are fundamental properties of fermionic coherent states and are also true for bosons. It is important to note that in addition to the space-time variables, fermionic fields require a spin index $\alpha=\uparrow,\downarrow$. So, if

$$|\xi\rangle=\exp\left\{-\sum_\alpha \xi_\alpha\hat{\psi}_\alpha^\dagger\right\}|0\rangle=\prod_\alpha\left(1-\xi_\alpha\hat{\psi}_\alpha^\dagger\right)|0\rangle \tag{252}$$

then we again prove relations (250) and (251):

$$\hat{\psi}_\alpha\left(1-\xi_\alpha\hat{\psi}^\dagger_\alpha\right)|0\rangle = \hat{\psi}_\alpha|0\rangle - \hat{\psi}_\alpha\xi_\alpha|1\rangle = \xi_\alpha\hat{\psi}_\alpha|1\rangle = \xi_\alpha|0\rangle = (\xi_\alpha - 0)|0\rangle = \left(\xi_\alpha - \xi^2_\alpha\hat{\psi}^\dagger_\alpha\right)|0\rangle = \xi_\alpha\left(1-\xi_\alpha\hat{\psi}^\dagger_\alpha\right)|0\rangle \tag{253}$$

So,

$$\hat{\psi}_\alpha|\xi\rangle = \hat{\psi}_\alpha\prod_\beta\left(1-\xi_\beta\hat{\psi}^\dagger_\beta\right)|0\rangle = \hat{\psi}_\alpha\prod_{\beta\neq\alpha}\left(1-\xi_\beta\hat{\psi}^\dagger_\beta\right)\left(1-\xi_\alpha\hat{\psi}^\dagger_\alpha\right)|0\rangle = \prod_{\beta\neq\alpha}\left(\hat{\psi}_\alpha + \xi_\beta\hat{\psi}_\alpha\hat{\psi}^\dagger_\beta\right)\left(1-\xi_\alpha\hat{\psi}^\dagger_\alpha\right)|0\rangle \tag{254}$$

Swapping the positions of the operators $\hat{\psi}_\alpha$ and $\hat{\psi}^\dagger_\beta$ in the first term of the product in the last expression, there is a sign change, and we have

$$\begin{aligned} \hat{\psi}_\alpha|\xi\rangle &= \prod_{\beta\neq\alpha}\left(\hat{\psi}_\alpha - \xi_\beta\hat{\psi}^\dagger_\beta\hat{\psi}_\alpha\right)\left(1-\xi_\alpha\hat{\psi}^\dagger_\alpha\right)|0\rangle = \prod_{\beta\neq\alpha}\left(1-\xi_\beta\hat{\psi}^\dagger_\beta\right)\hat{\psi}_\alpha\left(1-\xi_\alpha\hat{\psi}^\dagger_\alpha\right)|0\rangle = \\ &= \prod_{\beta\neq\alpha}\left(1-\xi_\beta\hat{\psi}^\dagger_\beta\right)\xi_\alpha\left(1-\xi_\alpha\hat{\psi}^\dagger_\alpha\right)|0\rangle = \xi_\alpha\prod_{\beta\neq\alpha}\left(1-\xi_\beta\hat{\psi}^\dagger_\beta\right)\left(1-\xi_\alpha\hat{\psi}^\dagger_\alpha\right)|0\rangle = \xi_\alpha\prod_\alpha\left(1-\xi_\alpha\hat{\psi}^\dagger_\alpha\right)|0\rangle = \xi_\alpha|\xi\rangle \end{aligned} \tag{255}$$

Therefore $|\xi\rangle$ is an eigenstate, which implies that it is a coherent state. From the coherent state, its adjoint is a left eigenstate of the creation operator:

$$\begin{aligned} \langle\xi|\hat{\psi}^\dagger_\alpha &= \langle 0|\exp\left\{-\sum_\beta\hat{\psi}_\beta\xi^*_\beta\right\}\hat{\psi}^\dagger_\alpha = \langle 0|\prod_\beta\left(1-\hat{\psi}_\beta\xi^*_\beta\right)\hat{\psi}^\dagger_\alpha = \langle 0|\left(1-\hat{\psi}_\alpha\xi^*_\alpha\right)\hat{\psi}^\dagger_\alpha\prod_{\beta\neq\alpha}\left(1-\hat{\psi}_\beta\xi^*_\beta\right) = \\ &= \langle 0|\hat{\psi}_\alpha\hat{\psi}^\dagger_\alpha\xi^*_\alpha\prod_{\beta\neq\alpha}\left(1-\hat{\psi}_\beta\xi^*_\beta\right) = \langle 0|\prod_\beta\left(1-\hat{\psi}_\beta\xi^*_\beta\right)\xi^*_\alpha = \langle\xi|\xi^*_\alpha \end{aligned} \tag{256}$$

So

$$\langle\xi|\hat{\psi}^\dagger_\alpha = \langle\xi|\xi^*_\alpha \tag{257}$$

It is useful to note that these coherent states are not orthonormal.

Similarly, we consider the application of a creation operator on a coherent state as for bosons:

$$\hat{\psi}^\dagger_\alpha|\xi\rangle = \hat{\psi}^\dagger_\alpha\exp\left\{-\sum_\beta\xi_\beta\hat{\psi}^\dagger_\beta\right\}|0\rangle = \hat{\psi}^\dagger_\alpha\prod_\beta\left(1-\xi_\beta\hat{\psi}^\dagger_\beta\right)|0\rangle = \hat{\psi}^\dagger_\alpha\left(1-\xi_\alpha\hat{\psi}^\dagger_\alpha\right)\prod_{\beta\neq\alpha}\left(1-\xi_\beta\hat{\psi}^\dagger_\beta\right)|0\rangle \tag{258}$$

Consider that $\left(\hat{\psi}^\dagger_\alpha\right)^2 = 0$ in the last expression, then

$$\hat{\psi}^\dagger_\alpha|\xi\rangle = \left(\hat{\psi}^\dagger_\alpha + \xi_\alpha\left(\hat{\psi}^\dagger_\alpha\right)^2\right)\prod_{\beta\neq\alpha}\left(1-\xi_\beta\hat{\psi}^\dagger_\beta\right)|0\rangle = \hat{\psi}^\dagger_\alpha\prod_{\beta\neq\alpha}\left(1-\xi_\beta\hat{\psi}^\dagger_\beta\right)|0\rangle = -\frac{\partial}{\partial\xi_\alpha}\left(1-\xi_\alpha\hat{\psi}^\dagger_\alpha\right)\prod_{\beta\neq\alpha}\left(1-\xi_\beta\hat{\psi}^\dagger_\beta\right)|0\rangle \tag{259}$$

or

$$\hat{\psi}^\dagger_\alpha|\xi\rangle = -\frac{\partial}{\partial\xi_\alpha}\left(1-\xi_\alpha\hat{\psi}^\dagger_\alpha\right)\prod_{\beta\neq\alpha}\left(1-\xi_\beta\hat{\psi}^\dagger_\beta\right)|0\rangle = -\frac{\partial}{\partial\xi_\alpha}\prod_\beta\left(1-\xi_\beta\hat{\psi}^\dagger_\beta\right)|0\rangle = -\frac{\partial}{\partial\xi_\alpha}|\xi\rangle \tag{260}$$

So,

$$\hat{\psi}_{\alpha}^{\dagger}|\xi\rangle = -\frac{\partial}{\partial\xi_{\alpha}}|\xi\rangle \tag{261}$$

Similarly,

$$\langle\xi|\hat{\psi}_{\alpha} = \langle 0|\prod_{\beta\neq\alpha}\left(1-\hat{\psi}_{\beta}\xi_{\beta}^{*}\right)\left(1-\hat{\psi}_{\alpha}\xi_{\alpha}^{*}\right)\hat{\psi}_{\alpha} = \frac{\partial}{\partial\xi_{\alpha}}\langle 0|\prod_{\beta}\left(1-\hat{\psi}_{\beta}\xi_{\beta}^{*}\right) = \frac{\partial}{\partial\xi_{\alpha}}\langle\xi| \tag{262}$$

or

$$\langle\xi|\hat{\psi}_{\alpha} = \frac{\partial}{\partial\xi_{\alpha}}\langle\xi| \tag{263}$$

Also,

$$\langle\xi|\hat{\psi}_{\alpha}^{\dagger} = \langle 0|\exp\left\{-\sum_{\beta}\xi_{\beta}^{*}\hat{\psi}_{\beta}\right\}\hat{\psi}_{\alpha}^{\dagger} = \langle 0|\prod_{\beta}\left(1-\xi_{\beta}^{*}\hat{\psi}_{\beta}\right)\hat{\psi}_{\alpha}^{\dagger} = \langle 0|\left(1-\xi_{\alpha}^{*}\hat{\psi}_{\alpha}\right)\hat{\psi}_{\alpha}^{\dagger}\prod_{\beta\neq\alpha}\left(1-\xi_{\beta}^{*}\hat{\psi}_{\beta}\right) \tag{264}$$

From where

$$\langle\xi|\hat{\psi}_{\alpha}^{\dagger} = \langle 0|\hat{\psi}_{\alpha}\hat{\psi}_{\alpha}^{\dagger}\xi_{\alpha}^{*}\prod_{\beta\neq\alpha}\left(1-\xi_{\beta}^{*}\hat{\psi}_{\beta}\right) = \langle 0|\prod_{\beta}\left(1-\xi_{\beta}^{*}\hat{\psi}_{\beta}\right)\xi_{\alpha}^{*} = \langle\xi|\xi_{\alpha}^{*} \tag{265}$$

Here, we considered that

$$\langle 0|\left[\hat{\psi}_{\alpha},\hat{\psi}_{\alpha}^{\dagger}\right]_{+} = \langle 0| \tag{266}$$

The inner product of two coherent states:

$$\begin{aligned}\langle\xi|\xi'\rangle &= \langle 0|\exp\left\{\sum_{\alpha}\xi_{\alpha}^{*}\hat{\psi}_{\alpha}\right\}\exp\left\{-\sum_{\alpha}\xi_{\alpha}'\hat{\psi}_{\alpha}^{\dagger}\right\}|0\rangle = \langle 0|\prod_{\alpha}\left(1+\xi_{\alpha}^{*}\hat{\psi}_{\alpha}\right)\prod_{\alpha}\left(1-\xi_{\alpha}'\hat{\psi}_{\alpha}^{\dagger}\right)|0\rangle = \\ &= \langle 0|\prod_{\alpha}\left(1+\xi_{\alpha}^{*}\hat{\psi}_{\alpha}\right)\left(1-\xi_{\alpha}'\hat{\psi}_{\alpha}^{\dagger}\right)|0\rangle = \langle 0|\prod_{\alpha}\left(1-\xi_{\alpha}'\hat{\psi}_{\alpha}^{\dagger}+\xi_{\alpha}^{*}\hat{\psi}_{\alpha}-\xi_{\alpha}^{*}\hat{\psi}_{\alpha}\xi_{\alpha}'\hat{\psi}_{\alpha}^{\dagger}\right)|0\rangle = \\ &= \prod_{\alpha}\left(\langle 0|0\rangle-\xi_{\alpha}'\langle 0|\alpha\rangle+\langle 0|\xi_{\alpha}^{*}\hat{\psi}_{\alpha}|0\rangle+\xi_{\alpha}^{*}\xi_{\alpha}'\langle 0|\hat{\psi}_{\alpha}\hat{\psi}_{\alpha}^{\dagger}|0\rangle\right)\end{aligned} \tag{267}$$

From

$$\hat{\psi}_{\alpha}|0\rangle = 0 \quad , \quad \langle 0|\hat{\psi}_{\alpha}\hat{\psi}_{\alpha}^{\dagger}|0\rangle = 1 \tag{268}$$

then

$$\langle\xi|\xi'\rangle = \prod_{\alpha}\left(1+\xi_{\alpha}^{*}\xi_{\alpha}'\right) = \exp\left\{\sum_{\alpha}\xi_{\alpha}^{*}\xi_{\alpha}'\right\} \tag{269}$$

Using the overcompleteness relation, we represent a state of the extended Fock space in terms of coherent states as follows:

$$|\psi\rangle = \int \prod_{\alpha} d\xi_{\alpha}^{*} d\xi_{\alpha} \exp\left\{-\sum_{\alpha} \xi_{\alpha}^{*}\xi_{\alpha}\right\} |\xi\rangle\langle\xi|\psi\rangle = \int \prod_{\alpha} d\xi_{\alpha}^{*} d\xi_{\alpha} \exp\left\{-\sum_{\alpha} \xi_{\alpha}^{*}\xi_{\alpha}\right\} \psi(\xi^{*})|\xi\rangle \tag{270}$$

where

$$\langle\xi|\psi\rangle = \psi(\xi^{*}) \tag{271}$$

2.4.1 Coherent State Overcompleteness Relation Proof

We show also that the overcompleteness relation for a coherent state basis is reflected in the closure relation:

$$\begin{aligned}\int \prod_{\alpha} d\xi_{\alpha}^{*} d\xi_{\alpha} \exp\left\{-\sum_{\alpha} \xi_{\alpha}^{*}\xi_{\alpha}\right\} |\xi\rangle\langle\xi| &= \prod_{\alpha} \int d\xi_{\alpha}^{*}\, d\xi_{\alpha} \left(1-\xi_{\alpha}^{*}\xi_{\alpha}\right)\left(1-\xi_{\alpha}\hat{\psi}_{\alpha}^{\dagger}\right)|0\rangle\langle 0|\left(1-\hat{\psi}_{\alpha}\xi_{\alpha}^{*}\right) \\ &= \prod_{\alpha} \left(|0\rangle\langle 0| + |1\rangle\langle 1|\right) = \hat{1}\end{aligned} \tag{272}$$

Therefore,

$$\int \prod_{\alpha} d\xi_{\alpha}^{*} d\xi_{\alpha} \exp\left\{-\sum_{\alpha} \xi_{\alpha}^{*}\xi_{\alpha}\right\} |\xi\rangle\langle\xi| = \hat{1} \tag{273}$$

and this also holds for bosons when

$$|\xi\rangle = \exp\left\{\chi \sum_{\alpha} \xi_{\alpha}\hat{\psi}_{\alpha}^{\dagger}\right\}|0\rangle \tag{274}$$

In (273), $\hat{1}$ is the Fock space identity operator. As shown in (273), because the coherent states generally are not orthogonal, the set of coherent states is **overcomplete**. We can then rewrite the **overcompleteness relation** in (273) as follows

$$\int d[\xi_{\alpha}^{*}]\, d[\xi_{\alpha}] \exp\left\{-\sum_{\alpha} \xi_{\alpha}^{*}\xi_{\alpha}\right\} |\xi\rangle\langle\xi| = \hat{1} \;,\quad N = \begin{cases} 2\pi i &, \text{ Bosons} \\ 1 &, \text{ Fermions} \end{cases} \tag{275}$$

so as to include the bosonic case. The integration measure in this case:

$$\int d[\xi_{\alpha}^{*}]\, d[\xi_{\alpha}] \equiv \int \prod_{\alpha} \frac{d\xi_{\alpha}^{*} d\xi_{\alpha}}{N} \tag{276}$$

Relation (273) can be proven exactly as in the case of bosons via Schur's lemma.

We examine another proof of relation (273) by considering the matrix element of its left-hand side between two arbitrary states of the Fock space by defining the operator:

$$\mathbf{A} = \int d\xi_\alpha^* \, d\xi_\alpha \exp\left\{-\sum_\alpha \xi_\alpha^* \xi_\alpha\right\} |\xi\rangle\langle\xi| \tag{277}$$

For sufficiency, we prove that for any vectors of the basis of Fock space,

$$\langle \alpha_1 \cdots \alpha_n | \mathbf{A} | \beta_1 \cdots \beta_m \rangle = \langle \alpha_1 \cdots \alpha_n | \beta_1 \cdots \beta_m \rangle \tag{278}$$

Considering equation (273), we have

$$\langle \alpha_1 \cdots \alpha_n | \xi \rangle = \langle 0 | \hat{\psi}_{\alpha_n} \cdots \hat{\psi}_{\alpha_1} | \xi \rangle = \xi_{\alpha_n} \cdots \xi_{\alpha_1} \quad , \quad \langle \xi | \beta_1 \cdots \beta_m \rangle = \langle \xi | \hat{\psi}^\dagger_{\beta_1} \cdots \hat{\psi}^\dagger_{\beta_m} | 0 \rangle = \xi^*_{\beta_1} \cdots \xi^*_{\beta_m} \tag{279}$$

But $\langle 0 | \xi \rangle = 1$. Therefore, we deduce

$$\langle \alpha_1 \cdots \alpha_n | \mathrm{A} | \beta_1 \cdots \beta_m \rangle = \int \prod_\alpha d\xi_\alpha^* d\xi_\alpha \exp\left\{-\sum_\alpha \xi_\alpha^* \xi_\alpha\right\} \langle \alpha_1 \cdots \alpha_n | \xi \rangle \langle \xi | \beta_1 \cdots \beta_m \rangle \tag{280}$$

or

$$\langle \alpha_1 \cdots \alpha_n | \mathrm{A} | \beta_1 \cdots \beta_m \rangle = \int \prod_\alpha d\xi_\alpha^* d\xi_\alpha \prod_\alpha \left(1 - \xi_\alpha^* \xi_\alpha\right) \xi_{\alpha_n} \cdots \xi_{\alpha_1} \xi^*_{\beta_1} \cdots \xi^*_{\beta_m} \tag{281}$$

Four types of integrals appear in (281):

$$\int d\xi_\alpha^* \, d\xi_\alpha \left(1 - \xi_\alpha^* \xi_\alpha\right) \begin{bmatrix} \xi_\alpha \xi_\alpha^* \\ \xi_\alpha \\ \xi_\alpha^* \\ 1 \end{bmatrix} = \begin{bmatrix} 1 \\ 0 \\ 0 \\ 1 \end{bmatrix} \tag{282}$$

So, the integral in (281) is nonvanishing only if each state α is either occupied in both $\langle \alpha_1 \cdots \alpha_n|$ and $|\beta_1 \cdots \beta_m\rangle$ or unoccupied in both states. This requires that $n = m$ and $\{\alpha_i\}$ is some permutation P of $\{\beta_i\}$ so that $\beta_i = \beta_{\mathrm{P}(i)}$. This permits us to easily evaluate the integral:

$$\int \prod_\alpha d\xi_\alpha^* d\xi_\alpha \prod_\alpha \left(1 - \xi_\alpha^* \xi_\alpha\right) \xi_{\alpha_n} \cdots \xi_{\alpha_1} \xi^*_{\alpha_{\mathrm{P}(1)}} \cdots \xi^*_{\alpha_{\mathrm{P}(n)}} = \int \prod_\alpha d\xi_\alpha^* d\xi_\alpha \prod_\alpha \left(1 - \xi_\alpha^* \xi_\alpha\right) (-1)^{\mathrm{P}} \xi_{\alpha_n} \cdots \xi_{\alpha_1} \xi^*_{\alpha_1} \cdots \xi^*_{\alpha_m} \tag{283}$$

Note that an even number of anticommutations is required to bring the integral over each state into the form of equation (282). So the value of the integral is as follows:

$$\int \prod_\alpha d\xi_\alpha^* d\xi_\alpha \prod_\alpha \left(1 - \xi_\alpha^* \xi_\alpha\right) (-1)^{\mathrm{P}} \xi_{\alpha_n} \cdots \xi_{\alpha_1} \xi^*_{\alpha_1} \cdots \xi^*_{\alpha_m} = (-1)^{\mathrm{P}} \tag{284}$$

We compare this with the following relation

$$\langle \alpha_1 \cdots \alpha_n | \beta_1 \cdots \beta_m \rangle = \delta_{nm} \sum_{\mathrm{P}} (-1)^{\mathrm{P}} \delta_{\alpha_i \beta_{\mathrm{P}(i)}} \tag{285}$$

that proves the closure relation (273).

From the aforementioned, we may note that fermion coherent states are not physical states of the system and do not belong to the Fock space. We consider, for example, the expectation value of the particle number operator:

$$\hat{n} = \sum_{\alpha} \xi_{\alpha}^{\dagger} \xi_{\alpha} \tag{286}$$

in the coherent state $|\xi\rangle$ is not a real number:

$$\frac{\langle \xi | \hat{n} | \xi \rangle}{\langle \xi | \xi \rangle} = \sum_{\alpha} \xi_{\alpha}^{*} \xi_{\alpha} \tag{287}$$

To conclude, we consider the contrast in the physical significance of bosonic and fermionic coherent states. The bosonic coherent states are physical states that naturally arise when the classical limit of quantum mechanics or quantum field theory is reached. Considering equations (8) through (12), for the classical limit, if the field operators are assumed to commute, then the classical field $\phi(\vec{r})$ at each point of space imitates the system's coherent state

$$\begin{aligned} |\xi\rangle &= \exp\left\{\chi \sum_{\alpha} \xi_{\alpha} \hat{\psi}_{\alpha}^{\dagger}\right\}|0\rangle = \exp\left\{\chi \sum_{\alpha} \xi_{\alpha}\left(\int d\vec{r}\, \xi_{\alpha}(\vec{r})\hat{\psi}^{\dagger}(\vec{r})\right)\right\}|0\rangle = \\ &= \exp\left\{\chi \int d\vec{r} \left(\sum_{\alpha} \xi_{\alpha} \xi_{\alpha}(\vec{r})\right)\hat{\psi}^{\dagger}(\vec{r})\right\}|0\rangle = \exp\left\{\chi \int d\vec{r}\, \phi(\vec{r})\hat{\psi}^{\dagger}(\vec{r})\right\}|0\rangle \end{aligned} \tag{288}$$

This example is a classical electromagnetic field viewed as a coherent state of photons. In contrast, fermionic coherent states do not belong to fermionic Fock space and are not physical observables as well as classical fermionic fields. Nonetheless, fermionic coherent states are very useful unifying factors in many fermionic and bosonic problems.

2.4.2 Trace of a Physical Quantity

The completeness relation provides a suitable expression for the trace of an operator as seen earlier for the case of bosons. Because the matrix elements $\langle \psi_k | \xi \rangle$ and $\langle \xi | \psi_k \rangle$ between states $|\psi_k\rangle$ in the Fock space and coherent states contain Grassmann numbers, then it follows from anticommutation relations that

$$\langle \psi_k | \xi \rangle \langle \xi | \psi_l \rangle = -\langle \xi | \psi_l \rangle \langle \psi_k | \xi \rangle \tag{289}$$

Defining a complete set of states $\{|n\rangle\}$ in Fock space, note that the inner products yield

$$\begin{aligned} \langle n|\xi\rangle\langle \xi|m\rangle &= \left(\xi_{\alpha_P} \cdots \xi_{\alpha_1}\right)\left(\xi_{\alpha_1}^{*} \cdots \xi_{\alpha_P}^{*}\right) = \chi^{P^2}\left(\xi_{\alpha_1}^{*} \cdots \xi_{\alpha_P}^{*}\right)\left(\xi_{\alpha_P} \cdots \xi_{\alpha_1}\right) \\ &= \chi^{P}\left(\xi_{\alpha_1}^{*} \cdots \xi_{\alpha_P}^{*}\right)\left(\xi_{\alpha_P} \cdots \xi_{\alpha_1}\right) = \langle \chi\xi|m\rangle\langle n|\xi\rangle \end{aligned} \tag{290}$$

We note that $\chi^{P^2} = \chi^P$. So, for a complete set of states in the Fock space, and making use of the sign factor χ, the trace of an operator can be written as follows:

$$\operatorname{Tr} A = \sum_n \langle n|\mathbf{A}|n\rangle = \int d[\xi_\alpha^*]d[\xi_\alpha]\exp\left\{-\sum_\alpha \xi_\alpha^*\xi_\alpha\right\}\sum_n \langle n|\xi\rangle\langle\xi|\mathbf{A}|n\rangle =$$
$$= \int d[\xi_\alpha^*]d[\xi_\alpha]\exp\left\{-\sum_\alpha \xi_\alpha^*\xi_\alpha\right\}\sum_n \langle\chi\xi|\mathbf{A}|n\rangle\langle n|\xi\rangle = \int d[\xi_\alpha^*]d[\xi_\alpha]\exp\left\{-\sum_\alpha \xi_\alpha^*\xi_\alpha\right\}\langle\chi\xi|\mathbf{A}|\xi\rangle \tag{291}$$

or

$$\operatorname{Tr}\mathbf{A} = \int d[\xi_\alpha^*]d[\xi_\alpha]\exp\left\{-\sum_\alpha \xi_\alpha^*\xi_\alpha\right\}\langle\chi\xi|\mathbf{A}|\xi\rangle \tag{292}$$

Here, the integration measure $d[\xi_\alpha^*]d[\xi_\alpha]$ defined in (276) considers fermions as well as bosons. It is instructive to note that contrasts to bosonic coherent states are quantum mechanical states. Where a quantum mechanical system is near the classical limit, the fermionic coherent states in the Fock space are not, but they are in the extended Fock space. In addition, there is no classical fermionic field that is observable.

The overcompleteness relation can be used to represent a state of the extended Fock space in terms of coherent states:

$$|\hat{\psi}\rangle = \int d[\xi_\alpha^*]d[\xi_\alpha]\exp\left\{-\sum_\alpha \xi_\alpha^*\xi_\alpha\right\}|\xi\rangle\langle\xi|\hat{\psi}\rangle = \int d[\xi_\alpha^*]d[\xi_\alpha]\exp\left\{-\sum_\alpha \xi_\alpha^*\xi_\alpha\right\}\hat{\psi}(\xi^*)|\xi\rangle \;,\; \langle\xi|\hat{\psi}\rangle = \hat{\psi}(\xi^*) \tag{293}$$

As seen in the case of bosons, the matrix element of a **normal-ordered operator** $A[\hat{\psi}^\dagger,\hat{\psi}]$ between two coherent states can be evaluated as

$$\langle\xi|\mathbf{A}[\hat{\psi}^\dagger,\hat{\psi}]|\xi'\rangle = \langle\xi|\xi'\rangle\mathbf{A}[\xi^*,\xi'] = \exp\{\xi^*\xi'\}\mathbf{A}[\xi^*,\xi'] \tag{294}$$

This implies that the matrix element of normal-ordered Hamiltonians is given by the polynomial representing the operator. Also, all creation operators are put on the left of all annihilation operators without using the commutation relations (for fermions, we must keep track of minus signs). It is instructive to note that the coherent state representation of the system is a quantum version of the phase space representation.

2.4.3 Functional Integral Time-Ordered Property

We show the time-ordered property of the functional integral by examining the path integral time-ordered property of n operators evaluated at different times:

$$t_1, t_2, \cdots, t_n \tag{295}$$

We show the path integral automatically orders the n operators in the right manner as expected from the time-order operator $\hat{T}$ defined as follows:

$$\hat{T}\xi_{\alpha|_1^n}\left(t|_1^n\right) = \chi^P \xi_{\alpha|_{P(1)1}^{P(n)}}\left(t|_{P(1)}^{P(n)}\right) \tag{296}$$

where

$$\xi_{\alpha|_1^n}\left(t|_1^n\right)=\xi_{\alpha_1}(t_1)\cdots\xi_{\alpha_n}(t_n) \quad , \quad \xi_{\alpha|_{P(1)1}^{P(n)}}\left(t|_{P(1)}^{P(n)}\right)=\xi_{\alpha P(1)}\left(t_{P(1)}\right)\cdots\xi_{\alpha P(n)}\left(t_{P(n)}\right) \tag{297}$$

From our standard convention, χ is -1 or $+1$ for fermions or bosons, respectively, and P is the permutation of $\{1,2,\cdots,n\}$ that orders the times chronologically with the latest time to the left:

$$t_{\alpha_{P(1)}} > t_{\alpha_{P(2)}} > \cdots > t_{\alpha_{P(n)}} \tag{298}$$

This orders creation operators to the left of annihilation operators (normal order) at equal times. We tailor the time-ordered operator for fermions and bosons to achieve anticommutation for the fermionic case due to time ordering. Besides, the time-ordering process has to guarantee simultaneous normal ordering of operators on the condition that we simultaneously order the creation and annihilation operators.

We first evaluate the matrix element of the evolution operator $\hat{U}$ between an initial coherent state $|\xi_i\rangle$ with components ξ_{α_i} and a final state $\langle\xi_f|$ with components $\xi^*_{\alpha_f}$. We time-slice the interval $[t_i,t_f]$ into n equal parts $\varepsilon=\frac{t_f-t_i}{n}$. In the matrix element, we insert in the k time slice of the closure relation

$$\int\prod_{\alpha}\frac{d\xi^*_{\alpha_k}d\xi_{\alpha_k}}{N}\exp\left\{-\sum_{\alpha}\xi^*_{\alpha_k}\xi_{\alpha_k}\right\}|\xi_{\alpha_k}\rangle\langle\xi_{\alpha_k}|=1 \tag{299}$$

So, the matrix element of the evolution operator $\hat{U}$:

$$U\left(\xi^*_{\alpha_f}t_f;\xi_{\alpha_i}t_i\right)\equiv U\left(t_f,t_i\right)=\left\langle\xi_{\alpha_f}\right|\exp\left\{-i\hat{H}\left(t_f-t_i\right)\right\}\left|\xi_{\alpha_i}\right\rangle\equiv\left\langle\xi_{\alpha_f}\right|\hat{U}\left(t_f,t_i\right)\left|\xi_{\alpha_i}\right\rangle \tag{300}$$

or

$$U\left(t_f,t_i\right)=\int\prod_{k=1}^{n-1}\prod_{\alpha}\frac{d\xi^*_{\alpha_k}d\xi_{\alpha_k}}{N}\exp\left\{-\sum_{k=1}^{n-1}\sum_{\alpha}\xi^*_{\alpha_k}\xi_{\alpha_k}\right\}\prod_{k=1}^{n}\left\langle\xi_{\alpha_k}\right|\exp\left\{-i\varepsilon\hat{H}\left[\hat{\psi}^\dagger_{\alpha_k},\hat{\psi}_{\alpha_k}\right]\right\}\left|\xi_{\alpha_{k-1}}\right\rangle \tag{301}$$

Because $\hat{H}\left[\hat{\psi}^\dagger_{\alpha_k},\hat{\psi}_{\alpha_k}\right]$ is a normal-ordered operator, then

$$\left\langle\xi_{\alpha_k}\right|\hat{H}\left[\hat{\psi}^\dagger_{\alpha_k},\hat{\psi}_{\alpha_k}\right]\left|\xi_{\alpha_{k-1}}\right\rangle=\left\langle\xi_{\alpha_k}|\xi_{\alpha_{k-1}}\right\rangle\hat{H}\left[\xi_{\alpha_k},\xi_{\alpha_{k-1}}\right]=\exp\left\{\xi^*_{\alpha_k}\xi_{\alpha_{k-1}}\right\}\hat{H}\left[\xi_{\alpha_k},\xi_{\alpha_{k-1}}\right] \tag{302}$$

Considering

$$S_\alpha\left[\xi^*_{\alpha_k},\xi_{\alpha_{k-1}}\right]=\xi^*_{\alpha_k}\xi_{\alpha_{k-1}}-i\varepsilon\hat{H}\left[\xi^*_{\alpha_k},\xi_{\alpha_{k-1}}\right] \tag{303}$$

then,

$$U\left(t_f,t_i\right)=\int\prod_{k=1}^{n-1}\left[\prod_{\alpha}\frac{d\xi^*_{\alpha_k}d\xi_{\alpha_k}}{N}\exp\left\{-\xi^*_{\alpha_k}\xi_{\alpha_k}\right\}\right]\exp\left\{\sum_{k=1}^{n-1}\sum_{\alpha}S_\alpha\left[\xi^*_{\alpha_k},\xi_{\alpha_{k-1}}\right]\right\} \tag{304}$$

For the integration measure

$$\int d\left[\xi^*_{\alpha_k}\right]d\left[\xi_{\alpha_k}\right]=\int\prod_{\alpha}\frac{d\xi^*_{\alpha_k}d\xi_{\alpha_k}}{N} \tag{305}$$

then,

$$U\left(t_f,t_i\right)=\prod_{k=1}^{n-1}\int d\big[\xi_{\alpha_k}^*\big]d\big[\xi_{\alpha_k}\big]\exp\big\{-\xi_{\alpha_k}^*\xi_{\alpha_k}\big\}\exp\left\{\sum_{k=1}^{n-1}\sum_{\alpha}S_{\alpha}\big[\xi_{\alpha_k}^*,\xi_{\alpha_{k-1}}\big]\right\} \tag{306}$$

or

$$U\left(t_f,t_i\right)=\prod_{k=1}^{n-1}\int d\big[\xi_{\alpha_k}^*\big]d\big[\xi_{\alpha_k}\big]\exp\big\{-\xi_{\alpha_k}^*\xi_{\alpha_k}\big\}\exp\left\{\sum_{\alpha}S_{\alpha}\big[\xi_{\alpha_k}^*,\xi_{\alpha_{k-1}}\big]\right\} \tag{307}$$

For definiteness, consider $\hat{\tilde{\psi}}$ also as being a creation (annihilation) operator; then from (296):

$$\prod_{k=1}^{n-1}\int d\big[\xi_{\alpha_k}^*\big]d\big[\xi_{\alpha_k}\big]\exp\big\{-\xi_{\alpha_k}^*\xi_{\alpha_k}\big\}\xi_{\alpha|_1^n}\big(t|_1^n\big)\xi_{\alpha|_{n+1}^{2n}}^*\big(t|_{n+1}^{2n}\big)\prod_{k=1}^{n}\exp\left\{\sum_{\alpha}S_{\alpha}\big[\xi_{\alpha_k}^*,\xi_{\alpha_{k-1}}\big]\right\}=$$

$$=\chi^{P(2n)}\prod_{k=1}^{n-1}\int d\big[\xi_{\alpha_k}^*\big]d\big[\xi_{\alpha_k}\big]\exp\big\{-\xi_{\alpha_k}^*\xi_{\alpha_k}\big\}\tilde{\xi}_{\alpha|_{P(1)}^{P(2n)}}\big(t|_{P(1)}^{P(2n)}\big)\prod_{k=1}^{n}\exp\left\{\sum_{\alpha}S_{\alpha}\big[\xi_{\alpha_k}^*,\xi_{\alpha_{k-1}}\big]\right\}\equiv\langle\xi\rangle \tag{308}$$

where

$$\xi_{\alpha|_{n+1}^{2n}}^*\big(t|_{n+1}^{2n}\big)=\xi_{\alpha_{n+1}}^*\big(t_{n+1}\big)\cdots\xi_{\alpha_{2n}}^*\big(t_{2n}\big)\quad,\quad\tilde{\xi}_{\alpha|_{P(1)}^{P(2n)}}\big(t|_{P(1)}^{P(2n)}\big)=\tilde{\xi}_{\alpha_{P(1)}}\big(t_{P(1)}\big)\cdots\tilde{\xi}_{\alpha_{P(2n)}}\big(t_{P(2n)}\big) \tag{309}$$

Depending on whether we place a creation (or annihilation) operator, we find the corresponding time-slice element:

$$\langle\xi\rangle=\chi^{P(2n)}\prod_{k=1}^{n-1}\int d\big[\xi_{\alpha_k}^*\big]d\big[\xi_{\alpha_k}\big]\exp\big\{-\xi_{\alpha_k}^*\xi_{\alpha_k}\big\}\tilde{\xi}_{\alpha|_{P(1)}^{P(2n)}}\big(\tilde{t}|_m^l\big)\prod_{k=1}^{n}\exp\left\{\sum_{\alpha}S_{\alpha}\big[\xi_{\alpha_k}^*,\xi_{\alpha_{k-1}}\big]\right\} \tag{310}$$

We split the time-slice elements considering the place of location of the evaluated fields and relocate them to the left or right of the time evolution:

$$\langle\xi\rangle=\chi^{P(2n)}\prod_{k=1}^{n-1}\int d\big[\xi_{\alpha_k}^*\big]d\big[\xi_{\alpha_k}\big]\exp\big\{-\xi_{\alpha_k}^*\xi_{\alpha_k}\big\}\prod_{k=m+1}^{n}\exp\left\{\sum_{\alpha}S_{\alpha}\big[\xi_{\alpha_k}^*,\xi_{\alpha_{k-1}}\big]\right\}\exp\left\{\sum_{\alpha}S_{\alpha}\big[\xi_{\alpha_m}^*,\xi_{\alpha_{m-1}}\big]\right\}\times$$

$$\times\tilde{\xi}_{\alpha|_{P(1)}^{P(2n)}}\big(\tilde{t}|_m^l\big)\exp\left\{\sum_{\alpha}\big(\xi_{\alpha_l}^*\xi_{\alpha_{l-1}}-i\varepsilon\hat{H}\big[\xi_{\alpha_n}^*,\xi_{\alpha_{n-1}}\big]\big)\right\}\prod_{k=1}^{l-1}\exp\left\{\sum_{\alpha}S_{\alpha}\big[\xi_{\alpha_k}^*,\xi_{\alpha_{k-1}}\big]\right\} \tag{311}$$

We temporarily apply the operators $\hat{U},\hat{U}^{\dagger}$, considering the positions of either $\hat{\psi}$ or $\hat{\psi}^{\dagger}$ and then

$$\langle\xi\rangle=\chi^{P(2n)}\prod_{k=1}^{n-1}\int d\big[\xi_{\alpha_k}^*\big]d\big[\xi_{\alpha_k}\big]\exp\big\{-\xi_{\alpha_k}^*\xi_{\alpha_k}\big\}\prod_{k=m+1}^{n}\langle\xi_{\alpha_k}|\hat{U}\big(t_k,t_{k-1}\big)|\xi_{\alpha_{k-1}}\rangle\langle\xi_{\alpha_m}|\hat{U}\big(t_m,t_{m-1}\big)\hat{\psi}_{\alpha_{P(1),m-1}}|\xi_{\alpha_{m-1}}\rangle\times$$

$$\times\cdots\times\langle\xi_{\alpha_l}|\hat{\psi}_{\alpha_{P(2n)}}^{\dagger}\hat{U}\big(t_l,t_{l-1}\big)|\xi_{\alpha_{l-1}}\rangle\prod_{k=1}^{l-1}\langle\xi_{\alpha_k}|\hat{U}\big(t_k,t_{k-1}\big)|\xi_{\alpha_{k-1}}\rangle \tag{312}$$

or

$$\langle\xi\rangle = \chi^{P(2n)}\langle\xi_{\alpha_f}|\hat{U}(t_n,t_{n-1})\cdots\hat{U}(t_{m+1},t_m)\hat{U}(t_m,t_{m-1})\hat{\psi}_{\alpha_{P(1)},m-1}\cdots\hat{\psi}^{\dagger}_{\alpha_{P(2n)}}\hat{U}^{\dagger}(t_l,t_{l-1})\cdots\hat{U}(t_2,t_1)\hat{U}(t_1,t_0)|\xi_{\alpha_i}\rangle \tag{313}$$

We examine two operators acting at the same time τ_k. To be consistent with the time-ordering operator definition, we bring the operators to the normal order at equal times. This permits two coherent states to be brought to one time evolution operator, and so the matrix element

$$\langle\xi_{\alpha_k}|\hat{\psi}^{\dagger}_{\alpha_{P(2)},k-1}\hat{U}(\tau_k,\tau_{k-1})\hat{\psi}_{\alpha_{P(1)},k-1}|\xi_{\alpha_{k-1}}\rangle \tag{314}$$

with the factor $\chi^{P(2n)}$ is in agreement with the time-ordering operator definition. This implies that when creation and annihilation operators act simultaneously, the creation operator is evaluated one time step later than the corresponding annihilation operator. Consequently, the identity

$$\langle\xi_{\alpha_f}(t_f)|\hat{T}\hat{\psi}_{\alpha_1}(t_1)\cdots\hat{\psi}_{\alpha_n}(t_n)\hat{\psi}^{\dagger}_{\alpha_{n+1}}(t_{n+1})\cdots\hat{\psi}^{\dagger}_{\alpha_{2n}}(t_{2n})|\xi_{\alpha_i}(t_i)\rangle \equiv \langle\xi\rangle \tag{315}$$

2.5 Gaussian Integrals

2.5.1 Multidimensional Gaussian Integral

We introduce integrals frequently encountered when evaluating matrix elements of operators in coherent states. These integrals will tend toward exponential functions that are polynomials in complex variables or Grassmann variables. For quadratic forms, these are generalizations of the familiar Gaussian integrals. Hence, for future reference, we present several useful integrals here. For convenience, we assume a real $n\times n$ matrix $\mathbf{A}$ symmetric and positive definite. Thus, there exists an orthogonal transformation $\mathbf{M}$ with $\mathbf{M}^{T}\mathbf{M} = \mathbf{M}\mathbf{M}^{T} = \hat{\mathbf{1}}$:

$$\mathbf{M}^{T}\mathbf{A}\mathbf{M} = \mathrm{diag}(\Delta_1,\Delta_2,\cdots,\Delta_n) = \mathbf{B} \tag{316}$$

Here, Δ_k are the eigenvalues of the matrix $\mathbf{A}$. So,

$$\int d\xi' \exp\{-\xi'^{T}\mathbf{A}\xi'\} = \int d\xi' \exp\{-\xi'^{T}(\mathbf{M}\mathbf{M}^{T})\mathbf{A}(\mathbf{M}\mathbf{M}^{T})\xi'\} = \int d\xi' \exp\{-(\xi'^{T}\mathbf{M})\mathbf{M}^{T}\mathbf{A}\mathbf{M}(\mathbf{M}^{T}\xi')\} =$$

$$= \int d\xi \exp\{-\xi^{T}\mathbf{B}\xi\} = \prod_{k=1}^{n}\left[\int d\xi_k \exp\{-\Delta_k\xi_k^2\}\right] = \prod_{k=1}^{n}\left[\sqrt{\frac{\pi}{\Delta_k}}\right] = \pi^{\frac{n}{2}}\prod_{k=1}^{n}\frac{1}{\sqrt{\Delta_k}} = \pi^{\frac{n}{2}}\frac{1}{\sqrt{\prod_{k=1}^{n}\Delta_k}} = \pi^{\frac{n}{2}}\frac{1}{\sqrt{\det(\mathbf{A})}} \tag{317}$$

Hence, for the multidimensional Gaussian integral, we shift the origin of integration via:

$$\xi = \xi' + \mathbf{A}^{-1}\eta \tag{318}$$

and imply

$$\xi' = \xi - \mathbf{A}^{-1}\eta \tag{319}$$

with

$$\xi'^{T} = \xi^{T} - \eta^{T}\left(\mathbf{A}^{-1}\right)^{T} \tag{320}$$

So,

$$-\xi'^{\mathrm{T}}\mathbf{A}\xi'-2\eta^{\mathrm{T}}\xi'=-\left[\xi^{\mathrm{T}}-\eta^{\mathrm{T}}\left(\mathbf{A}^{-1}\right)^{\mathrm{T}}\right]\mathbf{A}\left[\xi-\mathbf{A}^{-1}\eta\right]-2\left[\eta^{\mathrm{T}}\xi-\eta^{\mathrm{T}}\mathbf{A}^{-1}\eta\right]=-\xi^{\mathrm{T}}\mathbf{A}\xi+\xi^{\mathrm{T}}\mathbf{A}\mathbf{A}^{-1}\eta+\eta^{\mathrm{T}}\left(\mathbf{A}^{-1}\right)^{\mathrm{T}}\mathbf{A}\xi-$$
$$-\eta^{\mathrm{T}}\left(\mathbf{A}^{-1}\right)^{\mathrm{T}}\mathbf{A}\mathbf{A}^{-1}\eta-2\eta^{\mathrm{T}}\xi+2\eta^{\mathrm{T}}\mathbf{A}^{-1}\eta=-\xi^{\mathrm{T}}\mathbf{A}\xi+\xi^{\mathrm{T}}\eta+\eta^{\mathrm{T}}\left(\mathbf{A}^{-1}\right)^{\mathrm{T}}\mathbf{A}\xi-\eta^{\mathrm{T}}\left(\mathbf{A}^{-1}\right)^{\mathrm{T}}\eta-2\eta^{\mathrm{T}}\xi+2\eta^{\mathrm{T}}\mathbf{A}^{-1}\eta \tag{321}$$

Because the inverse of a symmetric matrix is a symmetric matrix, then

$$\left(\mathbf{A}^{-1}\right)^{\mathrm{T}}=\mathbf{A}^{-1} \tag{322}$$

and

$$-\xi'^{\mathrm{T}}\mathbf{A}\xi'-2\eta^{\mathrm{T}}\xi'=-\left[\xi^{\mathrm{T}}-\eta^{\mathrm{T}}\mathbf{A}^{-1}\right]\mathbf{A}\left[\xi-\mathbf{A}^{-1}\eta\right]-2\left[\eta^{\mathrm{T}}\xi-\eta^{\mathrm{T}}\mathbf{A}^{-1}\eta\right]=-\xi^{\mathrm{T}}\mathbf{A}\xi+\xi^{\mathrm{T}}\mathbf{A}\mathbf{A}^{-1}\eta+\eta^{\mathrm{T}}\mathbf{A}^{-1}\mathbf{A}\xi-$$
$$-\eta^{\mathrm{T}}\mathbf{A}^{-1}\mathbf{A}\mathbf{A}^{-1}\eta-2\eta^{\mathrm{T}}\xi+2\eta^{\mathrm{T}}\mathbf{A}^{-1}\eta=-\xi^{\mathrm{T}}\mathbf{A}\xi+\xi^{\mathrm{T}}\eta+\eta^{\mathrm{T}}\mathbf{A}^{-1}\mathbf{A}\xi-\eta^{\mathrm{T}}\mathbf{A}^{-1}\eta-2\eta^{\mathrm{T}}\xi+2\eta^{\mathrm{T}}\mathbf{A}^{-1}\eta \tag{323}$$

This permits us to solve the following Gaussian integral:

$$\int d\xi'\exp\left\{-\xi'^{\mathrm{T}}\mathbf{A}\xi'-2\eta^{\mathrm{T}}\xi'\right\}=\exp\left\{\eta^{\mathrm{T}}\mathbf{A}^{-1}\eta\right\}\int d\xi\exp\left\{-\xi^{\mathrm{T}}\mathbf{A}\xi\right\}=\pi^{\frac{n}{2}}\frac{1}{\sqrt{\det(\mathbf{A})}}\exp\left\{\eta^{\mathrm{T}}\mathbf{A}^{-1}\eta\right\} \tag{324}$$

So,

$$\int d\xi'\exp\left\{-\xi'^{\mathrm{T}}\mathbf{A}\xi'\pm2\eta^{\mathrm{T}}\xi'\right\}=\pi^{\frac{n}{2}}\frac{1}{\sqrt{\det(\mathbf{A})}}\exp\left\{\eta^{\mathrm{T}}\mathbf{A}^{-1}\eta\right\} \tag{325}$$

2.5.2 Multidimensional Complex Gaussian Integral

We evaluate multidimensional complex Gaussian integrals and assume the complex matrix $\mathbf{A}$ to be a positive definite Hermitian matrix. This implies that $\mathbf{A}=\mathbf{A}^{\dagger}$, and there exists a unitary transformation $\mathbf{S}$ that can diagonalize $\mathbf{A}$:

$$\mathbf{S}\mathbf{S}^{\dagger}=\mathbf{S}^{\dagger}\mathbf{S}=\hat{\mathbf{1}} \tag{326}$$

$$\mathbf{S}^{\dagger}\mathbf{A}\mathbf{S}=\mathrm{diag}(\Delta_1,\Delta_2,\cdots,\Delta_n)=\mathbf{B} \tag{327}$$

where Δ_k are the complex eigenvalues of the matrix $\mathbf{A}$. Therefore, we calculate the following complex Gaussian integral:

$$\int d\xi' d\xi'^{\dagger}\exp\left\{-\xi'^{\dagger}\mathbf{A}\xi'\right\}=\int d\xi' d\xi'^{\dagger}\exp\left\{-\xi'^{\dagger}\left(\mathbf{S}\mathbf{S}^{\dagger}\right)\mathbf{A}\left(\mathbf{S}\mathbf{S}^{\dagger}\right)\xi'\right\}=$$
$$=\int d\xi' d\xi'^{\dagger}\exp\left\{-\left(\xi'^{\dagger}\mathbf{S}\right)\mathbf{S}^{\dagger}\mathbf{A}\mathbf{S}\left(\mathbf{S}^{\dagger}\xi'\right)\right\}=\int d\xi d\xi^{\dagger}\exp\left\{-\xi^{\dagger}\mathbf{B}\xi\right\}=\prod_{k=1}^{n}\left[\int d\xi_k d\xi_k^{\dagger}\exp\left\{-\xi_k^{\dagger}\Delta_k\xi_k\right\}\right]$$
$$=\prod_{k=1}^{n}\frac{\pi}{\Delta_k}=\pi^n\frac{1}{\prod_{k=1}^{n}\Delta_k}=\pi^n\frac{1}{\det(\mathbf{A})}=\pi^n\exp\left\{-\mathrm{Tr}\left[\ln\mathbf{A}\right]\right\} \tag{328}$$

We shift the origin of integration via:

$$\xi'^{\dagger}=\xi^{\dagger}+\eta^{\dagger}\mathbf{A}^{-1} \quad , \quad \xi'=\xi+\mathbf{A}^{-1}\eta \tag{329}$$

then

$$-\xi'^{\dagger}\mathbf{A}\xi'+\eta^{\dagger}\xi'+\xi'^{\dagger}\eta=-\left[\xi^{\dagger}+\eta^{\dagger}\mathbf{A}^{-1}\right]\mathbf{A}\left[\xi+\mathbf{A}^{-1}\eta\right]+\eta^{\dagger}\left[\xi+\mathbf{A}^{-1}\eta\right]+\left[\xi^{\dagger}+\eta^{\dagger}\mathbf{A}^{-1}\right]\eta=-\xi^{\dagger}\mathbf{A}\xi-\xi^{\dagger}\mathbf{A}\mathbf{A}^{-1}\eta-$$
$$-\eta^{\dagger}\mathbf{A}^{-1}\mathbf{A}\xi-\eta^{\dagger}\mathbf{A}^{-1}\mathbf{A}\xi-\eta^{\dagger}\mathbf{A}^{-1}\mathbf{A}\mathbf{A}^{-1}\eta+\eta^{\dagger}\xi+\eta^{\dagger}\mathbf{A}^{-1}\eta+\xi^{\dagger}\eta+\eta^{\dagger}\mathbf{A}^{-1}\eta=-\xi^{\dagger}\mathbf{A}\xi-\xi^{\dagger}\eta-\eta^{\dagger}\xi-\eta^{\dagger}\mathbf{A}^{-1}\eta+$$
$$+\eta^{\dagger}\xi+\eta^{\dagger}\mathbf{A}^{-1}\eta+\xi^{\dagger}\eta+\eta^{\dagger}\mathbf{A}^{-1}\eta=-\xi^{\dagger}\mathbf{A}\xi+\eta^{\dagger}\mathbf{A}^{-1}\eta \tag{330}$$

This transformation permits us to calculate the following multidimensional complex Gaussian integral:

$$\int d\xi' d\xi'^{\dagger}\exp\left\{-\xi'^{\dagger}\mathbf{A}\xi'+\eta^{\dagger}\xi'+\xi'^{\dagger}\eta\right\}=\int d\xi d\xi^{\dagger}\exp\left\{-\xi^{\dagger}\mathbf{A}\xi+\eta^{\dagger}\mathbf{A}^{-1}\eta\right\}=$$
$$=\exp\left\{\eta^{\dagger}\mathbf{A}^{-1}\eta\right\}\int d\xi d\xi^{\dagger}\exp\left\{-\xi^{\dagger}\mathbf{A}\xi\right\}=\pi^{n}\frac{1}{\det(\mathbf{A})}\exp\left\{\eta^{\dagger}\mathbf{A}^{-1}\eta\right\} \tag{331}$$

2.5.3 Multidimensional Grassmann Gaussian Integral

Note that as the square of a Grassmann number vanishes, the rules for Grassmann algebra and Berezin integration may be sufficient to define Grassmann integration. Consider ξ_i and $\xi_i^{\dagger}$ to be independent Grassmann variables and the complex Hermitian matrix $\mathbf{A}$ with $i,j=1,2,\cdots,\infty$. These definitions will permit us to determine Gaussian Grassmann integrals. This implies that $\mathbf{A}=\mathbf{A}^{\dagger}$, and there exists a unitary transformation

$$\mathbf{S}: \mathbf{S}\mathbf{S}^{\dagger}=\mathbf{S}^{\dagger}\mathbf{S}=\hat{1} \tag{332}$$

$$\mathbf{S}^{\dagger}\mathbf{A}\mathbf{S}=\mathrm{diag}(\Delta_1,\Delta_2,\cdots,\Delta_n)=\mathbf{B} \tag{333}$$

Here, Δ_k are the eigenvalues of the matrix $\mathbf{A}$. Therefore, we calculate the following Grassmann Gaussian integral in the same manner as the complex Gaussian integral:

$$\int d\xi'^{\dagger} d\xi'\exp\left\{-\xi'^{\dagger}\mathbf{A}\xi'\right\}=\int d\xi' d\xi'^{\dagger}\exp\left\{-\xi'^{\dagger}\left(\mathbf{S}\mathbf{S}^{\dagger}\right)\mathbf{A}\left(\mathbf{S}\mathbf{S}^{+}\right)\xi'\right\}=\int d\xi' d\xi'^{\dagger}\exp\left\{-\left(\xi'^{\dagger}\mathbf{S}\right)\mathbf{S}^{\dagger}\mathbf{A}\mathbf{S}\left(\mathbf{S}^{\dagger}\xi'\right)\right\}=$$
$$=\int d\xi d\xi^{\dagger}\exp\left\{-\xi^{\dagger}\mathbf{B}\xi\right\}=\prod_{k=1}^{n}\left[\int d\xi_k^{\dagger} d\xi_k\exp\left\{-\Delta_k\xi_k^{\dagger}\xi_k\right\}\right]=\prod_{k=1}^{n}\int d\xi_k^{\dagger} d\xi_k\left(1-\Delta_k\xi_k^{\dagger}\xi_k\right)=\prod_{k=1}^{n}\Delta_k=\det(\mathbf{A}) \tag{334}$$

Shifting the origin of integration via:

$$\xi'^{\dagger}=\xi^{\dagger}+\eta^{\dagger}\mathbf{A}^{-1} \quad , \quad \xi'=\xi+\mathbf{A}^{-1}\eta \tag{335}$$

or

$$\xi_k'^{\dagger}=\xi_k^{\dagger}+\eta_k^{\dagger}\mathbf{A}^{-1} \quad , \quad \xi_k'=\xi_k+\mathbf{A}^{-1}\eta_k \tag{336}$$

then

$$-\xi'^{\dagger}\mathbf{A}\xi'+\eta^{\dagger}\xi'+\xi'^{\dagger}\eta=-\left[\xi^{\dagger}+\eta^{\dagger}\left(\mathbf{A}^{-1}\right)^{\dagger}\right]\mathbf{A}\left[\xi+\mathbf{A}^{-1}\eta\right]+\eta^{\dagger}\left[\xi+\mathbf{A}^{-1}\eta\right]+\left[\xi^{\dagger}+\eta^{\dagger}\left(\mathbf{A}^{-1}\right)^{\dagger}\right]\eta=-\xi^{\dagger}\mathbf{A}\xi-$$
$$-\xi^{\dagger}\mathbf{A}\mathbf{A}^{-1}\eta-\eta^{\dagger}\mathbf{A}^{-1}\mathbf{A}\xi-\eta^{\dagger}\left(\mathbf{A}^{-1}\right)^{\dagger}\mathbf{A}\xi-\eta^{\dagger}\left(\mathbf{A}^{-1}\right)^{\dagger}\eta+\eta^{\dagger}\xi+\eta^{\dagger}\mathbf{A}^{-1}\eta+\xi^{\dagger}\eta+\eta^{\dagger}\left(\mathbf{A}^{-1}\right)^{\dagger}\eta=-\xi^{\dagger}\mathbf{A}\xi-\xi^{\dagger}\eta-$$
$$-\eta^{\dagger}\xi-\eta^{\dagger}\mathbf{A}^{-1}\eta+\eta^{\dagger}\xi+\eta^{\dagger}\mathbf{A}^{-1}\eta+\xi^{\dagger}\eta+\eta^{\dagger}\mathbf{A}^{-1}\eta=-\xi^{\dagger}\mathbf{A}\xi+\eta^{\dagger}\mathbf{A}^{-1}\eta \tag{337}$$

or

$$-\xi'^{\dagger}\mathbf{A}\xi'+\eta^{\dagger}\xi'+\xi'^{\dagger}\eta=-\xi^{\dagger}\mathbf{A}\xi+\eta^{\dagger}\mathbf{A}^{-1}\eta \tag{338}$$

This transformation permits us to calculate the following multidimensional Grassmann Gaussian integral:

$$\int d\xi'^{\dagger}\,d\xi'\exp\left\{-\xi'^{\dagger}\mathbf{A}\xi'+\eta^{\dagger}\xi'+\xi'^{\dagger}\eta\right\}=\prod_{k=1}^{n}\int d\xi_k^{\dagger}d\xi_k\exp\left\{-\sum_{k,l}\xi_k^{\dagger}(\mathbf{A})_{kl}\xi_l+\eta_k^{\dagger}\xi_k+\xi_k^{\dagger}\eta_k\right\}=$$
$$=\int d\xi^{\dagger}\,d\xi\exp\left\{-\xi^{\dagger}\mathbf{A}\xi+\eta^{\dagger}\mathbf{A}^{-1}\eta\right\}=\exp\left\{\eta^{\dagger}\mathbf{A}^{-1}\eta\right\}\int d\xi^{\dagger}\,d\xi\exp\left\{-\xi^{\dagger}\mathbf{A}\xi\right\}=\exp\left\{\eta^{\dagger}\mathbf{A}^{-1}\eta\right\}\det(\mathbf{A}) \tag{339}$$

This result will permit us to obtain Green's functions or multipoint functions from functional derivatives over η. It should be noted that

$$\det\mathbf{A}=\exp\left\{\mathrm{Tr}\left[\ln\mathbf{A}\right]\right\} \tag{340}$$

We also consider that the determinant and the trace are both basis independent.

EXERCISE

Question: Compute the Grassmann Gaussian integral

$$\int d\xi^{*}\int d\xi\exp\left\{-\xi^{*}\mathrm{A}\xi\right\} \tag{341}$$

Answer: Considering the properties of Grassmann variables, we expand the exponential function in the integrand:

$$\int d\xi^{*}\int d\xi\left(1-\xi^{*}\mathrm{A}\xi\right)=-\mathrm{A}\int d\xi^{*}\int d\xi\xi^{*}\xi=\mathrm{A}\int d\xi^{*}\int d\xi\xi\xi^{*}=\mathrm{A}\int d\xi^{*}\xi^{*}=\mathrm{A} \tag{342}$$

2.6 Wick Theorem for Multidimensional Grassmann Integrals

We express a multidimensional integral $\mathbf{Z}\left[\eta,\eta^{*}\right]$, a **so-called generating function**, via source variables $\eta_{\alpha}^{*}(\tau)$ and $\eta_{\alpha}(\tau)$ that are c-numbers for bosons and anticommuting variables for fermions coupled linearly to the fields $\xi_{\alpha}^{*}(\tau)$ and $\xi_{\alpha}(\tau)$:

$$\mathbf{Z}\left[\eta,\eta^{*}\right]=\frac{\int d\left[\xi_{\alpha}^{*}\right]d\left[\xi_{\alpha}\right]\exp\left\{\sum_{\alpha,\alpha'}\xi_{\alpha}^{*}\mathbf{A}_{\alpha,\alpha'}\xi_{\alpha'}+\sum_{\alpha}\left(\eta_{\alpha}^{*}\xi_{\alpha}+\eta_{\alpha}\xi_{\alpha}^{*}\right)\right\}}{\int d\left[\xi_{\alpha}^{*}\right]d\left[\xi_{\alpha}\right]\exp\left\{\sum_{\alpha,\alpha'}\xi_{\alpha}^{*}\mathbf{A}_{\alpha,\alpha'}\xi_{\alpha'}\right\}} \tag{343}$$

For fermions, the source variables $\eta^*_\alpha(\tau)$ and $\eta_\alpha(\tau)$ are Grassmann variables and c-numbers for bosons. We shift the argument of the exponential function via:

$$\xi_\alpha = \xi'_\alpha + \mathbf{A}^{-1}_{\alpha,\alpha'}\eta_{\alpha'} \quad , \quad \xi^*_\alpha = \xi'^*_\alpha + \eta^*_{\alpha'}\mathbf{A}^{-1}_{\alpha,\alpha'} \tag{344}$$

then

$$\mathbf{Z}\left[\eta,\eta^*\right] = \frac{\int d\left[\xi^*_\alpha\right] d\left[\xi_\alpha\right] \exp\left\{\sum_{\alpha,\alpha'}\left(\xi'^*_\alpha \mathbf{A}_{\alpha,\alpha'}\xi'_{\alpha'} + \eta^*_\alpha \mathbf{A}^{-1}_{\alpha,\alpha'}\eta_{\alpha'}\right)\right\}}{\int d\left[\xi^*_\alpha\right] d\left[\xi_\alpha\right] \exp\left\{\sum_{\alpha,\alpha'}\xi'^*_\alpha \mathbf{A}_{\alpha,\alpha'}\xi'_{\alpha'}\right\}} = \exp\left\{\sum_{\alpha,\alpha'}\eta^*_\alpha \mathbf{A}^{-1}_{\alpha,\alpha'}\eta_{\alpha'}\right\} \tag{345}$$

2.6.1 Wick Theorem

From expressing $\mathbf{Z}\left[\eta,\eta^*\right]$ in (345), we find

$$\frac{\partial^{(n)}\mathbf{Z}\left[\eta,\eta^*\right]}{\partial\eta_{\alpha_n}(\tau_n)\cdots\partial\eta_{\alpha_1}(\tau_1)} = \chi^n\left(\sum_\alpha \eta^*_\alpha \mathbf{A}^{-1}_{\alpha,\alpha_n}\right)\cdots\left(\sum_\alpha \eta^*_\alpha \mathbf{A}^{-1}_{\alpha,\alpha_2}\right)\left(\sum_\alpha \eta^*_\alpha \mathbf{A}^{-1}_{\alpha,\alpha_1}\right)\exp\left\{\sum_{\alpha,\alpha'}\eta^*_\alpha \mathbf{A}^{-1}_{\alpha,\alpha'}\eta_{\alpha'}\right\} \tag{346}$$

For Grassmann numbers, only terms survive that contain each η_{α_k} only once. Because the derivative is carried n times, then we have n^n terms and expression (346) is rewritten

$$\frac{\partial^{(n)}\mathbf{Z}\left[\eta,\eta^*\right]}{\partial\eta_{\alpha_n}(\tau_n)\cdots\partial\eta_{\alpha_1}(\tau_1)} = \chi^n\sum_l^n\prod_{\mathrm{P}(l)\neq l}\eta^*_{\alpha_{\mathrm{P}(l)}}\mathbf{A}^{-1}_{\alpha_{\mathrm{P}(l)},\alpha_l}\exp\left\{\sum_{\alpha,\alpha'}\eta^*_\alpha \mathbf{A}^{-1}_{\alpha,\alpha'}\eta_{\alpha'}\right\} \tag{347}$$

It is viewed as a block of $n\times n$ summands and, for Grassmann numbers, all terms with more than one η_{α_k} vanish. So, of all the n^n terms, only $n!$ terms survive, and among these terms, all permutations of η_{α_k} are present. The derivative over all η_{α_k} leaves only terms independent of η^*_α and η_α out of the exponent in addition to terms containing η_α as a result of the product rule. Besides, letting $\eta=\eta^*=0$, then only the $n!$ permutations independent of η^*_α and η_α out of the exponential are conserved. For bosons, all the terms in equation 347 are present. However, for differentiation over all η_α, only terms with each η_α survive, and thus we have **Wick theorem**:

$$\left.\frac{\partial^{(2n)}\mathbf{Z}\left[\eta,\eta^*\right]}{\partial\eta^*_{\alpha_1}(\tau_1)\cdots\partial\eta^*_{\alpha_n}(\tau_n)\partial\eta_{\alpha'_n}(\tau'_n)\cdots\partial\eta_{\alpha'_1}(\tau'_1)}\right|_{\eta,\eta^*=0} = \chi^n\sum_{\mathrm{P}}\chi^{\mathrm{P}}\mathbf{A}^{-1}_{\alpha_{\mathrm{P}(n)},\alpha_n}\cdots\mathbf{A}^{-1}_{\alpha_{\mathrm{P}(1)},\alpha_1} \tag{348}$$

To write **Wick theorem in standard form, we define the so-called contractions: the process of identifying pairs of initial and final states in the n-particle Green's function**. This can be done by reexamining the integral

$$\mathbf{Z}\left[\eta,\eta^*\right] = \frac{\int d\left[\xi^*_\alpha\right] d\left[\xi_\alpha\right] \exp\left\{\sum_{\alpha,\alpha'}\xi^*_\alpha \mathbf{A}_{\alpha,\alpha'}\xi_{\alpha'} + \sum_\alpha\left(\eta^*_\alpha\xi_\alpha + \eta_\alpha\xi^*_\alpha\right)\right\}}{\int d\left[\xi^*_\alpha\right] d\left[\xi_\alpha\right] \exp\left\{\sum_{\alpha,\alpha'}\xi^*_\alpha \mathbf{A}_{\alpha,\alpha'}\xi_{\alpha'}\right\}} \tag{349}$$

and

$$\left\langle \xi_{\alpha_1}(\tau_1)\cdots\xi_{\alpha_n}(\tau_n)\xi^*_{\alpha'_n}(\tau'_n)\cdots\xi^*_{\alpha'_1}(\tau'_1)\right\rangle = \frac{\int d\left[\xi^*_\alpha\right]d\left[\xi_\alpha\right]\xi_{\alpha_1}\cdots\xi_{\alpha_n}\xi^*_{\alpha'_n}\cdots\xi^*_{\alpha'_1}\exp\left\{\sum_{\alpha,\alpha'}\xi^*_\alpha \mathbf{A}_{\alpha,\alpha'}\xi_{\alpha'}\right\}}{\int d\left[\xi^*_\alpha\right]d\left[\xi_\alpha\right]\exp\left\{\sum_{\alpha,\alpha'}\xi^*_\alpha \mathbf{A}_{\alpha,\alpha'}\xi_{\alpha'}\right\}} \tag{350}$$

or

$$\left\langle \xi_{\alpha_1}(\tau_1)\cdots\xi_{\alpha_n}(\tau_n)\xi^*_{\alpha'_n}(\tau'_n)\cdots\xi^*_{\alpha'_1}(\tau'_1)\right\rangle = \left.\frac{\partial^{(2n)}\mathbf{Z}\left[\eta,\eta^*\right]}{\partial\eta^*_{\alpha_1}(\tau_1)\cdots\partial\eta^*_{\alpha_n}(\tau_n)\partial\eta_{\alpha'_n}(\tau'_n)\cdots\partial\eta_{\alpha'_1}(\tau'_1)}\right|_{\eta,\eta^*=0}$$
$$= \sum_{\mathrm{P}}\chi^{\mathrm{P}}\mathbf{A}^{-1}_{\alpha_{\mathrm{P}(n)},\alpha_n}\cdots\mathbf{A}^{-1}_{\alpha_{\mathrm{P}(1)},\alpha_1} \tag{351}$$

or

$$\left\langle \xi_{\alpha_1}(\tau_1)\cdots\xi_{\alpha_n}(\tau_n)\xi^*_{\alpha'_n}(\tau'_n)\cdots\xi^*_{\alpha'_1}(\tau'_1)\right\rangle = \sum_{\mathrm{P}}\chi^{\mathrm{P}}\mathbf{A}^{-1}_{\alpha_{\mathrm{P}(n)},\alpha_n}\cdots\mathbf{A}^{-1}_{\alpha_{\mathrm{P}(1)},\alpha_1} = \sum_{\mathrm{P}}\chi^{\mathrm{P}}\prod_k\left\langle \xi^*_{\alpha_k \mathrm{P}(k)\alpha'_k}\xi_{\alpha_k \mathrm{P}(k)\alpha'_k}\right\rangle \tag{352}$$

This is **Wick theorem**, where we **sum over** P (**all possible Wick contractions)**. In particular,

$$\text{contractions} = \left\langle \xi_{\alpha_k}(\tau_k)\xi^*_{\alpha'_k}(\tau'_k)\right\rangle = \frac{\int d\left[\xi^*_\alpha\right]d\left[\xi_\alpha\right]\xi_{\alpha_k}\xi^*_{\alpha'_k}\exp\left\{\sum_{\alpha,\alpha'}\xi^*_\alpha \mathbf{A}_{\alpha,\alpha'}\xi_{\alpha'}\right\}}{\int d\left[\xi^*_\alpha\right]d\left[\xi_\alpha\right]\exp\left\{\sum_{\alpha,\alpha'}\xi^*_\alpha \mathbf{A}_{\alpha,\alpha'}\xi_{\alpha'}\right\}} \tag{353}$$

Therefore, Wick theorem considers the average of a product of fields with Gaussian action given as the sum of all possible Wick contractions. These contractions also correspond to bare field propagators.

The process of identifying pairs of initial and final states in the *k*-particle Green's function is often referred to as a contraction.

Note that a complete contraction is a configuration in which each $\xi_\alpha(\tau)$ is contracted with a $\xi^*_\alpha(\tau)$ and the overall sign is specified by χ^{P}. The permutation P is such that $\xi_{\alpha_i}(\tau_i)$ is contracted with $\xi^*_{\alpha_{\mathrm{P}(i)}}\left(\tau_{\alpha_{\mathrm{P}(i)}}\right)$. The effect of the creation operator $\xi^*_{\alpha'}(\tau')$ is to put the particle into the state α'. The system has to be back to the ground state before the final operator of $\langle\ |$ so one of the destruction operators $\xi_\alpha(\tau)$ should destroy the state α' and $\alpha=\alpha'$ for some α. For example,

$$\overleftrightarrow{\xi_\alpha(\tau)\xi_{\alpha'}}(\tau') = \left\langle \xi_\alpha(\tau)\xi_{\alpha'}(\tau')\right\rangle = 0\quad,\quad \overleftrightarrow{\xi^*_\alpha(\tau)\xi^*_{\alpha'}}(\tau') = \left\langle \xi^*_\alpha(\tau)\xi^*_{\alpha'}(\tau')\right\rangle = 0 \tag{354}$$

unless $\alpha=\alpha'$.

Therefore, within the pairing, a pairing bracket, the labels α and α', must be the same and denote eigenstates so that the creation and destruction operators refer to the same state.

3

Fermionic and Bosonic Path Integrals

Introduction

Having obtained a complete coherent state basis for the creation and annihilation operators, we could proceed by constructing path integrals for fermionic as well as bosonic systems. Because our emphasis is on the application of coherent states, it is more convenient to begin the application of coherent states formalism with the development of the path integral representation for the grand-canonical partition function of many-particle systems. The path integral formalism pioneered by Feynman [12–14] has proven to be an extremely useful tool for understanding and handling quantum mechanics, quantum field theory, and statistical mechanics. Apart from giving a global view of the entire system, the path integral offers:

1. An alternative to the descriptions based on differential equations such as the (nonlinear) Schrödinger equation and thus is often the only viable approach to many-body systems;
2. The advantage that position and momentum need not be expressed as (noncommuting) operators and that the covariance is directly established;
3. An ideal way of obtaining the classical limit of quantum mechanics;
4. A unified description of quantum dynamics and equilibrium quantum statistical mechanics;
5. A powerful influence and functional method for studying the dynamics of a low-dimensional system coupled to a harmonic bath.

3.1 Coherent State Path Integrals

From the previous interlude, we now have all the background to set up a unified path-integration for bosonic and fermionic systems. The path integral approach is a powerful tool that considers nonperturbative calculations for the investigation of many-particle systems. Our interest is in determining the equilibrium/nonequilibrium properties of a quantum fluid at some temperature T.

In our textbook, *Statistical Thermodynamics* [13], we derive the path integral formula starting from the time evolution operator and use its composition law n times, while afterward using its property of unitarity $(n-1)$ times. Further, we calculate the partition function of a particle or system of particles

$$Z = \mathrm{Tr}\exp\left\{-\beta\hat{H}\right\} \equiv \mathrm{Tr}\hat{\rho}(\beta) \tag{355}$$

with $\hat{H}$ being the Hamiltonian of the system and β the inverse temperature. This is the so-called imaginary time or Euclidean path integral that is closely related to the original Feynman path integral over the so-called Wick rotation. Essentially, this is an analytical continuation with a variable transformation

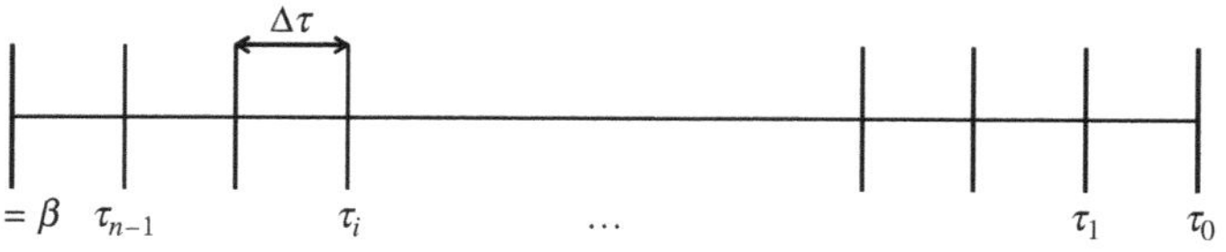

FIGURE 3.1 Time slice of the interval $[0, \beta]$ by $n-1$ intermediate points.

$t=-i\hbar\tau$, $0\le\tau\le\beta$ [13, 14]. So, to derive the path integral representation of the partition function Z, we time-slice the interval $[0,\beta]$ by $(n-1)$ intermediate points as shown in Figure 3.1. Hence, we set $\hbar, 2m, \kappa_F, \kappa_B = 1$, where κ_F is the Fermi wave number and κ_B the Boltzmann constant.

It is worth noting that while the path integral approach seems to be quite cumbersome in quantum mechanics, it does provide a powerful tool in quantum field theory.

In quantum statistics, it is very convenient to consider a system of a variable number of particles. Therefore, the ground state of the given system at $T=0$ can then be defined as the state having the lowest eigenvalue of the operator (the grand canonical Hamiltonian ensemble of a system that is **normal-ordered** with respect to some reference state $|0\rangle$ of a system of **fermions** or **bosons**):

$$\hat{H}' = \hat{H} - \mu\hat{N} \tag{356}$$

In this chapter, we investigate a quantum gas in the grand-canonical ensemble with the grand canonical Hamiltonian ensemble (356). The path integral formalism will be very appropriate for this purpose. In the grand-canonical ensemble, the total number of particles is not conserved. This involves a field theoretical approach given by reformulating nonrelativistic quantum mechanics in a field theory over the single-particle wave functions known as second quantization. It is possible to derive a path integral formulation for the partition function via coherent states. One deals with states with an indefinite number of particles because these coherent states form an overcomplete set in the Fock space.

We calculate the partition function in the grand-canonical ensemble of a many-particle system, which contains all information about the thermodynamic equilibrium properties of that system.

We consider the fact that $\hat{H}'$ is a normal-ordered operator, and the partition function in the grand-canonical ensemble is given as the trace of the density operator $\hat{\rho}$ with the help of (275) and (292). With $\{|n\rangle\}$ being a complete set of states in Fock space, noting from (290) the inner products $\langle n|\xi\rangle\langle\xi|m\rangle = \langle\chi\xi|m\rangle\langle n|\xi\rangle$, and making use of the sign factor χ, the trace of the density operator (thermal weighting factor) $\hat{\rho}(\beta) = \exp\left\{-\beta\left(\hat{H} - \mu\hat{N}\right)\right\}$ describing the partition function can be written as follows:

$$\begin{aligned}
Z = \mathrm{Tr}\hat{\rho} &= \sum_n \langle n|\hat{\rho}|n\rangle = \int d[\xi^*_\alpha]d[\xi_\alpha]\exp\left\{-\sum_\alpha \xi^*_\alpha\xi_\alpha\right\}\sum_n \langle n|\xi_\alpha\rangle\langle\xi_\alpha|\hat{\rho}|n\rangle = \\
&= \int d[\xi^*_\alpha]d[\xi_\alpha]\exp\left\{-\sum_\alpha \xi^*_\alpha\xi_\alpha\right\}\sum_n \langle\chi\xi_\alpha|\hat{\rho}|n\rangle\langle n|\xi_\alpha\rangle \\
&= \int d[\xi^*_\alpha]d[\xi_\alpha]\exp\left\{-\sum_\alpha \xi^*_\alpha\xi_\alpha\right\}\langle\chi\xi_\alpha|\hat{\rho}|\xi_\alpha\rangle
\end{aligned} \tag{357}$$

Here, $|\xi_\alpha\rangle$ is a coherent state; ξ_α is a c-number for bosons and a Grassmann variable for fermions. We now divide the imaginary time β into $n=\frac{\beta}{\varepsilon}$ steps and insert $n-1$ times the closure relation (276); (257) can now be rewritten

$$Z=\int d[\xi_\alpha^*]d[\xi_\alpha]\exp\left\{-\sum_\alpha \xi_\alpha^*\xi_\alpha\right\}\langle\chi\xi_\alpha|\hat\rho|\xi_\alpha\rangle= \\ =\int d[\xi_\alpha^*]d[\xi_\alpha]\exp\left\{-\sum_\alpha \xi_\alpha^*\xi_\alpha\right\}\langle\chi\xi_\alpha|\hat\rho(\beta,\tau_{n-1})\hat\rho(\tau_{n-1},\tau_{n-2})\cdots\hat\rho(\tau_2,\tau_1)\hat\rho(\tau_1,\tau_0)|\xi_\alpha\rangle \tag{358}$$

or

$$Z=\mathrm{Tr}\left[\hat\rho(\beta,\tau_{n-1})\hat\rho(\tau_{n-1},\tau_{n-2})\cdots\hat\rho(\tau_2,\tau_1)\hat\rho(\tau_1,\tau_0)\right] \\ =\int d[\xi_\alpha^*]d[\xi_\alpha]\exp\left\{-\sum_\alpha \xi_\alpha^*\xi_\alpha\right\}\langle\chi\xi_\alpha^*|\hat\rho(\beta)|\xi_\alpha\rangle \tag{359}$$

We observe that we are faced with the task of calculating the matrix elements

$$\langle\chi\xi_\alpha^*|\prod_{k=1}^{n-1}\hat\rho(\tau_k,\tau_{k-1})|\xi_\alpha\rangle \tag{360}$$

with the periodic (antiperiodic) boundary condition $\xi_{\alpha_0}=\xi_\alpha$ and $\xi_{\alpha_n}^*=\chi\xi_\alpha^*$ so,

$$Z=\mathrm{Tr}\exp\left\{-\beta\hat H'\right\}=\int d[\xi_\alpha^*]d[\xi_\alpha]\exp\left\{\chi\xi_\alpha^*\xi_\alpha\right\}\langle\chi\xi_\alpha^*|\prod_{k=1}^{n-1}\hat\rho(\tau_k,\tau_{k-1})|\xi_\alpha\rangle \tag{361}$$

Now, in equation (361), instead of inserting a complete set of states at each intermediate time τ_k, we insert an overcomplete set of coherent states $\left\{|\xi_{\alpha_k}\rangle\right\}$ at each time τ_k through the insertion of the resolution of the identity

$$\int d[\xi_{\alpha_k}^*]d[\xi_{\alpha_k}]\exp\left\{-\sum_\alpha \xi_{\alpha_k}^*\xi_{\alpha_k}\right\}|\xi_{\alpha_k}\rangle\langle\xi_{\alpha_k}|=1 \tag{362}$$

If we consider that the Hamiltonian $\hat H'\left[\hat\psi_{\alpha_k}^\dagger,\hat\psi_{\alpha_k}\right]$ is normal ordered, then

$$\langle\xi_{\alpha_k}|\hat H'\left[\hat\psi_{\alpha_k}^\dagger,\hat\psi_{\alpha_k}\right]|\xi_{\alpha_{k-1}}\rangle=\langle\xi_{\alpha_k}|\xi_{\alpha_{k-1}}\rangle\hat H'\left[\xi_{\alpha_k},\xi_{\alpha_{k-1}}\right]=\exp\left\{\xi_{\alpha_k}^*\xi_{\alpha_{k-1}}\right\}\hat H'\left[\xi_{\alpha_k},\xi_{\alpha_{k-1}}\right] \tag{363}$$

and the inner product

$$\langle\xi_{\alpha_k}|\xi_{\alpha_{k-1}}\rangle=\exp\left\{\xi_{\alpha_k}^*\xi_{\alpha_{k-1}}\right\} \tag{364}$$

Hence, we have the products of matrix elements of the form:

$$\begin{aligned}\left\langle \xi^*_{\alpha_k} \middle| \hat{\rho}(\tau_k,\tau_{k-1}) \middle| \xi_{\alpha_{k-1}} \right\rangle &\cong \left\langle \xi^*_{\alpha_k} \middle| 1-(\tau_k-\tau_{k-1})\hat{H}' \middle| \xi_{\alpha_{k-1}} \right\rangle \\ &= \left\langle \xi^*_{\alpha_k} \middle| \xi_{\alpha_{k-1}} \right\rangle - (\tau_k-\tau_{k-1})\left\langle \xi^*_{\alpha_k} \middle| \hat{H}' \middle| \xi_{\alpha_{k-1}} \right\rangle = \\ &= \exp\left\{\xi^*_{\alpha_k}\xi_{\alpha_{k-1}}\right\} - (\tau_k-\tau_{k-1})\hat{H}'\left[\xi^*_{\alpha_k},\xi_{\alpha_{k-1}}\right]\exp\left\{\xi^*_{\alpha_k}\xi_{\alpha_{k-1}}\right\} \\ &= \exp\left\{\xi^*_{\alpha_k}\xi_{\alpha_{k-1}}\right\}\left(1-(\tau_k-\tau_{k-1})\hat{H}'\left[\xi^*_{\alpha_k},\xi_{\alpha_{k-1}}\right]\right) \cong \\ &\cong \exp\left\{\xi^*_{\alpha_k}\xi_{\alpha_{k-1}} - (\tau_k-\tau_{k-1})\hat{H}'\left[\xi^*_{\alpha_k},\xi_{\alpha_{k-1}}\right]\right\}\end{aligned} \tag{365}$$

or

$$\left\langle \xi^*_{\alpha_k} \middle| \hat{\rho}(\tau_k,\tau_{k-1}) \middle| \xi_{\alpha_{k-1}} \right\rangle \cong \exp\left\{\xi^*_{\alpha_k}\xi_{\alpha_{k-1}} - (\tau_k-\tau_{k-1})\hat{H}'\left[\xi^*_{\alpha_k},\xi_{\alpha_{k-1}}\right]\right\} \tag{366}$$

The partition function can now be expressed in the form

$$Z = \prod_{k=1}^{n} \int d\left[\xi^*_{\alpha_k}\right] d\left[\xi_{\alpha_k}\right] \exp\left\{\left(\xi^*_{\alpha_k}\xi_{\alpha_{k-1}} - \xi^*_{\alpha_k}\xi_{\alpha_k} - \varepsilon\left(\hat{H}\left[\xi^*_{\alpha_k},\xi_{\alpha_{k-1}} - \mu\xi^*_{\alpha_k}\xi_{\alpha_{k-1}}\right]\right)\right)\right\} \tag{367}$$

But

$$\left(\xi^*_{\alpha_k}\xi_{\alpha_{k-1}} - \xi^*_{\alpha_k}\xi_{\alpha_k} - \varepsilon\left(\hat{H}\left[\xi^*_{\alpha_k},\xi_{\alpha_{k-1}} - \mu\xi^*_{\alpha_k}\xi_{\alpha_{k-1}}\right]\right)\right) = -\varepsilon\left(\xi^*_{\alpha_k}\frac{\xi_{\alpha_k}-\xi_{\alpha_{k-1}}}{\varepsilon} + \hat{H}\left[\xi^*_{\alpha_k},\xi_{\alpha_{k-1}} - \mu\xi^*_{\alpha_k}\xi_{\alpha_{k-1}}\right]\right) \tag{368}$$

Also

$$-\varepsilon\left(\xi^*_{\alpha_k}\frac{\xi_{\alpha_k}-\xi_{\alpha_{k-1}}}{\varepsilon} + \hat{H}\left[\xi^*_{\alpha_k},\xi_{\alpha_{k-1}} - \mu\xi^*_{\alpha_k}\xi_{\alpha_{k-1}}\right]\right) = -\varepsilon\left(\xi^*_{\alpha_k}\left(\frac{\xi_{\alpha_k}-\xi_{\alpha_{k-1}}}{\varepsilon} - \mu\xi_{\alpha_{k-1}}\right) + \hat{H}\left[\xi^*_{\alpha_k},\xi_{\alpha_{k-1}}\right]\right) \tag{369}$$

Letting the action functional be

$$S\left[\xi^*_{\alpha_k},\xi_{\alpha_{k-1}}\right] = \varepsilon\left(\xi^*_{\alpha_k}\left(\frac{\xi_{\alpha_k}-\xi_{\alpha_{k-1}}}{\varepsilon} - \mu\xi_{\alpha_{k-1}}\right) + \hat{H}\left[\xi^*_{\alpha_k},\xi_{\alpha_{k-1}}\right]\right) \tag{370}$$

then

$$Z = \prod_{k=1}^{n} \int d\left[\xi^*_{\alpha_k}\right] d\left[\xi_{\alpha_k}\right] \exp\left\{-S\left[\xi^*_{\alpha_k},\xi_{\alpha_{k-1}}\right]\right\} = \prod_{k=1}^{n} \int d\left[\xi^*_{\alpha_k}\right] d\left[\xi_{\alpha_k}\right] \exp\left\{-\sum_{k=1}^{n}\sum_{\alpha} S\left[\xi^*_{\alpha_k},\xi_{\alpha_{k-1}}\right]\right\} \tag{371}$$

We consider the cycling property of the trace

$$\mathrm{Tr}\left[\hat{\rho}(t_n,t_{n-1})\hat{\rho}(t_{n-1},t_{n-2})\cdots\hat{\rho}(t_1,t_0)\right] = \mathrm{Tr}\left[\hat{\rho}(t_1,t_0)\hat{\rho}(t_n,t_{n-1})\cdots\hat{\rho}(t_2,t_1)\right] \tag{372}$$

that yields the periodic (antiperiodic) boundary condition, $\chi\xi_{\alpha_n} = \xi_{\alpha_0}$. This emphasizes the equivalence of the interior and exterior coherent state intervals. Within the limit of an infinite number of time slices, this allows us to rewrite the following partition function:

$$Z = \lim_{n\to\infty} \prod_{k=1}^{n} \int \prod_{\alpha} \frac{d\xi^*_{\alpha_k} d\xi_{\alpha_k}}{N} \exp\left\{-S^n\left[\xi^*,\xi\right]\right\} \tag{373}$$

where the action functional

$$\begin{aligned} S^n\left[\xi^*,\xi\right] = \varepsilon \sum_{k=2}^{n} \sum_{\alpha} \left[\xi^*_{\alpha_k} \left(\frac{\xi_{\alpha_k} - \xi_{\alpha_{k-1}}}{\varepsilon} - \mu\xi_{\alpha_{k-1}} \right) + \hat{H}\left[\xi^*_{\alpha_k}, \xi_{\alpha_{k-1}}\right] \right] + \\ + \varepsilon \sum_{\alpha} \left[\xi^*_{\alpha_1} \left(\frac{\xi_{\alpha_1} - \chi\xi_{\alpha_n}}{\varepsilon} - \mu\chi\xi_{\alpha_n} \right) + \hat{H}\left[\xi^*_{\alpha_1}, \chi\xi_{\alpha_n}\right] \right] \end{aligned} \tag{374}$$

Considering the trajectory notation, we then symbolically write

$$\xi^*_{\alpha_k} \frac{\xi_{\alpha_k} - \xi_{\alpha_{k-1}}}{\varepsilon} \to \xi^*_{\alpha}(\tau) \frac{\partial \xi_{\alpha}(\tau)}{\partial \tau} \quad , \quad \xi_{\alpha_{k-1}} \to \xi_{\alpha}(\tau) \quad , \quad \hat{H}\left[\xi^*_{\alpha_k}, \xi_{\alpha_{k-1}}\right] \to \hat{H}\left[\xi^*_{\alpha}(\tau), \xi_{\alpha}(\tau)\right] \tag{375}$$

Therefore, from path integration, we have

$$Z = \int_{\xi_\alpha(\beta)=\chi\xi_\alpha(0)} d\left[\xi^*\right] d\left[\xi\right] \exp\left\{-S\left[\xi^*,\xi\right]\right\} \tag{376}$$

where we define the integration measure by

$$\int_{\xi_\alpha(\beta)=\chi\xi_\alpha(0)} d\left[\xi^*\right] d\left[\xi\right] \equiv \lim_{n\to\infty} \prod_{k=1}^{n} \int \prod_{\alpha} \frac{d\xi^*_{\alpha_k} d\xi_{\alpha_k}}{N} \tag{377}$$

and

$$S\left[\xi^*,\xi\right] = \int_0^\beta d\tau \sum_{\alpha} \left\{ \xi^*_{\alpha}(\tau) \left(\frac{\partial}{\partial \tau} - \mu \right) \xi_{\alpha}(\tau) + H\left[\xi^*_{\alpha}(\tau), \xi_{\alpha}(\tau)\right] \right\} \tag{378}$$

is the Euclidean action functional of the system.

Note that the trajectory form of path integration is simply a symbolic form of the discrete definition (373) that is confirmed from the trajectory notation from (375), which is indeed relevant for bosons and may not make sense for Grassmann numbers. For fermions, the notation $\frac{\partial \xi_{\alpha}(\tau)}{\partial \tau}$ is purely symbolic because there is no instance for which $\xi_{\alpha_k} - \xi_{\alpha_{k-1}}$ is small. The symbol should then be understood as $\lim_{\varepsilon\to 0} \frac{\xi_{\alpha_k} - \xi_{\alpha_{k-1}}}{\varepsilon}$. Nonetheless, the trajectory notation conveniently describes the path integration formulation of coherent states and equally can be rewritten via fields in space coordinates as seen in equation (288). As noted previously, all properties of fermionic coherent states are analogous to bosonic ones provided ξ^*, ξ imitates Grassmann variables. Therefore, the path integral for a fermionic system will imitate (formally) that of the bosonic system except that one must integrate on paths in Grassmann space that are antiperiodic:

$$Z=\int_{\xi_\alpha(\beta)=\chi\xi_\alpha(0)} d[\xi^*]d[\xi]\exp\{-S[\xi^*,\xi]\} \tag{379}$$

where now

$$S[\xi^*,\xi]=\int_0^\beta d\tau\sum_\alpha\int d\vec{r}\left\{\xi^*_\alpha(\tau,\vec{r})\frac{\partial}{\partial\tau}\xi_\alpha(\tau,\vec{r})+\mathrm{H}'[\xi^*_\alpha(\tau),\xi_\alpha(\tau)]\right\} \quad , \quad \alpha=\uparrow,\downarrow \tag{380}$$

is the action functional in Grassmann fields. It is instructive to note that for quadratic action functionals, the path integration approach can yield only analytical results. Suitable transformations and approximations can bring the action functional to a quadratic form in the desired functional integral over the complex field $\xi(\tau)$ with the boundary conditions $\xi_\alpha(0)=\xi_{0,\alpha}$, $\xi^*_{n,\alpha}(\beta)=\chi\xi^*_\alpha$. (This implies that the fields are periodic in $[0,\beta]$ for bosons and antiperiodic for fermions.) The field theory analogue of the Feynman path integral is a very versatile device that has become the main tool in field theory investigation [15]. To obtain the partition function Z, we set $\xi_{0,\alpha}=\xi_{n,\alpha}$ and then perform integration over ξ^* and ξ, considering the periodic boundary conditions

$$\xi_\alpha(0)=\xi_{0,\alpha} \quad , \quad \xi^*_{n,\alpha}(\beta)=\chi\xi^*_\alpha \tag{381}$$

and then

$$Z=\int d[\xi^*]d[\xi]\exp\{-S[\xi^*,\xi]\} \tag{382}$$

So far, we have related the partition function with a functional integral and found that the path integral for a fermionic system is identical (formally) to that of a bosonic system except for the fact that one must integrate on paths in Grassmann space, such as $\{\xi^*(\vec{r},\tau),\xi(\vec{r},\tau)\}$ that are antiperiodic. All possible information on the macroscopic states of a many-body system can be derived in principle from partition function Z.

3.2 Noninteracting Particles

3.2.1 Bare Partition Function

We now study the general many-particle Hamiltonian operator by first computing the partition function for a system of noninteracting particles with the grand canonical one-body Hamiltonian ensemble:

$$\hat{\mathrm{H}}_0=\sum_\alpha \epsilon_\alpha\hat{\psi}^\dagger_\alpha\hat{\psi}_\alpha \quad , \quad \epsilon_\alpha=\tilde{\epsilon}_\alpha-\mu \tag{383}$$

where $\tilde{\epsilon}_\alpha$ are the single-particle eigenvalues. This procedure allows us to express the Green's function via path integration and then to thread the interaction via the perturbation theory. This object proves to be a **reference** in the development of weakly interacting theories. In addition, the field integral representation of the bare partition function is an important operational building block for subsequent analysis of interacting problems.

We consider that $\hat{\psi}_\alpha(\tau)$ is periodic in time and can be expanded as a Fourier series:

$$\hat{\psi}_{\alpha,k'} = \frac{1}{\sqrt{n}} \sum_{k=1}^{n} \xi_{\alpha,k} \exp\{-i\omega_k \tau_{k'}\} \quad , \quad \hat{\psi}^*_{\alpha,k'} = \frac{1}{\sqrt{n}} \sum_{k=1}^{n} \xi^*_{\alpha,k} \exp\{i\omega_k \tau_{k'}\} \quad , \quad \tau_k = k\varepsilon \tag{384}$$

The $\xi_{\alpha,k}$ and $\xi^*_{\alpha,k}$ are complex variables for bosons and Grassmann numbers for fermions. Considering that the transformation

$$\hat{\psi}_{\alpha,k} \to \xi_{\alpha,k} \quad , \quad \hat{\psi}^*_{\alpha,k} \to \xi^*_{\alpha,k} \tag{385}$$

has a Jacobian equals to unity, then the discrete expression for action functional

$$\begin{aligned} S[\xi^*,\xi] = \varepsilon \sum_{k=2}^{n} \sum_{\alpha} \left[\xi^*_{\alpha,k} \left(\frac{\xi_{\alpha,k} - \xi_{\alpha,k-1}}{\varepsilon} - \mu \xi_{\alpha,k-1} \right) + \hat{H}\left[\xi^*_{\alpha,k}, \xi_{\alpha,k-1}\right] \right] \\ + \varepsilon \sum_{\alpha} \left[\xi^*_{\alpha,1} \left(\frac{\xi_{\alpha,1} - \chi \xi_{\alpha,n}}{\varepsilon} - \mu \chi \xi_{\alpha,n} \right) + \hat{H}\left[\xi^*_{\alpha,1}, \chi \xi_{\alpha,n}\right] \right] \end{aligned} \tag{386}$$

Expressing the Hamiltonian $\hat{H}$ explicitly via complex variables for bosons and via Grassmann numbers for fermions, then

$$\begin{aligned} S[\xi^*,\xi] = \varepsilon \sum_{k=2}^{n} \sum_{\alpha} \left[\xi^*_{\alpha,k} \left(\frac{\xi_{\alpha,k} - \xi_{\alpha,k-1}}{\varepsilon} - \mu \xi_{\alpha,k-1} \right) + \epsilon_\alpha \, \xi^*_{\alpha,k} \xi_{\alpha,k-1} \right] \\ + \varepsilon \sum_{\alpha} \left[\xi^*_{\alpha,1} \left(\frac{\xi_{\alpha,1} - \chi \xi_{\alpha,n}}{\varepsilon} - \mu \chi \xi_{\alpha,n} \right) + \chi \epsilon_\alpha \, \xi^*_{\alpha,1} \xi_{\alpha,n} \right] \end{aligned} \tag{387}$$

Rearrangement of the terms in the summands gives

$$\begin{aligned} S[\xi^*,\xi] &= \sum_{k=2}^{n} \sum_{\alpha} \left[\xi^*_{\alpha,k} \left(\xi_{\alpha,k} - \xi_{\alpha,k-1} - \varepsilon\mu \xi_{\alpha,k-1} \right) + \varepsilon \epsilon_\alpha \, \xi^*_{\alpha,k} \xi_{\alpha,k-1} \right] \\ &+ \sum_{\alpha} \left[\xi^*_{\alpha,1} \left(\xi_{\alpha,1} - \chi \xi_{\alpha,n} - \varepsilon\mu\chi \xi_{\alpha,n} \right) + \varepsilon\chi \epsilon_\alpha \, \xi^*_{\alpha,1} \xi_{\alpha,n} \right] = \\ &= \sum_{k=2}^{n} \sum_{\alpha} \left[\left(\xi^*_{\alpha,k}\xi_{\alpha,k} - \xi^*_{\alpha,k}\xi_{\alpha,k-1} - \varepsilon\mu \xi^*_{\alpha,k}\xi_{\alpha,k-1} \right) + \varepsilon \epsilon_\alpha \, \xi^*_{\alpha,k} \xi_{\alpha,k-1} \right] \\ &+ \sum_{\alpha} \left[\left(\xi^*_{\alpha,1}\xi_{\alpha,1} - \chi \xi^*_{\alpha,1}\xi_{\alpha,n} - \varepsilon\mu\chi \xi^*_{\alpha,1}\xi_{\alpha,n} \right) + \varepsilon\chi \epsilon_\alpha \, \xi^*_{\alpha,1} \xi_{\alpha,n} \right] \end{aligned} \tag{388}$$

or

$$S[\xi^*,\xi] = \sum_{k=2}^{n} \sum_{\alpha} \left[\xi^*_{\alpha,k}\xi_{\alpha,k} - (1 - \varepsilon \epsilon_\alpha) \xi^*_{\alpha,k}\xi_{\alpha,k-1} \right] + \sum_{\alpha} \left[\xi^*_{\alpha,1}\xi_{\alpha,1} - \chi (1 - \varepsilon \epsilon_\alpha) \xi^*_{\alpha,1}\xi_{\alpha,n} \right] \tag{389}$$

Letting

$$a_\alpha = 1 - \varepsilon \epsilon_\alpha \tag{390}$$

then

$$S\left[\xi^*,\xi\right]=\sum_{k=2}^{n}\sum_{\alpha}\left[\xi^*_{\alpha,k}\xi_{\alpha,k}-a_\alpha\xi^*_{\alpha,k}\xi_{\alpha,k-1}\right]+\sum_{\alpha}\left[\xi^*_{\alpha,1}\xi_{\alpha,1}-\chi a_\alpha\xi^*_{\alpha,1}\xi_{\alpha,n}\right] \tag{391}$$

We now introduce $\xi_{\alpha,k}$ and $\xi^*_{\alpha,k}$, which are complex variables for bosons and Grassmann numbers for fermions:

$$\xi_\alpha=\begin{bmatrix}\xi_{\alpha,1}\\ \xi_{\alpha,2}\\ \vdots\\ \vdots\\ \xi_{\alpha,n-1}\\ \xi_{\alpha,n}\end{bmatrix},\quad \xi^*_\alpha=\begin{bmatrix}\xi^*_{\alpha,1} & \xi^*_{\alpha,2} & \cdots & \cdots & \xi^*_{\alpha,n-1} & \xi^*_{\alpha,n}\end{bmatrix} \tag{392}$$

as well as the $n\times n$ matrix $\mathbf{S}^{(\alpha)}$:

$$\mathbf{S}^{(\alpha)}=\begin{bmatrix}1 & 0 & \cdots & \cdots & 0 & -\chi a_\alpha\\ -a_\alpha & 1 & 0 & & & 0\\ 0 & -a_\alpha & 1 & \ddots & & \vdots\\ \vdots & \ddots & \ddots & 1 & \ddots & \vdots\\ \vdots & & & -a_\alpha & 1 & 0\\ 0 & \cdots & \cdots & 0 & -a_\alpha & 1\end{bmatrix} \tag{393}$$

From

$$\mathbf{S}^{(\alpha)}\cdot\xi_\alpha=\begin{bmatrix}\xi_{\alpha,1}-\chi a_\alpha\xi_{\alpha,n}\\ -a_\alpha\xi_{\alpha,1}+\xi_{\alpha,2}\\ \vdots\\ \vdots\\ -a_\alpha\xi_{\alpha,n-2}+\xi_{\alpha,n-1}\\ -a_\alpha\xi_{\alpha,n-1}+\xi_{\alpha,n}\end{bmatrix} \tag{394}$$

then

$$\begin{aligned}\xi^*_\alpha\cdot\mathbf{S}^{(\alpha)}\cdot\xi_\alpha &= \xi^*_{\alpha,1}\xi_{\alpha,1}-\chi a_\alpha\xi^*_{\alpha,1}\xi_{\alpha,n}-a_\alpha\xi^*_{\alpha,2}\xi_{\alpha,1}+\xi^*_{\alpha,2}\xi_{\alpha,2}+\xi^*_{\alpha,3}\xi_{\alpha,3}-\\ &-\cdots-a_\alpha\xi^*_{\alpha,n-1}\xi_{\alpha,n-2}+\xi^*_{\alpha,n-1}\xi_{\alpha,n-1}-a_\alpha\xi^*_{\alpha,n}\xi_{\alpha,n-1}+\xi^*_{\alpha,n}\xi_{\alpha,n}\end{aligned} \tag{395}$$

or

$$\xi^*_\alpha\cdot\mathbf{S}^{(\alpha)}\cdot\xi_\alpha=\sum_{k=2}^{n}\left[\xi^*_{\alpha,k}\xi_{\alpha,k}-a_\alpha\xi^*_{\alpha,k}\xi_{\alpha,k-1}\right]+\xi^*_{\alpha,1}\xi_{\alpha,1}-\chi a_\alpha\xi^*_{\alpha,1}\xi_{\alpha,n} \tag{396}$$

Comparing this with (391) we rewrite the following action in the Gaussian form where the field components decouple while the time does not:

$$S\left[\xi^*,\xi\right]=\sum_{\alpha}\sum_{k,k'=1}^{n}\xi^*_{\alpha,k}\mathbf{S}^{(\alpha)}_{k,k'}\xi_{\alpha,k'} \tag{397}$$

Because there is no interaction, this renders the action matrix almost diagonal and permits us to solve the partition function exactly:

$$Z_0 = \lim_{n\to\infty} \int d[\xi^*]d[\xi]\prod_\alpha \exp\{-\xi^*_{\alpha,k}\mathbf{S}^{(\alpha)}_{kk'}\xi_{\alpha,k'}\} \tag{398}$$

where the integration measure is

$$\int d[\xi^*]d[\xi] = \prod_{k,k'=1}^{n}\int\prod_\alpha \frac{d\xi^*_{\alpha,k}d\xi_{\alpha,k}}{N} \tag{399}$$

Then

$$Z_0 = \lim_{n\to\infty}\prod_\alpha \left[\det\mathbf{S}^{(\alpha)}\right]^{-\chi} \tag{400}$$

We compute the determinant of $\mathbf{S}^{(\alpha)}$ by expanding by minors along the first row:

$$\lim_{n\to\infty}\det\mathbf{S}^{(\alpha)} = \lim_{n\to\infty}\left[1+(-\chi a_\alpha)(-1)^{n-1}(-a_\alpha)^{n-1}\right] = \lim_{n\to\infty}\left[1+(-1)^{n-1}\chi(-a_\alpha)^n\right] = \lim_{n\to\infty}\left[1-\chi\left(1-\frac{\beta}{n}\epsilon_\alpha\right)^n\right] \tag{401}$$

or

$$\lim_{n\to\infty}\det\mathbf{S}^{(\alpha)} = 1-\chi\exp\{-\beta\epsilon_\alpha\} \tag{402}$$

This permits us to compute the familiar expression for the bare partition function for noninteracting particles:

$$Z_0 = \prod_\alpha\left[1-\chi\exp\{-\beta\epsilon_\alpha\}\right]^{-\chi} \tag{403}$$

From here, the grand thermodynamic potential can be computed

$$\Omega_0 = -\frac{1}{\beta}\ln Z_0 = \frac{\chi}{\beta}\sum_\alpha \ln\left[1-\chi\exp\{-\beta\epsilon_\alpha\}\right] \tag{404}$$

and the mean number of particles:

$$N = -\left.\frac{\partial\Omega_0}{\partial\mu}\right|_T = \sum_\alpha \frac{1}{\exp\{\beta\epsilon_\alpha\}-\chi} \equiv \sum_\alpha n_\chi(\epsilon_\alpha) \tag{405}$$

Here, $n_+(\epsilon_\alpha) = n_{\mathrm{BE}}(\epsilon_\alpha)$ is the Bose-Einstein distribution function, and $n_-(\epsilon_\alpha) = n_{\mathrm{FD}}(\epsilon_\alpha)$ is the Fermi-Dirac distribution function. Considering the grand thermodynamic potential, we can also compute the mean energy

$$E = \left.\frac{\partial\beta\Omega_0}{\partial\beta}\right|_\mu + \mu N = \sum_\alpha \tilde{\epsilon}_\alpha\, n_\chi(\epsilon_\alpha) \tag{406}$$

and the entropy

$$S=-\left.\frac{\partial\Omega_0}{\partial T}\right|_{\mu}=-\beta\sum_{\alpha}\left(\epsilon_\alpha n_\chi(\epsilon_\alpha)-\frac{\chi}{\beta}\ln\left[1-\chi\exp\{-\beta\epsilon_\alpha\}\right]\right) \tag{407}$$

or

$$S=-\sum_{\alpha}\left(n_\chi(\epsilon_\alpha)\ln n_\chi(\epsilon_\alpha)-\chi\left[1+\chi n_\chi(\epsilon_\alpha)\right]\ln\left[1+\chi n_\chi(\epsilon_\alpha)\right]\right) \tag{408}$$

3.2.2 Inverse Matrix of $\mathbf{S}^{(\alpha)}$

We find the inverse matrix of $\left[\mathbf{S}^{(\alpha)}\right]^{-1}$ that relates the bare Green's function:

$$\left[\mathbf{S}^{(\alpha)}\right]^{-1}=\begin{bmatrix}\eta_{11} & \eta_{21} & \cdots & \cdots & \eta_{n1}\\ \eta_{12} & \eta_{22} & & & \eta_{n2}\\ \vdots & \vdots & & & \vdots\\ \vdots & & & & \\ \eta_{1n} & \eta_{2n} & \cdots & \cdots & \eta_{nn}\end{bmatrix} \tag{409}$$

and

$$\mathbf{S}^{(\alpha)}\left[\mathbf{S}^{(\alpha)}\right]^{-1}=\hat{1} \tag{410}$$

or

$$\begin{bmatrix}\eta_{11}-\chi a_\alpha\eta_{1n} & \eta_{21}-\chi a_\alpha\eta_{2n} & \cdots & \cdots & \eta_{n1}-\chi a_\alpha\eta_{nn}\\ -a_\alpha\eta_{11}+\eta_{12} & -a_\alpha\eta_{21}+\eta_{22} & \cdots & \cdots & -a_\alpha\eta_{n1}+\eta_{n2}\\ \vdots & \vdots & \cdots & \cdots & \vdots\\ \vdots & \vdots & \cdots & \cdots & \vdots\\ -a_\alpha\eta_{1,n-2}+\eta_{1,n-1} & -a_\alpha\eta_{2,n-2}+\eta_{2,n-1} & & & -a_\alpha\eta_{n,n-2}+\eta_{n,n-1}\\ -a_\alpha\eta_{1,n-1}+\eta_{1n} & -a_\alpha\eta_{2,n-1}+\eta_{2n} & \cdots & \cdots & -a_\alpha\eta_{n,n-1}+\eta_{nn}\end{bmatrix}=\hat{1} \tag{411}$$

The solution may be obtained by equating the elements at the same positions on the left and right sides of equation (411). For convenience, we consider matrix elements in positions $(1,k)$:

$$\eta_{11}-\chi a_\alpha\eta_{1n}=1\quad,\quad -a_\alpha\eta_{11}+\eta_{12}=0\quad,\quad -a_\alpha\eta_{1,2}+\eta_{13}=0\quad,\quad\cdots \tag{412}$$

From the first equation of (412), we have

$$\eta_{11}=1+\chi a_\alpha\eta_{1n} \tag{413}$$

Substituting for η_{11} in the second equation of (412), we then have

$$\eta_{12}=a_\alpha\eta_{11}=a_\alpha\left(1+\chi a_\alpha\eta_{1n}\right)=a_\alpha+\chi a_\alpha^2\eta_{1n} \tag{414}$$

Substituting also for η_{12} in the third equation of (412), we then have

$$\eta_{13} = a_\alpha \eta_{12} = a_\alpha\left(a_\alpha + \chi a_\alpha^2 \eta_{1n}\right) = a_\alpha^2 + \chi a_\alpha^3 \eta_{1n} \tag{415}$$

From equations (413) to (415), we arrive at the matrix element

$$\eta_{1k} = a_\alpha^{k-1} + \chi a_\alpha^k \eta_{1n} \tag{416}$$

where

$$\eta_{1n} = a_\alpha^{n-1} + \chi a_\alpha^n \eta_{1n} \tag{417}$$

Consequently,

$$\eta_{1n}\left(1 - \chi a_\alpha^n\right) = a_\alpha^{n-1} \tag{418}$$

and

$$\eta_{1n} = \frac{a_\alpha^{n-1}}{1 - \chi a_\alpha^n} \tag{419}$$

Substituting (419) into expression (416), then the matrix element

$$\eta_{1k} = a_\alpha^{k-1} + \chi a_\alpha^k \frac{a_\alpha^{n-1}}{1 - \chi a_\alpha^n} = \frac{a_\alpha^{k-1}}{1 - \chi a_\alpha^n} \tag{420}$$

We do the same thing for the matrix elements at positions $(2,k)$ and find that

$$\eta_{21} = \chi a_\alpha \eta_{2n}\,,\ \eta_{22} = 1 + a_\alpha \eta_{21} = 1 + \chi a_\alpha^2 \eta_{2n}\,,\ \eta_{23} = a_\alpha \eta_{22} = a_\alpha\left(1 + \chi a_\alpha^2 \eta_{2n}\right) = a_\alpha + \chi a_\alpha^3 \eta_{2n}\,,\cdots \tag{421}$$

Hence follows the recursion relation

$$\eta_{2k} = a_\alpha^{k-2} + \chi a_\alpha^k \eta_{2n} \tag{422}$$

from where

$$\eta_{2n} = a_\alpha^{n-2} + \chi a_\alpha^n \eta_{2n} \tag{423}$$

and

$$\eta_{2n}\left(1 - \chi a_\alpha^n\right) = a_\alpha^{n-2} \tag{424}$$

then

$$\eta_{2n} = \frac{a_\alpha^{n-2}}{1 - \chi a_\alpha^n} \tag{425}$$

and next follows the matrix element

$$\eta_{2k} = a_\alpha^{k-2} + \chi a_\alpha^k \frac{a_\alpha^{n-2}}{1 - \chi a_\alpha^n} = \frac{a_\alpha^{k-2}}{1 - \chi a_\alpha^n} \tag{426}$$

Now, we do the same for the matrix elements at positions $(3,k)$ and find that

$$\eta_{31} = \chi a_\alpha \eta_{3n} \,,\, \eta_{32} = a_\alpha \eta_{31} \,,\, \eta_{33} = 1 + a_\alpha \eta_{32} \,,\, \eta_{34} = a_\alpha \eta_{33} \,,\, \eta_{35} = a_\alpha \eta_{34} \,,\cdots \tag{427}$$

From here, we arrive at the recursion relation

$$\eta_{3k} = a_\alpha^{k-3} + \chi a_\alpha^k \eta_{3n} \tag{428}$$

where

$$\eta_{3n} = a_\alpha^{n-3} + \chi a_\alpha^n \eta_{3n} \tag{429}$$

and

$$\eta_{3n}\left(1 - \chi a_\alpha^n\right) = a_\alpha^{n-3} \tag{430}$$

or

$$\eta_{3n} = \frac{a_\alpha^{n-3}}{1 - \chi a_\alpha^n} \tag{431}$$

Then the matrix element

$$\eta_{3k} = a_\alpha^{k-3} + \chi a_\alpha^k \frac{a_\alpha^{n-3}}{1 - \chi a_\alpha^n} = \frac{a_\alpha^{k-3}}{1 - \chi a_\alpha^n} \tag{432}$$

From equations (420) to (432), we have the general expression for the matrix element:

$$\eta_{lk} = a_\alpha^{k-l} + \chi a_\alpha^k \frac{a_\alpha^{n-l}}{1 - \chi a_\alpha^n} = \frac{a_\alpha^{k-l}}{1 - \chi a_\alpha^n} \tag{433}$$

This permits us to arrive at the following inverse matrix:

$$\left[\mathbf{S}^{(\alpha)}\right]^{-1} = \frac{1}{1 - \chi a_\alpha^n} \begin{bmatrix} 1 & \chi a_\alpha^{n-1} & \chi a_\alpha^{n-2} & \cdots & \chi a_\alpha \\ a_\alpha & 1 & \chi a_\alpha^{n-1} & \cdots & \chi a_\alpha^2 \\ a_\alpha^2 & a_\alpha & 1 & & \\ & a_\alpha^2 & a_\alpha & & \\ & & a_\alpha^2 & & \\ a_\alpha^{n-3} & & & & \\ a_\alpha^{n-2} & a_\alpha^{n-3} & & & \chi a_\alpha^{n-1} \\ a_\alpha^{n-1} & a_\alpha^{n-2} & a_\alpha^{n-3} & \cdots & 1 \end{bmatrix} \tag{434}$$

3.3 Bare Green's Function via Generating Functional

We show that the single-particle Green's function $\mathbf{G}_0$ is indeed a Green's function in the mathematical sense of being a solution to a differential equation, with a delta distribution or an inhomogeneous local source term. We calculate the single-particle Green's function via a discrete path integral. Letting

$$\tau_q \equiv \tau = q\frac{\beta}{n} \quad , \quad \tau_r \equiv \tau' = r\frac{\beta}{n} \tag{435}$$

where q and r are integers so that

$$\mathbf{G}_0(\sigma_1\tau_q,\sigma_2\tau_r)=\frac{1}{Z_0}\int d[\xi_\alpha^*]d[\xi_\alpha]\xi_{\sigma_1}(\tau_q)\xi_{\sigma_2}^*(\tau_r)\exp\left\{-\int_0^\beta d\tau\left[\sum_\alpha \xi_\alpha^*(\tau)\left(\frac{\partial}{\partial\tau}-\mu\right)\xi_\alpha(\tau)+\hat{H}[\xi_\alpha^*(\tau),\xi_\alpha(\tau)]\right]\right\} \tag{436}$$

or

$$\mathbf{G}_0(\sigma_1\tau_q,\sigma_2\tau_r)=\frac{1}{Z_0}\lim_{n\to\infty}\prod_{k,k'=1}^{n}\left[\int\prod_\alpha \frac{d\xi_{\alpha,k}^*d\xi_{\alpha,k}}{N}\xi_{\sigma_1,q}\xi_{\sigma_2,r}^*\prod_\alpha \exp\left\{-\xi_{\alpha,k}^*\mathbf{S}_{kk'}^{(\alpha)}\xi_{\alpha,k'}\right\}\right] \tag{437}$$

or

$$\mathbf{G}_0(\sigma_1\tau_q,\sigma_2\tau_r)=\frac{1}{Z_0}\lim_{n\to\infty}\int d[\xi^*]d[\xi]\,\xi_{\sigma_1,q}\xi_{\sigma_2,r}^*\prod_\alpha \exp\left\{-\xi_{\alpha,k}^*\mathbf{S}_{kk'}^{(\alpha)}\xi_{\alpha,k'}\right\} \tag{438}$$

Here Z_0 is the discrete form of the partition function. Considering the fact that the action does not couple to different field components, integrating an odd function over a symmetric interval yields zero. Therefore,

$$\mathbf{G}_0(\sigma_1\tau_q,\sigma_2\tau_r)=\lim_{n\to\infty}\delta_{\sigma_1\sigma_2}\frac{\int d[\xi^*]d[\xi]\,\xi_{\sigma_1,q}\xi_{\sigma_2,r}^*\prod_\alpha \exp\left\{-\xi_{\alpha,k}^*\mathbf{S}_{kk'}^{(\alpha)}\xi_{\alpha,k'}\right\}}{\int d[\xi^*]d[\xi]\prod_\alpha \exp\left\{-\xi_{\alpha,k}^*\mathbf{S}_{kk'}^{(\alpha)}\xi_{\alpha,k'}\right\}} \tag{439}$$

3.3.1 Generating Functional

We express the Green's function via source variables $\eta_{\alpha,k}$ and $\eta_{\alpha,k}^*$, which have no physical significance and serve as c-numbers for bosons and as anticommuting variables for fermions and are coupled linearly to the fields $\xi_{\alpha,k}^*$ and $\xi_{\alpha,k}$. This is possible via the generating functional

$$\mathbf{Z}[\eta,\eta^*]=\frac{\int d[\xi^*]d[\xi]\prod_\alpha \exp\left\{-\xi_{\sigma_2,k}^*\mathbf{S}_{kk'}^{(\alpha)}\xi_{\sigma_1,k'}+\eta_k^*\xi_{\sigma_1,k}+\xi_{\sigma_2,k}^*\eta_k\right\}}{\int d[\xi^*]d[\xi]\prod_\alpha \exp\left\{-\xi_{\sigma_2,k}^*\mathbf{S}_{kk'}^{(\alpha)}\xi_{\sigma_1,k'}\right\}} \tag{440}$$

and with the bare Green's function being

$$\mathbf{G}_0(\sigma_1\tau_q,\sigma_2\tau_r)=\lim_{n\to\infty}\delta_{\sigma_1\sigma_2}\frac{\chi\partial^2}{\partial\eta_q^*\partial\eta_r}\mathbf{Z}[\eta,\eta^*]\Big|_{\eta=\eta^*=0} \tag{441}$$

We displace the argument of the exponential function in $\mathbf{Z}[\eta,\eta^*]$ by the following change of variables:

$$\xi_{\alpha,k}=\xi_{\sigma_1,k}+\left[\mathbf{S}_{kk'}^{(\alpha)}\right]^{-1}\eta_{k'}\quad,\quad \xi_{\alpha,k}^*=\xi_{\sigma_2,k}^*+\eta_k^*\left[\mathbf{S}_{kk'}^{(\alpha)}\right]^{-1} \tag{442}$$

then

$$\mathbf{Z}\left[\eta,\eta^*\right]=\frac{\int\prod_{\alpha}\prod_{k,k'=1}^{n} d\xi_{\alpha,k}^* d\xi_{\alpha,k}\exp\left\{\left[-\xi_{\sigma_2,k}^*\mathbf{S}_{kk'}^{(\alpha)}\xi_{\sigma_1,k'}+\boldsymbol{\eta}_k^*\left[\mathbf{S}_{kk'}^{(\alpha)}\right]^{-1}\eta_{k'}\right]\right\}}{\int\prod_{\alpha}\prod_{k,k'=1}^{n} d\xi_{\alpha,k}^* d\xi_{\alpha,k}\exp\left\{-\xi_{\sigma_2,k}^*\mathbf{S}_{kk'}^{(\alpha)}\xi_{\sigma_1,k'}\right\}}=\prod_{k,k'=1}^{n}\exp\left\{\boldsymbol{\eta}_k^*\left[\mathbf{S}_{kk'}^{(\alpha)}\right]^{-1}\boldsymbol{\eta}_{k'}\right\} \tag{443}$$

and

$$\mathbf{G}_0\left(\sigma_1\tau_q,\sigma_2\tau_r\right)=\lim_{n\to\infty}\delta_{\sigma_1\sigma_2}\frac{\chi\partial^2}{\partial\eta_q^*\partial\eta_r}\mathbf{Z}\left[\eta,\eta^*\right]\Bigg|_{\eta=\eta^*}=\lim_{n\to\infty}\delta_{\sigma_1\sigma_2}\frac{\chi\partial^2}{\partial\eta_q^*\partial\eta_r}\prod_{k,k'=1}^{n}\exp\left\{\boldsymbol{\eta}_k^*\left[\mathbf{S}_{kk'}^{(\alpha)}\right]^{-1}\boldsymbol{\eta}_{k'}\right\}\Bigg|_{\eta=\eta^*=0} \tag{444}$$

or

$$\mathbf{G}_0\left(\sigma_1\tau_q,\sigma_2\tau_r\right)=\lim_{n\to\infty}\delta_{\sigma_1\sigma_2}\left[\mathbf{S}_{qr}^{(\alpha)}\right]^{-1} \tag{445}$$

We observe that the matrix $\left[\mathbf{S}_{qr}^{(\alpha)}\right]^{-1}$ describes the discrete form of the Green's function $\mathbf{G}_0\left(\alpha\tau,\alpha'\tau'\right)$, which is definitely a Green's function that is a (discrete) solution to the differential equation

$$\left(\frac{\partial}{\partial\tau}+\epsilon_\alpha-\mu\right)\mathbf{G}\left(\alpha\tau,\alpha\tau'\right)=\delta\left(\tau-\tau'\right) \tag{446}$$

Similarly, the differential equation can be solved by considering the following boundary conditions:

$$\mathbf{G}_0\left(\alpha\tau,\alpha\tau'\right)=\chi\mathbf{G}_0\left(\alpha 0,\alpha\tau'\right) \tag{447}$$

From the previous procedure, it is obvious that the Green's function $\mathbf{G}_0\left(\alpha\tau,\alpha'\tau'\right)$ gives the expectation value of a system, where a particle is inserted or created in a state $\xi_{\alpha K}$ at a time τ travels through the medium to a time τ' and is removed or annihilated there.

We see that the single-particle Green's function can be evaluated via the inverse matrix (434) and where its diagonal represents the case when $\tau_q=\tau_r$. In the lower triangle, $\tau_q\geq\tau_r$ and in the upper triangle, $\tau_q\leq\tau_r$. The upper triangle is used in similar cases. For $q<r$ then,

$$\begin{aligned}\lim_{n\to\infty}\left[\mathbf{S}_{qr}^{(\alpha)}\right]^{-1}&=\lim_{n\to\infty}\frac{a_\alpha^{q-r}}{1-\chi a_\alpha^n}=\lim_{n\to\infty}a_\alpha^{q-r}\left(1+\frac{\chi a_\alpha^n}{1-\chi a_\alpha^n}\right)=\lim_{n\to\infty}a_\alpha^{q-r}\left(1-\frac{1}{a_\alpha^{-n}-\chi}\right)=\\&=\lim_{n\to\infty}\left(1-\varepsilon\left(\tilde{\epsilon}_\alpha-\mu\right)\right)^{q-r}\left(1-\frac{1}{\left(1-\varepsilon\left(\tilde{\epsilon}_\alpha-\mu\right)\right)^{-n}-\chi}\right)\end{aligned} \tag{448}$$

or

$$\lim_{n\to\infty}\left[\mathbf{S}_{qr}^{(\alpha)}\right]^{-1}=\lim_{n\to\infty}\left(1-\frac{\beta}{n}\left(\epsilon_\alpha-\mu\right)\right)^{q-r}\left(1-\frac{1}{\left(1-\frac{\beta}{n}\left(\tilde{\epsilon}_\alpha-\mu\right)\right)^{-n}-\chi}\right)=\exp\left\{-\left(\epsilon_\alpha-\mu\right)\left(\tau_q-\tau_r\right)\right\}\left(1-\chi n_\alpha\right) \tag{449}$$

where

$$n_\alpha = \frac{1}{\exp\{\beta(\tilde{\epsilon}_\alpha - \mu)\} - \chi} \tag{450}$$

are the well-known bosonic and fermionic occupation numbers with the chemical potential, μ, which can vary with temperature and concentration. For the case of phonons as well as photons, $\mu = 0$ because the excitations do not conserve particle number.

Then, for $q \le r$,

$$\begin{aligned}\lim_{n\to\infty}\left[\mathbf{S}_{qr}^{(\alpha)}\right]^{-1} &= \lim_{n\to\infty}\frac{\chi a_\alpha^{n+q-r}}{1-\chi a_\alpha^n} = \lim_{n\to\infty} a_\alpha^{q-r}\frac{\chi a_\alpha^n}{1-\chi a_\alpha^n} = \lim_{n\to\infty} a_\alpha^{q-r}\frac{\chi}{a_\alpha^{-n}-\chi} = \\ &= \lim_{n\to\infty}\left(1-\varepsilon(\epsilon_\alpha - \mu)\right)^{q-r}\frac{\chi}{\left(1-\varepsilon(\tilde{\epsilon}_\alpha-\mu)\right)^{-n}-\chi}\end{aligned} \tag{451}$$

or

$$\lim_{n\to\infty}\left[\mathbf{S}_{qr}^{(\alpha)}\right]^{-1} = \lim_{n\to\infty}\left(1-\frac{\beta}{n}(\tilde{\epsilon}_\alpha-\mu)\right)^{\frac{n}{\beta}(\tau_q-\tau_r)}\frac{\chi}{\left(1-\frac{\beta}{n}(\tilde{\epsilon}_\alpha-\mu)\right)^{-n}-\chi} = \exp\{-(\tilde{\epsilon}_\alpha-\mu)(\tau_q-\tau_r)\}\chi n_\alpha \tag{452}$$

For $q \ge r$,

$$\lim_{n\to\infty}\left[\mathbf{S}_{qr}^{(\alpha)}\right]^{-1} = \lim_{n\to\infty}\frac{a_\alpha^{q-r}}{1-\chi a_\alpha^n} = \lim_{n\to\infty}\left(1-\varepsilon(\tilde{\epsilon}_\alpha-\mu)\right)^{q-r}\left(1+\frac{\chi}{\left(1-\varepsilon(\tilde{\epsilon}_\alpha-\mu)\right)^{-n}-\chi}\right) \tag{453}$$

or

$$\lim_{n\to\infty}\left[\mathbf{S}_{qr}^{(\alpha)}\right]^{-1} = \exp\{-(\tilde{\epsilon}_\alpha-\mu)(\tau_q-\tau_r)\}(1+\chi n_\alpha) \tag{454}$$

3.4 Single-Particle Green's Function

3.4.1 Matsubara Green's Function

When considering systems that are homogeneous in space, such as a liquid or gas, it is convenient to examine the given quantities in momentum representation. Similarly, for quantities that are homogeneous in time, it is convenient to Fourier transform from time to frequency.

Let us now combine the previous results for single-particle Green's function:

$$\mathbf{G}_0(\alpha\tau,\alpha'\tau') = \left\langle \mathrm{T}\hat{\psi}_\alpha(\tau)\hat{\psi}_\alpha^\dagger(\tau')\right\rangle = \delta_{\alpha\alpha'}\exp\{-(\tilde{\epsilon}_\alpha-\mu)(\tau-\tau')\}\left[\theta(\tau-\tau'-\delta)(1+\chi n_\alpha)+\chi\theta(\tau'-\tau+\delta)n_\alpha\right] \tag{455}$$

or

$$\overbrace{\hat{\psi}_\alpha(\tau)\hat{\psi}_\alpha^\dagger(\tau')} = \mathbf{G}_0(\alpha\tau,\alpha'\tau') = \delta_{\alpha\alpha'}\mathbf{G}_\alpha(\tau-\tau'-\delta) \tag{456}$$

and also

$$\overline{\hat{\psi}_{\alpha}^{\dagger}(\tau')\hat{\psi}_{\alpha}(\tau)} = \chi\delta_{\alpha\alpha'}\mathbf{G}_{\alpha}(\tau-\tau'-\delta) \tag{457}$$

Here, the infinitesimal δ serves as a reminder that the second term in (455) contributes at equal times. A convenient reminder for the δ is that the time τ' associated with the creation operator is always shifted one time step later. In the previous example, the occupation number n_{α} is given by the Fermi-Dirac function n_F for fermions and the Bose-Einstein function n_B for bosons. In frequency space,

$$\mathbf{G}_0(\alpha, i\omega_n) = \int_0^{\beta} d(\tau-\tau')\exp(i\omega_n(\tau-\tau'))\mathbf{G}_0(\alpha\tau, \alpha'\tau') \tag{458}$$

Letting

$$(\tau-\tau') = \tilde{\tau} \to \tau \tag{459}$$

then

$$\mathbf{G}_0(\alpha, i\omega_n) = \delta_{\alpha\alpha'}\int_0^{\beta} d\tau\exp\{(i\omega_n - (\tilde{\epsilon}_{\alpha} - \mu))\tau\}\left[\theta(\tau-\delta)(1+\chi n_{\alpha}) + \chi\theta(-\tau+\delta)n_{\alpha}\right] \tag{460}$$

or

$$\mathbf{G}_0(\alpha, i\omega_n) = \delta_{\alpha\alpha'}\frac{\exp\{(i\omega_n - (\tilde{\epsilon}_{\alpha} - \mu))\beta\} - 1}{i\omega_n - (\tilde{\epsilon}_{\alpha} - \mu)}\frac{\exp\{(\tilde{\epsilon}_{\alpha} - \mu)\beta\}}{\exp\{(\tilde{\epsilon}_{\alpha} - \mu)\beta\} - \chi} \tag{461}$$

Considering that

$$\exp\{i\omega_n\beta\} = \chi \tag{462}$$

then

$$\mathbf{G}_0(\alpha, i\omega_n) = \delta_{\alpha\alpha'}\frac{\chi - \exp\{(\tilde{\epsilon}_{\alpha} - \mu)\beta\}}{\left[i\omega_n - (\tilde{\epsilon}_{\alpha} - \mu)\right]\left[\exp\{(\tilde{\epsilon}_{\alpha} - \mu)\beta\} - \chi\right]} \tag{463}$$

and

$$\mathbf{G}_0(\alpha, i\omega_n) = \delta_{\alpha\alpha'}\frac{-1}{i\omega_n - \tilde{\epsilon}_{\alpha} + \mu} \tag{464}$$

From the analytic continuation, we observe the following for the retarded Green's function:

$$\mathbf{G}_0^R(\alpha, \omega) = \mathbf{G}_0(\alpha, i\omega_n \to \omega + i\delta) \tag{465}$$

This result is also valid for interacting systems. The retarded Green's function can therefore be deduced from the Matsubara Green's function (finite temperature) for interacting systems. A convenient mnemonic for δ is that the time τ' associated with the creation operator is always shifted one time step later. The entire aforementioned derivation is calculated in order to justify the evaluation of the path integral at equal times.

3.5 Noninteracting Green's Function

We evaluate the Green's function **G** for noninteracting particles via path integrals where we express the creation $\hat{\psi}^{\dagger}_{\alpha}(\tau)$ and annihilation $\hat{\psi}_{\alpha}(\tau)$ operators as functions of time τ permitting the time-ordering operator that approximately interlaces operators with no explicit time dependence τ. In addition, the evolution operator is represented via a functional integral, and the operators $\left\{\hat{\psi}^{\dagger}_{\alpha}(\tau),\hat{\psi}_{\alpha}(\tau)\right\}$ within the time slice τ are replaced by the coherent state variables $\left\{\xi^{*}_{\alpha}(\tau),\xi_{\alpha}(\tau)\right\}$. The time-ordered product manipulations are facilitated by writing the thermal higher-order imaginary Green's function in the form

$$\mathbf{G}^{(2n)}(\alpha_1\tau_1,\cdots,\alpha_n\tau_n,\alpha'_n\tau'_n,\cdots,\alpha'_1\tau'_1)=\left\langle \mathrm{T}\hat{\psi}_{\alpha_1}(\tau_1)\cdots\hat{\psi}_{\alpha_n}(\tau_n)\hat{\psi}^{\dagger}_{\alpha'_n}(\tau'_n)\cdots\hat{\psi}^{\dagger}_{\alpha'_1}(\tau'_1)\right\rangle \tag{466}$$

with operators in the Heisenberg representation:

$$\hat{\psi}_{\alpha}(\tau)=\hat{\rho}(-\tau)\hat{\psi}_{\alpha}\hat{\rho}(\tau)\quad,\quad \hat{\psi}^{\dagger}_{\alpha}(\tau)=\hat{\rho}(-\tau)\hat{\psi}^{\dagger}_{\alpha}\hat{\rho}(\tau)\quad,\quad \hat{\mathrm{H}}'=\hat{\mathrm{H}}-\mu\hat{N} \tag{467}$$

Note that the statistical or thermal average of an operator **F** can be defined as:

$$\langle \mathbf{F}\rangle=\frac{1}{Z_0}\mathrm{Tr}\left[\hat{\rho}(\beta)\mathbf{F}\right]\quad,\quad Z_0=\mathrm{Tr}\left[\hat{\rho}(\beta)\right] \tag{468}$$

Therefore, the higher-order Green's functions via path integration while using the property of time-ordering operators can be rewritten

$$\begin{aligned}&\mathbf{G}^{(2n)}(\alpha_1\tau_1,\cdots,\alpha_n\tau_n,\alpha'_n\tau'_n,\cdots,\alpha'_1\tau'_1)=\left\langle \mathrm{T}\hat{\psi}_{\alpha_1}(\tau_1)\cdots\hat{\psi}_{\alpha_n}(\tau_n)\hat{\psi}^{\dagger}_{\alpha'_n}(\tau'_n)\cdots\hat{\psi}^{\dagger}_{\alpha'_1}(\tau'_1)\right\rangle=\\&=\frac{1}{Z_0}\mathrm{Tr}\left[\hat{\rho}(\beta)\mathrm{T}\hat{\psi}_{\alpha_1}(\tau_1)\cdots\hat{\psi}_{\alpha_n}(\tau_n)\hat{\psi}^{\dagger}_{\alpha'_n}(\tau'_n)\cdots\hat{\psi}^{\dagger}_{\alpha'_1}(\tau'_1)\right]=\\&=\frac{1}{Z_0}\mathrm{Tr}\left[\hat{\rho}(\beta)\chi^{\mathrm{P}}\tilde{\psi}_{\alpha_{\mathrm{P}(1)}}\left(\tau_{\mathrm{P}(1)}\right)\tilde{\psi}_{\alpha_{\mathrm{P}(2)}}\left(\tau_{\mathrm{P}(2)}\right)\cdots\tilde{\psi}^{\dagger}_{\alpha_{\mathrm{P}(2n)}}\left(\tau_{\mathrm{P}(2n)}\right)\right]\end{aligned} \tag{469}$$

or

$$\begin{aligned}&\mathbf{G}^{(2n)}(\alpha_1\tau_1,\cdots,\alpha_n\tau_n,\alpha'_n\tau'_n,\cdots,\alpha'_1\tau'_1)=\frac{1}{Z_0}\chi^{\mathrm{P}}\,\mathrm{Tr}\left[\hat{\rho}(\beta)\tilde{\psi}_{\alpha_{\mathrm{P}(1)}}\left(\tau_{\mathrm{P}(1)}\right)\tilde{\psi}_{\alpha_{\mathrm{P}(2)}}\left(\tau_{\mathrm{P}(2)}\right)\cdots\tilde{\psi}^{\dagger}_{\alpha_{\mathrm{P}(2n)}}\left(\tau_{\mathrm{P}(2n)}\right)\right]=\\&=\frac{1}{Z_0}\chi^{\mathrm{P}}\,\mathrm{Tr}\begin{bmatrix}\hat{\rho}(\beta)\hat{\rho}\left(-\tau_{\mathrm{P}(1)}\right)\tilde{\psi}_{\alpha_{\mathrm{P}(1)}}\left(\tau_{\mathrm{P}(1)}\right)\hat{\rho}\left(\tau_{\mathrm{P}(1)}\right)\hat{\rho}\left(-\tau_{\mathrm{P}(2)}\right)\tilde{\psi}_{\alpha_{\mathrm{P}(2)}}\left(\tau_{\mathrm{P}(2)}\right)\hat{\rho}\left(\tau_{\mathrm{P}(2)}\right)\times\\ \times\cdots\hat{\rho}\left(-\tau_{\mathrm{P}(2n)}\right)\tilde{\psi}^{\dagger}_{\alpha_{\mathrm{P}(2n)}}\left(\tau_{\mathrm{P}(2n)}\right)\hat{\rho}\left(\tau_{\mathrm{P}(2n)}\right)\end{bmatrix}\end{aligned} \tag{470}$$

Here, the permutation P arranges the times in chronological order. From

$$\exp\left\{-\int_{\tau_{\mathrm{P}(1)}}^{\beta}d\tau'\hat{\mathrm{H}}'\right\}=\hat{\rho}(\beta)\hat{\rho}\left(-\tau_{\mathrm{P}(1)}\right)\quad,\quad \exp\left\{-\int_{\tau_{\mathrm{P}(2)}}^{\tau_{\mathrm{P}(1)}}d\tau'\hat{\mathrm{H}}'\right\}=\hat{\rho}\left(\tau_{\mathrm{P}(1)}\right)\hat{\rho}\left(-\tau_{\mathrm{P}(2)}\right) \tag{471}$$

and

$$\exp\left\{-\int_{\tau_{\mathrm{P}(-1)}}^{\tau_{\mathrm{P}(k)}}d\tau'\hat{\mathrm{H}}'\right\}=\hat{\rho}\left(\tau_{\mathrm{P}(k)}\right)\hat{\rho}\left(-\tau_{\mathrm{P}(k-1)}\right) \tag{472}$$

then

$$\mathbf{G}^{(2n)}(\alpha_1\tau_1,\cdots,\alpha_n\tau_n,\alpha_n'\tau_n',\cdots,\alpha_1'\tau_1')=$$

$$=\frac{1}{Z_0}\chi^{P}\,\mathrm{Tr}\left[\begin{array}{l}\exp\left\{-\int_{\tau_{P(1)}}^{\beta}d\tau'\hat{H}'\right\}\hat{\psi}_{\alpha_{P(1)}}\left(\tau_{P(1)}\right)\exp\left\{-\int_{\tau_{P(2)}}^{\tau_{P(1)}}d\tau'\hat{H}'\right\}\hat{\psi}_{\alpha_{P(2)}}\left(\tau_{P(2)}\right)\exp\left\{-\int_{\tau_{P(3)}}^{\tau_{P(2)}}d\tau'\hat{H}'\right\}\times\\ \times\cdots\exp\left\{-\int_{\tau_{P(2n)}}^{\tau_{P(2n-1)}}d\tau'\hat{H}'\right\}\hat{\psi}_{\alpha_{P(2n)}}\left(\tau_{P(2n)}\right)\exp\left\{-\int_{\tau_{P(2n+1)}}^{\tau_{P(2n)}}d\tau'\hat{H}'\right\}\end{array}\right] \tag{473}$$

and, consequently,

$$\mathbf{G}^{(2n)}(\alpha_1\tau_1,\cdots,\alpha_n\tau_n,\alpha_n'\tau_n',\cdots,\alpha_1'\tau_1')=\frac{1}{Z_0}\mathrm{Tr}\left[\mathrm{T}\exp\left\{-\int_0^{\beta}d\tau'\hat{H}'\right\}\hat{\psi}_{\alpha_1}(\tau_1)\hat{\psi}_{\alpha_2}(\tau_2)\cdots\hat{\psi}^{\dagger}_{\alpha_{2n}}(\tau_{2n})\right]=$$

$$=\frac{1}{Z_0}\int d\left[\xi^*_\alpha\right]d\left[\xi_\alpha\right]\xi_{\alpha_1}(\tau_1)\xi_{\alpha_2}(\tau_2)\cdots\xi^*_{\alpha_{2n}}(\tau_{2n})\exp\left\{-\int_0^{\beta}d\tau\left[\sum_\alpha\xi^{\dagger}_\alpha(\tau)\left(\frac{\partial}{\partial\tau}-\mu\right)\xi_\alpha(\tau)+\hat{H}\left[\xi^*_\alpha(\tau),\xi_\alpha(\tau)\right]\right]\right\} \tag{474}$$

The formula shows integration over all field configurations of the system with different phases given by the Hamiltonian determinant and multiplied by the fixed fields, $\xi_{\alpha_k}(\tau_k)$ and $\xi^*_{\alpha_k}(\tau_k)$. The Green's function changes the system, when one particle with field $\xi_\alpha(\tau)$ is present at time τ, absent with field $\xi^*_\alpha(\tau')$, and absent at time τ'.

3.6 Average Value of a Functional

Perturbation Theory

We now consider a two-body system conveniently described by the following Hamiltonian interaction:

$$\hat{H}=\sum_\alpha\tilde{\epsilon}_\alpha\,\hat{\psi}^{\dagger}_\alpha(\tau)\hat{\psi}_\alpha(\tau)+\frac{1}{2}\sum_{\alpha\beta\gamma\delta}\left(\alpha\beta\middle|\hat{U}\middle|\lambda\delta\right)\hat{\psi}^{\dagger}_\alpha(\tau)\hat{\psi}^{\dagger}_\beta(\tau)\hat{\psi}_\gamma(\tau)\hat{\psi}_\delta(\tau) \tag{475}$$

This Hamiltonian interaction is partitioned into the sum of the one-body (noninteracting) operator $\hat{H}_0$ (the first summand in equation [475]) and the residual (interacting) operator $\hat{U}$ (the second summand in equation [475]). This residual operator generally may contain a one-body interaction as well as a many-body interaction serving as a perturbation. We express the quantities via the creation $\hat{\psi}^{\dagger}_\alpha(\tau)$ and annihilation $\hat{\psi}_\alpha(\tau)$ operators as functions of time τ denoting the time slice defining them. The operators $\left\{\hat{\psi}^{\dagger}_\alpha(\tau),\hat{\psi}_\alpha(\tau)\right\}$ within the time slice τ are then replaced by the coherent state variables $\left\{\xi^*_\alpha(\tau),\xi_\alpha(\tau)\right\}$ in the functional integral as seen earlier. We compute the grand partition function via a path integral with the starting point of the thermal average of the functional

$$\mathbf{F}\equiv\mathbf{F}\left[\xi^*_\alpha(\tau_i),\cdots,\xi_\gamma(\tau_k),\cdots\right] \tag{476}$$

in the interaction representation via the following path integral

$$\langle\mathbf{F}\rangle=\frac{1}{Z_0}\int_{\xi(\beta)=\chi\xi(0)}d\left[\xi^*\right]d\left[\xi\right]\exp\left\{-S_0\left[\xi^*,\xi\right]\right\}\mathbf{F}\left[\xi^*_\alpha(\tau_i),\cdots,\xi_\gamma(\tau_k),\cdots\right] \tag{477}$$

where Z_0 is the partition function of the noninteracting system:

$$Z_0 = \int_{\xi(\beta)=\chi\xi(0)} d\left[\xi^*\right] d\left[\xi\right] \exp\left\{-S_0\left[\xi^*,\xi\right]\right\} \tag{478}$$

and the action functional of the bare system

$$S_0\left[\xi^*,\xi\right] = \int_0^\beta d\tau \sum_\alpha \xi_\alpha^*(\tau) \mathbf{G}_{\alpha\alpha}^{-1} \xi_\alpha(\tau) \quad , \quad \mathbf{G}_{\alpha\alpha}^{-1} = \frac{\partial}{\partial\tau} + \tilde{\epsilon}_\alpha - \mu \tag{479}$$

It is instructive to note that the time ordering that was explicit for operators is implicit for (477) because the functional integral always represents time-ordered products.

Suppose we have an interacting system and then the partition function of the entire system:

$$Z = \int_{\xi(\beta)=\chi\xi(0)} d\left[\xi^*\right] d\left[\xi\right] \exp\left\{-S_0\left[\xi^*,\xi\right] - S_{\text{int}}\left[\xi^*,\xi\right]\right\} \tag{480}$$

or

$$Z = \int_{\xi(\beta)=\chi\xi(0)} d\left[\xi^*\right] d\left[\xi\right] \exp\left\{-S_0\left[\xi^*,\xi\right]\right\} \exp\left\{-S_{\text{int}}\left[\xi^*,\xi\right]\right\} = Z_0 \left\langle \exp\left\{-S_{\text{int}}\left[\xi^*,\xi\right]\right\}\right\rangle \tag{481}$$

For the perturbation theory, we use series expansion of the exponential function in (481):

$$\frac{Z}{Z_0} = \left\langle \sum_{n=0}^{\infty} \frac{(-1)^n}{n!} \left[S_{\text{int}}\left[\xi^*,\xi\right]\right]^n \right\rangle \tag{482}$$

or

$$\frac{Z}{Z_0} = \left\langle \sum_{n=0}^{\infty} \frac{(-1)^n}{n!} S_{\text{int}}\left[\xi_{\alpha_1}^*, \xi_{\alpha_1}\right] \times \cdots \times S_{\text{int}}\left[\xi_{\alpha_k}^*, \xi_{\alpha_k}\right] \times \cdots \times S_{\text{int}}\left[\xi_{\alpha_n}^*, \xi_{\alpha_n}\right] \right\rangle \tag{483}$$

This equation can be rewritten in the form:

$$\frac{Z}{Z_0} = \sum_{n=0}^{\infty} \frac{(-1)^n}{n!} \prod_{k=1}^{n} \left\langle S_{\text{int}}\left[\xi_{\alpha_k}^*, \xi_{\alpha_k}\right]\right\rangle \tag{484}$$

For the two-particle interaction then,

$$\frac{Z}{Z_0} = \sum_{n=0}^{\infty} \frac{(-1)^n}{2^n n!} \prod_{k=1}^{n} \sum_{\alpha_k\beta_k\gamma_k\delta_k} \left(\alpha_k\beta_k \middle| \hat{U} \middle| \gamma_k\delta_k\right) \int_0^\beta d\tau_k \left\langle \xi_{\alpha_k}^*(\tau_k) \xi_{\beta_k}^*(\tau_k) \xi_{\delta_k}(\tau_k) \xi_{\gamma_k}(\tau_k) \right\rangle \tag{485}$$

In equation (485), every term of the series is an average of a chronological product of particle field operators in the interaction representation. The **expectation value can be expanded via Wick theorem in all the contractions.**

Let us examine the example of an atomic Fermi gas by considering an electron gas [16]. The **Fermi liquid** concept is the deviation of the electron gas from the ideal gas behavior. This concept was introduced by Lev Davidovich Landau [17] and provides an effective description of interacting fermions that can be derived from microscopic physics either by resummation of quantum mechanical perturbation expansions or by renormalization group techniques.

The partition function Z_0 of a quantum ideal gas is an ideal test for the field-theoretical method. A detailed knowledge of a quantum ideal gas is an important step in understanding experiments with trapped atomic gases. The calculation of the retarded Green's functions and associated spectral functions is highly nontrivial and can only be approximated in most interacting systems. In next chapter, we use the so-called diagram technique to calculate some quantities. This technique has a great advantage over the ordinary form of the perturbation theory.

4

Perturbation Theory and Feynman Diagrams

Introduction

We now turn our attention to interacting systems described by non-Gaussian actions where the functional integral generally cannot be calculated exactly. Hence, it is necessary to adopt various approximation schemes for studying these interactions. The most common of these schemes is the perturbation theory in higher-order terms: when interactions among particles are weak, then they can be perturbatively taken into account with respect to kinetic energy. Generally, perturbation theory in many-particle systems is based on Wick theorem formulated in terms of Feynman diagrams, which are useful for providing an insight into the physical process that they represent. It is in the case of interacting particles and fields that the power of quantum field theory and Feynman diagrams indeed comes into play.

4.1 Representation as Diagrams

We now will introduce Feynman diagrams, which are nothing but a way of keeping track of the contractions mentioned earlier and, in particular, in relation (353). This correspondence is produced by following the **Feynman diagrammatic rules**:

1. For every Green's function

$$\mathbf{G}_0(\alpha_k\tau_k,\alpha'_k\tau'_k)=\delta_{\alpha_k\alpha'_k}\mathbf{G}_{\alpha_k}(\tau_k-\tau'_k) \tag{486}$$

with

$$\delta_{\alpha_k\alpha'_k}\mathbf{G}_{\alpha_k}(\tau_k-\tau'_k)=\delta_{\alpha_k\alpha'_k}\exp\{-(\epsilon_{\alpha_k}-\mu)(\tau_k-\tau'_k)\}\left[(1+\chi n_{\alpha_k})\theta(\tau_k-\tau'_k)+\chi n_{\alpha_k}\theta(\tau'_k-\tau_k)\right] \tag{487}$$

we have a directed line

$$\mathbf{G}_0(\alpha_k\tau_k,\alpha'_k\tau'_k)=\delta_{\alpha_k\alpha'_k}\mathbf{G}_{\alpha_k}(\tau_k-\tau'_k)= \tag{488}$$

that starts at $(\alpha_k\tau_k)$, the creation of a particle in the single-particle state α_k at time τ_k, and ends at $(\alpha'_k\tau'_k)$, the annihilation of a particle in the single-particle state α'_k at time τ'_k;

2. For every interaction $\left(\alpha_k\beta_k\left|\hat{U}\right|\gamma_k\delta_k\right)$, there is a wiggly line that does not require an arrow

$$U= \tag{489}$$

3. For each interaction, there is a vertex (vertex = the meeting of two directed and one wiggly line) with two incoming lines corresponding to $\xi_{\delta_k}(\tau_k)\xi_{\gamma_k}(\tau_k)$ and two outgoing lines corresponding to $\xi^*_{\alpha_k}(\tau_k)\xi^*_{\beta_k}(\tau_k)$. The n interaction in (485) can then be represented by n vertices with two outgoing lines α_k,β_k and two incoming lines γ_k,δ_k acting at time τ_k and corresponding to the factor

$$\left(\alpha_k\beta_k\middle|\hat{U}\middle|\gamma_k\delta_k\right) = \tag{490}$$

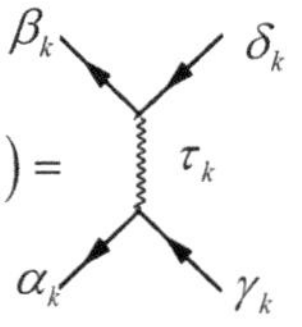

4.2 Generating Functionals

The cornerstone of the functional techniques is the concept of generating functionals sufficient to derive all propagators. Usually, generating functionals are obtained by coupling the field operators to one or more external fields. These are called the source fields $\eta_\alpha(\tau)$ and $\eta^*_\alpha(\tau)$, which are c-numbers for bosons and anticommuting variables for fermions. The source fields may be thought of as probes inside the quantum system; the propagators are probed by varying the source fields. In the current study, we express Green's functions (connected Green's functions) via source variables $\eta_\alpha(\tau)$ and $\eta^*_\alpha(\tau)$ coupled bilinearly to the fields $\xi^*_\alpha(\tau)$ and $\xi_\alpha(\tau)$:

$$Z\left[\eta,\eta^*\right] = \int d\left[\xi^*\right]d\left[\xi\right]\exp\left\{-S_0\left[\xi^*,\xi\right]+S\left[\eta^*,\eta,\xi\right]\right\} \tag{491}$$

where the bare action functional

$$S_0\left[\xi^*,\xi\right] = \sum_{\alpha\alpha'}\int_0^\beta d\tau d\tau'\xi^*_\alpha(\tau)\mathbf{G}^{-1}(\alpha\tau;\alpha'\tau')\xi_{\alpha'}(\tau') \tag{492}$$

and the source fields couple to the fermionic or bosonic operators as:

$$S\left[\eta^*,\eta,\xi\right] = \sum_{\alpha}\int_0^\beta d\tau\left(\xi^*_\alpha(\tau)\eta_\alpha(\tau)+\eta^*_\alpha(\tau)\xi_\alpha(\tau)\right) \tag{493}$$

Displacing the argument of the exponential function in equation (491) by the following change of variables:

$$\xi_\alpha = \xi'_\alpha + \mathbf{G}_{\alpha,\alpha'}\eta_{\alpha'} \quad , \quad \xi^*_\alpha = \xi'^*_\alpha + \eta^*_{\alpha'}\mathbf{G}_{\alpha,\alpha'} \tag{494}$$

equation (491) then becomes

$$Z\left[\eta,\eta^*\right] = Z[0,0]\exp\left\{\int d\tau d\tau'\sum_{\alpha\alpha'}\eta^*_\alpha(\tau)\mathbf{G}_{\alpha,\alpha'}\eta_{\alpha'}(\tau')\right\} \tag{495}$$

where $Z[0,0]$ imitates the partition function of the noninteracting system.

4.3 Wick Theorem

We now evaluate the following thermal average

$$\left\langle \hat{T}\xi_{\alpha_1}(\tau_1)\cdots\xi_{\alpha_n}(\tau_n)\xi^*_{\alpha_n'}(\tau_n')\cdots\xi^*_{\alpha_1'}(\tau_1')\right\rangle = \frac{1}{Z[0,0]}\int d\left[\xi^*\right]d\left[\xi\right]\xi_{\alpha_1}\cdots\xi_{\alpha_n}\xi^*_{\alpha_n'}\cdots\xi^*_{\alpha_1'}\exp\left\{-S_0\left[\xi^*,\xi\right]\right\} \tag{496}$$

that equally may be expressed as a functional derivative:

$$\left\langle \hat{T}\xi_{\alpha_1}(\tau_1)\cdots\xi_{\alpha_n}(\tau_n)\xi^*_{\alpha_n'}(\tau_n')\cdots\xi^*_{\alpha_1'}(\tau_1')\right\rangle = \frac{1}{Z[0,0]}\frac{\chi^n\partial^{(2n)}Z\left[\eta,\eta^*\right]}{\partial\eta^*_{\alpha_1}(\tau_1)\cdots\partial\eta^*_{\alpha_n}(\tau_n)\partial\eta_{\alpha_n'}(\tau_n')\cdots\partial\eta_{\alpha_1'}(\tau_1')}\Bigg|_{\eta,\eta^*=0} \tag{497}$$

or

$$\left\langle \hat{T}\xi_{\alpha_1}(\tau_1)\cdots\xi_{\alpha_n}(\tau_n)\xi^*_{\alpha_n'}(\tau_n')\cdots\xi^*_{\alpha_1'}(\tau_1')\right\rangle = \frac{\chi^n\partial^{(2n)}\mathbf{G}\left[\eta,\eta^*\right]}{\partial\eta^*_{\alpha_1}(\tau_1)\cdots\partial\eta^*_{\alpha_n}(\tau_n)\partial\eta_{\alpha_n'}(\tau_n')\cdots\partial\eta_{\alpha_1'}(\tau_1')}\Bigg|_{\eta,\eta^*=0} \tag{498}$$

or as the Feynman higher-order Green's functions

$$\mathbf{G}^{(2n)}\left(\alpha_1\tau_1,\cdots,\alpha_n\tau_n,\alpha_n'\tau',\cdots,\alpha_1'\tau_1'\right) = \left\langle \hat{T}\xi_{\alpha_1}(\tau_1)\cdots\xi_{\alpha_n}(\tau_n)\xi^*_{\alpha_n'}(\tau_n')\cdots\xi^*_{\alpha_1'}(\tau_1')\right\rangle \tag{499}$$

or simply,

$$\mathbf{G}^{(2n)}\left(\alpha_1\tau_1,\cdots,\alpha_n\tau_n,\alpha_n'\tau',\cdots,\alpha_1'\tau_1'\right) \equiv \left\langle \xi_{\alpha_1}(\tau_1)\cdots\xi_{\alpha_n}(\tau_n)\xi^*_{\alpha_n'}(\tau_n')\cdots\xi^*_{\alpha_1'}(\tau_1')\right\rangle \tag{500}$$

This last expression in equation (500) is introduced as a simple convention where $\xi_\alpha(\tau)$ and $\xi^*_\alpha(\tau)$ denote the imaginary time annihilation and creation operators, respectively, in the Heisenberg representation. From equations (496) through (500), **the functional derivative of $\mathbf{G}\left[\eta,\eta^*\right]$ is observed to yield vacuum expectation values of the time-ordered product of field operators (with such products providing an appropriate order for the field operators along the trajectories in the path integral):**

$$\left\langle \xi_{\alpha_1}(\tau_1)\cdots\xi_{\alpha_n}(\tau_n)\xi^*_{\alpha_n'}(\tau_n')\cdots\xi^*_{\alpha_1'}(\tau_1')\right\rangle = \sum_{\mathrm{P}}\chi^{\mathrm{P}}\left(\left\langle\xi^*_{\alpha_1'}(\tau_1')\xi_{\alpha\mathrm{P}(1)}\left(\tau_{\mathrm{P}(1)}\right)\right\rangle\cdots\left\langle\xi^*_{\alpha_n'}(\tau_n')\xi_{\alpha\mathrm{P}(n)}\left(\tau_{\mathrm{P}(n)}\right)\right\rangle\right) \tag{501}$$

or

$$\left\langle \xi_{\alpha_1}(\tau_1)\cdots\xi_{\alpha_n}(\tau_n)\xi^*_{\alpha_n'}(\tau_n')\cdots\xi^*_{\alpha_1'}(\tau_1')\right\rangle = \sum_{\mathrm{P}}\chi^{\mathrm{P}}\prod_k\left\langle\xi^*_{\alpha_k'}(\tau_k')\xi_{\alpha\mathrm{P}(k)}\left(\tau_{\mathrm{P}(k)}\right)\right\rangle \tag{502}$$

This is **Wick theorem: The thermal average is the sum over all complete sets of contractions whereby a complete contraction is a configuration in which each ξ_α is contracted with a ξ^*_α and the sign is specified by χ^{P} with P being the permutation such that $\xi^*_{\alpha_k}$ is contracted with $\xi_{\alpha\mathrm{P}(k)}$, that is, the factor χ^{P} takes into account that for fermions (minus sign) it costs a sign change every time a pair of operators $\xi^*_{\alpha_k}$ and $\xi_{\alpha\mathrm{P}(k)}$ are contracted.**

Wick theorem holds for imaginary as well as real times where,

$$\text{contractions} = \left\langle \xi^*_{\alpha'_n}(\tau'_n)\xi_{\alpha_n}(\tau_n)\right\rangle = \overbrace{\xi^*_{\alpha'_n}(\tau'_n)\xi_{\alpha_n}(\tau_n)} = \delta_{\alpha'_n\alpha_n}\mathbf{G}_{\alpha'_n}(\tau'_n-\tau_n-\delta) \tag{503}$$

So,

$$\overbrace{\xi_{\alpha'}(\tau')\xi^*_{\alpha}(\tau)} = \delta_{\alpha'\alpha}\mathbf{G}_{\alpha'}(\tau'-\tau) \quad , \quad \overbrace{\xi^*_{\alpha}(\tau)\xi_{\alpha'}(\tau')} = \chi\delta_{\alpha\alpha'}\mathbf{G}_{\alpha}(\tau-\tau') \tag{504}$$

Note that there are terms that do not pair creation and annihilation operators with such terms leading to matrix elements between orthogonal states:

$$\overbrace{\xi_{\alpha}(\tau)\xi_{\alpha'}(\tau')} = \left\langle \xi_{\alpha}(\tau)\xi_{\alpha'}(\tau')\right\rangle = 0 \quad , \quad \overbrace{\xi^*_{\alpha}(\tau)\xi^*_{\alpha'}(\tau')} = \left\langle \xi^*_{\alpha}(\tau)\xi^*_{\alpha'}(\tau')\right\rangle = 0 \tag{505}$$

unless $\alpha = \alpha'$. So, within the pairing, a pairing bracket (the labels α and α') must be the same and denote eigenstates where the creation and annihilation operators refer to the same state. Considering the expectation value of a product of an unequal number of ξ_α and ξ^*_α, then it still would be equal to the sum of all contractions because the complete expectation value would vanish, and at least one contraction in each complete set of contractions would also vanish. From (503) and (501) then,

$$\left\langle \xi_{\alpha_1}(\tau_1)\cdots\xi_{\alpha_n}(\tau_n)\xi^*_{\alpha'_n}(\tau'_n)\cdots\xi^*_{\alpha'_1}(\tau'_1)\right\rangle = \sum_{\mathrm{P}}\chi^{\mathrm{P}}\prod_{k=1}^{n}\delta_{\alpha_{\mathrm{P}(k)},\alpha'_k}\mathbf{G}_{\alpha'_k}\left(\tau_{\alpha_{\mathrm{P}(k)}}-\tau_{\alpha'_k}\right) \tag{506}$$

As a **consequence of Wick theorem, the n-particle Green's function for a noninteracting system is the sum of all permutations of the product of single-particle Green's functions.**

Note that the superscript $(2n)$ in (499) and (500) corresponds to the number of operators rather than the number n of particles propagating in the system. In normal systems, the only nonvanishing Green's functions are of the kind in (499) and (500) with n incoming particles and n outgoing particles. From (506), each contraction joins some $\xi^*_{\alpha_k}(\tau_k)$ to some $\xi_{\alpha'_k}(\tau'_k)$, resulting in the Green's function $\delta_{\alpha'_k\alpha_k}\mathbf{G}_{\alpha'_k}(\tau_k-\tau'_k)$ that diagrammatically is represented by a directed line in (488). This is interpreted as a particle introduced at τ_k with α_k that travels directly and then arrives at τ'_k with α'_k without interacting with (scattering off) any other particles. From Wick theorem, we see that once the single-particle Green's function $\mathbf{G}_\alpha$ is known, then higher-order Green's functions are also known, in a noninteracting system.

EXERCISE

From (499), write the **two-particle (four-point) Green's function** with its diagrammatic representation.

$$\mathbf{G}^{(2\times2)}_{\alpha_1\alpha_2}(\alpha_1\tau_1,\alpha_2\tau_2;\alpha'_2\tau'_2,\alpha'_1\tau'_1) = \left\langle \mathrm{T}\xi^*_{\alpha_1}(\tau_1)\xi^*_{\alpha_2}(\tau_2)\xi_{\alpha'_2}(\tau'_2)\xi_{\alpha'_1}(\tau'_1)\right\rangle \tag{507}$$

But

$$\left\langle \xi^*_{\alpha_1}(\tau_1)\xi^*_{\alpha_2}(\tau_2)\xi_{\alpha'_2}(\tau'_2)\xi_{\alpha'_1}(\tau'_1)\right\rangle = \overbrace{\xi^*_{\alpha_1}(\tau_1)\overbrace{\xi^*_{\alpha_2}(\tau_2)\xi_{\alpha'_2}(\tau'_2)}\xi_{\alpha'_1}(\tau'_1)} + \overbrace{\xi^*_{\alpha_1}(\tau_1)\underbrace{\xi^*_{\alpha_2}(\tau_2)\xi_{\alpha'_2}(\tau'_2)}\xi_{\alpha'_1}(\tau'_1)} \tag{508}$$

or

$$\left\langle \xi^*_{\alpha_1}(\tau_1)\xi^*_{\alpha_2}(\tau_2)\xi_{\alpha'_2}(\tau'_2)\xi_{\alpha'_1}(\tau'_1)\right\rangle = \left\langle \xi^*_{\alpha_1}(\tau_1)\xi_{\alpha'_1}(\tau'_1)\right\rangle\left\langle \xi^*_{\alpha_2}(\tau_2)\xi_{\alpha'_2}(\tau'_2)\right\rangle + \left\langle \xi^*_{\alpha_1}(\tau_1)\xi_{\alpha'_2}(\tau'_2)\right\rangle\left\langle \xi_{\alpha'_1}(\tau'_1)\xi^*_{\alpha_2}(\tau_2)\right\rangle \tag{509}$$

or

$$\left\langle \xi^*_{\alpha_1}(\tau_1)\xi^*_{\alpha_2}(\tau_2)\xi_{\alpha_2'}(\tau_2')\xi_{\alpha_1'}(\tau_1')\right\rangle = \delta_{\alpha_1\alpha_1'}\mathbf{G}_{\alpha_1}(\tau_1-\tau_1')\delta_{\alpha_2\alpha_2'}\mathbf{G}_{\alpha_2}(\tau_2-\tau_2')+\chi\delta_{\alpha_1\alpha_2'}\delta_{\alpha_1'\alpha_2}\mathbf{G}_{\alpha_1}(\tau_1-\tau_2')\mathbf{G}_{\alpha_1'}(\tau_1'-\tau_2) \tag{510}$$

Here, the two particles created in the states α_1 and α_2 at the time moments τ_1 and τ_2 propagate independently. The second term stems from the indiscernibility of the particles with the factor χ imitating bosonic or fermionic statistics. We represent the two-particle Green's function $\mathbf{G}^{(4)}_{\alpha_1\alpha_2}(\alpha_1\tau_1,\alpha_2\tau_2;\alpha_2'\tau_2',\alpha_1'\tau_1')$ diagrammatically, that is, the first and second summands in (510) represented by the first diagram and second diagrams, respectively, in equation (511):

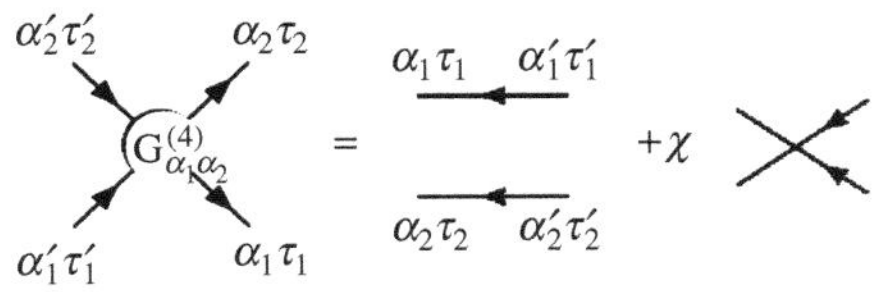

(511)

In this book, we denote the single-particle Green's function $\mathbf{G}^{(2)}_{\alpha\alpha}$ merely as $\mathbf{G}^{(2)}_{\alpha\alpha} \equiv \mathbf{G}^{(0)}_{\alpha}$. As noted earlier, we therefore observe that path integration always yields vacuum expectation values of time-ordered products of operators, and $\mathbf{Z}\left[\eta,\eta^*\right]$ is viewed as the generating functional of the propagators.

4.4 Perturbation Theory

Interacting Green's Function

We next consider perturbation theory and take into account the Hamiltonian determinant, which is the sum of a one-body operator $\hat{H}_0$ and the residual $\hat{U}$ Hamiltonian. This residual Hamiltonian generally may contain a single-particle as well as a many-particle interaction. The residual interaction will facilitate expressing the interacting Green's function via the perturbation theory and consequently calculate the self-energy given by the Dyson equation, which will be discussed later in this book. We now consider a many-particle system with a two-particle interaction and write the partition function:

$$Z = \int_{\xi(\beta)=\chi\xi(0)} d\left[\xi^*\right]d\left[\xi\right]\exp\left\{-S_0\left[\xi^*,\xi\right]-S_{\text{int}}\left[\xi^*,\xi\right]\right\} \tag{512}$$

where the bare action functional

$$S_0\left[\xi^*,\xi\right] = \sum_{\alpha\alpha'}\int_0^\beta d\tau d\tau'\xi^*_\alpha(\tau)\mathbf{G}^{-1}(\alpha\tau;\alpha'\tau')\xi_{\alpha'}(\tau') \tag{513}$$

and the two-particle interaction action functional

$$S_{\text{int}}\left[\xi^*,\xi\right] = \frac{1}{2}\int_0^\beta d\tau \sum_{\alpha_1\alpha_2\alpha_1'\alpha_2'}\left(\alpha_1\alpha_2\middle|\hat{U}\middle|\alpha_1'\alpha_2'\right)\xi^*_{\alpha_1}(\tau)\xi^*_{\alpha_2}(\tau)\xi_{\alpha_2'}(\tau)\xi_{\alpha_1'}(\tau) \tag{514}$$

The partition function in (512) can be rewritten

$$Z = \int_{\xi(\beta)=\chi\xi(0)} d\left[\xi^*\right]d\left[\xi\right]\exp\left\{-S_0\left[\xi^*,\xi\right]-S_{\text{int}}\left[\xi^*,\xi\right]\right\} = Z_0\left\langle\exp\left\{-S_{\text{int}}\left[\xi^*,\xi\right]\right\}\right\rangle \tag{515}$$

From this expression, we proceed to derive some form of Wick theorem that will facilitate evaluating the thermal averages of the products ξ^* and ξ as well as help in developing a set of rules for constructing Feynman diagrams. To apply the perturbation theory, we perform series expansion of the exponential function in the partition function (515) in powers of the two-body interaction $\left(\alpha_k\beta_k\middle|\hat{U}\middle|\gamma_k\delta_k\right)$:

$$\frac{Z}{Z_0}=\left\langle \exp\left\{-S_{\text{int}}\left[\xi^*,\xi\right]\right\}\right\rangle=\sum_{n=0}^{\infty}\frac{(-1)^n}{n!}\left\langle S_{\text{int}}^n\left[\xi^*,\xi\right]\right\rangle \tag{516}$$

where

$$\frac{Z}{Z_0}=\sum_{n=0}^{\infty}\frac{(-1)^n}{2^n n!}\prod_{k=1}^{n}\sum_{\alpha_k\beta_k\gamma_k\delta_k}\left(\alpha_k\beta_k\middle|\hat{U}\middle|\gamma_k\delta_k\right)\int_0^{\beta}d\tau_k\left\langle \xi_{\alpha_k}^*(\tau_k)\xi_{\beta_k}^*(\tau_k)\xi_{\delta_k}(\tau_k)\xi_{\gamma_k}(\tau_k)\right\rangle \tag{517}$$

and the partition function of the noninteracting system

$$Z_0=\int_{\xi(\beta)=\chi\xi(0)}d\left[\xi^*\right]d\left[\xi\right]\exp\left\{-S_0\left[\xi^*,\xi\right]\right\}$$

We note the time ordering that is explicit for operators is implicit here because the functional integral always represents the chronological product of field operators in the interaction representation. It is evident that any order in the perturbation expansion is an ensemble average evaluated over noninteracting states of a time-ordered product of creation and annihilation operators. In this manner, we can relate Wick theorem connecting these averages with contractions of operators. Though Wick theorem is an exact operator identity, the finite temperature Wick theorem is only valid for thermally averaged operators, as shown in equation (517), where every term of the series has an average of a chronological product of particle field operators in the interaction representation. Equation (517) shows Wick theorem applied in series expansion of $\frac{Z}{Z_0}$ where the average of a chronological product of particle field operators in the interaction representation connects the matrix elements of the potential $\left(\alpha_k\beta_k\middle|\hat{U}\middle|\gamma_k\delta_k\right)$ in all possible ways. So, from Wick theorem, it is now possible to justify **Feynman diagrammatic rules for series expansion in (517)**:

1. The expectation values are taken with respect to the noninteracting ground state.
2. All operators are in the interaction representation, which is really simply the Heisenberg representation for the noninteracting problem.
3. The interacting propagator is written as a sum of noninteracting expectation values. The nth term in the series contains $\left(\alpha_k\beta_k\middle|\hat{U}\middle|\gamma_k\delta_k\right)$ precisely n times.
4. From Wick theorem, each term in the series expansion can be written via noninteracting Green's functions. This is the general principle behind perturbation theory in quantum field theory.
5. Basically, the rules governing the constructing of the diagrams depend on the exact form of $\left(\alpha_k\beta_k\middle|\hat{U}\middle|\gamma_k\delta_k\right)$.
6. From (517), each contraction joins some $\xi_{\alpha_k}^*(\tau_k)$ to some $\xi_{\alpha_k'}(\tau_k')$ and resulting in the Green's function

$$\delta_{\alpha_k\alpha_k'}\mathbf{G}_{\alpha_k}(\tau_k-\tau_k')=\delta_{\alpha_k\alpha_k'}\exp\left\{-(\epsilon_{\alpha_k}-\mu)(\tau_k-\tau_k')\right\}\left[(1+\chi n_{\alpha_k})\theta(\tau_k-\tau_k'-\delta)+\chi n_{\alpha_k}\theta(\tau_k'-\tau_k+\delta)\right] \tag{518}$$

that is represented diagrammatically by a directed line.

$$\delta_{\alpha_k\alpha_k'}\mathbf{G}_{\alpha_k}(\tau_k-\tau_k')=\overset{\alpha_k\alpha_k}{}\longrightarrow\overset{\alpha_k'\alpha_k'}{} \tag{519}$$

7. Each interaction yields a vertex with two incoming lines corresponding to $\xi_{\delta_k}(\tau_k)\xi_{\gamma_k}(\tau_k)$ and outgoing lines corresponding to $\xi^*_{\alpha_k}(\tau_k)\xi^*_{\beta_k}(\tau_k)$. Therefore, the corresponding interactions' vertices $\left(\alpha_k\beta_k\middle|\hat{U}\middle|\gamma_k\delta_k\right)$ in (517) will be represented by vertices with two outgoing lines α_k, β_k and two incoming lines δ_k, γ_k acting at time τ_k:

$$\left(\alpha_k\beta_k\middle|\hat{U}\middle|\gamma_k\delta_k\right) = \tag{520}$$

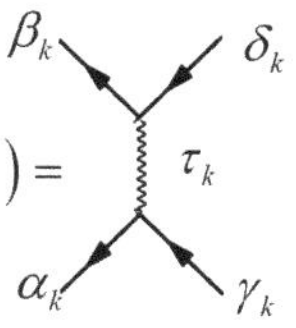

We observe from equation (517) that each summand of order n has n vertices and $2n$ fields ξ contracting with $2n$ fields ξ^* yielding $(2n)!$ contractions (or diagrams). Because the diagrams are seen as a representation of the contributing terms, they are not uniquely defined. If the diagrams can be transformed smoothly into each other then they are merely equal. This implies conserving the arrows and labels of the diagram. The overall prefactor $\frac{(-1)^n}{2^n n!}$ must be considered for each diagram but, as seen in (457), for fermions each contraction is accompanied by a χ factor. Therefore, we have to find the right sign. We compute some terms in the perturbation expansion in (517) and use Wick contractions to draw diagrams that represent them. For all possible Wick contractions, some of the diagrams look identical to each other. Therefore, we have to look for a technique that considers the diagram only once and that implies we work out the symmetry factor and the number of ways the number of diagrams have to be reduced. For the perturbative expansion of the order $n=1$, there are two diagrams that correspond to two contractions contributing to (517):

$$-\frac{1}{2}\sum_{\alpha\beta\gamma\delta}\left(\alpha\beta\middle|\hat{U}\middle|\gamma\delta\right)\int_0^\beta d\tau\left\langle\xi^*_\alpha(\tau)\xi^*_\beta(\tau)\xi_\delta(\tau)\xi_\gamma(\tau)\right\rangle \tag{521}$$

or

$$\begin{aligned}&-\frac{1}{2}\sum_{\alpha\beta\gamma\delta}\left(\alpha\beta\middle|\hat{U}\middle|\gamma\delta\right)\int_0^\beta d\tau\,\xi^*_\alpha(\tau)\xi^*_\beta(\tau)\xi_\delta(\tau)\xi_\gamma(\tau)-\\&-\frac{1}{2}\sum_{\alpha\beta\gamma\delta}\left(\alpha\beta\middle|\hat{U}\middle|\gamma\delta\right)\int_0^\beta d\tau\,\xi^*_\alpha(\tau)\xi^*_\beta(\tau)\xi_\delta(\tau)\xi_\gamma(\tau)\end{aligned} \tag{522}$$

or

$$-\frac{1}{2}\sum_{\alpha\beta\gamma\delta}\left(\alpha\beta\middle|\hat{U}\middle|\gamma\delta\right)\int_0^\beta d\tau\left[\chi\mathbf{G}(\alpha\tau,\gamma\tau)\chi\mathbf{G}(\beta\tau,\delta\tau)+\chi\mathbf{G}(\alpha\tau,\delta\tau)\mathbf{G}(\beta\tau,\gamma\tau)\right] \tag{523}$$

or

$$-\frac{1}{2}\sum_{\alpha\beta\gamma\delta}\left(\alpha\beta\middle|\hat{U}\middle|\gamma\delta\right)\int_0^\beta d\tau\left[\chi\delta_{\alpha\gamma}\mathbf{G}_\gamma(0)\chi\delta_{\beta\delta}\mathbf{G}_\delta(0)+\chi\delta_{\alpha\delta}\mathbf{G}_\delta(0)\delta_{\beta\gamma}\mathbf{G}_\gamma(0)\right] \tag{524}$$

or

$$-\frac{1}{2}\sum_{\gamma\delta}\int_0^\beta d\tau\left[\left(\gamma\delta\middle|\hat{U}\middle|\gamma\delta\right)\mathbf{G}_\gamma(0)\mathbf{G}_\delta(0)+\chi\left(\delta\gamma\middle|\hat{U}\middle|\gamma\delta\right)\mathbf{G}_\delta(0)\mathbf{G}_\gamma(0)\right] \tag{525}$$

We observe the two-particle interacting Green's function to be reduced to sum of two products with each product being single-particle Green's functions associated to an interaction vertex. The first product being

$$-\frac{1}{2}\sum_{\gamma\delta}(\gamma\delta|\hat{U}|\gamma\delta)\int_0^\beta d\tau \mathbf{G}_\gamma(0)\mathbf{G}_\delta(0) \equiv -\frac{1}{2}\sum_{\gamma\delta}(\gamma\delta|\hat{U}|\gamma\delta)\int_0^\beta d\tau \mathbf{G}(\gamma\tau,\gamma\tau)\mathbf{G}(\delta\tau,\delta\tau) \tag{526}$$

represented diagrammatically as

$$-\frac{1}{2}\sum_{\gamma\delta}(\gamma\delta|\hat{U}|\gamma\delta)\int_0^\beta d\tau \mathbf{G}(\gamma\tau,\gamma\tau)\mathbf{G}(\delta\tau,\delta\tau) \equiv \tag{527}$$

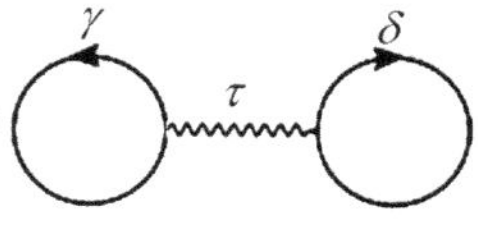

and the second

$$-\frac{1}{2}\chi\sum_{\gamma\delta}(\delta\gamma|\hat{U}|\gamma\delta)\int_0^\beta d\tau \mathbf{G}_\delta(0)\mathbf{G}_\gamma(0) = -\frac{1}{2}\chi\sum_{\gamma\delta}(\delta\gamma|\hat{U}|\gamma\delta)\int_0^\beta d\tau \mathbf{G}(\delta\tau,\delta\tau)\mathbf{G}(\gamma\tau,\gamma\tau) \tag{528}$$

represented diagrammatically as

$$-\frac{1}{2}\chi\sum_{\gamma\delta}(\delta\gamma|\hat{U}|\gamma\delta)\int_0^\beta d\tau \mathbf{G}(\delta\tau,\delta\tau)\mathbf{G}(\gamma\tau,\gamma\tau) \equiv \tag{529}$$

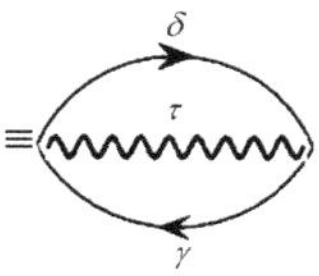

The two diagrams, (527) and (529), can be transformed smoothly into themselves by unity and exchange of the extremities of the vertex. Note each contraction yields a diagram in which single-particle propagators form some number of closed loops n_c. For the case of the first-order direct diagram in (527), we have $n_c = 2$, and for the first-order exchange diagram in (529), $n_c = 1$. For the diagrams (527) and (529), each contraction is accompanied by a $-\frac{1}{2}$ factor and $-\frac{1}{2}\chi$ factor, respectively.

Note that all terms in the partition function perturbation expansion can be represented by Feynman diagrams, which are often referred to as vacuum fluctuation graphs. The diagrammatic rules presented previously correctly consider all the contractions, propagators, and matrix elements. These rules must be augmented to consider the overall sign, factor, and a careful general definition of summation over all distinct diagrams. Two diagrams are distinct when they cannot be made to coincide with respect to topological structure, direction of arrows, and labels by some deformation. From here, summation over distinct labeled diagrams uniquely and correctly counts each contraction. We examine some examples of the second-order perturbative expansion in equation (517):

$$\frac{1}{2!2^2}\sum_{\alpha_1\cdots\delta_2}(\alpha_1\beta_1|\hat{U}|\gamma_1\delta_1)(\alpha_2\beta_2|\hat{U}|\gamma_2\delta_2)\int_0^\beta d\tau_1 d\tau_2 \langle \xi^*_{\alpha_1}(\tau_1)\cdots\xi_{\gamma_1}(\tau_1)\xi^*_{\alpha_2}(\tau_2)\cdots\xi_{\gamma_2}(\tau_2)\rangle \tag{530}$$

Because each line is associated with a propagator $\mathbf{G}^{(0)}_\alpha$ and each vertex with the matrix element $(\alpha_1\alpha_2|\hat{U}|\alpha'_1\alpha')$, we proceed to determine the combinatorial factor and sign of each diagram where, from the first-order diagram (527), we have

$$-\frac{1}{2}\langle \quad \rangle \rightarrow \tag{531}$$

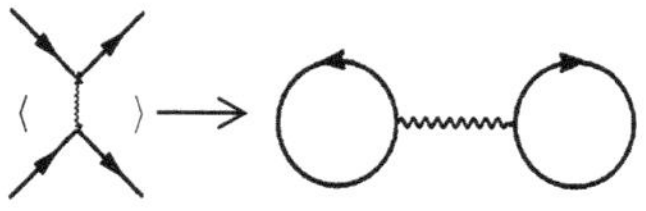

Here, the overall factor is 1/2 because there is a unique contraction that results in a diagram. We will now consider two second-order diagrams. We begin with the diagram:

$$\frac{1}{2!2^2}$$

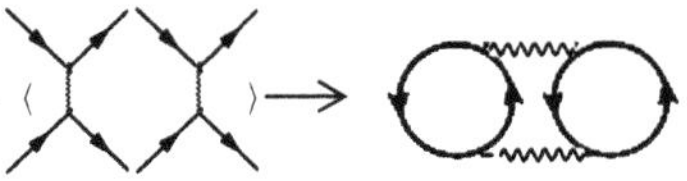

(532)

with the factor $\frac{1}{2!2^2}$. Two contractions are consistent with the diagram's topology, which can be seen by taking one of the outgoing lines of one of the two vertices and contacting one of the incoming line of the other vertex. This results in the combinatorial factor

$$\frac{1}{2!2^2}\times 2=\frac{1}{4} \tag{533}$$

Next, we consider another second-order diagram contributing to the single-particle propagator:

$$\frac{1}{2!2^2}$$

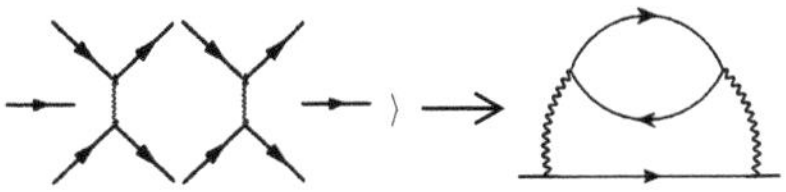

(534)

which imitates a particle interacting with itself via the particle-hole pair it created in the many-body system. The combinatorial factor is then

$$\frac{1}{2!2^2}\times\underbrace{2!}_{\text{interchange of vertices}}\times\underbrace{2}_{\text{connection of outgoing line to vertices}}\times\underbrace{2}_{\text{connection of incoming line to vertices}}=1 \tag{535}$$

This shows the combinatorial factor will always be unity for the n-particle propagator. But for the partition function, it is given by the number n_c of closed loops. This suffices to consider a particular contraction leading to the diagram so as to obtain the sign of the diagram, such as

$$\begin{aligned}
&\frac{1}{2!2^2}\sum_{\alpha_1\cdots\delta_2}\Gamma_{\alpha_1\cdots\delta_2}\int_0^\beta d\tau_1 d\tau_2 \xi^*_{\alpha_1}(\tau_1)\xi^*_{\beta_1}(\tau_1)\xi_{\delta_1}(\tau_1)\xi_{\gamma_1}(\tau_1)\xi^*_{\alpha_2}(\tau_2)\xi^*_{\beta_2}(\tau_2)\xi_{\delta_2}(\tau_2)\xi_{\gamma_2}(\tau_2)=\\
&=\frac{1}{2!2^2}\sum_{\alpha_1\cdots\delta_2}\Gamma_{\alpha_1\cdots\delta_2}\int_0^\beta d\tau_1 d\tau_2 \xi_{\gamma_2}(\tau_2)\xi^*_{\alpha_1}(\tau_1)\xi_{\delta_2}(\tau_2)\xi^*_{\beta_1}(\tau_1)\xi_{\delta_1}(\tau_1)\xi^*_{\beta_2}(\tau_2)\xi_{\gamma_1}(\tau_1)\xi^*_{\alpha_2}(\tau_2)=\\
&=\frac{1}{2!2^2}\sum_{\alpha_1\cdots\delta_2}\Gamma_{\alpha_1\cdots\delta_2}\int_0^\beta d\tau_1 d\tau_2 \delta_{\gamma_2\alpha_1}\mathbf{G}_{\gamma_2}(\tau_2-\tau_1)\delta_{\delta_2\beta_1}\mathbf{G}_{\delta_2}(\tau_2-\tau_1)\delta_{\delta_1\beta_2}\mathbf{G}_{\delta_1}(\tau_1-\tau_2)\delta_{\gamma_1\alpha_2}\mathbf{G}_{\gamma_1}(\tau_1-\tau_2)
\end{aligned} \tag{536}$$

where the interacting vertex is

$$\Gamma_{\alpha_1\cdots\delta_2}\equiv\left(\alpha_1\beta_1\middle|\hat{U}\middle|\gamma_1\delta_1\right)\left(\alpha_2\beta_2\middle|\hat{U}\middle|\gamma_2\delta_2\right) \tag{537}$$

We then have the second-order diagram contributing to the single-particle propagator:

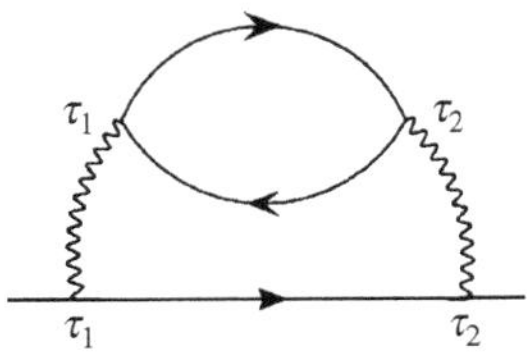

(538)

As our concern is only with complete contractions, all vertex connections form a closed loop with each such loop consisting of less than n vertices. Each vertex has an even number of fields with a particular closed loop having a pair of the fields. Hence, even Grassmann fields commute with even Grassmann fields. The pairs of fields involved in the cycle are consistent with the given closed loop and may be reordered without affecting the sign. This is done until the cycle of contractions achieves the form:

$$(\cdots)\xi^*\,\xi^*\,\xi\,\xi(\cdots)\xi^*\,\xi^*\,\xi\,\xi(\cdots)\xi^*\,\xi^*\,\xi\,\xi(\cdots) \tag{539}$$

where $(\cdots)$ are vertices not involved in the closed loop, and each $(\cdots)$ contains an even number of fields. Therefore, it commutes with the fields explicitly written in (539). From (539), each closed loop relates to a cycle of interactions starting at the left (L) or right (R) of some vertex and connects to some side of yet another vertex and so forth until it returns finally to the initial side of the vertex such that diagrammatically it may be rewritten:

$$\xi_L^*\,\xi_R^*\,\xi_R\,\xi_L\quad \xi_L^*\,\xi_R^*\,\xi_R\,\xi_L\quad \xi_L^*\,\xi_R^*\,\xi_R\,\xi_L \tag{540}$$

Here, we precisely specify the variables corresponding to the left and right sides of each vertex with all other labels suppressed because our concern is only with the signs. Because ξ_L is separated from ξ_L^* by two variables at each vertex, then $\xi_L^*\xi_R^*\xi_R\xi_L$ may be rewritten $\xi_L^*\xi_L\xi_R^*\xi_R$ such that the closed loop achieves the form:

$$\xi_L^*\,\xi_L\,\xi_R^*\,\xi_R\quad \xi_L^*\,\xi_L\,\xi_R^*\,\xi_R\quad \xi_L^*\,\xi_L\,\xi_R^*\,\xi_R \tag{541}$$

Considering (539) confirms that the pairs involved in the cycle are consistent with the given closed loop and may be reordered without affecting the sign until the cycle of contractions achieves the form:

$$\xi^*\xi\,\xi^*\,\xi\,\xi^*\,\xi(\cdots)\xi^*\,\xi \tag{542}$$

The sign associated with this closed loop is now apparent. From equation (542), the inner contractions $\xi\,\xi^*$ each yield the factor $+\mathbf{G}_{\alpha'}(\tau'-\tau)$, while the outer contraction $\xi^*\,\xi$ yields the factor $\chi\mathbf{G}_\alpha(\tau-\tau')$. Because the remaining loops consist of even Grassmann fields, they can be reordered to the same form without altering the sign. So, each closed loop gives the overall factor of $(-1)^n\chi^{n_c}$ with n_c being the number of closed loops and $(-1)^n$ resulting from the expansion of $\exp\{-S_{\text{int}}\}$.

From here, we then summarize the derived Feynman diagrammatic rules for constructing diagrams and providing a faithful representation of the complete set of contractions contributing to the nth order perturbation of the expansion of $\frac{Z}{Z_0}$:

1. Draw all distinct diagrams with n vertices

$$(\alpha\beta|\hat{U}|\gamma\delta) = \tag{543}$$

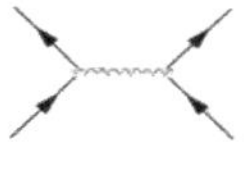

connected by directed lines . Two diagrams are distinct when not deformable so as to coincide completely, including the direction of arrows on propagators.

2. Assign a time label τ to each vertex and a single-particle index to each directed line associated with the quantity $\delta_{\alpha\alpha'}\mathbf{G}_{\alpha'}(\tau-\tau')$. An equal-time propagator with both ends connected to the same vertex is interpreted as

$$\delta_{\alpha\alpha'}\mathbf{G}_{\alpha'}(\tau-\tau') = \overrightarrow{\alpha'\tau' \quad \alpha\tau} \tag{544}$$

or

$$\delta_{\alpha\alpha'}\mathbf{G}_{\alpha'}(\tau-\tau') = \delta_{\alpha\alpha'}\exp\{-(\epsilon_{\alpha'}-\mu)(\tau-\tau')\}\left[(1+\chi n_{\alpha'})\theta(\tau-\tau'-\delta)+\chi n_{\alpha'}\theta(\tau'-\tau+\delta)\right] \tag{545}$$

where the infinitesimal quantity δ is included in the θ functions to indicate that the second term is to be used at equal times.

3. With each vertex, associate the factor

$$(\alpha\beta|\hat{U}|\gamma\delta) = \tag{546}$$

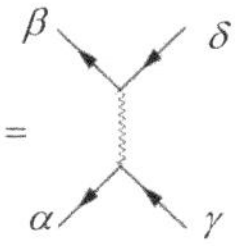

4. Sum over all single-particle indices α and integrate over all times τ over the interval $[0,\beta]$.
5. Multiply the result by the prefactor $\frac{(-1)^n}{2^n n!}\chi^{n_c}$, where n_c is the number of closed loops of single-particle propagators in the diagram.

Note that we can significantly reduce the number of diagrams by analyzing which diagrams contribute the same results. This can be done by defining the symmetry factor, which essentially is the investigation of the cases in which the time integration and the spatial integration within the interaction overlap matrix can be interchanged. The thermodynamics of the given system can be computed from $\ln Z$ rather than Z. In the next section, we examine the **linked cluster theorem**, which states that $\ln\frac{Z}{Z_0}$ is given by the sum of all fully connected Feynman diagrams.

4.4.1 Linked Cluster Theorem

Note that the Green's function $\mathbf{G}^{(2n)}$ should be obtained from the sum of all fully connected Feynman diagrams. Because disconnected diagrams can be expressed in terms of lower-order Green's functions, $\mathbf{G}^{(2n-2)}$, $\mathbf{G}^{(2n-4)}$ and so on, it is convenient to deal with connected Green's functions, $\mathbf{G}_c^{(2n)}$, which are defined as the sum of all completely connected diagrams with the corresponding function being

$$\mathbf{W}[\eta,\eta^*] = \ln\frac{Z}{Z_0} = \ln\mathbf{G}[\eta,\eta^*] \quad , \quad \mathbf{G}[\eta,\eta^*] \equiv \frac{Z[\eta,\eta^*]}{Z[0,0]} \tag{547}$$

This is analogous to the free energy in statistical mechanics, and $Z[0,0]$ is the noninteracting partition function for vanishing source fields. We expect by analogy that $\mathbf{W}[0,0]$ will be proportional to the total space-time volume VT and that if the source fields η,η^* are localized to a finite region of space-time, then $\mathbf{W}[\eta,\eta^*]-\mathbf{W}[0,0]$ is finite for the limit $VT\to\infty$. So, the functional derivatives of $\mathbf{W}[\eta,\eta^*]$ with respect to η,η^* should also be finite. This yields the so-called connected correlation functions

$$\langle\xi_{\alpha_1}(\tau_1)\cdots\xi_{\alpha_n}(\tau_n)\xi^*_{\alpha'_n}(\tau'_n)\cdots\xi^*_{\alpha'_1}(\tau'_1)\rangle_c \equiv \mathbf{G}_c^{(2n)}(\alpha_1\tau_1,\cdots,\alpha_n\tau_n,\alpha'_n\tau'_n,\cdots,\alpha'_1\tau'_1) \tag{548}$$

The reason for this is apparent when writing these functions in terms of Feynman diagrams.

We show this using the **replica method** such as in the **linked cluster theorem**. It is instructive to recall that $\mathbf{Z}\left[\eta,\eta^*\right]$ is the sum of vacuum-to-vacuum Feynman diagrams, and $\mathbf{W}\left[\eta,\eta^*\right]$ is the sum of connected vacuum-to-vacuum diagrams. In this chapter, we observe that a given diagram contributing to $\mathbf{Z}\left[\eta,\eta^*\right]$ in general is made up of several different connected diagrams that will factor into the contributions from each of the diagrams. We derive a linked cluster theorem for the generator of connected $2n$-point Green's functions via the replica technique, which evaluates $\mathbf{G}^R$ for integer R by replicating the system R times with the result expanded as follows:

$$\mathbf{G}^R = \exp\left\{R\ln\mathbf{G}\right\} = 1 + R\ln\mathbf{G} + \sum_{n=2}^{\infty}\frac{\left(R\ln\mathbf{G}\right)^n}{n!} \tag{549}$$

Therefore, evaluating $\mathbf{G}^R$ for integer R in the perturbation theory, $\ln\mathbf{G}$ (the generator of the connected $2n$-point Green's function) is given by the coefficient of terms proportional to R. Applying a more general statement of the technique, we calculate $\mathbf{G}^R$ for integer R and continue the function to $R=0$ that is unique by Carlson theorem. Then, from the linked cluster theorem, we observe that the resulting Green's function diagrams shall have the property that all connected diagrams are proportional to R and all disconnected diagrams must contain at least two factors of R. The terms proportional to R are singled out by

$$\lim_{R\to 0}\frac{\partial}{\partial R}\mathbf{G}^R = \lim_{R\to 0}\frac{\partial}{\partial R}\exp\left\{R\ln\mathbf{G}\right\} = \ln\mathbf{G} = \ln Z - \ln Z_0 \tag{550}$$

Therefore, we first evaluate $\mathbf{G}^R$ for integer R via perturbation theory and from (549), and then $\ln\mathbf{G}$ is given by the coefficient of the diagrams proportional to R.

To proceed, we consider the **interacting Green's function**:

$$\mathbf{G}^{(2n)}\left(\alpha_1\tau_1,\cdots,\alpha_n\tau_n,\alpha_n'\tau',\cdots,\alpha_1'\tau_1'\right) = \frac{\left\langle \xi_{\alpha_1}(\tau_1)\cdots\xi_{\alpha_n}(\tau_n)\xi^*_{\alpha_n'}(\tau_n')\cdots\xi^*_{\alpha_1'}(\tau_1')\exp\left\{-S_{\text{int}}\left[\xi^*,\xi\right]\right\}\right\rangle}{\left\langle \exp\left\{-S_{\text{int}}\left[\xi^*,\xi\right]\right\}\right\rangle} \tag{551}$$

and introduce the Green's function $\mathbf{G}^R$ of R replica of the systems and write it as a functional integral over R distinct fields $\left\{\xi_r^*,\xi_r\right\}$ for the rth copy with $r\in[1,R]$:

$$\mathbf{G}^R = \frac{1}{Z_0^R}\int\prod_{r=1}^{R} d\left[\xi_r^*\right]d\left[\xi_r\right]\xi_{\alpha_1}(\tau_1)\cdots\xi_{\alpha_n}(\tau_n)\xi^*_{\alpha_n'}(\tau_n')\cdots\xi^*_{\alpha_1'}(\tau_1')\exp\left\{-S_0\left[\xi_r^*,\xi_r\right]-S_{\text{int}}\left[\xi_r^*,\xi_r\right]\right\} \tag{552}$$

where

$$S_0\left[\xi_r^*,\xi_r\right] = \sum_r\int_0^\beta d\tau\sum_\alpha \xi^*_{\alpha_r}\mathbf{G}^{-1}_{\alpha\alpha}\xi_{\alpha_r} \quad , \quad \mathbf{G}^{-1}_{\alpha\alpha} = \frac{\partial}{\partial\tau} + \tilde{\epsilon}_\alpha - \mu \tag{553}$$

Applying (549) to (551) and developing the nominator and denominator of (551) via the perturbation theory, we observe that in the denominator, only contractions between the interactions immerge and are not connected to the external fields $\xi^*_{\alpha_n'}$. However, the nominator does couple the interaction with the external fields. So, the parts that are disconnected from the external field cancel out the disconnected fields in the denominator (and that we show via the replica method). We recall that the Green's function describes the expectation value of the system when a particle is created or annihilated in the system at a time moment τ' and annihilated or created at a time moment τ. Performing the evaluations via the perturbation theory, the interacting Green's function is given as the bare Green's function added to some correction terms. From the two fields corresponding to one time τ that are an even number apart, we then separate the $r=1$ component from the $R-1$ other components in (552) and write:

$$\mathbf{G}^R = \left(\frac{1}{Z_0} \int \prod_{r=1}^{R} d\left[\xi_r^*\right] d\left[\xi_r\right] \xi_{\alpha_1}(\tau_1) \xi_{\alpha_1'}^*(\tau_1') \exp\left\{-S_0\left[\xi_r^*, \xi_r\right] - S_{\text{int}}\left[\xi_r^*, \xi_r\right]\right\} \right) \times$$
$$\times \left(\frac{1}{Z_0^{R-1}} \int \prod_{r=2}^{R} d\left[\xi_r^*\right] d\left[\xi_r\right] \xi_{\alpha_2}(\tau_2) \cdots \xi_{\alpha_n}(\tau_n) \xi_{\alpha_n'}^*(\tau_n') \cdots \xi_{\alpha_2'}^*(\tau_2') \exp\left\{-S_0\left[\xi_r^*, \xi_r\right] - S_{\text{int}}\left[\xi_r^*, \xi_r\right]\right\} \right) \tag{554}$$

or

$$\mathbf{G}^R = \left\langle \xi_{\alpha_1}(\tau_1) \xi_{\alpha_1'}^*(\tau_1') \exp\left\{-S_{\text{int}}\left[\xi_r^*, \xi_r\right]\right\} \right\rangle \left(\left\langle \xi_{\alpha_2}(\tau_2) \cdots \xi_{\alpha_n}(\tau_n) \xi_{\alpha_n'}^*(\tau_n') \cdots \xi_{\alpha_2'}^*(\tau_2') \exp\left\{-S_{\text{int}}\left[\xi_r^*, \xi_r\right]\right\} \right\rangle \right)^{R-1} \tag{555}$$

which is achieved by integrating out the fields ξ_r^*, ξ_r for $r \geq 2$. Considering (552), the required Green's function is obtained for $R=0$:

$$\mathbf{G}^{(2n)} = \lim_{R \to 0} \mathbf{G}^R \tag{556}$$

We then expand $\mathbf{G}^R$ as

$$\mathbf{G}^R = \frac{1}{Z_0^R} \int \prod_{r=1}^{R} d\left[\xi_r^*\right] d\left[\xi_r\right] \xi_{\alpha_1}(\tau_1) \cdots \xi_{\alpha_n}(\tau_n) \xi_{\alpha_n'}^*(\tau_n') \cdots \xi_{\alpha_1'}^*(\tau_1') \exp\left\{-S_0\left[\xi_r^*, \xi_r\right]\right\} \exp\left\{-S_{\text{int}}\left[\xi_r^*, \xi_r\right]\right\} \tag{557}$$

Similar to the earlier expansion, we obtain the perturbation expansion for $\mathbf{G}^R$ by expanding $\exp\left\{-S_{\text{int}}\left[\xi_r^*, \xi_r\right]\right\}$:

$$\exp\left\{-S_{\text{int}}\left[\xi_r^*, \xi_r\right]\right\} = \sum_{n=0}^{\infty} \frac{(-1)^n}{n!} \prod_{k=1}^{n} S_{\text{int}}\left[\xi_{r_k}^*, \xi_{r_k}\right] \tag{558}$$

then

$$\mathbf{G}^R = \sum_{n=0}^{\infty} \frac{(-1)^n}{n!} \prod_{k=1}^{n} \left\langle \xi_{\alpha_k}(\tau_k) \cdots \xi_{\alpha_k'}^*(\tau_k') S_{\text{int}}\left[\xi_{r_k}^*, \xi_{r_k}\right] \right\rangle \tag{559}$$

For the two-body interaction, we have

$$\mathbf{G}^R = \sum_{n=0}^{\infty} \frac{(-1)^n}{2^n n!} \prod_{r=1}^{n} \sum_{\alpha_r \beta_r \gamma_r \delta_r} \left(\alpha_r \beta_r \middle| \hat{U} \middle| \gamma_r \delta_r\right) \int_0^{\beta} d\tau_r \left\langle \xi_{\alpha_r}(\tau_r) \cdots \xi_{\alpha_r'}^*(\tau_r) \xi_{\alpha_r}^*(\tau_r) \xi_{\beta_r}^*(\tau_r) \xi_{\delta_r}(\tau_r) \xi_{\gamma_r}(\tau_r) \right\rangle \tag{560}$$

Note that the Feynman rules for $\mathbf{G}^R$ are the same as those for $\mathbf{G}$. However, the Feynman diagrams consider the propagator carrying a replica index $r \in [1, R]$. Also, all propagators connected to the same vertex have the same replica index value because the propagators are given by $\delta_{\alpha,\alpha'} \mathbf{G}_\alpha(\tau - \tau')$. The propagators corresponding to the external lines α_r and α_r' have the replica index value $r=1$. As we sum over r index in the interaction vertex, a diagram with n disconnected parts (i.e., n parts not connected to the external legs) is proportional to R^n and vanishes in the limit $R=0$ when $n \geq 1$. For the connected diagrams ($n = 0$), the integration of the fields ξ_r^* with replica index $r \in [2, R]$ in (560) yields the factor Z_0^{R-1} so that there are only connected diagrams. This implies connected diagrams with all parts connected to the external vertex should be retained:

$$\mathbf{G}^{(2n)}(\alpha_1\tau_1,\cdots,\alpha_n\tau_n,\alpha_n'\tau_n',\cdots,\alpha_1'\tau_1') = \frac{1}{Z_0}\int d[\xi_r^*]d[\xi_r]\xi_{\alpha_r}(\tau_r)\cdots\xi_{\alpha_r'}^*(\tau_r')\exp\{-S_{\text{int}}[\xi_r^*,\xi_r]\} \quad (561)$$

The combinatorial factor of a diagram is computed in the same manner as for the partition function by a direct evaluation of the number of contractions leading to the given diagram as, for example, in the case of the second order seen previously. Our concern is only with connected diagrams. Hence, the diagrams consist only of directed lines and closed loops connected by the vertex and not over a propagator. The propagators within a loop are directed. So, the one-time label is fixed and cannot be permuted within a connected loop. The sign related to this closed loop is now obvious (as also was seen earlier), and our task is to determine the prefactor sign of the Wick contractions. From the definition of the Green's function, all $\xi_{\alpha_k'}^*(\tau_k')$ are an even number of fields separated from $\xi_{\alpha_k}(\tau_k)$ and, as seen earlier, the contraction

$$\overbrace{\xi_{\alpha_k}(\tau_k)\xi_{\alpha_k'}^*(\tau_k')} = +\mathbf{G}_{\alpha_k'\alpha_k}(\tau_k' - \tau_k) \quad (562)$$

If we arbitrary fix one $\xi_{\alpha_k'}^*(\tau_k')$, we consider the contractions with a permutation of its counterpart

$$\overbrace{\xi_{\alpha_k'}^*(\tau_k')\xi_{\alpha_{P(k)}}(\tau_{\alpha_{P(k)}})} = \chi^{P}\mathbf{G}_{\alpha_k'\alpha_{P(k)}}(\tau_{\alpha_{P(k)}} - \tau_{\alpha_k}') \quad (563)$$

with P being the necessary permutation. The sign of the Green's function changes to χ^{P}. We can then add the other contractions to form a closed loop, and because the remaining loops consist of even Grassmann fields, they can be reordered to the same form with the same sign. So, each closed loop yields a prefactor of χ^{n_c} with n_c being the number of closed loops (as seen earlier). We are now ready to give the **Feynman rules for the interacting Green's function.**

The diagrammatic rules for computing the r-order contribution to the n-particle Green's function $\mathbf{G}^{(2n)}(\alpha_1\tau_1,\cdots,\alpha_n\tau_n,\alpha_n'\tau_n',\cdots,\alpha_1'\tau_1')$:

1. Draw all distinct connected diagrams with n incoming lines $(\alpha_n'\tau_n',\cdots,\alpha_1'\tau_1')$, n outgoing lines $(\alpha_1\tau_1,\cdots,\alpha_n\tau_n)$, and r interaction vertices. Diagrams are distinct if they cannot be transformed into each other by fixing the external points and propagator direction.
2. The external points are given by the calculated Green's function assigned first to the interaction vertices. Next, the free points have to be connected with propagators, and each propagator is given an index. Then, each propagator is associated with $\mathbf{G}_0(\alpha\tau,\cdots,\alpha'\tau')$ and represented by a directed line.
3. Associate the matrix element $(\alpha_k\beta_k|\hat{U}|\gamma_k\delta_k)$ with each vertex.
4. Sum over all internal indices α and integrate over all internal times τ within the interval $[0,\beta]$. The indices $(\alpha_i'\tau_i')$ and $(\alpha_i\tau_i)$ of incoming and outgoing lines should be held fixed.
5. Multiply the result by the factor $(-1)^r\chi^{n_c+P}$, where n_c is the number of closed loops and P is the permutation such that each incoming line $(\alpha_i'\tau_i')$ ends at $(\alpha_{P(i)}\tau_{P(i)})$.

4.4.2 Green's Function Generating Functional

It is useful to construct the modified generating functional that only directly produces connected Feynman diagrams. We have seen that we can calculate the n-particle Green's function over a generating function defined by

$$Z[\eta,\eta^*] = \int d[\xi^*]d[\xi]\exp\{-S_0[\xi^*,\xi] + S[\eta^*,\eta,\xi]\} \quad (564)$$

where the bare and source action functionals are, respectively,

$$S_0[\xi^*,\xi]=\sum_{\alpha\alpha'}\int_0^\beta d\tau d\tau'\xi^*_\alpha(\tau)\mathbf{G}^{-1}(\alpha\tau;\alpha'\tau')\xi_{\alpha'}(\tau') \tag{565}$$

$$S[\eta^*,\eta,\xi]=\sum_{\alpha}\int_0^\beta d\tau\left(\xi^*_\alpha(\tau)\eta_\alpha(\tau)+\eta^*_\alpha(\tau)\xi_\alpha(\tau)\right) \tag{566}$$

and the extra terms $\eta_\alpha(\tau)$ and $\eta^*_\alpha(\tau)$ are external source fields coupling linearly to the fields $\xi^*_\alpha(\tau)$ and $\xi_\alpha(\tau)$. From the change of variables:

$$\xi_\alpha=\xi'_\alpha+\mathbf{G}_{\alpha,\alpha'}\eta_{\alpha'}\quad,\quad \xi^*_\alpha=\xi'^*_\alpha+\eta^*_{\alpha'}\mathbf{G}_{\alpha,\alpha'}\quad,\quad \mathbf{G}_{\alpha,\alpha'}\equiv\mathbf{G}(\alpha\tau;\alpha'\tau') \tag{567}$$

$Z[\eta,\eta^*]$ then becomes

$$Z[\eta,\eta^*]=Z[0,0]\exp\left\{\int d\tau d\tau'\sum_{\alpha\alpha'}\eta^*_\alpha(\tau)\mathbf{G}_{\alpha,\alpha'}\eta_{\alpha'}(\tau')\right\} \tag{568}$$

or

$$\mathbf{G}[\eta,\eta^*]\equiv\frac{Z[\eta,\eta^*]}{Z[0,0]} \tag{569}$$

The average in (568) is now taken with respect to the source field, and the thermal n-point Green's function is given by

$$\mathbf{G}^{(2n)}(\alpha_1\tau_1,\cdots,\alpha_n\tau_n,\alpha'_n\tau',\cdots,\alpha'_1\tau'_1)=\left.\frac{\chi^n\partial^{(2n)}\mathbf{G}[\eta,\eta^*]}{\partial\eta^*_{\alpha_1}(\tau_1)\cdots\partial\eta^*_{\alpha_n}(\tau_n)\partial\eta_{\alpha'_n}(\tau'_n)\cdots\partial\eta_{\alpha'_1}(\tau'_1)}\right|_{\eta,\eta^*=0} \tag{570}$$

So far, we have derived Feynman rules for the Green's function. We find that the diagrams are connected to all external fields. However, the *diagrams* are not all connected. This implies the diagrams are built from lower-order Green's functions. These parts can be cut off once via the replica technique.

Note that the fully connected diagrams are proportional to R, whereas a diagram with n_c connected pieces is of order R^{n_c}. So, the connected diagrams may be obtained from the functional

$$\lim_{R\to0}\frac{\partial}{\partial R}\mathbf{G}^R=\lim_{R\to0}\frac{\partial}{\partial R}\exp\{R\mathbf{W}[\eta,\eta^*]\}=\mathbf{W}[\eta,\eta^*]=\ln\mathbf{Z}[\eta,\eta^*]-\ln\mathbf{Z}[0,0] \tag{571}$$

or

$$\mathbf{W}[\eta,\eta^*]=\sum_{n=0}^{\infty}\frac{(-1)^n}{n!}\left\langle S^n_{\text{int}}\right\rangle_c=\sum\text{connnected diagrams} \tag{572}$$

Here, $\left\langle S^n_{\text{int}}\right\rangle_c$ represents the connected part of $\left\langle S^n_{\text{int}}\right\rangle$. We observe that, physically, $\mathbf{W}=\ln\mathbf{G}[\eta,\eta^*]$ represents the difference of the natural logarithm of the grand canonical partition function as well as the difference of the grand thermodynamic canonical potential in the presence and absence of sources:

$$\mathbf{W}[\eta,\eta^*]=-\beta\left(\Omega[\eta,\eta^*]-\Omega[0,0]\right)=\sum\text{connected diagrams} \tag{573}$$

The thermodynamic potential (in the absence of sources) of the noninteracting system

$$\Omega[0,0] \equiv \Omega_0 = \frac{\chi}{\beta}\sum_{\alpha} \ln\left[1-\chi\exp\{-\beta(\tilde{\epsilon}_{\alpha}-\mu)\}\right] \tag{574}$$

So, we introduce the connected Green's functions as

$$\mathbf{G}_c^{(2n)}(\alpha_1\tau_1,\cdots,\alpha_n\tau_n,\alpha_n'\tau_n',\cdots,\alpha_1'\tau_1') = \left.\frac{\chi^n \partial^{(2n)}\mathbf{G}\left[\eta,\eta^*\right]}{\partial\eta_{\alpha_1}^*(\tau_1)\cdots\partial\eta_{\alpha_n}^*(\tau_n)\partial\eta_{\alpha_n'}(\tau_n')\cdots\partial\eta_{\alpha_1'}(\tau_1')}\right|_{\eta=\eta^*=0} \tag{575}$$

or

$$\mathbf{G}_c^{(2n)}(\alpha_1\tau_1,\cdots,\alpha_n\tau_n,\alpha_n'\tau',\cdots,\alpha_1'\tau_1') \equiv \left.\frac{\chi^n \partial^{(2n)}\mathbf{W}\left[\eta,\eta^*\right]}{\partial\eta_{\alpha_1}^*(\tau_1)\cdots\partial\eta_{\alpha_n}^*(\tau_n)\partial\eta_{\alpha_n'}(\tau_n')\cdots\partial\eta_{\alpha_1'}(\tau_1')}\right|_{\eta=\eta^*=0} \tag{576}$$

which correspond to connected Feynman diagrams. From the previous definition, this is **defined as the sum of all connected Feynman diagrams linked to the external points** $(\alpha_1\tau_1,\cdots,\alpha_n\tau_n)$ **and** $(\alpha_n'\tau',\cdots,\alpha_1'\tau_1')$. It is convenient to deal with connected Green's functions because the sum of all disconnected diagrams merely corresponds to combinations of products of fewer-particle Green's functions.

We can write the diagrams directly in terms of the propagators $\delta_{\alpha\alpha'}\mathbf{G}_{\alpha'}(i\omega_n)$ in the frequency space by considering the interaction action functional:

$$S_{\text{int}} = \frac{1}{2\beta}\sum_{\substack{\alpha_1\alpha_2\alpha_1'\alpha_2' \\ \omega_n\omega_n'\omega_\nu}} (\alpha_1\alpha_2|\hat{U}|\alpha_1'\alpha_2')\hat{\psi}_{\alpha_1}^{\dagger}(i\omega_n+i\omega_\nu)\hat{\psi}_{\alpha_2}^{\dagger}(i\omega_n'-i\omega_\nu)\hat{\psi}_{\alpha_2'}(i\omega_n')\hat{\psi}_{\alpha_1'}(i\omega_n) \tag{577}$$

The sum of the frequencies related to the propagators entering the vertex is conserved in the interaction process and implies frequency conservation. We therefore have the following modifications of the diagrammatic rules:

1. Associate $\delta_{\alpha\alpha'}\mathbf{G}_{\alpha'}(i\omega_n)$ with each directed solid line, considering frequency conservation at each vertex. When both ends of a propagator are connected to the same vertex, multiply the propagator by $\exp\{i\omega_n\delta\}$.
2. Sum over all indices α as well as Matsubara frequencies ω_n.
3. Multiply the result by $\frac{1}{\beta^n}$ with n being the number of vertices.

For translational invariant systems, it is also useful to use the momentum basis $\{|\vec{\kappa},\alpha\rangle\}$, where α is the spin of the particle, as well as other internal indices. For particles interacting via a two-particle interaction $U(\vec{r}-\vec{r}')$, the action functional

$$S_{\text{int}} = \frac{1}{2\beta V}\sum_{\substack{\vec{\kappa}\vec{\kappa}'\vec{q} \\ \alpha\alpha'}} U(\vec{q})\hat{\psi}_{\alpha}^{\dagger}(\vec{\kappa}+\vec{q})\hat{\psi}_{\alpha'}^{\dagger}(\vec{\kappa}'-\vec{q})\hat{\psi}_{\alpha'}(\vec{\kappa}')\hat{\psi}_{\alpha}(\vec{\kappa}) = \frac{1}{2}\sum_{\vec{q}} U(\vec{q})\hat{n}(-\vec{q})\hat{n}(\vec{q}) \tag{578}$$

Here, $U(\vec{q})$ is the Fourier transform of $U(\vec{r}-\vec{r}')$ and

$$\hat{n}(\vec{q}) = \frac{1}{\sqrt{V}}\int d\vec{r}\exp\{-i\vec{q}\vec{r}\}\hat{n}(\vec{r}) = \frac{1}{\sqrt{\beta V}}\sum_{\vec{\kappa}\alpha}\hat{\psi}_{\alpha}^{\dagger}(\vec{\kappa})\hat{\psi}_{\alpha}(\vec{\kappa}+\vec{q}) \tag{579}$$

is the Fourier transform of the density operator

$$\hat{n}(\vec{r}) = \sum_{\alpha} \hat{\psi}^{\dagger}_{\alpha}(\vec{r})\hat{\psi}_{\alpha}(\vec{r}) \tag{580}$$

The interaction vertex assumes the following diagrammatic representation

$U(\vec{\kappa}' - \vec{\kappa}'') =$ 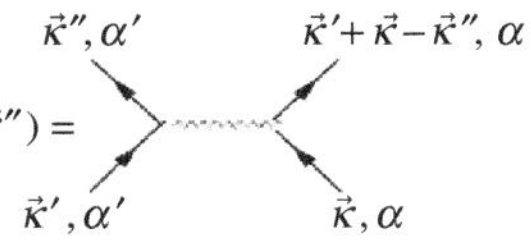

(581)

From this follows the modifications of the diagrammatic rules:

1. To each directed solid line, associate $\mathbf{G}^{(0)}_{\alpha}(\vec{\kappa}, i\omega_n)$ considering momentum, frequency, and spin conservation at each vertex. When both ends of a propagator are connected to the same vertex, multiply the propagator by $\exp\{i\omega_n\delta\}$.
2. With each vertex, associate

$U(\vec{\kappa}' - \vec{\kappa}'') =$ 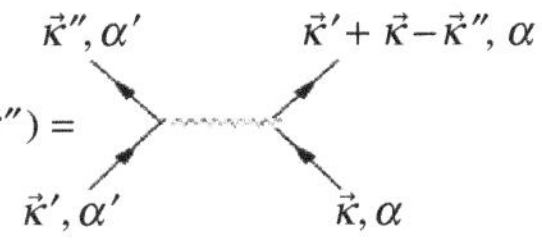

(582)

   where $\vec{q} = \vec{\kappa}' - \vec{\kappa}''$ is the transfer momentum in the interaction process.
3. With each vertex, associate

$U(\vec{\kappa}' - \vec{\kappa}'') + \chi U(\vec{\kappa} - \vec{\kappa}'') =$ 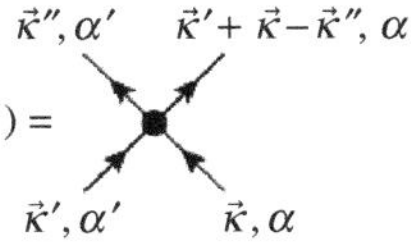

(583)

4. Sum over all momenta as well as Matsubara frequencies.
5. Multiply the result by $\frac{1}{(\beta V)^n}$ with n being the number of vertices.

Considering these rules, the first-order correction to the partition function $\frac{Z}{Z_0}$ reads

$$\begin{aligned} &-\frac{1}{2\beta V} U(\vec{q}=0) \sum_{\vec{\kappa}\vec{\kappa}'\alpha\alpha'} \mathbf{G}^{(0)}_{\alpha}(\vec{\kappa})\mathbf{G}^{(0)}_{\alpha'}(\vec{\kappa}')\exp\{i(\omega_n + \omega'_n)\delta\} \\ &-\frac{1}{2\beta V} \sum_{\vec{\kappa}\vec{\kappa}'\alpha} U(\vec{\kappa} - \vec{\kappa}')\mathbf{G}^{(0)}_{\alpha}(\vec{\kappa})\mathbf{G}^{(0)}_{\alpha}(\vec{\kappa}')\exp\{i(\omega_n + \omega'_n)\delta\} \end{aligned} \tag{584}$$

4.4.3 Green's Functions

Let us calculate the Green's functions perturbatively and various terms in the perturbative expansion represented by Feynman diagrams of the single-particle Green's function:

$$\mathbf{G}^{(2\times 1)}(\alpha_1\tau_1, \alpha_2\tau_2) \equiv \mathbf{G}(\alpha_1\tau_1, \alpha_2\tau_2) = \delta_{\alpha_1\alpha_2}\mathbf{G}_{\alpha_1}(\tau_1 - \tau_2) \tag{585}$$

or

$$\mathbf{G}(\alpha_1\tau_1,\alpha_2\tau_2)=-\left\langle\xi_{\alpha_1}(\tau_1)\xi^*_{\alpha_2}(\tau_2)\right\rangle=-\frac{Z_0}{Z}\left\langle\xi_{\alpha_1}(\tau_1)\xi^*_{\alpha_2}(\tau_2)\exp\{-S_{\text{int}}\}\right\rangle \tag{586}$$

We can evaluate **G** to any order in the interaction representation by expanding (586). From the expansion to the first order:

$$-\left\langle\xi_{\alpha_1}(\tau_1)\xi^*_{\alpha_2}(\tau_2)\exp\{-S_{\text{int}}\}\right\rangle=\mathbf{G}_0(\alpha_1\tau_1,\alpha_2\tau_2)+\left\langle\xi_{\alpha_1}(\tau_1)\xi^*_{\alpha_2}(\tau_2)S_{\text{int}}\right\rangle \tag{587}$$

where

$$\left\langle\xi_{\alpha_1}(\tau_1)\xi^*_{\alpha_2}(\tau_2)S_{\text{int}}\right\rangle=\frac{1}{2}\sum_{\beta_1,\cdots,\beta_2'}(\beta_1\beta_2|\hat{U}|\beta_1'\beta_2')\int_0^\beta d\tau\left\langle\xi_{\alpha_1}(\tau_1)\xi^*_{\alpha_2}(\tau_2)\xi^*_{\beta_1}(\tau_1)\xi^*_{\beta_2}(\tau_2)\xi_{\beta_2'}(\tau_2)\xi_{\beta_1'}(\tau_1)\right\rangle \tag{588}$$

4.4.3.1 Zeroth Order

For the zeroth order, we have

$$\mathbf{G}_0(\alpha_1\tau_1,\alpha_2\tau_2)= \tag{589}$$

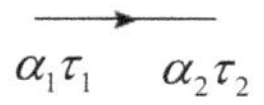

We interpret this as a particle with parameter α_1 introduced at time τ_1 travels directly and arrives at time τ_2 without interacting with (scattering off) any other particles. The first-order terms will then be when it scatters once, second-order when it scatters twice, and so on. From quantum mechanics, we sum up the appropriate weight of all of these methods of setting out from time τ_1 to arrive at time τ_2. This gives the total probability amplitude **G**.

4.4.3.2 First Order

The first-order correction to the **G** matrix:

$$\left\langle\xi_{\alpha_1}(\tau_1)\xi^*_{\alpha_2}(\tau_2)S_{\text{int}}\right\rangle=\frac{1}{2}\sum_{\beta_1,\cdots,\beta_2'}(\beta_1\beta_2|\hat{U}|\beta_1'\beta_2')\int_0^\beta d\tau\left\langle\xi_{\alpha_1}(\tau_1)\xi^*_{\alpha_2}(\tau_2)\xi^*_{\beta_1}(\tau_1)\xi^*_{\beta_2}(\tau_2)\xi_{\beta_2'}(\tau_2)\xi_{\beta_1'}(\tau_1)\right\rangle \tag{590}$$

It should be noted that in (590), $\left\langle\xi^*_{\beta_1}(\tau)\xi^*_{\beta_2}(\tau)\xi_{\beta_2'}(\tau)\xi_{\beta_1'}(\tau)\right\rangle$ is already normal ordered, so we have

$$\left\langle\xi_{\alpha_1}(\tau_1)\xi^*_{\alpha_2}(\tau_2)\right\rangle\left\langle\xi^*_{\beta_1}(\tau)\xi^*_{\beta_2}(\tau)\xi_{\beta_2'}(\tau)\xi_{\beta_1'}(\tau)\right\rangle \tag{591}$$

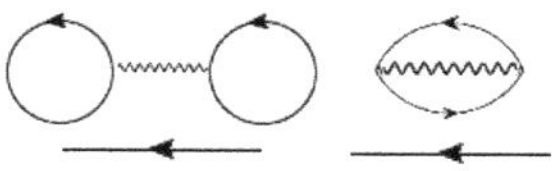

The quantity $\left\langle\xi^*_{\beta_1}(\tau)\xi^*_{\beta_2}(\tau)\xi_{\beta_2'}(\tau)\xi_{\beta_1'}(\tau)\right\rangle$ will cancel $\left\langle S(\infty,-\infty)\right\rangle$ in the denominator of (590), thereby implying that disconnected diagrams cancel the denominator, according to the linked cluster theorem.

From Wick theorem, we compute the first-order term (590) that yields six possible pairings (shown explicitly for convenience in Figure 4.1). For each possible pairing, by virtue of the anticommutation relations, the reordering of operators always brings the contracted ones together. Finally,

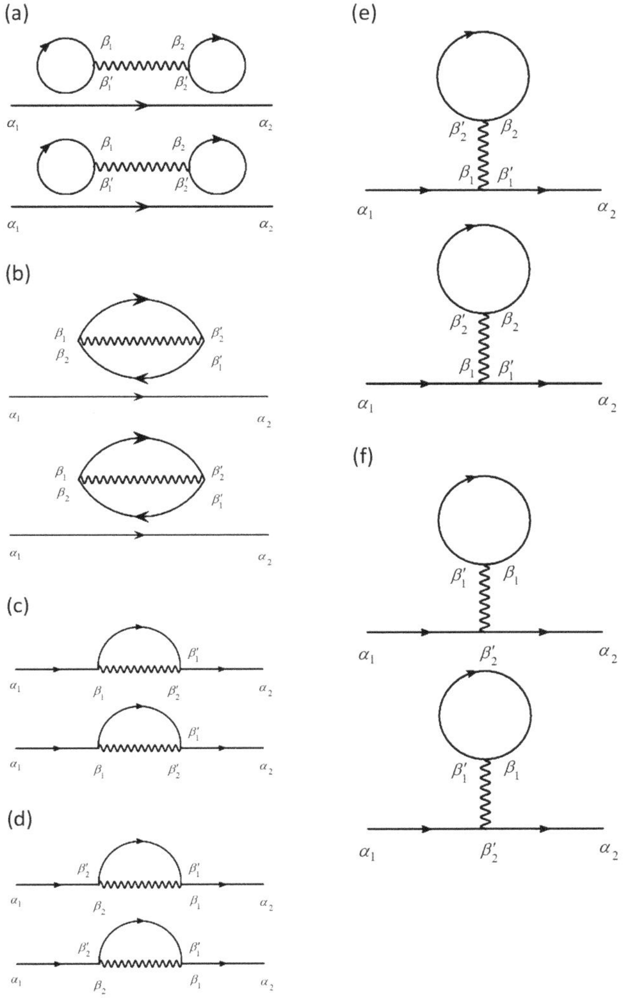

FIGURE 4.1 The six first-order diagrams, from the six Wick contractions.

inserting $|0\rangle\langle 0|$ between each pair of operators changes nothing because after each pair we return to the given state. In writing equation (590), the original time-ordering operator puts the six original operators in some order, including a relevant minus sign depending on the number of fermionic reordering. Our only worry will be which creation operators are to the right or left of which annihilation operators, as seen earlier. Regardless, it is easy to see that the time-ordering operators in each individual expectation value does the job correctly. We then can finally write down Wick theorem:

$$\langle \xi_{\alpha_1}(\tau_1)\xi^*_{\alpha_2}(\tau_2)\rangle\langle \xi^*_{\beta_1}(\tau)\xi_{\beta_1'}(\tau)\rangle\langle \xi^*_{\beta_2}(\tau)\xi_{\beta_2'}(\tau)\rangle = \mathbf{G}_0(\alpha_1\tau_1,\alpha_2\tau_2)\chi\mathbf{G}_0(\beta_1\tau,\beta_1'\tau)\chi\mathbf{G}_0(\beta_2\tau,\beta_2'\tau) \tag{592}$$

$$\langle \xi_{\alpha_1}(\tau_1)\xi^*_{\alpha_2}(\tau_2)\rangle\langle \xi^*_{\beta_1}(\tau)\xi_{\beta_2'}(\tau)\rangle\langle \xi^*_{\beta_2}(\tau)\xi_{\beta_1'}(\tau)\rangle = \mathbf{G}_0(\alpha_1\tau_1,\alpha_2\tau_2)\chi\mathbf{G}_0(\beta_1\tau,\beta_2'\tau)\chi\mathbf{G}_0(\beta_2\tau,\beta_1'\tau) \tag{593}$$

$$\langle \xi_{\alpha_1}(\tau_1)\xi^*_{\beta_1}(\tau)\rangle\langle \xi^*_{\alpha_2}(\tau_2)\xi_{\beta'_2}(\tau)\rangle\langle \xi^*_{\beta_2}(\tau)\xi_{\beta'_1}(\tau)\rangle = \mathbf{G}_0(\alpha_1\tau_1,\beta_1\tau)\chi\mathbf{G}_0(\alpha_2\tau_2,\beta'_2\tau)\chi\mathbf{G}_0(\beta_2\tau,\beta'_1\tau) \quad (594)$$

$$\langle \xi_{\alpha_1}(\tau_1)\xi^*_{\beta_2}(\tau)\rangle\langle \xi^*_{\alpha_2}(\tau_2)\xi_{\beta'_1}(\tau)\rangle\langle \xi^*_{\beta_1}(\tau)\xi_{\beta'_2}(\tau)\rangle = \mathbf{G}_0(\alpha_1\tau_1,\beta_2\tau)\chi\mathbf{G}_0(\alpha_2\tau_2,\beta'_1\tau)\chi\mathbf{G}_0(\beta_1\tau,\beta'_2\tau) \quad (595)$$

$$\langle \xi_{\alpha_1}(\tau_1)\xi^*_{\beta_1}(\tau)\rangle\langle \xi^*_{\alpha_2}(\tau_2)\xi_{\beta'_1}(\tau)\rangle\langle \xi^*_{\beta_2}(\tau)\xi_{\beta'_2}(\tau)\rangle = \mathbf{G}_0(\alpha_1\tau_1,\beta_1\tau)\chi\mathbf{G}_0(\alpha_2\tau_2,\beta'_1\tau)\chi\mathbf{G}_0(\beta_2\tau,\beta'_2\tau) \quad (596)$$

$$\langle \xi_{\alpha_1}(\tau_1)\xi_{\beta_2}(\tau)\rangle\langle \xi^*_{\alpha_2}(\tau_2)\xi_{\beta'_2}(\tau)\rangle\langle \xi^*_{\beta_1}(\tau)\xi_{\beta'_1}(\tau)\rangle = \mathbf{G}_0(\alpha_1\tau_1,\beta_2\tau)\chi\mathbf{G}_0(\alpha_2\tau_2,\beta'_2\tau)\chi\mathbf{G}_0(\beta_1\tau,\beta'_1\tau) \quad (597)$$

These six terms are shown in Figure 4.1 with diagrams (a) through (f) representing the six terms in the same order.

Note that there are two pairs of diagrams, (c,d) and (e,f) that look identical to each other, and the last two, (a) and (b) are **disconnected** and should not be taken into account when calculating the Green's function whereas **connected** diagrams contribute. So the contribution of (c,d) and (e,f) is half.

EXERCISE

Write down the remaining five Wick contractions of the first-order expression. Which of the expressions corresponds to which diagram in Figure 4.1? Put the momentum labels on diagrams (e) and (f).

4.4.3.3 Second Order

Considering the expansion to the second order:

$$-\langle \xi_{\alpha_1}(\tau_1)\xi^*_{\alpha_2}(\tau_2)\exp\{-S_{\text{int}}\}\rangle = \mathbf{G}_0(\alpha_1\tau_1,\alpha_2\tau_2)+\langle \xi_{\alpha_1}(\tau_1)\xi^*_{\alpha_2}(\tau_2)S_{\text{int}}\rangle-\frac{1}{2}\langle \xi_{\alpha_1}(\tau_1)\xi^*_{\alpha_2}(\tau_2)S^2_{\text{int}}\rangle \quad (598)$$

Here,

$$-\frac{1}{2}\langle \xi_{\alpha_1}(\tau_1)\xi^*_{\alpha_2}(\tau_2)S^2_{\text{int}}\rangle = \frac{1}{2}\sum_{\beta_1\beta_2\beta'_1\beta'_2}(\beta_1\beta_2|\hat{U}|\beta'_1\beta'_2)\int_0^\beta d\tau\langle \xi_{\alpha_1}(\tau_1)\xi^*_{\alpha_2}(\tau_2)\xi^*_{\gamma_1}(\tau_1)\xi^*_{\gamma_2}(\tau_2)\xi_{\gamma'_2}(\tau_2)\xi_{\gamma'_1}(\tau_1)\rangle \quad (599)$$

From (599), we may now rewrite (598) as follows:

$$\mathbf{G}^{(2\times1)}(\alpha_1\tau_1;\alpha_2\tau_2) \equiv \mathbf{G}(\alpha_1\tau_1;\alpha_2\tau_2) = \delta_{\alpha_1\alpha_2}\mathbf{G}_{\alpha_1}(\tau_1-\tau_2) - \sum_{\beta_1\beta_2\beta'_1\beta'_2}(\beta_1\beta_2|\hat{U}|\beta'_1\beta'_2)\times$$
$$\times\int_0^\beta d\tau\,\delta_{\alpha_1\alpha_2}\mathbf{G}_{\alpha_1}(\tau_1-\tau)\left[\chi\delta_{\beta'_1\alpha_2}\mathbf{G}_{\beta'_1}(\tau-\tau_2)\delta_{\beta'_2\beta_2}\mathbf{G}_{\beta'_2}(0)+\delta_{\beta'_2\alpha_2}\mathbf{G}_{\beta'_2}(\tau-\tau_2)\delta_{\beta'_1\beta_2}\mathbf{G}_{\beta'_1}(0)\right]+\cdots \quad (600)$$

This can be diagrammatically represented as

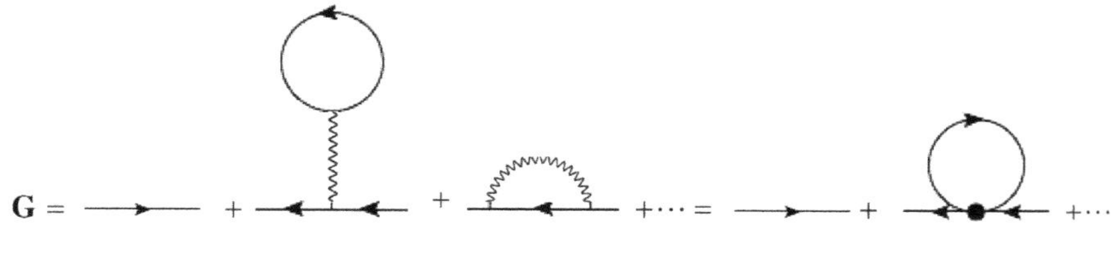

(601)

where,

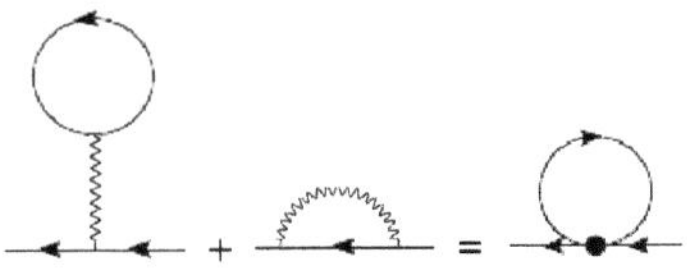

(602)

We observe that the correction to the bare particle Green's function is given by connected diagrams contributing to $-\left\langle \xi_{\alpha_1}(\tau_1)\xi^*_{\alpha_2}(\tau_2)\exp\{-S_{int}\}\right\rangle$. From here, we confirm by considering **only connected diagrams**, and we cancel the denominator. So, we conclude that the **Green's function is given by the sum of all connected diagrams**.

5

(Anti)Symmetrized Vertices

Introduction

The Feynman diagrams examined previously in this book treat direct and exchange matrix elements separately. It is simpler and more convenient for many purposes to combine them as a single (anti)symmetrized matrix element. Therefore, we formulate the perturbation theory via the (anti)symmetrized vertex introduced earlier for the residual Hamiltonian $\hat{U}$. For simplicity, we consider the following two-body action functional:

$$S_{\text{int}}\left[\hat{\psi}^{*},\hat{\psi}\right]=\frac{1}{4}\sum_{\alpha_1\alpha_2\alpha_1'\alpha_2'}\{\alpha_1\alpha_2|\hat{U}|\alpha_1'\alpha_2'\}\int_0^{\beta}d\tau\hat{\psi}^{\dagger}_{\alpha_1}(\tau)\hat{\psi}^{\dagger}_{\alpha_2}(\tau)\hat{\psi}_{\alpha_2'}(\tau)\hat{\psi}_{\alpha_1'}(\tau) \tag{603}$$

Its vertex is (anti)symmetrized under the exchange of two incoming or outgoing particles:

$$\{\alpha_1\alpha_2|\hat{U}|\alpha_1'\alpha_2'\}=\chi\{\alpha_1\alpha_2|\hat{U}|\alpha_2'\alpha_1'\}=\chi\{\alpha_2\alpha_1|\hat{U}|\alpha_1'\alpha_2'\}=(\alpha_1\alpha_2|\hat{U}|\alpha_1'\alpha_2')+\chi(\alpha_1\alpha_2|\hat{U}|\alpha_2'\alpha_1') \tag{604}$$

Because this vertex no longer distinguishes between direct and exchange diagrams of the two incoming particles, we then graphically represent it by a dot with two incoming and two outgoing lines, which considerably reduces the number of diagrams:

$$\{\alpha_1\alpha_2|\hat{U}|\alpha_1'\alpha_2'\}= \tag{605}$$

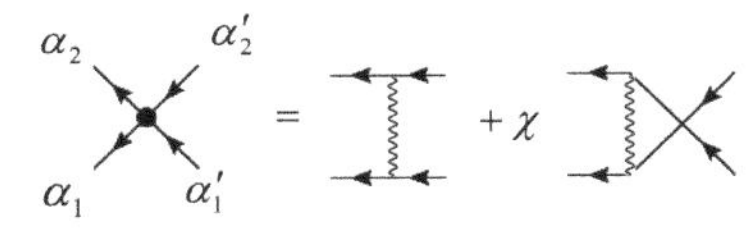

Using Wick theorem, the diagrams are constructed by drawing n vertices that connect all incoming lines with outgoing lines. From the (anti)symmetry of the vertex, two contractions that correspond with the exchange of the incoming or outgoing lines associated with a given vertex are equal:

$$\begin{aligned}&\sum_{\alpha_1'\alpha_2'}\{\alpha_1\alpha_2|\hat{U}|\alpha_1'\alpha_2'\}\hat{\psi}^{\dagger}_{\alpha_1}(\tau)\hat{\psi}^{\dagger}_{\alpha_2}(\tau)\hat{\psi}_{\alpha_2'}(\tau)\hat{\psi}_{\alpha_1'}(\tau)\cdots\hat{\psi}^{\dagger}_{\alpha_1'}(\tau)\hat{\psi}^{*}_{\alpha_2'}(\tau)=\\&=\sum_{\alpha_1'\alpha_2'}\{\alpha_1\alpha_2|\hat{U}|\alpha_1'\alpha_2'\}\hat{\psi}^{\dagger}_{\alpha_1}(\tau)\hat{\psi}^{\dagger}_{\alpha_2}(\tau)\hat{\psi}_{\alpha_2'}(\tau)\hat{\psi}_{\alpha_1'}(\tau)\cdots\hat{\psi}^{\dagger}_{\alpha_1'}(\tau)\hat{\psi}^{\dagger}_{\alpha_2'}(\tau)\end{aligned} \tag{606}$$

So, we will no longer need to distinguish the two incoming (or outgoing) lines of a vertex in contrast to the previous Feynman diagrams.

We note that a diagram that does not distinguish the two incoming (or two outgoing) lines as each vertex represents several sets of contractions.

How do we count the number of contractions associated with a given diagram? This can be done by defining an equivalent pair of lines as two directed propagators that begin at the same vertex, end at the same vertex, and point in the same direction. Because our concern is only with connected diagrams, the diagrams have only directed lines and closed loops connected by a vertex and are not connected over a propagator. For connected diagrams, the propagators within a loop are directed, while the time label is fixed and cannot be permuted within a loop. In a connected loop, one time label in a given loop fixes the time label and so, by recursion, all time labels are fixed and cannot be permuted. Our next task is to determine the prefactor sign of the Wick contractions.

We examine an example of the perturbative expansion of the order $n=1$ for $\frac{Z}{Z_0}$, where there are two diagrams that correspond to two contractions contributing to (517). So, from S_{int} considering (603), then

$$-\frac{1}{2}\sum_{\alpha_1\alpha_2\alpha_1'\alpha_2'}(\alpha_1\alpha_2|\hat{U}|\alpha_1'\alpha_2')\int_0^\beta d\tau\left[\mathbf{G}_0(\alpha_1'\tau,\alpha_1\tau)\mathbf{G}_0(\alpha_2'\tau,\alpha_2\tau)+\chi\mathbf{G}_0(\alpha_1'\tau,\alpha_2\tau)\mathbf{G}_0(\alpha_2'\tau,\alpha_1\tau)\right]=$$
$$=-\frac{1}{2}\sum_{\alpha_1\alpha_2\alpha_1'\alpha_2'}\{\alpha_1\alpha_2|\hat{U}|\alpha_1'\alpha_2'\}\int_0^\beta d\tau\mathbf{G}_0(\alpha_1'\tau,\alpha_1\tau)\mathbf{G}_0(\alpha_2'\tau,\alpha_2\tau) \tag{607}$$

As seen earlier, the two-particle Green's function is reduced to the sum of two product single-particle Green's functions:

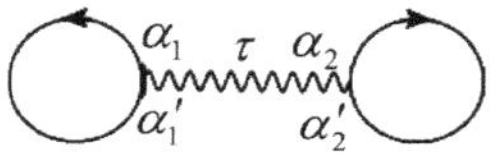

$$=-\frac{1}{2}\sum_{\alpha_1\alpha_2\alpha_1'\alpha_2'}(\alpha_1\alpha_2|\hat{U}|\alpha_1'\alpha_2')\int_0^\beta d\tau\mathbf{G}_0(\alpha_1'\tau,\alpha_1\tau)\mathbf{G}_0(\alpha_2'\tau,\alpha_2\tau)=$$

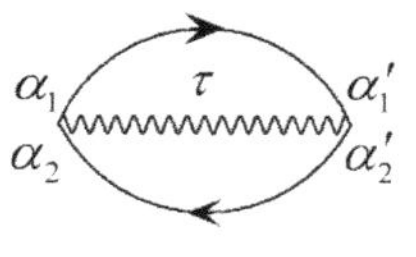

$$=-\frac{\chi}{2}\sum_{\alpha_1\alpha_2\alpha_1'\alpha_2'}(\alpha_1\alpha_2|\hat{U}|\alpha_1'\alpha_2')\int_0^\beta d\tau\mathbf{G}_0(\alpha_1'\tau,\alpha_2\tau)\mathbf{G}_0(\alpha_2'\tau,\alpha_1\tau) \tag{608}$$

The two diagrams can be represented by the following single diagram:

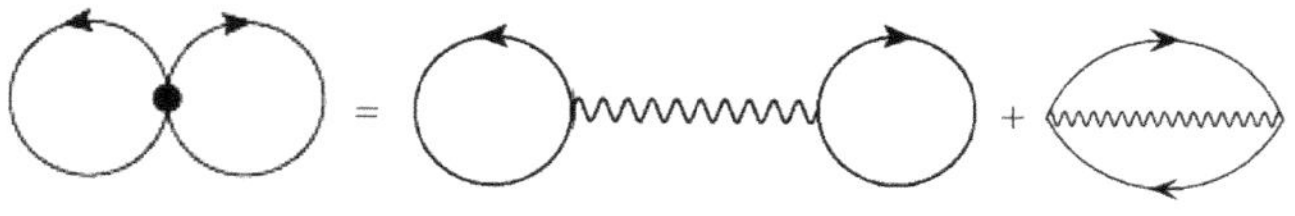

(609)

The second order has only three different diagrams, which are shown in Figure 5.1a, instead of eight as seen earlier when working with non-(anti)symmetrized vertices.

The interacting action functional S_{int} is then defined with a factor of $\frac{1}{4}$. We determine the combinatorial factor in a similar manner as seen earlier in the example of the second of the three diagrams in Figure 5.1b:

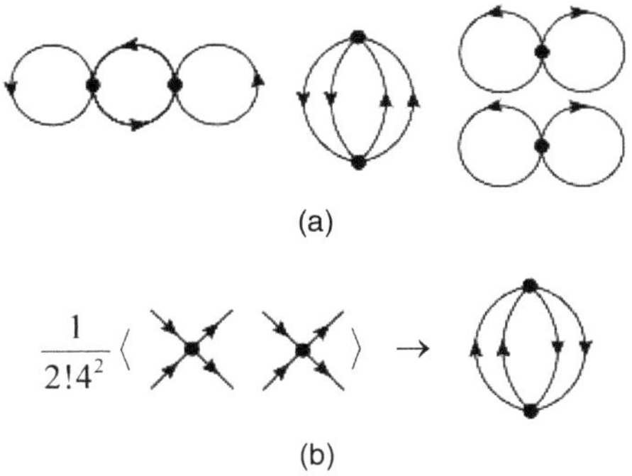

FIGURE 5.1 (a, b) Second-order diagrams.

This has a combinatorial factor of

$$\frac{1}{2!4^2}\times 4=\frac{1}{8} \tag{610}$$

where the factor “4” stems from the four ways to connect the two vertices in a manner consistent with the topology of the diagram. The sign of the diagram is determined by a particular contraction. We may as well use the fact that the diagram should have the same sign as one obtained via the replacement

$$\{\alpha_1\alpha_2|\hat{U}|\alpha_1'\alpha_2'\}\rightarrow(\alpha_1\alpha_2|\hat{U}|\alpha_1'\alpha_2') \tag{611}$$

An example would be to write the previous diagram as

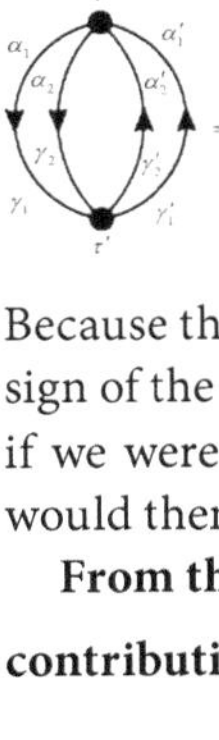

$$=\frac{1}{8}\sum_{\alpha_1\cdots\gamma_2'}\{\alpha_1\alpha_2|\hat{U}|\alpha_1'\alpha_2'\}\{\gamma_1\gamma_2|\hat{U}|\gamma_2'\gamma_1'\}\int_0^\beta d\tau d\tau'\mathbf{G}_0(\gamma_1\tau',\alpha_1\tau)\mathbf{G}_0(\gamma_2\tau',\alpha_2\tau)\mathbf{G}_0(\alpha_2'\tau,\gamma_2'\tau')\mathbf{G}_0(\alpha_1'\tau,\gamma_1'\tau') \tag{612}$$

Because there are two vertices and two closed loops, the sign is $(-1)^2\chi^2$. It is instructive to note that the sign of the diagram can be clearly defined only once it is expressed in terms of the vertices. For example, if we were to insert $\{\alpha_2\alpha_1|\hat{U}|\alpha_1'\alpha_2'\}$ in place of $\{\alpha_1\alpha_2|\hat{U}|\alpha_1'\alpha_2'\}$ in the previous equation, the diagram would then have a negative sign.

From these discussions, we arrive at the following diagrammatic rules for calculating the n-order contribution to the perturbation expansion of $\frac{Z}{Z_0}$ via the (anti)symmetrized vertices:

1. Associate noninteracting Green’s function to each directed line ↑:

$$\mathbf{G}_0(\alpha\tau,\alpha'\tau')=\delta_{\alpha\alpha'}\exp\{-(\epsilon_\alpha-\mu)(\tau-\tau')\}\left[(1+\chi n_\alpha)\theta(\tau-\tau'-\delta)+\chi n_\alpha\theta(\tau'-\tau+\delta)\right]=$$

$$\mathbf{G}_0(\alpha\tau,\alpha'\tau')=\underset{\alpha\tau}{}\longrightarrow\underset{\alpha'\tau'}{} \tag{613}$$

2. Associate

$$\{\alpha_1\alpha_2|\hat{U}|\alpha_1'\alpha_2'\}=(\alpha_1\alpha_2|\hat{U}|\alpha_1'\alpha_2')+\chi(\alpha_1\alpha_2|\hat{U}|\alpha_2'\alpha_1') \tag{614}$$

to each interaction vertex:

$$\{\alpha_1\alpha_2|\hat{U}|\alpha_1'\alpha_2'\} = \tag{615}$$

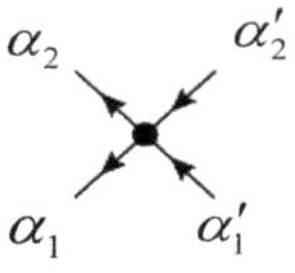

3. Sum over all internal single-particle indices α and integrate over all times τ where the integral runs over the interval $[0,\beta]$.
4. Multiply the diagram by the combinatorial factor and by $(-1)^n \chi^{n_c}$, where n is the number of vertices and n_c is the number of closed loops in the diagram obtained by replacing

$$\{\alpha_1\alpha_2|\hat{U}|\alpha_1'\alpha_2'\} = \tag{616}$$

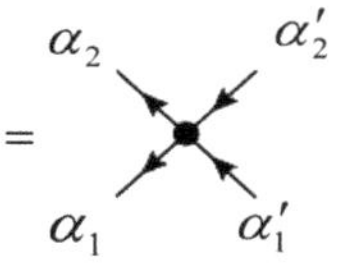

by a conventional vertex

$$(\alpha_1\alpha_2|\hat{U}|\alpha_1'\alpha_2') = \tag{617}$$

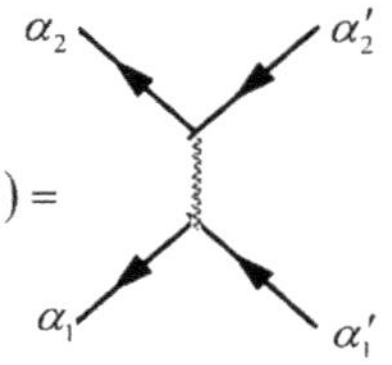

while counting the number of closed loops as for a Feynman diagram. Note that the order of the labels on the conventional vertex must match that of the matrix elements in the aforementioned rule 2.

It is important to generalize these rules to the case of the perturbative expansion for the n-particle Green's function $G^{(2n)}(\alpha_1\tau_1,\cdots,\alpha_n\tau_n,\alpha_n'\tau_n',\cdots,\alpha_1'\tau_1')$ where the last two rules are replaced by:

5. Sum over all indices α and integrate all times τ over the interval $[0,\beta]$, while the indices $(\alpha_i\tau_i)$ and $(\alpha_i'\tau_i')$ of the external lines are held fixed.
6. Multiply by the combinatorial factor and by $(-1)^n \chi^{n_c+P}$, where n is the number of vertices, and n_c is the number of closed loops in the diagram obtained by replacing all vertices

$$\{\alpha_1\alpha_2|\hat{U}|\alpha_1'\alpha_2'\} = \tag{618}$$

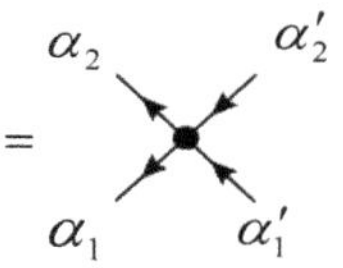

by a conventional vertex

$$(\alpha_1\alpha_2|\hat{U}|\alpha_1'\alpha_2') = \tag{619}$$

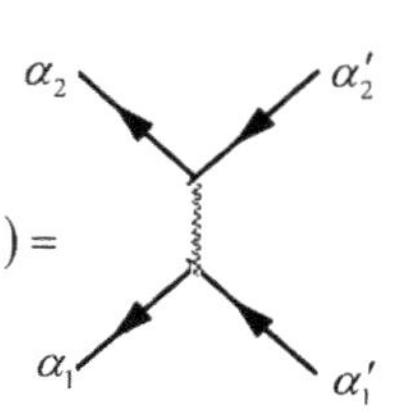

where P is the permutation of the incoming and outgoing lines.

It useful to recalculate via (anti)symmetrized vertices some of the contributions previously evaluated via Feynman diagrams. An example would be the first-order $n=1$ contribution to the single-particle Green's function $\mathbf{G}_0(\alpha\tau,\alpha'\tau')$ given by

$$-\sum_{\alpha_1\alpha_2\alpha_1'\alpha_2'}(\alpha_1\alpha_2|\hat{U}|\alpha_1'\alpha_2')\int_0^\beta d\tau''\mathbf{G}_0(\alpha\tau,\alpha_1\tau'')\left[\chi\mathbf{G}_0(\alpha_1'\tau'',\alpha'\tau')\mathbf{G}_0(\alpha_2'\tau'',\alpha_2\tau'')+\mathbf{G}_0(\alpha_2'\tau'',\alpha'\tau')\right.$$
$$\left.\mathbf{G}_0(\alpha_1'\tau'',\alpha_2\tau'')\right]=-\chi\sum_{\alpha_1\alpha_2\alpha_1'\alpha_2'}\int_0^\beta d\tau''\{\alpha_1\alpha_2|\hat{U}|\alpha_2'\alpha_1'\}\mathbf{G}_0(\alpha\tau,\alpha_1\tau'')\mathbf{G}_0(\alpha_1'\tau'',\alpha_2\tau'')\mathbf{G}_0(\alpha_2'\tau'',\alpha'\tau') \quad (620)$$

or equivalently by the diagram in (621):

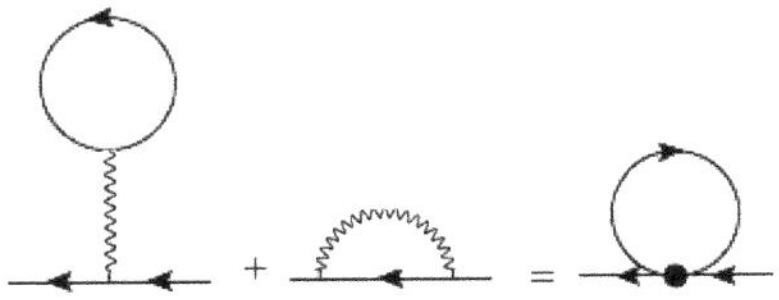

(621)

The first-order $n=1$ diagram with (anti)symmetrized vertex contributing to the single-particle Green's function.

In Figure 5.2, the second-order $n=2$ has three diagrams with (anti)symmetrized vertices contributing to the single-particle Green's function shown in (624):

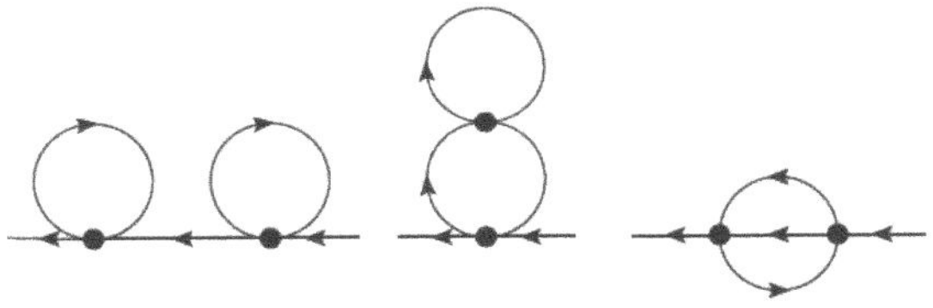

FIGURE 5.2 Second-order $n=2$ diagrams with (anti)symmetrized vertices contributing to the single-particle Green's function.

5.1 Fully (Anti)Symmetrized Vertices

We facilitate our discussions by introducing a uniform and compact notation for the creation and annihilation fermion operators. This is done by introducing a **charge index**, $c=\pm a$, to group the internal degrees of freedom α, and we define the following **charge explicit Grassmann new field**

$$\hat{\psi}_{\tilde{\alpha}}(\tau)=\begin{cases}\hat{\psi}_\alpha(\tau) & , \ c=- \\ \hat{\psi}^*_\alpha(\tau) & , \ c=+\end{cases} \quad (622)$$

Here $\tilde{\alpha}=(\alpha,c)$ and the **charge implicit Grassmann fields** $\hat{\psi}_{\alpha+}(\tau)$ and $\hat{\psi}_{\alpha-}(\tau)$ create a particle and a hole, respectively. This implies removing a particle from the system. This convenient notation allows us to treat $\hat{\psi}$ and $\hat{\psi}^*$ on an equal basis. We write the action functionals for the system as follows [3, 18, 19]:

$$S_0[\hat{\psi}]=\frac{1}{2}\int_0^\beta d\tau d\tau'\sum_{\tilde{\alpha}\tilde{\alpha}'}\hat{\psi}^{\mathrm{T}}_{\tilde{\alpha}}(\tau)\tilde{\mathbf{G}}_0^{-1}(\tilde{\alpha}\tau,\tilde{\alpha}'\tau')\hat{\psi}_{\tilde{\alpha}'}(\tau') \quad (623)$$

$$S_{\mathrm{int}}[\hat{\psi}]=\frac{1}{4!}\int_0^\beta d\tau\sum_{\tilde{\alpha}_1\tilde{\alpha}_2\tilde{\alpha}_3\tilde{\alpha}_4}U_{\tilde{\alpha}_1\tilde{\alpha}_2\tilde{\alpha}_3\tilde{\alpha}_4}\hat{\psi}_{\tilde{\alpha}_1}(\tau)\hat{\psi}_{\tilde{\alpha}_2}(\tau)\hat{\psi}_{\tilde{\alpha}_3}(\tau)\hat{\psi}_{\tilde{\alpha}_4}(\tau) \quad (624)$$

Here, $S_0\left[\hat{\psi}\right]$ and $S_{\text{int}}\left[\hat{\psi}\right]$ are, respectively, the bare and interaction parts of the action functional and the vector notation

$$\hat{\psi}^{\mathrm{T}}_{\tilde{\alpha}}(\tau)=\begin{bmatrix} \hat{\psi}_{\alpha}(\tau) & \hat{\psi}^{*}_{\tilde{\alpha}}(\tau) \end{bmatrix} \quad , \quad \hat{\psi}_{\tilde{\alpha}}(\tau)=\begin{bmatrix} \hat{\psi}_{\alpha}(\tau') \\ \hat{\psi}^{*}_{\alpha}(\tau') \end{bmatrix} \tag{625}$$

as well as the bare inverse matrix Green's function

$$\tilde{\mathbf{G}}_0^{-1}(\tilde{\alpha}\tau,\tilde{\alpha}'\tau')=\begin{bmatrix} 0 & -\mathbf{G}_0^{-1}(\tilde{\alpha}'\tau',\tilde{\alpha}\tau) \\ \mathbf{G}_0^{-1}(\tilde{\alpha}\tau,\tilde{\alpha}'\tau') & 0 \end{bmatrix} \tag{626}$$

By construction, permuting the arguments of the (anti)symmetric bare Green's function $\mathbf{G}_0$ we have:

$$\tilde{\mathbf{G}}_0(\tilde{\alpha}\tau,\tilde{\alpha}'\tau')=\chi\tilde{\mathbf{G}}_0(\tilde{\alpha}'\tau',\tilde{\alpha}\tau)=\begin{cases} -\mathbf{G}_0(\alpha\tau,\alpha'\tau') & , \quad c=-c'=- \\ 0 & , \quad c=c' \end{cases} \tag{627}$$

It is easy to verify that

$$\tilde{\mathbf{G}}_0^{-1}(\tilde{\alpha}\tau,\tilde{\alpha}'\tau')=\chi\tilde{\mathbf{G}}_0^{-1}(\tilde{\alpha}'\tau',\tilde{\alpha}\tau)=\begin{cases} -\mathbf{G}_0^{-1}(\alpha\tau,\alpha'\tau') & , \quad c=-c'=- \\ 0 & , \quad c=c' \end{cases} \tag{628}$$

The vertex $U_{\tilde{\alpha}_1\tilde{\alpha}_2\tilde{\alpha}_3\tilde{\alpha}_4}$ is (anti)symmetric under exchange of any two of its arguments:

$$U_{\tilde{\alpha}_1\tilde{\alpha}_2\tilde{\alpha}_3\tilde{\alpha}_4}=\chi U_{\tilde{\alpha}_2\tilde{\alpha}_1\tilde{\alpha}_3\tilde{\alpha}_4}=\chi U_{\tilde{\alpha}_1\tilde{\alpha}_2\tilde{\alpha}_4\tilde{\alpha}_3}=\chi U_{\tilde{\alpha}_3\tilde{\alpha}_2\tilde{\alpha}_1\tilde{\alpha}_4} \tag{629}$$

It satisfies the following conditions

$$U_{\tilde{\alpha}_1\tilde{\alpha}_2\tilde{\alpha}_3\tilde{\alpha}_4}=\begin{cases} \{\tilde{\alpha}_1\tilde{\alpha}_2|U|\tilde{\alpha}_4\tilde{\alpha}_3\} & , \quad c_1=c_2=-c_3=-c_4=+ \\ 0 & , \quad c_1+c_2+c_3+c_4\neq 0 \end{cases} \tag{630}$$

Wick theorem standard form then reads

$$\tilde{\mathbf{G}}_n(\tilde{\alpha}_1\tau_1,\cdots,\tilde{\alpha}_n\tau_n)\equiv\left\langle\hat{\psi}_{\tilde{\alpha}_1}(\tau_1)\cdots\hat{\psi}_{\tilde{\alpha}_n}(\tau_n)\right\rangle=\sum \text{all complete contractions} \tag{631}$$

We represent the various terms in the perturbation expansion of the partition function via Feynman diagrams. We represent the Green's function $\mathbf{G}_0(\tilde{\alpha}\tau,\tilde{\alpha}'\tau')$ by a **nondirected line** because we cannot distinguish between $\hat{\psi}_{\alpha}$ and $\hat{\psi}^{*}_{\alpha}$ as the difference is encoded in the charge index c of $\hat{\psi}_{\tilde{\alpha}}$:

The vertex $U_{\tilde{\alpha}_1\tilde{\alpha}_2\tilde{\alpha}_3\tilde{\alpha}_4}$ is represented by a dot:

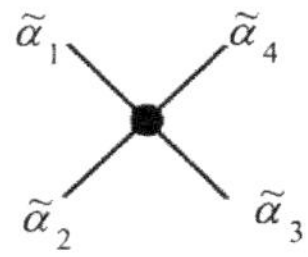

For the first-order correction to the partition function, considering the factor $\frac{1}{4!}$ in front of the interaction action functional (624) and the three ways to pair the lines of the vertex, the combinatorial factor is

$$\frac{1}{4!}\times 3=\frac{1}{8} \tag{632}$$

Considering the sign from a particular contraction, for example, say,

$$-\frac{1}{4!}\int_0^\beta d\tau \sum_{\tilde{\alpha}_1\tilde{\alpha}_2\tilde{\alpha}_3\tilde{\alpha}_4} U_{\tilde{\alpha}_1\tilde{\alpha}_2\tilde{\alpha}_3\tilde{\alpha}_4}\,\overbrace{\hat{\psi}_{\tilde{\alpha}_1}(\tau)\overbrace{\hat{\psi}_{\tilde{\alpha}_2}(\tau)\hat{\psi}_{\tilde{\alpha}_3}(\tau)}\hat{\psi}_{\tilde{\alpha}_4}(\tau)} \tag{633}$$

the combinatorial factor becomes

$$-\frac{1}{4!}\times 3=-\frac{1}{8} \tag{634}$$

So, the first-order diagrammatic correction can be represented by

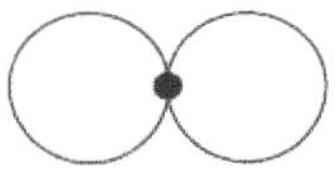

$$=-\frac{1}{8}\int_0^\beta d\tau \sum_{\tilde{\alpha}_1\tilde{\alpha}_2\tilde{\alpha}_3\tilde{\alpha}_4} U_{\tilde{\alpha}_1\tilde{\alpha}_2\tilde{\alpha}_3\tilde{\alpha}_4}\mathbf{G}_0(\tilde{\alpha}_1\tau,\tilde{\alpha}_4\tau)\mathbf{G}_0(\tilde{\alpha}_2\tau,\tilde{\alpha}_3\tau) \tag{635}$$

or equivalently by

$$\begin{aligned}
&-\frac{1}{8}\int_0^\beta d\tau \sum_{\tilde{\alpha}_1\tilde{\alpha}_2\tilde{\alpha}_3\tilde{\alpha}_4} U_{\tilde{\alpha}_1\tilde{\alpha}_2\tilde{\alpha}_3\tilde{\alpha}_4}\mathbf{G}_0(\tilde{\alpha}_1\tau,\tilde{\alpha}_4\tau)\mathbf{G}_0(\tilde{\alpha}_2\tau,\tilde{\alpha}_3\tau)=\\
&=-\frac{1}{8}\int_0^\beta d\tau \sum_{\substack{\alpha_1\alpha_2\alpha_3\alpha_4\\ c,c'=\pm}} U_{\alpha_1 c,\alpha_2 c',\alpha_3 -c',\alpha_4,-c}\mathbf{G}_0(\alpha_1,c,\tau;\alpha_4,-c,\tau)\mathbf{G}_0(\alpha_2.c',\tau;\alpha_3,-c',\tau)=\\
&=-\frac{1}{2}\int_0^\beta d\tau \sum_{\alpha_1\alpha_2\alpha_3\alpha_4} U_{\alpha_1+,\alpha_2+,\alpha_3-,\alpha_4-}\mathbf{G}_0(\alpha_4,-,\tau;\alpha_1,+,\tau)\mathbf{G}_0(\alpha_2.-,\tau;\alpha_3,+,\tau)
\end{aligned} \tag{636}$$

Similarly, we may do the same for the first-order contribution to the single-particle Green's function

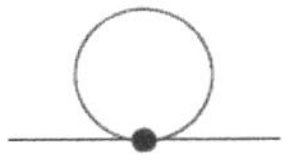

$$=-\frac{1}{2}\sum_{\tilde{\alpha}_1\cdots\tilde{\alpha}_4}\int_0^\beta d\tau'' U_{\tilde{\alpha}_1\tilde{\alpha}_2\tilde{\alpha}_3\tilde{\alpha}_4}\mathbf{G}_0(\tilde{\alpha}\tau,\tilde{\alpha}_1\tau'')\mathbf{G}_0(\tilde{\alpha}_3\tau'',\tilde{\alpha}_2,\tau'')\mathbf{G}_0(\tilde{\alpha}_4\tau'',\tilde{\alpha}'\tau') \tag{637}$$

For practical reasons, the usual Green's functions for superfluid or superconducting systems (where the global gauge symmetry is broken) can be easily found by assigning appropriate charges to the external points via the new field in (622). In such systems, the single-particle Green's function $\mathbf{G}(\tilde{\alpha}\tau,\tilde{\alpha}'\tau')$ has both **normal** $\mathscr{G}(\alpha,c,\tau,\alpha',-c,\tau')$ and **anomalous** $\mathscr{F}(\alpha,c,\tau,\alpha',c,\tau')$ components:

$$\tilde{\mathbf{G}}(\tilde{\alpha}\tau,\tilde{\alpha}'\tau')=\begin{bmatrix}\mathscr{F}(\alpha,+,\tau,\alpha',+,\tau') & \mathscr{G}(\alpha,+,\tau,\alpha',-,\tau')\\ \mathscr{G}(\alpha,-,\tau,\alpha',+,\tau') & \mathscr{F}^\dagger(\alpha,-,\tau,\alpha',-,\tau')\end{bmatrix} \tag{638}$$

or

$$\tilde{\mathbf{G}}(\tilde{\alpha}\tau,\tilde{\alpha}'\tau') = \tag{639}$$

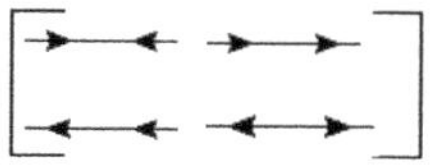

and

$$\mathscr{F}(\alpha,+,\tau,\alpha',+,\tau') = \left\langle \hat{T}\hat{\psi}^{\dagger}_{\alpha}(\tau)\hat{\psi}^{\dagger}_{\alpha}(\tau')\right\rangle \quad , \quad \mathscr{G}(\alpha,+,\tau,\alpha',-,\tau') = \left\langle \hat{T}\hat{\psi}^{\dagger}_{\alpha}(\tau)\hat{\psi}_{\alpha}(\tau')\right\rangle \tag{640}$$

$$\mathscr{F}^{\dagger}(\alpha,-,\tau,\alpha',-,\tau') = \left\langle \hat{T}\hat{\psi}_{\alpha}(\tau)\hat{\psi}_{\alpha}(\tau')\right\rangle \quad , \quad \mathscr{G}(\alpha,-,\tau,\alpha',+,\tau') = \left\langle \hat{T}\hat{\psi}_{\alpha}(\tau)\hat{\psi}^{\dagger}_{\alpha}(\tau')\right\rangle \tag{641}$$

It is important to note that superconducting transition is characterized by **spontaneous breaking of gauge symmetry**, corresponding to particle number (or charge) conservation. For the case of Bose-Einstein condensate, the electron pairs may **disappear** or **appear** from condensate, without changes in the macroscopic state of the system. It is instructive to note that the concept of spontaneous symmetry breaking plays the central role in the theory of second-order phase transitions, where the ground state of the **condensed** phase within the domain of temperature $T < T_c$ always possesses a symmetry **lower** than that of the Hamiltonian determinant describing the given phase transition.

It is instructive to note that, by definition,

$$\mathscr{G}(\alpha,+,\tau,\alpha',-,\tau') = \mathbf{G}(\alpha,\tau,\alpha',\tau') \tag{642}$$

and under exchange of any two of its arguments:

$$\mathscr{G}(\alpha,-,\tau,\alpha',+,\tau') = \chi\,\mathscr{G}(\alpha',+,\tau',\alpha,-,\tau) \quad , \quad \mathscr{G}(\alpha,+,\tau,\alpha',-,\tau') = \chi\,\mathscr{G}(\alpha',-,\tau',\alpha,+,\tau) \tag{643}$$

also

$$\mathscr{F}(\alpha,+,\tau,\alpha',+,\tau') = \chi\mathscr{F}(\alpha',+,\tau',\alpha,+,\tau) \quad , \quad \mathscr{F}^{\dagger}(\alpha,-,\tau,\alpha',-,\tau') = \chi\mathscr{F}^{\dagger}(\alpha',-,\tau',\alpha,-,\tau) \tag{644}$$

We will see later in this book that in the bosonic superfluid systems, the Green's functions $\mathbf{G}^{(2n+1)}$ with an odd number of legs do not vanish, and

$$\mathbf{G}^{(1)}(\tilde{\alpha},\tau) = \left\langle \psi_{\tilde{\alpha}}(\tau)\right\rangle \tag{645}$$

is the order parameter of the superfluid phase. It is important to note that superfluid or superconducting systems cannot be described within a perturbation expansion starting from the noninteracting limit. This is because for any system where symmetry is spontaneously broken, it is necessary to reorganize the perturbation theory about a broken-symmetry state.

6

Generating Functionals

Introduction

In this chapter, we define **connected Green's functions** with **one-particle (1PI) and two-particle (2PI) irreducible vertices.** As we have seen earlier, this allows reformulating and simplifying of the perturbation expansion. One cornerstone of the functional techniques is the concept of generating functionals that are obtained by coupling the field operators to one or more external fields, called the source fields. These source fields are thought of as our probes inside the quantum system. The introduction of the source term is artificial and serves us for technical reasons because it permits us to write higher-order correlation functions via derivatives. Once the derivative is taken, we set the source to zero. However, there are existing systems with source terms having a physical meaning, such as open systems. In such cases, the source cannot be set to zero. In this chapter, we will focus on the full $2n$-point correlation function generated via a source field coupled to fermion operators. This theory exists if we are given all the $2n$-particle transition amplitudes (Green's functions).

6.1 Connected Green's Functions

We have seen in the previous chapter that it is convenient to define a generating functional by adding the field operators to one or more external fields (called the source fields) to the given physical action functional coupling:

$$S\left[\hat{\psi}^*,\hat{\psi},\eta\right]=\sum_{\alpha}\int_0^{\beta}d\tau\left(\hat{\psi}^*_{\alpha}(\tau)\eta_{\alpha}(\tau)+\eta^*_{\alpha}(\tau)\hat{\psi}_{\alpha}(\tau)\right) \tag{646}$$

where η,η^* are c-numbers for bosons and are anticommuting variables for fermions. The path integral provides a very convenient representation of the Green's functions via the generating functional $\mathbf{Z}\left[\eta,\eta^*\right]$:

$$\mathbf{Z}\left[\eta,\eta^*\right]=\int d\left[\hat{\psi}^*\right]d\left[\hat{\psi}\right]\exp\left\{-S\left[\hat{\psi}^*,\hat{\psi}\right]+S\left[\hat{\psi}^*,\hat{\psi},\eta\right]\right\} \tag{647}$$

that generates all full Green's functions. As we discussed previously, the $2n$-particle imaginary time Green's functions are generated via the following relation:

$$\mathbf{G}^{(2n)}\left(\alpha_1\tau_1,\cdots,\alpha_n\tau_n,\alpha'_n\tau'_n,\cdots,\alpha'_1\tau'_1\right)=\left\langle\hat{\psi}_{\alpha_1}(\tau_1)\cdots\hat{\psi}_{\alpha_n}(\tau_n)\hat{\psi}^*_{\alpha'_n}(\tau'_n)\cdots\hat{\psi}^*_{\alpha'_1}(\tau'_1)\right\rangle \tag{648}$$

or

$$\mathbf{G}^{(2n)}\left(\alpha_1\tau_1,\cdots,\alpha_n\tau_n,\alpha'_n\tau'_n,\cdots,\alpha'_1\tau'_1\right)=\chi^n\left.\frac{\partial^{(2n)}\mathbf{G}\left[\eta,\eta^*\right]}{\partial\eta^*_{\alpha_1}(\tau_1)\cdots\partial\eta^*_{\alpha_n}(\tau_n)\partial\eta_{\alpha'_n}(\tau'_n)\cdots\partial\eta_{\alpha'_1}(\tau'_1)}\right|_{\eta,\eta^*=0} \tag{649}$$

and represented diagrammatically as

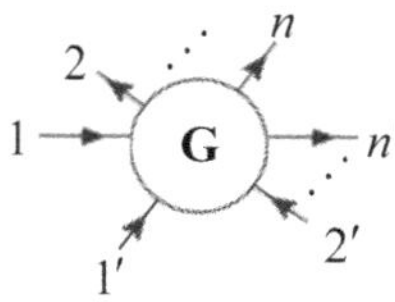

$$\equiv \mathbf{G}^{(2n)}\left(\alpha_1\tau_1,\cdots,\alpha_n\tau_n,\alpha_n'\tau_n',\cdots,\alpha_1'\tau_1'\right) \tag{650}$$

where $n \equiv (\alpha_n \tau_n)$ and $n' \equiv (\alpha_n' \tau_n')$ with

$$\mathbf{G}\left[\eta,\eta^*\right] \equiv \frac{Z\left[\eta,\eta^*\right]}{Z[0,0]} \tag{651}$$

being another generating functional because $Z[0,0]$ is independent of the external sources.

Though the Green's function $\mathbf{G}^{(2n)}$ can be obtained from the sum of all Feynman diagrams connected to the external legs, these diagrams have to be fully connected. Because the sum of all disconnected diagrams can be expressed via lower-order Green's functions $\mathbf{G}^{(2n-2)}$, $\mathbf{G}^{(2n-4)}$ and so on, it is convenient to deal with the **connected Green's function $\mathbf{G}_c^{(2n)}(1,\cdots,n,n',\cdots,1')$, which is defined as the sum of all fully connected diagrams linked to the external points $(1,\cdots,n)$, $(1',\cdots,n')$ and corresponding to the new generating functional $\mathbf{W}\left[\eta,\eta^*\right] = \ln \mathbf{G}\left[\eta,\eta^*\right]$. This implies that $\mathbf{W}\left[\eta,\eta^*\right]$ is the generating functional for connected Green's functions and is obtained from the generating functional $\mathbf{G}\left[\eta,\eta^*\right]$** via the replica method such as in the derivation of the linked cluster theorem [3], where the generating functional $\left(\mathbf{G}\left[\eta,\eta^*\right]\right)^R$ of the Green's functions of R replica of the systems can be written as a functional integral over R distinct fields $\left\{\hat{\psi}^*_{\alpha_r}(\tau), \hat{\psi}_{\alpha_r}(\tau)\right\}$ for the rth copy $r \in [1,R]$. The resulting Green's function diagrams will have the property that all connected diagrams will be proportional to R, whereas all disconnected diagrams will have at least two factors of R. The terms proportional to R are singled out by

$$\mathbf{W}\left[\eta,\eta^*\right] = \lim_{R\to 0}\frac{\partial}{\partial R}\left(\mathbf{G}\left[\eta,\eta^*\right]\right)^R = \lim_{R\to 0}\frac{\partial}{\partial R}\exp\left\{R\ln \mathbf{G}\left[\eta,\eta^*\right]\right\} \tag{652}$$

or

$$\mathbf{W}\left[\eta,\eta^*\right] = \ln \mathbf{G}\left[\eta,\eta^*\right] = \ln \mathbf{Z}\left[\eta,\eta^*\right] - \ln \mathbf{Z}[0,0] \tag{653}$$

and

$$\mathbf{G}^{(2n)}(1,\cdots,n,n',\cdots,1') = \chi^n \left.\frac{\partial^{(2n)}\mathbf{W}\left[\eta,\eta^*\right]}{\partial\eta^*_{\alpha_1}\cdots\partial\eta^*_{\alpha_n}\partial\eta_{\alpha_n'}\cdots\partial\eta_{\alpha_1'}}\right|_{\eta,\eta^*=0} \tag{654}$$

correspond to connected Feynman diagrams. According to the previous definition, they are called connected $2n$-point Green's functions $\mathbf{G}^{(2n)}$.

From (653), it is easy to show that

$$\mathbf{W}\left[\eta,\eta^*\right] = -\beta\left(\Omega\left[\eta,\eta^*\right] - \Omega[0,0]\right) \tag{655}$$

Therefore, physically, $\mathbf{W}$ should be the difference between the grand canonical potential in the presence and absence of sources as seen earlier. So, while the partition function provides the full correlation function, the functional $\mathbf{W}\left[\eta,\eta^*\right]$ gives only the connected Green's functions and their generating functional defined by

$$\mathbf{W}\left[\eta,\eta^*\right] = \ln \mathbf{Z}\left[\eta,\eta^*\right] \tag{656}$$

because the two-point correlation function or Green's function is connected in the case of fermions as there is no macroscopic fermionic field. So, in the absence of symmetry breaking, for the one-body connected Green's function, we have:

$$\mathbf{G}_c(\alpha_1\tau_1,\alpha_1'\tau_1') = \chi \left. \frac{\partial^{(2)} \mathbf{W}\left[\eta,\eta^*\right]}{\partial\eta^*_{\alpha_1}\,\partial\eta_{\alpha_1'}} \right|_{\eta,\eta^*=0} = \mathbf{G}(\alpha_1\tau_1,\alpha_1'\tau_1') \tag{657}$$

Hence, the functional $\mathbf{W}$ is the generating functional for the connected Green's function $\mathbf{G}$. We note that for fermionic fields, the two-point Green's function $\mathbf{G}$ is identical to the connected two-point function because the field expectation value always vanishes. However, higher derivatives of the generating functionals are quite different, and due to the chain rule, it is obvious that higher derivatives of the functional $\mathbf{W}\left[\eta,\eta^*\right]$ give more than one term. The extra terms subtract the unconnected parts. In mathematical terms, as seen earlier, the partition function generates moments while the $\mathbf{W}\left[\eta,\eta^*\right]$ functional generates cumulants.

Similarly, for the two-body connected Green's function in the absence of symmetry breaking, we have

$$\mathbf{G}_c^{(4)}(\alpha_1\tau_1,\alpha_2\tau_2,\alpha_2'\tau_2',\alpha_1'\tau_1') = \left. \frac{\partial^{(4)} \mathbf{W}\left[\eta,\eta^*\right]}{\partial\eta^*_{\alpha_1}\,\partial\eta^*_{\alpha_2}\,\partial\eta_{\alpha_2'}\,\partial\eta_{\alpha_1'}} \right|_{\eta,\eta^*=0} \tag{658}$$

or, equivalently,

$$\begin{aligned} &\mathbf{G}_c^{(4)}(\alpha_1\tau_1,\alpha_2\tau_2,\alpha_2'\tau_2',\alpha_1'\tau_1') = \\ &= \mathbf{G}^{(4)}(\alpha_1\tau_1,\alpha_2\tau_2,\alpha_2'\tau_2',\alpha_1'\tau_1') - \left[\mathbf{G}(\alpha_1\tau_1,\alpha_1'\tau_1')\mathbf{G}(\alpha_2\tau_2,\alpha_2'\tau_2') + \chi\mathbf{G}(\alpha_1\tau_1,\alpha_2'\tau_2')\mathbf{G}(\alpha_2\tau_2,\alpha_1'\tau_1')\right] \end{aligned} \tag{659}$$

It is instructive to note that the factor χ in the last summand stems from the fact that $\frac{\partial}{\partial\eta^*_{\alpha_2}}$ had to be commuted via an odd number of Grassmann variables. Equation (659) can be diagrammatically represented as follows:

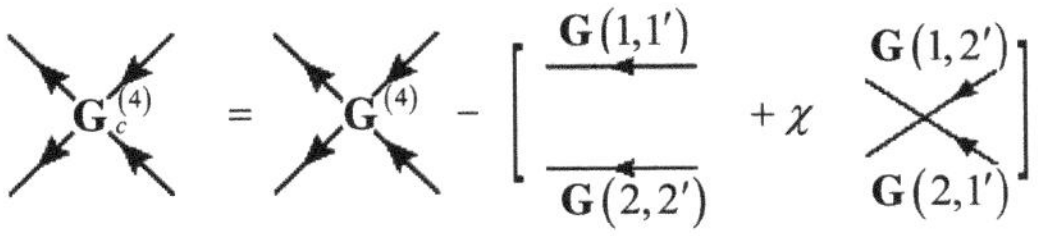

(660)

We note that the single-particle Green's function $\mathbf{G} \equiv \mathbf{G}^{(2)}$ and $\mathbf{G}_c^{(4)}$ is given by the sum of all fully connected diagrams.

6.2 General Case

We facilitate our discussions by deriving the general and compact notation by defining a charge implicit Grassmann field $\psi_{\tilde{\alpha}}(\tau)$, where $\tilde{\alpha} = (\alpha, c)$ and $c = \pm a$ is the charge index. For convenience, we group a pair of fermion arguments into an effective **bosonic** argument and use a single symbol to refer to it as $n \equiv (\tilde{\alpha}_n \tau_n)$ and $n' \equiv (\tilde{\alpha}_n' \tau_n')$. To proceed, we consider the following:

- **Care must be taken when calculating the functional derivatives over η and $G[\eta]$ as these quantities are antisymmetric and not all of their entries are independent variables.**
- **In using the chain rule for quantities implicitly dependent on η and $G[\eta]$, one must ensure only independent entries are varied in the chain rule to avoid double counting.**

From the generating functional

$$\mathbf{G}[\eta]=\frac{1}{\mathbf{Z}[0]}\int d[\psi]\exp\left\{-S[\psi]+\sum_{\tilde{\alpha}}\int_0^{\beta}d\tau\eta_{\tilde{\alpha}}(\tau)\psi_{\tilde{\alpha}}(\tau)\right\}\quad,\quad \mathbf{G}[\eta]=\frac{\mathbf{Z}[\eta]}{\mathbf{Z}[0]} \tag{661}$$

of the $2n$-point Green's function, we can directly calculate the two-point Green's function, as seen earlier:

$$\mathbf{G}^{(1)}(\tilde{\alpha}_1\tau_1)=\left.\frac{\partial \mathbf{G}[\eta]}{\partial \eta_{\tilde{\alpha}_1}(\tau_1)}\right|_{\eta=0}=\left\langle \psi_{\tilde{\alpha}_1}(\tau_1)\right\rangle \tag{662}$$

Taking additional functional derivatives over η yields higher-order correlators and can be shown by formally expanding $\mathbf{G}[\eta]$ about $\eta=0$:

$$\mathbf{G}[\eta]=1+\sum_{n=1}^{\infty}\frac{1}{n!}\int_0^{\beta}d\tau_1\cdots d\tau_n\sum_{\tilde{\alpha}_1\cdots\tilde{\alpha}_n}\mathbf{G}^{(n)}(\tilde{\alpha}_1\tau_1,\cdots,\tilde{\alpha}_n\tau_n)\eta_{\tilde{\alpha}_n}(\tau_n)\cdots\eta_{\tilde{\alpha}_1}(\tau_1) \tag{663}$$

Here,

$$\mathbf{G}^{(n)}(\tilde{\alpha}_1\tau_1,\cdots,\tilde{\alpha}_n\tau_n)=\left.\frac{\partial^{(n)}\mathbf{G}[\eta]}{\partial \eta_{\tilde{\alpha}_1}(\tau_1)\cdots\partial \eta_{\tilde{\alpha}_n}(\tau_n)}\right|_{\eta=0} \tag{664}$$

As seen earlier, we defined a two-connected diagram in the expansion of $\mathbf{G}[\eta]$ if it cannot be disentangled into the product of two disconnected pieces. We now find the generating functional for the two-connected $2n$-point Green's function using $\mathbf{G}[\eta]$ via the **linked cluster theorem**:

If $\mathbf{G}[\eta]$ is the generator of the $2n$-point Green's function, then

$$\mathbf{W}[\eta]=\ln\mathbf{G}[\eta] \tag{665}$$

is the generator of the two-connected $2n$-point Green's functions.

The proof of the **linked cluster theorem** can be revisited earlier in this text. From here on, the connected Green's functions and their generating functional are defined by

$$\mathbf{W}[\eta]=\ln\mathbf{Z}+\sum_{n=1}^{\infty}\frac{1}{n!}\int_0^{\beta}d\tau_1\cdots d\tau_n\sum_{\tilde{\alpha}_1\cdots\tilde{\alpha}_n}\mathbf{G}_c^{(n)}(\tilde{\alpha}_1\tau_1,\cdots,\tilde{\alpha}_n\tau_n)\eta_{\tilde{\alpha}_n}(\tau_n)\cdots\eta_{\tilde{\alpha}_1}(\tau_1) \tag{666}$$

with the connected functions obtained via functional derivatives according to:

$$\mathbf{G}_c^{(n)}(\tilde{\alpha}_1\tau_1,\cdots,\tilde{\alpha}_n\tau_n)=\left.\frac{\partial^{(n)}\mathbf{W}[\eta]}{\partial \eta_{\tilde{\alpha}_1}(\tau_1)\cdots\partial \eta_{\tilde{\alpha}_n}(\tau_n)}\right|_{\eta=0} \tag{667}$$

We compute the first few two-connected Green's functions to explicitly demonstrate the theorem:

$$\mathbf{G}_c^{(1)}(\tilde{\alpha}_1\tau_1)=\left.\frac{\partial \mathbf{W}[\eta]}{\partial \eta_{\tilde{\alpha}_1}(\tau_1)}\right|_{\eta=0}\equiv\mathbf{G}^{(1)}(\tilde{\alpha}_1\tau_1)=\left\langle \psi_{\tilde{\alpha}_1}(\tau_1)\right\rangle \tag{668}$$

$$\mathbf{G}_c^{(2)}(\tilde{\alpha}_1\tau_1,\tilde{\alpha}_2\tau_2)=\left.\frac{\partial^{(2)}\mathbf{W}[\eta]}{\partial \eta_{\tilde{\alpha}_1}(\tau_1)\partial \eta_{\tilde{\alpha}_2}(\tau_2)}\right|_{\eta=0}\equiv\mathbf{G}^{(2)}(\tilde{\alpha}_1\tau_1,\tilde{\alpha}_2\tau_2)-\left\langle \psi_{\tilde{\alpha}_1}(\tau_1)\right\rangle\left\langle \psi_{\tilde{\alpha}_2}(\tau_2)\right\rangle \tag{669}$$

and

$$\mathbf{G}_c^{(3)}(\tilde{\alpha}_1\tau_1,\tilde{\alpha}_2\tau_2,\tilde{\alpha}_3\tau_3)=\left.\frac{\partial^{(3)}\mathbf{W}[\eta]}{\partial\eta_{\tilde{\alpha}_1}(\tau_1)\partial\eta_{\tilde{\alpha}_2}(\tau_2)\partial\eta_{\tilde{\alpha}_3}(\tau_3)}\right|_{\eta=0} \tag{670}$$

or

$$\mathbf{G}_c^{(3)}(\tilde{\alpha}_1\tau_1,\tilde{\alpha}_2\tau_2,\tilde{\alpha}_3\tau_3)=\mathbf{G}^{(3)}(\tilde{\alpha}_1\tau_1,\tilde{\alpha}_2\tau_2,\tilde{\alpha}_3\tau_3)-\langle\psi_{\tilde{\alpha}_1}(\tau_1)\rangle\mathbf{G}_c^{(2)}(\tilde{\alpha}_2\tau_2,\tilde{\alpha}_3\tau_3)-$$
$$-\chi\langle\psi_{\tilde{\alpha}_2}(\tau_2)\rangle\mathbf{G}_c^{(2)}(\tilde{\alpha}_1\tau_1,\tilde{\alpha}_3\tau_3)-\langle\psi_{\tilde{\alpha}_3}(\tau_3)\rangle\mathbf{G}_c^{(2)}(\tilde{\alpha}_1\tau_1,\tilde{\alpha}_2\tau_2)-\langle\psi_{\tilde{\alpha}_1}(\tau_1)\rangle\langle\psi_{\tilde{\alpha}_2}(\tau_2)\rangle\langle\psi_{\tilde{\alpha}_3}(\tau_3)\rangle \tag{671}$$

Note that for a fermionic system, all Green's functions with an odd number of legs vanish. The previous expressions are also valid when $\eta\neq 0$.

6.3 Dyson-Schwinger Equations

We now examine another useful approach toward calculating the Green's functions via the so-called equation of motion method, where a differential equation is found for $\mathbf{G}$. Green's functions are total amplitudes for a given process and represent a sum of the amplitudes of all distinct ways in which this process can happen. This additive property is the central property of all quantum theories and mathematically is stated in terms of the Dyson-Schwinger equations (DSEs) developed by and named after Freeman J. Dyson and Julian S. Schwinger [20–22]. These are the coupled-integral equations of motion of the Green's functions of a quantum field theory that describe the propagation and the interaction of the fields. In addition, they permit us to gain some direct insight into the behavior of Green's functions.

One major problem for the DSEs is that there are an infinite number of them and also they are coupled. So, we can speak of an infinite system of DSEs. The full system of DSEs provides a complete nonperturbative description of the theory under study. Generally, it is impossible to solve the full/infinite system of equations. So, we have to truncate the system. Notwithstanding, the DSEs provide a continuum method for easily calculating Green's functions over many scales and allow us to gain some direct insight into the behavior of the Green's functions. It is important to note that untruncated DSEs are exact.

The most convenient way to derive the Dyson-Schwinger equations is via path integrals and to employ the approach of generating functions in the context of the path integral. Because the path integral is expressed in terms of formally unknown functions, the derivatives acting upon the generator are functional in nature, and so the generator is considered a generating functional:

$$\mathbf{G}[\eta]=\frac{1}{\mathbf{Z}[0]}\int d[\psi]\exp\left\{-S[\psi]+\sum_{\tilde{\alpha}}\int_0^\beta d\tau\eta_{\tilde{\alpha}}(\tau)\psi_{\tilde{\alpha}}(\tau)\right\} \tag{672}$$

We begin the derivation of the Schwinger-Dyson equations by noting that the expectation values produced by the generating functionals would remain invariant by an arbitrary shift in any of the field variables in the action functional:

$$\psi_{\tilde{\alpha}}(\tau)\to\psi'_{\tilde{\alpha}}(\tau)=\psi_{\tilde{\alpha}}(\tau)+\delta\psi_{\tilde{\alpha}}(\tau)\quad,\quad\delta\psi_{\tilde{\alpha}}(\tau)\to 0 \tag{673}$$

and

$$\mathbf{G}[\eta]=\frac{1}{\mathbf{Z}[0]}\int d[\psi]\exp\left\{-S[\psi_{\tilde{\alpha}}+\delta\psi_{\tilde{\alpha}}(\tau)]+\sum_{\tilde{\alpha}}\int_0^\beta d\tau\eta_{\tilde{\alpha}}(\tau)[\psi_{\tilde{\alpha}}(\tau)+\delta\psi_{\tilde{\alpha}}(\tau)]\right\} \tag{674}$$

where $\delta\psi_{\tilde{\alpha}}(\tau)$ is an arbitrary well decreasing function at infinity (so that it belongs to the class of functions to be integrated over in the path integral).

Let us make a variable substitution within the path integral for $\psi'_{\tilde{\alpha}}(\tau)$. This leaves the path integral measure $d[\psi]$ invariant (including the functional measure for Grassmann variables), and we recover the identical form for the generating functional defined in (672). If we allow $\delta\psi_{\tilde{\alpha}}(\tau)$ to be an infinitesimal shift in the fields, then the functional variation of $\mathbf{G}[\eta]$ with respect to any of the functional variables in the path integral is precisely zero. This is exactly what should be expected when integration is done over all possible variations in $\psi_{\tilde{\alpha}}(\tau)$. When coupled with the invariance of the path integral measure under such a transformation and the infinitesimal form, equality (674) reads:

$$\delta\mathbf{G}[\eta]=\frac{1}{\mathbf{Z}[0]}\int d[\psi]\frac{\delta}{\delta\psi_{\tilde{\alpha}}(\tau)}\exp\left\{-S[\psi]+\sum_{\tilde{\alpha}}\int_0^{\beta}d\tau\eta_{\tilde{\alpha}}(\tau)\psi_{\tilde{\alpha}}(\tau)\right\}=0 \tag{675}$$

And, after differentiation of the exponential function, then

$$\delta\mathbf{G}[\eta]=\frac{1}{\mathbf{Z}[0]}\int d[\psi]\exp\left\{-S[\psi_{\tilde{\alpha}}]+\sum_{\tilde{\alpha}}\int_0^{\beta}d\tau\eta_{\tilde{\alpha}}(\tau)\psi_{\tilde{\alpha}}(\tau)\right\}\sum_{\tilde{\alpha}}\int_0^{\beta}d\tau\delta\psi_{\tilde{\alpha}}(\tau)\left[\frac{\delta S}{\delta\psi_{\tilde{\alpha}}(\tau)}-\chi\eta_{\tilde{\alpha}}(\tau)\right]=0 \tag{676}$$

or

$$\delta\mathbf{G}[\eta]=\frac{1}{\mathbf{Z}[0]}\int d[\psi]\exp\left\{-S[\psi_{\tilde{\alpha}}]+\sum_{\tilde{\alpha}}\int_0^{\beta}d\tau\eta_{\tilde{\alpha}}(\tau)\psi_{\tilde{\alpha}}(\tau)\right\}\sum_{\tilde{\alpha}}\int_0^{\beta}d\tau\delta\psi_{\tilde{\alpha}}(\tau)\left[\frac{\delta S}{\delta\psi_{\tilde{\alpha}}(\tau)}-\chi\eta_{\tilde{\alpha}}(\tau)\right]=0 \tag{677}$$

We introduce χ depending on the particular field being considered. Because $\delta\psi_{\tilde{\alpha}}(\tau)$ is arbitrary, we can drop it together with the integral over $d\tau$:

$$\left[\frac{\delta S}{\delta\psi_{\tilde{\alpha}}(\tau)}-\chi\eta_{\tilde{\alpha}}(\tau)\right]\mathbf{G}[\eta]=0 \tag{678}$$

This is the basic Dyson-Schwinger variational equation for the Green's function generating functional [23].

Considering the action functionals (601) and (622), the equation of motion (678) becomes

$$\frac{1}{\mathbf{Z}[0]}\int d[\psi]\exp\left\{-S[\psi_{\tilde{\alpha}}]+\sum_{\tilde{\alpha}}\int_0^{\beta}d\tau\eta_{\tilde{\alpha}}(\tau)\psi_{\tilde{\alpha}}(\tau)\right\}\mathrm{F}[\psi]=0 \tag{679}$$

where

$$\mathrm{F}[\psi]=\int_0^{\beta}d\tau_1'\sum_{\tilde{\alpha}_1'}\mathbf{G}_0^{-1}(\tilde{\alpha}_1\tau_1,\tilde{\alpha}_1'\tau_1')\psi_{\tilde{\alpha}_1'}(\tau_1')+\frac{1}{3!}\sum_{\tilde{\alpha}_2'\tilde{\alpha}_3'\tilde{\alpha}_4'}U_{\tilde{\alpha}_1\tilde{\alpha}_2'\tilde{\alpha}_3'\tilde{\alpha}_4'}\psi_{\tilde{\alpha}_2'}(\tau_1)\psi_{\tilde{\alpha}_3'}(\tau_1)\psi_{\tilde{\alpha}'_4}(\tau_1)-\chi\eta_{\tilde{\alpha}_1}(\tau_1) \tag{680}$$

We find the equation of motion for the Green's function $\mathbf{G}^{(n)}$ by taking the $(n-1)$-order functional derivative with respect to η and at the end set $\eta=0$:

$$\int_0^{\beta}d\tau_1'\sum_{\tilde{\alpha}_1'}\mathbf{G}_0^{-1}(\tilde{\alpha}_1\tau_1,\tilde{\alpha}_1'\tau_1')\mathbf{G}^{(n)}(\tilde{\alpha}_n\tau_n,\cdots,\tilde{\alpha}_2\tau_2,\tilde{\alpha}_1'\tau_1')+\frac{1}{3!}\sum_{\tilde{\alpha}_2'\tilde{\alpha}_3'\tilde{\alpha}_4'}U_{\tilde{\alpha}_1\tilde{\alpha}_2'\tilde{\alpha}_3'\tilde{\alpha}_4'}\mathbf{G}^{(n+2)}(\tilde{\alpha}_n\tau_n,\cdots,\tilde{\alpha}_2\tau_2,\tilde{\alpha}_2'\tau_1,\tilde{\alpha}_3'\tau_1,\tilde{\alpha}_4'\tau_1)=$$

$$=\chi\sum_{k=2}^{n}\delta_{\tilde{\alpha}_k\tilde{\alpha}_1}\delta(\tau_k-\tau_1)\mathbf{G}^{(n-2)}(\tilde{\alpha}_n\tau_n,\cdots,\tilde{\alpha}_{k+1}\tau_{k+1},\tilde{\alpha}_{k-1}\tau_{k-1},\cdots,\tilde{\alpha}_2\tau_2) \tag{681}$$

The equation of motion for $\mathbf{G}^{(2)} = \mathbf{G}$ is as follows:

$$\int_0^\beta d\tau_1' \sum_{\tilde{\alpha}_1'} \mathbf{G}_0^{-1}(\tilde{\alpha}_1\tau_1, \tilde{\alpha}_1'\tau_1') \mathbf{G}(\tilde{\alpha}_1'\tau_1', \tilde{\alpha}_2\tau_2) + \frac{1}{3!} \sum_{\tilde{\alpha}_2'\tilde{\alpha}_3'\tilde{\alpha}_4'} U_{\tilde{\alpha}_1\tilde{\alpha}_2'\tilde{\alpha}_3'\tilde{\alpha}_4'} \mathbf{G}^{(4)}(\tilde{\alpha}_2'\tau_1, \tilde{\alpha}_3'\tau_1, \tilde{\alpha}_4'\tau_1, \tilde{\alpha}_2\tau_2) = \delta_{\tilde{\alpha}_1\tilde{\alpha}_2}\delta(\tau_1 - \tau_2) \tag{682}$$

This equation of motion is represented in (683).

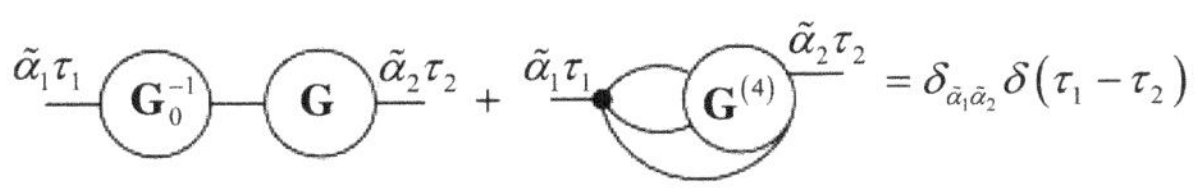

$$= \delta_{\tilde{\alpha}_1\tilde{\alpha}_2}\delta(\tau_1 - \tau_2) \tag{683}$$

Diagrammatic representation of the equation of motion for $\mathbf{G} = \mathbf{G}^{(2)}$ in the general case.

For a normal system, considering

$$\tilde{\alpha}_1 = (\alpha_1, +) \quad , \quad \tilde{\alpha}_2 = (\alpha_2, +) \tag{684}$$

Then,

$$\int_0^\beta d\tau_1' \sum_{\tilde{\alpha}_1'} \mathbf{G}_0^{-1}(\alpha_1\tau_1, \alpha_1'\tau_1') \mathbf{G}(\alpha_1'\tau_1', \alpha_2\tau_2) + \sum_{\alpha_2'\alpha_3'\alpha_4'} (\alpha_2'\alpha_1|U|\alpha_4'\alpha_3') \mathbf{G}^{(4)}(\alpha_3'\tau_1, \alpha_4'\tau_1, \alpha_2'\tau_1, \alpha_2\tau_2) = \delta_{\alpha_1\alpha_2}\delta(\tau_1 - \tau_2) \tag{685}$$

This equation of motion is represented in (686).

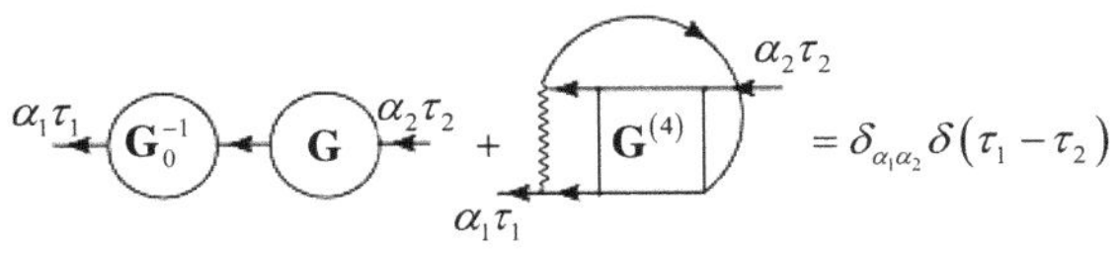

$$= \delta_{\alpha_1\alpha_2}\delta(\tau_1 - \tau_2) \tag{686}$$

Diagrammatic representation of the equation of motion for $\mathbf{G} = \mathbf{G}^{(2)}$, considering a normal system.

6.4 Effective Action For 1PI Green's Functions

One-particle irreducible (1PI) Green's functions play an important role in the renormalization of quantum field theories (particularly those theories with gauge invariance) in addition to nonperturbative calculations (the **so-called effective action method**). Therefore, it is desirable to construct another set of generating functionals for them that is particularly suitable for describing phase transitions and fields whose expectation value does not always vanish, such as the order parameter in a symmetry-broken phase.

The connected Green's functions can be composed of the so-called **one-particle irreducible (1PI) Green's functions'** vertex functions generated by the functional **(also called the effective action)** $\Gamma[\phi, \phi^*]$. Because all Green's functions can be constructed from the 1PI Green's functions, they fully characterize a quantum field theory. The generating functional for the 1PI Green's functions $\Gamma[\phi, \phi^*]$ is obtained from the generating functional for the connected Green's functions $\mathbf{W}[\eta, \eta^*]$ by the Legendre transformation [3], which we show in the next section.

6.4.1 Normal Systems

In the presence of external sources η, η^*, the fields $\psi_\alpha(\tau)$, and $\psi_\alpha^*(\tau)$, normal systems acquire nonzero expectation values as the derivative of the generator of the connected Green's functions:

$$\phi_\alpha(\tau)=\langle\psi_\alpha(\tau)\rangle_{\eta,\eta^*}=\frac{\int\prod_\alpha d\psi_\alpha d\psi^*_\alpha\psi_\alpha(\tau)\exp\left\{\sum_{\alpha,\alpha'}\psi^*_\alpha\mathbf{G}_{\alpha,\alpha'}\psi_{\alpha'}+\sum_\alpha\left(\eta^*_\alpha\psi_\alpha+\eta_\alpha\psi^*_\alpha\right)\right\}}{\int\prod_\alpha d\psi_\alpha d\psi^*_\alpha\exp\left\{\sum_{\alpha,\alpha'}\psi^*_\alpha\mathbf{G}_{\alpha,\alpha'}\psi_{\alpha'}+\sum_\alpha\left(\eta^*_\alpha\psi_\alpha+\eta_\alpha\psi^*_\alpha\right)\right\}}=-\frac{\partial\mathbf{W}\left[\eta,\eta^*\right]}{\partial\eta^*_\alpha(\tau)}\tag{687}$$

and its complex conjugate field

$$\phi^*_\alpha(\tau)=\langle\psi^*_\alpha(\tau)\rangle_{\eta,\eta^*}=\frac{\int\prod_\alpha d\psi_\alpha d\psi^*_\alpha\psi^*_\alpha(\tau)\exp\left\{\sum_{\alpha,\alpha'}\psi^*_\alpha\mathbf{G}_{\alpha,\alpha'}\psi_{\alpha'}+\sum_\alpha\left(\eta^*_\alpha\psi_\alpha+\eta_\alpha\psi^*_\alpha\right)\right\}}{\int\prod_\alpha d\psi_\alpha d\psi^*_\alpha\exp\left\{\sum_{\alpha,\alpha'}\psi^*_\alpha\mathbf{G}_{\alpha,\alpha'}\psi_{\alpha'}+\sum_\alpha\left(\eta^*_\alpha\psi_\alpha+\eta_\alpha\psi^*_\alpha\right)\right\}}=-\chi\frac{\partial\mathbf{W}\left[\eta,\eta^*\right]}{\partial\eta_\alpha(\tau)}\tag{688}$$

Here, $\phi_\alpha(\tau)$ and $\phi^*_\alpha(\tau)$ are c-numbers for bosonic fields and Grassmann variables for fermionic fields that are averaged fields. For the generating functional $\mathbf{W}\left[\eta,\eta^*\right]$, equation (688) for $\phi_\alpha\left[\eta,\eta^*\right]$ and $\phi^*_\alpha\left[\eta,\eta^*\right]$ is inverted to obtain the sources as functionals of the fields $\eta_\alpha\left[\phi_\alpha,\phi^*_\alpha\right]$ and $\eta^*_\alpha\left[\phi_\alpha,\phi^*_\alpha\right]$. We now introduce a new **effective action** $\Gamma\left[\phi,\phi^*\right]$ as the Legendre transform of $\mathbf{W}\left[\eta,\eta^*\right]$ in a functional sense [3]:

$$\Gamma\left[\phi,\phi^*\right]=-\mathbf{W}\left[\eta,\eta^*\right]-\sum_\alpha\int_0^\beta d\tau\left(\phi^*_\alpha(\tau)\eta_\alpha(\tau)+\eta^*_\alpha(\tau)\phi_\alpha(\tau)\right)\tag{689}$$

Here $\Gamma\left[\phi,\phi^*\right]$ depends explicitly on the new fields ϕ and ϕ^*, while the generating functional $\mathbf{W}\left[\eta,\eta^*\right]$ of the connected Green's functions depends only on the external sources η and η^*. It is instructive to note that $\Gamma\left[\phi,\phi^*\right]$ is the generating functional of the vertex functions or vertex functional that generates the 1PI Feynman diagrams. This implies the diagrams cannot be disconnected by cutting only one line. The advantage of this treatment lies in the fact that it is physically more transparent to formulate a nonequilibrium theory for the order parameter $\langle\psi\rangle$.

Taking the functional derivative of (689) by applying the chain rule to the effective action $\Gamma\left[\phi,\phi^*\right]$, we have the **reciprocity relation**, which plays the role of the generating functional of the 1PI vertices:

$$\frac{\partial\Gamma\left[\phi,\phi^*\right]}{\partial\phi^*_{\alpha'}(\tau)}=-\frac{\partial\mathbf{W}\left[\eta,\eta^*\right]}{\partial\phi^*_{\alpha'}(\tau)}-\sum_\alpha\int_0^\beta d\tau'\left(\chi\phi^*_\alpha(\tau')\frac{\partial\eta_\alpha(\tau')}{\partial\phi^*_{\alpha'}(\tau)}+\frac{\partial\phi^*_\alpha(\tau')}{\partial\phi^*_{\alpha'}(\tau)}\eta_\alpha(\tau')+\frac{\partial\eta^*_\alpha(\tau')}{\partial\phi^*_{\alpha'}(\tau)}\phi_\alpha(\tau')\right)\tag{690}$$

or

$$\begin{aligned}\frac{\partial\Gamma\left[\phi,\phi^*\right]}{\partial\phi^*_{\alpha'}(\tau)}=&-\sum_\alpha\int_0^\beta d\tau'\left(\frac{\partial\eta_\alpha(\tau')}{\partial\phi^*_{\alpha'}(\tau)}\frac{\partial\mathbf{W}\left[\eta,\eta^*\right]}{\partial\eta_\alpha(\tau')}+\frac{\partial\eta^*_\alpha(\tau')}{\partial\phi^*_{\alpha'}(\tau)}\frac{\partial\mathbf{W}\left[\eta,\eta^*\right]}{\partial\eta^*_\alpha(\tau')}\right)-\\&-\sum_\alpha\int_0^\beta d\tau'\left(\chi\phi^*_\alpha(\tau')\frac{\partial\eta_\alpha(\tau')}{\partial\phi^*_{\alpha'}(\tau)}+\frac{\partial\phi^*_\alpha(\tau')}{\partial\phi^*_{\alpha'}(\tau)}\eta_\alpha(\tau')+\frac{\partial\eta^*_\alpha(\tau')}{\partial\phi^*_{\alpha'}(\tau)}\phi_\alpha(\tau')\right)\end{aligned}\tag{691}$$

or

$$\begin{aligned}\frac{\partial\Gamma\left[\phi,\phi^*\right]}{\partial\phi^*_{\alpha'}(\tau)}=&-\sum_\alpha\int_0^\beta d\tau'\left(-\chi\frac{\partial\eta_\alpha(\tau')}{\partial\phi^*_{\alpha'}(\tau)}\phi^*_\alpha(\tau')-\frac{\partial\eta^*_\alpha(\tau')}{\partial\phi^*_{\alpha'}(\tau)}\phi_\alpha(\tau')\right)-\\&-\sum_\alpha\int_0^\beta d\tau'\left(\chi\phi^*_\alpha(\tau')\frac{\partial\eta_\alpha(\tau')}{\partial\phi^*_{\alpha'}(\tau)}+\frac{\partial\phi^*_\alpha(\tau')}{\partial\phi^*_{\alpha'}(\tau)}\eta_\alpha(\tau')+\frac{\partial\eta^*_\alpha(\tau')}{\partial\phi^*_{\alpha'}(\tau)}\phi_\alpha(\tau')\right)\end{aligned}\tag{692}$$

or

$$\frac{\partial\Gamma\left[\phi,\phi^*\right]}{\partial\phi^*_{\alpha'}(\tau)}=\sum_\alpha\int_0^\beta d\tau'\left(\chi\frac{\partial\eta_\alpha(\tau')}{\partial\phi^*_{\alpha'}(\tau)}\phi^*_\alpha(\tau')+\frac{\partial\eta^*_\alpha(\tau')}{\partial\phi^*_{\alpha'}(\tau)}\phi_\alpha(\tau')\right)-$$
$$-\sum_\alpha\int_0^\beta d\tau'\left(\chi\phi^*_\alpha(\tau')\frac{\partial\eta_\alpha(\tau')}{\partial\phi^*_{\alpha'}(\tau)}+\frac{\partial\phi^*_\alpha(\tau')}{\partial\phi^*_{\alpha'}(\tau)}\eta_\alpha(\tau')+\frac{\partial\eta^*_\alpha(\tau')}{\partial\phi^*_{\alpha'}(\tau)}\phi_\alpha(\tau')\right) \quad (693)$$

or

$$\frac{\partial\Gamma\left[\phi,\phi^*\right]}{\partial\phi^*_{\alpha'}(\tau)}=\sum_\alpha\int_0^\beta d\tau'\left(\chi\phi^*_\alpha(\tau')\frac{\partial\eta_\alpha(\tau')}{\partial\phi^*_{\alpha'}(\tau)}+\frac{\partial\eta^*_\alpha(\tau')}{\partial\phi^*_{\alpha'}(\tau)}\phi_\alpha(\tau')\right)-$$
$$-\sum_\alpha\int_0^\beta d\tau'\left(\chi\phi^*_\alpha(\tau')\frac{\partial\eta_\alpha(\tau')}{\partial\phi^*_{\alpha'}(\tau)}+\frac{\partial\phi^*_\alpha(\tau')}{\partial\phi^*_{\alpha'}(\tau)}\eta_\alpha(\tau')+\frac{\partial\eta^*_\alpha(\tau')}{\partial\phi^*_{\alpha'}(\tau)}\phi_\alpha(\tau')\right) \quad (694)$$

or

$$\frac{\partial\Gamma\left[\phi,\phi^*\right]}{\partial\phi^*_{\alpha'}(\tau)}=-\sum_\alpha\int_0^\beta d\tau'\eta_\alpha(\tau')\delta_{\alpha\alpha'}\delta(\tau-\tau')=-\eta_{\alpha'}(\tau) \quad (695)$$

We evaluate the companion equation as follows:

$$\frac{\partial\Gamma\left[\phi,\phi^*\right]}{\partial\phi_{\alpha'}(\tau)}=-\frac{\partial\mathbf{W}\left[\eta,\eta^*\right]}{\partial\phi_{\alpha'}(\tau)}-\sum_\alpha\int_0^\beta d\tau'\left(\chi\phi_\alpha(\tau')\frac{\partial\eta_\alpha(\tau')}{\partial\phi_{\alpha'}(\tau)}+\chi\eta^*_\alpha(\tau')\frac{\partial\phi_\alpha(\tau')}{\partial\phi_{\alpha'}(\tau)}+\frac{\partial\eta^*_\alpha(\tau')}{\partial\phi_{\alpha'}(\tau)}\phi_\alpha(\tau')\right) \quad (696)$$

or

$$\frac{\partial\Gamma\left[\phi,\phi^*\right]}{\partial\phi_{\alpha'}(\tau)}=-\sum_\alpha\int_0^\beta d\tau'\left(\frac{\partial\eta_\alpha(\tau')}{\partial\phi_{\alpha'}(\tau)}\frac{\partial\mathbf{W}\left[\eta,\eta^*\right]}{\partial\eta_\alpha(\tau')}+\frac{\partial\eta^*_\alpha(\tau')}{\partial\phi_{\alpha'}(\tau)}\frac{\partial\mathbf{W}\left[\eta,\eta^*\right]}{\partial\eta^*_\alpha(\tau')}\right)-$$
$$-\sum_\alpha\int_0^\beta d\tau'\left(\chi\phi_\alpha(\tau')\frac{\partial\eta_\alpha(\tau')}{\partial\phi_{\alpha'}(\tau)}+\chi\eta^*_\alpha(\tau')\frac{\partial\phi_\alpha(\tau')}{\partial\phi_{\alpha'}(\tau)}+\frac{\partial\eta^*_\alpha(\tau')}{\partial\phi_{\alpha'}(\tau)}\phi_\alpha(\tau')\right) \quad (697)$$

or

$$\frac{\partial\Gamma\left[\phi,\phi^*\right]}{\partial\phi_{\alpha'}(\tau)}=-\sum_\alpha\int_0^\beta d\tau'\left(-\chi\frac{\partial\eta_\alpha(\tau')}{\partial\phi_{\alpha'}(\tau)}\phi_\alpha(\tau')-\frac{\partial\eta^*_\alpha(\tau')}{\partial\phi_{\alpha'}(\tau)}\phi_\alpha(\tau')\right)-$$
$$-\sum_\alpha\int_0^\beta d\tau'\left(\chi\phi_\alpha(\tau')\frac{\partial\eta_\alpha(\tau')}{\partial\phi_{\alpha'}(\tau)}+\chi\eta^*_\alpha(\tau')\frac{\partial\phi_\alpha(\tau')}{\partial\phi_{\alpha'}(\tau)}+\frac{\partial\eta^*_\alpha(\tau')}{\partial\phi_{\alpha'}(\tau)}\phi_\alpha(\tau')\right) \quad (698)$$

or

$$\frac{\partial\Gamma\left[\phi,\phi^*\right]}{\partial\phi_{\alpha'}(\tau)}=\sum_\alpha\int_0^\beta d\tau'\left(\chi\frac{\partial\eta_\alpha(\tau')}{\partial\phi_{\alpha'}(\tau)}\phi_\alpha(\tau')+\frac{\partial\eta^*_\alpha(\tau')}{\partial\phi_{\alpha'}(\tau)}\phi_\alpha(\tau')\right)-$$
$$-\sum_\alpha\int_0^\beta d\tau'\left(\chi\phi_\alpha(\tau')\frac{\partial\eta_\alpha(\tau')}{\partial\phi_{\alpha'}(\tau)}+\chi\eta^*_\alpha(\tau')\frac{\partial\phi_\alpha(\tau')}{\partial\phi_{\alpha'}(\tau)}+\frac{\partial\eta^*_\alpha(\tau')}{\partial\phi_{\alpha'}(\tau)}\phi_\alpha(\tau')\right) \quad (699)$$

or

$$\frac{\partial\Gamma\left[\phi,\phi^*\right]}{\partial\phi_{\alpha'}(\tau)}=\sum_\alpha\int_0^\beta d\tau'\left(\chi\phi_\alpha(\tau')\frac{\partial\eta_\alpha(\tau')}{\partial\phi_{\alpha'}(\tau)}+\frac{\partial\eta_\alpha^*(\tau')}{\partial\phi_{\alpha'}(\tau)}\phi_\alpha(\tau')\right)-$$
$$-\sum_\alpha\int_0^\beta d\tau'\left(\chi\phi_\alpha(\tau')\frac{\partial\eta_\alpha(\tau')}{\partial\phi_{\alpha'}(\tau)}+\chi\eta_\alpha^*(\tau')\frac{\partial\phi_\alpha(\tau')}{\partial\phi_{\alpha'}(\tau)}+\frac{\partial\eta_\alpha^*(\tau')}{\partial\phi_{\alpha'}(\tau)}\phi_\alpha(\tau')\right) \tag{700}$$

or

$$\frac{\partial\Gamma\left[\phi,\phi^*\right]}{\partial\phi_{\alpha'}(\tau)}=-\chi\sum_\alpha\int_0^\beta d\tau'\eta_\alpha^*(\tau')\delta_{\alpha\alpha'}\delta(\tau-\tau')=-\chi\eta_{\alpha'}^*(\tau) \tag{701}$$

This explicitly defines η and η^* (for known $\Gamma\left[\phi,\phi^*\right]$) as a functional of $\phi_\alpha(\tau)$ and $\phi_\alpha^*(\tau)$. We consider the effective action as an analog to the Gibbs free energy seen in statistical thermodynamics [13].

If the sources are set equal to zero, equations (695) and (701) show that the effective action is stationary. This implies that, if we denote the fields in the absence of sources by $\bar{\phi}_\alpha(\tau)$ and $\bar{\phi}_\alpha^*(\tau)$, then

$$\frac{\partial\Gamma\left[\bar{\phi}_\alpha,\bar{\phi}_\alpha^*\right]}{\partial\bar{\phi}_\alpha(\tau)}=\frac{\partial\Gamma\left[\bar{\phi}_\alpha,\bar{\phi}_\alpha^*\right]}{\partial\bar{\phi}_\alpha^*(\tau)}=0 \tag{702}$$

It is instructive to note that in the case of Bose condensation, equation (702) has nonzero solutions $\bar{\phi}_\alpha(\tau)$ and $\bar{\phi}_\alpha^*(\tau)$. In the absence of symmetry breaking, the fields $\left\{\bar{\phi}_\alpha(\tau),\bar{\phi}_\alpha^*(\tau)\right\}$ are zero, and all Green's functions that have equal numbers of creation and annihilation operators vanish.

We observe from (689) that the Legendre transform implements a change in the active variable $(\eta,\eta^*)\to(\phi,\phi^*)$:

- The expectation value $\phi=\langle\psi\rangle_{\eta,\eta^*}$ is called the **classical field**. For ultracold bosons, it has a direct physical interpretation in terms of the condensate mean field.
- The effective action carries the same information as **Z** and **W**, only it is organized differently and generates the **one-particle irreducible (1PI)** correlation functions.

It is instructive to note that

- **The action principle is leveraged over to a full quantum status.**
- **Effective action can be understood as classical action plus fluctuations and provides a quasiclassical approximation (small fluctuations around a mean field).**
- **Symmetry principles are leveraged over from the classical action to a full quantum status.**

6.4.2 Self-Energy and Dyson Equation

While perturbation theory is fine in some cases, in many cases one needs to sum over whole classes of Feynman diagrams to give an elegant graphical representation of arbitrary contributions to perturbation series for Green's functions. We have observed in the previous chapter that specific diagrammatic rules for a given interacting system may be reduced to the study of (scattering) S-matrix perturbation expansion and the use of the Wick theorem [24, 25]. It is instructive to note that typical graphic elements of any diagram technique are Green's functions **lines** and interaction **vertices**, which are combined into Feynman diagrams of a certain **topology** depending on the type of interaction under consideration. In this book, we further examine these rules explicitly [24] for different types of interactions.

Consider the Feynman diagram technique that involves the possibility of performing **graphical summation** of infinite (sub)series of diagrams leading to the so-called Dyson equation [24, 25]. We find that the full Green's function is determined by the Dyson equation.

As discussed earlier, the generating function for vertex functions is an effective action. The **1PI** vertices are generated by the functional derivatives of the effective action $\Gamma\left[\phi,\phi^*\right]$:

$$\Gamma^{(2n)}\left(\alpha_1\tau_1,\cdots,\alpha_n\tau_n,\alpha_n'\tau_n',\cdots,\alpha_1'\tau_1'\right)=\left.\frac{\partial^{(2n)}\Gamma\left[\phi,\phi^*\right]}{\partial\phi_{\alpha_1}^*(\tau_1)\cdots\partial\phi_{\alpha_n}^*(\tau_n)\partial\phi_{\alpha_n'}(\tau_n')\cdots\partial\phi_{\alpha_1'}(\tau_1')}\right|_{\phi,\phi^*=0} \tag{703}$$

Some of the important properties of this functional are:

1. **They are one-particle irreducible (1PI) and therefore cannot be disconnected by removing a single internal propagator.**
2. **The connected Green's functions may be constructed from vertex functions using only tree diagrams. This implies diagrams with no closed propagator loops. This property is very useful in renormalization of field theories, where all the divergences arise from loop integrals that are isolated in the vertex functions Γ, and in the definition of consistent truncated expansions.**

These properties can be made more explicit by deriving a hierarchy of integral equations satisfied by the vertex functions as well as by the Green's functions. Inverting (703), we have the following vertex function

$$\Gamma\left[\phi,\phi^*\right]=\Gamma[0,0]+\sum_{n=1}^{\infty}\frac{1}{(n!)^2}\int_0^\beta d\tau_1\cdots d\tau_1'\sum_{\alpha_1\cdots\alpha_1'}\Gamma^{(2n)}\left(\alpha_1\tau_1,\cdots,\alpha_n\tau_n,\alpha_n'\tau_n',\cdots,\alpha_1'\tau_1'\right)\phi_{\alpha_1}^*(\tau_1)\cdots\phi_{\alpha_n}^*(\tau_n)\phi_{\alpha_n'}(\tau_n')\cdots\phi_{\alpha_1'}(\tau_1') \tag{704}$$

where,

$$\Gamma[0,0]=-\mathbf{W}[0,0]=\beta\Omega \tag{705}$$

To derive the Dyson equation, we require the following terms:

$$\begin{aligned}\frac{\partial\phi_{\alpha_3}(\tau_3)}{\partial\phi_{\alpha_1}(\tau_1)}&=\frac{\partial}{\partial\phi_{\alpha_1}(\tau_1)}\left[-\frac{\partial\mathbf{W}\left[\eta,\eta^*\right]}{\partial\eta_{\alpha_3}^*(\tau_3)}\right]\\&=-\sum_{\alpha_2}\int_0^\beta d\tau_2\left(\frac{\partial\eta_{\alpha_2}(\tau_2)}{\partial\phi_{\alpha_1}(\tau_1)}\frac{\partial^2\mathbf{W}\left[\eta,\eta^*\right]}{\partial\eta_{\alpha_2}(\tau_2)\partial\eta_{\alpha_3}^*(\tau_3)}+\frac{\partial\eta_{\alpha_2}^*(\tau_2)}{\partial\phi_{\alpha_1}(\tau_1)}\frac{\partial^2\mathbf{W}\left[\eta,\eta^*\right]}{\partial\eta_{\alpha_2}^*(\tau_2)\partial\eta_{\alpha_3}(\tau_3)}\right)\end{aligned} \tag{706}$$

or

$$\frac{\partial\phi_{\alpha_3}(\tau_3)}{\partial\phi_{\alpha_1}(\tau_1)}=-\sum_{\alpha_2}\int_0^\beta d\tau_2\left(-\frac{\partial^2\Gamma\left[\eta,\eta^*\right]}{\partial\phi_{\alpha_1}(\tau_1)\partial\phi_{\alpha_2}^*(\tau_2)}\frac{\partial^2\mathbf{W}\left[\eta,\eta^*\right]}{\partial\eta_{\alpha_2}(\tau_2)\partial\eta_{\alpha_3}^*(\tau_3)}-\chi\frac{\partial^2\Gamma\left[\eta,\eta^*\right]}{\partial\phi_{\alpha_1}(\tau_1)\partial\phi_{\alpha_2}(\tau_2)}\frac{\partial^2\mathbf{W}\left[\eta,\eta^*\right]}{\partial\eta_{\alpha_2}^*(\tau_2)\partial\eta_{\alpha_3}^*(\tau_3)}\right) \tag{707}$$

or

$$\frac{\partial\phi_{\alpha_3}(\tau_3)}{\partial\phi_{\alpha_1}(\tau_1)}=\sum_{\alpha_2}\int_0^\beta d\tau_2\left(\frac{\partial^2\Gamma\left[\eta,\eta^*\right]}{\partial\phi_{\alpha_1}(\tau_1)\partial\phi_{\alpha_2}^*(\tau_2)}\frac{\partial^2\mathbf{W}\left[\eta,\eta^*\right]}{\partial\eta_{\alpha_2}(\tau_2)\partial\eta_{\alpha_3}^*(\tau_3)}+\chi\frac{\partial^2\Gamma\left[\eta,\eta^*\right]}{\partial\phi_{\alpha_1}(\tau_1)\partial\phi_{\alpha_2}(\tau_2)}\frac{\partial^2\mathbf{W}\left[\eta,\eta^*\right]}{\partial\eta_{\alpha_2}^*(\tau_2)\partial\eta_{\alpha_3}^*(\tau_3)}\right) \tag{708}$$

also

$$\frac{\partial\phi^*_{\alpha_3}(\tau_3)}{\partial\phi^*_{\alpha_1}(\tau_1)} = \frac{\partial}{\partial\phi_{\alpha_1}(\tau_1)}\left[-\chi\frac{\partial\mathbf{W}[\eta,\eta^*]}{\partial\eta^*_{\alpha_3}(\tau_3)}\right]$$
$$= -\chi\sum_{\alpha_2}\int_0^\beta d\tau_2\left(\frac{\partial\eta^*_{\alpha_2}(\tau_2)}{\partial\phi^*_{\alpha_1}(\tau_1)}\frac{\partial^2\mathbf{W}[\eta,\eta^*]}{\partial\eta^*_{\alpha_2}(\tau_2)\partial\eta_{\alpha_3}(\tau_3)} + \frac{\partial\eta_{\alpha_2}(\tau_2)}{\partial\phi_{\alpha_1}(\tau_1)}\frac{\partial^2\mathbf{W}[\eta,\eta^*]}{\partial\eta_{\alpha_2}(\tau_2)\partial\eta_{\alpha_3}(\tau_3)}\right) \tag{709}$$

or

$$\frac{\partial\phi^*_{\alpha_3}(\tau_3)}{\partial\phi^*_{\alpha_1}(\tau_1)} = -\chi\sum_{\alpha_2}\int_0^\beta d\tau_2\left(-\chi\frac{\partial^2\Gamma[\eta,\eta^*]}{\partial\phi^*_{\alpha_1}(\tau_1)\partial\phi_{\alpha_2}(\tau_2)}\frac{\partial^2\mathbf{W}[\eta,\eta^*]}{\partial\eta^*_{\alpha_2}(\tau_2)\partial\eta_{\alpha_3}(\tau_3)}\right.$$
$$\left. - \frac{\partial^2\Gamma[\eta,\eta^*]}{\partial\phi^*_{\alpha_1}(\tau_1)\partial\phi^*_{\alpha_2}(\tau_2)}\frac{\partial^2\mathbf{W}[\eta,\eta^*]}{\partial\eta_{\alpha_2}(\tau_2)\partial\eta_{\alpha_3}(\tau_3)}\right) \tag{710}$$

or

$$\frac{\partial\phi^*_{\alpha_3}(\tau_3)}{\partial\phi^*_{\alpha_1}(\tau_1)} = \sum_{\alpha_2}\int_0^\beta d\tau_2\left(\frac{\partial^2\Gamma[\eta,\eta^*]}{\partial\phi^*_{\alpha_1}(\tau_1)\partial\phi_{\alpha_2}(\tau_2)}\frac{\partial^2\mathbf{W}[\eta,\eta^*]}{\partial\eta^*_{\alpha_2}(\tau_2)\partial\eta_{\alpha_3}(\tau_3)} + \chi\frac{\partial^2\Gamma[\eta,\eta^*]}{\partial\phi^*_{\alpha_1}(\tau_1)\partial\phi^*_{\alpha_2}(\tau_2)}\frac{\partial^2\mathbf{W}[\eta,\eta^*]}{\partial\eta_{\alpha_2}(\tau_2)\partial\eta_{\alpha_3}(\tau_3)}\right) \tag{711}$$

then

$$\frac{\partial\phi_{\alpha_3}(\tau_3)}{\partial\phi^*_{\alpha_1}(\tau_1)} = \frac{\partial}{\partial\phi_{\alpha_1}(\tau_1)}\left[-\frac{\partial\mathbf{W}[\eta,\eta^*]}{\partial\eta^*_{\alpha_3}(\tau_3)}\right]$$
$$= -\sum_{\alpha_2}\int_0^\beta d\tau_2\left(\frac{\partial\eta^*_{\alpha_2}(\tau_2)}{\partial\phi^*_{\alpha_1}(\tau_1)}\frac{\partial^2\mathbf{W}[\eta,\eta^*]}{\partial\eta^*_{\alpha_2}(\tau_2)\partial\eta^*_{\alpha_3}(\tau_3)} + \frac{\partial\eta_{\alpha_2}(\tau_2)}{\partial\phi^*_{\alpha_1}(\tau_1)}\frac{\partial^2\mathbf{W}[\eta,\eta^*]}{\partial\eta_{\alpha_2}(\tau_2)\partial\eta^*_{\alpha_3}(\tau_3)}\right) \tag{712}$$

or

$$\frac{\partial\phi_{\alpha_3}(\tau_3)}{\partial\phi^*_{\alpha_1}(\tau_1)} = -\sum_{\alpha_2}\int_0^\beta d\tau_2\left(-\chi\frac{\partial^2\Gamma[\eta,\eta^*]}{\partial\phi^*_{\alpha_1}(\tau_1)\partial\phi_{\alpha_2}(\tau_2)}\frac{\partial^2\mathbf{W}[\eta,\eta^*]}{\partial\eta^*_{\alpha_2}(\tau_2)\partial\eta^*_{\alpha_3}(\tau_3)} - \frac{\partial^2\Gamma[\eta,\eta^*]}{\partial\phi^*_{\alpha_1}(\tau_1)\partial\phi^*_{\alpha_2}(\tau_2)}\frac{\partial^2\mathbf{W}[\eta,\eta^*]}{\partial\eta_{\alpha_2}(\tau_2)\partial\eta^*_{\alpha_3}(\tau_3)}\right) \tag{713}$$

or

$$\frac{\partial\phi_{\alpha_3}(\tau_3)}{\partial\phi^*_{\alpha_1}(\tau_1)} = \sum_{\alpha_2}\int_0^\beta d\tau_2\left(\chi\frac{\partial^2\Gamma[\eta,\eta^*]}{\partial\phi^*_{\alpha_1}(\tau_1)\partial\phi_{\alpha_2}(\tau_2)}\frac{\partial^2\mathbf{W}[\eta,\eta^*]}{\partial\eta^*_{\alpha_2}(\tau_2)\partial\eta^*_{\alpha_3}(\tau_3)} + \frac{\partial^2\Gamma[\eta,\eta^*]}{\partial\phi^*_{\alpha_1}(\tau_1)\partial\phi^*_{\alpha_2}(\tau_2)}\frac{\partial^2\mathbf{W}[\eta,\eta^*]}{\partial\eta_{\alpha_2}(\tau_2)\partial\eta^*_{\alpha_3}(\tau_3)}\right) \tag{714}$$

and

$$\frac{\partial\phi^*_{\alpha_3}(\tau_3)}{\partial\phi_{\alpha_1}(\tau_1)} = \frac{\partial}{\partial\phi_{\alpha_1}(\tau_1)}\left[-\chi\frac{\partial\mathbf{W}[\eta,\eta^*]}{\partial\eta_{\alpha_3}(\tau_3)}\right]$$
$$= -\chi\sum_{\alpha_2}\int_0^\beta d\tau_2\left(\frac{\partial\eta^*_{\alpha_2}(\tau_2)}{\partial\phi_{\alpha_1}(\tau_1)}\frac{\partial^2\mathbf{W}[\eta,\eta^*]}{\partial\eta^*_{\alpha_2}(\tau_2)\partial\eta_{\alpha_3}(\tau_3)} + \frac{\partial\eta_{\alpha_2}(\tau_2)}{\partial\phi_{\alpha_1}(\tau_1)}\frac{\partial^2\mathbf{W}[\eta,\eta^*]}{\partial\eta_{\alpha_2}(\tau_2)\partial\eta_{\alpha_3}(\tau_3)}\right) \tag{715}$$

or

$$\frac{\partial \phi^*_{\alpha_3}(\tau_3)}{\partial \phi_{\alpha_1}(\tau_1)} = -\chi \sum_{\alpha_2} \int_0^\beta d\tau_2 \left(-\chi \frac{\partial^2 \Gamma[\eta,\eta^*]}{\partial \phi_{\alpha_1}(\tau_1) \partial \phi_{\alpha_2}(\tau_2)} \frac{\partial^2 \mathbf{W}[\eta,\eta^*]}{\partial \eta^*_{\alpha_2}(\tau_2) \partial \eta_{\alpha_3}(\tau_3)} - \frac{\partial^2 \Gamma[\eta,\eta^*]}{\partial \phi_{\alpha_1}(\tau_1) \partial \phi^*_{\alpha_2}(\tau_2)} \frac{\partial^2 \mathbf{W}[\eta,\eta^*]}{\partial \eta_{\alpha_2}(\tau_2) \partial \eta_{\alpha_3}(\tau_3)} \right) \tag{716}$$

or

$$\frac{\partial \phi^*_{\alpha_3}(\tau_3)}{\partial \phi_{\alpha_1}(\tau_1)} = \sum_{\alpha_2} \int_0^\beta d\tau_2 \left(\frac{\partial^2 \Gamma[\eta,\eta^*]}{\partial \phi_{\alpha_1}(\tau_1) \partial \phi_{\alpha_2}(\tau_2)} \frac{\partial^2 \mathbf{W}[\eta,\eta^*]}{\partial \eta^*_{\alpha_2}(\tau_2) \partial \eta_{\alpha_3}(\tau_3)} + \chi \frac{\partial^2 \Gamma[\eta,\eta^*]}{\partial \phi_{\alpha_1}(\tau_1) \partial \phi^*_{\alpha_2}(\tau_2)} \frac{\partial^2 \mathbf{W}[\eta,\eta^*]}{\partial \eta_{\alpha_2}(\tau_2) \partial \eta_{\alpha_3}(\tau_3)} \right) \tag{717}$$

This can be represented diagramatically by

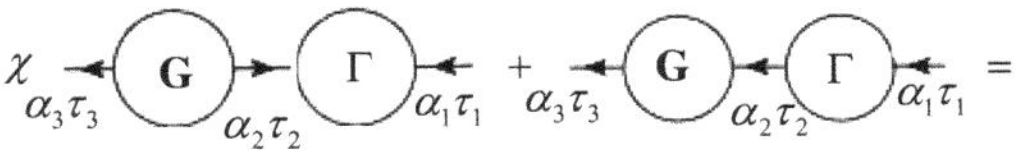

$$= \delta_{\alpha_3\alpha_1} \delta(\tau_3 - \tau_1) \tag{718}$$

From equations (708) to (717), we have the following matrix equation:

$$\sum_{\alpha_2} \int_0^\beta d\tau_2 \Gamma_{\phi\phi^*}(1,2) \mathbf{G}_c^{(2)}(2,3) = \delta_{\alpha_1\alpha_3} \delta(\tau_3 - \tau_1) \begin{bmatrix} 1 & 0 \\ 0 & 1 \end{bmatrix} \tag{719}$$

where

$$\Gamma_{\phi\phi^*}(1,2) = \begin{bmatrix} \dfrac{\partial^2 \Gamma[\eta,\eta^*]}{\partial \phi_{\alpha_1}(\tau_1) \partial \phi^*_{\alpha_2}(\tau_2)} & \dfrac{\partial^2 \Gamma[\eta,\eta^*]}{\partial \phi^*_{\alpha_1}(\tau_1) \partial \phi^*_{\alpha_2}(\tau_2)} \\ \dfrac{\partial^2 \Gamma[\eta,\eta^*]}{\partial \phi_{\alpha_1}(\tau_1) \partial \phi_{\alpha_2}(\tau_2)} & \dfrac{\partial^2 \Gamma[\eta,\eta^*]}{\partial \phi^*_{\alpha_1}(\tau_1) \partial \phi_{\alpha_2}(\tau_2)} \end{bmatrix},$$

$$\mathbf{G}_c^{(2)}(2,3) = \begin{bmatrix} \dfrac{\partial^2 \mathbf{W}[\eta,\eta^*]}{\partial \eta_{\alpha_2}(\tau_2) \partial \eta^*_{\alpha_3}(\tau_3)} & \chi \dfrac{\partial^2 \mathbf{W}[\eta,\eta^*]}{\partial \eta^*_{\alpha_2}(\tau_2) \partial \eta^*_{\alpha_3}(\tau_3)} \\ \chi \dfrac{\partial^2 \mathbf{W}[\eta,\eta^*]}{\partial \eta_{\alpha_2}(\tau_2) \partial \eta_{\alpha_3}(\tau_3)} & \dfrac{\partial^2 \mathbf{W}[\eta,\eta^*]}{\partial \eta^*_{\alpha_2}(\tau_2) \partial \eta_{\alpha_3}(\tau_3)} \end{bmatrix} \tag{720}$$

We observe from this equation that the matrix $\Gamma_{\phi\phi^*}(1,2)$ is the inverse of the matrix $\mathbf{G}_c^{(2)}(2,3)$, which is the connected Green's function. The matrix $\Gamma_{\phi\phi^*}(1,2)$ has the following matrix elements:

$$\Gamma_{\phi\phi} = \frac{\partial^2 \Gamma}{\partial \phi \partial \phi}, \quad \Gamma_{\phi\phi^*} = \frac{\partial^2 \Gamma}{\partial \phi \partial \phi^*}, \quad \Gamma_{\phi^*\phi} = \frac{\partial^2 \Gamma}{\partial \phi^* \partial \phi}, \quad \Gamma_{\phi^*\phi^*} = \frac{\partial^2 \Gamma}{\partial \phi^* \partial \phi^*} \tag{721}$$

This permits us to write $\Gamma_{\phi\phi^*}(1,2)$ as

$$\begin{bmatrix} \Gamma_{\phi\phi^*} & \Gamma_{\phi^*\phi^*} \\ \Gamma_{\phi\phi} & \Gamma_{\phi^*\phi} \end{bmatrix} = \chi \begin{bmatrix} \langle \psi\psi^* \rangle & \langle \psi^*\psi^* \rangle \\ \langle \psi\psi \rangle & \langle \psi^*\psi \rangle \end{bmatrix}^{-1} \tag{722}$$

Important property: The second derivative of the effective action is the inverse Green's function.

To understand the properties and physical significance of (721), we consider the absence of symmetry breaking. In this case, the Green's functions that consist of an unequal number of ψ and ψ^* vanish and the previous equations are reduced to

$$\sum_{\alpha_2}\int_0^\beta d\tau_2 \mathbf{G}_c^{(2)}(\alpha_1\tau_1,\alpha_2\tau_2)\Gamma_{\phi_{\alpha 2}^*\phi_{\alpha 3}} = \sum_{\alpha_2}\int_0^\beta d\tau_2 \Gamma_{\phi_{\alpha 1}^*\phi_{\alpha 2}} \mathbf{G}_c^{(2)}(\alpha_2\tau_2,\alpha_3\tau_3) = \delta_{\alpha_3\alpha_1}\delta(\tau_3-\tau_1) \tag{723}$$

Here, we establish a relationship between the interacting two-point connected Green's function and the two-point 1PI Green's function. This implies the link between connected and 1PI diagrams. Similarly, any higher connected Green's functions can be expressed via 1PI functions, with the diagrams of the connected Green's functions being constructed from 1PI parts linked by lines in a manner that cutting any of these lines translates the diagrams to disconnected ones. We now have the inverse connected Green's function

$$\Gamma_{\phi_{\alpha 1}^*\phi_{\alpha 2}} = \left[\mathbf{G}_c^{(2)}\right]^{-1}(\alpha_1\tau_1,\alpha_2\tau_2) = \left[\mathbf{G}^{(2)}\right]^{-1}(\alpha_1\tau_1,\alpha_2\tau_2) \tag{724}$$

in a matrix sense.

6.4.2.1 Self-Energy and Dyson Equation

We find that the Dyson series is a way to sum up infinite classes of diagrams. It is instructive to note that $\mathbf{G}_c^{(2)} = \mathbf{G}$ in the absence of broken symmetry. We may now conveniently express $\Gamma_{\phi_{\alpha 1}\phi_{\alpha 2}^*}$ in terms of the self-energy Σ, which is defined as the difference between the vertex function or inverse Green's function of the interacting system and the noninteracting system. This implies that

$$\Gamma_{\phi_{\alpha 1}\phi_{\alpha 2}^*} = \Gamma^{(0)}_{\phi_{\alpha 1}\phi_{\alpha 2}^*} + \Sigma_{\phi_{\alpha 1}\phi_{\alpha 2}^*} \tag{725}$$

The **exact self-energy** constitutes all **irreducible diagrams**. This implies **a disconnected diagram if a single propagator line is cut**. Equation (725) also may be written as follows

$$\Gamma^{(2)} = \mathbf{G}_0^{-1} + \Sigma \tag{726}$$

or

$$\mathbf{G}^{-1} = \mathbf{G}_0^{-1} + \Sigma \tag{727}$$

Here, Σ is the 1PI self-energy, and we retrieve the Dyson equation from the previous considerations. If this equation is multiplied by $\mathbf{G}$ from the right and $\mathbf{G}_0$ from the left, then we have

$$\mathbf{G}_0 = \mathbf{G} + \mathbf{G}_0\Sigma\mathbf{G} \tag{728}$$

or

$$\mathbf{G} = \mathbf{G}_0 - \mathbf{G}_0\Sigma\mathbf{G} \tag{729}$$

which is the Dyson equation:

$$\mathbf{G} = \mathbf{G}_0 + (-\mathbf{G}_0\Sigma)\mathbf{G} = \mathbf{G}_0 + (-\mathbf{G}_0\Sigma)\mathbf{G}_0 + (-\mathbf{G}_0\Sigma)\mathbf{G}_0(-\mathbf{G}_0\Sigma)\Sigma\mathbf{G}_0\cdots \tag{730}$$

or

$$\mathbf{G} = \mathbf{G}_0 \sum_{n=0}^{\infty} \left[(-\mathbf{G}_0 \Sigma) \right]^n = \frac{\mathbf{G}_0}{1 - \mathbf{G}_0 \Sigma} = \frac{1}{\mathbf{G}_0^{-1} - \Sigma} \tag{731}$$

This equation summarizes, in a particularly compact form, the various contributions to the exact one-particle connected Green's function. This implies that the sum of all topologically inequivalent and connected diagrams may be expressed in terms of the noninteracting Green's function plus an irreducible part.

Equation (731) can be read as a matrix equation, including summation over all indices and integration over internal indices. The Dyson equation can be explicitly written as follows:

$$\mathbf{G}(\alpha\tau, \alpha'\tau') = \mathbf{G}_0(\alpha\tau, \alpha'\tau') - \sum_{\alpha_1 \alpha_2} \int_0^{\beta} d\tau_1\, d\tau_2 \mathbf{G}_0(\alpha\tau, \alpha_1\tau_1) \Sigma(\alpha_1\tau_1, \alpha_2\tau_2) \mathbf{G}(\alpha_2\tau_2, \alpha'\tau') \tag{732}$$

Iterating this equation, we obtain the full perturbation series for the Green's function, and after Fourier transformation, the Dyson equation is reduced to an algebraic equation. Consequently, from the form of $\mathbf{G}_0$, we have

$$\mathbf{G}(\vec{\kappa}, \omega) = \frac{1}{\mathbf{G}_0^{-1}(\vec{\kappa}, \omega) - \Sigma(\vec{\kappa}, \omega) + i\delta \operatorname{sgn}(|\vec{\kappa}| - \kappa_F)} \tag{733}$$

The function $\Sigma(\vec{\kappa}, \omega)$ is known as the **exact self-energy** and is denoted by the circles in the diagrams. It is instructive to note that the self-energy $\Sigma(\vec{\kappa}, \omega)$ represents the compact form of all changes in a particle motion as a result of its interaction with other particles of the system. It is also the **effective field** or **potential** that the particle in state $\vec{\kappa}$ sees due to its interaction with all the other particles of the system. Certainly, this field is considerably more complicated than the Hartree-Fock field due to its ω-dependence, which describes the motion of the quasiparticle cloud. Equation (733) is completely general in as much as we define the self-energy in a general way:

Self-Energy

The self-energy $\Sigma(\vec{\kappa}, \omega)$ is defined as the sum of all diagrams that cannot be split into two by breaking a single fermion line. It is instructive to note that there must be momentum $\vec{\kappa}$ and energy ω coming in from the left and going out to the right of the self-energy diagrams, although the Green's function lines that carry this energy and momentum to the first vertex and away from the last vertex are not included. The graphical expansion of the self-energy Σ is evident from expressing the Dyson equation in (730) and its series expansion in diagrams such as in (738).

In diagrammatic terms, we write the total single-particle propagator $\mathbf{G}$, which is the sum of the amplitudes for all possible ways the particle can propagate through the given system:

(734)

or

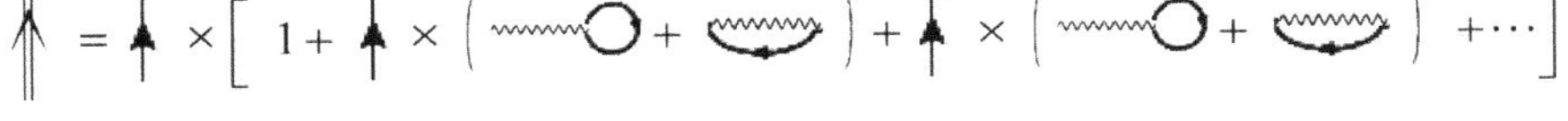
(735)

or

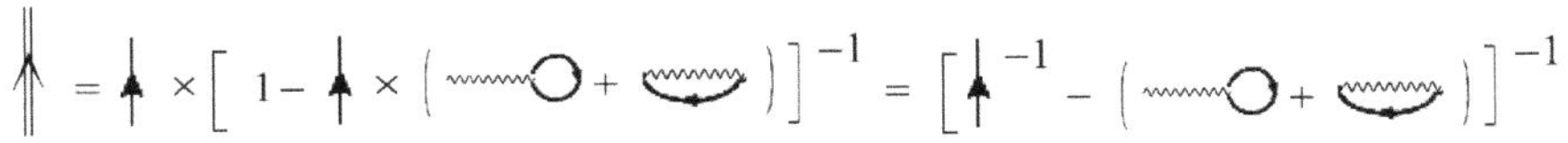
(736)

or

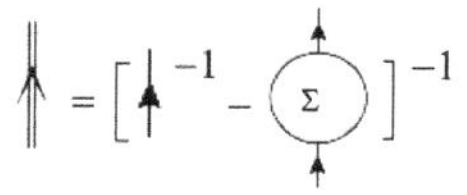

(737)

where the **irreducible self-energy** that is the sum of all proper (irreducible) self-energy parts is as follows:

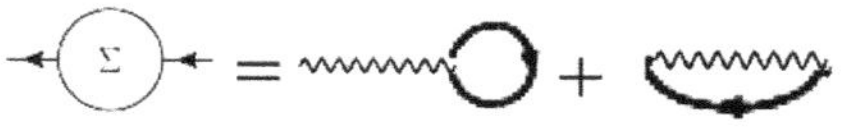

(738)

We have the following important consequences of the Dyson series:

- **Fewer independent diagrams to calculate**
- **The interacting G at the moment is like the Green's function of noninteracting fermions but with a renormalized energy spectrum:**

$$\varsigma(\vec{\kappa}) \to \varsigma(\vec{\kappa}) - \Sigma(\vec{\kappa}, \omega) \tag{739}$$

- **Hence, the term *self-energy***

If Σ is dependent only on $\vec{\kappa}$, then this would be a renormalized energy spectrum. Further, we examine the consequences of Σ being a function of energy.

The Dyson equation also can be diagrammatically represented as

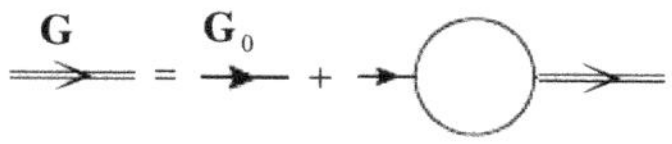

(740)

Diagrammatic representation of the Dyson equation connecting the Green's function and self-energy, where the double directed line stands for the full Green's function G and the single directed line, the bare Green's function G_0. The circles represent the exact self-energy function.

or

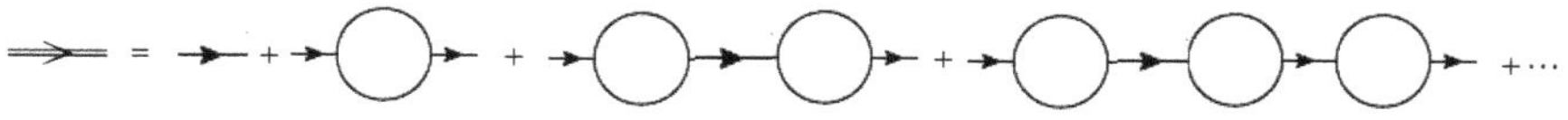

(741)

or

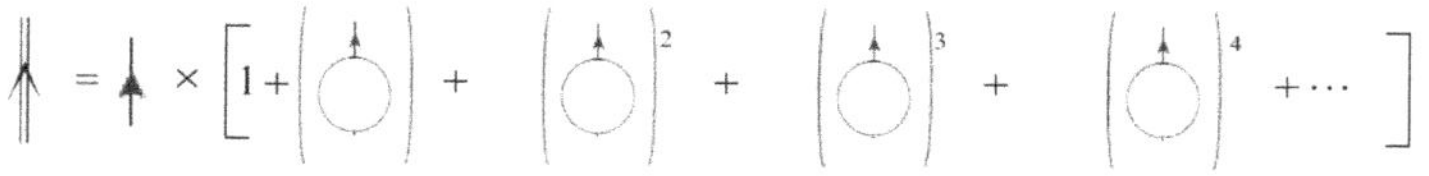

(742)

or

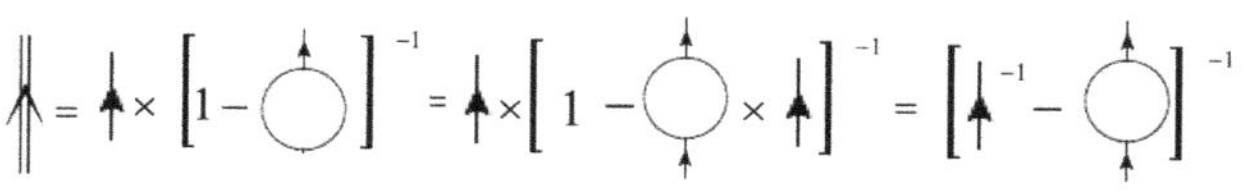

(743)

Diagrammatic representation of the Dyson equation, where the double directed line stands for the full Green's function G and the single directed line, the bare Green's function G_0. The circles represent the exact self-energy function.

We see from (740) that the Dyson equation is dressed. This implies the full Green's function stands on both sides of the given equation and in (740) is denoted by the double arrow. The circles denote the exact self-energy function $-\Sigma$. In (743), the full Green's function has been inserted, and only the first terms are considered. So, (743) corresponds to a perturbative description. Hence, if we calculate the self-energy, we get the correction of the noninteracting Green's function to the interacting Green's function. Here, the interacting Green's function is reduced to the connected Green's function. It is important to see how to express the self-energy in terms of the diagrammatic perturbation theory derived for the interacting Green's function. Therefore, two standard definitions are essential:

1. **A diagram is n-particle irreducible if it cannot be separated into two or more disconnected pieces by cutting n internal lines.**
2. **An amputated diagram has no propagator attached to the external legs $(\alpha_i\tau_i)$. Hence, each leg connects directly to an interaction vertex.**

Considering these definitions, we easily see that the self-energy $-\Sigma(\alpha_1\tau_1,\alpha_2\tau_2)$ is given by the sum of 1PI amputated diagrams connecting the legs $(\alpha_1\tau_1)$ and $(\alpha_2\tau_2)$. One easily verifies that all diagrams for the Green's function **G** are indeed obtained from the Dyson equation. For instance, the following self-energy diagram

$\Sigma =$ 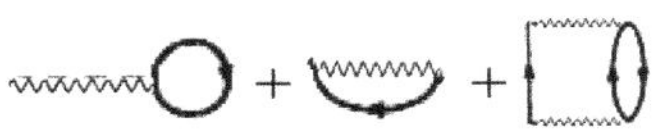 (744)

generates an infinite number of diagrams for the Green's function,

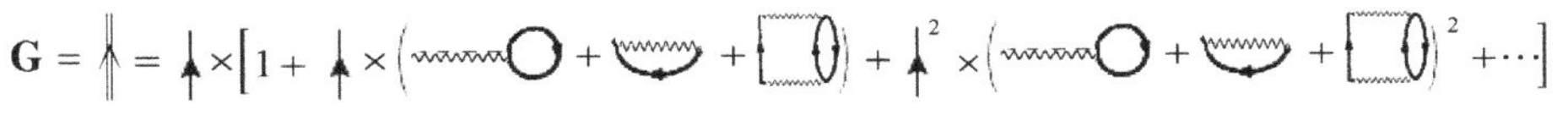 (745)

or

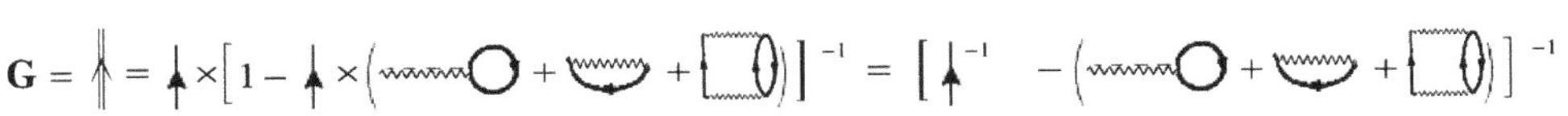 (746)

This constitutes a geometric series. It is instructive to note that the external legs in the self-energy diagrams are not part of the self-energy, although they are often drawn for clarity.

The diagrammatic rules for the self-energy follow directly from those corresponding to the one-particle Green's function. Calculating the nth-order contribution to the self-energy can be summarized by the following rules:

1. Draw all distinct 1PI amputated connected (two-leg) diagrams with n interaction vertices

   $\left(\alpha\beta\middle|\hat{U}\middle|\alpha'\beta'\right) =$ 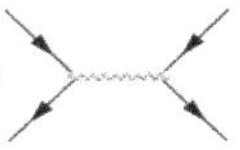 (747)

   The ingoing line is labeled as (α,β) and the outgoing line as (α',β'), and all the inner vertices are connected by propagators

   Two diagrams are equal when they can be transformed into each other by fixing the external legs (α,β) and (α',β') with the direction of the propagators.
2. Associate $\mathbf{G}_\alpha(\tau-\tau')$ to each directed line, where τ and τ' denote either internal or external times.

3. To each vertex, associate the matrix

$$(\alpha\alpha'|\hat{U}|\alpha_1\alpha_1') = $$ 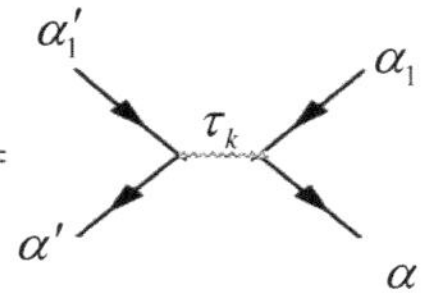 (748)

4. Sum over all internal indices α and integrate over all internal times τ_k over the interval $[0,\beta]$.
5. Multiply the result by the factor $(-1)^{n-1}\chi^{n_c}$, where n_c is the number of closed propagator loops.

These rules can be illustrated by considering the first- and second-order diagrams contributing to the self-energy as follows:

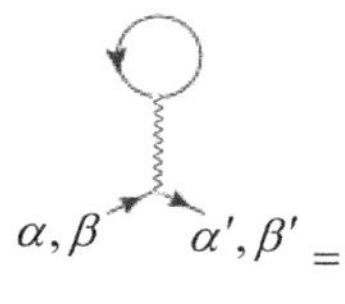

$$= -\chi\delta(\beta-\beta')\sum_{\alpha}(\alpha\gamma|\hat{U}|\alpha'\gamma)\mathbf{G}_{\gamma}(0) \tag{749}$$

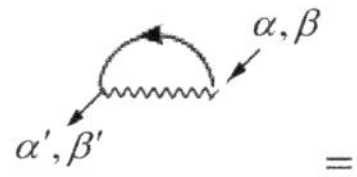

$$= \chi\delta(\beta-\beta')\sum_{\alpha}(\alpha\gamma|\hat{U}|\gamma\alpha')\mathbf{G}_{\gamma}(0) \tag{750}$$

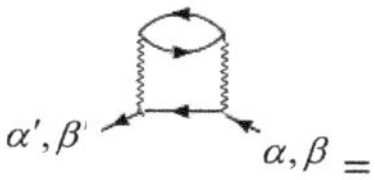

$$= -\chi\sum_{\gamma_1\gamma_2\gamma_3}(\alpha_1\gamma_3|\hat{U}|\gamma_1\gamma_2)(\gamma_1\gamma_2|\hat{U}|\alpha'\gamma_3)\mathbf{G}_{\gamma_1}(\beta-\beta')\mathbf{G}_{\gamma_2}(\beta-\beta')\mathbf{G}_{\gamma_3}(\beta'-\beta) \tag{751}$$

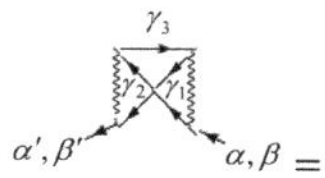

$$= -\chi\sum_{\gamma_1\gamma_2\gamma_3}(\alpha_1\gamma_3|\hat{U}|\gamma_2\gamma_1)(\gamma_1\gamma_2|\hat{U}|\alpha'\gamma_3)\mathbf{G}_{\gamma_1}(\beta-\beta')\mathbf{G}_{\gamma_2}(\beta-\beta')\mathbf{G}_{\gamma_3}(\beta'-\beta) \tag{752}$$

Therefore, the sum over all possible repeated irreducible self-energy parts is

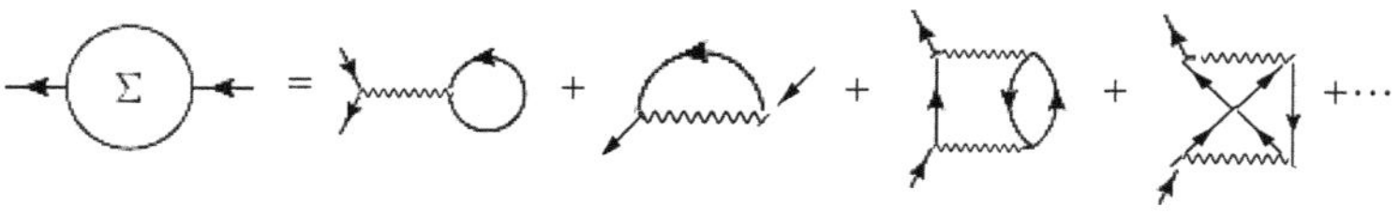 (753)

or

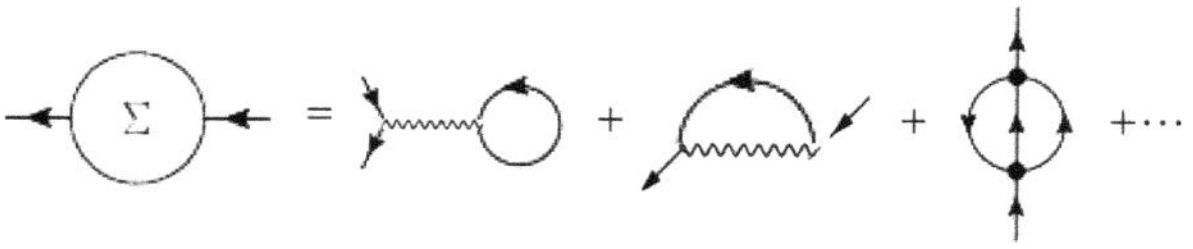 (754)

where

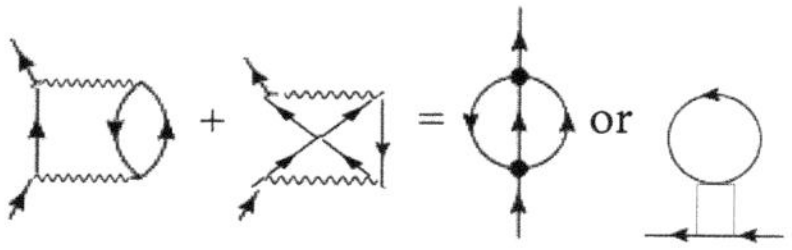 (755)

and the Dyson equation, which is the basis for the Green function

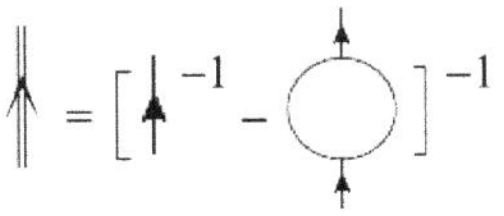

(756)

or

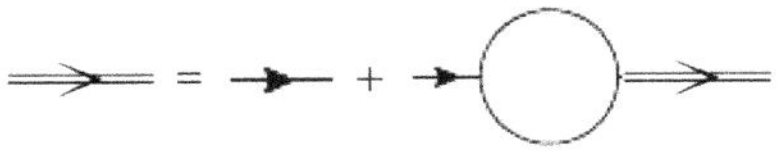

(757)

6.4.3 Higher-Order Vertices

We note that the essential features that make Σ differ from the one-particle vertex function $\Gamma_{\phi^*\phi}$ only by the trivial term $\Gamma^{(0)}_{\phi^*\phi}$ are

- **It is one-particle irreducible.**
- **The full one-particle Green's function, if obtained from equation (731), involves no loop integrals.**

We show that these two features generalize to n-particle vertex functions. If we take the functional derivatives of (719) with respect to η^* and η, we can relate the higher-order connected Green's functions $\mathbf{G}_c^{(2n)}$ with $n \geq 2$ to the 1PI vertices. Considering equation (664), we then have the following diagram

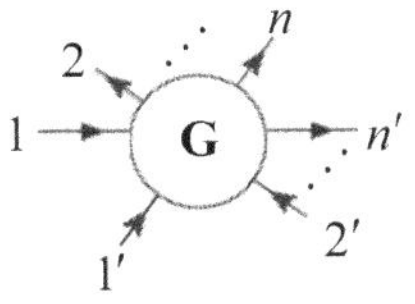

$$\equiv \mathbf{G}^{(2n)}\left(\alpha_1\tau_1,\cdots,\alpha_n\tau_n,\alpha'_n\tau'_n,\cdots,\alpha'_1\tau'_1\right) \tag{758}$$

that is equivalently represented by the following formula

$$\frac{\chi^n \partial^{(2n)} \mathbf{W}\left[\phi,\phi^*\right]}{\partial\eta^*(\tau_1)\cdots\partial\eta^*(\tau_n)\partial\eta(\tau'_{n'})\cdots\partial\eta(\tau'_{1'})} \tag{759}$$

and

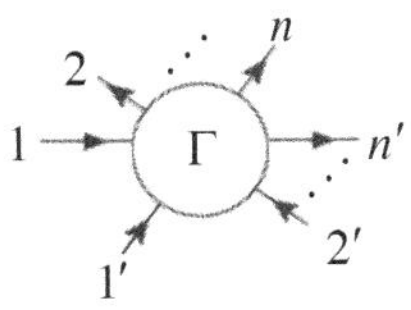

$$\equiv \Gamma^{(2n)}\left(\alpha_1\tau_1,\cdots,\alpha_n\tau_n,\alpha'_n\tau'_n,\cdots,\alpha'_1\tau'_1\right) \tag{760}$$

is equivalently represented by the formula

$$\Gamma^{(2n)}\left(\alpha_1\tau_1,\cdots,\alpha_n\tau_n,\alpha'_n\tau'_n,\cdots,\alpha'_1\tau'_1\right) = \frac{\partial^{(2n)}\Gamma\left[\phi,\phi^*\right]}{\partial\phi^*_{\alpha_1}(\tau_1)\cdots\partial\phi^*_{\alpha_n}(\tau_n)\partial\phi_{\alpha'_n}(\tau'_n)\cdots\partial\phi_{\alpha'_1}(\tau'_1)} \tag{761}$$

This permits us to write equation (719) in the following form:

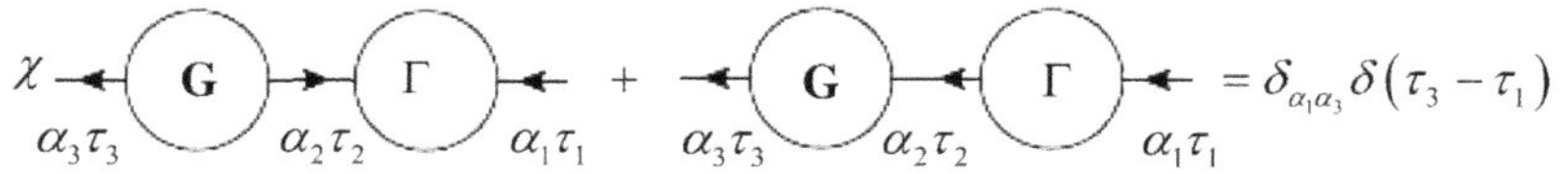

(762)

We further condense the notation by omitting signs and disregarding the directed lines. We do this by letting $\frac{\partial}{\partial\phi}$ represent either $\frac{\partial}{\partial\phi(\tau_i)}$ or $\frac{\partial}{\partial\phi^*(\tau_i)}$ and by letting $\frac{\partial}{\partial\eta}$ represent either $\frac{\partial}{\partial\eta(\tau_i)}$ or $\frac{\partial}{\partial\eta^*(\tau_i)}$. So, the functional derivative $\frac{\partial}{\partial\phi}$ applied to $\frac{\partial^{(n)}\Gamma}{\partial\phi^n}$ increases the number of legs by one:

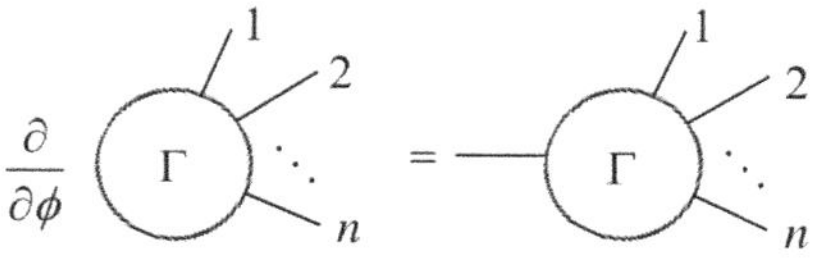

(763)

With $\frac{\partial}{\partial\phi} = \frac{\partial\eta}{\partial\phi}\frac{\partial}{\partial\eta}$, the functional derivative $\frac{\partial}{\partial\phi}$ applied to $\frac{\partial^{(n)}\mathbf{W}}{\partial\phi^n}$ then adds a leg containing $\frac{\partial\eta}{\partial\phi} = \frac{\partial^2\Gamma}{\partial\phi^2}$:

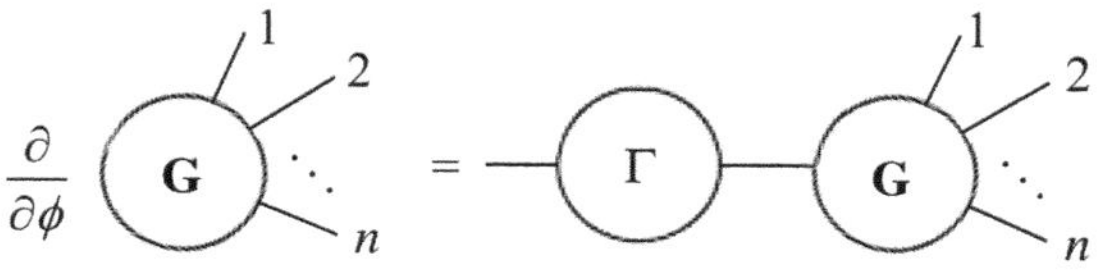

(764)

Considering this compact notation, the evaluation of

$$
\frac{\partial^{(n)}}{\partial\phi^n}[\phi] = \frac{\partial^{(n)}}{\partial\phi^n}\left[\frac{\partial\mathbf{W}}{\partial\eta}\right] \tag{765}
$$

for the successive values of n produces the desired hierarchy of equations. For $n=1$, we recover the abbreviated form of equation (762):

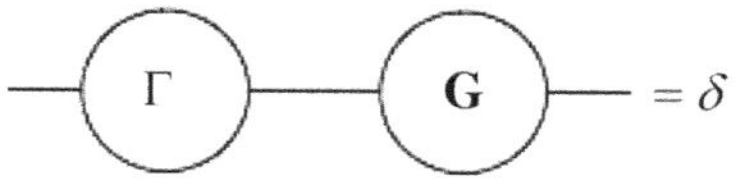

(766)

With successive derivatives, by letting $\frac{\partial}{\partial\phi}$ act on Γ s via (763) or on **G** s via (764) and considering no symmetry and by taking into account directed lines, we then arrive at

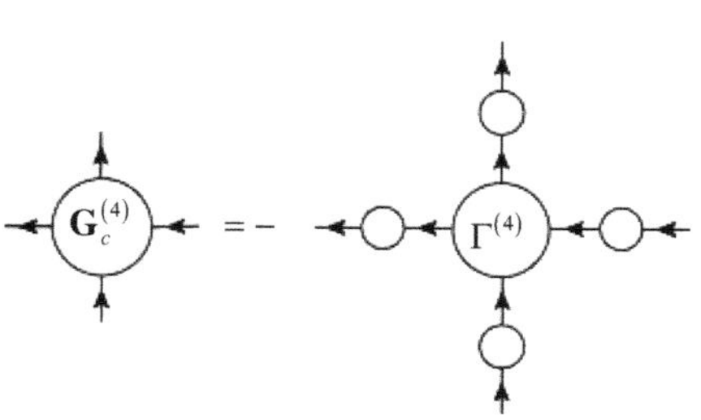

(767)

This can be represented explicitly by

$$\mathbf{G}_c^{(4)}(\alpha_1\tau_1,\alpha_2\tau_2,\alpha_1'\tau_1',\alpha_2'\tau_2') = -\int_0^\beta d\tilde{\tau}_1 d\tilde{\tau}_2 d\tilde{\tau}_2' d\tilde{\tau}_1' \sum_{\gamma_1\gamma_2\gamma_2'\gamma_1'} \mathbf{G}(\alpha_1\tau_1,\gamma_1\tilde{\tau}_1')\mathbf{G}(\alpha_2\tau_2,\gamma_2\tilde{\tau}_2') \times$$
$$\times \Gamma^{(4)}(\gamma_1\tilde{\tau}_1,\gamma_2\tilde{\tau}_2,\gamma_2'\tilde{\tau}_2',\gamma_1'\tilde{\tau}_1')\mathbf{G}(\gamma_2'\tilde{\tau}_2',\alpha_2'\tau_2')\mathbf{G}(\gamma_1'\tilde{\tau}_1',\alpha_1'\tau_1') \quad (768)$$

The sign of the diagrams is written explicitly with the one-particle Green's function $\mathbf{G}$ represented by an empty circle and $\Gamma^{(4)}(\gamma_1\tilde{\tau}_1,\gamma_2\tilde{\tau}_2,\gamma_2'\tilde{\tau}_2',\gamma_1'\tilde{\tau}_1')$ and the four-leg (or two-particle) 1PI vertex seen as the effective interaction vertex between two particles propagating in a many-particle medium. The two-particle Green's function is obtained from a tree diagram composed of Green's function and vertex functions of the same and lower order. The lower-order $\Gamma^{(4)}$ can be given by the bare vertex $\{\gamma_1\gamma_2|\hat{U}|\gamma_1'\gamma_2'\}$, and its perturbation expansion can be diagrammatically represented by the following:

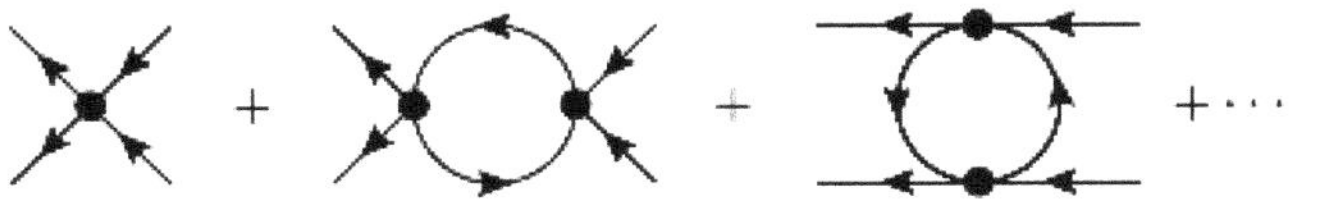

(769)

Because $\Gamma^{(4)}$ has the same symmetry properties as $\mathbf{G}_c^{(4)}$, it is (anti)symmetric under the exchange of the two incoming or outgoing particles,

$$\Gamma^{(4)}(\alpha_1\tau_1,\alpha_2\tau_2,\alpha_2'\tau_2',\alpha_1'\tau_1') = \chi\Gamma^{(4)}(\alpha_2\tau_2,\alpha_1\tau_1,\alpha_2'\tau_2',\alpha_1'\tau_1') = \chi\Gamma^{(4)}(\alpha_1\tau_1,\alpha_2\tau_2,\alpha_1'\tau_1',\alpha_2'\tau_2') \quad (770)$$

6.4.4 General Case

We generalize the grand canonical partition function by introducing an external source η that couples to two fields. This is different from coupling external sources to just one field, which results in 1PI formalism and brings about the 2PI nature of the functionals. We derive the fully symmetric 1PI vertices via sources coupled to field operators. The partition function may be written as follows:

$$\mathbf{Z}[\eta] = \int d[\psi]\exp\left\{-S[\psi] + \sum_{\tilde{\alpha}}\int_0^\beta d\tau\eta_{\tilde{\alpha}}(\tau)\psi_{\tilde{\alpha}}(\tau)\right\} = \exp\{\mathbf{W}[\eta]\} \quad (771)$$

Note that we coupled a term that was bilinear in the fermion fields to the external sources $\eta_{\tilde{\alpha}}$. From equation (771), we obtain the generating functional of the connected density correlation functions

$$\mathbf{W}[\eta] = \ln \mathbf{Z}[\eta] \quad (772)$$

We introduce fields $\phi_{\tilde{\alpha}}$ defined as the derivative of the functional $\mathbf{W}[\eta]$ with respect to the sources $\eta_{\tilde{\alpha}}$:

$$\phi_{\tilde{\alpha}}(\tau) = \langle\psi_{\tilde{\alpha}}(\tau)\rangle_\eta = \frac{\partial \mathbf{W}[\eta]}{\partial \eta_{\tilde{\alpha}}(\tau)} \quad (773)$$

Here, the fields $\phi_{\tilde{\alpha}}$ are not only functions of τ but also functionals of the sources $\eta_{\tilde{\alpha}}$.

Performing a Legendre transformation of $\mathbf{W}[\eta]$ with respect to the sources $\eta_{\tilde{\alpha}}$, we obtain the effective action $\Gamma[\phi]$:

$$\Gamma[\phi] = -\mathbf{W}[\eta] + \sum_{\tilde{\alpha}}\int_0^\beta d\tau\eta_{\tilde{\alpha}}(\tau)\phi_{\tilde{\alpha}}(\tau) \quad (774)$$

This effective action $\Gamma[\phi]$ contains the complete dynamics of the many-body system. Phenomenologically, this transformation to the 2PI effective action is viewed as a bosonization of the theory. This is because we have traded (in the fermion fields) the classical action S for the composite bosonic fields $\phi_{\tilde{\alpha}}$ in the effective action $\Gamma[\phi]$. The effective action $\Gamma[\phi]$ satisfies the following reciprocity relation and plays the role of the generating functional of the 1PI vertices:

$$\frac{\partial\Gamma[\phi]}{\partial\phi_{\tilde{\alpha}}(\tau)} = \chi\eta_{\tilde{\alpha}}(\tau) \tag{775}$$

If we consider the absence of external sources ($\eta = 0$), then the effective action is stationary. If the stationary value $\bar{\phi}$ is nonzero, then the gauge symmetry is broken.

We define the 1PI vertices in the state $\bar{\phi}$ by

$$\Gamma^{(n)}(\tilde{\alpha}_1\tau_1,\cdots,\tilde{\alpha}_n\tau_n) = \left.\frac{\partial^{(n)}\Gamma[\phi]}{\partial\phi_{\tilde{\alpha}_n}(\tau_n)\cdots\partial\phi_{\tilde{\alpha}_1}(\tau_1)}\right|_{\phi=\bar{\phi}} \tag{776}$$

And, by inverting (776), we have the following vertex function expended around $\bar{\phi}$:

$$\Gamma[\phi] = \Gamma[\bar{\phi}] + \sum_{n=1}^{\infty}\frac{1}{n!}\sum_{\tilde{\alpha}_1\cdots\tilde{\alpha}_n}\int_0^\beta d\tau_1\cdots d\tau_n\Gamma^{(n)}(\tilde{\alpha}_1\tau_1,\cdots,\tilde{\alpha}_n\tau_n)\left[\phi_{\tilde{\alpha}_1}(\tau_1)-\bar{\phi}_{\tilde{\alpha}_1}(\tau_1)\right]\cdots\left[\phi_{\tilde{\alpha}_n}(\tau_n)-\bar{\phi}_{\tilde{\alpha}_n}(\tau_n)\right] \tag{777}$$

These are (anti)symmetric under the exchange of two particles,

$$\Gamma^{(n)}(\tilde{\alpha}_1\tau_1\cdots\tilde{\alpha}_k\tau_k,\cdots,\tilde{\alpha}_l\tau_l,\cdots,\tilde{\alpha}_n\tau_n) = \chi\Gamma^{(n)}(\tilde{\alpha}_1\tau_1\cdots\tilde{\alpha}_l\tau_l,\cdots,\tilde{\alpha}_k\tau_k,\cdots,\tilde{\alpha}_n\tau_n) \tag{778}$$

From (773), we arrive at

$$\frac{\partial^{(2)}\Gamma[\phi]}{\partial\phi\partial\phi} = \chi\left(\frac{\partial^{(2)}\mathbf{W}[\eta]}{\partial\eta\partial\eta}\right)^{-1} \tag{779}$$

Considering $\eta = 0$ and $\phi = \bar{\phi}$, we then have

$$\Gamma^{(2)} = \left(\mathbf{G}_c^{(2)}\right)^{-1} \tag{780}$$

and

$$\Gamma^{(2)} = \mathbf{G}_0^{-1} + \Sigma \tag{781}$$

or

$$\mathbf{G}_c^{-1} = \mathbf{G}_0^{-1} + \Sigma \tag{782}$$

This is a generalization of the Dyson equation of systems with broken gauge symmetry and is also known as the **Dyson-Beliaev equation.** The self-energy has a 2×2 matrix structure with respect to the charge index c:

$$\Sigma(\alpha\tau,\alpha'\tau') = \begin{bmatrix} \Sigma(\alpha,+,\tau;\alpha',+,\tau') & \Sigma(\alpha,+,\tau;\alpha',-,\tau') \\ \Sigma(\alpha,-,\tau;\alpha',+,\tau') & \Sigma(\alpha,-,\tau;\alpha',-,\tau') \end{bmatrix} \tag{783}$$

or

$$\Sigma(\alpha\tau,\alpha'\tau') = \qquad (784)$$

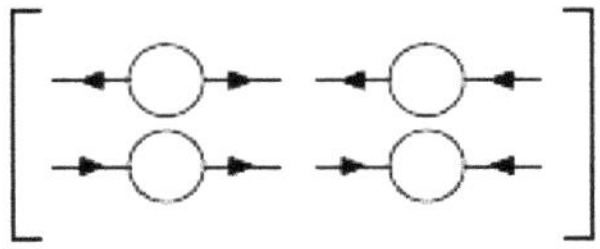

Considering (779) and relations

$$\frac{\partial}{\partial\eta_{\tilde{\alpha}_{n+1}}(\tau_{n+1})}\mathbf{G}_c^{(n)}(\tilde{\alpha}_1\tau_1,\cdots,\tilde{\alpha}_n\tau_n) = \chi^n\mathbf{G}_c^{(n+1)}(\tilde{\alpha}_1\tau_1,\cdots,\tilde{\alpha}_{n+1}\tau_{n+1}) \qquad (785)$$

$$\frac{\partial}{\partial\eta_{\tilde{\alpha}_{n+1}}(\tau_{n+1})}\Gamma^{(n)}(\tilde{\alpha}_1\tau_1,\cdots,\tilde{\alpha}_n\tau_n) = \int_0^\beta d\tau\sum_{\tilde{\alpha}}\mathbf{G}_c^{(2)}(\tilde{\alpha}_{n+1}\tau_{n+1},\tilde{\alpha}\tau)\,\Gamma^{(n+1)}(\tilde{\alpha}_1\tau_1,\cdots,\tilde{\alpha}_n\tau_n,\tilde{\alpha}\tau) \qquad (786)$$

This then permits us to express higher-order connected Green's functions $\mathbf{G}_c^{(n)}$ in terms of $\Gamma^{(m\leq n)}$ and $\mathbf{G}_c^{(m<n)}$. For example, let us say,

$$\mathbf{G}_c^{(3)}(\tilde{\alpha}_1\tau_1,\tilde{\alpha}_2\tau_2,\tilde{\alpha}_3\tau_3) = -\chi\int_0^\beta d\tilde{\tau}_1\,d\tilde{\tau}_2\,d\tilde{\tau}_3\sum_{\gamma_1\gamma_2\gamma_3}\mathbf{G}_c(\tilde{\alpha}_1\tau_1,\gamma_1\tilde{\tau}_1)\mathbf{G}_c(\tilde{\alpha}_2\tau_2,\gamma_2\tilde{\tau}_2)\mathbf{G}_c(\tilde{\alpha}_3\tau_3,\gamma_3\tilde{\tau}_3)\Gamma^{(3)}(\gamma_1\tilde{\tau}_1,\gamma_2\tilde{\tau}_2,\gamma_3\tilde{\tau}_3) \qquad (787)$$

Or, graphically, we can say

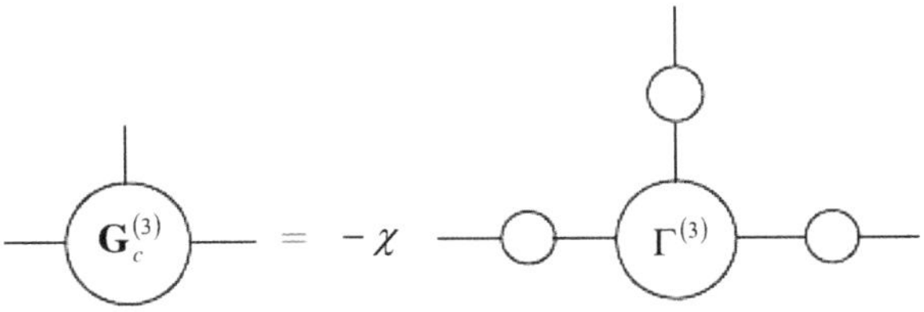

(788)

The empty circles represent the connected one-particle Green's function and identify $\Gamma^{(3)}$ as the three-leg 1PI vertex.

6.4.5 Luttinger-Ward Functional and 2PI Vertices

In this section, we introduce the Luttinger-Ward functional and the 2PI(**Two-Particle Irreducible)** vertices. These are used to set up approximations that satisfy conservation laws as well as nonperturbative approaches. The Luttinger-Ward functional (LWF)formalism has a strong resemblance to the least action principle in Lagrangian classical mechanics [13]. In the least action principle, the dynamical equations are obtained by requiring the stationarity of the classical action. For the Luttinger-Ward functional, we find a functional $\Gamma[\mathbf{G}]$ such that it becomes stationary at the exact $\mathbf{G} = \bar{\mathbf{G}}$.

6.4.5.1 Normal Systems

A cornerstone of the functional techniques is the concept of generating functionals. We generalize the grand canonical partition function by introducing an external source η that couples to two fields. This is different from coupling external sources to just one field, which yields the 1PI formalism and results in the 2PI nature of the functionals. So,

$$Z[\eta]=\exp\{W[\eta]\}=\int d[\psi]d[\psi^*]\exp\left\{-S[\psi,\psi^*]+\frac{1}{2}\sum_{\alpha,\alpha'}\int_0^\beta d\tau d\tau'\psi^*_{\tilde{\alpha}}(\tau)\eta_{\alpha\alpha'}(\tau,\tau')\psi_{\alpha'}(\tau')\right\} \tag{789}$$

The external source η coupled to two fields is decisive for obtaining 2PI quantities [3] and is defined as having the same (anti-)symmetry as $\mathbf{G}_0^{-1}$, that is,

$$\eta_{\alpha'\alpha}=\chi\eta_{\alpha\alpha'} \tag{790}$$

We take the generating functional derivative with respect to the source field η. The variations of $\mathbf{W}[\eta]$ with respect to η yield $\mathbf{G}$ via the relation:

$$\mathbf{W}^{(1)}\equiv\mathbf{G}_c(\gamma_1)=-\chi\frac{\partial\mathbf{W}[\eta]}{\partial\eta_{\gamma_1}}=\mathbf{G}(\gamma_1) \tag{791}$$

also

$$\mathbf{W}^{(2)}\equiv\mathbf{G}_c^{(2)}(\gamma_1,\gamma_2)=\frac{\partial^2\mathbf{W}[\eta]}{\partial\eta_{\gamma_1}\partial\eta_{\gamma_2}}=\mathbf{G}^{(2)}(\gamma_1,\gamma_2)-\mathbf{G}(\gamma_1)\mathbf{G}(\gamma_2) \tag{792}$$

and

$$\mathbf{W}^{(3)}\equiv\mathbf{G}_c^{(3)}(\gamma_1,\gamma_2,\gamma_3)=\frac{\partial^3\mathbf{W}[\eta]}{\partial\eta_{\gamma_1}\partial\eta_{\gamma_2}\partial\eta_{\gamma_3}} \tag{793}$$

or

$$\mathbf{G}_c^{(3)}(\gamma_1,\gamma_2,\gamma_3)=\mathbf{G}^{(3)}(\gamma_1,\gamma_2,\gamma_3)-\mathbf{G}(\gamma_1)\mathbf{G}_c^{(2)}(\gamma_2,\gamma_3)-\mathbf{G}(\gamma_2)\mathbf{G}_c^{(2)}(\gamma_1,\gamma_3)-\mathbf{G}(\gamma_3)\mathbf{G}_c^{(2)}(\gamma_1,\gamma_2)-\mathbf{G}(\gamma_1)\mathbf{G}(\gamma_2)\mathbf{G}(\gamma_3) \tag{794}$$

Then we can trade η with $\mathbf{G}$ via a Legendre transform of $-\ln\mathbf{Z}[\eta]$, which is a functional of the Green's function $\mathbf{G}$.

The higher derivatives

$$\mathbf{W}^{(n)}_{\gamma_1\cdots\gamma_n}=\frac{\partial^{(n)}\mathbf{W}[\eta]}{\partial\eta_{\gamma_1}\cdots\partial\eta_{\gamma_n}} \tag{795}$$

are the $2n$-point Green's functions, which are connected if each pair $(\tilde{\alpha}\tau,\tilde{\alpha}'\tau')$ is considered intrinsically connected. Considering the fact that $\mathbf{G}_\gamma$ is (anti)symmetric for the exchange $(\tilde{\alpha}\tau)\leftrightarrow(\tilde{\alpha}'\tau')$, then

$$\mathbf{W}^{(n)}(\gamma_1\cdots\gamma_i\cdots\gamma_j\cdots\gamma_n)=\mathbf{W}^{(n)}(\gamma_1\cdots\gamma_j\cdots\gamma_i\cdots\gamma_n)\quad,\quad\mathbf{W}^{(n)}(\gamma_1\cdots\gamma_i\cdots\gamma_n)=\chi\mathbf{W}^{(n)}(\gamma_1\cdots\bar{\gamma}_i\cdots\gamma_n) \tag{796}$$

The Legendre transformation of $\mathbf{W}[\eta]$ yields a functional of the free variable $\mathbf{G}$, that is, the **so-called 2PI effective action**

$$\Gamma(\mathbf{G})=\{-\mathbf{W}[\eta]-\eta\cdot\mathbf{G}\}\Big|_{\eta[\mathbf{G}]} \tag{797}$$

where the dot-product notation:

$$\eta\cdot\mathbf{G}=\frac{1}{2}\sum_{\gamma}\eta_\gamma\mathbf{G}_\gamma=\frac{1}{2}\sum_{\alpha\alpha'}\eta_{\alpha\alpha'}\mathbf{G}_{\alpha\alpha'}=\frac{\chi}{2}\mathrm{Tr}\,\eta\mathbf{G}=\mathbf{G}\cdot\eta \tag{798}$$

Here the factor $\frac{1}{2}$ is related to the (anti-)symmetry of the factors under exchange of α and α'. Because the quadratic part of the action functional is given by $\left[\mathbf{G}_0^{-1}+\eta\right]\cdot(\psi\psi)$, it then follows that

$$\Gamma[\mathbf{G}]=\Gamma[\mathbf{G},\mathbf{G}_0] \tag{799}$$

is explicitly dependent on $\mathbf{G}_0$ due to the term

$$\mathbf{G}_0^{-1}\cdot\mathbf{G}=\frac{\chi}{2}\mathrm{Tr}\mathbf{G}_0^{-1}\mathbf{G} \tag{800}$$

For the noninteracting case, we have

$$\mathbf{W}_0[\eta]=\ln\det\left[-\mathbf{G}_0^{-1}-\eta\right]^{-\frac{\chi}{2}}=-\frac{\chi}{2}\mathrm{Tr}\ln\left[-\mathbf{G}_0^{-1}-\eta\right] \tag{801}$$

$$\Gamma_0[\mathbf{G}]=-\frac{\chi}{2}\left(\mathrm{Tr}\ln[-\mathbf{G}]-\mathrm{Tr}\left[\mathbf{G}_0^{-1}\mathbf{G}-1\right]\right) \tag{802}$$

In the previous relations, the trace and logarithm functions must be interpreted in a functional sense.

We define the **Luttinger-Ward functional [26]** that is independent of $\mathbf{G}_0$ as:

$$\Phi[\mathbf{G}]=\Gamma[\mathbf{G}]-\Gamma_0[\mathbf{G}] \tag{803}$$

and we show that it is a **generating functional of the self-energy Σ's being the sum of all amputated one-particle 2PI diagrams with full Green's functions in place of bare Green's functions.** This will be the central quantity for all our 2PI considerations. We can find that the 2PI effective action satisfies the following relation:

$$\frac{\partial\Gamma(\mathbf{G})}{\partial\mathbf{G}(\gamma)}=-\chi\eta_{\alpha\alpha'}(\gamma) \tag{804}$$

This is typical for Legendre transforms, and we note that the actual Green's function $\mathbf{G}$ is obtained for vanishing external sources corresponding to the stationary value of the functional $\Gamma[\mathbf{G}]$. If we consider a noninteracting system, then $\Gamma[\mathbf{G}]$ is Gaussian and can be calculated exactly.

6.4.5.2 The Self-Consistent Dyson Equation

Considering equation (804), we determine the following **Dyson equation**:

$$\frac{\partial\Gamma(\mathbf{G})}{\partial\mathbf{G}(\gamma)}=-\chi\mathbf{G}^{-1}(\gamma)+\chi\mathbf{G}_0^{-1}(\gamma)+\frac{\partial\Phi[\mathbf{G}]}{\partial\mathbf{G}(\gamma)}=-\chi\eta_{\alpha\alpha'}(\gamma) \tag{805}$$

Physical quantities are obtained at vanishing external source $\eta(\bar{\mathbf{G}})=0$ where the quantities are marked by bars over them. So,

$$\bar{\eta}=0\quad,\quad\bar{\mathbf{Z}}=\mathbf{Z}[\bar{\eta}]\quad,\quad\bar{\mathbf{G}}=\mathbf{G}[\bar{\eta}]\quad,\quad\cdots \tag{806}$$

For the Legendre transformed functions of the free variable **G**, the physical state is equivalently defined by [27]:

$$\left.\frac{\partial \Gamma[\mathbf{G}]}{\partial \mathbf{G}}\right|_{\mathbf{G}=\bar{\mathbf{G}}} \equiv \Gamma^{(1)}\left[\bar{\mathbf{G}}\right]=0 \tag{807}$$

and we write

$$\Gamma[\mathbf{G}]\big|_{\mathbf{G}=\bar{\mathbf{G}}} \equiv \bar{\Gamma}=\Gamma\left[\bar{\mathbf{G}}\right] \quad , \quad \left.\frac{\partial \Phi[\mathbf{G}]}{\partial \mathbf{G}}\right|_{\mathbf{G}=\bar{\mathbf{G}}} \equiv \bar{\Phi}^{(1)}=\Phi^{(1)}\left[\bar{\mathbf{G}}\right] \quad , \quad \cdots \tag{808}$$

Therefore, using the source field to vanish $\eta(\bar{\mathbf{G}})=0$, then the **self-energy (identified as the 1PI self-energy) $\Sigma(\gamma)$ is the first-order functional derivative of the Luttinger-Ward functional (LWF)with respect to a Green's function (i.e., by breaking a line in the vacuum diagrams):**

$$\Sigma(\gamma)=-\chi\left.\frac{\partial \Phi[\mathbf{G}]}{\partial \mathbf{G}(\gamma)}\right|_{\mathbf{G}=\bar{\mathbf{G}}} \tag{809}$$

Here, $\bar{\mathbf{G}}$ is the solution of the **stationarity condition [27]:**

$$\frac{\partial \Gamma[\mathbf{G}]}{\partial \mathbf{G}}=0$$

and, consequently, follows the equation:

$$\mathbf{G}^{-1}(\gamma)=\mathbf{G}_0^{-1}(\gamma)-\Sigma(\gamma) \tag{810}$$

Equation (810) is the Dyson equation for the one-particle Green's function in the differential form, with $\Sigma(\gamma)$ identified as the 1PI self-energy. Because $\Sigma(\gamma)$ is the sum of all amputated one-particle 2PI diagrams with full propagator lines, then $\Phi[\mathbf{G}]$ is the sum of all connected 2PI vacuum Feynman diagrams (which are all 3PI) with full Green's functions G in place of bare Green's functions $\mathbf{G}_0$.

The Green's function **G** is obtained by inverting equation (810). We find this by multiplying equation (810) from the left and right by $\mathbf{G}(\gamma)$ and $\mathbf{G}_0(\gamma)$, respectively:

$$\mathbf{G}_0(\gamma)=\mathbf{G}(\gamma)-\mathbf{G}_0(\gamma)\Sigma(\gamma)\mathbf{G}(\gamma)$$

We represent the Green's functions and self-energy considering (809) in the 2×2 operator charge space:

$$\mathbf{G}(\tilde{\alpha}\tau,\tilde{\alpha}'\tau')=\begin{bmatrix} \mathscr{F}(\alpha,+,\tau,\alpha',+,\tau') & \mathscr{G}(\alpha,+,\tau,\alpha',-,\tau') \\ \mathscr{G}(\alpha,-,\tau,\alpha',+,\tau') & \mathscr{F}^{\dagger}(\alpha,-,\tau,\alpha',-,\tau') \end{bmatrix} \tag{811}$$

or

$\mathbf{G}(\tilde{\alpha}\tau,\tilde{\alpha}'\tau')=$ 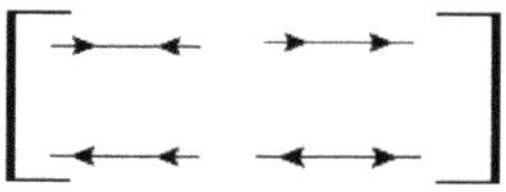 (812)

and

$$\Sigma(\alpha\tau,\alpha'\tau')=\begin{bmatrix}\Sigma(\alpha,+,\tau;\alpha',+,\tau') & \Sigma(\alpha,+,\tau;\alpha',-,\tau')\\ \Sigma(\alpha,-,\tau;\alpha',+,\tau') & \Sigma(\alpha,-,\tau;\alpha',-,\tau')\end{bmatrix} \tag{813}$$

or

$\Sigma(\alpha\tau,\alpha'\tau')=$ 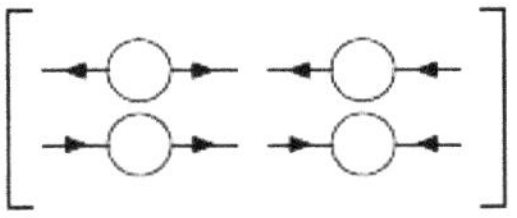 (814)

It is instructive to note that due to the antisymmetry of Σ in the superfluid, we have:

$$\Sigma(\alpha,+,\tau,\alpha',-,\tau')=\Sigma(\alpha,\tau,\alpha',\tau')=-\Sigma(\alpha',\tau',\alpha,\tau) \tag{815}$$

Also,

$$\mathscr{F}(\alpha,+,\tau,\alpha',+,\tau')=-\mathscr{F}(\alpha',+,\tau',\alpha,+,\tau)\quad,\quad \mathscr{F}^{\dagger}(\alpha,-,\tau,\alpha',-,\tau')=-\mathscr{F}^{\dagger}(\alpha',-,\tau',\alpha,-,\tau) \tag{816}$$

$$\mathscr{G}(\alpha,+,\tau,\alpha',-,\tau')=-\mathscr{G}(\alpha',-,\tau',\alpha,+,\tau) \tag{817}$$

The most general bilinear coupling is considered so that $\mathbf{G}[\eta]$ can also be used to calculate the anomalous Green's functions in the superconducting states where the gauge symmetry is spontaneously broken.

6.4.5.3 Diagrammatic Interpretation of LWF

We use equation **(809)** to find a diagrammatic interpretation of $\Phi[\mathbf{G}]$. **From the self-energy Σ, defined in (809) as the sum of all amputated one-particle 2PI diagrams with full propagator lines, we can deduce the Luttinger-Ward functional $\Phi[\mathbf{G}]$ to have a simple diagrammatic interpretation as the sum of all connected 2PI vacuum Feynman diagrams (these are called skeletons diagrams because they do not contain self-energy insertions) with full Green's functions G in place of bare Green's functions $\mathbf{G}_0$, as seen earlier and as shown in equation (818). Incidentally, all diagrams that are 2PI are also 3PI as seen earlier.**

$\Phi[\mathbf{G}]=$  (818)

Luttinger-Ward functional (LWF) $\Phi[\mathbf{G}]$ taken as the sum of all connected 2PI vacuum Feynman diagrams with full Green's functions G in place of bare Green's functions $\mathbf{G}_0$.

This result is represented formally as follows:

$$\Phi[\mathbf{G}]=\left[\ln\int d[\psi]d[\psi^*]\exp\left\{-S_{\text{int}}[\psi,\psi^*]\right\}\right]\Bigg|_{2\text{PI},\mathbf{G}_0\to\mathbf{G},} \tag{819}$$

Here $S_{\text{int}}[\psi,\psi^*]$ is the interaction part of the action functional that is at least cubic in the field operators. When applied to normal systems, all lines are the usual $\mathbf{G}$ functions. The anomalous Green's functions must be included when applying to superconducting states.

The diagrammatic expansion rules for $\Phi[\mathbf{G}]$ are similar to those of the thermodynamic potential. The perturbation expansion of the full Green's function $\mathbf{G}$ with respect to the bare Green's function $\mathbf{G}_0$ is depicted in (820), while that of Σ is depicted in (821).

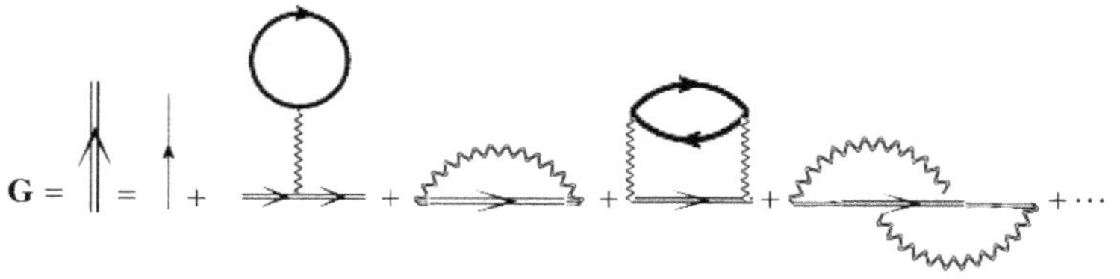

(820)

The perturbation expansion of the full Green's function $\mathbf{G}$ represented by a double directed line with respect to the bare Green's function $\mathbf{G}_0$ represented by a single directed line. The terms after the first single directed line is the perturbation expansion of the exact self-energy Σ with respected to $\mathbf{G}_0$ and represented as an expansion with respect the full Green's function $\mathbf{G}$.

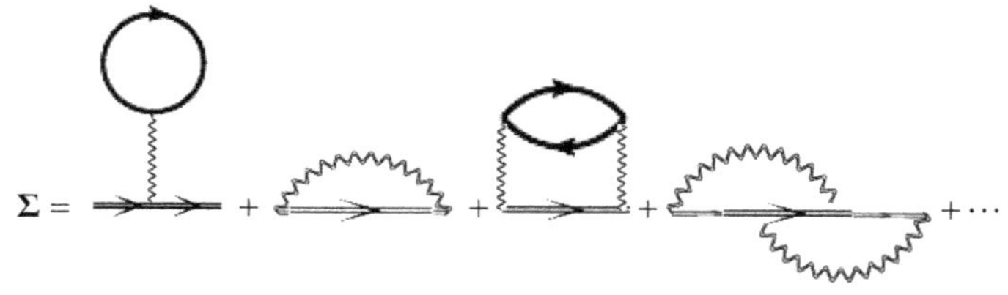

(821)

The perturbation expansion of the exact self-energy Σ with respected to the bare Green's function $\mathbf{G}_0$ represented as an expansion with respect to the full Green's function $\mathbf{G}$.

It is instructive to note that as the functional derivation with respect to $\mathbf{G}$ entails removing a Green's function $\mathbf{G}$ from the diagrams contributing to the Luttinger-Ward functional, then the diagrammatic definition of $\Phi[\mathbf{G}]$ reproduces the expansion of Σ in terms of $\mathbf{G}$. Another useful expression for $\Phi[\mathbf{G}]$ is the definition originally given by Luttinger and Ward [26]:

$$\Phi[\mathbf{G}] = \sum_{n,k} \frac{1}{2n} \int d\gamma \mathbf{G}(\gamma) \Sigma_k^{(n)}(\gamma) \tag{822}$$

Here, $\Sigma_k^{(n)}$ denotes the 1PI self-energy diagrams with n interaction lines and k runs over the topologically distinct diagrams. It is also worth noting that a two-particle reducible diagram for $\Phi[\mathbf{G}]$ would give a contribution to the self-energy that is not 1PI. The diagrammatic rules for $\Phi[\mathbf{G}]$ are the same as those for $-\ln \mathbf{Z}$ and also are consistent with the diagrammatic rules for the self-energy. For example, for the first-order of $\Phi[\mathbf{G}]$, then

$$\begin{aligned} \Phi[\mathbf{G}] = \frac{1}{2} \sum_{\alpha_1 \cdots \alpha_2'} \left(\alpha_1\alpha_2 \middle| \hat{U} \middle| \alpha_1'\alpha_2'\right) \int_0^\beta d\tau \mathbf{G}(\alpha_1'\tau, \alpha_1\tau) \mathbf{G}(\alpha_2'\tau, \alpha_2\tau) + \\ + \frac{\chi}{2} \sum_{\alpha_1 \cdots \alpha_2'} \left(\alpha_1\alpha_2 \middle| \hat{U} \middle| \alpha_1'\alpha_2'\right) \int_0^\beta d\tau \mathbf{G}(\alpha_1'\tau, \alpha_2\tau) \mathbf{G}(\alpha_2'\tau, \alpha_1\tau) \end{aligned} \tag{823}$$

The first-order functional derivative $\Phi^{(1)}$ reproduces the lowest-order contribution to the self-energy, with $\mathbf{G}_0$ replaced by $\mathbf{G}$.

So far, we have treated the Green's functions and the self-energy as matrices in the operator charge space. Though such a matrix notation is convenient for derivations, it is more apparent to make the charge structure explicit in practice. We can make the charge matrix structure explicit by writing out the matrix elements of the Dyson equation. Let us find the integral form of the Dyson equation in (810):

$$\mathbf{G}^{-1}(\gamma)=\mathbf{G}_0^{-1}(\gamma)-\Sigma(\gamma) \tag{824}$$

We multiply this equation by $\mathbf{G}$ and $\mathbf{G}_0^{-1}$ from left and right:

$$\mathbf{G}(\tilde{\alpha}\tau,\tilde{\alpha}'\tau')=\mathbf{G}_0(\tilde{\alpha}\tau,\tilde{\alpha}'\tau')+\sum_{\alpha_1\alpha_2}\int_0^{\beta} d\tau_1\, d\tau_2 \mathbf{G}_0(\tilde{\alpha}\tau,\tilde{\alpha}_1\tau_1)\Sigma(\tilde{\alpha}_1\tau_1,\tilde{\alpha}_2\tau_2)\mathbf{G}(\tilde{\alpha}_2\tau_2,\tilde{\alpha}'\tau') \tag{825}$$

The Green's functions and self-energy are treated as matrices in the 2×2 operator charge space, and a matrix product is implied everywhere. If we write the matrix products explicitly, then in (825), we have (as shown earlier):

$$\mathbf{G}(\tilde{\alpha}\tau,\tilde{\alpha}'\tau')=\begin{bmatrix} \mathscr{F}(\alpha,+,\tau,\alpha',+,\tau') & \mathscr{G}(\alpha,+,\tau,\alpha',-,\tau') \\ \mathscr{G}(\alpha,-,\tau,\alpha',+,\tau') & \mathscr{F}^{\dagger}(\alpha,-,\tau,\alpha',-,\tau') \end{bmatrix} \tag{826}$$

$$\Sigma(\tilde{\alpha}\tau,\tilde{\alpha}'\tau')=\begin{bmatrix} \Sigma(\alpha,+,\tau;\alpha',+,\tau') & \Sigma(\alpha,+,\tau;\alpha',-,\tau') \\ \Sigma(\alpha,-,\tau;\alpha',+,\tau') & \Sigma(\alpha,-,\tau;\alpha',-,\tau') \end{bmatrix} \tag{827}$$

and

$$\mathbf{G}_0(\tilde{\alpha}\tau,\tilde{\alpha}'\tau')=\begin{bmatrix} 0 & \mathscr{G}_0(\alpha,+,\tau,\alpha',-,\tau') \\ \mathscr{G}_0(\alpha,-,\tau,\alpha',+,\tau') & 0 \end{bmatrix} \tag{828}$$

6.4.5.4 2PI Vertices and Bethe-Salpeter Equation

Earlier in this chapter, we inferred the diagrammatic interpretation of $\Phi[\mathbf{G}]$ via the Dyson equation. The 2PI structure of $\Phi[\mathbf{G}]$ can be shown more directly via the generating functional of the two connected vacuum diagrams $\mathbf{W}[\eta]$. The Luttinger-Ward functional $\Phi[\mathbf{G}]$ is given by the sum of the 2PI diagrams with the $2n$-point 2PI vertices $\Phi^{(n)}$:

$$\Phi^{(n)}(\gamma_1,\cdots,\gamma_n)=\left.\frac{\chi^n\partial^{(n)}\Phi[\mathbf{G}]}{\partial\mathbf{G}(\gamma_1)\cdots\partial\mathbf{G}(\gamma_n)}\right|_{\mathbf{G}=\bar{\mathbf{G}}} \tag{829}$$

and n external **bosonic** legs $\gamma_i=(\alpha_i\tau_i,\alpha_i'\tau_i')$. It is instructive to note that it is 2PI in the sense that it cannot be separated into two disconnected pieces by cutting only two internal lines. The external leg $\gamma=(\tilde{\alpha}\tau,\tilde{\alpha}'\tau')$ corresponds to a particle-hole pair when $c=-c'$ and to a particle-particle pair when $c=c'$. Considering the fact that $\mathbf{G}_\gamma$ is (anti)symmetric for the exchange $(\tilde{\alpha}\tau)\leftrightarrow(\tilde{\alpha}'\tau')$, then

$$\Phi^{(n)}(\gamma_1\cdots\gamma_i\cdots\gamma_j\cdots\gamma_n)=\Phi^{(n)}(\gamma_1\cdots\gamma_j\cdots\gamma_i\cdots\gamma_n)\quad,\quad \Phi^{(n)}(\gamma_1\cdots\gamma_i\cdots\gamma_n)=\chi\Phi^{(n)}(\gamma_1\cdots\bar{\gamma}_i\cdots\gamma_n) \tag{830}$$

Normally, all vertices obtained from $\Phi[\mathbf{G}]$ are 2PI and cannot be separated into two disconnected pieces by cutting only two internal lines if each external bosonic leg is taken as a single piece. If we consider the leg γ_i as corresponding to a particle-hole pair, then the vertices $\Phi^{(n)}$ are irreducible in the particle-hole channel. This is a consequence of the choice of $\eta_{\alpha\alpha'}(\gamma)$ as an external source that couples to a particle-hole pair. The Green's functions in the 2PI formalism are defined as follows:

$$\mathbf{W}^{(n)}(\gamma_1,\cdots,\gamma_n)=\left.\frac{\chi^n\partial^{(n)}\ln\mathbf{Z}[\eta]}{\partial\eta_{\alpha_1'\alpha_1}(\gamma_1)\cdots\partial\eta_{\alpha_n'\alpha_n}(\gamma_n)}\right|_{\eta=0} \tag{831}$$

From here, considering $\gamma_i=(\alpha_i\tau_i,\alpha_i'\tau_i')$, we find that

$$\mathbf{W}^{(1)}(\gamma_1)=\mathbf{G}^{(2)}(\gamma_1)\equiv\mathbf{G}(\gamma_1) \tag{832}$$

$$\mathbf{W}^{(2)}(\gamma_1,\gamma_2)=\mathbf{G}^{(4)}(\gamma_1,\gamma_2)-\mathbf{G}(\gamma_1)\mathbf{G}(\gamma_2) \tag{833}$$

Here, $\mathbf{W}^{(2)}$ is a **connected** Green's function with respect to the external bosonic legs (γ_1) and (γ_2). The $\mathbf{W}^{(2)}$ has a role similar to that of the connected Green's function $\mathbf{G}_c^{(4)}$ in the 1PI formalism.

Letting

$$\gamma=(\alpha\tau,\alpha'\tau'),\quad \sum_\gamma=\int_0^\beta d\tau d\tau'\sum_{\alpha,\alpha'} \tag{834}$$

and relating $\mathbf{W}^{(2)}$ to $\Phi^{(2)}$, we then start with relation (804), and we have

$$\frac{\partial^2\Gamma(\mathbf{G})}{\partial\mathbf{G}(\gamma_1)\partial\mathbf{G}(\gamma_2)}=-\chi\frac{\partial\eta_{\alpha\alpha'}(\gamma_1)}{\partial\mathbf{G}(\gamma_2)} \tag{835}$$

Also, the last term in equation (835) can be related to the inverse of the two-connected two-particle propagator by first considering that:

$$-\chi\frac{\partial\mathbf{G}(\gamma_2)}{\partial\eta_{\alpha\alpha'}(\gamma_1)}=-\chi\frac{\partial}{\partial\eta_{\alpha\alpha'}(\gamma_1)}\left[\frac{\partial\ln\mathbf{Z}[\eta]}{\partial\eta_{\alpha\alpha'}(\gamma_2)}\right]\equiv\mathbf{G}_c^{(4)}(\gamma_1,\gamma_2) \tag{836}$$

The multiplication is defined as a **bosonic** matrix product between two antisymmetric four-point functions:

$$(\mathrm{AB})(\gamma_1,\gamma_2)=\sum_{\gamma_3}\mathrm{A}(\gamma_1,\gamma_3)\mathrm{B}(\gamma_3,\gamma_2) \tag{837}$$

and an antisymmetric bosonic four-point identity matrix $\mathrm{I}_{\gamma_1\gamma_2}$:

$$\mathrm{I}_{\gamma_1\gamma_2}=\frac{\partial\eta_{\gamma_2}}{\partial\eta_{\gamma_1}}=\delta_{\bar{\alpha}_1\bar{\alpha}_2}\delta_{\bar{\alpha}_1'\bar{\alpha}_2'}\delta(\tau_1-\tau_2)\delta(\tau_1'-\tau_2')+\chi\delta_{\bar{\alpha}_1\bar{\alpha}_2'}\delta_{\bar{\alpha}_1'\bar{\alpha}_2}\delta(\tau_1-\tau_2')\delta(\tau_1'-\tau_2) \tag{838}$$

If we use equation (803), then equation (835) is written as follows:

$$\Pi^{-1}(\gamma_1,\gamma_2)+\Phi^{(2)}(\gamma_1,\gamma_2)=-\left[\mathbf{G}_c^{(4)}(\gamma_1,\gamma_2)\right]^{-1} \tag{839}$$

where the inverse two-particle propagator is given by:

$$\Pi^{-1}(\tilde{\alpha}_1\tau_1,\tilde{\alpha}_1'\tau_1',\tilde{\alpha}_2\tau_2,\tilde{\alpha}_2'\tau_2')\equiv\mathbf{G}^{-1}(\tilde{\alpha}_1\tau_1,\tilde{\alpha}_2\tau_2)\mathbf{G}^{-1}(\tilde{\alpha}_1'\tau_1',\tilde{\alpha}_2'\tau_2')+\chi\mathbf{G}^{-1}(\tilde{\alpha}_1\tau_1,\tilde{\alpha}_2'\tau_2')\mathbf{G}^{-1}(\tilde{\alpha}_1'\tau_1',\tilde{\alpha}_2\tau_2) \quad (840)$$

In a particle number conserving system, Π denotes either a particle-hole or a particle-particle pair propagator, depending on the charge indices. For example, the charge indices $c_1=-c_2=c_3=-c_4$ renders $\Pi(\gamma_1,\gamma_2)$ a particle-hole propagator, while $c_1=c_2=-c_3=-c_4$ renders it a particle-particle propagator.

Equation (839) implies that $\Phi^{(2)}$ is the sum of all diagrams connected to γ_1 and γ_2 with the two-particle reducible graphs removed; consequently, it justifies the terminology 2PI vertex for $\Phi^{(2)}$. These same methods can be used to show that all $\Phi^{(n)}$ is the sum of all 2PI diagrams pinned to n bosonic external legs.

6.4.5.5 Bethe-Salpeter Equation

Relation (839) permits us to rewrite relation (835) in the form of an integral equation for $\mathbf{W}^{(2)}$:

$$\mathbf{W}^{(2)}=\left[\Pi^{-1}+\Phi^{(2)}\right]^{-1}=\Pi-\Pi\Phi^{(2)}\mathbf{W}^{(2)} \quad (841)$$

We drop the shared (γ_1,γ_2) argument for brevity; Π is the inverse of Π^{-1} and has the interpretation of the bare two-particle Green's function:

$$\Pi(\tilde{\alpha}_1\tau_1,\tilde{\alpha}_1'\tau_1',\tilde{\alpha}_2\tau_2,\tilde{\alpha}_2'\tau_2')\equiv\mathbf{G}(\tilde{\alpha}_1\tau_1,\tilde{\alpha}_2\tau_2)\mathbf{G}(\tilde{\alpha}_1'\tau_1',\tilde{\alpha}_2'\tau_2')+\chi\mathbf{G}(\tilde{\alpha}_1\tau_1,\tilde{\alpha}_2'\tau_2')\mathbf{G}(\tilde{\alpha}_1'\tau_1',\tilde{\alpha}_2\tau_2) \quad (842)$$

Relation (841) is known as the Bethe-Salpeter equation for the two-particle Green's function $\mathbf{W}^{(2)}$ relating to the 2PI vertex $\Phi^{(2)}$. This is represented diagrammatically as follows:

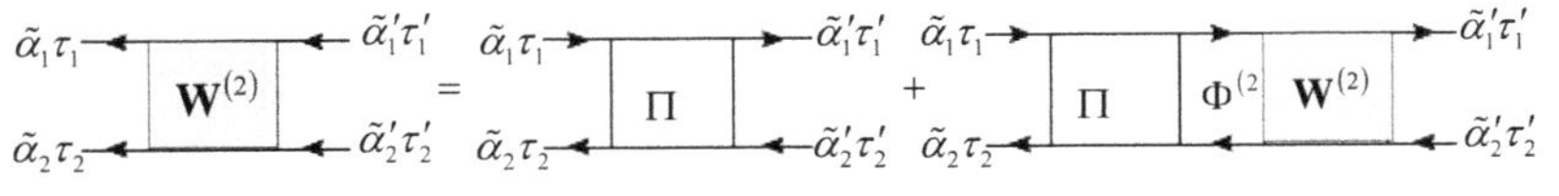

(843)

Diagrammatic representation of the Bethe-Salpeter equation for the 2-particle Green's function $\mathbf{W}^{(2)}$ relating the 2PI vertex $\Phi^{(2)}$ and Π representing the bare two-particle Green's function. Here, $\Phi^{(2)}$ and Π play the role of the self-energy Σ and the bare Green's function $\mathbf{G}_0$ respectively.

This is equivalent to the Dyson equation (782) for the two-particle Green's function. Here, $\Phi^{(2)}$ and Π play the role of the self-energy Σ and the bare Green's function $\mathbf{G}_0$, respectively.

The Bethe-Salpeter equation for $\mathbf{W}^{(2)}$ provides another interpretation of the diagrammatic expansion of the Luttinger-Ward functional where, from definition, $\mathbf{W}^{(2)}$ is the sum of connected diagrams pinned to two external bosonic legs. The 1PI diagrams can be removed by replacing $\mathbf{G}_0$ with $\mathbf{G}$ in the perturbation expansion for $\mathbf{W}^{(2)}$. The Bethe-Salpeter equation removes the 2PI diagrams so that $\Phi^{(2)}$ is the sum of all 2PI diagrams pinned to two bosonic vertices. Because $\Phi^{(2)}$ is obtained from Φ by breaking two fermionic lines and converting them into bosonic vertices, we again arrive at Φ being the sum of all 2PI vacuum diagrams, such as in equation (819). This argument is true for $\mathbf{G}_c^{(4)}$ considering how $\mathbf{W}^{(2)}$ relates with $\mathbf{G}_c^{(4)}$:

$$\mathbf{W}^{(2)}(\gamma_1,\gamma_2)=\mathbf{G}_c^{(4)}(\tilde{\alpha}_1\tau_1,\tilde{\alpha}_2\tau_2,\tilde{\alpha}_2'\tau_2',\tilde{\alpha}_1'\tau_1')+\chi\mathbf{G}(\tilde{\alpha}_1'\tau_1',\tilde{\alpha}_2'\tau_2')\mathbf{G}(\tilde{\alpha}_2\tau_2,\tilde{\alpha}_1'\tau_1') \quad (844)$$

From equation (844), we relate $\mathbf{G}_c^{(4)}$ and the 1PI vertex $\Gamma^{(4)}$. This permits us to obtain the relation between $\Gamma^{(4)}$ and $\Phi^{(2)}$:

$$\Gamma^{(4)}\left(\tilde{\alpha}_1\tau_1,\tilde{\alpha}_2\tau_2,\tilde{\alpha}_2'\tau_2',\tilde{\alpha}_1'\tau_1'\right)=\Phi^{(2)}\left(\tilde{\alpha}_1'\tau_1',\tilde{\alpha}_1\tau_1,\tilde{\alpha}_2'\tau_2',\tilde{\alpha}_2\tau_2\right)$$
$$\left[1+\Pi\left(\tilde{\alpha}_1'\tau_1',\tilde{\alpha}_1\tau_1,\tilde{\alpha}_2'\tau_2',\tilde{\alpha}_2\tau_2\right)\Phi^{(2)}\left(\tilde{\alpha}_1'\tau_1',\tilde{\alpha}_1\tau_1,\tilde{\alpha}_2'\tau_2',\tilde{\alpha}_2\tau_2\right)\right]^{-1} \quad (845a)$$

or

$$\Gamma^{(4)}=\left[\Phi^{(2)-1}+\Pi\right]^{-1}=\Phi^{(2)}-\Phi^{(2)}\Pi\Gamma^{(4)} \quad (845b)$$

This can be represented diagrammatically as follows:

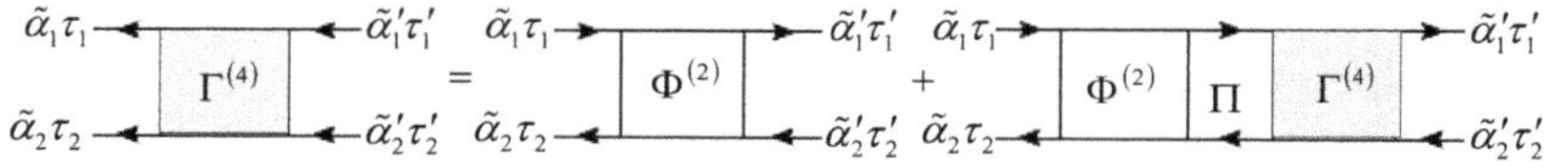

(846)

Diagrammatic representation of the Bethe-Salpeter equations relating the 1PI vertex $\Gamma^{(4)}$ to $\Phi^{(2)}$ (representing the sum of all 2PI diagrams) and Π, representing the bare 2-particle Green's function.

In a normal system, $c_1+c_1'+c_2+c_2'=0$, the Bethe-Salpeter equation can be projected onto the particle-hole $c_1+c_1'=c_2+c_2'=0$ and onto the particle-particle $c_1=c_1'=-c_2=-c_2'$ channels. The corresponding vertices, $\Phi_{\text{ph}}^{(2)}$ and $\Phi_{\text{pp}}^{(2)}$, are 2PI in the particle-hole and particle-particle channels, respectively.

7

Random Phase Approximation (RPA)

Introduction

In this section of the book, we will examine different problems that apply the field theory in condensed matter physics. We will consider bosonic as well as fermionic systems and apply the techniques of the quantum field theory we have studied thus far.

In this chapter, we will study a model describing electrons in a metal known as the electron gas model that considers a system of electrons interacting with each other via the instantaneous Coulomb force $\frac{e^2}{r}$ (**Jellium model**). There are N electrons in a large volume V, with an average density $n_0 = \frac{N}{V}$. The **Jellium model** assumes the global neutrality of the system that is maintained by the presence of a **uniform positive charge background** of density n_0 in a volume V. This is an idealization of the positive metal ions distributed inside the metal and consequently cancels the infinite static Coulomb potential. Therefore, the static uniform positive charge background does not contribute in any way to the change in energy and momentum of the electrons.

The electron gas model provides a framework from which we introduce standard many-body approaches such as the **Gell-Mann-Bruckner theory [28]** or **random phase approximation** (RPA) (time-dependent, Hartree-Fock theory) and show how they are formulated within diagrammatic or functional integral techniques. Our discussions will be within the RPA: plasmon mode, screening, Fermi-liquid behavior, and so on. For real crystals, the positive charge (ions) cannot be uniform and have their own dynamics (lattice vibrations).

Consider the potential energy (PE) of the system:

$$\mathrm{PE} \approx \left(\frac{3}{4\pi a_{\mathrm{B}}^3}\right)^{\frac{1}{3}} \frac{e^2}{\varepsilon_0 r_s} \tag{847}$$

where

$$r_s = \left(\frac{3}{4\pi n_0 a_{\mathrm{B}}^3}\right)^{\frac{1}{3}} \tag{848}$$

is the radius in atomic units of a sphere that encloses one-unit electron gas and is used universally to describe the density of an electron gas; a_{B}, is the Bohr radius. So, the Coulomb energy per particle is e^2 divided by the characteristic length.

We relate the density n_0 to the Fermi wave vector:

$$n_0 = 2\int \frac{d\vec{\kappa}}{(2\pi)^3} n(\vec{\kappa}) = \frac{1}{\pi^2}\int_0^{\kappa_F} \kappa^2\, d\kappa = \frac{\kappa_{\mathrm{F}}^3}{3\pi^2} \tag{849}$$

The Fermi wave vector κ_F and energy are related to r_s:

$$\kappa_F a_B = \left(3\pi^2 n_0\right)^{\frac{1}{3}} a_B = \left(\frac{9\pi}{4}\right)^{\frac{1}{3}} \left(\frac{4\pi n_0 a_B^3}{3}\right)^{\frac{1}{3}} = \left(\frac{9\pi}{4}\right)^{\frac{1}{3}} \frac{1}{r_s} R_y \tag{850}$$

Then, the kinetic energy:

$$\epsilon_F = \frac{\hbar^2 \kappa_F^2}{2m} = \left(\kappa_F a_B\right)^2 \left(\frac{\hbar^2}{2m a_B^2}\right) = \left(\frac{9\pi}{4}\right)^{\frac{2}{3}} \frac{1}{r_s^2} R_y \tag{851}$$

Similarly, the plasma frequency:

$$\omega_p = \hbar \left(\frac{4\pi e^2 n_0}{m}\right)^{\frac{1}{2}} = \left(\frac{12}{r_r^3}\right)^{\frac{1}{2}} R_y \tag{852}$$

In the previous relations, the Rydberg energy R_y:

$$R_y = \frac{e^2}{2a_B} \tag{853}$$

So, in a homogenous electron gas, the average kinetic energy of the electrons is proportional to κ_F^2 that, by dimensional analysis, is inversely proportional to the square of the characteristic length r_s^2. We estimate the domain of validity of a perturbative treatment of the Coulomb interaction by considering the ratio

$$\frac{PE}{KE} \approx \frac{1}{n_0^{\frac{1}{3}} a_B} \tag{854}$$

$$\frac{1}{n_0} = \frac{3\pi^2}{\kappa_F^3} = \frac{4}{3}\pi\left(r_s a_B\right)^3 \tag{855}$$

then

$$\frac{PE}{KE} \approx r_s = \frac{r_0}{a_B} \tag{856}$$

This measures electron separation r_0, which may be obtained from

$$\frac{4}{3}\pi r_0^3 \equiv \frac{1}{n_0} \tag{857}$$

Therefore, when the electron gas has sufficiently high density (i.e., $r_s << 1$), the kinetic energy will be larger than the potential energy. Consequently, the electrons will behave as free particles because the potential energy is a perturbation on the dominant (Fermi pressure) kinetic energy, and the interactions are not very important. In the high-density limit, the free-particle picture is expected to be valid. Therefore, the RPA is expected to be justified at high-density electron gas when $r_s << 1$. Because typical metals correspond to a range $r_s \approx 2 \div 5$, then the RPA results have limited applicability even for conventional metals.

In this chapter, we will see how to go beyond the RPA. In the opposite limit, $r_s >> 1$, the electrons are expected to be localized and form a Wigner crystal because at very low density, the zero-point kinetic energy associated with localizing the electrons eventually becomes negligible in comparison with the electrostatic energy of a classical lattice.

We apply the path integral formalism that is one of the most feasible approaches for the calculation of ground-state properties of heavy fermions. A microscopic treatment of large fermions via many-body wave functions is difficult as the number of possible configurations rapidly increases with increasing particle numbers. The path integral formalism may be a good description of the fermionic properties via Green's functions and scales more favorably to heavy fermions.

7.1 Path Integral Formalism

The most important feature responsible for superconductivity is the effective attraction among electrons due to **overscreening** by the ions of the metal. The ions respond to the motion of the electrons and, in certain cases, produce an effective interaction among electrons that is attractive. This was discovered by Fröhlich [29], and later, by Cooper [30], who demonstrated that the attractive interaction could produce a two-electron bound state in the presence of a Fermi sphere of Bloch electrons. The bound pairs have properties similar to those of bosons. At exceedingly low temperatures, these bosons condense into the superconducting state.

7.1.1 Quantum Three-Dimensional Coulomb Gas

We now consider a quantum three-dimensional coulomb gas and apply the field-theoretic techniques to systems of interacting electrons. We make a detour into the formalism so that we can handle systems of fermions. The quantum Jellium model in three-dimensional position space is defined by the following Hamiltonian determinant in second quantized form:

$$\hat{H} = \hat{H}_0 + \hat{H}_{int} \tag{858}$$

where, $\hat{H}_0$ is the kinetic energy, and the Coulomb interaction, $\hat{H}_{int}$, is written in second-quantized form as

$$\hat{H}_0 = \sum_{\sigma} \int d\vec{r}\, \hat{\psi}^{\dagger}_{\sigma}(\vec{r}) \left(-\frac{\hbar^2 \nabla^2}{2m} \right) \hat{\psi}_{\sigma}(\vec{r}) \tag{859}$$

and

$$\hat{H}_{int} = \frac{1}{2} \sum_{\sigma\sigma'} \int d\vec{r}\, d\vec{r}' \hat{\psi}^{\dagger}_{\sigma}(\vec{r}) \hat{\psi}^{\dagger}_{\sigma'}(\vec{r}') U(\vec{r}' - \vec{r}) \hat{\psi}_{\sigma'}(\vec{r}') \hat{\psi}_{\sigma}(\vec{r}) \tag{860}$$

respectively, with $\sigma = \uparrow, \downarrow$ being spin indices and the Coulomb interaction

$$U(\vec{r}' - \vec{r}) = \frac{e^2}{|\vec{r}' - \vec{r}|} \tag{861}$$

We write the grand-canonical partition function describing the system as:

$$Z = \mathrm{Tr}\left[\exp\left\{ -\beta \hat{H} \right\} \right] \tag{862}$$

Because the exact number of electrons is not accessible experimentally in a macroscopic piece of metal, we choose the grand-canonical ensemble instead of the canonical ensemble to describe the Jellium model. We consider first a system of noninteracting spinless fermions with chemical potential μ. Therefore, we modify the Hamiltonian of the system:

$$\hat{H}' = \hat{H} - \mu\hat{N} \tag{863}$$

and the grand-canonical partition function describing the system:

$$Z = \mathrm{Tr}\exp\left\{-\beta\hat{H}'\right\} = \int d\left[\hat{\psi}^{\dagger}\right]d\left[\hat{\psi}\right]\exp\left\{-S\left[\hat{\psi}^{\dagger},\hat{\psi}\right]\right\} \tag{864}$$

where

$$S\left[\hat{\psi}^{\dagger},\hat{\psi}\right] = S_0\left[\hat{\psi}^{\dagger},\hat{\psi}\right] + S_{\mathrm{int}}\left[\hat{\psi}^{\dagger},\hat{\psi}\right] \tag{865}$$

Here,

$$S_0\left[\hat{\psi}^{\dagger},\hat{\psi}\right] = \int d\vec{r}d\tau\sum_{\sigma}\hat{\psi}_{\sigma}^{\dagger}(\vec{r},\tau)\left[-\mathbf{G}_0^{-1}\right]\hat{\psi}_{\sigma}(\vec{r},\tau) \tag{866}$$

with the bare fermionic operator being

$$-\mathbf{G}_0^{-1} = \partial_{\tau} - \frac{\nabla^2}{2m} - \mu \tag{867}$$

The Grassmann variables obey antiperiodic boundary conditions in imaginary time:

$$\hat{\psi}_{\sigma}^{\dagger}(\vec{r},\tau+\beta) = -\hat{\psi}_{\sigma}^{\dagger}(\vec{r},\tau) \quad , \quad \hat{\psi}_{\sigma}(\vec{r},\tau+\beta) = -\hat{\psi}_{\sigma}(\vec{r},\tau) \tag{868}$$

The interaction is shown schematically in Figure 7.1, which represents the scattering of two electrons with opposite spins and total momentum $\vec{q}$ (conserved in the process).

It is instructive to note that because the coherent states form a basis of eigenvectors of the annihilation operators in Fock space, the Feynman path integral technique is a suitable framework to treat general many-particle Hamiltonians expressed in second quantized form. Furthermore, in addition to recovering previously known results within a single formalism, one is able to go beyond and gain further insight into the physics of the problem. In particular, by including fluctuations about the saddle-point solutions, the method leads to the determination of the collective modes of the system, which become extremely important at finite temperatures.

Let us Fourier transform the field operators in (859):

$$\hat{\psi}_{\sigma}(\vec{r}) = \frac{1}{\sqrt{V}}\sum_{\vec{\kappa}}\exp\left\{i\vec{\kappa}\vec{r}\right\}\hat{\psi}_{\vec{\kappa}\sigma} \quad , \quad \hat{\psi}_{\sigma}^{\dagger}(\vec{r}) = \frac{1}{\sqrt{V}}\sum_{\vec{\kappa}}\exp\left\{-i\vec{\kappa}\vec{r}\right\}\hat{\psi}_{\vec{\kappa}\sigma}^{\dagger} \tag{869}$$

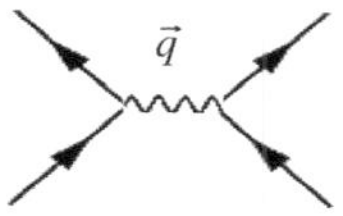

FIGURE 7.1 Scattering of two electrons with opposite spins and total momentum $\vec{q}$.

So, the kinetic energy $\hat{H}_0$ becomes

$$\hat{H}_0 = \frac{1}{V}\sum_{\vec{\kappa}\vec{\kappa}'\sigma}\int d\vec{r}\,\hat{\psi}^{\dagger}_{\vec{\kappa}\sigma}\frac{\hbar^2\vec{\kappa}'^2}{2m}\hat{\psi}_{\vec{\kappa}'\sigma}\exp\{i\vec{r}(\vec{\kappa}-\vec{\kappa}')\} = \frac{1}{V}\sum_{\vec{\kappa}\vec{\kappa}'\sigma}\delta_{\vec{\kappa}-\vec{\kappa}'}\hat{\psi}^{\dagger}_{\vec{\kappa}\sigma}\frac{\hbar^2\vec{\kappa}'^2}{2m}\hat{\psi}_{\vec{\kappa}'\sigma} \tag{870}$$

or

$$\hat{H}_0 = \sum_{\vec{\kappa}\sigma}\frac{\hbar^2\vec{\kappa}^2}{2m}\hat{\psi}^{\dagger}_{\vec{\kappa}\sigma}\hat{\psi}_{\vec{\kappa}'\sigma} \tag{871}$$

7.1.2 Translationally Invariant System

For the translationally invariant system, the interaction between two particles at $\vec{r}'$ and $\vec{r}$ is dependent only on the difference $(\vec{r}'-\vec{r})$ and not on $\vec{r}'$ and $\vec{r}$ separately. So, for the Coulomb interaction

$$U(\vec{r}',\vec{r}) \equiv U(\vec{r}'-\vec{r}) \tag{872}$$

Therefore, if the two particles within the system at positions $\vec{r}'$ and $\vec{r}$ are translated by the same vector $\vec{R}'$:

$$\vec{R} = \frac{\vec{r}'+\vec{r}}{2} \quad , \quad \vec{R}' = \vec{r}'-\vec{r} \tag{873}$$

their interaction energy does not change, and the Coulomb interaction Hamiltonian $\hat{H}_{int}$ in (860) is rewritten as

$$\hat{H}_{int} = \frac{1}{2V^2}\sum_{\vec{\kappa}\vec{\kappa}'\vec{\kappa}''\vec{\kappa}'''\sigma\sigma'}\int d\vec{r}d\vec{r}'\exp\{i(-\vec{\kappa}\vec{r}-\vec{\kappa}'\vec{r}'+\vec{\kappa}''\vec{r}'+\vec{\kappa}'''\vec{r})\}\hat{\psi}^{\dagger}_{\vec{\kappa}\sigma}\hat{\psi}^{\dagger}_{\vec{\kappa}'\sigma'}U(\vec{r}'-\vec{r})\hat{\psi}_{\vec{\kappa}''\sigma'}\hat{\psi}_{\vec{\kappa}'''\sigma} \tag{874}$$

or

$$\hat{H}_{int} = \frac{1}{2V^2}\sum_{\vec{\kappa}\vec{\kappa}'\vec{\kappa}''\vec{\kappa}'''\sigma\sigma'}\int d\vec{R}d\vec{R}'\exp\{i(-\vec{\kappa}-\vec{\kappa}'+\vec{\kappa}''+\vec{\kappa}''')\vec{R}\}\exp\left\{i(\vec{\kappa}-\vec{\kappa}'+\vec{\kappa}''-\vec{\kappa}''')\frac{\vec{R}'}{2}\right\}\hat{\psi}^{\dagger}_{\vec{\kappa}\sigma}\hat{\psi}^{\dagger}_{\vec{\kappa}'\sigma'}U(\vec{R}')\hat{\psi}_{\vec{\kappa}''\sigma'}\hat{\psi}_{\vec{\kappa}'''\sigma} \tag{875}$$

We can now perform the integral over $\vec{R}$:

$$\hat{H}_{int} = \frac{1}{2V}\sum_{\vec{\kappa}\vec{\kappa}'\vec{\kappa}''\vec{\kappa}'''\sigma\sigma'}\int d\vec{R}'\delta_{\vec{\kappa}+\vec{\kappa}'-\vec{\kappa}''-\vec{\kappa}'''}\exp\left\{i(\vec{\kappa}-\vec{\kappa}'+\vec{\kappa}''-\vec{\kappa}''')\frac{\vec{R}'}{2}\right\}\hat{\psi}^{\dagger}_{\vec{\kappa}\sigma}\hat{\psi}^{\dagger}_{\vec{\kappa}'\sigma'}U(\vec{R}')\hat{\psi}_{\vec{\kappa}''\sigma'}\hat{\psi}_{\vec{\kappa}'''\sigma} \tag{876}$$

or

$$\hat{H}_{int} = \frac{1}{2V}\sum_{\vec{\kappa}\vec{\kappa}'\vec{\kappa}''\sigma\sigma'}\int d\vec{R}'\exp\{-i(\vec{\kappa}'-\vec{\kappa}'')\vec{R}'\}\hat{\psi}^{\dagger}_{\vec{\kappa}\sigma}\hat{\psi}^{\dagger}_{\vec{\kappa}'\sigma'}U(\vec{R}')\hat{\psi}_{\vec{\kappa}''\sigma'}\hat{\psi}_{\vec{\kappa}+\vec{\kappa}'-\vec{\kappa}'',\sigma} \tag{877}$$

Substituting new momentum variables

$$\vec{\kappa}_1 = \vec{\kappa}+\vec{\kappa}'-\vec{\kappa}'' \quad , \quad \vec{\kappa}_2 = \vec{\kappa}'' \quad , \quad \vec{q} = \vec{\kappa}'-\vec{\kappa}'' \tag{878}$$

we then obtain

$$\hat{H}_{int} = \frac{1}{2V} \sum_{\vec{\kappa}_1 \vec{\kappa}_2 \vec{q} \sigma \sigma'} U(\vec{q}) \hat{\psi}^{\dagger}_{\vec{\kappa}_1 - \vec{q}, \sigma} \hat{\psi}^{\dagger}_{\vec{\kappa}_2 + \vec{q}, \sigma'} \hat{\psi}_{\vec{\kappa}_2 \sigma'} \hat{\psi}_{\vec{\kappa}_1 \sigma} \tag{879}$$

with the Fourier transform of the Coulomb interaction:

$$U(\vec{q}) = \int d\vec{R}' \exp\{-i\vec{q}\vec{R}'\} U(\vec{R}') \tag{880}$$

Each term in the summation in (879) represents a scattering process. So, the interaction Hamiltonian $\hat{H}_{int}$ should be the sum over all processes in which two electrons come in with momenta $\vec{\kappa}_1$ and $\vec{\kappa}_2$, a momentum of $\vec{q}$ is transferred from one to the other via the Coulomb interaction, and the electrons scatter with momenta $\vec{\kappa}_1 - \vec{q}$ and $\vec{\kappa}_2 + \vec{q}$ (see Figure 7.2).

We can obtain the Coulomb potential $U(\vec{q})$ most easily via the Fourier transforming of the Poisson equation for a point charge where the electronic interaction is mediated by the electrostatic potential $\phi(\vec{r})$:

$$\nabla^2 \phi(\vec{r}) = -4\pi\rho(\vec{r}) = -4\pi e \delta(\vec{r}) \tag{881}$$

It follows that

$$\int d\vec{r} \exp\{-i\vec{q}\vec{r}\} \nabla^2 \phi(\vec{r}) = -4\pi e \tag{882}$$

Conducting integration by parts on the left-hand side of equation (882), we have

$$\int d\vec{r} (-i\vec{q})^2 \exp\{-i\vec{q}\vec{r}\} \phi(\vec{r}) = -4\pi e \tag{883}$$

This follows that

$$q^2 \phi(\vec{q}) = 4\pi e \tag{884}$$

or

$$\phi(\vec{q}) = \frac{4\pi e}{q^2} \tag{885}$$

So, the bare pairwise Coulomb interaction assumes the form

$$U(\vec{q}) = \frac{4\pi e^2}{q^2} \tag{886}$$

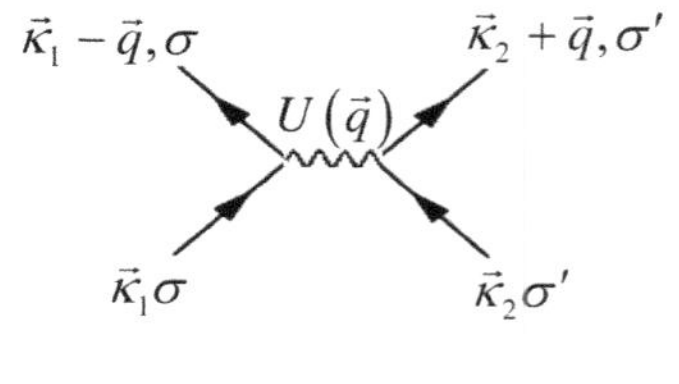

FIGURE 7.2 Schematic representation of the interaction of two elections. The two electrons come in with momenta $\vec{\kappa}_1$ and $\vec{\kappa}_2$, a momentum of $\vec{q}$ is transferred from one to the other via the Coulomb interaction, and the electrons scatter with momenta $\vec{\kappa}_1 - \vec{q}$ and $\vec{\kappa}_2 + \vec{q}$.

which is strongly repulsive. However, if we (indirectly) measure the interaction among charges in a metal, we do not find $U(\vec{q})$ but a reduced interaction. First of all, there will be a dielectric function ε from the polarizability of the ion cores:

$$U(\vec{q}) \to \frac{4\pi e^2}{\varepsilon q^2} \tag{887}$$

Nonetheless, this only leads to a quantitative change, not a qualitative one. For now, unless otherwise stated, we absorb the factor $\frac{1}{\varepsilon}$ into e^2:

$$\frac{e^2}{\varepsilon} \to e^2 \tag{888}$$

It is instructive to note that a test charge in a metal is **screened** by a cloud of opposite charge so that from far away, the effective charge is strongly reduced.

Next, we perform the denotation

$$\mathbf{D}_{0q} = -\frac{4\pi}{q^2} \tag{889}$$

Hence, the pairwise Coulomb interaction assumes the form

$$U(\vec{q}) = -e^2\mathbf{D}_{0q} \tag{890}$$

The interaction action functional of our system can therefore be rewritten as

$$S_{\text{int}}\left[\hat{\psi}^{\dagger},\hat{\psi}\right] = -\frac{1}{2V}\sum_{\vec{\kappa}_1\vec{\kappa}_2\vec{q}\sigma\sigma'} e^2\mathbf{D}_{0q}\hat{\psi}^{\dagger}_{\vec{\kappa}_1-\vec{q},\sigma}\hat{\psi}^{\dagger}_{\vec{\kappa}_2+\vec{q},\sigma'}\hat{\psi}_{\vec{\kappa}_2\sigma'}\hat{\psi}_{\vec{\kappa}_1\sigma} \tag{891}$$

Writing the Fourier transform of the electronic density operator as

$$\rho_{\vec{q}} = \sum_{\vec{\kappa}\sigma}\hat{\psi}^{\dagger}_{\vec{\kappa}+\vec{q},\sigma}\hat{\psi}_{\vec{\kappa}\sigma} \tag{892}$$

then

$$S_{\text{int}}\left[\hat{\psi}^{\dagger},\hat{\psi}\right] = -\frac{1}{2V}\sum_{\vec{\kappa}_1\vec{\kappa}_2\vec{q}\sigma\sigma'} e^2\rho_{\vec{q}}\mathbf{D}_{0q}\rho_{-\vec{q}} \tag{893}$$

From (890), we observe that the contribution of small q to the Coulomb potential is important because e^2 is small. Since (891) diverges, simple perturbation theory cannot be applied; therefore, the problem is definitely nontrivial. So, we develop the RPA theory by introducing the diagrammatic method where a corresponding infinitesimal series of diagrams is added for the divergent parts. The functional integral method is convenient because this method is deduced in a very compact form.

7.2 RPA Functional Integral

Auxiliary Field Method

We treat the previous problem via the perturbation theory known as **Random Phase Approximation** (RPA) or via the mean field theory (Hubbard Stratonovich transformation). In RPA, we write the partition function

$$Z = \int d\left[\hat{\psi}^{\dagger}\right]d\left[\hat{\psi}\right]\exp\left\{-S\left[\hat{\psi}^{\dagger},\hat{\psi}\right]\right\} \tag{894}$$

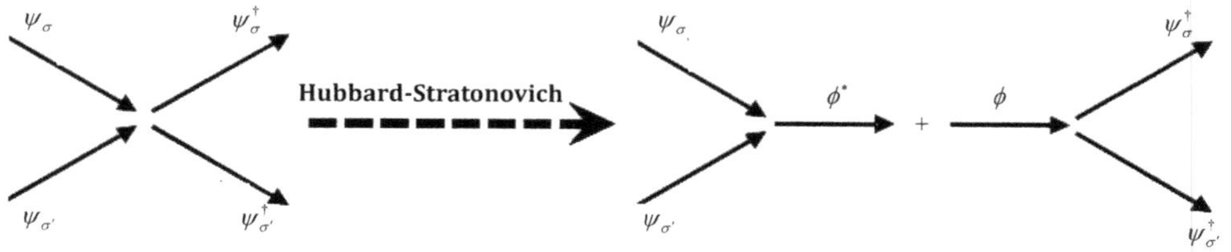

FIGURE 7.3 The Hubbard-Stratonovich transformation illustration with the interaction term originally composed of quartic term fermion fields decomposed into quadratic term fermion fields.

and the action functional of the interacting fermions:

$$S\left[\hat{\psi}^{\dagger},\hat{\psi}\right]=\int_{0}^{\beta}d\tau\left[\sum_{\vec{\kappa},\sigma}\psi_{\vec{\kappa},\sigma}^{\dagger}(\tau)\left[-\mathbf{G}_{0}^{-1}\right]\psi_{\vec{\kappa},\sigma}(\tau)+\frac{1}{2V}\sum_{\vec{q}\neq 0}\rho_{-\vec{q}}(\tau)\left[(ie)^{2}\mathbf{D}_{0q}\right]\rho_{\vec{q}}(\tau)\right] \tag{895}$$

In the absence of interaction, the second term in (895) is absent and the functional integral (894) becomes

$$Z_{0}=\int d\left[\hat{\psi}^{\dagger}\right]d\left[\hat{\psi}\right]\exp\left\{-S_{0}\left[\hat{\psi}^{\dagger},\hat{\psi}\right]\right\}=\det\left[-\mathbf{G}_{0}^{-1}\right] \tag{896}$$

The last term in (895) is quartic in the fermionic variables. Hence, the functional integral appearing in the expression of the partition function (894) cannot be computed exactly. To deal with this last term in (895), the quartic term may be conveniently treated via a Hubbard-Stratonovich transformation that serves to decouple the interaction by introducing auxiliary **Hubbard-Stratonovich c-field** $\phi_{\vec{q}}(\tau)$ [31, 32, 9], which renders the action functional quadratic in the fermion fields (Grassmann variables) $\hat{\psi}_{\sigma}(\vec{r},\tau)$ and $\hat{\psi}_{\sigma}^{\dagger}(\vec{r},\tau)$ (Figure 7.3).

Then, considering interaction, we introduce the collective degrees of freedom via the so-called **Hubbard-Stratonovich transformation**, which decouples the interaction by introducing auxiliary c-field $\phi_{\vec{q}}(\tau)$. Therefore, we observe that the **Hubbard–Stratonovich transformation** entails the transformation of a system of interacting particles into a system of **free** particles interacting with a bosonic field. In this case, the transformation from a four-fermion-field interaction to two–fermion–fields plus an auxiliary–boson–field one. It is instructive to note that our physics is dominated by **charge density fluctuations (plasmons)** and so we rewrite the action functional in terms of **densities**.

7.2.1 Gaussian Fluctuations

We resolve the four-product of the fields in the second term of (895) in equation (894), which is a Gaussian integral in terms of the densities $\rho_{\vec{q}}(\tau)$ and $\rho_{-\vec{q}}(\tau)$, at the expense of introducing an additional functional integration:

$$\exp\left\{-S_{\text{int}}\left[\hat{\psi}^{\dagger},\hat{\psi}\right]\right\}=\exp\left\{\int_{0}^{\beta}d\tau\frac{1}{2V}\sum_{\vec{q}\neq 0}\rho_{-\vec{q}}(\tau)\left[(ie)^{2}\mathbf{D}_{0q}\right]\rho_{\vec{q}}(\tau)\right\}=N\int d[\phi]\exp\left\{-\mathrm{F}[\phi]\right\} \tag{897}$$

where

$$\mathrm{F}[\phi]=\frac{1}{2}\int_{0}^{\beta}d\tau\left(\sum_{\vec{q}\neq 0}\left[\phi_{-\vec{q}}(\tau)\left[\mathbf{D}_{0q}\right]^{-1}\phi_{\vec{q}}(\tau)+\frac{ie}{\sqrt{V}}\left[\rho_{-\vec{q}}(\tau)\phi_{\vec{q}}(\tau)+\rho_{\vec{q}}(\tau)\phi_{-\vec{q}}(\tau)\right]\right]\right) \tag{898}$$

From here, the normalization factor N is absorbed into the integration measure $d[\phi]$; then the partition function (894) may be rewritten

$$Z=\int d\left[\hat{\psi}^{\dagger}\right]d\left[\hat{\psi}\right]d[\phi]\exp\left\{-S\left[\hat{\psi}^{\dagger},\hat{\psi},\phi\right]\right\} \tag{899}$$

where the action functional now becomes

$$S\left[\hat{\psi}^{\dagger},\hat{\psi},\phi\right]=S_0\left[\psi^{*},\psi\right]+\frac{1}{2}\sum_{\vec{q}\neq 0}\left[\phi_{-\vec{q}}(\tau)\left[\mathbf{D}_{0q}\right]^{-1}\phi_{\vec{q}}(\tau)+\frac{ie}{\sqrt{V}}\left[\rho_{-\vec{q}}(\tau)\phi_{\vec{q}}(\tau)+\rho_{\vec{q}}(\tau)\phi_{-\vec{q}}(\tau)\right]\right] \tag{900}$$

Physically, the c-field ϕ imitates the scalar potential related to charge fluctuations; hence, charge neutrality is realized by the following condition:

$$\phi_{\vec{q}=0}(\tau)=\frac{1}{\sqrt{V}}\int d\vec{r}\phi(\vec{r},\tau)=0 \tag{901}$$

Introducing the Matsubara expansion of the bosonic and fermionic fields:

$$\phi(\vec{r},\tau)=\frac{1}{\sqrt{\beta V}}\sum_{\vec{q},l}\exp\left\{-i\left(\omega_l\tau-\vec{q}\vec{r}\right)\right\}\phi_{\vec{q},\omega_l} \quad , \quad \phi^{*}(\vec{r},\tau)=\frac{1}{\sqrt{\beta V}}\sum_{\vec{q},l}\exp\left\{i\left(\omega_n\tau-\vec{q}\vec{r}\right)\right\}\phi^{*}_{\vec{q},\omega_l} \tag{902}$$

$$\hat{\psi}_{\sigma}(\vec{r},\tau)=\frac{1}{\sqrt{\beta V}}\sum_{\vec{\kappa},n}\exp\left\{-i\left(\omega_n\tau-\vec{\kappa}\vec{r}\right)\right\}\hat{\psi}_{\vec{\kappa}\omega_n\sigma} \quad , \quad \hat{\psi}^{\dagger}_{\sigma}(\vec{r},\tau)=\frac{1}{\sqrt{\beta V}}\sum_{\vec{\kappa},n}\exp\left\{i\left(\omega_n\tau-\vec{\kappa}\vec{r}\right)\right\}\hat{\psi}^{\dagger}_{\vec{\kappa}\omega_n\sigma} \tag{903}$$

respectively, the grand-canonical partition function now becomes

$$Z=N'\int d[\phi]\int d\left[\hat{\psi}^{\dagger}\right]d\left[\hat{\psi}\right]\exp\left\{-\tilde{S}\left[\hat{\psi}^{\dagger},\hat{\psi},\phi\right]\right\} \tag{904}$$

where

$$\tilde{S}\left[\hat{\psi}^{\dagger},\hat{\psi},\phi\right]=\int_0^{\beta}d\tau\int d\vec{r}\left[\frac{1}{8\pi}\left(-\phi(\vec{r},\tau)\Delta\phi(\vec{r},\tau)\right)+\sum_{\sigma}\psi^{\dagger}_{\sigma}(\vec{r},\tau)\left[-\mathbf{G}^{-1}\right]\psi_{\sigma}(\vec{r},\tau)\right] \tag{905}$$

or

$$\tilde{S}\left[\hat{\psi}^{\dagger},\hat{\psi},\phi\right]=\sum_{\vec{q},l}\left[\frac{1}{2}\phi_{-\vec{q},-\omega_l}\left[-\mathbf{D}_{0q}\right]^{-1}\phi_{\vec{q},\omega_l}+\sum_{\vec{\kappa},n,\sigma}\psi^{\dagger}_{\vec{\kappa}+\vec{q},\omega_n+\omega_l,\sigma}\left(\left[-\mathbf{G}_0^{-1}\right]\delta_{\vec{q},0}\delta_{\omega_l,0}+\frac{ie}{\sqrt{\beta V}}\phi_{\vec{q},\omega_l}\right)\psi_{\vec{\kappa},\omega_n,\sigma}\right] \tag{906}$$

with

$$-\mathbf{G}^{-1}=-\mathbf{G}_0^{-1}+\hat{\mathbf{M}}_1 \quad , \quad \hat{\mathbf{M}}_1=ie\phi(\vec{r},\tau) \tag{907}$$

7.2.1.1 Integration over Grassmann Variables

We introduce a new auxiliary field via a gauge of the c-field by

$$\phi(\vec{r},\tau)\to\phi(\vec{r},\tau)+ie\int\frac{d\vec{r}'}{\left|\vec{r}-\vec{r}'\right|}\rho(\vec{r}',\tau) \tag{908}$$

and also consider the Poisson equation

$$\Delta\frac{1}{|\vec{r}-\vec{r}'|}=-4\pi\delta(\vec{r}-\vec{r}') \quad (909)$$

Then the functional in (905) becomes

$$\begin{aligned}\tilde{S}\left[\hat{\psi}^{\dagger},\hat{\psi},\phi\right]=&\int_0^\beta d\tau\int d\vec{r}\hat{\psi}^{\dagger}(\vec{r},\tau)\left[-\mathbf{G}^{-1}\right]\hat{\psi}(\vec{r},\tau)+\\&+\frac{1}{8\pi}\int_0^\beta d\tau\int d\vec{r}\left(\phi(\vec{r},\tau)+ie\int\frac{d\vec{r}'}{|\vec{r}-\vec{r}'|}\rho(\vec{r}',\tau)\right)\left(\Delta\phi(\vec{r},\tau)-ie4\pi\rho(\vec{r},\tau)\right)\end{aligned} \quad (910)$$

It is instructive to note that the introduction of the new auxiliary c-field ϕ leaves the partition function invariant:

$$Z=\int d\left[\hat{\psi}^{\dagger}\right]d\left[\hat{\psi}\right]\exp\left\{-S\left[\hat{\psi}^{\dagger},\hat{\psi}\right]\right\}=\int d\left[\hat{\psi}^{\dagger}\right]d\left[\hat{\psi}\right]d[\varphi]\exp\left\{-\tilde{S}\left[\hat{\psi}^{\dagger},\hat{\psi},\phi\right]\right\} \quad (911)$$

where

$$\tilde{S}\left[\hat{\psi}^{\dagger},\hat{\psi},\phi\right]=\int_0^\beta d\tau\int d\vec{r}\left[\hat{\psi}^{\dagger}(\vec{r},\tau)\left[-\mathbf{G}^{-1}\right]\hat{\psi}(\vec{r},\tau)-\frac{1}{8\pi}\nabla\varphi(\vec{r},\tau)\nabla\varphi(\vec{r},\tau)\right] \quad (912)$$

We examine the following action functional in relation (912):

$$\tilde{S}[\phi]=-\frac{1}{8\pi}\int_0^\beta d\tau\int d\vec{r}\,\nabla\phi(\vec{r},\tau)\nabla\phi(\vec{r},\tau) \quad (913)$$

We perform integration by parts and then

$$\tilde{S}[\phi]=-\frac{1}{8\pi}\int_0^\beta d\tau\left[\phi(\vec{r},\tau)\nabla\phi(\vec{r},\tau)\Big|_{\text{boundary}}+\int d\vec{r}\phi(\vec{r},\tau)\Delta\phi(\vec{r},\tau)\right] \quad (914)$$

Considering boundary conditions, the first term is zero; hence,

$$\tilde{S}[\phi]=-\frac{1}{8\pi}\int_0^\beta d\tau\int d\vec{r}\,\nabla\phi(\vec{r},\tau)\nabla\phi(\vec{r},\tau)=\frac{1}{8\pi}\int_0^\beta d\tau\int d\vec{r}\phi(\vec{r},\tau)\Delta\phi(\vec{r},\tau) \quad (915)$$

Therefore,

$$S\left[\hat{\psi}^{\dagger},\hat{\psi},\phi\right]=\int_0^\beta d\tau\int d\vec{r}\left[\hat{\psi}^{\dagger}(\vec{r},\tau)\left[-\mathbf{G}^{-1}\right]\hat{\psi}(\vec{r},\tau)+\frac{1}{8\pi}\phi(\vec{r},\tau)\Delta\phi(\vec{r},\tau)\right] \quad (916)$$

Hence, with the approximate normalization of the measure $d[\phi]$,

$$\int d[\phi]\exp\left\{-S[\phi]\right\}=1 \quad (917)$$

From here, we evaluate the following partition function by integrating over Grassmann variables $\hat{\psi}$. So,

$$Z[\phi] = \int d\left[\hat{\psi}^{\dagger}\right] d\left[\hat{\psi}\right] \exp\left\{-S\left[\hat{\psi}^{\dagger}, \hat{\psi}, \phi\right]\right\} = \exp\left\{-S_{\text{eff}}[\phi]\right\} \tag{918}$$

where

$$S^{\text{eff}}[\phi] = \frac{1}{8\pi} \int_0^{\beta} d\tau \int d\vec{r} \left(-\phi(\vec{r},\tau)\Delta\phi(\vec{r},\tau)\right) - 2\,\text{Tr}\ln\left[-\mathbf{G}^{-1}\right] \tag{919}$$

We obtain the functional trace Tr when exponentiating the fermionic determinant:

$$\left[\det\left[-\mathbf{G}^{-1}\right]\right]^2 = \exp\left\{2\ln\det\left[-\mathbf{G}^{-1}\right]\right\} = \exp\left\{2\,\text{Tr}\ln\left[-\mathbf{G}^{-1}\right]\right\} \tag{920}$$

The prefactor of twice multiplying the trace is due to the spin degeneracy.

7.2.1.2 Fermionic Determinant Gaussian Expansion

We examine the last term $\text{Tr}\ln\left[-\mathbf{G}^{-1}\right]$ in (919):

$$\text{Tr}\ln\left[-\mathbf{G}^{-1}\right] = \text{Tr}\ln\left[-\mathbf{G}_0^{-1} + \mathbf{M}_1\right] = \text{Tr}\ln\left[-\mathbf{G}_0^{-1}\left[1 - \mathbf{G}_0\mathbf{M}_1\right]\right] = \text{Tr}\ln\left[-\mathbf{G}_0^{-1}\right] + \text{Tr}\ln\left[1 - \mathbf{G}_0\mathbf{M}_1\right] \tag{921}$$

We then perform the following expansion to the expected order:

$$\text{Tr}\ln\left[-\mathbf{G}^{-1}\right] = \text{Tr}\ln\left[-\mathbf{G}_0^{-1}\right] - \sum_{n=1}^{\infty} \frac{1}{n}\text{Tr}\left[\mathbf{G}_0\mathbf{M}_1\right]^n \tag{922}$$

The first term corresponds to the free particle and the second to the interaction of the particle and the field. The RPA limits this expansion to second order. Also

$$\left[\mathbf{G}_0\right]_{\nu\nu'} = \left[i\omega_n - \frac{\vec{\kappa}^2}{2m} + \mu\right]^{-1} \delta_{\nu\nu'} \equiv \left[i\omega_n - \xi_{\vec{\kappa}}\right]^{-1} \delta_{\nu\nu'} \equiv \mathbf{G}_{0\nu}\delta_{\nu\nu'} \quad , \quad \left[\mathbf{M}_1\right]_{\nu\nu'} = \frac{ie}{\sqrt{\beta V}}\phi_{\nu-\nu'} \quad , \quad \nu = \left(\vec{\kappa}, \omega_n\right) \tag{923}$$

where $\mathbf{G}_{0\nu}$ is the single-particle Green's function of free fermions.

The following diagrammatic notations are very important for the given problem:

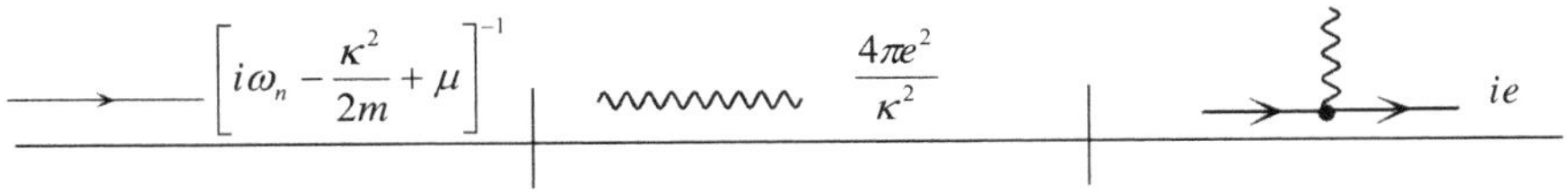

FIGURE 7.4 Some diagrammatic notations of Gaussian expansion. The first being the single-particle fermionic Green's function with the directed arrow indicating the electric charge flow, the second, the interacting field and the third, the vertex (field interaction with the electric charge). Momenta and energies meeting at a vertex obey momentum and energy conservation.

We examine the contribution of the first order:

$$-\text{Tr}\left[\mathbf{G}_0\mathbf{M}_1\right] = -\sum_{\nu\nu'}\left[\mathbf{G}_0\right]_{\nu\nu'}\left[\mathbf{M}_1\right]_{\nu'\nu} = -\sum_{\nu\nu'}\mathbf{G}_{0\nu}\delta_{\nu\nu'}\left[\mathbf{M}_1\right]_{\nu'\nu} = -\sum_{\nu}\mathbf{G}_{0\nu}\left[\mathbf{M}_1\right]_{\nu\nu} \tag{924}$$

and the second-order

$$-\frac{1}{2}\mathrm{Tr}[\mathbf{G}_0\mathbf{M}_1]^2=-\frac{1}{2}\sum_{\nu}\mathbf{G}_{0\nu}[\mathbf{M}_1]_{\nu\nu'}\mathbf{G}_{0\nu'}[\mathbf{M}_1]_{\nu'\nu}=-\frac{1}{2}\sum_{\nu q}\mathbf{G}_{0\nu}[\mathbf{M}_1]_{\nu(\nu+q)}\mathbf{G}_{0(\nu+q)}[\mathbf{M}_1]_{(\nu+q)\nu} \tag{925}$$

where we observe the expansion of the fermionic logarithm in $S^{\mathrm{eff}}[\phi]$ in powers of $\frac{\mathbf{M}_1}{[-\mathbf{G}_0^{-1}]}$ to yield the first nonvanishing contribution to the second order, which is only Gaussian:

$$-\frac{1}{2}\mathrm{Tr}[\mathbf{G}_0\mathbf{M}_1]^2=\frac{e^2}{4}\sum_{q}\phi_q\Pi_q^{\mathrm{RPA}}\phi_{-q} \tag{926}$$

while the first-order contribution vanishes due to the charge-neutrality condition $\phi_{q=0}(\tau)=0$.

Here, the **polarization function**

$$\Pi_q^{\mathrm{RPA}}=\frac{2}{\beta V}\sum_{\nu}\mathbf{G}_{0\nu}\mathbf{G}_{0(\nu+q)} \tag{927}$$

results from the integration over the fermionic variables within the RPA. An important building block is entering the calculation of the **Cooper pair propagator in superconductors**. The factor of two is due to the spin. It is the **(causal) density-density propagator** for a noninteracting Fermi gas and contains information about the actual excitation spectrum of the Fermi-gas, which has a fixed number of particles. The function Π_q^{RPA} is a property of the occupied states of a Fermi gas at temperature $T=\beta^{-1}$, implying that all states within a small energy domain $\frac{T}{\epsilon_{\mathrm{F}}}<<1$ around the Fermi energy $\epsilon_{\mathrm{F}}=\frac{\kappa_{\mathrm{F}}^2}{2m}$. This is at sufficiently low temperatures relative to the Fermi energy. The density-density propagator Π_q^{RPA} also gives the field ϕ some nontrivial dynamics, implying that ϕ is no longer instantaneous. If we ignore the coupling to the electron gas, ϕ represents a field with a **bare propagator** proportional to q^{-2}; so, there exists a divergence due to the singularity of Coulomb interaction at small momentum transfers q. This implies long-range correlation mediated by an electric field in a vacuum. This problem is solved by performing an infinite series of diagram summation describing the **screening** of Coulomb potential by free electrons.

The truncation of the expansion to the second order yields the RPA to the Jellium model where the grand-canonical partition function:

$$Z=N'\left[\det\left[-\mathbf{G}_0^{-1}\right]\right]^2\int d[\phi]\exp\left\{-S^{\mathrm{RPA}}[\phi]\right\} \tag{928}$$

and where the RPA action functional in the grand-canonical ensemble:

$$S^{\mathrm{RPA}}[\phi]=\frac{1}{2}\sum_{q=(\vec{q},\omega_l)}\phi_q\left[-\mathbf{D}_q^{\mathrm{RPA}}\right]^{-1}\phi_{-q} \tag{929}$$

with

$$\mathbf{D}_q^{\mathrm{RPA}}=\frac{1}{\left[\mathbf{D}_{0q}^{\mathrm{RPA}}\right]^{-1}+e^2\Pi_q^{\mathrm{RPA}}}=\frac{\mathbf{D}_{0q}^{\mathrm{RPA}}}{1+e^2\mathbf{D}_{0q}^{\mathrm{RPA}}\Pi_q^{\mathrm{RPA}}}=\mathbf{D}_{0q}^{\mathrm{RPA}}\sum_{n=0}^{\infty}\left[(ie)^2\Pi_q^{\mathrm{RPA}}\mathbf{D}_{0q}^{\mathrm{RPA}}\right]^n\equiv\mathbf{D}_{0q}^{\mathrm{RPA}}\sum_{n=0}^{\infty}\left[\Sigma_q\mathbf{D}_{0q}^{\mathrm{RPA}}\right]^n \tag{930}$$

This is an effective interaction (known as random phase approximation, or RPA) that is frequency ω-dependent. The modification of the interaction results from the polarization of the medium,

which is an example of screening that generally is frequency-dependent, corresponding to the accounting of retardation effects due to characteristic time of electron response to instantaneous Coulomb interaction.

The quantity Σ_q defines what is known as the **self-energy of the collective field** ϕ:

$$\Sigma_q = (ie)^2 \Pi_q^{\mathrm{RPA}} \tag{931}$$

The coupling of the field to a medium yields a **self-energy** proportional to Π_q and translates the long-range power law correlation into an exponential screening law that we will study further in this chapter.

The RPA polarization function Π_q^{RPA} can be represented as follows:

$$\Pi_q^{\mathrm{RPA}} = \frac{\Pi_{0q}^{\mathrm{RPA}}}{1 + e^2 \mathbf{D}_{0q}^{\mathrm{RPA}} \Pi_{0q}^{\mathrm{RPA}}} = \Pi_{0q}^{\mathrm{RPA}} \sum_{n=0}^{\infty} \left[(ie)^2 \Pi_{0q}^{\mathrm{RPA}} \mathbf{D}_{0q}^{\mathrm{RPA}} \right]^n \tag{932}$$

and diagrammatically can be represented as

Π_q^{RPA} = 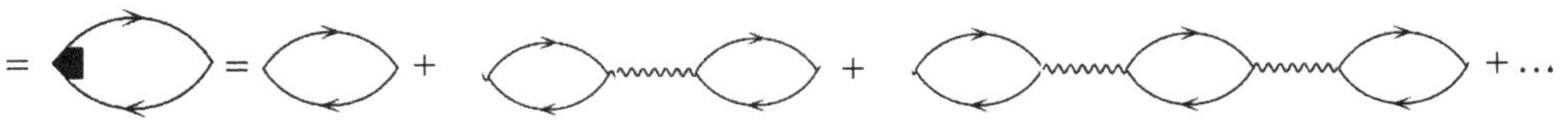 (933)

or

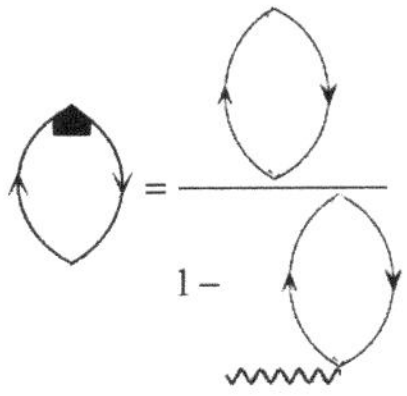 (934)

FIGURE 7.5 Diagrammatic representation of the density-density response function Π_q^{RPA} in the RPA.

The **RPA** is an equality that highlights the fact that the irreducible vertex, $U(\vec{q}) = e^2 \mathbf{D}_{0q}^{\mathrm{RPA}}$, plays the role of an irreducible self-energy in the particle-hole channel. From here, we observe that

- **RPA is based on the assumption that the phase relation among different particle-hole excitations entering the perturbation series that yields the aforementioned relation (932) are random such that interference terms vanish on the average.**
- **The polarization propagator is expressed in terms of the sum over irreducible polarization parts Π_{0q}^{RPA}.**

In the previous expansion, we consider $\left| e^2 \mathbf{D}_{0q}^{\mathrm{RPA}} \Pi_{0q}^{\mathrm{RPA}} \right| < 1$. The mean-field theory, which considers fluctuations within the RPA, is not a systematic method to solve interacting many-body problems but rather a practical method that can be justified a posteriori by comparison with experiments.

From equation (930), the limit $e \to 0$ confirms the absence of Coulomb interaction. So the Euclidean propagator $\mathbf{D}_{0q}^{\mathrm{RPA}}$ is just the bare Coulomb potential in Fourier space that is instantaneous because it is independent of the Matsubara frequency ω_l.

Dyson Equation

Equation (930) is an approximate solution to the following Dyson equation:

$$\mathbf{D}_q = \mathbf{D}_{0q} + \mathbf{D}_{0q} \Sigma_q \mathbf{D}_q \tag{935}$$

Assume that the Coulombic part of the effective ion-ion interaction is reduced by the electronic dielectric constant, which has implications for normal modes of short wavelength:

$$\mathbf{D}_{0q} = \left(1 - \mathbf{D}_{0q}\Sigma_q\right)\mathbf{D}_q = \varepsilon_q \mathbf{D}_q \tag{936}$$

Here, ε_q is the (generalized Euclidean) dielectric constant that is a proportionality constant between the bare $\mathbf{D}_{0q}$ and renormalized propagators $\mathbf{D}_q$ in Dyson equation. From (936), we also have

$$\mathbf{D}_q = \frac{1}{1 - \mathbf{D}_{0q}\Sigma_q}\mathbf{D}_{0q} \tag{937}$$

or

$$\mathbf{D}_q = \left\langle \phi_q \phi_{-q} \right\rangle = \frac{\int d[\phi]\phi_q\phi_{-q} \exp\left\{-S^{\text{eff}}[\phi]\right\}}{\int d[\phi]\exp\left\{-S^{\text{eff}}[\phi]\right\}} \tag{938}$$

with

$$S^{\text{eff}}[\phi] = \frac{1}{8\pi}\int_0^\beta d\tau \int d\vec{r}\left(-\phi(\vec{r},\tau)\Delta\phi(\vec{r},\tau)\right) - 2\,\text{Tr}\ln\left[-\mathbf{G}^{-1}\right] \tag{939}$$

From this, and from (930), we determine some diagrammatic definitions:

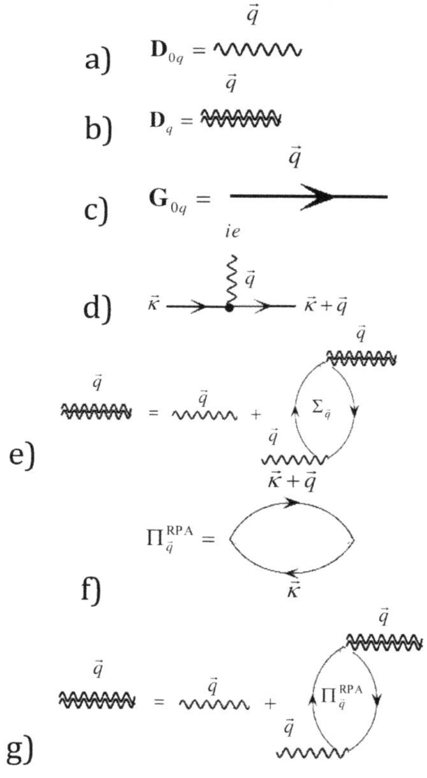

FIGURE 7.6 Feynman diagrams: (a) Unperturbed bosonic propagator $\mathbf{D}_{0q}$; (b) Exact bosonic propagator $\mathbf{D}_q$; (c) Unperturbed fermionic propagator $\mathbf{G}_{0q}$; (e) Dyson equation; and (f) RPA electron-hole polarization bubble. The (g) RPA propagator follows from (e) with the substitution of the self-energy Σ_q by the electron-hole bubble $(ie)^2\,\Pi_q^{\text{RPA}}$.

7.2.1.3 Diagrammatic Interpretation of the RPA

We derive the RPA for the Jellium model, which amounts to expanding the logarithm of the fermionic determinant up to quadratic order in the electron charge in the effective action $S^{\text{eff}}[\phi]$ for the collective field ϕ. This collective field ϕ is introduced via a Hubbard-Stratonovich transformation and couples to local electronic charge fluctuations such as the scalar potential does in electrodynamics. Therefore, it imitates an effective **scalar potential**, and the RPA trades the fermionic partition function for the Jellium model in the grand-canonical ensemble in favor of the effective bosonic, partition function.

The polarization function $\Pi^{\text{RPA}}(q)$ translates the effects of the Coulomb interaction within RPA where the limit $e \to 0$ implies that if we insert two static **infinitely heavy** unit point charges at positions $\vec{r}$ and $\vec{r}'$ in the noninteracting Fermi gas, then they will interact via the **instantaneous** bare Coulomb potential $\frac{\delta(\tau-\tau')}{|\vec{r}-\vec{r}'|}$ renormalized by the response of the Fermi sea to switching on e.

From (929), we consider the RPA sum of polarization bubbles as an effective interaction that we represent diagrammatically by a double wiggly line and interpret as a **screened** interaction between two particles:

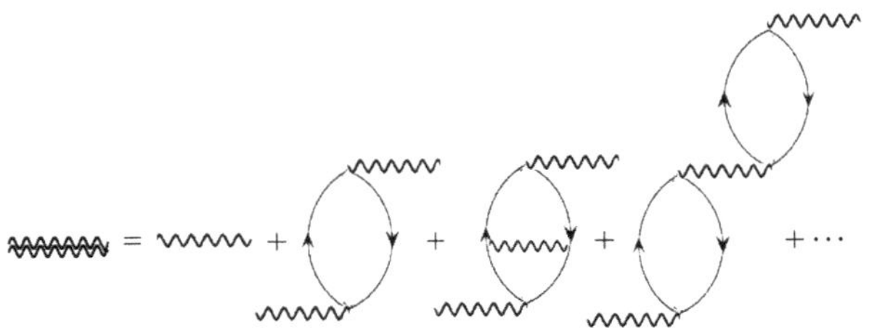
(940)

or

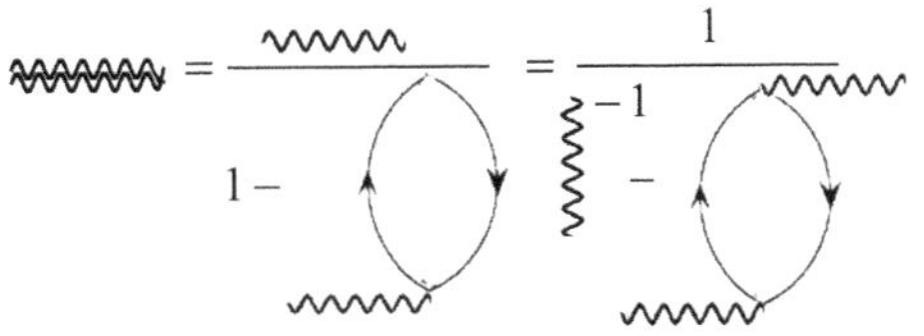
(941)

or

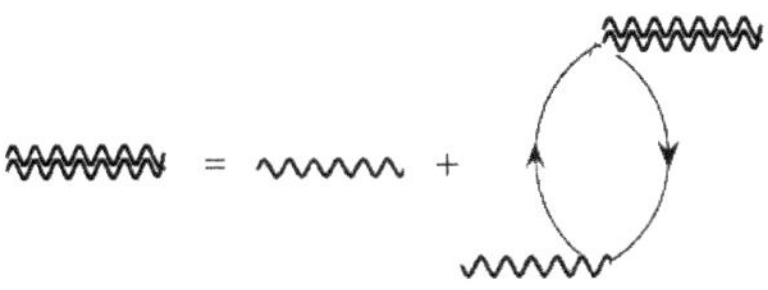
(942)

FIGURE 7.7 Diagrammatic representation of the effective interaction in the random phase approximation.

These are **polarization diagrams** because they show how the interaction causes the medium to become **virtually polarized** in all possible ways. The diagrams from the second and upward after the equal sign in (940), which have one interaction line entering and one leaving, are called **polarization** diagrams and show the possible ways how the interaction causes the medium to become **virtually** polarized. We observe that **random phase approximation** entails taking the noninteracting polarizability function for **polarization bubbles** in the diagrammatic expansion. Physically, it is obvious that two particles could interact via the direct interaction. That is to say, the interaction due to one particle induces a density fluctuation in the electron sea that, in turn, interacts with the second particle. Alternatively, they could interact via two intermediate density fluctuations, and so on.

7.2.1.4 Saddle-Point Approximation

We see that the exact effective action $S^{\text{eff}}[\phi]$ in (929) for the order parameter ϕ is expanded up to quadratic order within the RPA. **Because this expansion takes place in the argument of an exponential, this is obviously not second-order perturbation theory in the electric charge** e.

The terminology of the order parameter for the field ϕ is to indicate the treatment of the Coulomb interaction within a mean-field theory with the inclusion of Gaussian fluctuations around the **mean-field value** of the order parameter. The **stationary phase (saddle-point or mean-field) approximation or method of steepest descent** permits us to reorganize the perturbation expansion as a **loop expansion** about a broken-symmetry state.

The saddle-point equation can be obtained from equation (904):

$$\frac{\delta \ln Z}{\delta \phi_q} = \frac{N'}{Z}\int d[\phi]\int d[\hat{\psi}^\dagger]d[\hat{\psi}]\frac{\delta \tilde{S}}{\delta \phi_q}\exp\{-\tilde{S}[\hat{\psi}^\dagger,\hat{\psi},\phi]\} = \left\langle -[\mathbf{D}_{0q}]^{-1}\phi_{-q} + \frac{ie}{\sqrt{\beta V}}\rho_{-q}\right\rangle \tag{943}$$

For the saddle-point,

$$\frac{\delta \tilde{S}}{\delta \phi_q} = 0 \tag{944}$$

then

$$\langle \phi_{-q}\rangle = \frac{ie}{\sqrt{\beta V}}\mathbf{D}_{0q}\langle \rho_{-q}\rangle \tag{945}$$

If the mean field

$$\phi_{-q} = \sqrt{\beta V}\,\delta_{-q,0}\phi_0 \tag{946}$$

is chosen static and uniform with $\phi = \phi_0$,

$$\langle \delta_{-q,0}\phi_0\rangle = \frac{ie}{\beta V}\mathbf{D}_{0q}\langle \rho_{-q}\rangle \tag{947}$$

From here, it is obvious that ϕ_0 describes quantized charge density fluctuations (plasmons). It is instructive to note that ϕ_0 is purely imaginary and ensures that the mean-field action functional is real.

Equation (945) also suggests that

$$\mathbf{D}_q = \langle \phi_q\phi_{-q}\rangle = -\mathbf{D}_{0q}^2\frac{(ie)^2}{\beta V}\langle \rho_{-q}\rho_q\rangle \tag{948}$$

Examining the diagrammatic counterparts to equation (935) given in Figure 7.6 (e) and (g), in terms of the original electrons, $\mathbf{D}_q$ closely imitates the **density-density correlation function** $\langle \rho_q\rho_{-q}\rangle$ as seen from equation (948). The poles of the propagator $\mathbf{D}_q$ are interpreted as collective excitations.

7.2.1.5 Lindhard Function and Plasmon Oscillations

We examine the dynamic density-density propagator or polarization function again:

$$\Pi^{\text{RPA}}_{q,\omega_l} = \frac{2}{\beta V}\sum_{\vec{\kappa},n}\frac{1}{(i\omega_n - \xi_{\vec{\kappa}})(i\omega_n + i\omega_l - \xi_{\vec{\kappa}+\vec{q}})} \tag{949}$$

This can be represented diagrammatically as follows:

$$\Pi^{\text{RPA}}_{q,\omega_l} = \tag{950}$$

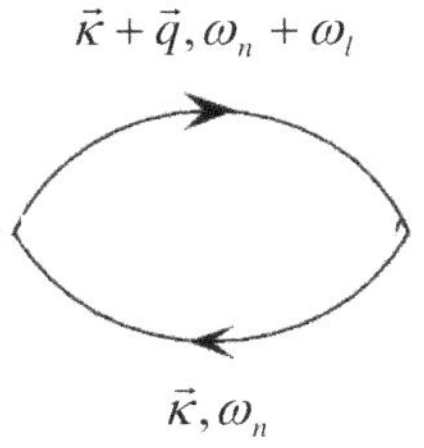

FIGURE 7.8 Polarization function.

The self-energy is responsible for renormalizing the propagator and the polarization loop is responsible for renormalizing the interaction (since two interaction lines can be attached to the vertices). Relation (949) is known as **the Lindhard function** and has information about the probability (and phases) of all possible virtual processes as seen earlier. It is also essential when we go beyond the noninteracting system because from the principles of quantum mechanics, all possible paths must be added—not just the **on-shell** (i.e., energy conserving) ones.

Let us perform the summation over fermionic Matsubara frequencies ω_n. Considering the Fermi-Dirac distribution function,

$$n_{\text{FD}}(\epsilon) = \frac{1}{\exp\{\beta \epsilon\} + 1} \tag{951}$$

This function has equidistant first-order poles at $\epsilon_n = i\omega_n$ with residues

$$\text{Res}\, n_{\text{FD}}(\epsilon)\big|_{i\omega_n} = -\frac{1}{\beta} \tag{952}$$

For given $\vec{\kappa}$, let $C_{\vec{\kappa}}$ be the path running antiparallel to the imaginary axis infinitesimally close to its left and parallel to the imaginary axis infinitesimally close to its right. This implies that it goes around the imaginary axis in a counterclockwise manner. From the residue theorem,

$$\Pi^{\text{RPA}}_{q,\omega_l} = -\frac{2}{V}\sum_{\vec{\kappa}}\int_{C_{\vec{\kappa}}}\frac{d\epsilon}{2\pi i}\frac{n_{\text{FD}}(\epsilon)}{(\epsilon - \xi_{\vec{\kappa}})(\epsilon + i\omega_l - \xi_{\vec{\kappa}+\vec{q}})} \tag{953}$$

If $\vec{q} \neq 0$, then the integrand (for given $\vec{\kappa}$) has two isolated first-order poles along the imaginary axis at

$$\epsilon_{\vec{\kappa}} = \xi_{\vec{\kappa}} \quad , \quad \epsilon_{\vec{\kappa}+\vec{q},\omega_l} = \xi_{\vec{\kappa}+\vec{q}} - i\omega_l \tag{954}$$

with residues

$$\frac{1}{2\pi i}\frac{n_{\text{FD}}(\xi_{\vec{\kappa}})}{\xi_{\vec{\kappa}} - \xi_{\vec{\kappa}+\vec{q}} + i\omega_l} \tag{955}$$

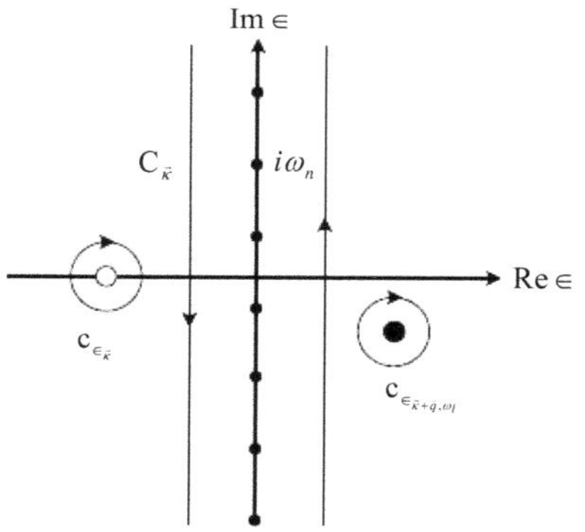

FIGURE 7.9 Poles of Euclidean polarization function on and off imaginary $\in$-axis in the representation of equation (953) arising from an arbitrarily chosen $\vec{\kappa}$ contribution. The hole- (particle-) like poles are off the imaginary axis and are denoted by an empty (filled) circle. Poles on the imaginary axis at the Matsubara frequencies ω_n are denoted by smaller filled circles. The quantities $C_{\vec{\kappa}}$, $c_{\in_{\vec{\kappa}}}$, and $c_{\in_{\vec{\kappa}+\vec{q},\omega l}}$ are closed integration paths.

and

$$\frac{1}{2\pi i}\frac{n_{\text{FD}}\left(\xi_{\vec{\kappa}+\vec{q}}-i\omega_l\right)}{\xi_{\vec{\kappa}+\vec{q}}-\xi_{\vec{\kappa}}-i\omega_l}=-\frac{1}{2\pi i}\frac{n_{\text{FD}}\left(\xi_{\vec{\kappa}+\vec{q}}\right)}{\xi_{\vec{\kappa}}-\xi_{\vec{\kappa}+\vec{q}}+i\omega_l} \tag{956}$$

The given two first-order poles combine into a second-order pole for $\omega_l = 0$ and $\vec{q}\to 0$. From the Cauchy's theorem, we deform the contour of integration into two small circles $c_{\in_{\vec{\kappa}}}$ and $c_{\in_{\vec{\kappa}+\vec{q},\omega l}}$ encircling $\in_{\vec{\kappa}}$ and $\in_{\vec{\kappa}+\vec{q},\omega_l}$, respectively, in a clockwise manner as in Figure 7.9.

$$C_{\vec{\kappa}}\to c_{\in_{\vec{\kappa}}}\cup c_{\in_{\vec{\kappa}+\vec{q},\omega l}} \tag{957}$$

A second application of the residue theorem yields

$$\Pi^{\text{RPA}}_{q,\omega_l}=(-1)^2\frac{2}{V}\sum_{\vec{\kappa}}\frac{n_{\text{FD}}\left(\xi_{\vec{\kappa}}\right)-n_{\text{FD}}\left(\xi_{\vec{\kappa}+\vec{q}}\right)}{\xi_{\vec{\kappa}}-\xi_{\vec{\kappa}+\vec{q}}+i\omega_l}=\frac{2}{V}\sum_{\vec{\kappa}}\frac{n_{\text{FD}}\left(\xi_{\vec{\kappa}-\frac{\vec{q}}{2}}\right)-n_{\text{FD}}\left(\xi_{\vec{\kappa}+\frac{\vec{q}}{2}}\right)}{\xi_{\vec{\kappa}-\frac{\vec{q}}{2}}-\xi_{\vec{\kappa}+\frac{\vec{q}}{2}}+i\omega_l} \tag{958}$$

It is instructive to note that the change of variable is valid in the thermodynamic limit $V\to\infty$. Considering equation (958) and the fact that, at zero temperature, the Fermi-Dirac distribution becomes the Heaviside step function,

$$\lim_{\beta\to\infty} n_{\text{FD}}(\xi)=\theta(-\xi)=\begin{cases}0 & ,\ \xi>0\\ 1 & ,\ \text{otherwise}\end{cases} \tag{959}$$

This implies that all single-particle states with an energy smaller than the chemical potential (Fermi energy) are occupied in the Fermi sea at zero temperature.

Also,

$$\lim_{\beta\to\infty}\left[n_{\text{FD}}\left(\xi_{\vec{\kappa}-\frac{\vec{q}}{2}}\right)-n_{\text{FD}}\left(\xi_{\vec{\kappa}+\frac{\vec{q}}{2}}\right)\right]\neq 0 \tag{960}$$

which is equivalent to

$$\xi_{\vec{\kappa}-\frac{\vec{q}}{2}}\xi_{\vec{\kappa}+\frac{\vec{q}}{2}} < 0 \tag{961}$$

This implies that for the difference of the Fermi-Dirac distributions of two single-particle energies to be nonvanishing at zero temperature, one of the two single-particle states must be above the Fermi energy, while the other single-particle state must be below it. The difference in the Fermi-Dirac distributions selects an electron-hole pair that we find represented by the polarization bubble diagram in Figure 7.8.

At low temperatures, the polarization function (958) is thus controlled by the geometrical properties of the Fermi sea, which is the unperturbed ground state of the Fermi gas. We will need some characteristic scales of the Fermi sea. The Fermi wave vector κ_F is defined by filling up all available single-particle energy levels:

$$\frac{N}{V} = \frac{1}{V}\sum_{\vec{\kappa}\sigma}\theta(-\xi_{\vec{\kappa}}) = \frac{2}{V}\sum_{\vec{\kappa}\sigma}\theta\left(\frac{\vec{\kappa}_F^2}{2m} - \frac{\vec{\kappa}^2}{2m}\right) = 2\frac{1}{(2\pi)^3}\frac{4\pi}{3}\kappa_F^3 = \frac{\kappa_F^3}{3\pi^2} \tag{962}$$

We have the Fermi velocity

$$v_F = \frac{\kappa_F}{m} \approx \left(\frac{N}{V}\right)^{\frac{1}{3}} \tag{963}$$

and Fermi energy

$$\epsilon_F = \frac{\vec{\kappa}_F^2}{2m} \approx \left(\frac{N}{V}\right)^{\frac{2}{3}} \tag{964}$$

If we examine relation (958) again and consider both the zero-temperature and infinite-volume limits:

$$\frac{1}{V}\sum_{\vec{\kappa}} \to = \int\frac{d\vec{\kappa}}{(2\pi)^3} \quad , \quad \frac{1}{V}\sum_{\vec{q}}\frac{1}{\beta}\sum_{\epsilon} \to = \int\frac{d\vec{\kappa}}{(2\pi)^3}\int\frac{d\epsilon}{2\pi} \tag{965}$$

then

$$\Pi^{RPA}_{q,\omega_l} = 2\int\frac{d\vec{\kappa}}{(2\pi)^3}\frac{n_{FD}(\xi_{\vec{\kappa}}) - n_{FD}(\xi_{\vec{\kappa}+\vec{q}})}{\xi_{\vec{\kappa}} - \xi_{\vec{\kappa}+\vec{q}} + i\omega_l} \tag{966}$$

It is worth noting that similar to defining the self-energy, we can define the exact irreducible polarization bubble, Π^{RPA}_{q,ω_l} as the sum over all bubble diagrams that cannot be cut in two by splitting a single interaction line:

$\Pi^{RPA}_{q,\omega_l} =$  (967)

or

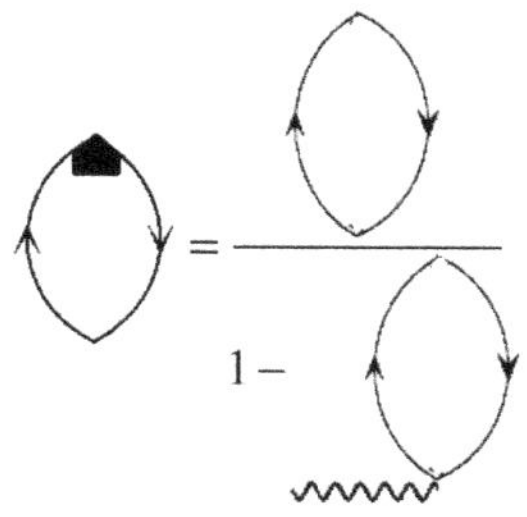

(968)

The RPA resummation then becomes exact (the generic expression for the propagator in terms of self-energy) yielding the **effective (screened) potential** or

$$U_{\text{eff}}^{\text{RPA}}(\vec{q},\omega_l) = U(q) + U(q)\Pi(\vec{q},\omega_l)U(q) + U(q)\Pi(\vec{q},\omega_l)U(q)\Pi(\vec{q},\omega_l)U(q) + \cdots =$$
$$= U(q)\left[1 + \Pi(\vec{q},\omega_l)U(q) + \Pi(\vec{q},\omega_l)U(q)\Pi(\vec{q},\omega_l)U(q) + \cdots\right] \quad (969)$$

This geometric series can sum as follows

$$U_{\text{eff}}^{\text{RPA}}(\vec{q},\omega_l) = \frac{U(q)}{1 - U(q)\Pi(\vec{q},\omega_l)} \quad (970)$$

This expression (970) shows the **effective interaction that, in general, is dependent on frequency** ω_l, **corresponding to the account of retardation effects due to characteristic time of electron response to instantaneous Coulomb interaction.**

Using equation (966), let us determine the analytic continuation $i\omega_l \to \omega + i\delta$ where

$$\Pi_{q,\omega}^{\text{RPA}} = 2\int \frac{d\vec{\kappa}}{(2\pi)^3} \frac{n_{\text{FD}}(\xi_{\vec{\kappa}}) - n_{\text{FD}}(\xi_{\vec{\kappa}+\vec{q}})}{\xi_{\vec{\kappa}} - \xi_{\vec{\kappa}+\vec{q}} + \omega + i\delta} \quad (971)$$

If a symmetric change of variables $\vec{\kappa} \equiv -\vec{\kappa}' - \vec{q}$ and $\omega + i\delta \to -\omega - i\delta$ is introduced in the second integrand of relation (971):

$$\frac{n_{\text{FD}}(\xi_{\vec{\kappa}+\vec{q}})}{\xi_{\vec{\kappa}} - \xi_{\vec{\kappa}+\vec{q}} + \omega + i\delta} = \frac{n_{\text{FD}}(\xi_{-\vec{\kappa}'-\vec{q}+\vec{q}})}{\xi_{-\vec{\kappa}'-\vec{q}} - \xi_{-\vec{\kappa}'-\vec{q}+\vec{q}} - \omega - i\delta} = \frac{n_{\text{FD}}(\xi_{-\vec{\kappa}'})}{\xi_{-\vec{\kappa}'-\vec{q}} - \xi_{-\vec{\kappa}'} - \omega - i\delta} \quad (972)$$

Then, changing the primes into the unprimed and taking the inversion of the momentum space, we obtain the **dynamic polarization function**

$$\Pi_{q,\omega}^{\text{RPA}} = 4\int \frac{d\vec{\kappa}}{(2\pi)^3} \frac{n_{\text{FD}}(\xi_{\vec{\kappa}})}{\xi_{\vec{\kappa}} - \xi_{\vec{\kappa}+\vec{q}} + \omega + i\delta} \quad (973)$$

This function shows that relevant excitations originating from it are particle-hole excitations. One electron with momentum $\vec{\kappa}$ is removed from the ground state of a completely filled Fermi sea and is inserted again outside the Fermi sea in some state with momentum $\vec{\kappa} + \vec{q}$.

7.2.1.6 Particle-Hole Pair Excitation

The excited states of the Fermi gas consist of many-body states. Here, one of the electrons is excited to a higher single-particle energy state. Because electrons are conserved, excited electrons emanate from the Fermi sea, leaving a hole in its initial position. Relative to the ground state, such an electron and hole each carry positive energy because each corresponds to a missing electron below the Fermi energy $\in_F$.

Examining the expression of $\Pi_{q,\omega}^{RPA}$, we observe that the step functions are written as the particle and hole distribution functions

$$n_{\text{particle}} = n_{\text{FD}}\left(\xi_{\vec{\kappa}}\right) = \theta\left(-\xi_{\vec{\kappa}}\right) \quad , \quad n_{\text{hole}} = 1 - n_{\text{FD}}\left(\xi_{\vec{\kappa}}\right) = \theta\left(\xi_{\vec{\kappa}}\right) \tag{974}$$

The first equation imitates the propagator of a particle-hole pair, with $\left|\vec{\kappa}+\vec{q}\right| > \kappa_F$. The $n_{\text{FD}}\left(\xi_{\vec{\kappa}}\right)$ shows the excitation of a hole with momentum $\vec{\kappa}$. This implies that the excitation energy $\in\left(\vec{\kappa}\right) = \xi_{\vec{\kappa}+\vec{q}} - \xi_{\vec{\kappa}}$ is positive and that the plasmon becomes strongly damped (**Landau damping**). This no longer is a well-defined excitation of the system. The second equation has the momenta of the particle and hole reversed. It imitates the first equation after a time reversal operation. This implies that the causal Green's function (a two-particle causal Green's function as well as the single-particle Green's function) is a superposition of the advanced and retarded Green's functions. For $\left(1 - n_{\text{FD}}\left(\xi_{\vec{\kappa}}\right)\right)$, we have an excitation of a particle with momentum $\vec{\kappa}+\vec{q}$ and a particle transfer to a higher energy level that creates a particle-hole pair. Therefore, we are involved in the formation of all possible particle-hole pairs that carry momentum $\vec{q}$ and that satisfy energy conservation. This implies that we sort for the spectrum of the real excitations of the Fermi gas (i.e., a particle being transferred to a higher energy that creates a particle-hole pair).

So, we see that the dynamic polarization function $\Pi_{q,\omega}^{RPA}$ also contains effects of electron-electron interaction and includes information not only about the renormalization of potentials but about the excitation spectrum of the metal as well.

7.2.1.7 Lindhard Formula

Dynamic Polarization

Consider again the **dynamic polarization function**

$$\Pi_{q,\omega}^{RPA} = 4\int \frac{d\vec{\kappa}}{(2\pi)^3} \frac{n_{\text{FD}}\left(\xi_{\vec{\kappa}}\right)}{\xi_{\vec{\kappa}} - \xi_{\vec{\kappa}+\vec{q}} + \omega + i\delta} \tag{975}$$

From

$$\frac{1}{s+i\delta} = \frac{\text{P}}{s} - i\pi\delta(s) \tag{976}$$

The real part of $\Pi_{q,\omega}^{RPA}$ is then:

$$\Pi_{q,\omega}^{0} = \text{Re}\,\Pi_{q,\omega}^{RPA} = 4\int \frac{d\vec{\kappa}}{(2\pi)^3} \frac{n_{\text{FD}}\left(\xi_{\vec{\kappa}}\right)}{\xi_{\vec{\kappa}} - \xi_{\vec{\kappa}+\vec{q}} + \omega} \tag{977}$$

This is known as the **Lindhard formula** (the density-density response function or polarization operator) [33]. This is the **irreducible** part of the function $\Pi_{q,\omega}^{RPA}$ and implies that the part cannot be split into two disconnected pieces by cutting a single Coulomb line $U\left(\vec{q}\right)$. The Lindhard function has information about the probability (and phases) of all possible virtual processes. It is also essential when we go beyond the noninteracting system—as we know from the principles of quantum mechanics, all possible paths must be added and not just the **on-shell** (i.e., energy conserving) ones.

Considering

$$n_{\text{FD}}\left(\xi_{\vec{\kappa}}\right) = \theta\left(\kappa_F - \left|\vec{\kappa}\right|\right) \tag{978}$$

from (977), we then have

$$\Pi^{0}_{q,\omega}=4\int\frac{d\vec{\kappa}}{(2\pi)^{3}}\frac{n_{\mathrm{FD}}(\xi_{\vec{\kappa}})}{\xi_{\vec{\kappa}}-\xi_{\vec{\kappa}+\vec{q}}+\omega} \tag{979}$$

The excitation energy of a particle-hole pair is given by

$$\xi_{\vec{\kappa}+\vec{q}}-\xi_{\vec{\kappa}}=\frac{q^{2}}{2m}+\frac{\kappa q}{m}\cos\theta \tag{980}$$

where θ is the angle between $\vec{\kappa}$ and $\vec{q}$. So,

$$\Pi^{0}_{q,\omega}=-\frac{1}{\pi^{2}}\int_{0}^{\kappa_{\mathrm{F}}}\kappa^{2}d\kappa\int_{-1}^{1}\frac{d(\cos\theta)}{\omega-\frac{q^{2}}{2m}-\frac{\kappa q}{m}\cos\theta} \tag{981}$$

and introducing the following dimensionless variables

$$\kappa'=\frac{|\vec{\kappa}|}{\kappa_{\mathrm{F}}}\quad,\quad q'=\frac{|\vec{q}|}{2\kappa_{\mathrm{F}}}\quad,\quad \varpi=\frac{\omega}{v_{\mathrm{F}}|\vec{q}|} \tag{982}$$

we have

$$\Pi^{0}_{q,\omega}=-\frac{m\kappa_{\mathrm{F}}^{2}}{\pi^{2}|\vec{q}|}\int_{0}^{1}\kappa'^{2}d\kappa'\int_{0}^{1}\frac{d(\cos\theta)}{\varpi-\kappa'\cos\theta-q'}=-\frac{N_{\mathrm{F}}}{q'}\int_{0}^{1}\kappa'^{2}d\kappa'\ln\left[\frac{\varpi+\kappa'-q'}{\varpi-\kappa'-q'}\right]=\frac{N_{\mathrm{F}}}{q'}zh(z) \tag{983}$$

So,

$$\Pi^{0}_{q,\omega}=\frac{N_{\mathrm{F}}}{q'}zh(z) \tag{984}$$

where

$$z=\varpi-q' \tag{985}$$

and

$$h(z)=1+\frac{1-z^{2}}{2z}\ln\left|\frac{1+z}{1-z}\right| \tag{986}$$

is the **unrenormalized Lindhard function** [33]. This function has information on the probability and phases of all possible virtual processes. The quantity $N_{\mathrm{F}}=\frac{m\kappa_{\mathrm{F}}}{2\pi^{2}}$ is the noninteracting density of states per spin at the Fermi surface $\xi_{\vec{\kappa}}=0$. From relation (927), it can be inferred that the effective screening length increases with the momentum transfer $|\vec{q}|$. It is more difficult to make electrons screen out potentials on shorter wavelengths.

Static Screening

We examine relation (927) for convenience and first for the static screening (three-dimensional case). This implies that we examine the effective interaction in the limit $\omega=0$ [33]. The static limit $\omega\to 0$ of the **polarization function** yields the following **unrenormalized Lindhard function** $h(z)$ multiplied by the factor $-N_F$:

$$\Pi^0_{q,\omega=0} = -N_F h(q') \tag{987}$$

The presence of logarithmic singularities when $|\vec{q}| = 2\kappa_F$ are responsible for the so-called Friedel oscillations in the context of the Jellium model or the Ruderman-Kittel-Kasuya-Yosida oscillations of the static spin susceptibility induced by a magnetic impurity in a free electron gas.

Consider again

$$\Pi^{RPA}_{q,\omega} = 2\int \frac{d\vec{\kappa}}{(2\pi)^3} \frac{n_{FD}(\xi_{\vec{\kappa}}) - n_{FD}(\xi_{\vec{\kappa}+\vec{q}})}{\xi_{\vec{\kappa}} - \xi_{\vec{\kappa}+\vec{q}} + \omega + i\delta} \tag{988}$$

and the symmetric change of variables $\vec{\kappa} \equiv -\vec{\kappa}' - \vec{q}$ and $\omega + i\delta \to -\omega - i\delta$ in the second integrand:

$$\frac{n_{FD}(\xi_{\vec{\kappa}+\vec{q}})}{\xi_{\vec{\kappa}} - \xi_{\vec{\kappa}+\vec{q}} + \omega + i\delta} = \frac{n_{FD}(\xi_{-\vec{\kappa}'-\vec{q}+\vec{q}})}{\xi_{-\vec{\kappa}'-\vec{q}} - \xi_{-\vec{\kappa}'-\vec{q}+\vec{q}} - \omega - i\delta} = \frac{n_{FD}(\xi_{-\vec{\kappa}'})}{\xi_{-\vec{\kappa}'-\vec{q}} - \xi_{-\vec{\kappa}'} - \omega - i\delta} \tag{989}$$

from (976), we then obtain the imaginary part of $\Pi^{RPA}_{q,\omega}$:

$$\mathrm{Im}\,\Pi^{RPA}_{q,\omega} = 2\pi \int \frac{d\vec{\kappa}}{(2\pi)^3} \left\{ (1 - n_{FD}(\xi_{\vec{\kappa}})) n_{FD}(\xi_{\vec{\kappa}+\vec{q}}) \delta(\omega - \xi_{\vec{\kappa}+\vec{q}} + \xi_{\vec{\kappa}}) + n_{FD}(\xi_{\vec{\kappa}}) (`1 - n_{FD}(\xi_{\vec{\kappa}+\vec{q}})) \delta(\omega - \xi_{\vec{\kappa}+\vec{q}} + \xi_{\vec{\kappa}}) \right\} \tag{990}$$

or

$$A_{q,\omega} \equiv \mathrm{Im}\,\Pi^{RPA}_{q,\omega} = 2\pi \int \frac{d\vec{\kappa}}{(2\pi)^3} \left\{ \left(n_{FD}(\xi_{\vec{\kappa}}) + n_{FD}(\xi_{\vec{\kappa}+\vec{q}}) - 2 n_{FD}(\xi_{\vec{\kappa}}) n_{FD}(\xi_{\vec{\kappa}+\vec{q}}) \right) \right\} \delta(\omega - \xi(\vec{\kappa}+\vec{q}) + \xi(\vec{\kappa})) \tag{991}$$

We do the change of variables

$$\vec{\kappa} = -\vec{\kappa}' - \vec{q} \quad , \quad \vec{\kappa} + \vec{q} = -\vec{\kappa}' \tag{992}$$

then

$$A_{q,\omega} = 2\pi \int \frac{d\vec{\kappa}}{(2\pi)^3} n_{FD}(\xi_{\vec{\kappa}}) (1 - n_{FD}(\xi_{\vec{\kappa}})) \left\{ \delta(\omega - \xi_{\vec{\kappa}+\vec{q}} + \xi_{\vec{\kappa}}) + \delta(\omega + \xi_{\vec{\kappa}+\vec{q}} - \xi_{\vec{\kappa}}) \right\} \tag{993}$$

The excitation spectrum is visible from $A_{q,\omega}$ and relates to the absorption of energy by the electrons subject to a time-dependent external perturbation. From this, we find that $A_{q,\omega}$ is proportional to the Dirac delta functions that describe the energy conservation law of the system. If there is a particle-hole excitation with momentum $\vec{q}$ in the noninteracting Fermi gas, then $A_{q,\omega}$ is nonzero for a wide range of energies. This implies that the excitation could have a wide range of energies. We call this an incoherent excitation. For the three-dimensional free Fermi gas, all particle-hole excitations are incoherent, and this strongly contrasts with the form of the spectral function for the free Fermi gas

$$A_{q,\omega} = \delta(\omega - \xi_q) \tag{994}$$

which shows single-particle excitations are coherent. Apart from the fact that the Dirac delta functions enforce conservation of energy, the function $A_{q,\omega}$ also contains information about real processes.

We note that $A_{q,\omega}$ corresponds to Fermi golden rule, which is known from time-dependent perturbation theory. This implies the transition rate from the ground state to an excited state of energy ω and momentum $\vec{q}$.

7.2.1.8 Spectral Function

Let us reexamine relation (993):

$$A_{q,\omega} = 2\pi \int \frac{d\vec{\kappa}}{(2\pi)^3} n_{FD}(\xi_{\vec{\kappa}})(1 - n_{FD}(\xi_{\vec{\kappa}}))\{\delta(\omega - E_{\vec{\kappa}+\vec{q}}) + \delta(\omega + E_{\vec{\kappa}+\vec{q}})\} \tag{995}$$

where,

$$E_{\vec{\kappa}+\vec{q}} = \xi_{\vec{\kappa}+\vec{q}} - \xi_{\vec{\kappa}} = \frac{q^2}{2m} + \frac{|\vec{\kappa}||\vec{q}|\cos\theta}{m} \tag{996}$$

So,

$$\delta(\omega - E_{\vec{\kappa}+\vec{q}}) = \delta\left(\omega - \frac{q^2}{2m} - \frac{\vec{\kappa}\vec{q}}{m}\right) = m\delta\left(m\omega - \vec{\kappa}\vec{q} - \frac{q^2}{2}\right) = \frac{m}{\kappa_F^2}\delta\left(\frac{m\omega}{\kappa_F^2} - \frac{\vec{\kappa}}{\kappa_F}\frac{\vec{q}}{\kappa_F} - \frac{q^2}{2\kappa_F^2}\right) \tag{997}$$

and also is represented as follows

$$\delta(\omega - E_{\vec{\kappa}+\vec{q}}) = \frac{2m}{\kappa_F^2}\delta\left(\frac{2m\omega}{\kappa_F^2} - \frac{2\vec{\kappa}}{\kappa_F}\frac{\vec{q}}{2\kappa_F} - \frac{2q^2}{4\kappa_F^2}\right) = \frac{m}{\kappa_F^2}\delta(\nu - \vec{\kappa}'\vec{q}' - q'^2) \tag{998}$$

where

$$\kappa' = \frac{|\vec{\kappa}|}{\kappa_F} \quad , \quad q' = \frac{|\vec{q}|}{2\kappa_F} \quad , \quad \nu = \frac{m\omega}{\kappa_F^2} \tag{999}$$

Hence,

$$A_{q,\omega} = \frac{2\pi m\kappa_F^3}{\kappa_F^2}\int \frac{d\vec{\kappa}'}{(2\pi)^3}\theta(1-\kappa')\theta(|\vec{\kappa}' + \vec{q}'| - 1)\{\delta(\nu - \vec{\kappa}'\vec{q}' - q'^2) + \delta(\nu + \vec{\kappa}'\vec{q}' + q'^2)\} \tag{1000}$$

or

$$A_{q,\omega} = \frac{N_F}{2}\int \kappa'^2 d\kappa' \int_0^{2\pi} d\phi \sin\phi\, \theta(1-\kappa')\theta(|\vec{\kappa}' + \vec{q}'| - 1)\{\delta(\nu - \kappa' q'\cos\phi - q'^2) + \delta(\nu + \kappa' q'\cos\phi + q'^2)\} \tag{1001}$$

or

$$A_{q,\omega} = N_F\int \kappa'^2 d\kappa' \int_0^1 dx\, \theta(1-\kappa')\theta(|\vec{\kappa}' + \vec{q}'| - 1)\{\delta(\nu - \kappa' q' x - q'^2) + \delta(\nu + \kappa' q' x + q'^2)\} \tag{1002}$$

From this function, it should be noted that

$$1 - \kappa' > 0 \quad , \quad |\vec{\kappa}' + \vec{q}'| - 1 > 0 \tag{1003}$$

and implies that

$$\kappa' < 1 \quad , \quad |\vec{\kappa}' + \vec{q}'| = |\vec{\kappa}'|^2 + |\vec{q}'|^2 + 2\vec{\kappa}'\vec{q}' > 1 \tag{1004}$$

So,

$$\nu = \frac{m\omega}{\kappa_F^2} = \kappa' q' + q'^2 = \frac{|\vec{\kappa}|}{\kappa_F}\frac{|\vec{q}|}{2\kappa_F} + \left(\frac{|\vec{q}|}{2\kappa_F}\right)^2 \tag{1005}$$

or

$$2\frac{\omega}{v_F|\vec{q}|} \equiv 2\varpi = \frac{|\vec{\kappa}|}{\kappa_F} + \frac{|\vec{q}|}{2\kappa_F} \quad , \quad \varpi = \frac{\omega}{v_F|\vec{q}|} \tag{1006}$$

Therefore, $|\vec{\kappa}'|^2 > 1 - 2\varpi$ and for $z = \varpi - q'$, then for $q' \leq 1$, the spectral function $A_{q,\omega}$ is linear at low energy and corresponds to an arc of parabola at higher energy:

$$A_{q,\omega} = \pi N_F \begin{cases} \varpi & , \quad 0 \leq \omega \leq |\omega_{\min}(\vec{q})| \\ \dfrac{1-z^2}{4q'} & , \quad |\omega_{\min}(\vec{q})| \leq \omega \leq \omega_{\max}(\vec{q}) \end{cases} \tag{1007}$$

Here,

$$\omega_{\min}(\vec{q}) = \frac{q^2}{2m} - v_F|\vec{q}| \quad , \quad \omega_{\max}(\vec{q}) = \frac{q^2}{2m} + v_F|\vec{q}| \tag{1008}$$

We see that the particle-hole continuum extends from $\omega = 0$ up to $\omega = \omega_{\max}(\vec{q})$. For $q' \geq 1$, the linear part is no longer present, and the spectral function $A_{q,\omega}$ reduces to an arc of parabola at higher energy:

$$A_{q,\omega} = \frac{\pi N_F}{4q'}(1 - z^2) \quad , \quad \omega_{\min}(\vec{q}) \leq \omega \leq \omega_{\max}(\vec{q}) \tag{1009}$$

In this case, there is no excitation at low energy, and the particle-hole continuum extends from $\omega_{\min}(\vec{q})$ up to $\omega_{\max}(\vec{q})$ (Figure 7.10).

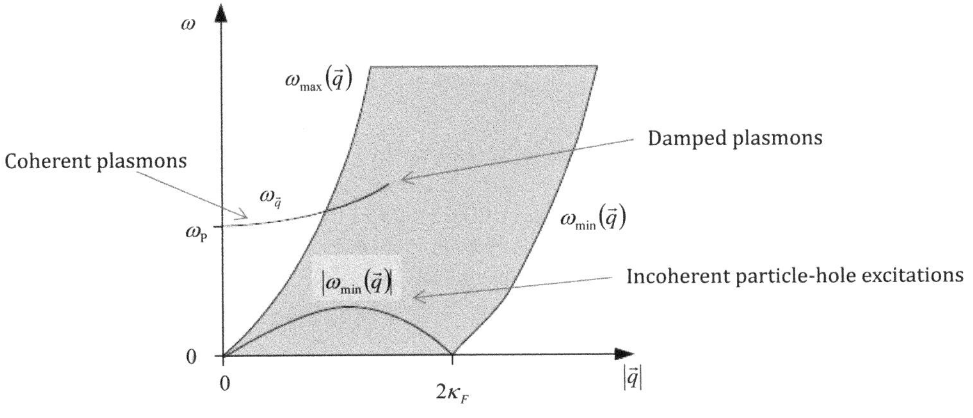

FIGURE 7.10 Plasmon mode dispersion $\omega_{\vec{q}}$ in the electron liquid. The shaded area shows the particle-hole excitation continuum. The coherent, weakly dispersive plasmon mode at small momenta and large energy becomes damped as it enters the incoherent particle-hole continuum.

In the region of large ω and small q, plasmon excitations are stable; for larger q, the excitation crosses the boundary of the incoherent particle-hole background. Here, the plasmon is a bound state of a particle, and a hole decays into the given background. Thus, the plasmons acquire a finite lifetime.

7.2.1.9 Plasma Oscillations And Landau Damping

Due to the Coulomb interaction, there arises a collective excitation known as **plasma oscillation**. For a long-ranged interaction such as the Coulomb interaction, this oscillation appears at a finite frequency for small momenta q. This may be observed when energetic electrons scatter from a metallic crystal. If an energetic electron strikes a metal, it may excite a plasmon (whose energy may be about 10eV) and the scattered electron would then be downshifted in energy by an equal amount relative to the incident electron. This will be derived here via the dynamic polarization function $\Pi_{q,\omega}^{\mathrm{RPA}}$ assuming $q << \kappa_{\mathrm{F}}$. So, from

$$\Pi_{q,\omega}^{\mathrm{RPA}} = 2\int \frac{d\vec{\kappa}}{(2\pi)^3} \frac{n_{\mathrm{FD}}\left(\xi_{\vec{\kappa}}\right) - n_{\mathrm{FD}}\left(\xi_{\vec{\kappa}+\vec{q}}\right)}{\xi_{\vec{\kappa}} - \xi_{\vec{\kappa}+\vec{q}} + \omega + i\delta} \tag{1010}$$

then

$$\mathrm{A}_{q,\omega} = \mathrm{Im}\,\Pi_{q,\omega}^{\mathrm{RPA}} = 2\pi \int \frac{d\vec{\kappa}}{(2\pi)^3} \left(n_{\mathrm{FD}}\left(\xi_{\vec{\kappa}}\right) - n_{\mathrm{FD}}\left(\xi_{\vec{\kappa}+\vec{q}}\right)\right) \delta\left(\xi_{\vec{\kappa}} - \xi_{\vec{\kappa}+\vec{q}} + \omega\right) \tag{1011}$$

Considering

$$n_{\mathrm{FD}}\left(\xi_{\vec{\kappa}+\vec{q}}\right) - n_{\mathrm{FD}}\left(\xi_{\vec{\kappa}}\right) = n_{\mathrm{FD}}\left(\xi_{\vec{\kappa}}\right) - n_{\mathrm{FD}}\left(\xi_{\vec{\kappa}}\right) + \frac{\partial n_{\mathrm{FD}}\left(\xi_{\vec{\kappa}}\right)}{\partial \xi_{\vec{\kappa}}} \frac{\partial \xi_{\vec{\kappa}}}{\partial \vec{\kappa}} \vec{q} = \vec{q}\vec{v} \frac{\partial n_{\mathrm{FD}}\left(\xi_{\vec{\kappa}}\right)}{\partial \xi_{\vec{\kappa}}} \equiv \vec{q}\nabla_{\vec{\kappa}} n_{\mathrm{FD}}\left(\xi_{\vec{\kappa}}\right) \tag{1012}$$

Then the spectral function

$$\mathrm{A}_{q,\omega} = \mathrm{Im}\,\Pi_{q,\omega}^{\mathrm{RPA}} = 2\pi \int \frac{d\vec{\kappa}}{(2\pi)^3} \delta\left(\xi_{\vec{\kappa}} - \xi_{\vec{\kappa}+\vec{q}} + \omega\right) \vec{q}\nabla_{\vec{\kappa}} n_{\mathrm{FD}}\left(\xi_{\vec{\kappa}}\right) \tag{1013}$$

permits us to find the spectrum of the density fluctuations of the particle-hole pair excitations. This imitates the particle-hole pair excitations of the noninteracting electron gas.

The contribution to the collective plasmon mode at the frequency $\omega_{\vec{q}}$ may be obtained via the **generalized dielectric function**

$$\varepsilon_{\vec{q},\omega}^{\mathrm{RPA}} \equiv 1 - U\left(\vec{q}\right)\Pi_{q,\omega}^{\mathrm{RPA}} \equiv 1 - \mathrm{F}_{\omega_l,q}^{\mathrm{RPA}} \tag{1014}$$

Note that in the interacting medium, the interaction among the fermions now become frequency dependent, indicating that the interactions among the particles are therefore **retarded**. The dielectric properties of the medium arise due to the polarization of the medium by the field. The equation (940) is the sum of the diagrams representing the polarization of the electron gas by the field of one of the electrons in the gas itself. At zero temperature, (1014) is the excitation in the Jellium model with momentum q and real-time frequency ω_q that shows up as a zero of the analytic continuation of the dielectric function to the negative imaginary axis:

$$\lim_{\omega \to -i\omega_q + \delta} \varepsilon_{q,\omega} = 0 \tag{1015}$$

The infinitesimal frequency δ ensures the perturbation on the Jellium model to be switched slowly on adiabatically. The physical interpretation of (1015) is that a harmonic perturbation with arbitrary **small**

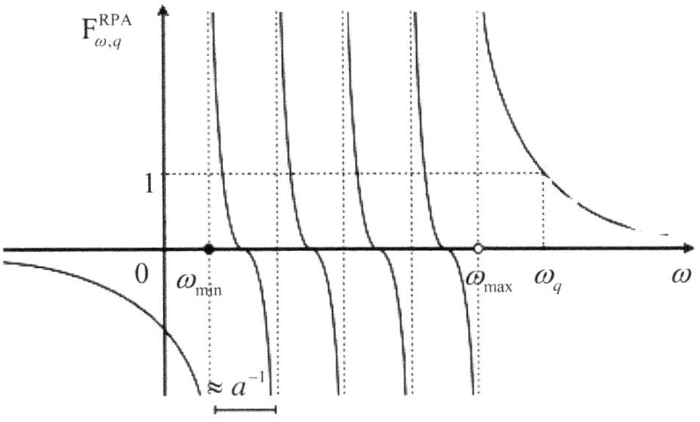

FIGURE 7.11 Plot of $F^{RPA}_{\omega,q}$ versus real-time frequency ω at fixed momentum $q > 2\kappa_F$.

amplitude induces a nonvanishing response of the Fermi sea in the form of a nonvanishing renormalization of the Coulomb potential. So, for any $q \neq 0$, a pole of $\mathbf{D}_q$ implies a zero of ε_q for any $q \neq 0$. The harmonic perturbation may arise by introducing a charge fluctuation in the electron gas that varies periodically in space and time with wave vector q and real-time frequency ω_q, respectively.

The Jellium model supports free oscillation modes with the dispersion ω_q because these oscillations may not be forced by an external probe to the electronic system. The excitation spectrum can be obtained within the RPA via the following equation:

$$\lim_{\omega \to -i\omega_q + \delta} \varepsilon^{RPA}_{q,\omega} = 1 - F^{RPA}_{\omega,q} = 0 \tag{1016}$$

We observe from Figure 7.11 that the polarization function decays as $\frac{1}{\omega^2}$ for $|\omega| >> |\omega_{min}|, |\omega_{max}|$. This may also be proven via the following formula:

$$\Pi^0_{q,\omega} = \frac{2}{\omega} \int \frac{d\vec{\kappa}}{(2\pi)^3} \frac{n_{FD}(\xi_{\vec{\kappa}}) - n_{FD}(\xi_{\vec{\kappa}+\vec{q}})}{1 + \frac{\xi_{\vec{\kappa}} - \xi_{\vec{\kappa}+\vec{q}}}{\omega}} \tag{1017}$$

or

$$\Pi^0_{q,\omega} = \frac{2}{\omega} \int \frac{d\vec{\kappa}}{(2\pi)^3} \left(n_{FD}(\xi_{\vec{\kappa}}) - n_{FD}(\xi_{\vec{\kappa}+\vec{q}})\right) - \frac{2}{\omega} \int \frac{d\vec{\kappa}}{(2\pi)^3} \left(n_{FD}(\xi_{\vec{\kappa}}) - n_{FD}(\xi_{\vec{\kappa}+\vec{q}})\right) \frac{\xi_{\vec{\kappa}} - \xi_{\vec{\kappa}+\vec{q}}}{\omega} \tag{1018}$$

But

$$\xi_{\vec{\kappa}+\vec{q}} - \xi_{\vec{\kappa}} = \xi_{\vec{\kappa}} - \xi_{\vec{\kappa}} + \frac{\partial \xi_{\vec{\kappa}}}{\partial \vec{\kappa}} \vec{q} = \vec{v}\vec{q} \tag{1019}$$

where $\vec{v} = \frac{\partial \xi_{\vec{\kappa}}}{\partial \vec{\kappa}}$ is the group velocity, then

$$\Pi^0_{q,\omega} = -\frac{2}{\omega} \int \frac{d\vec{\kappa}}{(2\pi)^3} \vec{q}\vec{v} \frac{\partial n_{FD}(\xi_{\vec{\kappa}})}{\partial \xi_{\vec{\kappa}}} - \frac{2}{\omega^2} \int \frac{d\vec{\kappa}}{(2\pi)^3} (\vec{q}\vec{v})^2 \frac{\partial n_{FD}(\xi_{\vec{\kappa}})}{\partial \xi_{\vec{\kappa}}} \tag{1020}$$

or

$$\Pi^0_{q,\omega} = \frac{2}{\omega} \int \frac{2\pi\kappa^2 d\kappa}{(2\pi)^3} qv \frac{\partial n_{FD}(\xi_{\vec{\kappa}})}{\partial \xi_{\vec{\kappa}}} \int_{-1}^{1} \cos\theta d(\cos\theta) + \frac{2}{\omega^2} \int \frac{2\pi\kappa^2 d\kappa}{(2\pi)^3} (qv)^2 \frac{\partial n_{FD}(\xi_{\vec{\kappa}})}{\partial \xi_{\vec{\kappa}}} \int_{-1}^{1} \cos^2\theta d(\cos\theta) \tag{1021}$$

Since

$$-\frac{\partial n_{\mathrm{FD}}(\xi_{\vec{\kappa}})}{\partial \xi_{\vec{\kappa}}} = \delta(\xi_{\vec{\kappa}} - \xi_{\kappa_{\mathrm{F}}}) \tag{1022}$$

then

$$\Pi^0_{q,\omega} = -\frac{2}{\omega^2}\frac{2\pi}{(2\pi)^3}(qv_{\mathrm{F}})^2\frac{2}{3}m\int \kappa d\xi_{\vec{\kappa}}\delta(\xi_{\vec{\kappa}} - \xi_{\kappa_{\mathrm{F}}}) = -\frac{2}{\omega^2}\frac{2\pi}{(2\pi)^3}(qv_{\mathrm{F}})^2\frac{2}{3}m\kappa_{\mathrm{F}} \tag{1023}$$

Performing the analytic continuation for $\omega \to -i\omega_q + \delta$ onto the negative imaginary axis yields the **dielectric function**

$$\varepsilon^{\mathrm{RPA}}_{\vec{q},\omega} \equiv 1 - U(\vec{q})\Pi^{\mathrm{RPA}}_{q,\omega} \equiv 1 - \frac{4\pi e^2}{q^2}\frac{2}{\omega_q^2}\frac{2\pi}{(2\pi)^3}(qv_{\mathrm{F}})^2\frac{2}{3}m\kappa_{\mathrm{F}} = 1 - \frac{\omega_{\mathrm{P}}^2}{\omega_q^2} \tag{1024}$$

where the square of the **plasmon frequency**:

$$\omega_{\mathrm{P}}^2 = \frac{(2ev_F)^2 m\kappa_{\mathrm{F}}}{3\pi} \tag{1025}$$

which are the density oscillations of a free electron gas. Because the given system has a tendency to stay neutral everywhere, electrostatic forces will tend to bring spontaneous electronic density fluctuations back toward the uniform state. However, there is overshooting due to the electron inertia. Therefore, oscillations arise at this particular natural frequency, called the **plasma frequency**.

The number of intercepts between $F^{\mathrm{RPA}}_{\omega_l,q}$ and the constant line at 1 for $\omega_{\min} \le \omega \le \omega_{\max}$ scales as the inverse of the level spacing $\approx \frac{1}{a}$ (i.e., as $a = V^{\frac{1}{3}}$). There is one more intercept between $F^{\mathrm{RPA}}_{\omega,q}$ and the constant line at 1 for $\omega_{q_{\max}} < \omega$. This intercept takes place at the plasma frequency ω_{P}. When the plasmon dispersion merges with the electron-hole continuum, it is damped (**Landau damping**) due to the allowed decay into electron-hole excitations. This yields a finite lifetime of the plasmons within the electron-hole continuum corresponding to a finite width of the resonance of the collective excitation:

$$\tau = \frac{1}{\mathrm{A}_{q,\omega}} \tag{1026}$$

where

$$\mathrm{A}_{q,\omega} = 2\pi\int \frac{d\vec{\kappa}}{(2\pi)^3}\delta(\xi_{\vec{\kappa}} - \xi_{\vec{\kappa}+\vec{q}} + \omega_{\vec{q}})\vec{q}\nabla_{\vec{\kappa}} n_{\mathrm{FD}}(\xi_{\vec{\kappa}}) \tag{1027}$$

In the RPA, the plasmon infinite lifetime exists provided that $\mathrm{A}_{q,\omega} \neq 0$, and (this also implies) provided it does not overlap with the particle-hole pair continuum. The delta function, $\delta(\xi_{\vec{\kappa}} - \xi_{\vec{\kappa}+\vec{q}} + \omega)$, selects electron velocities $\vec{v} = \frac{\vec{\kappa}}{m}$ close to the phase velocity $\frac{\omega_{\vec{q}}}{|\vec{q}|}$ of the plasmon density wave in that $\frac{\vec{\kappa}\vec{q}}{m} = \omega_{\vec{q}}$. So, there exists a small range of electron velocities where electrons are able to surf the plasmon wave. Electrons initially moving slightly more slowly than the plasmon wave pump energy from the plasmon wave as they are accelerated up to the wave speed by the wave's leading edge. Conversely, electrons moving initially faster than the plasmon wave give up energy to the plasmon wave as they are decelerated down to the wave speed by the wave's trailing edge. Because the electron velocity distribution $\vec{q}\nabla_{\vec{\kappa}} n_{\mathrm{FD}}(\xi_{\vec{\kappa}})$ is skewed in favor of low electron energy, **the net effect is to damp the wave.**

This damping is called Landau damping. Of course, it is no longer a well-defined excitation of the system. The absence of damping at small q is the property of the RPA. The multipair excitation, which is not taken into account in the RPA, provides the main damping mechanism when the Landau damping is ineffective.

7.2.1.10 Thomas-Fermi Screening

We analyze the potential U felt by the electrons exposed to a static field $(\omega \to 0)$ via some **limiting cases of the unrenormalized Lindhard function** $h(z)$:

$$h(z \to 0) = 1 + \frac{1-z^2}{2z}\left(2z + 2\frac{z^3}{3} + \cdots\right) \approx 2 - 2\frac{z^2}{3} + \cdots \tag{1028}$$

From (970), we then have

$$-\mathbf{D}^{\text{RPA}}_{q,\omega\to 0} \equiv U_{\text{eff}}\left(q,\omega \to 0\right) = \frac{U(q)}{1 - U(q)\Pi^0_{q,\omega\to 0}} \tag{1029}$$

This implies that

$$-\mathbf{D}^{\text{RPA}}_{q,\omega\to 0} \equiv U_{\text{eff}}(q) = \frac{U(q)}{1 - U(q)(-N_{\text{F}})} = \frac{4\pi e^2}{q^2 + q^2_{\text{TF}}} = \frac{4\pi e^2}{q^2 + \chi^2} = \frac{4\pi e^2}{q^2 \varepsilon_{\text{TF}}(q)} \tag{1030}$$

and its position-space Fourier transformation yields the **Yukawa potential**, which corresponds with the **screened Coulomb interaction** that is exponentially suppressed beyond the screening length $\frac{1}{\chi} \equiv \frac{1}{q_{TF}}$ (this length is of the order of $10^{-8}\,\text{m} = 10\,\text{nm}$ in typical metals):

$$U(\vec{r}) = \frac{e^2}{|\vec{r}|}\exp\{-\chi|\vec{r}|\} = \frac{e^2}{|\vec{r}|}\exp\{-q_{TF}|\vec{r}|\} \tag{1031}$$

The potential is screened by a rearrangement of the electrons, and this changes the long-ranged Coulomb potential into a **Yukawa potential** with exponential decay where

$$q^2_{\text{TF}} = \frac{4m\kappa_F e^2}{\pi} \tag{1032}$$

is the square of the **Thomas-Fermi screening wave vector**. For ordinary metals, q_{TF} is typically of the same order of magnitude as κ_{F}. This implies that the screening length is of the order $5\,\text{Å}$ and is comparable to the distance between neighboring atoms. Consequently, external electric fields cannot penetrate the metal but are screened on this length q^{-1}_{TF}. This legitimates one of the basic assumptions used in electrostatics with metals.

In the previous relations,

$$U(q) = \frac{4\pi e^2}{q^2} \tag{1033}$$

is the Fourier transform of the Coulomb potential,

$$N_{\text{F}} = \frac{m\kappa_{\text{F}}}{2\pi^2} \tag{1034}$$

is the density of states at the Fermi level, and

$$\chi^2 = \frac{4m\kappa_F e^2}{\pi} \tag{1035}$$

is the square of the Debye-Hükkel inverse radius. Considering that

$$\kappa_F^3 = 3\pi^2\left(\frac{N}{V}\right) \tag{1036}$$

then

$$\chi^2 = \frac{4m\kappa_F e^2}{\pi} = 4me^2\left(3\pi\left(\frac{N}{V}\right)\right)^{\frac{1}{3}} \tag{1037}$$

and the Debye radius

$$R_D = e^{-1}(4m)^{-\frac{1}{2}}\left(3\pi\left(\frac{N}{V}\right)\right)^{-\frac{1}{6}} \tag{1038}$$

and the Thomas-Fermi dielectric function

$$\varepsilon_{TF}(q) = 1 + \frac{q_{TF}^2}{q^2} \tag{1039}$$

This function has a simple form that makes it easy to use in several calculations. At small distances $|\vec{q}| >> q_{TF}$, the impurity charge is unscreened because $\varepsilon_{TF}(q) \cong 1$, while at large distances, the screening is very effective. A necessary condition for the quasiclassical Thomas-Fermi theory to be justified is $q_{TF} << \kappa_F$. The dependence on e^2 in the Thomas-Fermi screening length is nonanalytic in the vicinity of $e^2 = 0$.

In expression (1031), the bare Coulomb interaction is thus profoundly modified by the Fermi sea. This expression is exponentially suppressed at length scales larger than the screening length q_{TF}^{-1}. The Fermi sea is characterized by a continuum of particle-hole excitations causing a nonvanishing lifetime of the field ϕ at nonvanishing frequencies and screening in the static limit. We see that in the presence of the electron sea, the long-range Coulomb interaction turns out to be short range. This has the following physical interpretation: If there is a point of higher than average density (i.e., negative charge), then the electrons in the neighborhood will be repelled from the given region. This creates a region (screening region of positive charge) around the given point of lower-than-average density. For long distances, a different electron feels the initial negative charge in addition to the screening region around it. Each almost compensates the other, thus making the Coulomb interaction effectively short ranged. It is important to note that with a screened potential, the Hartree-Fock calculation yields a nonvanishing effective mass m^*. In this case, the metals are no longer superconductors. It is instructive to note that for long-distance effects, applying the limit $q \to 0$ yields the correct long-distance behavior of the effective Coulomb interaction.

7.2.1.11 Friedel Oscillations

We examine the screening effect again by considering once more

$$\mathrm{D}_{q,\omega_l}^{\mathrm{RPA}} \equiv \frac{U(q)}{1 - U(q)\Pi_{q,\omega_l}^{\mathrm{RPA}}} \tag{1040}$$

and we show that there is still a weak singularity in the polarization that induces a long-range oscillatory interaction among the particles. But,

$$U_{q,\omega_l \to 0}^{\mathrm{RPA}} = \frac{U(q)}{1-U(q)\Pi_{0q}} \equiv \frac{U(q)}{\varepsilon(q)} \tag{1041}$$

then

$$U_{\mathrm{RPA}}(\vec{r}) = \int \frac{d\vec{q}}{(2\pi)^3} \exp\{i\vec{q}\vec{r}\} U_{\mathrm{RPA}}(\vec{q}) = \int \frac{d\vec{q}}{(2\pi)^3} \exp\{i\vec{q}\vec{r}\} \frac{U(\vec{q})}{\varepsilon(\vec{q})} \tag{1042}$$

or

$$U_{\mathrm{RPA}}(\vec{r}) = \int \frac{d\vec{q}}{(2\pi)^3} \exp\{i\vec{q}\vec{r}\} \frac{U(\vec{q})}{\varepsilon(\vec{q})} = \int_0^\infty 2\pi \frac{q^2 dq}{(2\pi)^3} \frac{U(\vec{q})}{\varepsilon(\vec{q})} \int_{-1}^{1} dx \exp\{iqrx\} = \int_0^\infty \frac{q dq}{2\pi^2 r} \sin(qr) \frac{U(\vec{q})}{\varepsilon(\vec{q})} \tag{1043}$$

From where

$$U_{\mathrm{RPA}}(\vec{r}) = \frac{1}{2\pi^2 r} \left\{ -\frac{1}{r} \cos(qr) \frac{qU(\vec{q})}{\varepsilon(\vec{q})} \Bigg|_0^\infty + \frac{1}{r} \int_0^\infty dq \cos(qr) \left(\frac{qU(\vec{q})}{\varepsilon(\vec{q})} \right)' \right\} \tag{1044}$$

or

$$U_{\mathrm{RPA}}(\vec{r}) = \frac{1}{2\pi^2 r} \left\{ \frac{1}{r^2} \sin(qr) \frac{d}{dq} \left(\frac{qU(\vec{q})}{\varepsilon(\vec{q})} \right) \Bigg|_0^\infty - \frac{1}{r^2} \int_0^\infty dq \sin(qr) \left(\frac{qU(\vec{q})}{\varepsilon(\vec{q})} \right)'' \right\} \tag{1045}$$

$$\left(\frac{qU(\vec{q})}{\varepsilon(\vec{q})} \right)'' \Bigg|_{2p_F} \approx -2\kappa_F U(2\kappa_F) \frac{\varepsilon''}{\varepsilon^2} \tag{1046}$$

So,

$$\int_0^\infty d(qr) \frac{\sin(qr)}{(q-2\kappa_F)r} = \int_{-2p_F r}^\infty dy \frac{\sin(y+2\kappa_F r)}{y} \Bigg|_{2\kappa_F r >> 1} = \int_{-\infty}^\infty dy \frac{\sin y \cos(2\kappa_F r) + \cos y \sin(2\kappa_F r)}{y} = \pi \cos(2\kappa_F r) \tag{1047}$$

We see that at $q = 2\kappa_F$, the dielectric constant is not analytic, and this singularity is known as the Kohn anomaly [5]. This is a consequence of the sharpness of the Fermi surface in κ-space. It gives rise to an oscillating term in the RPA screened potential:

$$U_{\mathrm{RPA}}(\vec{r}) \approx \frac{\cos(2\kappa_F |\vec{r}| + \phi)}{|\vec{r}|^3} \tag{1048}$$

where ϕ is a phase. **This oscillatory behavior is known as the Friedel oscillation in the context of the Jellium model or as the Ruderman-Kittel-Kasuya-Yosida oscillation of the static spin susceptibility induced by a magnetic impurity in a free electron gas.**

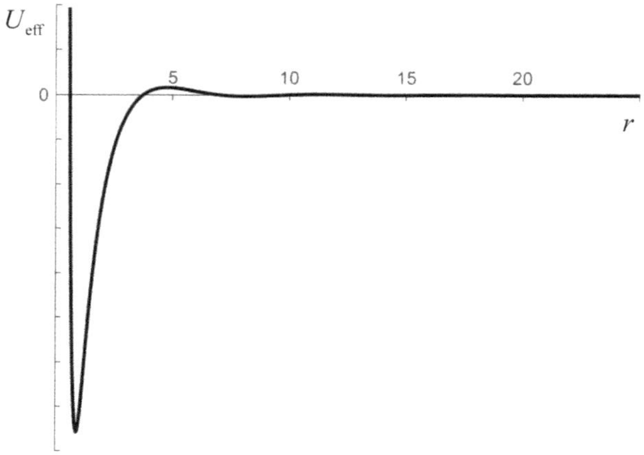

FIGURE 7.12 Screened potential versus the oscillatory radius. There is a substantial weakly decaying oscillatory behavior (Friedel oscillation) for smaller radii.

So, the Fourier transform to position space of the Lindhard function amounts to the replacement

$$U_{\text{RPA}}(\vec{r}) \approx \frac{\cos\left(2\kappa_F|\vec{r}|+\phi\right)}{|\vec{r}|^3} \tag{1049}$$

Consequently, at large distances, the screened potential $U_{\text{RPA}}(\vec{r})$ of a point charge at $\omega_l = 0$ has the form (1049). So, screening at large distances has a more noticeable structure than a simple Yukawa potential, as predicted by the Thomas-Fermi theory, with a substantial weakly decaying oscillatory term.

7.2.1.12 Dynamic Polarization Function

So far, we have discussed the dynamic polarization function for a three-dimensional parabolic band. We perform similar calculations for one- and two-dimensional systems. We also use the following limit for the unrenormalized Lindhard function when $z \to \infty$

$$h(z) = 1 + \frac{1-z^2}{2z}\ln\left|\frac{1+z^{-1}}{1-z^{-1}}\right| \approx 1 + \frac{1-z^2}{2z}\left(\frac{z}{z} + \frac{2}{3}\frac{1}{z^3} + \cdots\right) \approx \frac{2}{3}\frac{1}{z^2} + O\left(\frac{1}{z^4}\right) \tag{1050}$$

Let us consider a two-dimensional case:

$$\Pi^0_{q,\omega=0} = 4\int \frac{d\vec{\kappa}}{(2\pi)^2}\frac{n_{\text{FD}}(\xi_{\vec{\kappa}})}{\xi_{\vec{\kappa}} - \xi_{\vec{\kappa}+\vec{q}}} = -4\int_0^{\kappa_F}\frac{\kappa d\kappa}{(2\pi)^2}\int_0^{2\pi}\frac{d\theta}{\frac{q^2}{2m} + \frac{\kappa q}{m}\cos\theta} = -\frac{2}{\pi^2}\frac{m}{q}\int_0^{\kappa_F}\kappa d\kappa \ln\left|\frac{q+2\kappa}{q-2\kappa}\right| \tag{1051}$$

Then

$$\Pi_{\omega_l=0,q} = -\frac{m}{\pi q}\int_0^{\kappa_F} d\kappa \ln\left|\frac{q+2\kappa}{q-2\kappa}\right| = -\frac{m}{2\pi^2}\left[\ln\left(1-z^2\right) + z\ln\left(\frac{1+z}{1-z}\right)\right], \quad z = \frac{2\kappa_F}{q} \tag{1052}$$

For a one-dimensional case,

$$\Pi_{\omega=0,q} = -4\int_0^{\kappa_F}\frac{d\kappa}{2\pi}\frac{1}{\frac{q^2}{2m} + \frac{\kappa q}{m}} = -\frac{8}{\pi}\frac{m}{q^2}\int_0^{\kappa_F} d\kappa \frac{1}{1+\frac{2\kappa}{q}} = -\frac{2m}{\pi q}\int_0^z \frac{dy}{1+y} = -\frac{2m}{\pi q}\ln(1+z), \quad z = \frac{2\kappa_F}{q} \tag{1053}$$

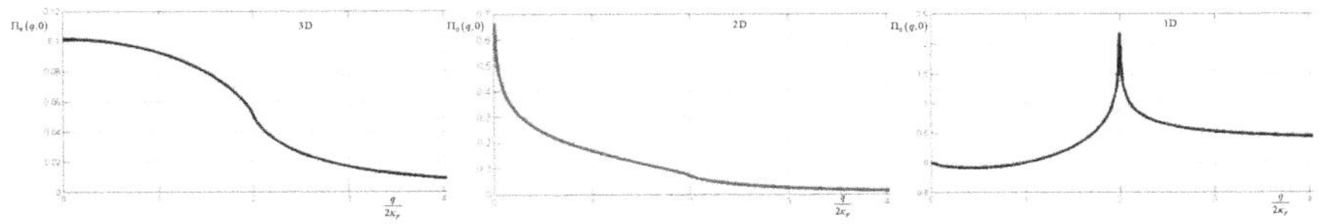

FIGURE 7.13 Lindhard functions for different dimensions. The lower the dimension, the stronger the singularity at $q = 2\kappa_F$. There is a logarithmic divergence in one-dimension while in two-dimension, a kink, and in three-dimension only the derivative diverges.

As $q \to 0$ then

$$\Pi_{\omega=0,q\to 0} = -\frac{2m}{\pi q}\ln z \tag{1054}$$

It is interesting to see that $\Pi_{\omega=0,q}$ has a singularity at $q = 2\kappa_F$ in all dimensions. This singularity becomes weaker as the dimensionality is increased. In one dimension, there is a logarithmic divergence, while in two dimensions there is a kink, and in three dimensions only the derivative diverges. The one-dimensional case is always unstable.

From the aforementioned, it is worth noting that in the presence of the polarizable medium (**the electron sea**), the long-range Coulomb interaction essentially achieves the short-range regime. **How can we interpret this physically?** Suppose there is a point of higher than average density, **that is, a negative charge**. In this case, the electrons in the nearby vicinity are repelled from the given region, creating a region around the given point of lower-than-average density. This implies a screening region of positive charge. At long distances, another electron sees not only the initial negative charge but also the screening region around it, and one nearly cancels out the other. Therefore, this means the Coulomb interaction is effectively short ranged.

7.2.1.13 Ground-State Energy in the RPA

We calculate the thermodynamic potential as a function of the interaction energy via a coupling constant integration. The Hamiltonian determinant of the system is written:

$$\hat{H} = \hat{H}_0 + \lambda \hat{H}_{\text{int}} \tag{1055}$$

This allows us to calculate the partition function and the thermodynamic potential:

$$Z = \text{Tr}\exp\left\{-\beta\hat{H}\right\} \quad , \quad \frac{\partial\Omega}{\partial\lambda} = \frac{1}{\lambda}\left\langle \lambda\hat{H}_{\text{int}}\right\rangle_\lambda \tag{1056}$$

where the average $\langle\cdots\rangle_\lambda$ is taken with the Hamiltonian $\hat{H}$. We express the potential energy as a function of the density-density response function:

$$\left\langle \hat{H}_{\text{int}}\right\rangle_\lambda = -\frac{1}{2}\sum_{\vec{q}\neq 0} e^2 D_{0q}\left\langle \rho_{-\vec{q}}\rho_{\vec{q}} - \frac{\hat{N}}{V}\right\rangle_\lambda = \frac{1}{2}\sum_{\vec{q}\neq 0} e^2 D_{0q}\left[\frac{1}{\beta}\sum_{\omega_\nu}\Pi_{\vec{q}}^{\text{RPA}\,\lambda}\exp\{i\omega_\nu\delta\} + n^\lambda\right] \tag{1057}$$

Here, $\Pi_{\vec{q}}^{\mathrm{RPA}\lambda}$, as well as $n^{\lambda} = \frac{\langle \hat{N} \rangle_{\lambda}}{V}$, depend on λ. From (932) and (1030), we have

$$\Omega = \Omega_0 + \int_0^1 \frac{d\lambda}{\lambda} \frac{1}{2} \sum_{\vec{q}\neq 0} \lambda e^2 \mathbf{D}_{0q} \left[\frac{1}{\beta} \sum_{\omega_\nu} \Pi_{\vec{q}}^{\mathrm{RPA}\lambda} \exp\{i\omega_\nu \delta\} + n^{\lambda} \right] \tag{1058}$$

The quantity $\Omega(\lambda = 0) = \Omega_0$ is the thermodynamic potential of the noninteracting electron gas.

In the RPA, we have

$$\Pi_{\vec{q}}^{\mathrm{RPA}\lambda} = \frac{\Pi_{0q}}{1 - \lambda \Sigma_q \mathbf{D}_{0q}} = \Pi_{0q} + \lambda \Sigma_q \mathbf{D}_{0q} \frac{\Pi_{0q}}{1 - \lambda \Sigma_q \mathbf{D}_{0q}} \quad , \quad \Sigma_q = (ie)^2 \Pi_{0q} \tag{1059}$$

Substituting (1059) into (1058) and summing over λ, we have the ground state energy:

$$E_{\mathrm{RPA}} = \Omega(T = 0) + \mu N = E_0 + E_{\mathrm{exchange}} + E_{\mathrm{corr}}^{\mathrm{RPA}} \tag{1060}$$

where the correlation energy in the RPA is expressed as:

$$E_{\mathrm{corr}}^{\mathrm{RPA}} = \frac{1}{2\beta} \sum_{\vec{q}\neq 0} \left[-\Sigma_q \mathbf{D}_{0q} + \ln\left[1 - \Sigma_q \mathbf{D}_{0q}\right] \right] \tag{1061}$$

The exchange (Fock) energy is expressed as:

$$E_{\mathrm{exchange}} = \frac{1}{2} \sum_{\vec{q}\neq 0} e^2 \mathbf{D}_{0q} \left[\frac{1}{\beta} \sum_{\omega_\nu} \Pi_{0q} \exp\{i\omega_\nu \delta\} + n \right] \tag{1062}$$

An elementary calculation yields

$$E_{\mathrm{exchange}} \equiv E_x = \frac{1}{V} \sum_{\vec{\kappa},\vec{q}\neq 0} e^2 \mathbf{D}_{0q} n_{\vec{\kappa}+\vec{q}}^0 n_{\vec{\kappa}}^0 \quad , \quad n_{\vec{\kappa}}^0 = \theta\left(\kappa_F - |\vec{\kappa}|\right) \tag{1063}$$

This is the first-order correction $\langle \hat{\mathrm{H}}_{\mathrm{int}} \rangle$ to the energy E_0 of the noninteracting electron gas. This energy stems from the antisymmetry of the ground state with respect to the exchange of two particles. In diagrammatic form, it is represented as:

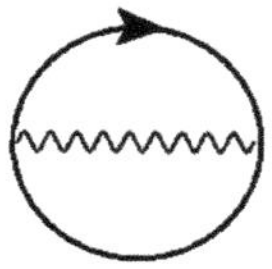

FIGURE 7.14 First-order correction $\langle \hat{\mathrm{H}}_{\mathrm{int}} \rangle$ to the energy E_0. For a uniform electron gas, the Coulomb interaction correction to the ground state energy comes only through the exchange interaction that is a purely quantum effect.

The total exchange energy per electron in terms of the parameter r_s:

$$\frac{E_{\text{exchange}}}{N} \equiv \frac{1}{V}\sum_{\vec{\kappa},\vec{q}\neq 0} e^2 \mathbf{D}_{0q} n^0_{\vec{\kappa}+\vec{q}} n^0_{\vec{\kappa}} = -\frac{1}{n}\int \frac{d\vec{q}}{(2\pi)^3}\int \frac{d\vec{\kappa}}{(2\pi)^3}\frac{4\pi e^2}{|\vec{q}-\vec{\kappa}|^2}\theta(\kappa_F - |\vec{\kappa}|)\theta(\kappa_F - |\vec{q}|) \tag{1064}$$

or

$$\frac{E_{\text{exchange}}}{N} = -\frac{e^2\kappa_F}{\pi n}\int \frac{d\vec{\kappa}}{(2\pi)^3}\theta(\kappa_F - |\vec{\kappa}|)h\left(\frac{\kappa}{\kappa_F}\right) = -\frac{3e^2\kappa_F}{4\pi} = -\frac{3}{2\pi}\left(\frac{9\pi}{4}\right)^{\frac{1}{3}}\frac{1}{r_s}\mathrm{R}_y \tag{1065}$$

Here

$$n^0_{\vec{\kappa}} = \theta(\kappa_F - |\vec{\kappa}|) \tag{1066}$$

We see from relation (1065) that for a uniform electron gas, the Coulomb interaction correction to the ground state energy is negative and comes only through the exchange interaction that is a purely quantum effect. We also see from relation (1065) that the average exchange energy per electron is $\frac{3}{2}$ of the value at the Fermi energy. The additional factor of $\frac{1}{2}$ enters the ground state energy to account for the fact that the exchange energy is a pair interaction. The kinetic energy for a single particle is expressed as $\epsilon_\kappa = \frac{\kappa^2}{2m}$. To determine the contribution to the ground state energy, we then sum over all the particles in the ground state:

$$E_0 = \sum_{\vec{\kappa}\sigma} \epsilon_\kappa n(\kappa) = 2V\int \frac{d\vec{\kappa}}{(2\pi)^3}\frac{\kappa^2}{2m}n(\kappa) = \frac{N}{n_0}\frac{1}{2\pi^2 m}\int_0^{\kappa_F}\kappa^4 d\kappa = \frac{3}{5}\frac{\hbar^2\kappa_F^2}{2m}N = \frac{3}{5}\epsilon_F N = \frac{3}{5}\left(\frac{9\pi}{4}\right)^{\frac{2}{3}}\frac{1}{r_s^2}\mathrm{R}_y N \tag{1067}$$

Then, the kinetic energy per electron:

$$\frac{E_0}{N} = \frac{3}{5}\epsilon_F = \frac{3}{5}\left(\frac{9\pi}{4}\right)^{\frac{2}{3}}\frac{1}{r_s^2}\mathrm{R}_y \tag{1068}$$

So far, we have found two terms for the energy of the particle:

$$E(\kappa) = \frac{\hbar^2\kappa^2}{2m} + \Sigma_x(\kappa) + \cdots \tag{1069}$$

where the exchange self-energy may be written as follows:

$$\Sigma_x = \frac{1}{V}\sum_{\vec{q}} e^2 \mathbf{D}_{0q} n^0_{\vec{\kappa}+\vec{q}} \tag{1070}$$

The corresponding two terms for the ground state energy per particle are:

$$E_g = \frac{3}{5}\left(\frac{9\pi}{4}\right)^{\frac{2}{3}}\frac{1}{r_s^2}\mathrm{R}_y - \frac{3}{2\pi}\left(\frac{9\pi}{4}\right)^{\frac{1}{3}}\frac{1}{r_s}\mathrm{R}_y + \cdots \tag{1071}$$

The ground state energy appears to be a power series of r_s. Usually, it is unsafe to extrapolate from just two terms such as for the given case. From here, we can only say the ground state energy is a power series of r_s. The next term will be of order $O(r_s \ln r_s)$. The zero order could be interpreted as either a constant or as $\ln r_s$. However, both of these terms are present. The series, therefore, takes the following form:

$$E_g = \frac{3}{5}\left(\frac{9\pi}{4}\right)^{\frac{2}{3}}\frac{1}{r_s^2}\mathrm{R}_y - \frac{3}{2\pi}\left(\frac{9\pi}{4}\right)^{\frac{1}{3}}\frac{1}{r_s}\mathrm{R}_y + 0.0622\ln r_s - 0.094 + O(r_s \ln r_s)\mathrm{R}_y \tag{1072}$$

The terms from the third onward can be found in reference [34]. The first term is called the **Hartree term**, and the first two terms are called the **Hartree-Fock terms.** In typical metals, $r_s \approx 2 \div 5$ indicating that electronic interactions need to be accounted for to calculate the energy of a metal with any hope of precision. The RPA gives the next two leading corrections to the expansion in powers of r_s [35], though from a computational sense, the relevance of such an expansion to metals is uncertain.

The energy terms beyond Hartree-Fock are known as the correlation energy [34]:

$$\frac{E_{\text{corr}}}{N} = 0.0622\ln r_s - 0.094 + O(r_s \ln r_s)\mathrm{R}_y \tag{1073}$$

The name is applied both to the additional energy terms in the self-energy of an electron of wave vector $\vec{\kappa}$ as well as to the ground state energy obtained by averaging over all of the electrons. So, the validity of the RPA can be improved by adding the second-order contributions to the total ground state energy per particle:

$$E_g = \frac{3}{5}\left(\frac{9\pi}{4}\right)^{\frac{2}{3}}\frac{1}{r_s^2}\mathrm{R}_y - \frac{3}{2\pi}\left(\frac{9\pi}{4}\right)^{\frac{1}{3}}\frac{1}{r_s}\mathrm{R}_y + \frac{E_{\text{corr}}}{N} \tag{1074}$$

The correlation energy result is accurate in the limit of $r_s \to 0$. There is some uncertainty regarding the maximum value of r_s for which the few terms provide an accurate description. The radius of convergence of the power series is about $r_s \le 1$. Actual metals have values of r_s up to about six. This series does not give sensible numbers at these low densities.

Besides the total ground state energy, the term correlation energy is often applied to other quantities. An example is the correlation energy of a particle of wave vector $\vec{\kappa}$ that is a term beyond Hartree-Fock:

$$E(\kappa) = \frac{\hbar^2\kappa^2}{2m} + \Sigma_x(\kappa) + \Sigma_{\text{corr}}(\kappa, i\kappa_n) \tag{1075}$$

The self-energy, apart from correlation $\Sigma_{\text{corr}}(\kappa, i\kappa_n)$, depends upon the particle energy $i\kappa_n$. The energy and the wave vector can be averaged to obtain the contribution E_{corr} to the ground state correlation energy.

7.2.1.14 Compressibility

Suppose

$$\in(r_s) \equiv \frac{E}{N} = \frac{3}{5}\left(\frac{9\pi}{4}\right)^{\frac{2}{3}}\frac{1}{r_s^2} - \frac{3}{2\pi}\left(\frac{9\pi}{4}\right)^{\frac{1}{3}}\frac{1}{r_s} + 0.0622\ln r_s - 0.094 + O(r_s \ln r_s)\mathrm{R}_y \tag{1076}$$

then we compute the pressure

$$P = -\left.\frac{\partial E}{\partial V}\right|_N = n^2\frac{\partial \in}{\partial n} = -\frac{n r_s}{3}\frac{\partial \in}{\partial r_s} \tag{1077}$$

and the inverse compressibility

$$\frac{1}{\kappa} = -V\frac{\partial E}{\partial V}\bigg|_N = n\frac{\partial P}{\partial n} = \frac{\rho r_s}{3}\left[-\frac{2}{3}\frac{\partial \epsilon}{\partial r_s} + \frac{r_s}{3}\frac{\partial^2 \epsilon}{\partial r_s^2}\right] \tag{1078}$$

We assume in the calculations that the background of positive charge adjusts itself to maintain neutrality at no energy cost.

7.2.1.15 One-Particle Property: Hartree-Fock Theory

We have calculated the collective excitations of the electron liquid assuming the one-electron properties imitate the noninteracting electron gas. The assumption is qualitatively correct if the electron liquid is a Fermi liquid. Otherwise, the self-energy corrections in the one-particle propagator would invalidate the calculation of the density-density response function. In this chapter, we show that the electron liquid is indeed a Fermi liquid.

Investigating the RPA in the functional integral formalism, the Hartree self-energy at the saddle-point (or mean-field) approximation was obtained. The Hartree self-energy vanishes in the Jellium model due to the neutrality of the system. We will study the failure of the Hartree-Fock theory in the electron liquid. This can be done by relating the self-energy to the density-density response function via the general formula

$$\frac{1}{\beta V}\sum_{\vec{\kappa}} \Sigma(\vec{\kappa})\mathbf{G}_{0\vec{\kappa}} \exp\{i\omega_n\delta\} = \frac{1}{2V}\sum_{\vec{q}} e^2\mathbf{D}_{0q}\left[\frac{1}{\beta}\sum_{\omega_\nu} \Pi_{\vec{q}} \exp\{i\omega_\nu\delta\} + n\right] \tag{1079}$$

From here, we easily deduce that the Hartree-Fock approximation $\Sigma \equiv \Sigma_x$ corresponds to the density-density response function

$$\tilde{\Pi}_{0\vec{q}} = \frac{2}{\beta V}\sum_{\vec{\kappa}} \mathbf{G}_{0\vec{\kappa}}\mathbf{G}_{0(\vec{\kappa}+\vec{q})} \tag{1080}$$

Relation (1080) can be represented in diagrammatic form as follows:

$$\tilde{\Pi}^{\text{RPA}}_{0q} =$$ 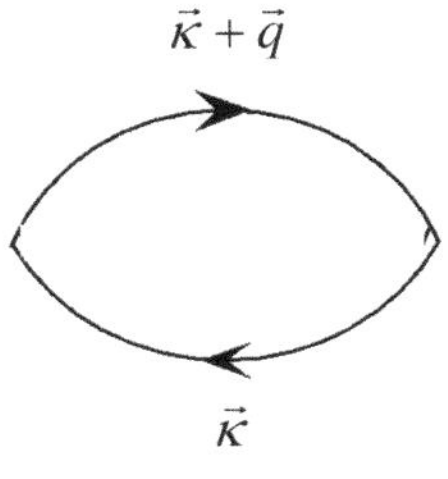 (1081)

This is obtained from $\Pi_{0\vec{q}}$ by replacing the bare propagator $\mathbf{G}_0$ by $\mathbf{G} = \left[\mathbf{G}_0^{-1} - \Sigma_x\right]^{-1}$.

It is instructive to note that for a single-particle Green's function in (1080), we add an extra particle and let it propagate. Here, we make a density fluctuation. That is to say, a particle-hole pair, and let it propagate. So, the quantity $\tilde{\Pi}_{0\vec{q}}$ will contain information about the actual excitation spectrum of the Fermi gas that has a fixed number of particles. In $\tilde{\Pi}_{0\vec{q}}$, the external momentum $\vec{q}$ is within the diagram, while the internal momentum $\vec{\kappa}$ is summed over. There is also an internal spin within the given diagram. The sum also is done over the internal spin within the diagram. We think of a hole as an electron propagating backward in time. This is known as a particle-hole, or a polarization loop.

The equation in (1080) does not include screening. Therefore, it is a very poor approximation of the density-density response function of the electron liquid. As the RPA does include screening and gives us a reasonable satisfying account of density fluctuations, a better approximation for the self-energy can be as follows:

$$\Sigma(\vec{\kappa})=\frac{1}{\beta V}\sum_{\vec{q}}\frac{e^2\mathbf{D}_{0q}}{1-\tilde{\Sigma}_q\mathbf{D}_{0q}}\mathbf{G}_{0(\vec{\kappa}+\vec{q})} \quad , \quad \tilde{\Sigma}_q=(ie)^2\tilde{\Pi}_{0\vec{q}} \tag{1082}$$

The corresponding expression for the density-density response function has the RPA form with all one-particle propagators dressed with the self-energy in (1082). To simplify its computation, we relinquish self-consistency and replace $\mathbf{G}$ by $\mathbf{G}_0$; this yields the following RPA self-energy

$$\Sigma_{\text{RPA}}(\vec{\kappa})=\frac{1}{\beta V}\sum_{\vec{q}}\frac{e^2\mathbf{D}_{0q}}{1-\Sigma_q\mathbf{D}_{0q}}\mathbf{G}_{0(\vec{\kappa}+\vec{q})}=\frac{1}{\beta V}\sum_{\vec{q}}\frac{e^2\mathbf{D}_{0q}}{\varepsilon_{\text{RPA}}(\vec{q})}\mathbf{G}_{0(\vec{\kappa}+\vec{q})} \tag{1083}$$

This imitates the exchange self-energy with the Coulomb interaction $e^2\mathbf{D}_{0q}$ replaced by the effective interaction $\frac{e^2\mathbf{D}_{0q}}{\varepsilon_q^{\text{RPA}}}$ and without the self-consistency; then

$$\varepsilon_q^{\text{RPA}}=1-\Sigma_q\mathbf{D}_{0q} \tag{1084}$$

is the RPA dielectric function. We can rewrite the RPA self-energy:

$$\Sigma_{\text{RPA}}(\vec{\kappa})=\frac{1}{\beta V}\sum_{\vec{q}}e^2\mathbf{D}_{0q}\mathbf{G}_{0(\vec{\kappa}+\vec{q})}\left(1+\Sigma_q\mathbf{D}_{0q}\right) \tag{1085}$$

This can be represented diagrammatically:

(1086)

We imagine that initially in the Fock diagram, we replace the bare interaction by the effective interaction (**dressed interaction**). The improvement over the Hartree theory amounts to including the Fock term (diagram) Σ_x. In this way, we replace the **bare Green's function** line by the true Green's function including self-energy corrections:

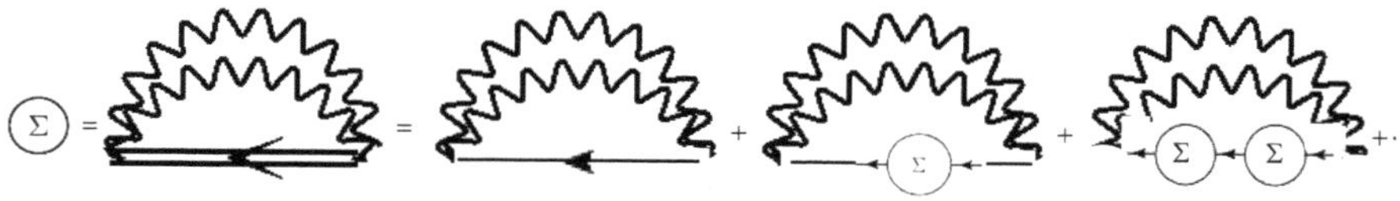

(1087)

This can be solved self-consistently (**self-consistent Hartree-Fock technique**) or we can go even further:

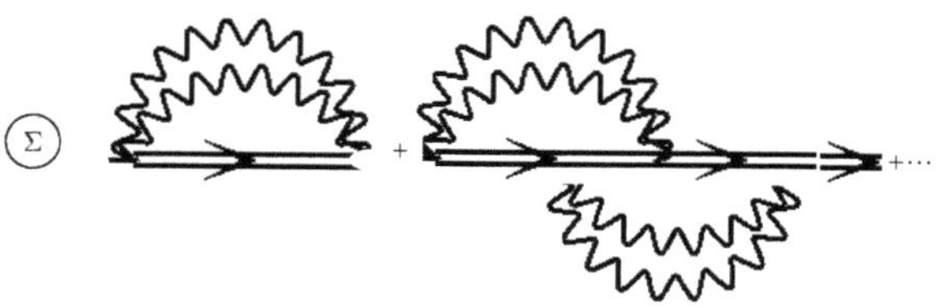

(1088)

It is instructive to note that the use of the **doubled** line that denotes the following Green's function

$$\mathbf{G} = \left[\mathbf{G}_0^{-1} - \Sigma_x\right]^{-1} \quad , \quad \Sigma_{\text{exchange}}(\kappa) \equiv \Sigma_x(\kappa) \tag{1089}$$

significantly reduces the number of diagrams. However, we must be very careful to make sure no diagram is double counted.

We compute the Fock term that is the exchange self-energy:

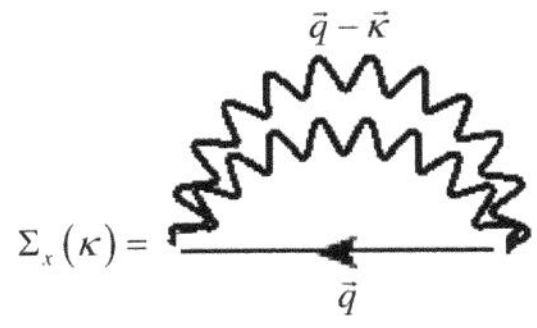

(1090)

or

$$\Sigma_x(\kappa) = -\frac{1}{\beta V}\sum_{\vec{q}} U(\vec{q}-\vec{\kappa})\mathbf{G}_{0q}\exp\{i\omega_l\delta\} = -\frac{1}{\beta V}\sum_{\vec{q}} U(\vec{q}-\vec{\kappa})\frac{\exp\{i\omega_l\delta\}}{i\omega_l - \xi(\vec{q}) - \Sigma_x(\vec{q})} \tag{1091}$$

or

$$\Sigma_x(\kappa) = -\frac{1}{V}\sum_{\vec{q}} U(\vec{q}-\vec{\kappa}) n_{\text{F}}\left[\xi(\vec{q}) + \Sigma_x(\vec{q})\right] \tag{1092}$$

The self-energy is the only contribution with one Coulomb line. This self-energy depends only on the magnitude of the wave vector $|\vec{\kappa}|$ of the particle and not on the frequency. So, the Green's function imitates the noninteracting one, with

$$\xi(\vec{\kappa}) \to \xi(\vec{\kappa}) + \Sigma(\kappa) \tag{1093}$$

At zero temperature, this relation becomes

$$\Sigma_x(\kappa) = -\frac{1}{V}\sum_{\vec{q}} U(\vec{q}-\vec{\kappa})\theta\left(\kappa_{\text{F}} - |\vec{q}|\right) \tag{1094}$$

Here, the Fermi wave vector is defined by

$$\xi(\kappa_{\text{F}}) + \Sigma_x(\kappa_{\text{F}}) = 0 \tag{1095}$$

and

$$n = \frac{2}{\beta V}\sum_{\vec{\kappa}} \mathbf{G}_{0\kappa}\exp\{i\omega_n\delta\} = \frac{2}{V}\sum_{\vec{\kappa}} n_{\text{F}}\left[\xi(\vec{\kappa}) + \Sigma_x(\kappa)\right] = \frac{2}{V}\sum_{\vec{\kappa}} \theta\left(\kappa_{\text{F}} - |\vec{\kappa}|\right) \tag{1096}$$

Therefore, κ_{F} is the same as in the noninteracting electron gas. Therefore, we have the **Luttinger theorem:**

The interactions do not change the volume of the Fermi surface.

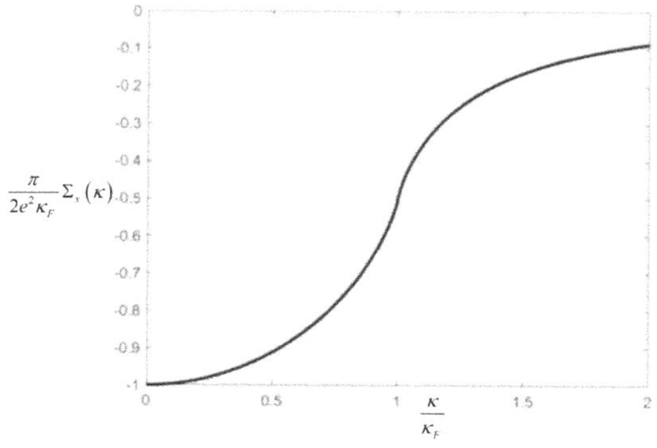

FIGURE 7.15 Plot of $\frac{\pi}{2e^2\kappa_F}\Sigma_x(\kappa)$ versus $z = \frac{\kappa}{\kappa_F}$.

For a frequency-independent self-energy, the theorem is trivially satisfied. From (1094), we have

$$\Sigma_x(\kappa) = -\frac{1}{V}\sum_{\vec{q}\neq 0} U(\vec{q}-\vec{\kappa})\theta(\kappa_F - |\vec{q}|) = -\int \frac{d\vec{q}}{(2\pi)^3}\frac{4\pi e^2}{|\vec{q}-\vec{\kappa}|^2}n(\vec{q}) = -\frac{e^2}{\pi}\int_0^{\kappa_F} q^2\,dq\int_{-1}^{1}\frac{d(\cos\theta)}{\kappa^2+q^2-2q\kappa\cos\theta} \tag{1097}$$

or

$$\Sigma_x(\kappa) = -\frac{e^2}{\pi\kappa}\int_0^{\kappa_F} q\,dq\ln\left|\frac{\kappa+q}{\kappa-q}\right| = -\frac{e^2\kappa_F}{\pi}h(z) \quad , \quad z = \frac{\kappa}{\kappa_F} \tag{1098}$$

Here, $h(z)$ is the unrenormalized Lindhard function, which gives the wave dependence of the exchange energy.

From the plot in Figure 7.15, the value of the exchange energy at $z = 0$ is:

$$\Sigma_x(0) = -\frac{2e^2\kappa_F}{\pi} \tag{1099}$$

In the vicinity of $z = 1$, the curve of the self-energy rises steeply and achieves zero at large values of z. At the Fermi energy, when $\kappa = \kappa_F$, the value is

$$\Sigma_x = -\frac{e^2\kappa_F}{\pi} \tag{1100}$$

This point is interesting because the self-energy has an infinite slope. The derivative of the self-energy:

$$\frac{d\Sigma_x(z)}{dz} = \frac{e^2\kappa_F}{\pi}\frac{1}{2z}\left(\frac{1+z^2}{z}\ln\left|\frac{1+z}{1-z}\right| - 2\right) \quad , \quad z = \frac{\kappa}{\kappa_F} \tag{1101}$$

This derivative has a logarithmic divergence as $z \to 1$. This predicts the effective mass to achieve the value zero at the Fermi surface, and all the metals are superconductors. This unphysical result is due to the fact that the screening effect is not taken into account. It is well established that an electric field cannot penetrate the interior of a metal. So, the effective interaction among the electrons cannot be given by a long-range Coulomb force. Considering that the exchange self-energy is independent of frequency, the effective mass is given as follows:

$$\frac{m}{m^*}=1+\frac{\partial\Sigma_x(z)}{\partial\in_\kappa}=-\frac{e^2m}{\pi\kappa}\frac{dh(z)}{dz}=-\frac{e^2m}{2\pi\kappa_F}\frac{1}{z^2}\left(\frac{1+z^2}{z}\ln\left|\frac{1+z}{1-z}\right|-2\right) \tag{1102}$$

This diverges at the Fermi energy as $z\to 1$. Considering that the Fermi velocity is defined as $\frac{d\in}{d\kappa}$ at $\kappa=\kappa_F$, this implies that within the Hartree-Fock approximation, there is an infinite Fermi velocity. Therefore, the inverse effective mass really diverges at the Fermi surface. This would have numerous observable consequences experimentally. The electron gas would be unstable, while the specific heat C_V would diverge at low temperatures. Several metals at low temperatures have unusual properties. For example, some are superconducting while others magnetic. However, simple metals such as alkalis are stable at low temperatures, so that this exchange instability is regarded as being absent. The examination of further terms in the perturbation theory produces another divergence in the effective mass, which exactly cancels that due to exchange. The effective mass and specific heat are not divergent. So, the first-order perturbation theory is never going to be effective enough.

It is instructive to note that the self-energy function has an infinite number of terms. We will examine some **low-order** terms and find which terms contribute to the constant and $\ln r_s$ terms. The term **low-order** refers to the order of the self-energy diagram that is the number of internal Coulomb lines. The correlation energy is the sum of all contributions with two or more Coulomb lines.

8

Phase Transitions and Critical Phenomena

Introduction

Regimes of behavior in which a given macroscopic variable of thermodynamic significance (**ordered phase**) changes from being zero in the **disordered phase** to a nonzero value in the ordered phase are separated by phase **transitions.** Near T_c **(critical temperature)**, the O.P. **(order parameter)** is close to a phase transition. So, an order parameter is a thermodynamic variable that is zero on one side of the transition and nonzero on the other; we can ferromagnets as an example, where the magnetization $\mathcal{M}$ is the order parameter. In the ferromagnetic phase, the magnetic moments undergo a spontaneous transition from a disordered phase with no net magnetization to an ordered phase with a nonzero net magnetization. Here, the magnetic moments are aligned in a definite direction leading to a degenerate ground state.

Because one of the system's symmetries is spontaneously broken when the system moves from a disordered to an ordered state, the order parameter is defined as the measure of the degree of order across the phase boundaries. When the spins (angular momenta) have a sign change during time reversal, the spontaneous magnetization in a ferromagnet breaks the time-reversal symmetry. The order parameter in a one-dimensional crystal is the local displacement, while the order parameter in a ferromagnetic material is the local magnetization.

Many observable properties should display interesting behavior as a function of $T - T_c$ (see, for example, Figure 8.1), where the order parameter is zero on the high temperature, disordered side and nonzero in the ordered, low-temperature side of the phase transition.

A disordered liquid crystallizing to form a solid crystal of long-range order is another example of a phase transition that breaks the continuous translational symmetry because each point in a crystal does not have the same properties as those observed in a fluid.

Similarly, the superconducting phase transition of conventional superconductors can also be explained in terms of symmetry breaking, where Cooper pairs are due to electrons experiencing phonon-mediated interactions' attractive force. The many single-electron wave functions are now transformed into a collective wave function representing the condensate while breaking the global phase U(1) symmetry, with the pair density acting as the order parameter.

Superconductivity is characterized by a vanishing static electrical resistivity and an exclusion of the magnetic field from the interior of a sample [36]. This relates the phenomena of **superfluidity** (in helium-3 and helium-4) and **Bose-Einstein condensation** (in weakly interacting boson systems). Microscopically, superfluidity in helium-3 (He-3) relates to superconductivity most closely because both phenomena involve the condensation of **fermions**, while in helium-4 (He-4) and

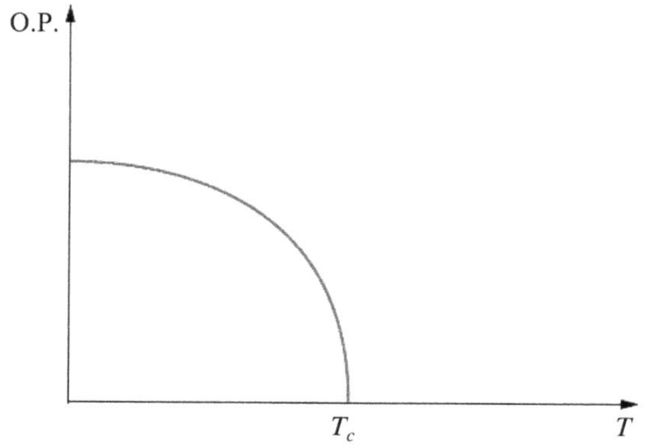

FIGURE 8.1 Variation of the order parameter (O.P.) with temperature T.

Bose-Einstein condensates, it is **bosons** that condense. In this chapter, we will examine a few of interesting examples:

1. **Magnetization:** As T increases toward T_c, the spontaneous magnetization must vanish as

$$\mathcal{M}(T) \approx (T_c - T)^{\beta} \text{ with } \beta > 0 \tag{1103}$$

2. **Susceptibility:** As T decreases toward T_c in the paramagnetic state, the magnetic susceptibility $\chi(T)$ must diverge as

$$\chi(T) \approx (T_C - T)^{-\gamma} \text{ with } \gamma > 0 \tag{1104}$$

3. **Specific heat:** As T decreases toward T_c in the paramagnetic state, the specific heat has a characteristic of singularity given by

$$C(T) \approx (T_C - T)^{-\alpha} \text{ with } \alpha > 0 \tag{1105}$$

In mean field theory, where the interactions are replaced by their values in the presence of a self-consistently determined average magnetization, we find $\beta = \frac{1}{2}$ and $\gamma = 1$ for all dimensions. The mean field values do not agree with experiments or with several exactly solvable theoretical models for T very close to T_c, for example:

1. $\beta = \frac{1}{8}$ in the two-dimensional Ising model.
2. $\beta \approx \frac{1}{3}$ in the three-dimensional Heisenberg model.
3. $\gamma \approx 1.25$ for most three-dimensional phase transitions instead of the mean field predictions of $\gamma = 1$.

In the early 1970s, K.G. Wilson [37] developed the renormalization group theory of phase transitions to describe the behavior of systems in the region $T \approx T_c$. In all examples of phase transition, the order parameter (O.P.):

$$\text{O.P.} = \begin{cases} \neq 0 & , \quad T < T_c \\ = 0 & , \quad T \geq T_c \end{cases} \tag{1106}$$

a. Scalar O.P.
b. Vector O.P.
c. In principle, complex possible

In the free energy $F = F(p,T)$, the pressure p and temperature T characterize the mechanical and thermal equilibrium, respectively, in the volume V.

Consider the example when ϕ is a scalar O.P. such that $F = F(\phi)$. This may permit us to examine the following Landau theory of phase transition [38].

8.1 Landau Theory of Phase Transition

The concept of the **order parameter** was introduced by Landau to describe phase transitions. This theory neglects fluctuations, implying that the order parameter is assumed constant in time and space. Therefore, the Landau theory is a **mean-field theory**.

We examine the quantitative theory of phase transition that entails considering the thermodynamic quantities of the body for given deviation from the symmetrical state. That is to say, we consider given values of the O.P. ϕ. We represent the thermodynamic potential of the body, for example, as a function of p, T, and ϕ. It is instructive to note that if the thermodynamic potential is, for example, the Helmholtz **free energy** $F = F(p,T,\phi)$, the variable ϕ is not necessarily situated in the same dimension as the variables p and T. However, the variables p and T can be arbitrarily specified and the value of ϕ can be determined from the thermal equilibrium condition when $F = F(p,T,\phi)$ has its extremal value for given p and T. The continuity of the change of state in phase transitions demands the O.P. ϕ take arbitrary small values in the neighborhood of the transition point. For that reason, we examine the Taylor expansion of $F = F(\phi)$:

$$F(\phi) = F_0 + F_1\phi + F_2\phi^2 + F_3\phi^3 + F_4\phi^4 \tag{1107}$$

We could do the same for vector or complex O.P.

1. $F_1 = 0$
2. It is instructive to note that the coefficient $F_2(p,T)$ in the second-order term should vanish at the transition point $F_2(p,T) = 0$. When can this be true? It is true, when in the symmetrical phase, the zero value of the O.P. must correspond to the stable state (i.e., to the minimum of F) when $F_2(p,T) > 0$. On the other side of the transition point, in the unsymmetrical phase, the nonzero values of the O.P. must also correspond to the stable state (i.e., to the minimum of F) when only $F_2(p,T) < 0$, and this corresponds to the so-called **Mexican-hat potential**. Because $F_2(p,T)$ is positive on one side of the transition point and negative on the other, it must vanish at the transition point $F_2(p,T) = 0$ and, **in Landau theory, this corresponds to the phase transition.**
3. What about $F_3(p,T)$? If the transition point is a stable state, that is, if the function $F = F(p,T,\phi)$ is a minimum at $\phi = 0$, then it is necessary that the third-order term $F_3 = 0$ and the fourth-order term is positive:

$$F(\phi) = F_0 + F_2\phi^2 + F_4\phi^4 \tag{1108}$$

If F_4 is positive at the transition point, then it should be positive in the neighborhood of that point.

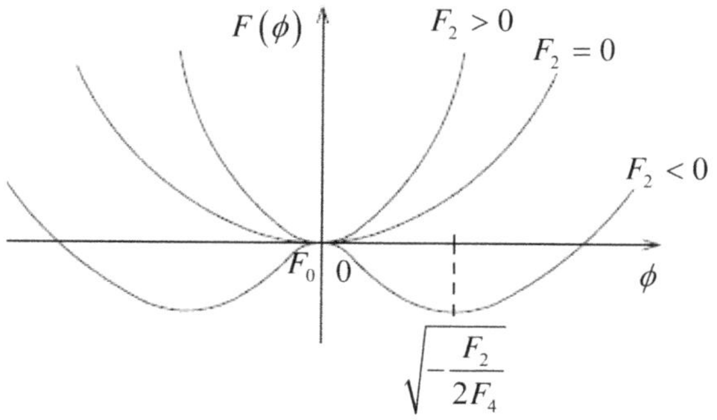

FIGURE 8.2 Variation of the free energy $F(\phi)$ versus the order parameter ϕ. In the second-order term $F_3(p,T)\equiv 0$ vanish at the transition point $F_2(p,T)=0$. In the symmetrical phase, the zero value of O.P. correspond to stable state when $F_2(p,T)>0$ while in the unsymmetrical phase, the nonzero values of O.P. correspond to stable state when only $F_2(p,T)<0$ and corresponding to the Mexican-hat potential. Because $F_2(p,T)$ is positive on one side and negative on the other of the transition point, it must vanish at the transition point $F_2(p,T)=0$ and, in Landau theory, corresponds to the phase transition.

Remark

We examine two particular cases. In one, we consider the third-order term identically zero due to the bodies' symmetry $F_3(p,T)\equiv 0$. This brings us to the only condition at the transition point $F_2(p,T)=0$ that determines the relation between p and T. This allows us to have a line of **phase transition points of the second kind** in the pT-plane. **Suppose that $F_3(p,T)\neq 0$, the transition points can be determined from two equations, $F_2(p,T)=0$ and $F_3(p,T)=0$. Therefore, the continuous phase transitions may exist only at isolated points.**

Unless otherwise stated, we consider the case where a line of continuous transition points is of interest. This case involves transitions resulting from the appearance or disappearance of a magnetic structure due to their symmetry under time reversal. It is instructive to note that this transformation cannot alter the thermodynamic potential while the magnetic moment (acting as O.P.) changes sign. This can be a motivation for the expansion of the thermodynamic potential not having odd-order terms. For $F_2(p,T_c)=0$ with $p=const$, then

$$F_2 = aV\frac{\tau}{2} \quad , \quad F_4 = b\frac{V}{4}>0 \quad , \quad \tau=\frac{T-T_c}{T_c} \tag{1109}$$

and

$$F(\phi)=F_0+aV\frac{\tau}{2}\phi^2+b\frac{V}{4}\phi^4 \tag{1110}$$

It should be noted that

$$F_2 = aV\frac{\tau}{2}>0 \tag{1111}$$

in the symmetrical phase and

$$F_2 = aV\frac{\tau}{2}<0 \tag{1112}$$

in the unsymmetrical phase, and the **transition points are determined from**

$$F_2 = aV\frac{\tau}{2} = 0 \tag{1113}$$

Neglecting the field contribution, we have

$$F(\phi) = F_0 + aV\frac{\tau}{2}\phi^2 + b\frac{V}{4}\phi^4$$

The dependence of ϕ on the temperature in the neighborhood of the transition point in the unsymmetrical phase can be determined from the condition that the free energy should be a minimum as a function of ϕ:

$$\frac{\partial}{\partial\phi}F(\phi) = aV\tau\phi + bV\phi^3 = 0 \tag{1114}$$

If $\tau > 0\,(T > T_c)$ (symmetrical phase), we have $\phi = 0$ for a stable minimum, and if $\tau < 0\,(T < T_c)$ (unsymmetrical phase), we have $\phi = 0$ for a maximum and $\phi_0 = \pm\left(\frac{a|\tau|}{b}\right)^{\frac{1}{2}}$ for minima.

8.2 Entropy and Specific Heat

From our knowledge of the mean-field free energy as a function of temperature, we can calculate further thermodynamic variables. For spontaneous symmetry breaking, the entropy

$$S = -\left.\frac{\partial F}{\partial T}\right|_{\text{on a line}} = -\left(\frac{\partial F}{\partial \phi}\right)_T \frac{d\phi}{dT} - \left(\frac{\partial F}{\partial T}\right)_\phi = -\left(\frac{\partial F}{\partial T}\right)_{\phi=const} \tag{1115}$$

or

$$S = S_0 - \frac{a}{2}V\frac{\phi^2}{T_c} = S_0 - \frac{a^2}{2b}V\frac{\tau}{T_c} = S_0 + \frac{a^2}{2b}V\frac{(T-T_c)}{T_c} \tag{1116}$$

At $T = T_c$ or $\tau = 0$, then $S = S_0$ (symmetrical phase) and the change in heat is

$$\Delta q = T(S_1 - S_2) = 0 \tag{1117}$$

Therefore, we see that at the transition point $T = T_c$, the entropy is **continuous** as expected. By definition, this means that **the phase transition is continuous** (i.e., not of first order). As the aforementioned expression for F includes only the contributions of superconductivity or superfluidity, the entropy calculated from it also will contain only these contributions.

The heat capacity at constant volume of the superconductor or superfluid can be obtained as follows

$$C_p = T\left(\frac{\partial S}{\partial T}\right)_p = C_{p0} + \frac{a^2V}{2bT_c} \tag{1118}$$

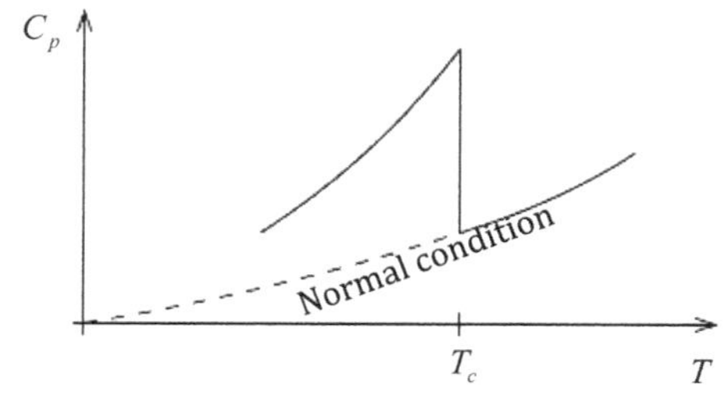

FIGURE 8.3 Variation of the heat capacity C_p versus the temperature T.

Thus, the heat capacity at $T = T_c$ has a **jump discontinuity** of

$$\Delta C_p = \frac{a^2 V}{2bT_c} \tag{1119}$$

We see that the heat capacity is discontinuous at the phase transition point of the second-order. Because $b > 0$ and $C_p > \Delta C_p$, at the transition point, the specific heat increases when going from the symmetrical to the unsymmetrical phase irrespective of their positions on the temperature scale.

Question

What are the discontinuity of C_V, the thermal expansion coefficient, and the compressibility?

8.3 External Field Effect on a Phase Transition

We now consider how the properties of a phase transition change when a body is subjected to an external field with a value depending on the O. P., ϕ. The field appears as a perturbation $-Vh\phi$ that is linear in the field strength $h \neq 0$ and in the O.P. it is:

$$F(\phi) = F_0 + aV\frac{\tau}{2}\phi^2 + b\frac{V}{4}\phi^4 - Vh\phi \tag{1120}$$

It is instructive to note that no matter how weak the O.P. is in any field, $\phi \neq 0$ for all temperatures. So, the field reduces the symmetry of the more symmetrical phase and thereby leads to the disappearance of the difference between the two phases. Consequently, the discrete phase transition point disappears, and the transition is "**smoothed out.**" We show that **instead of a sharp discontinuity in the specific heat, there is an anomalous spread over a temperature range**.

We now will do a quantitative investigation of the transition by writing the equilibrium condition:

$$\frac{\partial F}{\partial \phi} = \tau a V\phi + bV\phi^3 - Vh = 0 \quad , \quad h = \tau a\phi + b\phi^3 \tag{1121}$$

This permits us to see the dependence of the O.P. ϕ on the field h, which should be different from the temperature below and above the critical temperature T_c. Considering the expression for h in equation (1121), when $\tau > 0$, its right-hand side should increase monotonically with increased ϕ. We may foresee that the equation for any given value h has only one real root that vanishes when $h = 0$. Thus, the function $\phi(h)$ should be single-valued with the sign of h corresponding to that of ϕ.

When $\tau > 0$, then h is not a monotonous function of ϕ. Therefore, we resort to three different roots over a certain range of values of h. Consequently, $\phi(h)$ is no longer single-valued.

If $\tau = 0$, then $\phi = \left(\frac{h}{b}\right)^{\frac{1}{3}}$ is the O.P. in the strong fields. We can verify that in this limit, the specific heat C_p is independent of the field.

We may define the critical index (critical isotherm) δ as $\phi = h^{\frac{1}{\delta}}$ and then from the Landau theory, $\delta = 3$.

The susceptibility

$$\chi = \left(\frac{\partial \phi}{\partial h}\right)\bigg|_{h \to 0} \tag{1122}$$

and

$$\frac{\partial \phi}{\partial h}\left(\tau a + 3b\phi^2\right) = 1 \tag{1123}$$

It should be noted that

$$\phi^2 = \begin{cases} -\dfrac{\tau a}{b} & , \quad \tau < 0 \\ 0 & , \quad \tau > 0 \end{cases} \tag{1124}$$

If we consider that

$$\phi^2 = -\frac{\tau a}{b} \tag{1125}$$

Then

$$\frac{\partial \phi}{\partial h}(\tau a - 3\tau a) = 1 \tag{1126}$$

or

$$\chi = \frac{1}{2a|\tau|} \quad , \quad T < T_c \tag{1127}$$

And if

$$\phi^2 = 0 \tag{1128}$$

then

$$\frac{\partial \phi}{\partial h}(\tau a) = 1 \tag{1129}$$

or

$$\chi = \frac{1}{a\tau} \quad , \quad T > T_c \tag{1130}$$

Thus,

$$\chi = \begin{cases} \chi = \dfrac{1}{a\tau} & , \quad T > T_c \\ \dfrac{1}{2a|\tau|} & , \quad T < T_c \end{cases} \tag{1131}$$

and $\chi \approx |\tau|^{-\gamma}$ corresponds to the region of weak fields; the exponent $\gamma = 1$ also is assigned to this region. As already mentioned, the infinite value of χ for $\tau \to 0$ is a consequence of the fact that the minimum of $F(\phi)$ becomes steadily achieved as the transition point is approached. As a result of this, we observe that a slight perturbation has a considerable effect on the equilibrium value of ϕ. Then, from the Landau theory, $\gamma = 1$. An additional exponent can be obtained from $C_p \approx |\tau|^{-\alpha}$, and thus, $\alpha = 0$. Hence, the Landau theory gives the following critical exponents:

$$\alpha = 0 \, , \beta = \frac{1}{2} \, , \gamma = 1 \, , \delta = 3 \tag{1132}$$

From experimentation, this may not always be true due to fluctuation. So, we observe from the aforementioned that the critical point $T = T_c$ is the prominent instance where the Fermi liquid theory breaks down.

8.4 Ginzburg-Landau Theory

We should note that the essence of phase transition of the second order is a singularity of the thermodynamic functions for the body. What should be the physical nature of this singularity? This entails an anomalous increase in the fluctuations of the O.P. that relates the gentle excursion to the extremal value (minimum) of the thermodynamic potential about the transition point. In this section, we consider that the change in symmetry in the transition is described only by one parameter ϕ. **In order to also describe spatially nonuniform situations, Ginzburg and Landau went beyond the Landau description for a constant order parameter**.

Consider the volume V of the sample and that of the thermostat or environment V_0 such that

$$V + V_0 = const \tag{1133}$$

then the total energy fluctuation

$$\Delta E_{tot} = \Delta E + \Delta E_0 = \Delta E - p_0 \Delta V_0 + T_0 \Delta S_0 \tag{1134}$$

is needed to bring the system out of equilibrium for given constant values of pressure p_0 and temperature T_0. Here, ΔE is the fluctuation of the energy of the sample, and ΔE_0 is that of the environment. It should be noted that for mechanical stability,

$$\Delta V + \Delta V_0 = 0 \tag{1135}$$

and no heat flow:

$$\Delta S + \Delta S_0 = 0 \tag{1136}$$

then

$$\Delta E_{tot} = \Delta E + p_0 \Delta V + T_0 \Delta S = \Delta\left(E + p_0 V + T_0 S\right) = \Delta F \tag{1137}$$

The fluctuation probability for constant values of pressure and temperature:

$$W = \text{const} \times \exp\left\{-\frac{\Delta F}{T}\right\} \tag{1138}$$

Because the O.P. changes in space $\phi = \phi(r)$ due to fluctuation, F is a functional of the O.P.:

$$F[\phi] = F_0 + \int\left(\frac{1}{2}c(\nabla\phi)^2 + \frac{1}{2}a\tau\phi^2 + \frac{1}{4}b\phi^4 - h\phi\right)dV \tag{1139}$$

We assume fluctuation to be small with the main role played by long-wavelength fluctuations in which the fluctuating quantity varies slowly through the body:

$$\tilde{\phi} = \phi - \phi_0 = \Delta\phi << \phi_0 \quad , \quad \phi = \phi_0 + \tilde{\phi} \tag{1140}$$

We shall see later on that these fluctuations increase anomalously in the neighborhood of the transition point.

Let us do the Fourier transformation

$$\tilde{\phi} = \frac{1}{\sqrt{V}}\sum_{\vec{\kappa}} \phi_{\vec{\kappa}} \exp\{i\vec{\kappa}\vec{r}\} \tag{1141}$$

Since ϕ is supposed to be real, then

$$\phi_{-\vec{\kappa}} = \phi_{\vec{\kappa}}^* \tag{1142}$$

and

$$\phi^2 = \left(\tilde{\phi} + \phi_0\right)^2 = \phi_0^2 + 2\tilde{\phi}\phi_0 + \tilde{\phi}^2 \quad , \quad \phi^4 = \left(\tilde{\phi} + \phi_0\right)^4 = \phi_0^4 + 4\tilde{\phi}\phi_0^3 + 6\phi_0^2\tilde{\phi}^2 + 4\phi_0\tilde{\phi}^3 + \tilde{\phi}^4 \tag{1143}$$

We look at the particular case of the fluctuations in the symmetrical phase. We consider the absence of a field h. It should be noted that $\Delta\phi$ does not have a constant part and

$$\phi_{\vec{\kappa}=0} = 0 \, , \phi_{\vec{\kappa}=0}^3 = 0 \tag{1144}$$

and then

$$\Delta F = \sum_{\vec{\kappa}(\text{semisphere})} \left(c\kappa^2 + a\tau + 3b\phi_0^2\right)\left|\phi_{\vec{\kappa}}\right|^2 \tag{1145}$$

This permits us to evaluate the fluctuation probability

$$W = \text{const}\times\exp\left\{-\frac{\Delta F}{T}\right\} = \text{const}\times\prod_{\vec{\kappa}}\exp\left\{-\frac{\left(c\kappa^2+a\tau+3b\phi_0^2\right)}{T}\left|\phi_{\vec{\kappa}}\right|^2\right\} \tag{1146}$$

This shows the independence of fluctuations with different wave lengths.

Let us evaluate the fluctuation correction to the free energy:

$$F_{fl} = -T\ln\sum_{\vec{\kappa}}\exp\left\{\frac{\Delta F}{T}\right\} = -T\ln\prod_{\vec{\kappa}}\int\exp\left\{-\frac{\left(c\kappa^2+a\tau+3b\phi_0^2\right)}{T}\left|\phi_{\vec{\kappa}}\right|^2\right\}d\,\mathrm{Re}\phi_{\vec{\kappa}}\,d\,\mathrm{Im}\phi_{\vec{\kappa}} \tag{1147}$$

or

$$F_{fl} = -T\ln\sum_{\vec{\kappa}}\ln\int_0^{\infty}\exp\left\{-\frac{\left(c\kappa^2+a\tau+3b\phi_0^2\right)}{T}\left|\phi_{\vec{\kappa}}\right|^2\right\}2\pi\left|\phi_{\vec{\kappa}}\right|d\left|\phi_{\vec{\kappa}}\right| \tag{1148}$$

or

$$F_{fl} = -T\ln\sum_{\vec{\kappa}}\ln\int_0^{\infty}\exp\left\{-\frac{\left(c\kappa^2+a\tau+3b\phi_0^2\right)}{T}\left|\phi_{\vec{\kappa}}\right|^2\right\}\pi d\left|\phi_{\vec{\kappa}}\right|^2 = -T\sum_{\vec{\kappa}}\ln\left[\frac{\pi T}{c\kappa^2+a\tau+3b\phi_0^2}\right] \tag{1149}$$

or

$$F_{fl} = -T\sum_{\vec{\kappa}}\ln\left[\frac{\pi T}{c\kappa^2+a\tau+3b\phi_0^2}\right] \tag{1150}$$

Considering $\tau = \frac{T-T_c}{T_c}$, from

$$\frac{\partial}{\partial T} = \frac{\partial}{\partial\tau}\frac{\partial\tau}{\partial T} = \frac{1}{T_c}\frac{\partial}{\partial\tau} \quad , \quad \frac{\partial^2}{\partial T^2} = \frac{1}{T_c^2}\frac{\partial^2}{\partial\tau^2} \tag{1151}$$

and the fluctuation correction to the specific heat, then

$$C_{fl} = -\frac{1}{T_c^2}\frac{\partial^2 F_{fl}}{\partial\tau^2} = a^2\sum_{\vec{\kappa}}\frac{1}{\left(c\kappa^2+a\tau+3b\phi_0^2\right)^2} \tag{1152}$$

From

$$\sum_{\vec{\kappa}}(\cdots) = V\int\frac{d\vec{\kappa}}{(2\pi)^3}(\cdots) = V\int\frac{2\pi\kappa^2 d\kappa}{(2\pi)^3}(\cdots) \tag{1153}$$

then

$$C_{fl} = a^2 \sum_{\vec{\kappa}} \frac{1}{\left(c\kappa^2 + a\tau + 3b\phi_0^2\right)^2} = \frac{a^2 V}{4\pi^2} \int_0^\infty \frac{\kappa^2 d\kappa}{\left(c\kappa^2 + A\right)^2} = \frac{a^2 V}{16\pi c^{\frac{3}{2}}} \frac{1}{\left(a\tau + 3b\phi_0^2\right)^{\frac{1}{2}}} \tag{1154}$$

Considering the **Ornstein-Zernike theory** we then have

$$C_{fl} \approx \frac{1}{|\tau|^{\frac{1}{2}}} \text{ for } \phi_0 = 0 \tag{1155}$$

From the mean-field theory, C is a step at T_c. So, including fluctuations, we obtain a divergence of the form in (1155) for $T \to T_c$. This is a consequence of superfluid fluctuations in the normal state. The system **exhibits** superfluid states at relatively low energies, while the mean-field state is still normal. From

$$\frac{a^2 V}{16\pi c^{\frac{3}{2}}} \frac{1}{(a\tau)^{\frac{1}{2}}} << \frac{a^2 V}{3bT_c} \tag{1156}$$

or

$$1 >> \tau >> \left(\frac{bT_c}{\left(ac^3\right)^{\frac{1}{2}}} \right)^2 = \frac{b^2 T_c^2}{ac^3} \tag{1157}$$

then

$$\frac{b^2 T_c^2}{ac^3} << \tau << 1 \tag{1158}$$

where

$$G_i = \frac{b^2 T_c^2}{ac^3} \tag{1159}$$

This is called the **Ginzburg number**. The Ginzburg-Landau theory is **not applicable** for

$$\tau < G_i \tag{1160}$$

where fluctuations are strong. So, the condition

$$G_i << \tau << 1 \tag{1161}$$

should be satisfied.

For superconductors,

$$G_i \approx \left(\frac{\Delta}{\epsilon_F} \right)^2 \approx 10^{-8} << 1 \tag{1162}$$

This shows that **for superconductors, the Ginzburg-Landau theory is always applicable**. For magnetics, $G_i \approx 0.1 \div 0.2$. We can thus conclude that **the Ginzburg-Landau theory does not work within the region**

$$G_i > 1 \tag{1163}$$

Considering the superfluidity of helium four (He-4), the Ginzburg-Landau theory is not applicable at all. We now examine how to describe the behavior of the system for

$$\tau << G_i \tag{1164}$$

with the help of the scaling invariance.

Question

What is the jump in the compressibility $-\left(\frac{\partial V}{\partial p}\right)\Big|_T$, the jump in the specific heat ΔC_V, and the jump in the thermal expansion coefficient $\frac{1}{V}\left(\frac{\partial V}{\partial T}\right)\Big|_p$ at the second-order phase transition point:

$$\beta = -\frac{1}{V}\left(\frac{\partial V}{\partial p}\right)\Bigg|_T \tag{1165}$$

$$\alpha = \frac{1}{V}\left(\frac{\partial V}{\partial T}\right)\Bigg|_p \tag{1166}$$

Let us understand the Ginzburg criterion using a simple formula. We use the Landau theory and assume $\tau \approx 1$ for the purpose of estimation.

$$\frac{E}{V} \approx \frac{F}{V} \approx \in \frac{N}{V} \tag{1167}$$

where $\in$ is the energy of the particles and $\frac{N}{V}$ is the density of the particles:

$$\in \approx T_c \quad , \quad \frac{N}{V} \approx \frac{1}{r_0^3} \tag{1168}$$

and where r_0 is an average distance between the particles. On the other hand,

$$\frac{E}{V} \approx \frac{F}{V} \approx a\phi^2 \approx b\phi^4 \tag{1169}$$

Assuming $\tau \approx 1$ or $\phi^2 \approx \frac{a}{b}$, then

$$E \approx \frac{a^2}{b} \approx \frac{T_c}{r_0^3} \quad , \quad r_0^3 \approx \frac{T_c b}{a^2} \tag{1170}$$

On the other hand,

$$c(\nabla\phi)^2 \approx c\phi^2\kappa_c^2 \quad , \quad \kappa_c \approx \frac{1}{r_c} \tag{1171}$$

where r_c is the correlation radius of the fluctuation. This determines the order of magnitude of the distance at which the correlation decreases significantly.

$$c(\nabla\phi)^2 \approx a\tau\phi^2 \quad , \quad \frac{c}{r_c^2} \approx a\tau \to r_c \approx \left(\frac{c}{a\tau}\right)^{\frac{1}{2}} \tag{1172}$$

As the transition point is approached, the correlation radius increases as

$$r_c \approx \left(\frac{c}{a\tau}\right)^{\frac{1}{2}} \tag{1173}$$

then

$$\left(\frac{r_0}{r_c}\right)^6 \approx \frac{T_c^2 b^2}{a^4 \frac{c^3}{a^3}} \approx \frac{T_c^2 b^2}{ac^3} = G_i \tag{1174}$$

Here, $r_0 \approx a$ is the interatomic distance. Therefore, if $\left(\frac{r_0}{r_c}\right) >> 1$, then there is room for the Landau theory. If $\left(\frac{r_0}{r_c}\right) \approx 1$, then it cannot be smaller than the interatomic distance. There is no room for the Landau theory and

$$G_i << |\tau| << 1 \tag{1175}$$

This ensures that the fluctuations are sufficiently small and also that the Landau theory is applicable. We also observe that at the transition point itself, the correlation function decreases as $\frac{1}{r_c}$. If we find ourselves in the region $|\tau| << G_i$ (the critical region), then the fluctuations are dominant and we resort to scaling $r \to \lambda r$ with the hypothesis that asserts that all physical relations do not depend on the scale.

Let us define the critical exponents. We have seen previously that $\phi \approx \tau^{\frac{1}{2}}$ and, in general, $\phi \approx \tau^{\beta}$. From the Landau theory, $\beta = \frac{1}{2}$. It is instructive to note that the existence of phase transitions of the second-order is based on certain hypotheses that are entirely plausible. However, this is not rigorously proven. The results can be proven by experimentation and by numerical calculations for certain simple models. The results show that as $T \to T_c$, the quantity $\frac{\partial C_p}{\partial T}$ as well as C_p become infinite in many cases. This permits us to draw various conclusions on the behavior of certain thermodynamic quantities. When $C_p = T\left(\frac{\partial S}{\partial T}\right)_p$ becomes infinite, it follows that the entropy of the body can be written as

$$S = S(T, p - p_c(T)) \tag{1176}$$

with $p = p_c(T)$ describing the equation of the curve of points of the phase transition in the pT-plane.

The universality of the limiting laws for the behavior of a substance in the fluctuation range in the neighborhood of the transition point also implies the universality of the critical exponents. We expect their values to be the same for all transitions with a change of symmetry described by only one order parameter.

8.5 The Scaling Hypothesis

Here, we find the set of critical exponents $\{\alpha,\beta,\gamma,\delta,\cdots\}$ that characterizes the singular behavior at the vicinity of a second-order critical point. This power law dependence on thermodynamic quantities are a symptom of scaling behavior. Mean field estimates of the critical exponents are unreliable due to fluctuations. Nonetheless, since the various thermodynamic quantities are related, the critical exponents are not completely independent of each other. The aim of this section is to use the scaling hypothesis to establish relationships between these exponents.

The important consequence of the scaling hypothesis is the additional dilation symmetry of the critical system. Under a change of scale, the critical correlation functions behave as follows:

$$A(\lambda r)=\lambda^{-\Delta_A}A(r) \tag{1177}$$

This implies a scale invariance or self-similarity [39, 40].

Suppose a snapshot of the critical system is enlarged by a factor of λ. Notwithstanding the change in contrast $\left(\lambda^{-\Delta_A}\right)$, the resulting snapshot statistically imitates the original one. Let us apply this to the order parameter

$$\phi(\lambda r)=\lambda^{-\Delta_\phi}\phi(r) \quad , \quad \tau(\lambda r)=\lambda^{-\Delta_\tau}\tau(r) \quad , \quad \tau=\frac{T-T_c}{T_c} \tag{1178}$$

It should be noted that

$$\phi(\lambda r)=f\left(\tau(\lambda r)\right) \tag{1179}$$

Then

$$\phi(\lambda r)=\lambda^{-\Delta_\phi}\phi(r)=\lambda^{-\Delta_\phi}\tau^{\beta}(r) \quad , \quad \tau(\lambda r)=\lambda^{-\Delta_\tau}\tau(r) \tag{1180}$$

$$\tau(r)=\frac{\tau(\lambda r)}{\lambda^{-\Delta_\tau}} \quad , \quad \phi(\lambda r)=\lambda^{-\Delta_\phi}\lambda^{-\Delta_\tau\beta}\tau^{\beta}(\lambda r) \quad , \quad \Delta_\tau\beta-\Delta_\phi=0 \quad , \quad \beta=\frac{\Delta_\phi}{\Delta_\tau} \tag{1181}$$

8.6 Identities from the d-Dimensional Space

We now consider the generalized homogeneity assumption involving the d-dimensional space, which are known as hyperscaling relations:

$$V(\lambda r)=\lambda^{-\Delta_d}V(r)=\lambda^{d}V(r) \quad , \quad h(\lambda r)=\lambda^{-\Delta_h}h(r) \quad , \quad \phi(\lambda r)=\lambda^{-\Delta_\phi}\phi(r) \quad , \quad \Delta_\phi+\Delta_h=d \quad , \quad \Delta_d=-d \tag{1182}$$

We examine the susceptibility

$$\chi = \left.\frac{\partial\phi}{\partial h}\right|_{h\to 0} \quad , \quad \gamma = \frac{\Delta_h - \Delta_\phi}{\Delta_\tau} \quad , \quad \Delta_\chi = \Delta_\phi - \Delta_h \quad , \quad \delta = \frac{\Delta_h}{\Delta_\phi} \tag{1183}$$

Proof

$$\phi(\lambda r) = \lambda^{-\Delta_\phi}\phi(r) = \lambda^{-\Delta_\phi} h^{\frac{1}{\delta}}(r) \quad , \quad h(\lambda r) = \lambda^{-\Delta_h} h(r) \to h(r) = \frac{h(\lambda r)}{\lambda^{-\Delta_h}} \tag{1184}$$

$$\phi(\lambda r) = \lambda^{-\Delta_\phi + \frac{\Delta_h}{\delta}} h^{\frac{1}{\delta}}(\lambda r) \quad , \quad \Delta_\phi = \frac{\Delta_h}{\delta} \to \delta = \frac{\Delta_h}{\Delta_\phi} \tag{1185}$$

The specific heat

$$C = -\frac{1}{V}\frac{\partial^2 F[\phi]}{\partial\tau^2} \quad , \quad C_{(\lambda r)} = \lambda^{-\Delta_C} C(r) \tag{1186}$$

$$C(r) \approx |\tau|^{-\alpha} \quad , \quad \Delta_C = d - 2\Delta_\tau \tag{1187}$$

$$C(\lambda r) = \lambda^{-\Delta_C} C(r) = \lambda^{-\Delta_C} |\tau|^{-\alpha} \tag{1188}$$

$$\tau(\lambda r) = \lambda^{-\Delta_\tau}\tau(r) \to \tau(r) = \lambda^{\Delta_\tau}\tau(\lambda r) \tag{1189}$$

$$-\Delta_C - \alpha\Delta_\tau = 0 \quad , \quad 2\Delta_\tau - \alpha\Delta_\tau = 0 \tag{1190}$$

$$\alpha = 2 - \frac{d}{\Delta_\tau} \tag{1191}$$

Let us introduce new indices following the definition:

$$\Delta_\phi = \frac{1}{2}(d - 2 + \eta) \tag{1192}$$

The correlation length:

$$r_c(\tau) \approx \tau^{-\nu} \quad , \quad \nu = \frac{1}{\Delta_\tau} \tag{1193}$$

As proof:

$$r_c(\lambda r)=\lambda^{-\Delta_{r_c}}r_c(r)=\lambda\tau^{-\nu} \quad , \quad \Delta_{r_c}=-1 \tag{1194}$$

$$\tau(r)=\lambda^{\Delta_\tau}\tau(\lambda r) \quad , \quad 1-\nu\Delta_\tau=0 \tag{1195}$$

$$\nu=\frac{1}{\Delta_\tau} \tag{1196}$$

We exclude Δ_τ and Δ_ϕ, so the scaling equations:

$$\alpha=2-\frac{d}{\Delta_\tau}=2-d\nu \tag{1197}$$

$$2-\alpha=d\nu \tag{1198}$$

$$\gamma=\frac{\Delta_h-\Delta_\phi}{\Delta_\tau} \quad , \quad \Delta_h=\Delta_\phi\delta \tag{1199}$$

$$\gamma=\frac{\Delta_\phi}{\Delta_\tau}(\delta-1)=\beta(\delta-1) \tag{1200}$$

$$\beta+\gamma=\beta\delta \tag{1201}$$

$$\beta=\frac{\Delta_\phi}{\Delta_\tau}=\frac{1}{2}\nu(d-2+\eta)\to 2\beta=\nu(d-2+\eta) \tag{1202}$$

$$\Delta_\phi+\Delta_h=d\to(1+\delta)(d-2+\eta)=2d \tag{1203}$$

$$2\beta=\frac{2\nu d}{1+\delta}=\frac{2-\alpha}{1+\delta} \tag{1204}$$

From equation (1201), we have

$$1+\delta=\frac{2\beta+\gamma}{\beta} \tag{1205}$$

and

$$\alpha+4\gamma+2\beta=2 \tag{1206}$$

$$\nu d-\nu(2-\eta)=2\beta \tag{1207}$$

$$2-\alpha-2\beta=\nu(2-\eta) \tag{1208}$$

From equation (1204), we have

$$\delta = \frac{2-\alpha-\beta}{\beta} \tag{1209}$$

From equation (1201), we have

$$\beta\delta = 2-\alpha-\beta = \beta+\gamma \tag{1210}$$

$$\alpha+2\beta+\gamma = 2 \tag{1211}$$

$$\nu(2-\eta) = \gamma \tag{1212}$$

We observe from the aforementioned that two independent exponents are sufficient to describe all singular critical behavior:

$$\alpha+2\beta+\gamma = 2 \quad , \quad \beta+\gamma = \beta\delta \quad , \quad \nu(2-\eta) = \gamma \quad , \quad 2-\alpha = d\nu \tag{1213}$$

The first equation is credited to J. W. Essam and M. E. Fisher [41–44], and the second is credited to B. Widom [45, 46].

If we assume that $\alpha = 0 \quad , \quad \eta = 0$, then

$$2\beta+\gamma = 2 \quad , \quad \beta+\gamma = \beta\delta \quad , \quad \gamma = 2\nu \quad , \quad \nu = \frac{2}{d} \tag{1214}$$

For $d = 3$, then

$$\nu = \frac{2}{3} \quad , \quad \gamma = \frac{4}{3} \quad , \quad \beta = \frac{1}{2}\left(2-\frac{4}{3}\right) = \frac{1}{3} \quad , \quad \delta = \frac{\beta+\gamma}{\beta} = \frac{\frac{5}{3}}{\frac{1}{3}} = 5 \tag{1215}$$

Let us substitute Landau theory indices into the scaling equations in Table 8.2.

We see from Tables 8.1 and 8.2 that the relation between α and ν does not agree with the mean-field values, $\alpha = 0$ and $\nu = \frac{1}{2}$, which are valid for $d > 4$. Any theory of critical behavior consequently must account for the validity of this relation in low dimensions and the breakdown for $d > 4$. It follows from

TABLE 8.1 Scaling Theory and the Landau Theory Indices.

Theory	α	β	γ	δ	ν	η
Scaling Theory	0	1/2	1	3	1/2	0
Landau Theory	0	1/3	4/3	5	2/3	0

TABLE 8.2 Scaling Theory and the Landau Theory Indices.

Scaling Theory	$\alpha + 2\beta + \gamma = 2$	$\beta + \gamma = \beta\delta$	$(2-\eta)\nu = \gamma$	$2 - \alpha = d\nu$
Landau Theory	$0 + 2(1/2) + 1 = 2$	$1/2 + 1 = (1/2)3$	$2(1/2) = 1$	$d = 4$

here that the Landau theory is always valid for dimensionality d = 4. If d ≥ 4, the fluctuations are weak, and if d < 4, the fluctuations are strong. Thus, the heat capacity associated with fluctuation

$$C_{fl}^{d=4} \approx \int_0^\infty \frac{\kappa^3 d\kappa}{\left(c\kappa^2 + a\tau + 3b\phi_0^2\right)^2} \approx \ln|\tau| \tag{1216}$$

It follows from here that all fluctuation corrections to the specific heat ARE NOT power-law $|\tau|^n$ but $\left(\ln|\tau|\right)^n$. Examples are some ferro-electrons and some antiferromagnets.

8.7 Energy Fluctuation

Consider the volume V of the sample and that of the thermostat or environment V_0, then the total energy fluctuation

$$\Delta E_{\text{tot}} = \Delta E + \Delta E_0 = \Delta E - p_0 \Delta V_0 + T_0 \Delta S_0 \tag{1217}$$

Here ΔE is the fluctuation of the energy of the sample and ΔE_0 is that of the environment. It should be noted that for mechanical stability,

$$\Delta V + \Delta V_0 = 0 \tag{1218}$$

and for no heat flow,

$$\Delta S + \Delta S_0 = 0 \tag{1219}$$

then

$$\Delta E_{\text{tot}} = \Delta E + p_0 \Delta V - T_0 \Delta S = \Delta\left(E + p_0 V - T_0 S\right) = \Delta \Phi_{\text{Gibbs}} \tag{1220}$$

This permits us to calculate the fluctuation probability

$$W = \text{const} \times \exp\left\{-\frac{\Delta E_{\text{tot}}}{T}\right\} = \exp\left\{-\frac{\Delta \Phi_{\text{Gibbs}}}{T}\right\} \tag{1221}$$

Because the order parameter changes in space $\phi = \phi(r)$ due to fluctuation, then $\boldsymbol{\Phi}$ is a functional of the O.P.:

$$\Phi[\phi] = \Phi_0 + \int\left(\frac{1}{2}c(\nabla\phi)^2 + \frac{1}{2}a\tau\phi^2 + \frac{1}{4}b\phi^4\right)dV \tag{1222}$$

We assume fluctuations to be small, with the main role played by long-wave fluctuations:

$$\tilde{\phi} = \phi - \phi_0 = \Delta\phi << \phi_0 \quad , \quad \phi = \phi_0 + \tilde{\phi} \tag{1223}$$

Let us do the Fourier transformation

$$\tilde{\phi} = \frac{1}{\sqrt{V}} \sum_{\vec{\kappa}} \phi_{\vec{\kappa}} \exp\left\{i\vec{\kappa}\vec{r}\right\} \tag{1224}$$

Because $\tilde{\phi}$ is supposed to be real, then from

$$\sum_{\vec{\kappa}} \phi_{\vec{\kappa}}^* \exp\{-i\vec{\kappa}\vec{r}\} = \sum_{\vec{\kappa}} \phi_{\vec{\kappa}} \exp\{i\vec{\kappa}\vec{r}\} \tag{1225}$$

we have

$$\phi_{-\vec{\kappa}} = \phi_{\vec{\kappa}}^* \quad , \quad \phi^2 = \left(\tilde{\phi}+\phi_0\right)^2 = \phi_0^2 + 2\tilde{\phi}\phi_0 + \tilde{\phi}^2 \quad , \quad \phi^4 = \left(\tilde{\phi}+\phi_0\right)^4 = \phi_0^4 + 4\tilde{\phi}\phi_0^3 + 6\phi_0^2\tilde{\phi}^2 + 4\phi_0\tilde{\phi}^3 + \tilde{\phi}^4 \tag{1226}$$

It should be noted that

$$(\nabla\phi)^2 = \nabla\phi\cdot\nabla\phi = \nabla\sum_{\vec{\kappa}_1}\phi_{\vec{\kappa}_1}\exp\{i\vec{\kappa}_1\vec{r}\}\nabla\sum_{\vec{\kappa}_2}\phi_{\vec{\kappa}_2}\exp\{i\vec{\kappa}_2\vec{r}\} \underset{\vec{\kappa}_2\to-\vec{\kappa}_1}{=} \nabla\sum_{\vec{\kappa}_1}\phi_{\vec{\kappa}_1}\exp\{i\vec{\kappa}_1\vec{r}\}\nabla\sum_{\vec{\kappa}_2}\phi_{-\vec{\kappa}_2}\exp\{-i\vec{\kappa}_2\vec{r}\} \tag{1227}$$

or

$$(\nabla\phi)^2 = \sum_{\vec{\kappa}_1}\phi_{\vec{\kappa}_1}\phi_{-\vec{\kappa}}^*\vec{\kappa}_1\vec{\kappa}_2\exp\{i(\vec{\kappa}_1-\vec{\kappa}_2)\}\vec{r} \tag{1228}$$

But

$$\int d\vec{r}\exp\{i(\vec{\kappa}_1-\vec{\kappa}_2)\}\vec{r} = \delta_{\vec{\kappa}_1\vec{\kappa}_2} \tag{1229}$$

then

$$(\nabla\phi)^2 = \sum_{\vec{\kappa}}\vec{\kappa}^2|\phi_{\vec{\kappa}}|^2 \quad , \quad \tilde{\phi}^2 = \sum_{\vec{\kappa}}|\phi_{\vec{\kappa}}|^2 \tag{1230}$$

and

$$\phi_{\vec{\kappa}=0} = 0\,,\,\phi_{\vec{\kappa}=0}^3 = 0 \tag{1231}$$

then

$$\Delta\Phi = \sum_{\vec{\kappa}(\text{semisphere})}\left(c\kappa^2 + a\tau + 3b\phi_0^2\right)|\phi_{\vec{\kappa}}|^2 \tag{1232}$$

This allows us to evaluate the fluctuation probability

$$W = \text{const}\times\exp\left\{-\frac{\Delta\Phi}{T}\right\} = \text{const}\times\prod_{\vec{\kappa}}\exp\left\{-\frac{\left(c\kappa^2+a\tau+3b\phi_0^2\right)}{T}|\phi_{\vec{\kappa}}|^2\right\} \tag{1233}$$

This shows the independence of fluctuations with different wave lengths.

Let us evaluate the fluctuation correction to the specific heat:

$$\Phi_{fl} = -T\ln\sum_{\vec{\kappa}}\exp\left\{\frac{\Delta\Phi}{T}\right\} = -T\ln\prod_{\vec{\kappa}}\int\exp\left\{-\frac{\left(c\kappa^2+a\tau+3b\phi_0^2\right)}{T}|\phi_{\vec{\kappa}}|^2\right\}d\,\mathrm{Re}\,\phi_{\vec{\kappa}}\,d\,\mathrm{Im}\,\phi_{\vec{\kappa}} \tag{1234}$$

or

$$\Phi_{fl} = -T\ln\sum_{\vec{\kappa}}\ln\int_0^{\infty}\exp\left\{-\frac{\left(c\kappa^2 + a\tau + 3b\phi_0^2\right)}{T}\left|\phi_{\vec{\kappa}}\right|^2\right\}2\pi\left|\phi_{\vec{\kappa}}\right|d\left|\phi_{\vec{\kappa}}\right| \tag{1235}$$

or

$$\Phi_{fl} = -T\ln\sum_{\vec{\kappa}}\ln\int_0^{\infty}\exp\left\{-\frac{\left(c\kappa^2 + a\tau + 3b\phi_0^2\right)}{T}\left|\phi_{\vec{\kappa}}\right|^2\right\}\pi d\left|\phi_{\vec{\kappa}}\right|^2 = -T\sum_{\vec{\kappa}}\ln\left[\frac{\pi T}{c\kappa^2 + a\tau + 3b\phi_0^2}\right] \tag{1236}$$

or

$$\Phi_{fl} = -T\sum_{\vec{\kappa}}\ln\left[\frac{\pi T}{c\kappa^2 + a\tau + 3b\phi_0^2}\right] \tag{1237}$$

Considering

$$\tau = \frac{T - T_c}{T_c} \tag{1238}$$

then from

$$\frac{\partial}{\partial T} = \frac{\partial}{\partial \tau}\frac{\partial \tau}{\partial T} = \frac{1}{T_c}\frac{\partial}{\partial \tau} \quad , \quad \frac{\partial^2}{\partial T^2} = \frac{1}{T_c^2}\frac{\partial^2}{\partial \tau^2} \tag{1239}$$

and

$$C_{fl} = -\frac{1}{T_c^2}\frac{\partial^2 \Phi_{fl}}{\partial \tau^2} = a^2\sum_{\vec{\kappa}}\frac{1}{\left(c\kappa^2 + a\tau + 3b\phi_0^2\right)^2} \tag{1240}$$

$$\sum_{\vec{\kappa}}(\cdots) = V\int\frac{d\vec{\kappa}}{(2\pi)^3}(\cdots) = V\int\frac{2\pi\kappa^2 d\kappa}{(2\pi)^3}(\cdots) \tag{1241}$$

$$C_{fl} = a^2\sum_{\vec{\kappa}}\frac{1}{\left(c\kappa^2 + a\tau + 3b\phi_0^2\right)^2} = \frac{a^2 V}{4\pi^2}\int_0^{\infty}\frac{\kappa^2 d\kappa}{\left(c\kappa^2 + A\right)^2} = \frac{a^2 V}{16\pi c^{\frac{3}{2}}}\frac{1}{\left(a\tau + 3b\phi_0^2\right)^{\frac{1}{2}}} \tag{1242}$$

Then, considering the Ornstein-Zernike theory, we have

$$C_{fl} \approx \frac{1}{|\tau|^{\frac{1}{2}}} \text{ for } \phi_0 = 0 \tag{1243}$$

$$\frac{a^2 V}{16\pi c^{\frac{3}{2}}}\frac{1}{(a\tau)^{\frac{1}{2}}} << \frac{a^2 V}{3bT_c} \tag{1244}$$

or

$$1 >> \tau >> \left(\frac{bT_c}{\left(ac^3\right)^{\frac{1}{2}}} \right)^2 = \frac{b^2T_c^2}{ac^3} \tag{1245}$$

$$\Delta C = \frac{a^2V}{3bT_c} \quad , \quad C_{fl} << \Delta C \tag{1246}$$

$$\frac{b^2T_c^2}{ac^3} << \tau << 1 \tag{1247}$$

The number

$$G_i = \frac{b^2T_c^2}{ac^3} \approx \left(\frac{r}{r_c} \right)^6 \tag{1248}$$

is called the Ginzburg number. The Ginzburg-Landau theory is **not applicable** for $\tau < G_i$, where fluctuations are strong. Thus the condition

$$G_i << \tau << 1 \tag{1249}$$

should be satisfied.

For superconductors,

$$G_i \approx \left(\frac{\Delta}{\epsilon_F} \right)^2 \approx 10^{-8} << 1 \tag{1250}$$

This shows that for superconductors, the Ginzburg-Landau theory is **always applicable**. For magnetics, $G_i \approx 0.1 \div 0.2$. Thus, we can conclude that the Ginzburg-Landau theory does not work within the region $G_i > 1$. Considering the superfluidity of helium-four (He-4), the Ginzburg-Landau theory is not applicable at all.

In the next chapter, we will examine how to describe the behavior of the system for $\tau << G_i$ with the help of the scaling invariance.

9

Weakly Interacting Bose Gas

Introduction

Interactions in atomic gases are usually very weak, but they can have important effects. For example, superfluidity does not occur in an ideal Bose gas but exists in the weakly interacting Bose system. When particles in a Bose gas interact with each other, quantum field theory is needed to understand their behavior. Considering weak and short-range interactions, in the limit of low temperature and low density, the Bose condensation can be studied by a procedure outlined by Bogoliubov [47, 48].

Consider a superfluid He-4 poured into a container filled with a very fine glass powder that could act like a weakly interacting Bose gas, due to the diluting effect of the powder. The condensation process observed in such a system can be explained by the superfluid transition in an ensemble of very small micro grain fluid elements of He-4. The condition of being dilute permits one to consider only the s-wave in the scattering process. Consequently, the interaction can be approximated by a δ-function repulsion that acts only upon the s-wave.

We now consider the problem of two interacting electrons in the presence of a filled Fermi sea. The time-independent second-quantized grand-canonical energy has a free particle term:

$$\hat{H}_0 = \int d\vec{r}\left(\frac{1}{2m}\nabla\hat{\psi}_\sigma^\dagger(\vec{r})\nabla\hat{\psi}_\sigma(\vec{r}) - \mu\hat{\psi}_\sigma^\dagger(\vec{r})\hat{\psi}_\sigma(\vec{r})\right) \tag{1251}$$

or

$$\hat{H}_0 = \frac{1}{V}\sum_{\vec{\kappa}\vec{\kappa}'}\int d\vec{r}\left(\frac{1}{2m}(-i\vec{\kappa})(i\vec{\kappa}'')\hat{\psi}_{\vec{\kappa}'}^\dagger\hat{\psi}_{\vec{\kappa}} - \mu\hat{\psi}_{\vec{\kappa}'}^\dagger\hat{\psi}_{\vec{\kappa}}\right)\exp\left\{i(\vec{\kappa}-\vec{\kappa}')\vec{r}\right\} \tag{1252}$$

The interaction Hamiltonian determinant of the two-electron system has the form

$$\hat{H}_{\text{int}} = \frac{1}{2}\int d\vec{r}\int d\vec{r}'\hat{\psi}^\dagger(\vec{r})\hat{\psi}^\dagger(\vec{r}')U\left(\left|\vec{r}-\vec{r}'\right|\right)\hat{\psi}(\vec{r}')\hat{\psi}(\vec{r}) \tag{1253}$$

But

$$\frac{1}{V}\int d\vec{r}\exp\left\{i(\vec{\kappa}-\vec{\kappa}'')\vec{r}\right\} = \delta_{\vec{\kappa}\vec{\kappa}'} \quad , \quad \tilde{\epsilon}_0(\vec{\kappa}) = \frac{\hbar^2\vec{\kappa}^2}{2m} \tag{1254}$$

Then the free-particle Hamiltonian (1252) becomes

$$\hat{H}_0 = \sum_{\vec{\kappa}}\xi_{\vec{\kappa}}\hat{\psi}_{\vec{\kappa}}^\dagger\hat{\psi}_{\vec{\kappa}} \quad , \quad \xi_{\vec{\kappa}} = \tilde{\epsilon}_0(\vec{\kappa}) - \mu \tag{1255}$$

and we rewrite the interaction Hamiltonian determinant as

$$\hat{\mathrm{H}}_{\text{int}} = \sum_{\vec{\kappa},\vec{\kappa}',q\neq 0} \frac{U(\vec{q})}{2V} \hat{\psi}^{\dagger}_{\vec{\kappa}-\vec{q}} \hat{\psi}^{\dagger}_{\vec{\kappa}'+\vec{q}} \hat{\psi}_{\vec{\kappa}} \hat{\psi}_{\vec{\kappa}'} \tag{1256}$$

The sum over wave vectors is restricted to $|\vec{\kappa}| > \vec{\kappa}_F$; the Fermi wave vector, $\tilde{\in}_0(\vec{\kappa})$ is the single-particle energy and $\in(\vec{\kappa}) \equiv \xi_{\vec{\kappa}}$ is the energy in a grand-canonical ensemble. Considering only s-wave scattering, we restrict the potential to the repulsive δ-function potential

$$U(|\vec{r}-\vec{r}'|) = \lambda\delta(\vec{r}-\vec{r}') \tag{1257}$$

Then,

$$\hat{\mathrm{H}}_{\text{int}} = \frac{\lambda}{2V} \sum_{\vec{\kappa},\vec{\kappa}',q\neq 0} \hat{\psi}^{\dagger}_{\vec{\kappa}-\vec{q}} \hat{\psi}^{\dagger}_{\vec{\kappa}'+\vec{q}} \hat{\psi}_{\vec{\kappa}} \hat{\psi}_{\vec{\kappa}'} \tag{1258}$$

Here λ is the interaction strength.

We examine a simple example of a quantum liquid that is a weakly nonideal gas. In this case, the interaction among gas particles is relatively small. Thus, we examine a weakly interacting Bose gas at almost zero temperature. This requires that the scattering amplitude of the particles should be small compared to the parameter, which characterizes the range of the forces:

$$\frac{a}{r_0} << 1 \quad , \quad \frac{N}{V} a^3 << 1 \tag{1259}$$

where the scattering length is a, and r_0 characterizes the range of the forces (average distance between gas particles) [49].

It is useful to note that under the conditions of smallness of the momenta of the colliding particles, we consider only s-wave scattering to the first approximation. For the s-wave approximation, the interaction between every pair of particles is identical, and the scattering does not depend on the angle. The amplitude of the s-wave scattering is denoted by a. If the s-wave scattering contributes terms of the order $\frac{a}{r_0}$ and higher to the total energy, then the p-wave scattering contributes terms of an order no lower than $\left(\frac{a}{r_0}\right)^3$. Therefore, up to terms of first order, the scattering is regarded as isotropic. This permits us to neglect triple collisions.

In our derivation, we assume that the interactions among gas particles are repulsive. This implies that the scattering amplitude is positive. It is important to note that for the case of a Bose gas, the assumption is connected with the fact that even for infinitesimal attractive forces, a Bose gas cannot remain dilute at low temperatures. For the Fermi gas, the attraction among particles achieves superfluidity. We will develop a theory based on Bogoliubov [47, 48] for a weakly interacting Bose system. The dynamics are described by the following Hamiltonian determinant:

$$\hat{\mathrm{H}} = \sum_{\vec{\kappa}} \in(\vec{\kappa}) \hat{\psi}^{\dagger}_{\vec{\kappa}} \hat{\psi}_{\vec{\kappa}} + \frac{1}{2} \sum_{\vec{\kappa}_1 \vec{\kappa}_2 \vec{q}} \frac{U(q)}{V} \hat{\psi}^{\dagger}_{\vec{\kappa}_1 - q} \hat{\psi}^{\dagger}_{\vec{\kappa}_2 + q} \hat{\psi}_{\vec{\kappa}_1} \hat{\psi}_{\vec{\kappa}_2} \tag{1260}$$

where the Fourier transform of the interaction coupling energy:

$$U(\vec{q}) = \int d\vec{r}\, U(\vec{r}) \exp\{-i\vec{q}\vec{r}\} \quad (1261)$$

We apply the mean field approximation for our study. First, we consider the fact that the matrix elements of the Bose operators $\hat{\psi}_{\vec{\kappa}}$ are equal to $\sqrt{N_{\vec{\kappa}}}$, and then we need only take into account the interaction of particles in the condensate with each other and the interaction of **excited** particles with particles in the condensate, neglecting the interaction of **excited** particles with each other. For bosons, the following commutation relation is satisfied

$$\hat{\psi}_{\vec{\kappa}}\hat{\psi}^{\dagger}_{\vec{\kappa}'} - \hat{\psi}^{\dagger}_{\vec{\kappa}'}\hat{\psi}_{\vec{\kappa}} = \delta_{\vec{\kappa}\vec{\kappa}'} \quad (1262)$$

Considering the condition of orthonormality of the $|n_{\vec{\kappa}}\rangle$, we have

$$\begin{gathered} \hat{\psi}_{\vec{\kappa}}|n_{\vec{\kappa}}\rangle = \sqrt{n_{\vec{\kappa}}}\,|n_{\vec{\kappa}}-1\rangle \quad , \quad \hat{\psi}^{\dagger}_{\vec{\kappa}}|n_{\vec{\kappa}}\rangle = \sqrt{n_{\vec{\kappa}}+1}\,|n_{\vec{\kappa}}+1\rangle \\ \hat{\psi}^{\dagger}_{\vec{\kappa}}\hat{\psi}_{\vec{\kappa}}|n_{\vec{\kappa}}\rangle = n_{\vec{\kappa}}|n_{\vec{\kappa}}\rangle \quad , \quad \hat{\psi}_{\vec{\kappa}}\hat{\psi}^{\dagger}_{\vec{\kappa}}|n_{\vec{\kappa}}\rangle = (1+n_{\vec{\kappa}})|n_{\vec{\kappa}}\rangle \end{gathered} \quad (1263)$$

9.1 Bose-Einstein Condensation

At the temperature $T = 0$, the noninteracting bosons will have a tendency to accumulate at the ground state. We consider again the following action functional

$$S\left[\hat{\psi}^{\dagger},\hat{\psi}\right] = S_0\left[\hat{\psi}^{\dagger},\hat{\psi}\right] + S_{\text{int}}\left[\hat{\psi}^{\dagger},\hat{\psi}\right] \quad (1264)$$

From the Fourier transform of the field operators,

$$\hat{\psi}(\vec{r},\tau) = \frac{1}{\sqrt{\beta V}}\sum_{n,\vec{\kappa}} \exp\{-i(\omega_n\tau - \vec{\kappa}\vec{r})\}\hat{\psi}_{n,\vec{\kappa}} \quad , \quad \hat{\psi}^{\dagger}(\vec{r},\tau) = \frac{1}{\sqrt{\beta V}}\sum_{n,\vec{\kappa}} \exp\{i(\omega_n\tau - \vec{\kappa}\vec{r})\}\hat{\psi}^{\dagger}_{n,\vec{\kappa}} \quad (1265)$$

then

$$S_0\left[\hat{\psi}^{\dagger},\hat{\psi}\right] = \sum_{n,\vec{\kappa},\sigma} \hat{\psi}^{\dagger}_{n,\vec{\kappa},\sigma}(i\omega_n - \epsilon_\sigma)\hat{\psi}_{n,\vec{\kappa},\sigma} \quad , \quad \epsilon_\sigma = \tilde{\epsilon}_{\vec{\kappa},\sigma} - \mu \quad (1266)$$

and

$$S_{\text{int}}\left[\hat{\psi}^{\dagger},\hat{\psi}\right] = \frac{1}{2\beta V}\sum_{n,\vec{\kappa},\vec{\kappa}',\vec{q},\sigma} \hat{\psi}^{\dagger}_{n,\vec{\kappa}+\vec{q},\sigma}\hat{\psi}^{\dagger}_{n,\vec{\kappa}'-\vec{q},\sigma} U(\vec{q})\hat{\psi}_{n,\vec{\kappa}',\sigma}\hat{\psi}_{n,\vec{\kappa},\sigma} \quad (1267)$$

Here, ω_n are the discrete (fermionic or bosonic) Matsubara frequencies. The free partition function for noninteracting bosons:

$$Z_0 = \int d\left[\hat{\psi}^{\dagger}\right] d\left[\hat{\psi}\right] \exp\left\{\sum_{n,\vec{\kappa},\sigma} \hat{\psi}^{\dagger}_{n,\vec{\kappa},\sigma}(i\omega_n - \epsilon_\sigma)\hat{\psi}_{n,\vec{\kappa},\sigma}\right\} \quad (1268)$$

This permits us to have

$$Z_0 = \prod_{\sigma} \left[1 - \exp\{-\beta \in_\sigma\}\right]^{-1} \tag{1269}$$

The grand thermodynamic potential can be computed

$$\Omega_0 = -\frac{1}{\beta} \ln Z_0 = \frac{1}{\beta} \sum_{\sigma} \ln\left[1 - \exp\{-\beta \in_\sigma\}\right] \tag{1270}$$

From here, we compute the mean number of particles:

$$N(\mu) = -\left.\frac{\partial \Omega_0}{\partial \mu}\right|_T = \sum_{\sigma} \frac{1}{\exp\{\beta \in_\sigma\} - 1} \equiv \sum_{\sigma} n(\in_\sigma) \tag{1271}$$

Here, $n(\in_\sigma) = n_{BE}(\in_\sigma)$ is the Bose-Einstein distribution function. Considering the grand thermodynamic potential, we can also compute the mean energy

$$\mathrm{E} = \left.\frac{\partial \beta \Omega_0}{\partial \beta}\right|_{\mu} + \mu N = \sum_{\sigma} \tilde{\in}_\sigma \, n(\in_\sigma) \tag{1272}$$

and the entropy

$$S = -\left.\frac{\partial \Omega_0}{\partial T}\right|_{\mu} = -\beta \sum_{\sigma} \left(\in_\sigma n_{\mathrm{BE}}(\in_\sigma) - \frac{1}{\beta} \ln\left[1 - \exp\{-\beta \in_\sigma\}\right] \right) \tag{1273}$$

or

$$S = -\sum_{\sigma} \left(n_{\mathrm{BE}}(\in_\sigma) \ln n_{\mathrm{BE}}(\in_\sigma) - \left[1 + n_{\mathrm{BE}}(\in_\sigma)\right] \ln\left[1 + n_{\mathrm{BE}}(\in_\sigma)\right] \right) \tag{1274}$$

We observe from the aforementioned that the mean number of particles monotonically increases with μ:

$$\frac{\partial N(\mu)}{\partial \mu} = \sum_{\sigma} \frac{\beta \exp\{\beta \in_\sigma\}}{\left[\exp\{\beta \in_\sigma\} - 1\right]^2} > 0 \tag{1275}$$

The maximum value $N_{\max}$ is reached for $\mu = 0$. The value

$$N = N(\mu = 0) = N_{\max} = \sum_{\sigma} \left. n_{BE}(\in_\sigma) \right|_{\mu = 0} \tag{1276}$$

is the maximum number of particles that can be accumulated and that determines the critical temperature T_c:

$$T > T_c = T_c(N) \tag{1277}$$

We note that the critical temperature T_c is determined when the chemical potential μ formally changes sign. For $T < T_c$, the ground state has a macroscopic occupation number

$$N_0 = N - \sum_{\epsilon_\sigma > E} n_{BE}(\epsilon_\sigma) \tag{1278}$$

For the three-dimensional noninteracting bosons with energy $\tilde{\epsilon}_{\vec{\kappa},\sigma}$, we have the following [13]:

$$N_0 = N - N_{E>0} = N\left[1 - \left(\frac{T}{T_c}\right)^{\frac{3}{2}}\right], \qquad T < T_c = \frac{3.3}{g^{\frac{2}{3}}}\frac{\hbar^2}{m}\left(\frac{N}{V}\right)^{\frac{2}{3}} \tag{1279}$$

We study interacting bosons via the Hamiltonian formalism. We have just seen that for noninteracting bosons

$$\mathrm{E} = \sum_{\sigma} \tilde{\epsilon}_\sigma \, n(\epsilon_\sigma) \tag{1280}$$

All particles in the condensate are in the ground state of an ideal Bose gas. This is for the zero-state energy. In this case, the occupation number

$$N_{\vec{\kappa}=0} = N_0 \quad , \quad N_{\vec{\kappa}\neq 0} = 0 \tag{1281}$$

For a weakly ideal gas (in the ground and in weakly excited states), the numbers $N_{\vec{\kappa}}$ are different from zero. However, the numbers $N_{\vec{\kappa}}$ are small compared to the macroscopically large number N_0. This justifies our regarding the operators $\hat{\psi}_0^\dagger, \hat{\psi}_0$ as c-numbers by replacing them as follows:

$$\hat{\psi}_0^\dagger \to \sqrt{N_0} \quad , \quad \hat{\psi}_0 \to \sqrt{N_0} \tag{1282}$$

The commutators of these operators with one another or with other operators $\hat{\psi}_{\vec{\kappa}}^\dagger, \hat{\psi}_{\vec{\kappa}}$ are either 0 or 1. This implies that, in any event, they are small compared to the matrix elements of the operators $\hat{\psi}_0^\dagger, \hat{\psi}_0$.

The total number of particles in the system can be written in the form

$$N = \sum_{\vec{\kappa}\neq 0} \hat{\psi}_{\vec{\kappa}}^\dagger \hat{\psi}_{\vec{\kappa}} + \hat{\psi}_0^\dagger \hat{\psi}_0 \tag{1283}$$

Here, at very low temperature, the state with $\vec{\kappa} = 0$ will contain a macroscopic number of particles $\hat{\psi}_0^\dagger \hat{\psi}_0 = N_0$ that is much larger than the number of excited particles $N - N_0$. The interaction energy of the gas particles is given as follows:

$$\hat{\mathrm{H}}_{\mathrm{int}} = \frac{1}{2}\sum_{\vec{\kappa}_1\vec{\kappa}_2 q} \frac{U(q)}{V} \hat{\psi}_{\vec{\kappa}_1 - q}^\dagger \hat{\psi}_{\vec{\kappa}_2 + q}^\dagger \hat{\psi}_{\vec{\kappa}_1} \hat{\psi}_{\vec{\kappa}_2} \tag{1284}$$

Here, we consider the collision of two particles in the condensate. We can see that the Hamiltonian determinant (1284) depends on the particular value of the wave vector.

We represent this interaction diagrammatically as in relation (1285) (Figure 9.1):

$$\hat{\text{H}}_{\text{int}} = \frac{1}{2}\sum_{\vec{\kappa}_1\vec{\kappa}_2 q}\frac{U(q)}{V}\hat{\psi}^{\dagger}_{\vec{\kappa}_1-q}\hat{\psi}^{\dagger}_{\vec{\kappa}_2+q}\hat{\psi}_{\vec{\kappa}_1}\hat{\psi}_{\vec{\kappa}_2} \rightarrow \tag{1285}$$

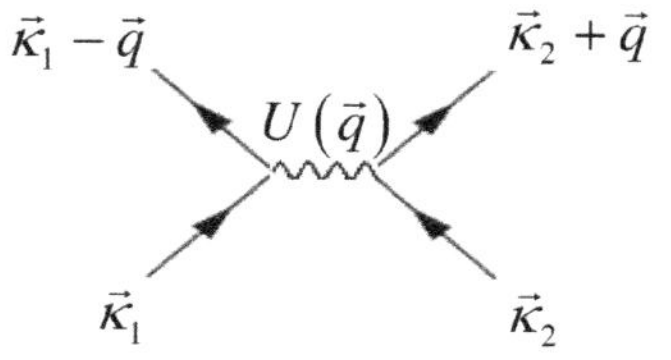

FIGURE 9.1 The interaction described by equation (1284) illustrated graphically, and showing a particle of momentum $\vec{\kappa}_1$ scattered into a new momentum state $\vec{\kappa}_1 - \vec{q}$ while another particle is scattered from $\vec{\kappa}_2$ into $\vec{\kappa}_2 + \vec{q}$.

In the two-particle interaction contribution in equations (1284) and (1285), we find the following combinations of wave vectors that yield the most important contributions by separating out the terms containing the zero modes and rewriting the interaction Hamiltonian determinant as follows:

If all those in the condensate state

1. $\vec{\kappa}_1 = \vec{\kappa}_2 = \vec{q} = 0$ then $\hat{\text{H}}_{\text{int}} = \frac{1}{2}\frac{U(0)}{V}N_0^2$ (1286)

$$U(\vec{r}) = \begin{cases} \text{const} & , \; r \leq a \\ 0 & , \; r > a \end{cases} \tag{1287}$$

If the interaction is short-ranged, then

$$U(q) = \int d\vec{r}\, U(\vec{r})\exp\{-i\vec{q}\vec{r}\} = 2\pi\int_{-1}^{1}dx\int_{0}^{\infty}r^2 dr U(r)\exp\{-iqrx\} = 2\pi\int_{0}^{\infty}r^2 dr\frac{U(r)}{qr}2\sin qr \approx U(0) = \frac{4\pi\hbar^2}{m}a \tag{1288}$$

Then $E_0 = \frac{1}{2}\frac{U(0)}{V}N_0^2$ is the ground state energy, and $\mu = \frac{\partial E_0}{\partial N_0} = \frac{U(0)}{V}N_0$ is the chemical potential.

2. $\vec{\kappa}_1 = \vec{\kappa}_2 = 0 \;\; , \;\; \vec{q} \neq 0$ then $\hat{\text{H}}_{\text{int}} = \frac{1}{2}\frac{U(0)}{V}N_0\sum_{\vec{q}}\hat{\psi}^{\dagger}_{-\vec{q}}\hat{\psi}^{\dagger}_{+\vec{q}}$ (1289)

3. $\vec{\kappa}_1 - \vec{q} = 0, \vec{\kappa}_2 + \vec{q} = 0$ then $\hat{\text{H}}_{\text{int}} = \frac{1}{2}\frac{U(0)}{V}N_0\sum_{\vec{q}}\hat{\psi}_{-\vec{q}}\hat{\psi}_{\vec{q}}$ (1290)

4. $\vec{\kappa}_1 - \vec{q} = 0, \vec{\kappa}_2 = 0$ then $\hat{\text{H}}_{\text{int}} = \frac{1}{2}\frac{U(0)}{V}N_0\sum_{\vec{q}}\hat{\psi}^{\dagger}_{\vec{q}}\hat{\psi}_{\vec{q}}$ (1291)

5. $\vec{\kappa}_1 = 0, \vec{\kappa}_2 + \vec{q} = 0$ then $\hat{H}_{int} = \frac{1}{2}\frac{U(0)}{V} N_0 \sum_q \hat{\psi}^\dagger_{-\vec{q}} \hat{\psi}_{-\vec{q}}$ (1292)

6. $\vec{q} = 0, \vec{\kappa}_2 = 0 \quad , \quad \vec{\kappa}_1 \neq 0$ then $\hat{H}_{int} = \frac{1}{2}\frac{U(0)}{V} N_0 \sum_{\vec{\kappa}_1} \hat{\psi}^\dagger_{\vec{\kappa}_1} \hat{\psi}_{\vec{\kappa}_1}$ (1293)

7. $\vec{q} = 0, \vec{\kappa}_1 = 0 \quad , \quad \vec{\kappa}_2 \neq 0$ then $\hat{H}_{int} = \frac{1}{2}\frac{U(q)}{V} N_0 \sum_{\vec{\kappa}_2} \hat{\psi}^\dagger_{\vec{\kappa}_2} \hat{\psi}_{\vec{\kappa}_2}$ (1294)

Due to the fact that the particle number N_0 is very large, the harmonic oscillator associated with the term $\hat{\psi}^\dagger_0, \hat{\psi}_0$ in the Hamiltonian operator behaves almost classically. So, we consider $\hat{\psi}^\dagger_0, \hat{\psi}_0$ approximately as c-numbers. This permits us to have the Hamiltonian

$$\hat{H} = E_0 + \sum_{\vec{\kappa}} \xi_{\vec{\kappa}} \hat{\psi}^\dagger_{\vec{\kappa}} \hat{\psi}_{\vec{\kappa}} + \beta \sum_{\vec{q} \neq 0} \left(\hat{\psi}^\dagger_{\vec{q}} \hat{\psi}^\dagger_{-\vec{q}} + \hat{\psi}_{\vec{q}} \hat{\psi}_{-\vec{q}} + 2 \hat{\psi}^\dagger_{\vec{q}} \hat{\psi}_{\vec{q}} \right) \tag{1295}$$

$$\beta = \frac{N_0}{2V} U(0) \quad , \quad E_0 = \frac{N_0^2}{2V} U(0) \tag{1296}$$

Here E_0 is the ground state energy of the gas, and μ is the chemical potential at $T = 0$. It is useful to note that because of the weakness of the interactions in a dilute gas, the ground state slightly differs from the ground state of an ideal gas. The same applies to the case of weakly excited states. Therefore, the number of particles with zero energy can be obtained from the formula

$$N_0 = N - \sum_{\vec{q} \neq 0} \hat{\psi}^\dagger_{\vec{q}} \hat{\psi}_{\vec{q}} \tag{1297}$$

This implies

$$N_0^2 = \left(N - \sum_{\vec{q} \neq 0} \hat{\psi}^\dagger_{\vec{q}} \hat{\psi}_{\vec{q}} \right)^2 \approx N^2 - 2N \sum_{\vec{q} \neq 0} \hat{\psi}^\dagger_{\vec{q}} \hat{\psi}_{\vec{q}} \tag{1298}$$

Then the Hamiltonian determinant takes the form

$$\hat{H} = E_0 + \sum_{\vec{q} \neq 0} \xi_{\vec{q}} \hat{\psi}^\dagger_{\vec{q}} \hat{\psi}_{\vec{q}} + \beta \sum_{\vec{q} \neq 0} \left(\hat{\psi}^\dagger_{\vec{q}} \hat{\psi}^\dagger_{-\vec{q}} + \hat{\psi}_{-\vec{q}} \hat{\psi}_{\vec{q}} \right) \tag{1299}$$

where

$$\xi_{\vec{q}} = \epsilon_{\vec{q}} + \frac{N}{V} U(0) \equiv \epsilon_{\vec{q}} + 2\beta \quad , \quad \beta = \frac{N}{2V} U(0) \tag{1300}$$

9.2 Bogoliubov Transformation

Equation (1299) is the Hamiltonian quadratic in the creation and annihilation operators and can therefore be diagonalized by means of a canonical transformation in order to find the energy levels. The motivation for the diagonalization is that, at the ground state, the particles of an ideal Bose gas occupy the lowest level, with zero energy. The gas is often said to be in the condensate or the condensed state. The diagonalization is done by introducing the canonical transformation known as a **Bogoliubov transformation [47, 48]**:

$$\begin{bmatrix} \hat{\Psi}_{\vec{q}}^{\dagger} \\ \hat{\Psi}_{-\vec{q}} \end{bmatrix} = \begin{bmatrix} u_{\vec{q}} & v_{\vec{q}} \\ v_{\vec{q}} & u_{\vec{q}} \end{bmatrix} \begin{bmatrix} \hat{b}_{\vec{q}}^{\dagger} \\ \hat{b}_{-\vec{q}} \end{bmatrix} \tag{1301}$$

or

$$\hat{\Psi}_{\vec{q}} = \hat{b}_{\vec{q}} u_{\vec{q}} + \hat{b}_{-\vec{q}}^{\dagger} v_{\vec{q}} \quad , \quad \hat{\Psi}_{\vec{q}}^{\dagger} = \hat{b}_{\vec{q}}^{\dagger} u_{\vec{q}} + \hat{b}_{-\vec{q}} v_{\vec{q}} \tag{1302}$$

From here, we have the following commutation relation

$$\hat{b}_{\vec{q}}\hat{b}_{\vec{q}}^{\dagger} - \hat{b}_{\vec{q}}^{\dagger}\hat{b}_{\vec{q}} = \left(u_{\vec{q}}\hat{\Psi}_{\vec{q}} - v_{\vec{q}}\hat{\Psi}_{-\vec{q}}^{\dagger}\right)\left(u_{\vec{q}}\hat{\Psi}_{\vec{q}}^{\dagger} - v_{\vec{q}}\hat{\Psi}_{-\vec{q}}\right) - \left(u_{\vec{q}}\hat{\Psi}_{\vec{q}}^{\dagger} - v_{\vec{q}}\hat{\Psi}_{-\vec{q}}\right)\left(u_{\vec{q}}\hat{\Psi}_{\vec{q}} - v_{\vec{q}}\hat{\Psi}_{-\vec{q}}^{\dagger}\right) \tag{1303}$$

or

$$\hat{b}_{\vec{q}}\hat{b}_{\vec{q}}^{\dagger} - \hat{b}_{\vec{q}}^{\dagger}\hat{b}_{\vec{q}} = u_{\vec{q}}^2\left(\hat{\Psi}_{\vec{q}}\hat{\Psi}_{\vec{q}}^{\dagger} - \hat{\Psi}_{\vec{q}}^{\dagger}\hat{\Psi}_{\vec{q}}\right) + v_{\vec{q}}^2\left(\hat{\Psi}_{-\vec{q}}^{\dagger}\hat{\Psi}_{-\vec{q}} - \hat{\Psi}_{-\vec{q}}\hat{\Psi}_{-\vec{q}}^{\dagger}\right) + u_{\vec{q}}v_{\vec{q}}\left(\hat{\Psi}_{\vec{q}}\hat{\Psi}_{-\vec{q}} - \hat{\Psi}_{-\vec{q}}^{\dagger}\hat{\Psi}_{\vec{q}}^{\dagger} - \hat{\Psi}_{-\vec{q}}\hat{\Psi}_{\vec{q}} - \hat{\Psi}_{\vec{q}}^{\dagger}\hat{\Psi}_{-\vec{q}}^{\dagger}\right) \tag{1304}$$

This yields

$$\hat{b}_{\vec{q}}\hat{b}_{\vec{q}}^{\dagger} - \hat{b}_{\vec{q}}^{\dagger}\hat{b}_{\vec{q}} = u_{\vec{q}}^2 - v_{\vec{q}}^2 = 1 \tag{1305}$$

We can now transform the Hamiltonian (1299):

$$\hat{\mathrm{H}} = \mathrm{E}_0 + \sum_{\vec{q}\neq 0} \xi_{\vec{q}}\left(\hat{b}_{\vec{q}}^{\dagger}u_{\vec{q}} + \hat{b}_{-\vec{q}}v_{\vec{q}}\right)\left(\hat{b}_{\vec{q}}u_{\vec{q}} + \hat{b}_{-\vec{q}}^{\dagger}v_{\vec{q}}\right) + \hat{\mathrm{H}}_{01} \tag{1306}$$

where

$$\hat{\mathrm{H}}_{01} = \beta\sum_{\vec{q}\neq 0}\left(\left(\hat{b}_{\vec{q}}^{\dagger}u_{\vec{q}} + \hat{b}_{-\vec{q}}v_{\vec{q}}\right)\left(\hat{b}_{-\vec{q}}^{\dagger}u_{-\vec{q}} + \hat{b}_{\vec{q}}v_{-\vec{q}}\right) + \left(\hat{b}_{-\vec{q}}u_{-\vec{q}} + \hat{b}_{\vec{q}}^{\dagger}v_{-\vec{q}}\right)\left(\hat{b}_{\vec{q}}u_{\vec{q}} + \hat{b}_{-\vec{q}}^{\dagger}v_{\vec{q}}\right)\right) \tag{1307}$$

It is instructive to note that

$$u_{\vec{q}} = u_{-\vec{q}} \quad , \quad v_{\vec{q}} = v_{-\vec{q}} \tag{1308}$$

and, considering (1305), we have

$$\hat{\mathrm{H}} = \mathrm{E}_0 + \hat{\mathrm{H}}_1 + \hat{\mathrm{H}}_2 \tag{1309}$$

with

$$\hat{H}_1 = \sum_{\vec{q}\neq 0}\left[\hat{b}^\dagger_{\vec{q}}\hat{b}_{\vec{q}}\left(\xi_{\vec{q}}u^2_{\vec{q}} + 2\beta u_{\vec{q}}v_{\vec{q}}\right) + \hat{b}_{-\vec{q}}\hat{b}^\dagger_{-\vec{q}}\left(\xi_{\vec{q}}v^2_{\vec{q}} + 2\beta u_{\vec{q}}v_{\vec{q}}\right)\right] \quad (1310)$$

$$\hat{H}_2 = \sum_{\vec{q}\neq 0}\left[\hat{b}^\dagger_{\vec{q}}\hat{b}^\dagger_{-\vec{q}}\left(\beta\left(u^2_{\vec{q}} + v^2_{\vec{q}}\right) + \xi_{\vec{q}}u_{\vec{q}}v_{\vec{q}}\right) + \hat{b}_{-\vec{q}}\hat{b}_{\vec{q}}\left(\beta\left(u^2_{\vec{q}} + v^2_{\vec{q}}\right) + \xi_{\vec{q}}u_{\vec{q}}v_{\vec{q}}\right)\right] \quad (1311)$$

The previously mentioned Hamiltonian can also be simplified as follows:

$$\hat{H} = E_0 + \sum_{\vec{q}\neq 0}\left[\left(\xi_{\vec{q}}v^2_{\vec{q}} + 2\beta u_{\vec{q}}v_{\vec{q}}\right) + \hat{b}^\dagger_{\vec{q}}\hat{b}_{\vec{q}}\left(\xi_{\vec{q}}\left(u^2_{\vec{q}} + v^2_{\vec{q}}\right) + 4\beta u_{\vec{q}}v_{\vec{q}}\right) + \left(\hat{b}^\dagger_{\vec{q}}\hat{b}^\dagger_{-\vec{q}} + \hat{b}_{-\vec{q}}\hat{b}_{\vec{q}}\right)\left(\xi_{\vec{q}}u_{\vec{q}}v_{\vec{q}} + \beta\left(u^2_{\vec{q}} + v^2_{\vec{q}}\right)\right)\right] \quad (1312)$$

For the Hamiltonian $\hat{H}$ to be diagonalized, we require that

$$\hat{b}^\dagger_{\vec{q}}\hat{b}^\dagger_{-\vec{q}} = \hat{b}_{-\vec{q}}\hat{b}_{\vec{q}} = 0 \quad (1313)$$

and

$$\beta\left(u^2_{\vec{q}} + v^2_{\vec{q}}\right) + \xi_{\vec{q}}u_{\vec{q}}v_{\vec{q}} = 0 \quad , \quad u^2_{\vec{q}} - v^2_{\vec{q}} = 1 \quad , \quad \beta > 0 \quad (1314)$$

Therefore, it follows that

$$u_{\vec{q}}v_{\vec{q}} < 0 \quad (1315)$$

From the first equation of (1314), we have

$$\left[\beta\left(u^2_{\vec{q}} + v^2_{\vec{q}}\right)\right]^2 = \left[\xi_{\vec{q}}u_{\vec{q}}v_{\vec{q}}\right]^2 \quad (1316)$$

Then, considering the second equation of (1314), we have

$$v^4_{\vec{q}} + v^2_{\vec{q}} - \frac{\beta^2}{\xi^2_{\vec{q}} - 4\beta^2} = 0 \quad (1317)$$

From here, we have

$$v^2_{\vec{q},1,2} = -\frac{1}{2}\left(1 \pm \frac{\xi_{\vec{q}}}{E_{\vec{q}}}\right) \quad , \quad E^2_{\vec{q}} = \xi^2_{\vec{q}} - 4\beta^2 \quad (1318)$$

Considering the second equation of (1314) and the fact that $v^2_{\vec{q}} > 0$,

$$v^2_{\vec{q}} = \frac{1}{2}\left(\frac{\xi_{\vec{q}}}{E_{\vec{q}}} - 1\right) \quad , \quad u^2_{\vec{q}} = \frac{1}{2}\left(\frac{\xi_{\vec{q}}}{E_{\vec{q}}} + 1\right) \quad (1319)$$

From here,

$$u^2_{\vec{q}}v^2_{\vec{q}} = \frac{1}{4}\left(\frac{\xi^2_{\vec{q}}}{E^2_{\vec{q}}} - 1\right) = \frac{\beta^2}{E^2_{\vec{q}}} \quad (1320)$$

From the inequality (1315), we have

$$u_{\vec{q}}v_{\vec{q}} = -\frac{\beta}{\mathrm{E}_{\vec{q}}} \tag{1321}$$

We can now rewrite the Hamiltonian of our system as

$$\hat{\mathrm{H}} = \mathrm{E}_0 + \sum_{\vec{q}\neq 0}\left[\left(\xi_{\vec{q}}\frac{1}{2}\left(\frac{\xi_{\vec{q}}}{\mathrm{E}_{\vec{q}}}-1\right)+2\beta\left(-\frac{\beta}{\mathrm{E}_{\vec{q}}}\right)\right)+\hat{b}_{\vec{q}}^{\dagger}\hat{b}_{\vec{q}}\left(\xi_{\vec{q}}\left(\frac{1}{2}\left(\frac{\xi_{\vec{q}}}{\mathrm{E}_{\vec{q}}}+1\right)+\frac{1}{2}\left(\frac{\xi_{\vec{q}}}{\mathrm{E}_{\vec{q}}}-1\right)\right)+4\beta\left(-\frac{\beta}{\mathrm{E}_{\vec{q}}}\right)\right)\right] \tag{1322}$$

or

$$\hat{\mathrm{H}} = \mathrm{E}_0 + \sum_{\vec{q}\neq 0}\left[\left(\frac{1}{2}\left(\frac{\xi_{\vec{q}}^2}{\mathrm{E}_{\vec{q}}}-\frac{4\beta^2}{\mathrm{E}_{\vec{q}}}\right)-\frac{\xi_{\vec{q}}}{2}\right)+\hat{b}_{\vec{q}}^{\dagger}\hat{b}_{\vec{q}}\left(\frac{\xi_{\vec{q}}^2}{\mathrm{E}_{\vec{q}}}-\frac{4\beta^2}{\mathrm{E}_{\vec{q}}}\right)\right] \tag{1323}$$

But from

$$\frac{\xi_{\vec{q}}^2}{\mathrm{E}_{\vec{q}}}-\frac{4\beta^2}{\mathrm{E}_{\vec{q}}} = \frac{\xi_{\vec{q}}^2-4\beta^2}{\mathrm{E}_{\vec{q}}} = \frac{\mathrm{E}_{\vec{q}}^2}{\mathrm{E}_{\vec{q}}} = \mathrm{E}_{\vec{q}} \tag{1324}$$

then

$$\hat{\mathrm{H}} = \mathrm{E}_0 + \sum_{\vec{q}\neq 0}\left[\frac{1}{2}\left(\mathrm{E}_{\vec{q}}-\xi_{\vec{q}}\right)+\hat{b}_{\vec{q}}^{\dagger}\hat{b}_{\vec{q}}\mathrm{E}_{\vec{q}}\right] \tag{1325}$$

It follows that for the case of weakly excited states of a dilute Bose gas, the model of elementary excitations can be described by the energy spectrum:

$$\mathrm{E}_{\vec{q}} = \sqrt{\xi_{\vec{q}}^2-4\beta^2} = \sqrt{\left(\epsilon_{\vec{q}}+\frac{N}{V}U(0)\right)^2-4\beta^2} = \sqrt{\left(\epsilon_{\vec{q}}+2\beta\right)^2-4\beta^2} = \sqrt{\epsilon_{\vec{q}}^2+4\beta\,\epsilon_{\vec{q}}} \tag{1326}$$

or

$$\mathrm{E}_{\vec{q}} = \sqrt{\left(\frac{\hbar^2\vec{q}^{\,2}}{2m}\right)^2+\frac{N}{V}U(0)\frac{\hbar^2\vec{q}^{\,2}}{m}} \tag{1327}$$

This is a phonon dispersion relation. In the limiting case of small momenta where $q \to 0$, we then have

$$\mathrm{E}_{\vec{q}} = \sqrt{\frac{N}{V}U(0)\frac{\hbar^2\vec{q}^{\,2}}{m}} = \hbar qc = \frac{\hbar^2}{m}\frac{q}{a}\sqrt{4\pi r_0} \tag{1328}$$

Here, c **is the velocity of second sound**:

$$c = \sqrt{\frac{N}{V}U(0)\frac{1}{m}} = \sqrt{\frac{N}{V}\frac{4\pi\hbar^2 a}{m}\frac{1}{m}} = \frac{\hbar}{ma}\sqrt{4\pi\frac{N}{V}a^3} = \frac{\hbar}{ma}\sqrt{4\pi r_0} \tag{1329}$$

and the gas parameter is r_0:

$$r_0 = \frac{N}{V}a^3 \tag{1330}$$

So,

$$E_{\vec{q}} = \begin{cases} c\hbar q & , \quad \hbar q << mc \\ \dfrac{\hbar^2 q^2}{2m} & , \quad \hbar q >> mc \end{cases} \tag{1331}$$

The first case is for small momenta that correspond to the phonon part of the spectrum of the Bose liquid due to interaction (indication of **superfluidity**), while the second case is for large momenta and corresponds to the energy of a free particle devoid of interaction (indication of **no superfluidity**). In the previous relations, c is the **velocity of second sound**. Sound in a condensate is an indication of superfluidity. The energies achieve the value zero for $q \to 0$ in accordance with the **Nambu-Goldstone theorem**, which states that a spontaneous breakdown of a continuous symmetry leads to excitation with such an energy-momentum relation (**Figure 9.2**). If we make a plot of relation (1327) versus the momentum, we observe that it has a minimum with respect to energy. This is referred to as the **roton energy**. The interpretation of these states was first given by Feynman in 1955 [50].

The number of gas particles with zero momentum and at absolute zero is found to be as follows

$$N_0 = N - \sum_{\vec{q}} N_{\vec{q}} = N - \frac{V}{(2\pi\hbar)^3}\int N_{\vec{q}} d\vec{q} = N\left[1 - \frac{8}{3}\sqrt{\frac{n_0 a^3}{\pi}}\right]_{T=0}, N_{\vec{q}} = \frac{m^2c^4}{2E_{\vec{q}}\left[E_{\vec{q}} + \epsilon_{\vec{q}} + mc^2\right]}, E_{\vec{q}} = \sqrt{c^2q^2 + \epsilon_q^2} \tag{1332}$$

or

$$N_0 = N\left[1 - \frac{8}{3}\sqrt{\frac{n_0 a^3}{\pi}}\right]_{T=0} \tag{1333}$$

We see that interaction changes the number of particles in the condensate. It is useful to note that not all particles are in the condensed state.

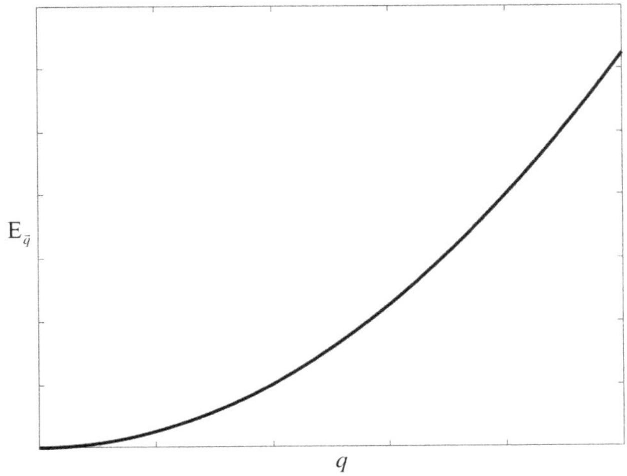

FIGURE 9.2 Plot of the energy $E_{\vec{q}}$ versus the momentum q. The large momenta correspond to a free particle energy devoid of interaction (indication of **no superfluidity**). The energy achieves the value zero for $q \to 0$ in accordance with the **Nambu-Goldstone theorem.**

At low temperatures, the quasiparticles are frozen out, and the ground state energy is given as follows:

$$\hat{H} = E_0 + \sum_{\vec{q}\neq 0} \frac{1}{2}\left[E_{\vec{q}} - \xi_{\vec{q}}\right] \tag{1334}$$

Particles with nonzero momentum $\vec{q}$ lie outside the condensate and constitute the uncondensed part of the liquid. Their number is given by the expectation value in the vacuum state of quasiparticles $|0\rangle$:

$$\left\langle \hat{\psi}_{\vec{q}}^{\dagger}\hat{\psi}_{\vec{q}} \right\rangle \equiv \langle 0|\left(u_{\vec{q}}\hat{b}_{\vec{q}}^{\dagger} + v_{\vec{q}}\hat{b}_{-\vec{q}}\right)\left(u_{\vec{q}}\hat{b}_{\vec{q}} + v_{\vec{q}}\hat{b}_{-\vec{q}}^{\dagger}\right)|0\rangle = v_{\vec{q}}^2 \tag{1335}$$

and

$$\left\langle \hat{\psi}_{\vec{q}}^{\dagger}\hat{\psi}_{-\vec{q}}^{\dagger} \right\rangle \equiv \langle 0|\left(u_{\vec{q}}\hat{b}_{\vec{q}}^{\dagger} + v_{\vec{q}}\hat{b}_{-\vec{q}}\right)\left(u_{\vec{q}}\hat{b}_{-\vec{q}}^{\dagger} + v_{\vec{q}}\hat{b}_{\vec{q}}\right)|0\rangle = u_{\vec{q}}v_{\vec{q}} \tag{1336}$$

The total number of uncondensed particles is given as follows:

$$N_u = \sum_{\vec{q}\neq 0} \left\langle \hat{\psi}_{\vec{q}}^{\dagger}\hat{\psi}_{\vec{q}} \right\rangle \equiv \sum_{\vec{q}\neq 0} v_{\vec{q}}^2 \tag{1337}$$

We calculate the density ρ_u of uncondensed particles by considering the expression of $v_{\vec{q}}^2$ from (1319):

$$\rho_u = \sum_{\vec{q}} v_{\vec{q}}^2 = \int \frac{d\vec{q}}{(2\pi\hbar)^3} v_{\vec{q}}^2 = -\frac{1}{2}\int \frac{d\vec{q}}{(2\pi\hbar)^3}\left(1 - \frac{\xi(\vec{q})}{\varepsilon(\vec{q})}\right) \tag{1338}$$

Let us measure all momenta in units of

$$k_{2\beta} = \left(\frac{4m\beta}{\hbar}\right)^{\frac{1}{2}} \tag{1339}$$

which is the characteristic wave number associated with s-wave scattering length:

$$q = \hbar k_{2\beta}\kappa \tag{1340}$$

Then the density of uncondensed particles in equation (1338) may be calculated as follows:

$$\rho_u = \frac{N_u}{V} = \frac{k_{2\beta}^3}{4\pi^2}\int_0^{\infty} d\kappa\kappa^2 \left(\frac{\kappa^2+1}{\sqrt{\left(\kappa^2+1\right)^2 - 1}} - 1\right) = \frac{k_{2\beta}^3}{4\pi^2}\frac{\sqrt{2}}{3} \tag{1341}$$

Equation (1341) shows that even at zero temperature, the repulsion scatters a small set of particles out of the condensate and consequently causes depletion to a fraction [52, 53].

$$\frac{\rho_0}{\rho} = 1 - \frac{\rho_u}{\rho} = 1 - \frac{N_u}{N} \tag{1342}$$

Here ρ is the particle density, and ρ_0 is the condensate density. Consider the strongly interacting superfluid He-4. For this case, only about 8% of the particles condense in the zero-temperature state [53].

However, all particles participate in the superfluid motion [102]. The superfluid density equals the total density, and the condensate and the normal fluid move together.

If we consider that the birth of the Bose gas is a result of charged particles, for the elementary excitation, we then have

$$E_{\vec{q}} = \sqrt{\left(\frac{\hbar^2\vec{q}^2}{2m}\right)^2 + \frac{N}{V}\frac{4\pi e^2}{q^2}\frac{\hbar^2\vec{q}^2}{m}} = \sqrt{\left(\frac{\hbar^2\vec{q}^2}{2m}\right)^2 + \frac{N}{V}\frac{4\pi e^2\hbar^2}{m}} \equiv \sqrt{\left(\frac{\hbar^2\vec{q}^2}{2m}\right)^2 + \Delta^2} \tag{1343}$$

where

$$\Delta^2 = \frac{N}{V}\frac{4\pi e^2\hbar^2}{m} = \omega_p^2 \tag{1344}$$

and Δ is the gap of the excitation spectrum that coincides with the plasmon frequency ω_p.

9.3 Nonideal Bose Gas Path Integral Formalism

Considering the nonideal Bose gas at absolute temperature, the following action functional contains the electron-electron interaction (that is simply a Coulomb potential):

$$S_{\text{int}}\left[\hat{\psi}^\dagger, \hat{\psi}\right] = -\frac{1}{4\beta V}\sum_{\nu_1+\nu_2=\nu_3+\nu_4}\left[U(\vec{\kappa}_1 - \vec{\kappa}_3) + U(\vec{\kappa}_1 - \vec{\kappa}_4)\right]\hat{\psi}^\dagger_{\nu_1}\hat{\psi}^\dagger_{\nu_2}\hat{\psi}_{\nu_3}\hat{\psi}_{\nu_4}$$

$= -\frac{1}{4\beta V}\sum_{\nu_1+\nu_2=\nu_3+\nu_4}\Big[$ 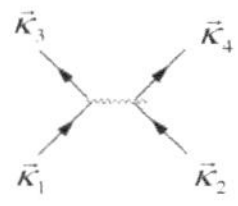 $+$ 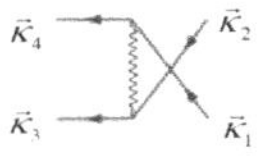 $\Big]\hat{\psi}^\dagger_{\nu_1}\hat{\psi}^\dagger_{\nu_2}\hat{\psi}_{\nu_3}\hat{\psi}_{\nu_4} = -\frac{1}{4\beta V}\sum_{\nu_1+\nu_2=\nu_3+\nu_4}\Big|$ 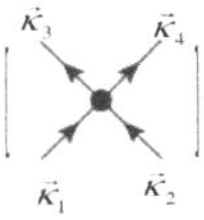 $\Big|\hat{\psi}^\dagger_{\nu_1}\hat{\psi}^\dagger_{\nu_2}\hat{\psi}_{\nu_3}\hat{\psi}_{\nu_4}$

(1345)

Here, the summation runs over the four-momenta $\nu = (i\omega_n, \vec{\kappa})$ comprised Matsubara and momentum components, and summation over repeated spin indices is implied. For weak and short-range interactions, in the limits of low temperature and low density, the Bose condensation can be studied by a procedure from Bogoliubov [47, 48, 54]. Note that the ground state of a system of noninteracting bosons is a Bose-Einstein condensate. Hence, we separate the operators corresponding to the annihilation and creation of particles in the state with zero momentum:

$$\hat{\psi}(\vec{r},\tau) \to \hat{\tilde{\psi}}(\vec{r},\tau) + \alpha \quad , \quad \hat{\psi}^\dagger(\vec{r},\tau) \to \hat{\tilde{\psi}}^\dagger(\vec{r},\tau) + \alpha^* \tag{1346}$$

$$\hat{\psi}_\nu = \hat{\phi}_\nu + \alpha(\beta V)^{\frac{1}{2}}\delta_{\nu,0} \quad , \quad \hat{\psi}^\dagger_\nu = \hat{\phi}^\dagger_\nu + \alpha^*(\beta V)^{\frac{1}{2}}\delta_{\nu,0} \tag{1347}$$

We substitute into the action functional S:

$$S_0 = \sum_\nu\left(i\omega_n - \frac{\kappa^2}{2m} + \mu\right)\hat{\phi}^\dagger_\nu\hat{\phi}_\nu + \mu\beta V\alpha^*\alpha + \mu(\beta V)^{\frac{1}{2}}\left(\alpha\hat{\phi}^\dagger_0 + \alpha^*\hat{\phi}_0\right) \tag{1348}$$

and

$$S_1 = -\frac{1}{4\beta V}\sum_{\nu_1+\nu_2=\nu_3+\nu_4}\left(U(\vec{\kappa}_1-\vec{\kappa}_3)+U(\vec{\kappa}_1-\vec{\kappa}_4)\right)F\left(\hat{\phi},\hat{\phi}^\dagger,\alpha,\alpha^*\right) \tag{1349}$$

where

$$F\left(\hat{\phi},\hat{\phi}^\dagger,\alpha,\alpha^*\right)=\left(\hat{\phi}^\dagger_{\nu_1}+\alpha^*(\beta V)^{\frac{1}{2}}\delta_{\nu_1,0}\right)\left(\hat{\phi}^\dagger_{\nu_2}+\alpha^*(\beta V)^{\frac{1}{2}}\delta_{\nu_2,0}\right)\left(\hat{\phi}_{\nu_3}+\alpha(\beta V)^{\frac{1}{2}}\delta_{\nu_3,0}\right)\left(\hat{\phi}_{\nu_4}+\alpha(\beta V)^{\frac{1}{2}}\delta_{\nu_4,0}\right) \tag{1350}$$

or

$$\begin{aligned}F\left(\hat{\phi},\hat{\phi}^\dagger,\alpha,\alpha^*\right)=&\left(\hat{\phi}^\dagger_{\nu_1}\hat{\phi}^\dagger_{\nu_2}+\hat{\phi}^\dagger_{\nu_1}\alpha^*(\beta V)^{\frac{1}{2}}\delta_{\nu_2,0}+\alpha^*(\beta V)^{\frac{1}{2}}\delta_{\nu_1,0}\hat{\phi}^\dagger_{\nu_2}+\left(\alpha^*\right)^2\beta V\delta_{\nu_1,0}\delta_{\nu_2,0}\right)\times\\&\times\left(\hat{\phi}_{\nu_4}\hat{\phi}_{\nu_3}+\alpha(\beta V)^{\frac{1}{2}}\delta_{\nu_4,0}\hat{\phi}_{\nu_3}+\alpha(\beta V)^{\frac{1}{2}}\delta_{\nu_3,0}\hat{\phi}_{\nu_4}+\alpha^2\beta V\delta_{\nu_3,0}\delta_{\nu_4,0}\right)\end{aligned} \tag{1351}$$

or

$$\begin{aligned}F\left(\hat{\phi},\hat{\phi}^\dagger,\alpha,\alpha^*\right)=&\hat{\phi}^\dagger_{\nu_1}\hat{\phi}^\dagger_{\nu_2}\hat{\phi}_{\nu_4}\hat{\phi}_{\nu_3}+|\alpha|^4(\beta V)^2\delta_{\nu_1,0}\delta_{\nu_2,0}\delta_{\nu_3,0}\delta_{\nu_4,0}+\alpha^2\beta V\delta_{\nu_3,0}\delta_{\nu_4,0}\hat{\phi}^\dagger_{\nu_1}\hat{\phi}^\dagger_{\nu_2}+\\&+\alpha(\beta V)^{\frac{1}{2}}\hat{\phi}^\dagger_{\nu_1}\hat{\phi}^\dagger_{\nu_2}\left(\delta_{\nu_4,0}\hat{\phi}_{\nu_3}+\delta_{\nu_3,0}\hat{\phi}_{\nu_4}\right)+\left(\alpha^*\right)^2\beta V\delta_{\nu_1,0}\delta_{\nu_2,0}\hat{\phi}_{\nu_4}\hat{\phi}_{\nu_3}+\alpha^*|\alpha|^2(\beta V)^{\frac{3}{2}}\delta_{\nu_1,0}\delta_{\nu_2,0}\left(\delta_{\nu_4,0}\hat{\phi}_{\nu_3}+\delta_{\nu_3,0}\hat{\phi}_{\nu_4}\right)+\\&+\alpha^*(\beta V)^{\frac{1}{2}}\hat{\phi}_{\nu_3}\hat{\phi}_{\nu_4}\left(\delta_{\nu_2,0}\hat{\phi}^\dagger_{\nu_1}+\delta_{\nu_1,0}\hat{\phi}^\dagger_{\nu_2}\right)+\alpha|\alpha|^2(\beta V)^{\frac{3}{2}}\delta_{\nu_3,0}\delta_{\nu_4,0}\left(\delta_{\nu_1,0}\hat{\phi}^\dagger_{\nu_2}+\delta_{\nu_2,0}\hat{\phi}^\dagger_{\nu_1}\right)+\\&+|\alpha|^2\beta V\left(\hat{\phi}^\dagger_{\nu_1}\delta_{\nu_2,0}+\hat{\phi}^\dagger_{\nu_2}\delta_{\nu_1,0}\right)\left(\hat{\phi}_{\nu_3}\delta_{\nu_4,0}+\hat{\phi}_{\nu_4}\delta_{\nu_3,0}\right)\end{aligned} \tag{1352}$$

Then,

$$S_1^{(0)} = -\frac{1}{4\beta V}\sum_{\nu_1+\nu_2=\nu_3+\nu_4}\left(U(\vec{\kappa}_1-\vec{\kappa}_3)+U(\vec{\kappa}_1-\vec{\kappa}_4)\right)\hat{\phi}^\dagger_{\nu_1}\hat{\phi}^\dagger_{\nu_2}\hat{\phi}_{\nu_4}\hat{\phi}_{\nu_3} \tag{1353}$$

$$S_1^{(1)} = -|\alpha|^2\sum_{\nu}\left(U(0)+U(\nu)\right)\hat{\phi}^\dagger_{\nu}\hat{\phi}_{\nu}\quad,\quad S_1^{(2)}=-\alpha|\alpha|^2(\beta V)^{\frac{1}{2}}\hat{\phi}^\dagger_0\quad,\quad S_1^{(3)}=-\alpha^*|\alpha|^2(\beta V)^{\frac{1}{2}}\hat{\phi}_0 \tag{1354}$$

$$S_1^{(4)} = -\frac{1}{2}\alpha^2\sum_{\nu}U(\vec{\kappa})\hat{\phi}^\dagger_{\nu}\hat{\phi}^\dagger_{-\nu}\quad,\quad S_1^{(5)}=-\frac{1}{2}\left(\alpha^*\right)^2\sum_{\nu}U(\vec{\kappa})\hat{\phi}_{\nu}\hat{\phi}_{-\nu} \tag{1355}$$

$$S_1^{(6)} = -\alpha(\beta V)^{\frac{1}{2}}\frac{1}{4\beta V}\sum_{\nu_1+\nu_2=\nu_3}\left(U(\vec{\kappa}_1)+U(\vec{\kappa}_2)\right)\hat{\phi}^\dagger_{\nu_1}\hat{\phi}^\dagger_{\nu_2}\hat{\phi}_{\nu_3} \tag{1356}$$

$$S_1^{(7)} = -\alpha^*(\beta V)^{\frac{1}{2}} \frac{2}{4\beta V} \sum_{\nu_1+\nu_2=\nu_3} \left(U(\vec{\kappa}_1)+U(\vec{\kappa}_2)\right)\hat{\phi}_{\nu_3}^\dagger \hat{\phi}_{\nu_2}\hat{\phi}_{\nu_1} \tag{1357}$$

$$S_1^{(8)} = -\frac{1}{2}|\alpha|^4 U(0) \tag{1358}$$

So,

$$S=\beta V\left(\mu|\alpha|^2 - \frac{1}{2}U(0)|\alpha|^4\right)-(\beta V)^{\frac{1}{2}}\left(\gamma^*\hat{\phi}_0+\gamma\hat{\phi}_0^\dagger\right)-|\alpha|^2\sum_\nu\left(U(0)+U(\nu)\right)\hat{\phi}_\nu^\dagger\hat{\phi}_\nu+\sum_\nu\left(i\omega_n-\frac{\vec{\kappa}^2}{2m}+\mu\right)\hat{\phi}_\nu^\dagger\hat{\phi}_\nu -$$

$$-\frac{1}{2}\sum_\nu U(\vec{\kappa})\left(\alpha^2\hat{\phi}_\nu^\dagger\hat{\phi}_{-\nu}^\dagger+\left(\alpha^*\right)^2\hat{\phi}_\nu\hat{\phi}_{-\nu}\right)-\frac{1}{4\beta V}\sum_{\nu_1+\nu_2=\nu_3+\nu_4}\left[\left(U(\vec{\kappa}_1-\vec{\kappa}_3)+U(\vec{\kappa}_1-\vec{\kappa}_4)\right)\hat{\phi}_{\nu_1}^\dagger\hat{\phi}_{\nu_2}^\dagger\hat{\phi}_{\nu_3}\hat{\phi}_{\nu_4}\right]-$$

$$-\frac{1}{2(\beta V)^{\frac{1}{2}}}\sum_{\nu_1+\nu_2=\nu_3}\left[\left(U(\vec{\kappa}_1)+U(\vec{\kappa}_2)\right)\left(\alpha\hat{\phi}_{\nu_1}^\dagger\hat{\phi}_{\nu_2}^\dagger\hat{\phi}_{\nu_3}+\alpha^*\hat{\phi}_{\nu_3}^\dagger\hat{\phi}_{\nu_2}\hat{\phi}_{\nu_1}\right)\right] \tag{1359}$$

Considering

$$S_0=\left(\mu|\alpha|^2-\frac{1}{2}U(0)|\alpha|^4\right)\beta V \tag{1360}$$

from the saddle-point approximation, then

$$\frac{\partial S_0}{\partial|\alpha|^2}=0 \tag{1361}$$

we have

$$\mu=U(0)|\alpha|^2=U(0)n_0\neq 0 \tag{1362}$$

The quantity $|\alpha|^2=n_0$ is the total of the Bose-Einstein condensate. The linear terms in $\hat{\phi}_0^\dagger,\hat{\phi}_0$ disappear. So,

$$S=\sum_{\nu\neq 0}\left(i\omega_n-\frac{\vec{\kappa}^2}{2m}+\mu\right)\hat{\phi}_\nu^\dagger\hat{\phi}_\nu-|\alpha|^2\sum_\nu\left(U(0)+U(\nu)\right)\hat{\phi}_\nu^\dagger\hat{\phi}_\nu-\frac{1}{2}\sum_\nu U(\vec{\kappa})\left(\alpha^2\hat{\phi}_\nu^\dagger\hat{\phi}_{-\nu}+\left(\alpha^*\right)^2\hat{\phi}_\nu\hat{\phi}_{-\nu}\right)-$$

$$-\frac{1}{4\beta V}\sum_{\nu_1+\nu_2=\nu_3+\nu_4}\left(U(\vec{\kappa}_1-\vec{\kappa}_3)+U(\vec{\kappa}_1-\vec{\kappa}_4)\right)\hat{\phi}_{\nu_1}^\dagger\hat{\phi}_{\nu_2}^\dagger\hat{\phi}_{\nu_3}\hat{\phi}_{\nu_4}-$$

$$-\frac{1}{2(\beta V)^{\frac{1}{2}}}\sum_{\nu_1+\nu_2=\nu_3}\left(U(\vec{\kappa}_1)+U(\vec{\kappa}_3)\right)\left(\alpha\hat{\phi}_{\nu_1}^\dagger\hat{\phi}_{\nu_2}^\dagger\hat{\phi}_{\nu_3}+\alpha^*\hat{\phi}_{\nu_3}^\dagger\hat{\phi}_{\nu_2}\hat{\phi}_{\nu_3}\right) \tag{1363}$$

The following diagrammatic notations are very important for the given problem:

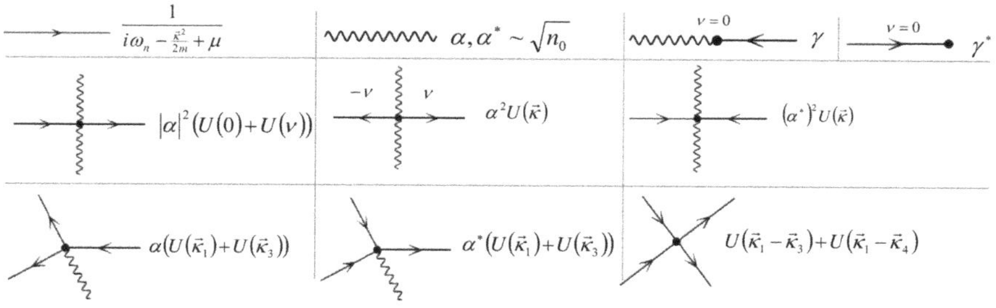

FIGURE 9.3 Some diagrammatic notations of some scattering mechanisms and different vertices. The diagram in first row, first column, the bare particle propagator; second diagram, second row, second column, 2-leg vertex with two virtual interactions; third diagram, third row, third column, 3-leg vertex with a virtual interaction; second diagram, first row, second column, virtual interaction; second diagram, second row, second column, 2-leg vertex with two virtual interactions; third diagram, third row, second column, 3-leg vertex with a virtual interaction; third diagram, first row, third column, field propagator and a virtual interaction; third diagram, second row, third column, 2-leg vertex with two virtual interactions; third diagram, third row, third column, 4-leg vertex; fourth diagram, first row, fourth column, bare field propagator.

We also introduce the Green's functions:
Normal Green's function:

$$\mathbf{G}(\nu) = -\langle \hat{\psi}_\nu \hat{\psi}_\nu^\dagger \rangle \tag{1364}$$

and the **anomalous Green's function:**

$$\mathbf{G}^<(\nu) = -\langle \hat{\psi}_\nu \hat{\psi}_{-\nu} \rangle \tag{1365}$$

$$\mathbf{G}^>(\nu) = -\langle \hat{\psi}_\nu^\dagger \hat{\psi}_{-\nu}^\dagger \rangle \tag{1366}$$

9.3.1 Beliaev-Dyson Equations

We select among all the diagrams that determine the self-energy Σ (with pair interaction between particles) those that represent various offshoots connected to the external lines by one wiggly line and denote their sum by $\Sigma_A \equiv \Sigma_{11}$. All such diagrams are present in one **skeleton diagram**. What are **skeleton diagrams?** They are diagrams made up of thick lines and blocks with each such diagram equivalent to a definite infinite set of ordinary diagrams of various orders. One such **skeleton diagram** is of the following form of the self-energy:

$\Sigma_A =$

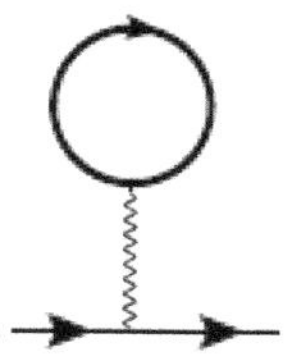

(1367)

Here, in this diagram, the loop corresponds to the exact density $n(\mu)$ of the system. It therefore follows from the definition in (1367) that

$$\Sigma_A = n(\mu) U(0) \quad , \quad U(0) \to \frac{4\pi}{m} a \tag{1368}$$

The remaining part of the exact self-energy Σ is denoted by $\Sigma_B \equiv \Sigma_{21}$:

$\Sigma_B =$ 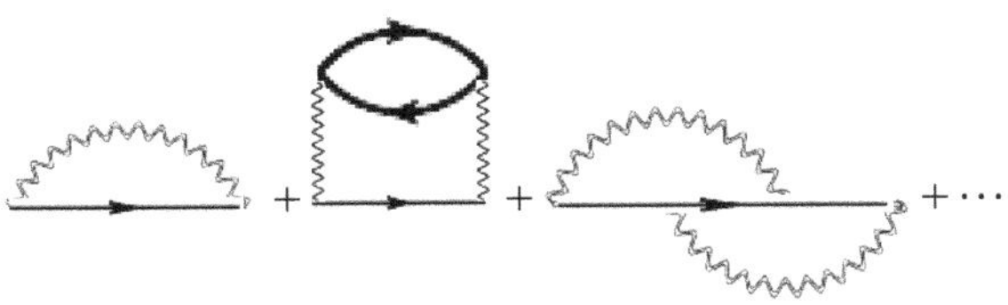 (1369)

So,

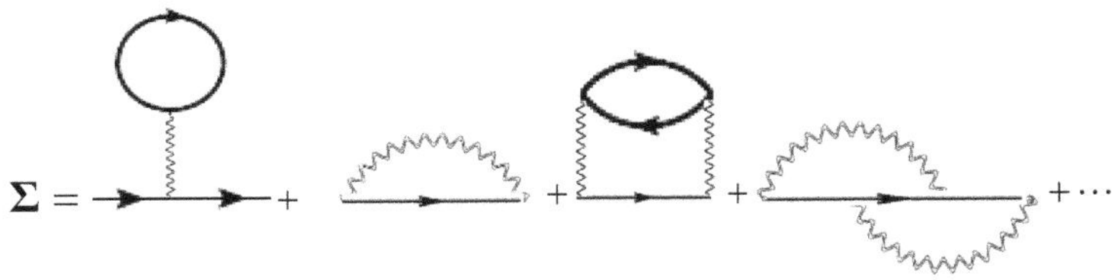 (1370)

and

$$\Sigma = n(\mu)U(0) + \Sigma_B \tag{1371}$$

Hence, only the diagrams in Σ_B need to be specially calculated.

The diagrammatic representation of the Green's function $\mathbf{G}$ in terms of the self-energy Σ and the bare particle Green's function $\mathbf{G}_0$ are shown as follows:

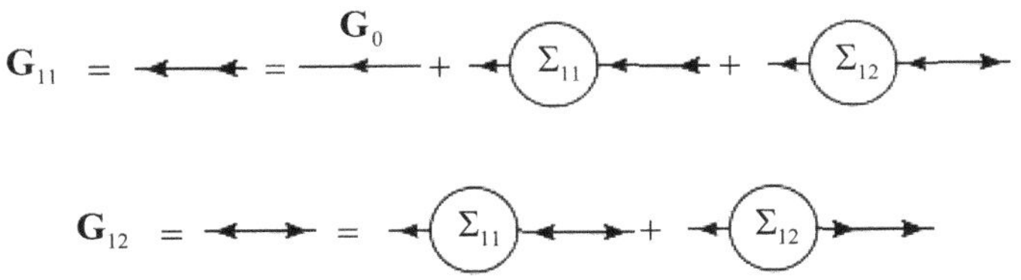

(1372)

(1373)

FIGURE 9.4 Diagrammatic representation of the Beliaev-Dyson equations. The double directed line for $\mathbf{G}$ and the single directed line for $\mathbf{G}_0$.

This is represented explicitly as follows:

$$\mathbf{G}_{11} \equiv \mathbf{G}(\vec{\kappa}) = \mathbf{G}_0(\vec{\kappa}) + \mathbf{G}_0(\vec{\kappa})\Sigma_{11}\mathbf{G}(\vec{\kappa}) + \mathbf{G}_0(\vec{\kappa})\Sigma_{12}(\vec{\kappa})\mathbf{G}^{>}(\vec{\kappa}) \tag{1374}$$

$$\mathbf{G}_{21} \equiv \mathbf{G}^{>}(\vec{\kappa}) = \mathbf{G}_0(-\vec{\kappa})\Sigma_{21}(-\vec{\kappa})\mathbf{G}(\vec{\kappa}) + \mathbf{G}_0(-\vec{\kappa})\Sigma_{11}(-\vec{\kappa})\mathbf{G}^{>}(\vec{\kappa}) \tag{1375}$$

From here, we write in a more convenient form:

$$\left[i\omega_n - \xi(\vec{\kappa}) - \Sigma_{11}(\vec{\kappa})\right]\mathbf{G}(\vec{\kappa}) - \Sigma_{12}(\vec{\kappa})\mathbf{G}^{>}(\vec{\kappa}) = 1, \left[-i\omega_n - \xi(\vec{\kappa}) - \Sigma_{11}(-\vec{\kappa})\right]\mathbf{G}^{>}(\vec{\kappa}) - \Sigma_{21}(-\vec{\kappa})\mathbf{G}(\vec{\kappa}) = 0 \tag{1376}$$

Suppose

$$\mathrm{A}_{\pm}(\vec{\kappa}) = \frac{\Sigma_{11}(\vec{\kappa}) \pm \Sigma_{11}(-\vec{\kappa})}{2} \tag{1377}$$

then we solve for $\mathbf{G}(\vec{\kappa})$ and $\mathbf{G}^{>}(\vec{\kappa})$ in (1376) and for

$$i\omega \to \omega + i\delta \quad , \quad \delta \to 0 \tag{1378}$$

then

$$\mathbf{G}(\vec{\kappa}) = \frac{\omega_n + \xi(\vec{\kappa}) + A_+(\vec{\kappa}) + A_-(\vec{\kappa})}{\left[\omega - A_-(\vec{\kappa})\right]^2 - \left[\xi(\vec{\kappa}) + A_+(\vec{\kappa})\right]^2 + \Sigma_{12}(\vec{\kappa})\Sigma_{21}(-\vec{\kappa})} \tag{1379}$$

and

$$\mathbf{G}^{>}(\vec{\kappa}) = -\frac{\Sigma_{21}(-\vec{\kappa})}{\left[\omega - A_-(\vec{\kappa})\right]^2 - \left[\xi(\vec{\kappa}) + A_+(\vec{\kappa})\right]^2 + \Sigma_{12}(\vec{\kappa})\Sigma_{21}(-\vec{\kappa})} \tag{1380}$$

These are the generalized expressions of the usual Green's functions for the one-particle function via the self-energy part. Because we see from equation (1380) that $\Sigma_{12}(\vec{\kappa})$ is an even function, then

$$\Sigma_{12} = \Sigma_{21} \tag{1381}$$

From (1380), we consider the following equation to determine the poles of the Green's functions:

$$\left[\omega - A_-(\vec{\kappa})\right]^2 - \left[\xi(\vec{\kappa}) + A_+(\vec{\kappa})\right]^2 + \Sigma_{12}(\vec{\kappa})\Sigma_{21}(-\vec{\kappa}) = 0 \tag{1382}$$

From physical interpretations, this equation has solutions for arbitrary small $\vec{\kappa}$ and ω. The class of possible solutions for the energy spectrum of the excitations for small $\vec{\kappa}$ must have the acoustic spectrum $\omega = c|\vec{\kappa}|$, with c being the velocity of sound. This implies the spectrum imitates the long-wavelength oscillations. From equation (1382), the energy spectrum becomes

$$\omega = A_-(\vec{\kappa}) \pm \sqrt{\left[\xi(\vec{\kappa}) + A_+(\vec{\kappa}) - \Sigma_{12}(\vec{\kappa})\right]\left[\xi(\vec{\kappa}) + A_+(\vec{\kappa}) + \Sigma_{12}(\vec{\kappa})\right]} \tag{1383}$$

In the presence of condensate, we have

$$\mu = \Sigma_{11} - \Sigma_{12} = n_0 U(0) \tag{1384}$$

So

$$\Sigma_{12} = n_0 U(\vec{\kappa}) \tag{1385}$$

and in diagrammatic form, we have

$\Sigma_{12} =$ 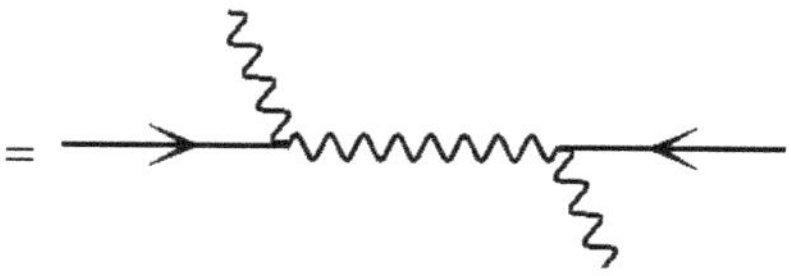 (1386)

Also

$$\Sigma_{11} = n_0\left(U(\vec{\kappa}) + U(0)\right) \tag{1387}$$

and, in diagrammatic form, we have

$\Sigma_{11} =$ 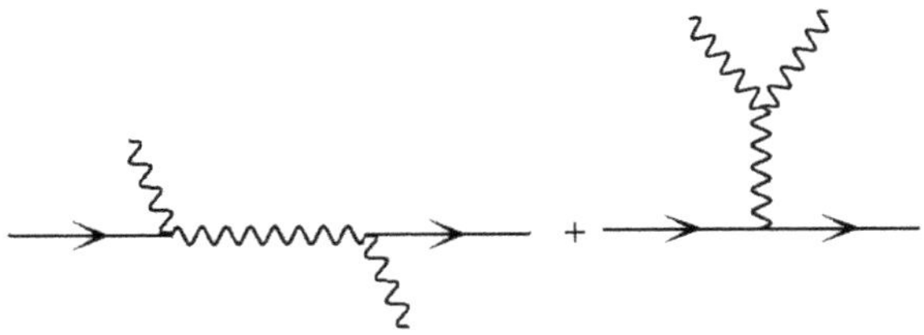 (1388)

Thus,

$$\Sigma_{11} + \Sigma_{12} - \mu = n_0 U(0) + 2\Sigma_{12} - \mu \equiv \Sigma - \mu = 2n_0 U(\vec{\kappa}) \tag{1389}$$

where

$$\Sigma = n_0 U(0) + 2\Sigma_{12} \tag{1390}$$

Here, n_0 is the condensate density.

From the RPA, we have

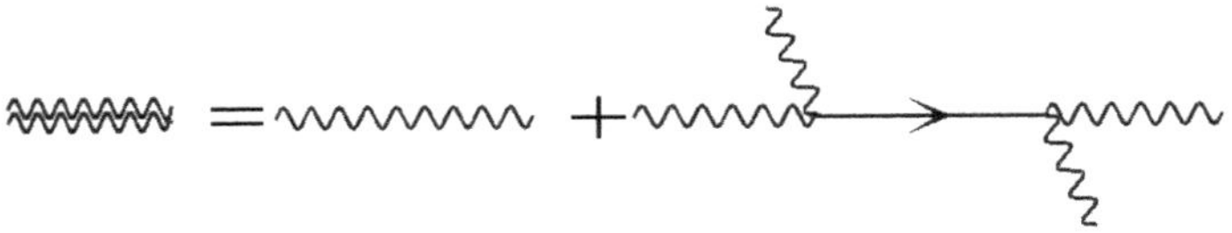 (1391)

$$U = U_{\text{eff}} = U_0 + U_0\left(G(\vec{\kappa}) + G^{<}(\vec{\kappa}) + G(-\vec{\kappa}) + G^{<}(\vec{\kappa})\right) n_0 U_0 \tag{1392}$$

or

$$U = \frac{4\pi e^2}{\kappa^2}\left(1 + \frac{\frac{\kappa^2}{m} n_0 \frac{4\pi e^2}{\kappa^2}}{\omega_n^2 + \epsilon_n^2}\right) = \frac{4\pi e^2}{\kappa^2}\frac{\omega_n^2 + \left(\frac{\kappa^2}{2m}\right)^2}{\omega_n^2 + \epsilon_n^2} \tag{1393}$$

For the static case $\omega_n = 0$, then

$$U_{\text{eff}} = \frac{\frac{4\pi e^2}{4m^2}\kappa^2}{\omega_p^2 + \left(\frac{\kappa^2}{2m}\right)^2} = 4\pi e^2 \frac{\kappa^2}{\kappa^4 + (2m\omega_p)^2} \tag{1394}$$

10

Superconductivity Theory

Introduction

In this chapter, we examine **Bardeen, Cooper and Schrieffer** (BCS) theory [54, 55] named for three famous American scientists who created this successful (semi-)microscopic theory of superconductivity [55]. This theory describes a collective quantum phenomenon that is contrary to the Ginzburg-Landau theory and does not depend on microscopic details but rather establishes a general phenomenological framework for the transition from the normal to the superconducting phase. This involves the phonon-mediated dynamic attraction between two electrons leading to their subsequent pairing (a so-called **Cooper pair**) [30], which is a loose quasi-boson entity comprised of two fermionic quasiparticles with oppositely directed spins and momenta, that is, between states $(\vec{\kappa},\uparrow)$ and $(-\vec{\kappa},\downarrow)$.

In the BCS theory, the basic feature is that pairing occurs between electrons in states with opposite momentum and opposite spins producing an effective attraction between the electrons. If this attractive interaction is strong enough, two electrons may form a bound state or composite boson (Cooper pair).

The two spins in the Cooper pair are combined into the spin state with $S = 0$ (**singlet state**). In BCS theory, this is selected because other choices of spin combination would lead to a **triplet state** with $S = 1$ and would imply that the superconducting state with magnetic properties is in effect absent for simple metals. The triplet state has smaller binding energy and so is less favored. However, triplet pairing is possible and may exist in heavy fermion solids such as UPt_3 and UBe_{13}. Therefore, the choice of $S = 0$ (**singlet state**) seems reasonable.

Superconductivity was first discovered by Kammerlingh Onnes in 1911 [56]. In 1957, a successful microscopic theory of the phenomenon finally was proposed by Bardeen, Cooper, and Schrieffer. It is known today as the **BCS** theory [55]. The essence of the microscopic explanation is that a very weak interaction between the conduction electrons of a metal can, at sufficiently low temperatures, yield quantum states characterized by a highly correlated motion of all the electrons. This stems from an arbitrarily small attractive interaction between two electrons above a filled Fermi sea sufficient to create a bound state. This yields electronic pairs (known as **Cooper pairs**) that could move without resistance in the presence of a driving electric field [30].

Note that the electrons near the Fermi energy interact with their pair on the opposite side of the Fermi sea. Mutual scattering produces a singularity in the scattering amplitude that leads to the binding of the electron pair. Indeed, all electron pairs simultaneously are involved in the process leading to a phase transition in the entire metal. The existence of the singularity depends on the peak (sharpness) at the Fermi level in the spectrum or Fermi distribution function. If all electrons near the Fermi energy become paired, then we must reconsider whether the sharp distribution still exists. A self-consistent superconductivity theory and, in particular, BCS theory can determine the properties of the bound electron pairs.

10.1 BCS Superconductivity Theory

10.1.1 Electron-Phonon Interaction in a Solid State

We consider lattice ions, their mutual interaction, and interaction between lattice ions and electrons. While the electron gas modifies the electron-electron interaction and yields an **effective** or **screened** interaction, the ionic lattice and interaction between electrons and phonons modifies the electron-electron interaction. This phonon-modified interaction is attractive and is the basis of superconductivity.

We examine superconductivity theory via the functional integral approach where the statistical model is built on classical field configurations. The formulation of the phonon mechanism of superconductivity is based on the description of the:

- **Statistical many-electron system with a variable number of particles.**
- **Acoustic long wavelength equilibrium crystal lattice vibration.**
- **Process of electron-phonon (or electron-lattice) interaction where the direct contribution of electron-electron interaction's playing an important role in the excitation spectrum of the lattice vibrations may be ignored.**

We consider the vibration of a crystal lattice and assume acoustic longitudinal long wavelength excitation where the characteristic vibration frequency ω:

$$\omega << \frac{1}{a} \tag{1346}$$

and where a is the lattice constant. The condition (1346) permits the solid state to imitate a continuous medium. That is, imitating the sound with crystal lattice vibration. This results to an ideal incompressible liquid moving with a longitudinal long wavelength sound wave. A small perturbation in an ideal liquid is described by the function:

$$\rho'(\vec{r},t) = \rho(\vec{r},t) - \nu_0 \tag{1347}$$

where ν_0 and $\rho(\vec{r},t)$ are the densities of the unperturbed liquid and that of the liquid subjected to the sound wave, respectively. The liquid flow potential is related to its field velocity potential:

$$\vec{\mathrm{v}}(\vec{r},t) = \nabla\varphi(\vec{r},t) \tag{1348}$$

and satisfies the continuity equation

$$\frac{\partial\rho'(\vec{r},t)}{\partial t} \cong -\nu_0 \nabla\varphi(\vec{r},t) \tag{1349}$$

The classical energy of the liquid subjected to the wave may be written:

$$\mathrm{E} = \int d\vec{r}\left[\frac{\nu_0 \mathrm{v}^2}{2} + \frac{u^2}{2\nu_0}\rho'^2\right] \tag{1350}$$

where u is the speed of sound in the liquid.

For a statistical equilibrium system, we do the analytical continuation by introducing imaginary time:

$$t \to t' = -i\tau \quad , \quad 0 \le \tau \le \beta \tag{1351}$$

The phonon action functional has the form:

$$S_{ph}[\rho'] = \int_0^{\beta} d\tau \int d\vec{r} \left[\frac{\nu_0}{2} \left(\nabla\varphi(\vec{r},\tau) \right)^2 + \frac{u^2}{2\nu_0} \rho'^2(\vec{r},\tau) \right] \tag{1352}$$

where the potential $\varphi(\vec{r},\tau)$ and the density $\rho'(\vec{r},\tau)$ are related in the $(\vec{r},\tau)$ space via equation (1349):

$$i\frac{\partial \rho'(\vec{r},\tau)}{\partial \tau} = -\nu_0 \nabla\varphi(\vec{r},\tau) \quad , \quad 0 \le \tau \le \beta \tag{1353}$$

while satisfying the periodic boundary conditions

$$\rho'(\vec{r},0) = \rho'(\vec{r},\beta) \quad , \quad \varphi(\vec{r},0) = \varphi(\vec{r},\beta) \tag{1354}$$

We study the interaction of the **phonon cloud** with the electron and consider a homogenous isotropic medium so that the deformation of the lattice yields a **deformation potential** in the first-order approximation of $\rho'(\vec{r},\tau)$:

$$U_{\text{deform}} \cong \frac{1}{\nu_0} \int d\vec{r}' \, W\left(\left|\vec{r}-\vec{r}'\right|\right) \rho'(\vec{r}') \tag{1355}$$

Here, the Coulomb potential $W\left(\left|\vec{r}-\vec{r}'\right|\right)$ is at the characteristic length scale of order of the lattice constant a. The condition of the sound wave vector:

$$\left|\vec{q}\right| << \frac{1}{a} \tag{1356}$$

So, the interaction between electrons is well described by the instantaneous contact interaction

$$W\left(\left|\vec{r}-\vec{r}'\right|\right) \simeq \lambda \delta(\vec{r}-\vec{r}') \tag{1357}$$

instead of a true and complicated potential and

$$U_{\text{deform}} \cong \lambda \frac{\rho'(\vec{r})}{\nu_0} \tag{1358}$$

Here, λ is the electron-phonon interaction constant.

Considering the interaction of the electrons with the deformation of the lattice, the Hamiltonian determinant is written in the form:

$$\hat{H}_{e-ph} = \sum_{\sigma} \int d\vec{r} \, U_{\text{deform}}(\vec{r}) \hat{\psi}_{\sigma}^{\dagger}(\vec{r}) \hat{\psi}_{\sigma}(\vec{r}) \cong \frac{\lambda}{\nu_0} \sum_{\sigma} \int d\vec{r} \, \rho'(\vec{r}) \hat{\psi}_{\sigma}^{\dagger}(\vec{r}) \hat{\psi}_{\sigma}(\vec{r}) \tag{1359}$$

This classical model Hamiltonian imitates the Fröhlich Hamiltonian and describes the electron-phonon interaction via second quantization. The action functional may therefore be written as

$$S_{e-ph}[\hat{\psi},\varphi] \equiv S_{\text{int}}[\hat{\psi},\rho'] = \frac{\lambda}{\nu_0} \int_0^{\beta} d\tau \int d\vec{r} \sum_{\sigma} \int d\vec{r} \, \rho'(\vec{r}) \hat{\psi}_{\sigma}^{\dagger}(\vec{r}) \hat{\psi}_{\sigma}(\vec{r}) \tag{1360}$$

From (1352), (1353), (1354), and (1360), the action functional for the entire system then is

$$S\left[\hat{\psi},\rho'\right]=S_e\left[\hat{\psi}^{\dagger},\hat{\psi}\right]+S_{ph}\left[\hat{\psi},\rho'\right]+S_{e-ph}\left[\hat{\psi},\rho'\right] \quad (1361)$$

and the partition function

$$Z=\int d\left[\hat{\psi}_\sigma\right]d\left[\hat{\psi}_\sigma^{\dagger}\right]d\left[\rho'\right]\exp\left\{-S\left[\hat{\psi},\rho'\right]\right\} \quad (1362)$$

considering the boundary conditions

$$\rho'(\vec{r},0)=\rho'(\vec{r},\beta) \quad , \quad \hat{\psi}_\sigma(\vec{r},0)=-\hat{\psi}_\sigma(\vec{r},\beta) \quad (1363)$$

The partition function (1362) may be conveniently evaluated by going to reciprocal space via the Matsubara expansion for the fermionic field:

$$\hat{\psi}_\sigma(\vec{r},\tau)=\left(\frac{V}{(2\pi)^3}\right)^{\frac{1}{2}}\sum_{\omega_n}\int d\vec{\kappa}\exp\left\{-i\left(\omega_n\tau-\vec{\kappa}\vec{r}\right)\right\}\hat{\psi}_{\omega_n\vec{\kappa}\sigma} \quad (1364)$$

Here, $\omega_n=\frac{2n+1}{\beta}\pi$ is the fermionic Matsubara frequency. For the phonon fields $\varphi(\vec{r},\tau)$, $\rho'(\vec{r},\tau)$, we have the Matsubara expansion

$$\varphi(\vec{r},\tau)=\left(\frac{V}{(2\pi)^3}\right)^{\frac{1}{2}}\sum_{\Omega_n}\int\left(\frac{\Omega_n}{\nu_0\kappa^2}\right)^{\frac{1}{2}}d\vec{\kappa}\exp\left\{-i\left(\Omega_n\tau-\vec{\kappa}\vec{r}\right)\right\}\varphi_{\vec{\kappa}\Omega_n} \quad (1365)$$

Considering the condition that the field is real, then

$$\varphi_{\vec{\kappa}\Omega_n}=\varphi_{-\vec{\kappa},-\Omega_n} \quad (1366)$$

Also,

$$\rho'(\vec{r},\tau)=\left(\frac{V}{(2\pi)^3}\right)^{\frac{1}{2}}\sum_{\Omega_n}\int\left(\frac{\Omega_n}{\nu_0\kappa^2}\right)^{\frac{1}{2}}d\vec{\kappa}\exp\left\{-i\left(\Omega_n\tau-\vec{\kappa}\vec{r}\right)\right\}\rho'_{\vec{\kappa}\Omega_n} \quad (1367)$$

with the bosonic Matsubara frequency being $\Omega_n=\frac{2n}{\beta}\pi$. So, from the equation of continuity of $\rho'(\vec{r},\tau)$ in space, we then find the following relation for the Fourier coefficients from (1365) and (1367):

$$\rho'_{\vec{\kappa}\Omega_n}=\frac{\nu_0\kappa^2}{\Omega_n}\varphi_{\vec{\kappa}\Omega_n} \quad (1368)$$

Hence, from (1364) to (1368), we find the following action functionals

$$S_e\left[\hat{\psi}\right]=\beta V\sum_{\omega_n}\int d\vec{\kappa}\hat{\psi}^{\dagger}_{\omega_n\vec{\kappa}\sigma}\left[i\omega_n+\frac{\vec{\kappa}^2}{2m}-\mu\right]\hat{\psi}_{\omega_n\vec{\kappa}\sigma} \tag{1369}$$

$$S_{ph}\left[\varphi\right]=\frac{\beta V}{2}\sum_{\Omega_n}\int d\vec{q}\varphi^{*}_{\vec{\kappa}\Omega_n}\left[\Omega_n\left(1+\frac{u^2q^2}{\Omega_n^2}\right)\right]\varphi_{\vec{\kappa}\Omega_n} \tag{1370}$$

$$S_{e-ph}\left[\hat{\psi},\varphi\right]=\lambda\beta\left(\frac{V}{2\pi}\right)^{\frac{3}{2}}\sum_{\omega_n}\sum_{\Omega_n}\int d\vec{q}d\vec{\kappa}\left(\frac{q^2}{\nu_0\Omega_n}\right)^{\frac{1}{2}}\sum_{\sigma}\left[\hat{\psi}^{\dagger}_{\omega_n+\Omega_n,\vec{\kappa}+\vec{q},\sigma}\hat{\psi}_{\omega_n\vec{\kappa}\sigma}\varphi_{\vec{q}\Omega_n}\hat{\psi}^{\dagger}_{\omega_n\vec{\kappa}\sigma}+\text{h.c.}\right] \tag{1371}$$

Here, h.c. denotes the Hermitian conjugate. The partition function of the given system is then rewritten:

$$Z=\int\prod_{\omega_n\vec{\kappa}}d\hat{\psi}^{\dagger}_{\omega_n\vec{\kappa}\sigma}d\hat{\psi}_{\omega_n\vec{\kappa}\sigma}\prod_{\vec{q}\Omega_n}d\varphi_{\vec{q}\Omega_n}\exp\left\{-\left(S_e\left[\hat{\psi}\right]+S_{ph}\left[\varphi\right]+S_{e-ph}\left[\hat{\psi},\varphi\right]\right)\right\} \tag{1372}$$

10.1.2 Effective Four-Fermion BCS Theory

The partition function in (1372) is Gaussian in the phonon fields, and so it can easily be integrated over phonon fields to give the effective four-dimensional fermion (4-fermion) partition function:

$$Z=N\int\prod_{\omega_n\vec{\kappa}}d\hat{\psi}^{\dagger}_{\omega_n\vec{\kappa}\sigma}d\hat{\psi}_{\omega_n\vec{\kappa}\sigma}\exp\left\{-\left(S_e\left[\hat{\psi}\right]-S^{\text{eff}}_{e-e}\left[\hat{\psi}\right]\right)\right\} \tag{1373}$$

where the effective 4-fermion interaction action functional $S^{\text{eff}}_{e-e}\left[\hat{\psi}\right]$:

$$S^{\text{eff}}_{e-e}\left[\hat{\psi}\right]=\frac{\lambda^2\beta V}{\nu_0(2\pi)^3}\sum_{\omega_n\omega_{n'}}\int d\vec{\kappa}d\vec{\kappa}'\sum_{\Omega_n}\int d\vec{q}\frac{q^2}{\Omega_n^2+u^2q^2}\sum_{\sigma\sigma'}\hat{\psi}^{\dagger}_{\omega_n+\Omega_n,\vec{\kappa}+\vec{q},\sigma}\hat{\psi}^{\dagger}_{\omega_{n'}-\Omega_n,\vec{\kappa}'-\vec{q},\sigma'}\hat{\psi}_{\omega_{n'}\vec{\kappa}'\sigma'}\hat{\psi}_{\omega_n\vec{\kappa}\sigma} \tag{1374}$$

From here, we easily move to the $(\vec{\kappa},\tau)$ space and, letting $i\equiv(\vec{r}_i,\tau_i)$, then

$$S^{\text{eff}}_{e-e}\left[\hat{\psi}\right]=S^{\text{eff}(1)}_{e-e}\left[\hat{\psi}\right]+S^{\text{eff}(2)}_{e-e}\left[\hat{\psi}\right] \tag{1375}$$

where

$$S^{\text{eff}(1)}_{e-e}\left[\hat{\psi}\right]=\frac{1}{\beta}\int d\tau_1\int_{\tau_1>\tau_2}d\tau_2\int d\vec{r}_1d\vec{r}_2\sum_{\sigma\sigma'}\hat{\psi}^{\dagger}_{\sigma}(1)\hat{\psi}^{\dagger}_{\sigma'}(2)u^{>}(1-2)\hat{\psi}_{\sigma'}(2)\hat{\psi}_{\sigma}(1) \tag{1376}$$

and

$$S^{\text{eff}(2)}_{e-e}\left[\hat{\psi}\right]=\frac{1}{\beta}\int d\tau_1\int_{\tau_1<\tau_2}d\tau_2\int d\vec{r}_1d\vec{r}_2\sum_{\sigma\sigma'}\hat{\psi}^{\dagger}_{\sigma}(1)\hat{\psi}^{\dagger}_{\sigma'}(2)u^{<}(1-2)\hat{\psi}_{\sigma'}(2)\hat{\psi}_{\sigma}(1) \tag{1377}$$

with the effective electron-phonon interaction potential being

$$u(1-2) \equiv u(\vec{r}_1-\vec{r}_2,\tau_1-\tau_2) = \frac{\lambda^2}{\nu_0}\sum_{\Omega_n}\int d\vec{q}\frac{q^2}{\Omega_n^2+u^2q^2}\exp\{i\vec{q}(\vec{r}_1-\vec{r}_2)-i\Omega_n(\tau_1-\tau_2)\} \tag{1378}$$

and

$$u^>(1-2) \equiv u(\vec{r}_1-\vec{r}_2,\tau_1-\tau_2>0) \quad , \quad u^<(1-2) \equiv u(\vec{r}_1-\vec{r}_2,\tau_1-\tau_2<0) \tag{1379}$$

We solve (1378) by first taking the Matsubara sum over Ω_n. But,

$$\sum_{\Omega_n}\frac{q^2}{\Omega_n^2+u^2q^2}\exp\{-i\Omega_n\tau\} = -\sum_{\Omega_n}\frac{q^2}{(i\Omega_n)^2-\epsilon^2}\exp\{-i\Omega_n\tau\} \tag{1380}$$

where

$$\epsilon(q) \equiv uq \tag{1381}$$

and

$$\sum_{\Omega_n}\frac{q^2}{(i\Omega_n)^2-\epsilon^2}\exp\{-i\Omega_n\tau\} = \frac{1}{2\epsilon}\sum_{\Omega_n}\exp\{i\Omega_n\,\tau|_{+0}\}\left[\frac{q^2}{i\Omega_n-\epsilon}-\frac{q^2}{i\Omega_n+\epsilon}\right] \tag{1382}$$

or

$$\sum_{\Omega_n}\frac{q^2}{(i\Omega_n)^2-\epsilon^2}\exp\{-i\Omega_n\tau\} = \frac{q}{2u}\sum_{\Omega_n}\exp\{i\Omega_n\,\tau|_{+0}\}\left[\frac{1}{i\Omega_n-\epsilon}-\frac{1}{i\Omega_n+\epsilon}\right] \tag{1383}$$

This can be evaluated by first considering the sum

$$\sum_{n=-\infty}^{\infty}\frac{\exp\{i\Omega_n 0^+\}}{i\Omega_n-\epsilon} \tag{1384}$$

and first introducing the integral

$$\mathrm{I} = \lim_{\eta\to 0^+}\int_C \frac{\exp\{\eta z\}}{z-\epsilon}\frac{dz}{\exp\{\beta z\}\pm 1} \tag{1385}$$

Here, C is a circle of infinite radius centered at $z=0$. As $|z|\to\infty$, if $\mathrm{Re}\,z>0$, then the absolute value of the integrand is of the order $\left(\frac{1}{|z|}\right)\exp\{-\beta\,\mathrm{Re}\,z\}$. If $\mathrm{Re}\,z<0$, then the absolute value of the integrand is of the order $\left(\frac{1}{|z|}\right)\exp\{\eta\,\mathrm{Re}\,z\}$. So, the integrand is exponentially small as $|z|\to\infty$; hence, $\mathrm{I}=0$.

For bosons,

$$\mathrm{I} = \lim_{\eta\to 0^+}\int_C \frac{\exp\{\eta z\}}{z-\epsilon}\frac{dz}{\exp\{\beta z\}-1} = 0 \tag{1386}$$

The poles of the integrand are at $z = \frac{2n\pi i}{\beta}$ and at $z = \in$ for imaginary n. So, from the residue theorem

$$\sum_{n=-\infty}^{\infty} \frac{\exp\{i\Omega_n 0^+\}}{i\Omega_n - \in} = -\beta f_{\mathrm{B}}(\in) \tag{1387}$$

where $\Omega_n = \frac{2n\pi}{\beta}$ and

$$f_{\mathrm{B}}(\in) = \frac{1}{\exp\{\beta \in\} - 1} \tag{1388}$$

For the case of fermions,

$$\mathrm{I} = \lim_{\eta \to 0^+} \int_C \frac{\exp\{\eta z\}}{z - \in} \frac{dz}{\exp\{\beta z\} + 1} = 0 \tag{1389}$$

So, the poles of the integrand are at $z = \frac{(2n+1)\pi i}{\beta}$ and at $z = \in$ for imaginary n. Therefore, from the residue theorem,

$$\sum_{n=-\infty}^{\infty} \frac{\exp\{i\Omega_n 0^+\}}{i\Omega_n - \in} = \beta f_{\mathrm{F}}(\in) \tag{1390}$$

where $\Omega_n = \frac{(2n+1)\pi}{\beta}$ and

$$f_{\mathrm{F}}(\in) = \frac{1}{\exp\{\beta \in\} + 1} \tag{1391}$$

Hence,

$$\sum_{n=-\infty}^{\infty} \frac{\exp\{i\Omega_n 0^+\}}{i\Omega_n - \in} = \begin{cases} \beta f_{\mathrm{F}}(\in) & \text{fermions} \\ -\beta f_{\mathrm{B}}(\in) & \text{bosons} \end{cases} \tag{1392}$$

Thus,

$$\frac{q}{2u} \sum_{\Omega_n} \exp\{i\Omega_n \tau|_{+0}\} \left[\frac{1}{i\Omega_n - \in} - \frac{1}{i\Omega_n + \in} \right] = \frac{\beta q}{2u} \left(f_{\mathrm{F}}(\in) - f_{\mathrm{F}}(-\in) \right) = \frac{\beta q}{2u} \left(2 f_{\mathrm{F}}(\in) - 1 \right) \tag{1393}$$

as

$$f_{\mathrm{F}}(-\in) = 1 - f_{\mathrm{F}}(\in) \tag{1394}$$

and the Matsubara sum over Ω_n in (1378), considering (1379), gives us

$$u^{>}(1-2) = \frac{\lambda^2 \beta}{2u\nu_0} \int \frac{d\vec{q}}{(2\pi)^3} \sum_{\chi=\pm} \chi q \frac{\exp\{-\chi u q(\tau_1 - \tau_2)\}}{1 - \exp\{-\chi \beta u q\}} \exp\{i\vec{q}(\vec{r}_1 - \vec{r}_2)\} \tag{1395}$$

and

$$u^{<}(1-2)=\frac{\lambda^2\beta}{2u\nu_0}\int\frac{d\vec{q}}{(2\pi)^3}\sum_{\chi=\pm}\chi q\frac{\exp\{\chi uq(\tau_1-\tau_2)\}}{1-\exp\{-\chi\beta uq\}}\exp\{i\vec{q}(\vec{r}_1-\vec{r}_2)\} \tag{1396}$$

Substituting (1395) and (1396) into (1375) and moving to the momentum-frequency representation yields the effective 4-fermion interaction action functional:

$$S_{e-e}^{\text{eff}}[\hat{\psi}]=\frac{\lambda^2V^2}{u^2\nu_0}\sum_{\omega_1\omega_2\omega_3\omega_4}\int d\vec{\kappa}_1\,d\vec{\kappa}_2\int\frac{d\vec{q}}{(2\pi)^3}\sum_{\sigma\sigma'}\hat{\psi}^{\dagger}_{\omega_1,\vec{\kappa}_1+\vec{q},\sigma}\hat{\psi}^{\dagger}_{\omega_2,\vec{\kappa}_2-\vec{q},\sigma'}\mathrm{V}(q,\omega_i)\hat{\psi}_{\omega_3\vec{\kappa}_2\sigma'}\hat{\psi}_{\omega_4\vec{\kappa}_1\sigma} \tag{1397}$$

where the effective electron-electron potential:

$$\mathrm{V}(q,\omega_i)=\left[\frac{\beta}{2}\sum_{\chi=\pm}\frac{\chi\in(q)}{i(\omega_1-\omega_4-i\chi\in)}\frac{1}{1-\exp\{-\chi\beta\in\}}+\begin{pmatrix}\omega_1\to\omega_2\\ \omega_4\to\omega_3\end{pmatrix}\right]\delta(\omega_1+\omega_2-\omega_3-\omega_4) \tag{1398}$$

In (1398), $\begin{pmatrix}\omega_1\to\omega_2\\ \omega_4\to\omega_3\end{pmatrix}$ implies we write the first term in (1398) and do a swap of the frequencies. In further evaluations, we consider $\mathrm{V}(q,\omega_i)$ to be the double count of the first term in (1398). Integrating over the phonon momenta in (1397), we consider the contribution of long wavelength phonons. Therefore, we set the upper limit of the energy to be the Debye frequency:

$$\in(q)=uq\le\omega_D \tag{1399}$$

From the BCS theory where the electro-phonon coupling constant λ is small, we then apply the perturbation theory with respect to λ when calculating the scattering amplitude of an electron on an electron. For the present model, this process occurs only due to the exchange of a virtual phonon. From the 4-fermion partition function in (1373), we write the following Feynman diagrams for the given process:

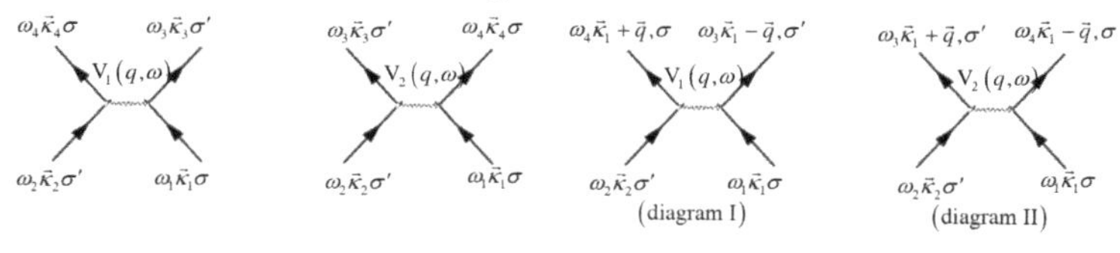

FIGURE 10.1 Feynman diagrams for given virtual phonon processes for given effective electron-electron potentials $\mathrm{V}_1(\vec{q},\omega)$ and $\mathrm{V}_2(\vec{q},\omega)$ and given spin indices σ and σ'. The diagrams show the scattering from one state $\omega_i\vec{\kappa}_i\sigma'$ to another $\omega_k\vec{\kappa}_k\sigma$.

Here,

$$\mathrm{V}_1(q,\omega)=-\frac{\in\beta}{i(\omega_4-\omega_1)+\in}\frac{1}{1-\exp\{\beta\in\}}\delta(\omega_4+\omega_3-\omega_1-\omega_2)\delta(\vec{\kappa}_4-\vec{\kappa}_1-\vec{q})\delta(\vec{\kappa}_3-\vec{\kappa}_2+\vec{q}) \tag{1400}$$

and

$$V_2(q,\omega) = \frac{\in\beta}{-i(\omega_4-\omega_1)+\in}\frac{1}{1-\exp\{\beta\in\}}\delta(\omega_4+\omega_3-\omega_1-\omega_2)\delta(\vec{\kappa}_4-\vec{\kappa}_1+\vec{q})\delta(\vec{\kappa}_3-\vec{\kappa}_2-\vec{q}) \quad (1401)$$

In the second series of diagrams in Figure 10.1, for example, the first diagram shows scattering of two particles from the states $\omega_2\vec{\kappa}_2\sigma'$, $\omega_4\vec{\kappa}_4\sigma$ to $\omega_1\vec{\kappa}_1\sigma$, $\omega_3\vec{\kappa}_3\sigma'$. The Coulomb potential $W(|\vec{r}-\vec{r}'|)$ conserves momentum and spin momentum (indicated by the Dirac delta functions in [1400] and [1401]) because it depends on only one $|\vec{r}-\vec{r}'|$ and cannot move the center of mass of the two colliding particles or flip their spins. From the lower-order approximation with respect to the electron-phonon constant, the vertex function is written as the sum of two diagrams:

$$\Gamma(q) = (\text{diagram I}) + (\text{diagram II}) \quad (1402)$$

Because the incoming and outgoing electronic states are real, so holds the law of conservation of energy and momentum. Therefore, for (diagram I) and (diagram II), respectively, we have

$$-i\omega_4 = \frac{(\vec{\kappa}_1+\vec{q})^2}{2m}-\mu \quad , \quad -i\omega_4 = \frac{(\vec{\kappa}_1-\vec{q})^2}{2m}-\mu \quad (1403)$$

We evaluate the scattering amplitude (1402) considering scattering of the electrons on or at the neighborhood of the Fermi surface. In our case, we have two physical quantities, that is, the Fermi energy

$$\in_F = \frac{\vec{\kappa}_F^2}{2m} = \mu \quad (1404)$$

and the Debye frequency ω_D. So, for any solid state, we have the condition

$$\in_F >> \omega_D \quad (1405)$$

This considers all electrons at the neighborhood of the Fermi surface. Consequently, the following condition is also satisfied

$$|\in(\kappa)-\in_F| = \left|\frac{\vec{\kappa}^2}{2m}-\mu\right| \le \omega_D \quad (1406)$$

So, at the neighborhood of the Fermi surface, we have

$$i\omega_1 = i\omega_2 = i\omega_3 = i\omega_4 \quad (1407)$$

and then (1402) becomes

$$\Gamma(q) = \frac{\lambda^2}{u^2 v_0}\sum_{\chi=\mp}\frac{\chi\in(q)}{\frac{\vec{\kappa}_4^2}{2m}-\frac{\vec{\kappa}_1^2}{2m}+\chi\in(q)}\frac{1}{1-\exp\{\chi\beta\in(q)\}} \cong \frac{\lambda^2}{u^2 v_0} \quad (1408)$$

Therefore, electrons at the neighborhood of the Fermi surface (1406) have the following scattering amplitude:

$$\Gamma(q) \cong \Gamma(0) = \frac{\lambda^2}{u^2 v_0} > 0 \quad (1409)$$

This implies that for such electrons, the effective 4-fermion potential is an attractive one. The BCS approximation therefore implies the swap of the effective 4-fermion interaction potential in (1397) with the vertex function in (1409). Physically, this implies searching for the contribution to the partition function of virtual electrons in the domain

$$\left|\frac{\vec{\kappa}^2}{2m}-\mu\right|\leq\omega_D \tag{1410}$$

For this condition, for small λ, the attraction between the quasiparticles is achieved and so

$$Z=\int\prod_{\omega_n\vec{\kappa}}d\hat{\psi}^{\dagger}_{\omega_n\vec{\kappa}\sigma}\,d\hat{\psi}_{\omega_n\vec{\kappa}\sigma}\exp\left\{-\left(S_e\left[\hat{\psi}\right]-\tilde{S}^{\text{eff}}_{e-e}\left[\hat{\psi}\right]\right)\right\} \tag{1411}$$

where the effective 4-fermion interaction action functional $S^{\text{eff}}_{e-e}\left[\hat{\psi}\right]$:

$$\tilde{S}^{\text{eff}}_{e-e}\left[\hat{\psi}\right]=\Gamma(0)\beta V^2\sum_{\omega_1\omega_2\omega_3\omega_4}\delta(\omega_4+\omega_3-\omega_1-\omega_2)\int d\vec{\kappa}_1\,d\vec{\kappa}_2\int\frac{d\vec{q}}{(2\pi)^3}\sum_{\sigma\sigma'}\hat{\psi}^{\dagger}_{\omega_4,\vec{\kappa}_1+\vec{q},\sigma}\hat{\psi}^{\dagger}_{\omega_3,\vec{\kappa}_2-\vec{q},\sigma'}\hat{\psi}_{\omega_2\vec{\kappa}_2\sigma'}\hat{\psi}_{\omega_1\vec{\kappa}_1\sigma} \tag{1412}$$

From here, we easily move to the $(\vec{\kappa},\tau)$ space and, consequently, the action functional for the local 4-fermion interaction

$$\tilde{S}^{\text{eff}}_{e-e}\left[\hat{\psi}\right]=\Gamma(0)\int_0^{\beta}d\tau\int d\vec{\kappa}\sum_{\sigma\sigma'}\hat{\psi}^{\dagger}_{\sigma}(\vec{\kappa},\tau)\hat{\psi}^{\dagger}_{\sigma'}(\vec{\kappa},\tau)\hat{\psi}_{\sigma'}(\vec{\kappa},\tau)\hat{\psi}_{\sigma}(\vec{\kappa},\tau) \tag{1413}$$

The integral over all $\vec{\kappa}$ space in (1413) is an illusion because the condition (1410) is equivalent to an insertion in the integral over space coordinates—the effective cutoff limit. The functional integral (1413) can be effectively evaluated by introducing an auxiliary functional integral via the auxiliary **Hubbard-Stratonovich c-field [31]** $\Delta^{*}_{\sigma\sigma'}(\vec{\kappa},\tau)$, $\Delta_{\sigma\sigma'}(\vec{\kappa},\tau)$, which serves to decouple the interaction and render the action functional quadratic in the fermionic fields $\hat{\psi}_{\sigma}(\vec{\kappa},\tau)$ and $\hat{\psi}^{\dagger}_{\sigma}(\vec{\kappa},\tau)$ (**Figure 10.2**).

So,

$$\exp\left\{\tilde{S}^{\text{eff}}_{e-e}\left[\hat{\psi}\right]\right\}=N_1\int d\left[\Delta^{*}_{\sigma\sigma'}\right]d\left[\Delta_{\sigma\sigma'}\right]\exp\left\{\mathbf{F}\left[\Delta^{*}_{\sigma\sigma'},\Delta_{\sigma\sigma'}\right]\right\} \tag{1414}$$

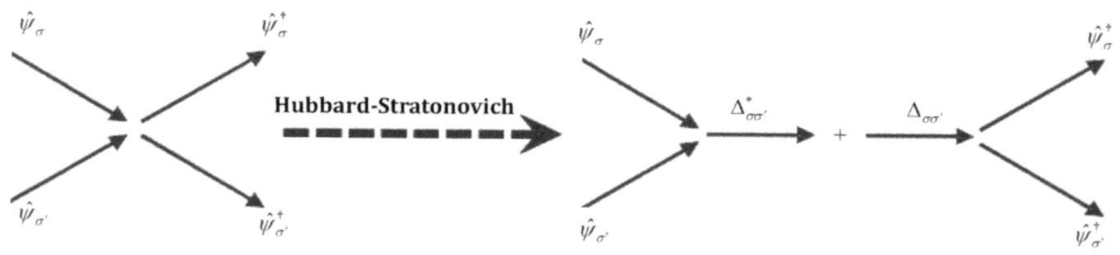

FIGURE 10.2 The Hubbard-Stratonovich transformation illustration with the interaction term originally composed of quartic term fermion fields decomposed into quadratic in the fermion fields.

where

$$\mathbf{F}\left[\Delta^*_{\sigma\sigma'},\Delta_{\sigma\sigma'}\right]=\int_0^\beta d\tau\int d\vec{\kappa}\left(\frac{\Delta^*_{\sigma\sigma'}(\vec{\kappa},\tau)\Delta_{\sigma\sigma'}(\vec{\kappa},\tau)}{\Gamma(0)}+\left[\begin{array}{cc}\Delta^*_{\sigma\sigma'}(\vec{\kappa},\tau) & \Delta_{\sigma\sigma'}(\vec{\kappa},\tau)\end{array}\right]\left[\begin{array}{c}A(\vec{\kappa},\tau)\\ A^\dagger(\vec{\kappa},\tau)\end{array}\right]\right) \tag{1415}$$

and the operators that act by raising or lowering, respectively, the particle number by 2:

$$A^\dagger_{\sigma\sigma'}(\vec{\kappa},\tau)=\hat{\psi}^\dagger_\sigma(\vec{\kappa},\tau)\hat{\psi}^\dagger_{\sigma'}(\vec{\kappa},\tau)\quad,\quad A_{\sigma'\sigma}(\vec{\kappa},\tau)=\hat{\psi}_{\sigma'}(\vec{\kappa},\tau)\hat{\psi}_\sigma(\vec{\kappa},\tau) \tag{1416}$$

The auxiliary fields $\Delta^*_{\sigma\sigma'}(\vec{\kappa},\tau)$, $\Delta_{\sigma\sigma'}(\vec{\kappa},\tau)$ are not dynamic; therefore, their equations of motion have the form:

$$\Delta_{\sigma\sigma'}(\vec{\kappa},\tau)=\Gamma(0)A_{\sigma,\sigma'}(\vec{\kappa},\tau) \tag{1417}$$

From (1417), $\Delta_{\sigma\sigma'}(\vec{\kappa},\tau)$ has the sense of a bosonic field linked to fermions (Cooper pairs).

10.1.3 Effective Action Functional

We rewrite the partition function (1411) considering (1414) via the Nambu spinors:

$$\psi=\left[\begin{array}{c}\psi_{\sigma'}\\ \psi^\dagger_\sigma\end{array}\right],\ \psi^\dagger=\left[\begin{array}{cc}\psi^\dagger_{\sigma'} & \psi_\sigma\end{array}\right] \tag{1418}$$

then

$$Z_{BCS}=\int d[\psi]d[\psi^\dagger]d[\Delta^*_{\sigma\sigma'}]d[\Delta_{\sigma\sigma'}]\exp\left\{\mathbf{F}\left[\Delta^*_{\sigma\sigma'},\Delta_{\sigma\sigma'}\right]\right\} \tag{1419}$$

where

$$\mathbf{F}\left[\Delta^*_{\sigma\sigma'},\Delta_{\sigma\sigma'}\right]=\int_0^\beta d\tau\int d\vec{\kappa}\left(-\frac{\Delta^*_{\sigma\sigma'}(\vec{\kappa},\tau)\Delta_{\sigma\sigma'}(\vec{\kappa},\tau)}{\Gamma(0)}+\psi^\dagger\left[-\mathbf{G}^{-1}\right]\psi\right) \tag{1420}$$

and

$$\left[-\mathbf{G}^{-1}\right]=\left[\begin{array}{cc}\mathbf{G}^{-1}_{01} & -\Delta(\vec{\kappa},\tau)\\ -\Delta^*(\vec{\kappa},\tau) & \mathbf{G}^{-1}_{02}\end{array}\right] \tag{1421}$$

with

$$\mathbf{G}^{-1}_{01}=-\frac{\partial}{\partial\tau}+\frac{\nabla^2_{\vec{\kappa}}}{2m}-\mu\quad,\quad \mathbf{G}^{-1}_{02}=\frac{\partial}{\partial\tau}-\frac{\nabla^2_{\vec{\kappa}}}{2m}+\mu \tag{1422}$$

Evaluating the Gaussian integral (1419) over fermionic variables (1418), we then have

$$Z_{\mathrm{BCS}} = \int d\left[\Delta^*_{\sigma\sigma'}\right] d\left[\Delta_{\sigma\sigma'}\right] \exp\left\{-S^{\mathrm{eff}}\left[\Delta^*_{\sigma\sigma'}, \Delta_{\sigma\sigma'}\right]\right\} \tag{1423}$$

where

$$S^{\mathrm{eff}}\left[\Delta^*_{\sigma\sigma'}, \Delta_{\sigma\sigma'}\right] = -\int_0^\beta d\tau \int d\vec{\kappa} \frac{\Delta^*_{\sigma\sigma'}(\vec{\kappa},\tau)\Delta_{\sigma\sigma'}(\vec{\kappa},\tau)}{\Gamma(0)} + \mathrm{Tr}\ln\left[-\mathbf{G}^{-1}\right] \tag{1424}$$

From equation (1423) or (1424), we can achieve all the results of the BCS theory and, in particular, the expression of the grand thermodynamic potential that permits us to find the excitation spectrum of Cooper pairs and the behavior of Cooper pairs at the neighborhood of phase transition temperature.

10.1.4 Critical Temperature

The saddle-point solution of (1424) can be obtained when $\Delta_{\sigma\sigma'} = \Delta$ via the equation

$$\left.\frac{\partial S^{\mathrm{eff}}\left[\Delta^*_{\sigma\sigma'}, \Delta_{\sigma\sigma'}\right]}{\partial \Delta^*_{\sigma\sigma'}}\right|_{\Delta_{\sigma\sigma'}=\Delta} = 0 \tag{1425}$$

or

$$-\frac{\beta V}{\Gamma(0)}\Delta + \mathrm{Tr}\left[\left[-\mathbf{G}\right]\frac{\partial}{\partial \Delta^*_{\sigma\sigma'}}\left[-\mathbf{G}^{-1}\right]\Big|_{\Delta_{\sigma\sigma'}=\Delta}\right] = 0 \tag{1426}$$

The inverse Green's function $\left[-\mathbf{G}^{-1}\right]$ in (1426) can be conveniently obtained by expressing (1421) via the momentum and frequency:

$$\begin{bmatrix} i\omega + \varsigma_{\vec{\kappa}} & -\Delta \\ -\Delta^* & i\omega - \varsigma_{\vec{\kappa}} \end{bmatrix} \mathbf{G}(\vec{\kappa},\omega) = \hat{\mathbf{1}} \tag{1427}$$

Here, $\hat{\mathbf{1}}$ is a unit matrix and

$$\mathbf{G}(\vec{\kappa},\omega) = \frac{-1}{\omega^2 + \mathrm{E}^2_{\vec{\kappa}}} \begin{bmatrix} i\omega + \varsigma_{\vec{\kappa}} & -\Delta \\ -\Delta^* & i\omega - \varsigma_{\vec{\kappa}} \end{bmatrix} \tag{1428}$$

with

$$\varsigma_{\vec{\kappa}} = \frac{\vec{\kappa}^2}{2m} - \mu \equiv \epsilon_{\vec{\kappa}} - \mu \tag{1429}$$

and

$$\mathrm{E}_{\vec{\kappa}} = \sqrt{\varsigma^2_{\vec{\kappa}} + |\Delta|^2} \tag{1430}$$

is the **Bogoliubov dispersion,** which is the quasi-particle energy.

Also from (1421), we have

$$\frac{\partial}{\partial \Delta^*_{\sigma\sigma'}}\left[-\mathbf{G}^{-1}\right] = -\begin{bmatrix} 0 & 0 \\ 1 & 0 \end{bmatrix} \equiv -\tau^- \tag{1431}$$

So, the second term in (1426) can now be represented as follows:

$$\mathrm{Tr}\left[\left[-\mathbf{G}\right]\frac{\partial}{\partial \Delta^*_{\sigma\sigma'}}\left[-\mathbf{G}^{-1}\right]\Bigg|_{\Delta_{\sigma\sigma'}=\Delta}\right] = \mathrm{Tr}\left[-\left[-\mathbf{G}\right]\tau^-\right] \tag{1432}$$

Substituting this into (1421), we then arrive at the following scaling equation

$$\frac{\Gamma(0)}{\beta V}\mathrm{Tr}\left[\frac{1}{\omega_n^2 + \mathrm{E}_{\vec{\kappa}}^2}\right] = 1 \tag{1433}$$

Also, rewriting the functional trace via the momentum-frequency representation, we arrive at the standard form of the scaling equation:

$$\frac{\Gamma(0)}{\beta}\int_{|\varsigma_{\vec{\kappa}}|\leq\omega_D}\frac{d\vec{\kappa}}{(2\pi)^2}\sum_{\omega_n}\frac{1}{\omega_n^2 + \mathrm{E}_{\vec{\kappa}}^2} = 1 \tag{1434}$$

So, the sum in (1434) over the Matsubara frequency ω_n yields

$$\frac{\Gamma(0)}{4\pi^2}\int_{|\varsigma_{\vec{\kappa}}|\leq\omega_D}\frac{\kappa^2 d\kappa}{\sqrt{(\in-\mu)^2+\Delta^2}}\tanh\frac{\beta}{2}\sqrt{(\in-\mu)^2+\Delta^2} = 1 \tag{1435}$$

This equation is approximately solved with the help of the following condition

$$\frac{\omega_D}{\mu} << 1 \quad , \quad T << \omega_D \tag{1436}$$

The critical (or BCS theory) temperature T_c is defined as one when the condensate of Cooper pairs "**disappears**" leading to the **quenching** of the superconducting phase:

$$\Delta(T_c) = 0 \tag{1437}$$

From condition (1436) for the critical temperature and from (1435), we have

$$\int_{|\varepsilon-\mu|\leq\omega_D}\frac{\kappa^2 d\kappa}{\in-\mu}\tanh\frac{\beta_c}{2}(\in-\mu) \cong 2m\sqrt{2m\mu}\int_0^{\frac{\omega_D}{2T_c}} d\eta\frac{\tanh\eta}{\eta} \cong 2m\sqrt{2m\mu}\ln\frac{4\gamma\omega_D}{2\pi T_c} \tag{1438}$$

From here, the critical temperature is found to be

$$T_c \cong \frac{2\gamma\omega_D}{\pi}\exp\left\{-\frac{1}{\Gamma(0)\nu_{\mathrm{F}}}\right\} \tag{1439}$$

where $\nu_{\mathrm{F}} = \frac{m\kappa_{\mathrm{F}}}{2\pi^2}$ is the density of states of particles on the Fermi surface and $\ln\gamma = C = 0.577\cdots$ (Euler constant).

In (1435), considering $T = 0$ and the change of variables,

$$\eta = \in - \mu \cong v_F(\kappa - \kappa_F) \quad , \quad v_F \equiv \frac{\kappa_F}{m} \quad , \quad \kappa_F = \sqrt{2m\mu} \tag{1440}$$

and with the help of

$$\Delta \cong T \quad , \quad T << \omega_D \tag{1441}$$

then

$$\Delta << \omega_D \tag{1442}$$

so,

$$\frac{\Gamma(0)}{4\pi^2} \int_{|\varepsilon-\mu|\le\omega_D} \frac{\kappa^2 d\kappa}{\sqrt{(\in - \mu)^2 + \Delta^2}} \cong \frac{1}{v_F^3} \int_{-\omega_D}^{\omega_D} d\eta \frac{(\eta - v_F\kappa_F)^2}{\sqrt{\eta^2 + \Delta^2}} \cong \frac{2\kappa_F}{v_F} \ln \frac{2\omega_D}{\Delta} \tag{1443}$$

and substituting this into (1435) then the saddle-point solution

$$\Delta \cong 2\omega_D \exp\left\{ -\frac{2}{\Gamma(0) v_F} \right\} \tag{1444}$$

We can therefore establish the relation between T_c and Δ:

$$T_c \cong \frac{\gamma}{\pi} \Delta \cong 0.57\Delta \tag{1445}$$

Expressions (1439) and (1445) completely coincide with the results obtained from the BCS theory of superconductivity.

10.2 Mean-Field Theory

We find the Matsubara single-particle Green's function with the help of the **equation of motion approach**. The basic feature of the BCS ground state wave function is that it is found in a superconducting position of electronic states containing different numbers of electrons. Therefore, such a formulation is possible in the grand canonical ensemble. Considering the Heisenberg representation, we write down the standard equations of motion for electronic operators, taking into account real time:

$$\frac{\partial \hat{\psi}_\sigma}{\partial t} = i\left[\hat{H}, \hat{\psi}_\sigma \right] \tag{1446}$$

The commutator in the right-hand side is calculated directly using commutation relations for operators $\hat{\psi}_\sigma(\vec{r}, t)$:

$$\hat{\rho}_{\sigma\sigma'}(\vec{r},\tau;\vec{r}',\tau') + \hat{\rho}_{\sigma'\sigma}(\vec{r}',\tau';\vec{r},\tau) = \delta_{\sigma\sigma'}\delta(\vec{r} - \vec{r}') \tag{1447}$$

$$\hat{\rho}_{\sigma\sigma'}(\vec{r},\tau;\vec{r}',\tau') + \hat{\rho}_{\sigma'\sigma}(\vec{r}',\tau';\vec{r},\tau) = A^{\dagger}_{\sigma\sigma'}(\vec{r},\tau;\vec{r}',\tau') + A_{\sigma'\sigma}(\vec{r}',\tau';\vec{r},\tau) = 0 \tag{1448}$$

where

$$\hat{\rho}_{\sigma\sigma'}(\vec{r},\tau;\vec{r}',\tau') = \hat{\psi}_\sigma(\vec{r},\tau)\hat{\psi}^\dagger_{\sigma'}(\vec{r}',\tau') \quad , \quad \hat{\rho}_{\sigma'\sigma}(\vec{r}',\tau';\vec{r},\tau) = \hat{\psi}^\dagger_{\sigma'}(\vec{r}',\tau)\hat{\psi}_\sigma(\vec{r},\tau) \tag{1449}$$

$$A^\dagger_{\sigma\sigma'}(\vec{r},\tau;\vec{r}',\tau') = \hat{\psi}^\dagger_\sigma(\vec{r},\tau)\hat{\psi}^\dagger_{\sigma'}(\vec{r},\tau;\vec{r}',\tau') \quad , \quad A_{\sigma'\sigma}(\vec{r}',\tau';\vec{r},\tau) = \hat{\psi}_{\sigma'}(\vec{r}',\tau')\hat{\psi}_\sigma(\vec{r},\tau) \tag{1450}$$

Let us introduce imaginary time $t = -i\tau$, then

$$\frac{\partial\hat{\psi}_\sigma}{\partial\tau} = \left[\hat{H}, \hat{\psi}_\sigma\right] \tag{1451}$$

From here, we therefore compute

$$\frac{\partial\hat{\psi}_\sigma}{\partial\tau} = \mathbf{M}_2\hat{\psi}_\sigma - \frac{|\lambda|}{2}\left\langle\left[\hat{H}_{\text{int}}, \hat{\psi}_\sigma\right]\right\rangle - \mu\sum_{\sigma'}\left[\hat{\psi}^\dagger_\sigma\hat{\psi}_\sigma, \hat{\psi}_{\sigma'}\right] \quad , \quad \mathbf{M}_2 = \frac{\nabla^2}{2m} + \mu \tag{1452}$$

The last commutator may be computed as follows

$$-\mu\left[\hat{\psi}^\dagger_\sigma\hat{\psi}_\sigma, \hat{\psi}_{\sigma'}\right] = -\mu\left[\hat{\psi}^\dagger_\sigma\hat{\psi}_\sigma\hat{\psi}_{\sigma'} - \hat{\psi}_{\sigma'}\hat{\psi}^\dagger_\sigma\hat{\psi}_\sigma\right] = -\mu\left[-\hat{\psi}^\dagger_\sigma\hat{\psi}_{\sigma'}\hat{\psi}_\sigma - \hat{\psi}_{\sigma'}\hat{\psi}^\dagger_\sigma\hat{\psi}_\sigma\right] = \mu\hat{\psi}_\sigma \tag{1453}$$

$$-\mu\left[\hat{\psi}^\dagger_\sigma\hat{\psi}_\sigma, \hat{\psi}^\dagger_{\sigma'}\right] = -\mu\left[\hat{\psi}^\dagger_\sigma\hat{\psi}_\sigma\hat{\psi}^\dagger_{\sigma'} - \hat{\psi}^\dagger_{\sigma'}\hat{\psi}^\dagger_\sigma\hat{\psi}_\sigma\right] = -\mu\left[\hat{\psi}^\dagger_\sigma\left(\delta_{\sigma\sigma'} + \hat{\psi}^\dagger_{\sigma'}\hat{\psi}_\sigma\right) - \hat{\psi}^\dagger_{\sigma'}\hat{\psi}^\dagger_\sigma\hat{\psi}_\sigma\right] = -\mu\hat{\psi}^\dagger_\sigma \tag{1454}$$

We consider the second term of the equation of motion by examining

$$\hat{H}^{\text{MF}}_{\text{int}} = -|\lambda|\left(\left\langle\hat{\psi}^\dagger_\sigma\hat{\psi}^\dagger_{\sigma'}\right\rangle\hat{\psi}_{\sigma'}\hat{\psi}_\sigma + \hat{\psi}^\dagger_\sigma\hat{\psi}^\dagger_{\sigma'}\left\langle\hat{\psi}_{\sigma'}\hat{\psi}_\sigma\right\rangle\right) \tag{1455}$$

$$\begin{aligned}\left[\hat{H}^{\text{MF}}_{\text{int}}, \hat{\psi}_{\sigma'}\right] &= -|\lambda|\left\langle\hat{\psi}_{\sigma'}\hat{\psi}_\sigma\right\rangle\left(\hat{\psi}^\dagger_\sigma\hat{\psi}^\dagger_{\sigma'}\hat{\psi}_{\sigma'} - \hat{\psi}_{\sigma'}\hat{\psi}^\dagger_\sigma\hat{\psi}^\dagger_{\sigma'}\right) = -|\lambda|\left\langle\hat{\psi}_{\sigma'}\hat{\psi}_\sigma\right\rangle\left(\hat{\psi}^\dagger_\sigma\left(1 - \hat{\psi}_{\sigma'}\hat{\psi}^\dagger_{\sigma'}\right) - \hat{\psi}_{\sigma'}\hat{\psi}^\dagger_\sigma\hat{\psi}^\dagger_{\sigma'}\right)\\ &= -|\lambda|\left\langle\hat{\psi}_{\sigma'}\hat{\psi}_\sigma\right\rangle\hat{\psi}^\dagger_\sigma\end{aligned} \tag{1456}$$

$$\begin{aligned}\left[\hat{H}^{\text{MF}}_{\text{int}}, \hat{\psi}^\dagger_\sigma\right] &= -|\lambda|\left\langle\hat{\psi}^\dagger_\sigma\hat{\psi}^\dagger_{\sigma'}\right\rangle\left(\hat{\psi}_{\sigma'}\hat{\psi}_\sigma\hat{\psi}^\dagger_\sigma - \hat{\psi}^\dagger_\sigma\hat{\psi}_{\sigma'}\hat{\psi}_\sigma\right) = -|\lambda|\left\langle\hat{\psi}^\dagger_\sigma\hat{\psi}^\dagger_\sigma\right\rangle\left(\hat{\psi}_{\sigma'}\left(1 - \hat{\psi}^\dagger_\sigma\hat{\psi}_{\sigma'}\right) - \hat{\psi}^\dagger_\sigma\hat{\psi}_{\sigma'}\hat{\psi}_\sigma\right)\\ &= -|\lambda|\left\langle\hat{\psi}^\dagger_\sigma\hat{\psi}^\dagger_{\sigma'}\right\rangle\hat{\psi}_{\sigma'}\end{aligned} \tag{1457}$$

So,

$$\frac{\partial\hat{\psi}_{\sigma'}}{\partial\tau} = \mathbf{M}_2\hat{\psi}_{\sigma'} - |\lambda|\left\langle\hat{\psi}_{\sigma'}\hat{\psi}_\sigma\right\rangle\hat{\psi}^\dagger_\sigma \quad , \quad \frac{\partial\hat{\psi}^\dagger_{\sigma'}}{\partial\tau} = -\mathbf{M}_2\hat{\psi}^\dagger_{\sigma'} + |\lambda|\left\langle\hat{\psi}^\dagger_\sigma\hat{\psi}^\dagger_{\sigma'}\right\rangle\hat{\psi}_\sigma \tag{1458}$$

$$\frac{\partial\hat{\psi}_\sigma}{\partial\tau} = \mathbf{M}_2\hat{\psi}_\sigma + |\lambda|\left\langle\hat{\psi}_{\sigma'}\hat{\psi}_\sigma\right\rangle\hat{\psi}^\dagger_{\sigma'} \quad , \quad \frac{\partial\hat{\psi}^\dagger_\sigma}{\partial\tau} = -\mathbf{M}_2\hat{\psi}^\dagger_\sigma - |\lambda|\left\langle\hat{\psi}^\dagger_\sigma\hat{\psi}^\dagger_{\sigma'}\right\rangle\hat{\psi}_{\sigma'} \tag{1459}$$

It is instructive to note that the ground state of a superconductor assumes that at $T = 0$ we have a **condensate** of Cooper pairs with a massive macroscopic number of particles. This state does not change at all physically if the number of pairs in the condensate is changed by unity. This can be expressed mathematically by the appearance of nonzero (in the limit of number of particles $N \to \infty$) values of matrix elements:

$$\lim_{N\to\infty} \langle m,N|\mathrm{A}_{\sigma'\sigma}(\vec{r}_2,\tau_2;\vec{r}_1,\tau_1)|m,N+2\rangle = \lim_{N\to\infty}\left(\langle m,N+2|\mathrm{A}^{\dagger}_{\sigma\sigma'}(\vec{r}_1,\tau_1;\vec{r}_2,\tau_2)|N,m\rangle\right)^{*} \neq 0 \tag{1460}$$

In this book, for convenience, we drop the symbol of limit and diagonal matrix index m, numbering **the same** states of the system with different numbers of particles. Therefore,

$$\langle N|\mathrm{A}_{\sigma'\sigma}(\vec{r}_2,\tau_2;\vec{r}_1,\tau_1)|N+2\rangle = \left(\langle N+2|\mathrm{A}^{\dagger}_{\sigma\sigma'}(\vec{r}_1,\tau_1;\vec{r}_2,\tau_2)|N\rangle\right)^{*} \neq 0 \tag{1461}$$

Therefore, it is important to note that the superconducting transition is characterized by **spontaneous breaking of gauge symmetry** that corresponds to particle number (or charge) conservation. The electron pairs may **disappear** in condensate or **appear** from condensate without a change in the macroscopic state of the system (for $N \to \infty$). It is instructive to note that the concept of spontaneous symmetry breaking plays a central role in the theory of second-order phase transitions. Here, the ground state of a **condensed** phase within the domain of temperature $T < T_c$ always possesses a symmetry **lower** than that of the Hamiltonian determinant that describes the given phase transition.

The propagation of particles (**Landau-quasiparticles**) through the many-body system becomes more complex compared to normal systems due to the presence of a condensate in the given system. Consider a particle presence with momentum $\vec{q}$ at time $\tau = 0$. Then, there exists two contributions to the propagator if we neglect finite lifetime effects of the particles at that point. We describe the pairing of electrons by introducing (in addition to **normal Green's function**):

$$\rightarrow\!\!\rightarrow \mathcal{G}_{\sigma\sigma'}(\vec{r},\tau;\vec{r}',\tau') = -\langle N|\mathrm{T}_{\tau}\hat{\rho}_{\sigma\sigma'}(\vec{r},\tau;\vec{r}',\tau')|N\rangle \tag{1462}$$

which is a new correlation function, **the anomalous Green's function:**

$$\langle N|\mathrm{T}_{\tau}\mathrm{A}|N+2\rangle = \mathcal{F} \quad , \quad \langle N+2|\mathrm{T}_{\tau}\mathrm{A}^{\dagger}|N\rangle = \mathcal{F}^{\dagger} \tag{1463}$$

$$\rightarrow\!\!\leftarrow \mathcal{F}_{\sigma\sigma'}(\vec{r},\tau;\vec{r}',\tau') = -\langle N|\mathrm{T}_{\tau}\mathrm{A}_{\alpha\beta}(\vec{r},\tau;\vec{r}',\tau')|N+2\rangle \tag{1464}$$

$$\longleftrightarrow \mathcal{F}^{\dagger}_{\sigma\sigma'}(\vec{r},\tau;\vec{r}',\tau') = -\langle N+2|\mathrm{T}_{\tau}\mathrm{A}^{\dagger}_{\sigma\sigma'}(\vec{r},\tau;\vec{r}',\tau')|N\rangle \tag{1465}$$

that satisfies the following symmetry properties, which follow directly from commutation relations for electron operators:

$$\mathcal{F}_{\sigma\sigma'}(\vec{r},\tau;\vec{r}',\tau') = -\mathcal{F}_{\sigma'\sigma}(\vec{r}',\tau';\vec{r},\tau) \quad , \quad \mathcal{F}^{\dagger}_{\sigma\sigma'}(\vec{r},\tau;\vec{r}',\tau') = -\mathcal{F}^{\dagger}_{\sigma'\sigma}(\vec{r}',\tau';\vec{r},\tau) \tag{1466}$$

These **anomalous Green's functions** destroy or create a Cooper pair in the superconducting ground state. Though the Green's function in (1462) has a different algebraic form in the superconducting state, it has the usual definition.

In the following, we consider only (spin) singlet Cooper pairing, which is realized in the majority of known metallic superconductors. We may separate spin dependence using the following representation:

$$\mathscr{F}_{\sigma\sigma'} = \mathbf{a}_{\sigma\sigma'}\mathscr{F} \quad , \quad \mathscr{F}^{\dagger}_{\sigma\sigma'} = \mathbf{b}_{\sigma\sigma'}\mathscr{F}^{\dagger} \tag{1467}$$

From the Pauli Exclusion Principle,

$$\left\langle \hat{\psi}_{\sigma}(\vec{r},\tau)\hat{\psi}_{\sigma}(\vec{r},\tau)\right\rangle = 0 \tag{1468}$$

Hence,

$$\mathscr{F}_{\sigma\sigma} = \mathscr{F}^{\dagger}_{\sigma\sigma} = 0 \tag{1469}$$

and, consequently,

$$\mathbf{a}_{\sigma\sigma} = \mathbf{b}_{\sigma\sigma} = 0 \tag{1470}$$

This allows us to have the following matrices

$$\mathbf{a} = \begin{bmatrix} 0 & a_1 \\ a_2 & 0 \end{bmatrix} \quad , \quad \mathbf{b} = \begin{bmatrix} 0 & b_1 \\ b_2 & 0 \end{bmatrix} \tag{1471}$$

For the singlet pairing,

$$\mathscr{F}(\vec{r}-\vec{r}') = \mathscr{F}(\vec{r}'-\vec{r}) \tag{1472}$$

from (1466), we then have

$$\mathbf{a}_{\sigma\sigma'} = -\mathbf{a}_{\sigma'\sigma} \quad , \quad \mathbf{b}_{\sigma\sigma'} = -\mathbf{b}_{\sigma'\sigma} \tag{1473}$$

Therefore,

$$\mathbf{a} = \begin{bmatrix} 0 & a \\ -a & 0 \end{bmatrix} \quad , \quad \mathbf{b} = \begin{bmatrix} 0 & b \\ -b & 0 \end{bmatrix} \tag{1474}$$

Considering, $\left(\mathscr{F}^{+}_{\sigma\sigma'}\right)^{*} = -\mathscr{F}_{\sigma\sigma'}$, then $\mathbf{b}^{*} = -\mathbf{a}$, and consequently,

$$\mathbf{a} = a\begin{bmatrix} 0 & 1 \\ -1 & 0 \end{bmatrix} \quad , \quad \mathbf{b} = -a^{*}\begin{bmatrix} 0 & 1 \\ -1 & 0 \end{bmatrix} \tag{1475}$$

Hence, the spin dependence of anomalous Green's functions reduces to a unit antisymmetric spinor of the second rank:

$$g_{\sigma\sigma'} = \begin{bmatrix} 0 & 1 \\ -1 & 0 \end{bmatrix} = i\sigma^{y}_{\sigma\sigma'} \quad , \quad \left[\hat{g}^{2}\right]_{\sigma\sigma'} = -\delta_{\sigma\sigma'} \tag{1476}$$

So, the anomalous Green's functions have the following form

$$\mathscr{F}_{\sigma\sigma'}(\vec{r},\tau;\vec{r}',\tau') = g_{\sigma\sigma'}\mathscr{F}(\vec{r}-\vec{r}';\tau-\tau') \quad , \quad \mathscr{F}^{\dagger}_{\sigma\sigma'}(\vec{r},\tau;\vec{r}',\tau') = -g_{\sigma\sigma'}\mathscr{F}^{\dagger}(\vec{r}-\vec{r}';\tau-\tau') \tag{1477}$$

and satisfy the **antiperiodicity** condition $\tau-\tau' \equiv \tilde{\tau} \to \tau$:

$$\mathscr{F}(\tau) = -\mathscr{F}(\tau+\beta) \quad , \quad \mathscr{F}^{\dagger}(\tau) = -\mathscr{F}^{\dagger}(\tau+\beta) \tag{1478}$$

Then, the Fourier series expansion of the given functions over τ has only odd Matsubara frequencies $\omega_n = \dfrac{(2n+1)\pi}{\beta}$. Also,

$$\mathscr{F}_{\sigma\sigma'}(0+) = g_{\sigma\sigma'}\lim_{\substack{\vec{r}\to\vec{r}' \\ \tau\to\tau'+0}}\mathscr{F}(\vec{r}-\vec{r}';\tau-\tau') \quad , \quad \mathscr{F}^{\dagger}_{\sigma\sigma'}(0+) = -g_{\sigma\sigma'}\lim_{\substack{\vec{r}\to\vec{r}' \\ \tau\to\tau'+0}}\mathscr{F}^{\dagger}(\vec{r}-\vec{r}';\tau-\tau') \tag{1479}$$

It is important to recall that spin dependence of the normal Green's function $\mathscr{G}_{\sigma\sigma'}(\vec{r},\tau;\vec{r}',\tau')$ (for a non-magnetic system) reduces to

$$\mathscr{G}_{\sigma\sigma'}(\vec{r},\tau;\vec{r}',\tau') = \delta_{\sigma\sigma'}\mathscr{G}(\vec{r},\tau;\vec{r}',\tau') \tag{1480}$$

We also recall that in a homogeneous and stationary system, we have the following dependence:

$$\mathscr{F}(\vec{r},\tau;\vec{r}',\tau') \equiv \mathscr{F}(\vec{r}-\vec{r}',\tau-\tau') \quad , \quad \mathscr{F}^{\dagger}(\vec{r},\tau;\vec{r}',\tau') \equiv \mathscr{F}^{\dagger}(\vec{r}-\vec{r}',\tau-\tau') \quad , \quad \mathscr{G}(\vec{r},\tau;\vec{r}',\tau') \equiv \mathscr{G}(\vec{r}-\vec{r}',\tau-\tau') \tag{1481}$$

We represent the Green's functions of the Fermi gas with attraction, defined by the Hamiltonian (1346) in the form:

$$\mathscr{G}_{\sigma\sigma'}(\vec{r},\tau;\vec{r}',\tau') = -\left\langle T_{\tau}\hat{\psi}_{\sigma}(\vec{r},\tau)\hat{\psi}^{\dagger}_{\sigma'}(\vec{r}',\tau')\right\rangle = -\theta(\tau-\tau')\hat{\psi}_{\sigma}(\vec{r},\tau)\hat{\psi}^{\dagger}_{\sigma'}(\vec{r}',\tau') + \theta(\tau'-\tau)\hat{\psi}^{\dagger}_{\sigma'}(\vec{r}',\tau')\hat{\psi}_{\sigma}(\vec{r},\tau) \tag{1482}$$

We find the equation of motion by first differentiating the Matsubara Green's function with respect to τ:

$$\frac{\partial}{\partial\tau}\mathscr{G}_{\sigma\sigma'}(\vec{r},\tau;\vec{r}',\tau') = -\left\langle T_{\tau}\frac{\partial\hat{\psi}_{\alpha}(\vec{r},\tau)}{\partial\tau}\hat{\psi}^{\dagger}_{\sigma'}(\vec{r}',\tau')\right\rangle - \delta_{\sigma\sigma'}\delta(\tau-\tau')\delta(\vec{r}-\vec{r}') \tag{1483}$$

Considering (1458) to (1483), we then have as follows:

$$-\mathbf{G}_0^{-1}\mathscr{G}_{\sigma\sigma'} - |\lambda|\left\langle\hat{\psi}_{\sigma}\hat{\psi}_{\sigma'}\right\rangle\mathscr{F}^{\dagger} = \delta_{\sigma\sigma'}\delta(\tau-\tau')\delta(\vec{r}-\vec{r}') \tag{1484}$$

$$\breve{\mathbf{G}}_0^{-1}\mathscr{F}^{\dagger} = -|\lambda|\left\langle\hat{\psi}^{\dagger}_{\sigma'}\hat{\psi}^{\dagger}_{\sigma}\right\rangle\mathscr{G}_{\sigma\sigma'} \tag{1485}$$

where

$$\mathbf{G}_0^{-1} = \partial_{\tau} - \frac{\Delta}{2m} - \mu \quad , \quad \breve{\mathbf{G}}_0^{-1} = \partial_{\tau} + \frac{\Delta}{2m} + \mu \tag{1486}$$

We introduce the notation

$$\Delta = |\lambda| \mathscr{F}(\vec{r},\tau;\vec{r}',\tau') = |\lambda| \mathscr{F}(0+) = -|\lambda| \langle \hat{\psi}_{\sigma'} \hat{\psi}_{\sigma} \rangle \tag{1487}$$

$$\Delta^* = |\lambda| \mathscr{F}^\dagger(\vec{r},\tau;\vec{r}',\tau') = |\lambda| \mathscr{F}^\dagger(0+) = -|\lambda| \langle \hat{\psi}^\dagger_{\sigma} \hat{\psi}^\dagger_{\sigma'} \rangle \tag{1488}$$

which plays the role of the **energy gap**, later in this book, in the spectrum of elementary excitations of a superconductor and, simultaneously, the **order parameter** for a superconducting transition, which distinguishes the superconducting state from the normal state. In this way, it plays the same role as the magnetization in the normal-metal-ferromagnet transition. Therefore, the following **Nambu-Gorkov equation in the differential form**:

$$\mathbf{G}^{-1} \hat{\mathscr{G}} = \hat{1} \tag{1489}$$

where

$$\mathbf{G}^{-1} = \begin{bmatrix} \mathbf{G}_0^{-1} & \Delta \\ \Delta^* & \breve{\mathbf{G}}_0^{-1} \end{bmatrix} , \quad \hat{\mathscr{G}} = \begin{bmatrix} \mathscr{G}(\vec{r}-\vec{r}';\tau-\tau') & \mathscr{F}(\vec{r}-\vec{r}';\tau-\tau') \\ \mathscr{F}^\dagger(\vec{r}-\vec{r}';\tau-\tau') & \mathscr{G}(\vec{r}'-\vec{r};\tau'-\tau) \end{bmatrix} \tag{1490}$$

$$\hat{1} = \begin{bmatrix} \delta(\tau-\tau')\delta(\vec{r}-\vec{r}') & 0 \\ 0 & \delta(\tau-\tau')\delta(\vec{r}-\vec{r}') \end{bmatrix} \tag{1491}$$

We therefore obtain the system of equations (1489), determining the Green's functions of a superconductor [57].

Let us write the Gorkov equations in the momentum representation by first introducing the **Nambu notations**:

$$\hat{\psi} = \begin{bmatrix} \hat{\psi}_\uparrow \\ \hat{\psi}^\dagger_\downarrow \end{bmatrix} , \quad \hat{\psi}^\dagger = \begin{bmatrix} \hat{\psi}^\dagger_\uparrow & \hat{\psi}_\downarrow \end{bmatrix} \tag{1492}$$

and the full Green's function, which is a 2×2 matrix in the fermion space:

$$\mathbf{G} = -\langle T_\tau \hat{\psi}\hat{\psi}^\dagger \rangle = \begin{bmatrix} -\langle T_\tau \hat{\psi}_\uparrow \hat{\psi}^\dagger_\uparrow \rangle & -\langle T_\tau \hat{\psi}_\uparrow \hat{\psi}_\downarrow \rangle \\ -\langle T_\tau \hat{\psi}^\dagger_\downarrow \hat{\psi}^\dagger_\uparrow \rangle & -\langle T_\tau \hat{\psi}^\dagger_\downarrow \hat{\psi}_\downarrow \rangle \end{bmatrix} = \hat{\mathscr{G}} \tag{1493}$$

It is important to note that in the temperature technique, all quantities are expanded in the Fourier series:

$$\mathscr{G}(\vec{r},\tau) = \frac{1}{\beta}\sum_n \int \frac{d\vec{p}}{(2\pi)^3} \exp\{-i(\omega_n\tau - \vec{p}\vec{r})\} \mathscr{G}(\vec{\kappa},\omega_n) \quad , \quad \vec{p} = \hbar\vec{\kappa} \tag{1494}$$

$$\mathscr{F}^\dagger(\vec{r},\tau) = \frac{1}{\beta}\sum_n \int \frac{d\vec{p}}{(2\pi)^3} \exp\{-i(\omega_n\tau - \vec{p}\vec{r})\} \mathscr{F}^\dagger(\vec{\kappa},\omega_n) \tag{1495}$$

Here, $\omega_n = \frac{(2n+1)\pi}{\beta}$, then

$$\mathbb{G}^{-1}\begin{bmatrix} \mathcal{G} \\ \mathcal{F}^{\dagger} \end{bmatrix} = \begin{bmatrix} 1 \\ 0 \end{bmatrix} \tag{1496}$$

where the Gorkov inverse Green's function

$$\mathbb{G}^{-1} = \mathbb{G}_0^{-1} + \Delta \quad , \quad \mathbb{G}_0^{-1} = \begin{bmatrix} \mathrm{G}_0^{-1} & 0 \\ 0 & \breve{\mathrm{G}}_0^{-1} \end{bmatrix} \quad , \quad \Delta = \begin{bmatrix} 0 & \Delta \\ \Delta^* & 0 \end{bmatrix} \tag{1497}$$

$$\mathrm{G}_0^{-1} = i\omega_n - \xi_{\vec{\kappa}} \quad , \quad \breve{\mathrm{G}}_0^{-1} = i\omega_n + \xi_{\vec{\kappa}} \quad , \quad \xi_{\vec{\kappa}} = \frac{\hbar^2\vec{\kappa}^2}{2m} - \mu \tag{1498}$$

and

$$\det \mathbb{G}^{-1} = (i\omega_n - \xi_{\vec{\kappa}})(i\omega_n + \xi_{\vec{\kappa}}) - |\Delta|^2 = -\omega_n^2 - \xi_{\vec{\kappa}}^2 - |\Delta|^2 \tag{1499}$$

For the energy spectrum,

$$\det \mathbb{G}^{-1} = -\omega_n^2 - \xi_{\vec{\kappa}}^2 - |\Delta|^2 = 0 \tag{1500}$$

or

$$\mathrm{E}(\vec{\kappa}) = \sqrt{\xi_{\vec{\kappa}}^2 + |\Delta|^2} \tag{1501}$$

which imitates the spectrum of elementary excitations of BCS theory. From here, we arrive at an excitation spectrum having a gap size Δ.

From equation (1496), we have the explicit expressions of the Green's functions in the momentum representation:

$$\mathcal{G} = -\frac{i\omega_n + \xi_{\vec{\kappa}}}{\omega_n^2 + \xi_{\vec{\kappa}}^2 + |\Delta|^2} = -\frac{i\omega_n + \xi_{\vec{\kappa}}}{\omega_n^2 + \mathrm{E}^2(\vec{\kappa})} \quad , \quad \mathcal{F}^{\dagger} = \frac{\Delta^*}{\omega_n^2 + \xi_{\vec{\kappa}}^2 + |\Delta|^2} = \frac{\Delta^*}{\omega_n^2 + \mathrm{E}^2(\vec{\kappa})} \tag{1502}$$

We do the diagrammatic representation

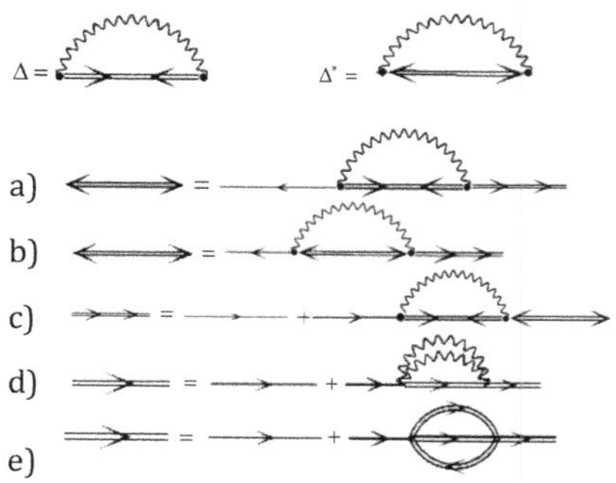

FIGURE 10.3 Self-energy due to pairing interaction built on the anomalous Green's, function (a) and (b) and another form of diagrammatic representation of Gorkov equations (c)–(e).

It is instructive to note that in the absence of an external field, the functions $\mathcal{F}^{\dagger}$ and $\mathcal{F}$ are equal, and the quantity Δ is real. This is related to the fact that the analytic properties of the functions in the temperature technique are uniquely defined.

10.3 Green's Function via Bogoliubov Coefficients

We perform the following transformation to have another representation for the Green's functions:

$$-\frac{i\omega_n+\xi_{\vec{\kappa}}}{\omega_n^2+\xi_{\vec{\kappa}}^2+|\Delta|^2}=\frac{i\omega_n+\xi_{\vec{\kappa}}}{\left(i\omega_n-E(\vec{\kappa})\right)\left(i\omega_n+E(\vec{\kappa})\right)}=\frac{u_{\vec{\kappa}}^2}{i\omega_n-E(\vec{\kappa})}+\frac{v_{\vec{\kappa}}^2}{i\omega_n+E(\vec{\kappa})} \tag{1503}$$

where $u_{\vec{\kappa}}$ and $v_{\vec{\kappa}}$ are the well-known Bogoliubov coefficients:

$$u_{\vec{\kappa}}^2=\frac{1}{2}\left(1+\frac{\xi_{\vec{\kappa}}}{E(\vec{\kappa})}\right)\quad,\quad v_{\vec{\kappa}}^2=\frac{1}{2}\left(1-\frac{\xi_{\vec{\kappa}}}{E(\vec{\kappa})}\right)\quad,\quad u_{\vec{\kappa}}v_{\vec{\kappa}}=\frac{\Delta_{\vec{\kappa}}}{2E(\vec{\kappa})}\quad,\quad |u_{\vec{\kappa}}|^2+|v_{\vec{\kappa}}|^2=1 \tag{1504}$$

Therefore, the Green's functions:

$$\mathscr{G}=\frac{u_{\vec{\kappa}}^2}{i\omega_n-E(\vec{\kappa})}+\frac{v_{\vec{\kappa}}^2}{i\omega_n+E(\vec{\kappa})}\quad,\quad \mathscr{F}^{\dagger}=u_{\vec{\kappa}}v_{\vec{\kappa}}\left(\frac{1}{i\omega_n+E(\vec{\kappa})}-\frac{1}{i\omega_n-E(\vec{\kappa})}\right) \tag{1505}$$

In the previous relations, $v_{\vec{\kappa}}^2$ is the probability that the pair $\vec{\kappa}$ is occupied. If we make a plot of $v_{\vec{\kappa}}^2$, we observe $v_{\vec{\kappa}}^2$ to be a rounded Fermi distribution. For small Δ, then

$$v_{\vec{\kappa}}^2=\begin{cases}1 & ,\ \xi_{\vec{\kappa}}<0 \quad \text{deep inside the Fermi sphere}\\ 0 & ,\ \xi_{\vec{\kappa}}>0 \quad \text{far outside the Fermi sphere}\end{cases} \tag{1506}$$

We find the gap equation of BCS theory via the following relation that, by definition,

$$\Delta^*=|\lambda|\mathscr{F}^{\dagger}(0+)=|\lambda|T\sum_{\omega_n}\int\frac{d\vec{\kappa}}{(2\pi)^3}\frac{\Delta^*}{\omega_n^2+\xi_{\vec{\kappa}}^2+|\Delta|^2} \tag{1507}$$

or

$$1=\frac{|\lambda|}{(2\pi)^3}T\int d\vec{\kappa}\sum_{\omega_n}\frac{1}{\omega_n^2+\xi_{\vec{\kappa}}^2+|\Delta|^2} \tag{1508}$$

This is the **gap equation of BCS theory**. Considering the summation

$$\sum_{n}\frac{1}{(2n+1)^2+x^2}=\frac{\pi}{2x}\tanh\frac{\pi x}{2}\quad,\quad x^2=\frac{|\Delta|^2+\xi_{\vec{\kappa}}^2}{\pi^2T^2} \tag{1509}$$

the divergence of integral in (1508) is cut off, as in the BCS approach. This is a consequence of electrons only from the layer of the width of $2\omega_D$ around the Fermi level attracting each other. Therefore,

$$1=\frac{m\kappa_F}{2\pi^2}|\lambda|\int_{-\omega_D}^{\omega_D}d\xi\frac{\tanh\left(\frac{\sqrt{\Delta^2+\xi_{\vec{\kappa}}^2}}{2T}\right)}{2\sqrt{\Delta^2+\xi_{\vec{\kappa}}^2}} \tag{1510}$$

This gives the temperature dependence of the gap $\Delta(T)$ of BCS theory. So, for the instability of the system $T = 0$, then

$$\tanh\left(\frac{\sqrt{\Delta^2 + \xi_{\vec{\kappa}}^2}}{2T}\right) = 1 \tag{1511}$$

and

$$1 = \frac{m\kappa_F}{2\pi^2}|\lambda| \int_0^{\omega_D} \frac{d\xi}{\sqrt{\Delta^2 + \xi_{\vec{\kappa}}^2}} \tag{1512}$$

Consider the integral

$$\int \frac{dx}{\sqrt{x^2 + a^2}} = \ln\left|x + \sqrt{x^2 + a^2}\right| = \text{Arcsinh}\frac{x}{a} \tag{1513}$$

Equation (1508) then yields

$$1 = \frac{m\kappa_F|\lambda|}{2\pi^2}\ln\frac{2\omega_D}{\Delta} \tag{1514}$$

which is the standard result of BCS theory:

$$\Delta = 2\omega_D \exp\left\{-\frac{2\pi^2}{|\lambda| m\kappa_F}\right\} \tag{1515}$$

For this relation to hold, the weak coupling condition must be satisfied:

$$\frac{|\lambda| m\kappa_F}{\pi^2} << 1 \tag{1516}$$

Relation (1515) permits us to find the energy gap:

$$\mathrm{E}_g = 2\Delta = 4\omega_D \exp\left\{-\frac{2\pi^2}{|\lambda| m\kappa_F}\right\} \tag{1517}$$

In order to find the superconducting transition temperature T_C defined by BCS expression [54] and then $\Delta = 0$:

$$1 = \frac{mp_F}{2\pi^2}|\lambda| \int_0^{\omega_D} \frac{\tanh\frac{\xi}{2T_C}}{\xi} d\xi \tag{1518}$$

From the change of variables $x = \frac{\xi}{2T_C}$, we then do integration by parts

$$1 = \frac{m\kappa_F}{2\pi^2}|\lambda| \int_0^{\frac{\omega_D}{2T_C}} \frac{\tanh x}{x} dx = \frac{m\kappa_F}{2\pi^2}|\lambda| \left\{ \ln x \tanh x\Big|_0^{\frac{\omega_D}{2T_C}} - \int_0^{\infty} \frac{\ln x}{\cosh^2 x} dx \right\} \tag{1519}$$

where in the remaining integral we have replaced the upper limit $x = \frac{\omega_D}{2T_c}$ by infinity, due to fast convergence and $T << \omega_D$. Then,

$$1 = \frac{m\kappa_F}{2\pi^2}|\lambda|\left\{\ln\frac{\omega_D}{2T_C} + \ln\frac{4\gamma}{\pi}\right\} \tag{1520}$$

Where $\ln\gamma = C \cong 0.577$ (Euler constant) and

$$1 = \frac{m\kappa_F}{2\pi^2}|\lambda|\ln\frac{2\omega_D \exp\{C\}}{\pi T_C} \tag{1521}$$

or

$$T_C = \frac{2\omega_D}{\pi}\exp\left\{C - \frac{2\pi^2}{|\lambda| m\kappa_F}\right\} = \frac{2\gamma\omega_D}{\pi}\exp\left\{-\frac{1}{|\lambda| n_F}\right\} \tag{1522}$$

Here, n_F is the density of states at the Fermi level for the single spin projection. The quantity $|\lambda|n_F$ determines the dimensionless coupling constant of the pairing interaction. It is essential to stress that dependence on this constant in (1522) is nonanalytic. Therefore, it cannot be expanded in powers of λ for $\lambda \to 0$. Thus comparing with (1522), we have:

$$\Delta = \frac{\pi}{\gamma}T_C \tag{1523}$$

We also can write the Gorkov equations as follows:

$$\mathscr{G}(\omega_n, \vec{\kappa}) = \frac{1}{i\omega_n - \xi_{\vec{\kappa}}} - \frac{1}{i\omega_n - \xi_{\vec{\kappa}}}\Delta.\mathscr{F}^\dagger(\omega_n, \vec{\kappa}) \tag{1524}$$

$$\mathscr{F}^\dagger(\omega_n, \vec{\kappa}) = \frac{1}{-i\omega_n - \xi_{-\vec{\kappa}}}\Delta^*\mathscr{G}(\omega_n, \vec{\kappa}) = -\frac{1}{i\omega_n + \xi_{-\vec{\kappa}}}\Delta^*\mathscr{G}(\omega_n, \vec{\kappa}) \tag{1525}$$

Substituting (1525) into (1524) many times, we then obtain for the normal Green's function perturbation theory expansion in powers of Δ:

$$\mathscr{G}(\omega_n, \vec{\kappa}) = \frac{1}{i\omega_n - \xi_{\vec{\kappa}}} + \frac{1}{i\omega_n - \xi_{\vec{\kappa}}}\frac{|\Delta|^2}{(i\omega_n + \xi_{\vec{\kappa}})(i\omega_n - \xi_{\vec{\kappa}})} + \frac{1}{i\omega_n - \xi_{\vec{\kappa}}}\frac{|\Delta|^4}{(i\omega_n + \xi_{\vec{\kappa}})^2(i\omega_n - \xi_{\vec{\kappa}})^2} + \cdots \tag{1526}$$

or

$$\mathscr{G}(\omega_n, \vec{\kappa}) = \frac{i\omega_n + \xi_{\vec{\kappa}}}{(i\omega_n)^2 - \xi_{\vec{\kappa}}^2 - |\Delta|^2} \tag{1527}$$

It is instructive to note that in the series for anomalous Green's function, the zero-order term is absent, which implies that no **bare** anomalous Green's function exists.

10.4 The BCS Ground State

We have observed that the Fermi sea becomes unstable due to the scattering of electrons in states $\left|\vec{\kappa},\uparrow\right\rangle$ and $\left|-\vec{\kappa},\downarrow\right\rangle$. Bardeen, Cooper, and Schrieffer (BCS) proposed an ansatz for the new ground state based on the idea that electrons from the states $\left|\vec{\kappa},\uparrow\right\rangle$ and $\left|-\vec{\kappa},\downarrow\right\rangle$ form the so-called Cooper pairs. The ground state is a superposition of states built up of such pairs with the ansatz being

$$\left|\hat{\Psi}_{\text{BCS}}\right\rangle = \prod_{\vec{\kappa}}\left(u_{\vec{\kappa}} + v_{\vec{\kappa}}\hat{\psi}^{\dagger}_{\vec{\kappa}\uparrow}\hat{\psi}^{\dagger}_{-\vec{\kappa}\downarrow}\right)|0\rangle \tag{1528}$$

where the coefficients $u_{\vec{\kappa}}$ and $v_{\vec{\kappa}}$ are variational parameters.

For the normalization condition

$$\begin{aligned}\left\langle\hat{\Psi}_{\text{BCS}}\middle|\hat{\Psi}_{\text{BCS}}\right\rangle &= \langle 0|\prod_{\vec{\kappa}}\left(u^{*}_{\vec{\kappa}} + v^{*}_{\vec{\kappa}}\hat{\psi}_{-\vec{\kappa}\downarrow}\hat{\psi}_{\vec{\kappa}\uparrow}\right)\prod_{\vec{\kappa}'}\left(u_{\vec{\kappa}'} + v_{\vec{\kappa}'}\hat{\psi}^{\dagger}_{\vec{\kappa}'\uparrow}\hat{\psi}^{\dagger}_{-\vec{\kappa}'\downarrow}\right)|0\rangle = \\ &= \langle 0|\prod_{\vec{\kappa}}\left(\left|u_{\vec{\kappa}}\right|^{2} + u^{*}_{\vec{\kappa}}v_{\vec{\kappa}}\hat{\psi}^{\dagger}_{\vec{\kappa}\uparrow}\hat{\psi}^{\dagger}_{-\vec{\kappa}\downarrow} + u_{\vec{\kappa}}v^{*}_{\vec{\kappa}}\hat{\psi}_{-\vec{\kappa}\downarrow}\hat{\psi}_{\vec{\kappa}\uparrow} + \left|v_{\vec{\kappa}}\right|^{2}\right)|0\rangle\end{aligned} \tag{1529}$$

to be satisfied then

$$\left|u_{\vec{\kappa}}\right|^{2} + \left|v_{\vec{\kappa}}\right|^{2} = 1 \tag{1530}$$

and so

$$\left\langle\hat{\Psi}_{\text{BCS}}\middle|\hat{\Psi}_{\text{BCS}}\right\rangle = \prod_{\vec{\kappa}}\left(\left|u_{\vec{\kappa}}\right|^{2} + \left|v_{\vec{\kappa}}\right|^{2}\right) = 1 \tag{1531}$$

The number of particles is not fixed in this state. However, it is not a problem in the thermodynamic limit as it is in statistical mechanics, the grand canonical ensemble produces the same thermodynamics as the canonical ensemble. It is instructive to note that the occupations of $\left|\vec{\kappa},\uparrow\right\rangle$ and $\left|-\vec{\kappa},\downarrow\right\rangle$ are maximally correlated where either both are occupied or both are empty. The state $\left|\hat{\Psi}_{\text{BCS}}\right\rangle$ is peculiar because it is a superposition of states with different total electron numbers. The superconductor is entangled with a much larger electron reservoir with the total electron number in the superconductor and the reservoir being fixed. Consequently, expressions containing unequal numbers of electronic creation and annihilation operators have nonvanishing expectation values. For example, say,

$$\left\langle\hat{\psi}^{\dagger}_{\vec{\kappa}'\uparrow}\hat{\psi}^{\dagger}_{-\vec{\kappa}'\downarrow}\right\rangle_{\text{BCS}} = \langle 0|\prod_{\vec{\kappa}'}\left(u^{*}_{\vec{\kappa}'} + v^{*}_{\vec{\kappa}'}\hat{\psi}_{-\vec{\kappa}'\downarrow}\hat{\psi}_{\vec{\kappa}'\uparrow}\right)\hat{\psi}^{\dagger}_{\vec{\kappa}'\uparrow}\hat{\psi}^{\dagger}_{-\vec{\kappa}'\downarrow}\prod_{\vec{\kappa}''}\left(u_{\vec{\kappa}''} + v_{\vec{\kappa}''}\hat{\psi}^{\dagger}_{\vec{\kappa}''\uparrow}\hat{\psi}^{\dagger}_{-\vec{\kappa}''\downarrow}\right)|0\rangle \tag{1532}$$

or

$$\left\langle\hat{\psi}^{\dagger}_{\vec{\kappa}'\uparrow}\hat{\psi}^{\dagger}_{-\vec{\kappa}'\downarrow}\right\rangle_{\text{BCS}} = \langle 0|v^{*}_{\vec{\kappa}}u_{\vec{\kappa}}\prod_{\vec{\kappa}'\neq\vec{\kappa}}\left(\left|u_{\vec{\kappa}'}\right|^{2} + \left|v_{\vec{\kappa}'}\right|^{2}\right)|0\rangle = v^{*}_{\vec{\kappa}}u_{\vec{\kappa}} \tag{1533}$$

Limiting ourselves to the so-called reduced Hamiltonian determinant that is relevant for both the Cooper problem and the BCS theory

$$\hat{\text{H}} = \sum_{\vec{\kappa}\sigma}\xi_{\vec{\kappa}}\hat{\psi}^{\dagger}_{\vec{\kappa}\sigma}\hat{\psi}_{\vec{\kappa}\sigma} + \hat{\text{H}}_{\text{int}} \quad , \quad \xi_{\vec{\kappa}} = \frac{\hbar^{2}\vec{\kappa}^{2}}{2m} - \mu \tag{1534}$$

then the coefficients $u_{\vec{\kappa}}$ and $v_{\vec{\kappa}}$ are chosen so as to minimize the expectation value of the energy, $\left\langle \hat{H} \right\rangle_{BCS}$, under the condition (1530) for all $\vec{\kappa}$ where $\left|\hat{\Psi}_{BCS}\right\rangle$ is a **variational ansatz**. Here,

$$\hat{H} = \sum_{\vec{\kappa}\sigma} \xi_{\vec{\kappa}} \hat{\Psi}^{\dagger}_{\vec{\kappa}\sigma} \hat{\Psi}_{\vec{\kappa}\sigma} + \hat{H}_{int} \quad , \quad \xi_{\vec{\kappa}} = \frac{\hbar^2 \vec{\kappa}^2}{2m} - \mu \tag{1535}$$

and

$$\hat{H}_{int} = \frac{1}{N} \sum_{\vec{\kappa}\vec{\kappa}'} V_{\vec{\kappa}\vec{\kappa}'} \hat{\Psi}^{\dagger}_{\vec{\kappa}\uparrow} \hat{\Psi}^{\dagger}_{-\vec{\kappa}\downarrow} \hat{\Psi}_{-\vec{\kappa}'\downarrow} \hat{\Psi}_{\vec{\kappa}'\uparrow} \tag{1536}$$

We consider that

- **Only electrons with energies $\left|\xi_{\vec{\kappa}}\right| \le \omega_D$ are relative to the Fermi energy.**
- **The instability is due to the scattering between electrons in single-particle states $\left|\vec{\kappa},\uparrow\right\rangle$ and $\left|-\vec{\kappa},\downarrow\right\rangle$.**

Here the Debye frequency ω_D is the characteristic scale of acoustic phonons, as seen earlier. For the Cooper problem, assuming $V_{\vec{\kappa}\vec{\kappa}'}$ in equation (1536) is attractive in a finite strip above the Fermi energy:

$$V_{\vec{\kappa}\vec{\kappa}'} = \begin{cases} -V_0 & , \quad \left|\xi_{\vec{\kappa}}\right| < \omega_D \quad , \quad \left|\xi_{\vec{\kappa}'}\right| < \omega_D \\ 0 & , \quad \text{otherwise} \end{cases} \tag{1537}$$

We consider that

- **Scattering without momentum transfer, $\vec{\kappa} = \vec{\kappa}'$, contributes negligibly compared to $\vec{\kappa} \neq \vec{\kappa}'$ because there are many more scattering channels for $\vec{\kappa} \neq \vec{\kappa}'$.**
- **The components $u_{-\vec{\kappa}} = u_{\vec{\kappa}}$ and $v_{-\vec{\kappa}} = v_{\vec{\kappa}}$ are restrictions of the variational ansatz.**

Therefore

$$\begin{aligned} \left\langle \hat{H} \right\rangle_{BCS} &= \sum_{\vec{\kappa}\sigma} \xi_{\vec{\kappa}} \langle 0| \prod_{\vec{q}} \left(u^*_{\vec{q}} + v^*_{\vec{q}} \hat{\Psi}_{-\vec{q}\downarrow} \hat{\Psi}_{\vec{q}\uparrow}\right) \hat{\Psi}^{\dagger}_{\vec{\kappa}\sigma} \hat{\Psi}_{\vec{\kappa}\sigma} \prod_{\vec{q}'} \left(u_{\vec{q}'} + v_{\vec{q}'} \hat{\Psi}^{\dagger}_{\vec{q}'\uparrow} \hat{\Psi}^{\dagger}_{-\vec{q}'\downarrow}\right) |0\rangle + \\ &+ \frac{1}{N} \sum_{\vec{\kappa}\vec{\kappa}'} V_{\vec{\kappa}\vec{\kappa}'} \langle 0| \prod_{\vec{q}} \left(u^*_{\vec{q}} + v^*_{\vec{q}} \hat{\Psi}_{-\vec{q}\downarrow} \hat{\Psi}_{\vec{q}\uparrow}\right) \hat{\Psi}^{\dagger}_{\vec{\kappa}\uparrow} \hat{\Psi}^{\dagger}_{-\vec{\kappa}\downarrow} \hat{\Psi}_{-\vec{\kappa}'\downarrow} \hat{\Psi}_{\vec{\kappa}'\uparrow} \prod_{\vec{q}'} \left(u_{\vec{q}'} + v_{\vec{q}'} \hat{\Psi}^{\dagger}_{\vec{q}'\uparrow} \hat{\Psi}^{\dagger}_{-\vec{q}'\downarrow}\right) |0\rangle \end{aligned} \tag{1538}$$

or

$$\begin{aligned} \left\langle \hat{H} \right\rangle_{BCS} &= \sum_{\vec{\kappa}} \xi_{\vec{\kappa}} \langle 0| \left|v_{\vec{\kappa}}\right|^2 \hat{\Psi}_{-\vec{\kappa}'\downarrow} \hat{\Psi}_{\vec{\kappa}'\uparrow} \hat{\Psi}^{\dagger}_{\vec{\kappa}\uparrow} \hat{\Psi}_{\vec{\kappa}'\uparrow} \hat{\Psi}^{\dagger}_{\vec{\kappa}\uparrow} \hat{\Psi}^{\dagger}_{-\vec{\kappa}\downarrow} |0\rangle + \sum_{\vec{\kappa}} \xi_{\vec{\kappa}} \langle 0| \left|v_{-\vec{\kappa}}\right|^2 \hat{\Psi}_{\vec{\kappa}'\downarrow} \hat{\Psi}_{-\vec{\kappa}'\uparrow} \hat{\Psi}^{\dagger}_{\vec{\kappa}\downarrow} \hat{\Psi}_{\vec{\kappa}'\downarrow} \hat{\Psi}^{\dagger}_{-\vec{\kappa}\uparrow} \hat{\Psi}^{\dagger}_{\vec{\kappa}\downarrow} |0\rangle + \\ &+ \frac{1}{N} \sum_{\vec{\kappa}\vec{\kappa}'} V_{\vec{\kappa}\vec{\kappa}'} \langle 0| v^*_{\vec{\kappa}} u^*_{\vec{\kappa}'} u_{\vec{\kappa}} v_{\vec{\kappa}'} \hat{\Psi}_{-\vec{\kappa}'\downarrow} \hat{\Psi}_{\vec{\kappa}'\uparrow} \hat{\Psi}^{\dagger}_{\vec{\kappa}\uparrow} \hat{\Psi}^{\dagger}_{-\vec{\kappa}\downarrow} \hat{\Psi}_{-\vec{\kappa}'\downarrow} \hat{\Psi}_{\vec{\kappa}'\uparrow} \hat{\Psi}^{\dagger}_{\vec{\kappa}'\uparrow} \hat{\Psi}^{\dagger}_{-\vec{\kappa}'\downarrow} |0\rangle \end{aligned} \tag{1539}$$

or

$$\left\langle \hat{H} \right\rangle_{BCS} = \sum_{\vec{\kappa}} 2\xi_{\vec{\kappa}} \left|v_{\vec{\kappa}}\right|^2 + \frac{1}{N} \sum_{\vec{\kappa}\vec{\kappa}'} V_{\vec{\kappa}\vec{\kappa}'} v^*_{\vec{\kappa}} u^*_{\vec{\kappa}'} u_{\vec{\kappa}} v_{\vec{\kappa}'} \equiv E_{BCS} \tag{1540}$$

This energy is minimized with respect to the $u_{\vec{\kappa}}$ and $v_{\vec{\kappa}}$. In order for E_{BCS} to be real, the phases of $u_{\vec{\kappa}}$ and $v_{\vec{\kappa}}$ must be the same. However, as E_{BCS} is invariant under

$$u_{\vec{\kappa}} \to u_{\vec{\kappa}} \exp\{i\phi_{\vec{\kappa}}\} \quad , \quad v_{\vec{\kappa}} \to v_{\vec{\kappa}} \exp\{i\phi_{\vec{\kappa}}\} \tag{1541}$$

we can select all $u_{\vec{\kappa}}$ and $v_{\vec{\kappa}}$ real. From normalization in (1530), we can parametrize the coefficients:

$$u_{\vec{\kappa}} = \cos\theta_{\vec{\kappa}} \quad , \quad v_{\vec{\kappa}} = \sin\theta_{\vec{\kappa}} \tag{1542}$$

So, from (1542) and (1540), we have

$$E_{BCS} = \sum_{\vec{\kappa}} 2\xi_{\vec{\kappa}}(1-\cos 2\theta_{\vec{\kappa}}) + \frac{1}{4N}\sum_{\vec{\kappa}\vec{\kappa}'} V_{\vec{\kappa}\vec{\kappa}'} \sin 2\theta_{\vec{\kappa}} \cos 2\theta_{\vec{\kappa}'} \tag{1543}$$

The stationary value is obtained from

$$\frac{\partial E_{BCS}}{\partial \theta_{\vec{q}}} = 2\xi_{\vec{q}} \sin 2\theta_{\vec{q}} + \frac{1}{2N}\sum_{\vec{\kappa}'} V_{\vec{q}\vec{\kappa}'} \sin 2\theta_{\vec{\kappa}'} \cos 2\theta_{\vec{q}} + \frac{1}{2N}\sum_{\vec{\kappa}} V_{\vec{\kappa}\vec{q}} \sin 2\theta_{\vec{\kappa}} \cos 2\theta_{\vec{q}} \tag{1544}$$

or

$$\frac{\partial E_{BCS}}{\partial \theta_{\vec{q}}} = 2\xi_{\vec{q}} \sin 2\theta_{\vec{q}} + \frac{1}{N}\sum_{\vec{\kappa}'} V_{\vec{q}\vec{\kappa}'} \sin 2\theta_{\vec{\kappa}'} \cos 2\theta_{\vec{q}} = 0 \tag{1545}$$

We replace $\vec{q}$ by $\vec{\kappa}$ and parameterize $\theta_{\vec{\kappa}}$ as follows

$$\sin 2\theta_{\vec{\kappa}} = \frac{\Delta_{\vec{\kappa}}}{E(\vec{\kappa})} \quad , \quad \cos 2\theta_{\vec{\kappa}} = \frac{\xi_{\vec{\kappa}}}{E(\vec{\kappa})} \quad , \quad E(\vec{\kappa}) = \sqrt{\xi_{\vec{\kappa}}^2 + |\Delta_{\vec{\kappa}}|^2} \tag{1546}$$

Substituting these in equation (1545), we then have

$$2\xi_{\vec{\kappa}} \frac{\Delta_{\vec{\kappa}}}{E(\vec{\kappa})} + \frac{1}{N}\sum_{\vec{\kappa}'} V_{\vec{\kappa}\vec{\kappa}'} \frac{\Delta_{\vec{\kappa}'}}{E(\vec{\kappa}')} \frac{\xi_{\vec{\kappa}}}{E(\vec{\kappa})} = 0 \tag{1547}$$

or

$$\Delta_{\vec{\kappa}} = -\frac{1}{2N}\sum_{\vec{\kappa}'} V_{\vec{\kappa}\vec{\kappa}'} \frac{\Delta_{\vec{\kappa}'}}{E(\vec{\kappa}')} \tag{1548}$$

As noted earlier, this is also known as the **BCS gap equation**. The variational parameters $u_{\vec{\kappa}}$ and $v_{\vec{\kappa}}$ in terms of $\Delta_{\vec{\kappa}}$ can now also be represented as in (1504). Therefore, the **relative** sign of $u_{\vec{\kappa}}$ and $v_{\vec{\kappa}}$ is the sign of $\Delta_{\vec{\kappa}}$ and the **absolute** sign. For example, $u_{\vec{\kappa}}$ is irrelevant due to the invariance of E_{BCS} under simultaneous phase rotations of $u_{\vec{\kappa}}$ and $v_{\vec{\kappa}}$. Considering our case

$$V_{\vec{\kappa}\vec{\kappa}'} = \begin{cases} -V_0 & , \ |\xi_{\vec{\kappa}}| < \omega_D \ , \ |\xi_{\vec{\kappa}'}| < \omega_D \\ 0 & , \ \text{otherwise} \end{cases} \tag{1549}$$

the BCS gap equation then becomes

$$V_{\vec{\kappa}\vec{\kappa}'} = \begin{cases} -\dfrac{1}{2N}\displaystyle\sum_{\vec{\kappa}',|\xi_{\vec{\kappa}'}|<\omega_D}(-V_0)\dfrac{\Delta_{\vec{\kappa}'}}{\mathrm{E}(\vec{\kappa}')} & , \quad |\xi_{\vec{\kappa}}|<\omega_D \\ 0 & , \quad \text{otherwise} \end{cases} \tag{1550}$$

This can be solved with the ansatz

$$\Delta_{\vec{\kappa}} = \begin{cases} \Delta_0 > 0 & , \quad |\xi_{\vec{\kappa}}|<\omega_D \\ 0 & , \quad \text{otherwise} \end{cases} \tag{1551}$$

So,

$$\Delta_0 = \frac{V_0}{2N}\sum_{\vec{\kappa}',|\xi_{\vec{\kappa}'}|<\omega_D}\frac{\Delta_0}{\mathrm{E}(\vec{\kappa}')} \tag{1552}$$

This implies that

$$1 = \frac{V_0}{2N}\sum_{\vec{\kappa}',|\xi_{\vec{\kappa}'}|<\omega_D}\frac{1}{\mathrm{E}(\vec{\kappa}')} = \frac{V_0}{2}\int_{-\omega_D}^{\omega_D} d\xi\, n_{\mathrm{F}}(\mu+\xi)\frac{1}{\sqrt{\xi^2+\Delta_0^2}} \tag{1553}$$

Here, $n_{\mathrm{F}}(\in)$ is the density of states per spin direction and per unit cell. If this density of states is approximately constant within $\pm\omega_D$ of the Fermi energy, we have

$$1 = \frac{V_0 n_{\mathrm{F}}(\in_{\mathrm{F}})}{2}\int_{-\omega_D}^{\omega_D}\frac{d\xi}{\sqrt{\xi^2+\Delta_0^2}} = V_0 n_{\mathrm{F}}(\in_{\mathrm{F}})\,\mathrm{Arcsinh}\frac{\omega_D}{\Delta_0} \tag{1554}$$

This implies that

$$\sinh\frac{1}{V_0 n_{\mathrm{F}}(\in_{\mathrm{F}})} = \frac{\omega_D}{\Delta_0} \tag{1555}$$

or

$$\Delta_0 = \frac{\omega_D}{\sinh\dfrac{1}{V_0 n_{\mathrm{F}}(\in_{\mathrm{F}})}} \tag{1556}$$

In the **weak-coupling limit** of small $V_0 n_{\mathrm{F}}(\in_{\mathrm{F}})$, this yields the same result as in (1418):

$$\Delta_0 \cong 2\omega_D \exp\left\{-\frac{1}{V_0 n_{\mathrm{F}}(\in_{\mathrm{F}})}\right\} \tag{1557}$$

The value of Δ_0 agrees with T_c for the **Cooper instability**, which we see in the following.

Now, substituting $u_{\vec{\kappa}}$ and $v_{\vec{\kappa}}$ into $\mathrm{E}_{\mathrm{BCS}}$ yields the energy gain due to the superconducting state (i.e., the **condensation energy**):

$$\mathrm{E}_{\mathrm{BCS}} = \sum_{\vec{\kappa}}\xi_{\vec{\kappa}}\left(1-\frac{\xi_{\vec{\kappa}}}{\mathrm{E}(\vec{\kappa})}\right) + \frac{1}{4N}\sum_{\vec{\kappa}\vec{\kappa}'}V_{\vec{\kappa}\vec{\kappa}'}\frac{\Delta_{\vec{\kappa}}}{\mathrm{E}(\vec{\kappa})}\frac{\Delta_{\vec{\kappa}'}}{\mathrm{E}(\vec{\kappa}')} \tag{1558}$$

Using the simple form of $V_{\vec{k}\vec{k}'}$ and assuming that $N_F(\in)$ is constant within $\pm\omega_D$ of the Fermi energy but not outside of this interval, we have

$$\begin{aligned}E_{BCS} &= N\int_{-\infty}^{-\omega_D} d\xi n_F(\mu+\xi)2\xi + N\int_{-\omega_D}^{\omega_D} d\xi n_F(\mu+\xi)\left(1-\frac{\xi}{\sqrt{\xi^2+\Delta_0^2}}\right)+ \\ &+N\int_{-\omega_D}^{\omega_D} d\xi\int_{-\omega_D}^{\omega_D} d\xi' n_F(\mu+\xi)n_F(\mu+\xi')(-V_0)\frac{\Delta_0^2}{4\sqrt{\xi^2+\Delta_0^2}\sqrt{\xi'^2+\Delta_0^2}}\end{aligned} \tag{1559}$$

or

$$E_{BCS} \cong 2N\int_{-\infty}^{-\omega_D} d\xi n_F(\mu+\xi)\xi - Nn_F(\in_F)\left[\omega_D\sqrt{\omega_D^2+\Delta_0^2}+\Delta_0^2\,\text{Arcsinh}\frac{\omega_D}{\Delta_0}\left(1+V_0 n_F(\in_F)\text{Arcsinh}\frac{\omega_D}{\Delta_0}\right)\right] \tag{1560}$$

From the gap equation (1554), we then have

$$E_{BCS} \cong 2N\int_{-\infty}^{-\omega_D} d\xi n_F(\mu+\xi)\xi - Nn_F(\in_F)\omega_D\sqrt{\omega_D^2+\Delta_0^2} \tag{1561}$$

The normal-state energy as well as the energy difference can be obtained by taking the limiting value $\Delta_0 \to 0$:

$$\Delta E_{BCS} = E_{BCS} - E_{BCS}\big|_{\Delta_0\to 0} = Nn_F(\in_F)\omega_D\left(\omega_D-\sqrt{\omega_D^2+\Delta_0^2}\right) \tag{1562}$$

For the weak-coupling limit $\frac{\Delta_0}{\omega_D} << 1$, the energy difference:

$$\Delta E_{BCS} \cong -\frac{1}{2}Nn_F(\in_F)\Delta_0^2 \tag{1563}$$

10.5 Gauge Invariance

Next, we study the role of a vector potential $\vec{A}(\vec{r})$ and gauge invariance in the Gorkov equations. Consider the superconductor is placed in an external field, for example, in an electromagnetic field. The electromagnetic field may be introduced into the Gorkov equations for the case of $\hat{\psi}$ and $\hat{\psi}^\dagger$, respectively, as

$$\nabla\hat{\psi} \to \left(\nabla - i\frac{e}{c}\vec{A}\right)\hat{\psi} \quad , \quad \nabla\hat{\psi}^\dagger \to \left(\nabla + i\frac{e}{c}\vec{A}\right)\hat{\psi}^\dagger \tag{1564}$$

Here, $\vec{A}$ is an external vector. Therefore, the equation for $\mathscr{G}$ and $\mathscr{F}^\dagger$ becomes

$$\left(i\frac{\partial}{\partial t}+\frac{1}{2m}\left(\nabla-i\frac{e}{c}\vec{A}\right)^2+\mu\right)\mathscr{G}(\vec{r},t;\vec{r}',t')+i\lambda\mathscr{F}(\vec{r},t;\vec{r}',t')\mathscr{F}^\dagger(\vec{r},t;\vec{r}',t')=\delta(t-t')\delta(\vec{r}-\vec{r}') \tag{1565}$$

$$\left(i\frac{\partial}{\partial t}-\frac{1}{2m}\left(\nabla+i\frac{e}{c}\vec{A}\right)^2-\mu\right)\mathscr{F}^\dagger(\vec{r},t;\vec{r}',t')+i\lambda\mathscr{F}^+(\vec{r},t;\vec{r}',t')\mathscr{G}(\vec{r},t;\vec{r}',t')=0 \tag{1566}$$

If we consider the vector potential undergoes the gauge transformation

$$\vec{A} \to \vec{A} + \frac{\partial \varphi}{\partial \vec{r}} = \vec{A} + \nabla \varphi \tag{1567}$$

then the field transforms the following as

$$\mathscr{G}(\vec{r},t;\vec{r}',t') \to \mathscr{G}(\vec{r},t;\vec{r}',t') \exp\{ie(\varphi - \varphi')\} \tag{1568}$$

$$\mathscr{F}(\vec{r},t;\vec{r}',t') \to \mathscr{F}(\vec{r},t;\vec{r}',t') \exp\{ie(\varphi + \varphi')\} \tag{1569}$$

$$\mathscr{F}^{\dagger}(\vec{r},t;\vec{r}',t') \to \mathscr{F}^{\dagger}(\vec{r},t;\vec{r}',t') \exp\{-ie(\varphi + \varphi')\} \tag{1570}$$

The charged electron fields (operators) are transformed as follows:

$$\hat{\psi}(\vec{r}) \to \hat{\psi} \exp\{ie\varphi(\vec{r})\} \quad , \quad \hat{\psi}^{\dagger}(\vec{r}) \to \hat{\psi}^{\dagger} \exp\{-ie\varphi(\vec{r})\} \tag{1571}$$

The gap functions

$$\Delta(\vec{r},t;\vec{r},t) \approx |\lambda| \mathscr{F}(\vec{r},t;\vec{r},t) \quad , \quad \Delta^{*}(\vec{r},t;\vec{r},t) \approx |\lambda| \mathscr{F}^{\dagger}(\vec{r},t;\vec{r},t) \tag{1572}$$

which in the external field are transformed as follows:

$$\mathscr{F}(\vec{r},t;\vec{r},t) \to \mathscr{F}(\vec{r},t;\vec{r},t) \exp\{2ie\varphi(\vec{r})\} \quad , \quad \mathscr{F}^{\dagger}(\vec{r},t;\vec{r},t) \to \mathscr{F}^{\dagger}(\vec{r},t;\vec{r},t) \exp\{-2ie\varphi(\vec{r})\} \tag{1573}$$

This implies that the order parameter of a superconducting (**gap**) Δ is a charged field with electric charge $2e$, that is, the double charge of an electron (Cooper pair condensate).

From the aforementioned, we observe that the Gorkov equations serve as a basis for further investigations of the superconducting state. It is observed also that results from the Gorkov equations agree with those obtained from the mean-field BCS theory.

10.6 Diagrammatic Approach to Superconductivity

10.6.1 Ladder Approximation

The one-particle Green's functions do not fully describe interacting quantum systems because they do not provide information about correlation effects. Quantities such as linear response functions are related to the **two-particle Green's function**. The two-particle Green's function is the sum over the probability amplitudes for all the ways two particles can enter the system, interact with each other and with the particles in the system, and then scatter again. The two-particle Matsubara Green's function is defined as

$$\mathbf{G}^{(4)}(\alpha_1\tau_1, \alpha_1'\tau_1', \alpha_2\tau_2, \alpha_2'\tau_2') = \left\langle \mathrm{T}\hat{\psi}^{*}_{\alpha_1}(\tau_1)\hat{\psi}^{*}_{\alpha_2}(\tau_2)\hat{\psi}_{\alpha_2'}(\tau_2')\hat{\psi}_{\alpha_1'}(\tau_1')\right\rangle \tag{1574}$$

Letting $\alpha_i\tau_i \equiv i$, $\alpha_i'\tau_i' \equiv i'$, then

$$\mathbf{G}^{(4)}(1,1',2;2') = \left\langle \mathrm{T}\hat{\psi}^{*}_{\alpha_1}(\tau_1)\hat{\psi}^{*}_{\alpha_2}(\tau_2)\hat{\psi}_{\alpha_2'}(\tau_2')\hat{\psi}_{\alpha_1'}(\tau_1')\right\rangle \tag{1575}$$

We express the full two-particle Green's function in terms of the four-leg (or two-particle) 1PI vertex $\Gamma^{(4)}$ defined as the 1PI part of the two-particle Green's function. This implies that any two-particle vertex contribution remains connected when any of its internal propagators is cut. This is represented diagrammatically in Figure 10.4, and the algebraic form of the relation is as follows:

$$-\mathbf{G}_c^{(4)}(\alpha_1\tau_1,\alpha_2\tau_2,\alpha_1'\tau_1',\alpha_2'\tau_2')=\mathbf{G}(1,3)\mathbf{G}(2,4)+\chi\mathbf{G}(1,4)\mathbf{G}(2,3)+\int_0^\beta d\tilde{\tau}_1 d\tilde{\tau}_2 d\tilde{\tau}_2' d\tilde{\tau}_1' \times$$
$$\times\sum_{\gamma_1\gamma_2\gamma_2'\gamma_1'}\mathbf{G}(\alpha_1\tau_1,\gamma_1\tilde{\tau}_1')\mathbf{G}(\alpha_2\tau_2,\gamma_2\tilde{\tau}_2')\Gamma^{(4)}(1,2,3,4)\mathbf{G}(\gamma_2'\tilde{\tau}_2',\alpha_2'\tau_2')\mathbf{G}(\gamma_1'\tilde{\tau}_1',\alpha_1'\tau_1') \tag{1576}$$

where

$$1=\alpha_1\tau_1 \quad , \quad 2=\alpha_2\tau_2 \quad , \quad 3=\alpha_1'\tau_1' \quad , \quad 4=\alpha_2'\tau_2' \tag{1577}$$

The two-body connected Green's function is presented graphically as follows:

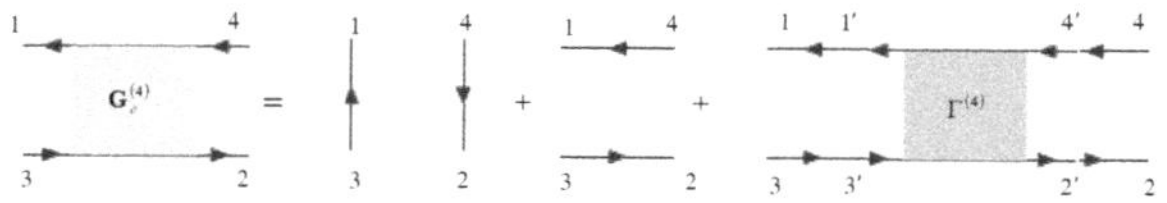

(1578)

FIGURE 10.4 The full two-particle Green's function $\mathbf{G}_c^{(4)}$ and the two-particle or 1PI vertex $\Gamma^{(4)}$ relation.

Here, all the one-particle propagators are full. In a sense, the 1PI vertex $\Gamma^{(4)}$ is the correlated part of the two-particle Green's function. It vanishes in the noninteracting system—unlike the full two-particle $\mathbf{G}_c^{(4)}$. The two-particle or 1PI vertex

$$\Gamma^{(4)}(1,2,3,4)\equiv\Gamma^{(4)}(\gamma_1\tilde{\tau}_1,\gamma_2\tilde{\tau}_2,\gamma_1'\tilde{\tau}_1',\gamma_2'\tilde{\tau}_2') \tag{1579}$$

as seen relates the effective 2PI irreducible vertex $\Phi^{(2)}$ and is represented diagrammatically as follows:

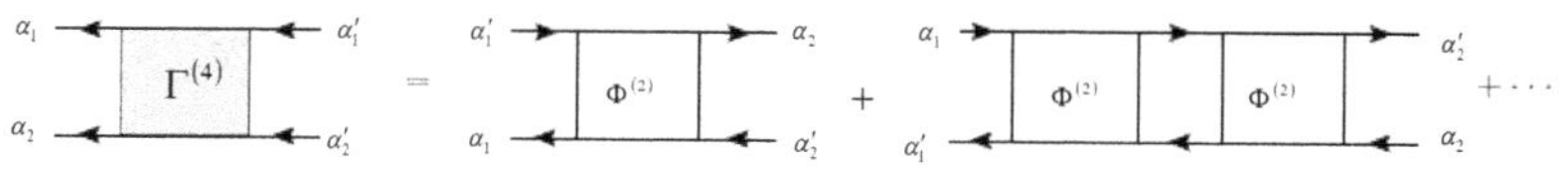

(1580)

or

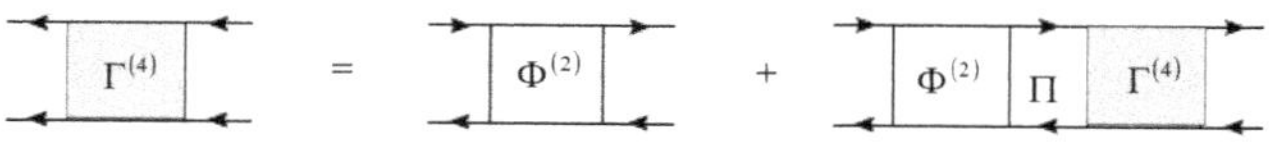

(1581)

FIGURE 10.5 Diagrammatic representation of the Bethe-Salpeter equation relating the 1PI vertex $\Gamma^{(4)}$ to $\Phi^{(2)}$ in the Ladder diagrams.

Analytically, the 1PI vertex can be represented as

$$\Gamma^{(4)}=\Phi^{(2)}-\Phi^{(2)}\Pi\Gamma^{(4)} \tag{1582}$$

Here, the effective 2PI irreducible vertex $\Phi^{(2)}$, which describes the propagation of two particles from the points $(\alpha_1\tau_1, \alpha_1'\tau_1')$ to $(\alpha_2\tau_2, \alpha_2'\tau_2')$, can be represented diagrammatically in Figure 10.6 as follows:

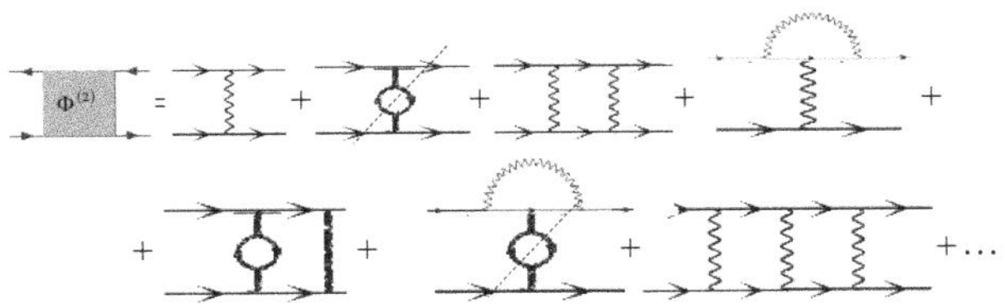

(1583)

FIGURE 10.6 Diagrammatic definition of the block (irreducible vertex) $\Phi^{(2)}$. Those diagrams that are crossed out can be cut by two lines of a particle and antiparticle.

But

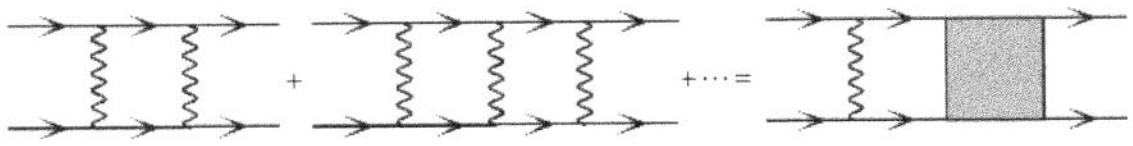

(1584)

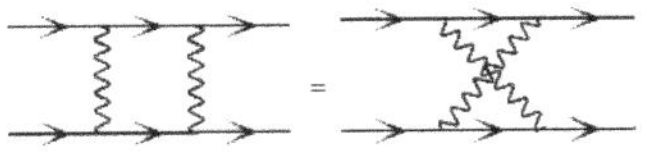

(1585)

FIGURE 10.7 Bethe-Salpeter equation and simplification of Ladder diagrams.

Then, $\Phi^{(2)}$ appears to be conceptually similar to the Dyson-like integral equation for the one-particle Green's function and relates the original potential:

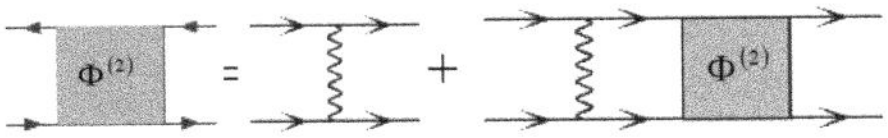

(1586)

FIGURE 10.8 Dyson-like integral equation for the one-particle Green's function.

10.6.2 Bethe-Salpeter Equation

Considering that the central part of the correlation function is described by a vertex, we rewrite the full vertex $\Gamma^{(4)}$ in equation (1582) and seeing the aforementioned can be written in diagrammatic form

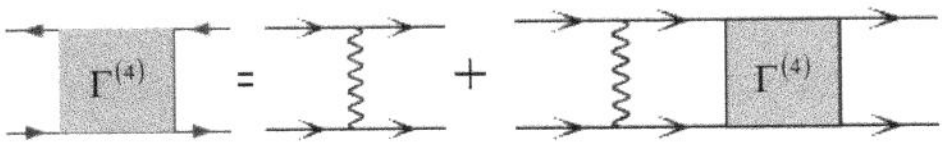

(1587)

FIGURE 10.9 The full vertex $\Gamma^{(4)}$ in equation (1582).

Translating from diagrammatic to an algebraic formulation, we obtain the following analytic form:

$$\Gamma^{(4)} = \Phi^{(2)} - \Phi^{(2)}\Pi\Gamma^{(4)} = \Phi^{(2)} + \Gamma^{(4)}(-\Pi)\Phi^{(2)} = \Phi^{(2)} + \Gamma^{(4)}\left(\mathrm{I} + \tilde{\mathrm{I}}\right)\Phi^{(2)} = \Phi^{(2)} + \Phi^{(2)}\left(\mathrm{I} + \tilde{\mathrm{I}}\right)\Gamma^{(4)} \tag{1588}$$

where I and $\tilde{I}$ will be calculated as follows via the Green's functions. Equation (1588) is obtained from the diagrammatic representation for $\Gamma^{(4)}$, if diagrams are summed in **inverse** order. Introducing the scattering amplitude λ, defined by the equation:

$$\lambda = \Phi^{(2)} + \Phi^{(2)}\tilde{I}\lambda = \Phi^{(2)} + \lambda\tilde{I}\Phi^{(2)} \tag{1589}$$

where

$$\lambda = \lim_{\omega \to 0} \Gamma^{(4)} \tag{1590}$$

The order of the limit is very important. Multiplying equation (1588) from the left side by $\left(1+\lambda\tilde{I}\right)$, we have

$$\Gamma^{(4)} = \lambda + \lambda I \Gamma^{(4)} = \lambda + \Gamma^{(4)} I \lambda \tag{1591}$$

This can be checked directly by:

$$\left(1+\lambda\tilde{I}\right)\Gamma^{(4)} = \left(1+\lambda\tilde{I}\right)\Phi^{(2)} + \left(1+\lambda\tilde{I}\right)\Phi^{(2)}\left(I+\tilde{I}\right)\Gamma^{(4)} = \lambda + \lambda\left(I+\tilde{I}\right)\Gamma^{(4)} = \lambda + \lambda I \Gamma^{(4)} + \lambda\tilde{I}\Gamma^{(4)} \tag{1592}$$

So, solving this equation for $\Gamma^{(4)}$, we arrive at

$$\Gamma \equiv \Gamma^{(4)} = \frac{\lambda}{1-\lambda I} \tag{1593}$$

10.6.3 Cooper Instability

We show that the attractive effective interaction leads to an instability of the normal state (i.e., of the Fermi sea). We consider scattering of two electrons due to the effective interaction. We consider that at $T=0$ the interacting electrons have a small **sum** of momenta (for example as in equation (1594) so that $\vec{\kappa}_1 + \vec{\kappa}_2 \approx 0$ (nearly opposite momenta). The transferred momentum $\vec{q} = \vec{\kappa}_2 - \vec{\kappa}_1$ is not small, and $\left|\vec{\kappa}_1\right| = \left|\vec{\kappa}_2\right| \approx \kappa_F$ or $\left|\vec{q}\right| \approx 2\kappa_F$. At the same time, for electrons that are close to the Fermi surface, we have $\epsilon_3 \approx \epsilon_1 \approx 0$ or $\epsilon_1 + \epsilon_3 = \Omega \approx 0$. This implies that

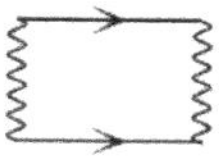

$= I\lambda^2$ (1594)

which corresponds to the appearance of electron–electron attraction. This confirms the general idea that electrons in metals with opposite momenta and spins (**Pauli Exclusion Principle**) attract each other due to phonon exchange, which is of basic importance to the BCS approach to superconductivity [36]. Taking the simplest approach via the BCS model Hamiltonian determinant, the real interaction due to phonon exchange may be replaced by an effective point (similar to attraction that is different from zero only for electrons from the layer of the width of $\approx 2\omega_D$ around the Fermi surface) [36].

We consider the scattering of two electrons due to the effective interaction with a single scattering event represented by the diagram in Figure 10.10:

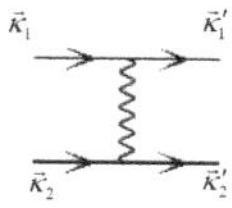

(1595)

FIGURE 10.10 Single scattering event of two electrons.

We also consider the scattering of two electrons due to the effective interaction with multiple scattering events represented by the diagram in Figure 10.11:

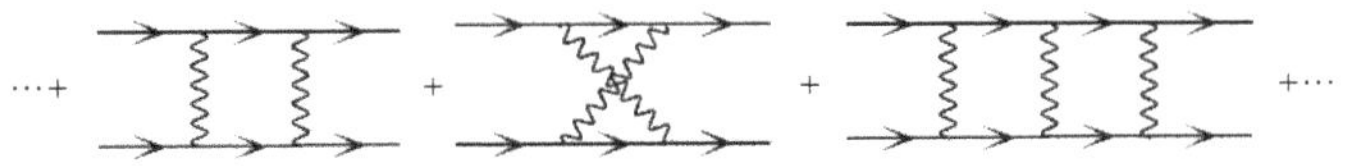

(1596)

FIGURE 10.11 Multiple scattering event of two electrons.

Instability occurs if the given series diverges and the scattering is infinitely strong. In this way, the perturbative expansion in the interaction strength U represented by the diagrams breaks down. This implies that the true equilibrium state cannot be achieved from the equilibrium state when $U = 0$ (i.e., the noninteracting Fermi gas), by perturbation theory. **A state perturbatively connected to the free Fermi gas is called a Landau Fermi liquid** and is an appropriate description for normal metals. Conversely, a scattering instability implies that the equilibrium state is no longer a Fermi liquid. As in RPA, it is sufficient to consider dominant diagrams at each order, which are the **ladder diagrams** and do not contain crossing interaction lines. In addition, instability first is achieved for the scattering of two electrons with opposite momentum, frequency, and spin. Therefore, our restriction will be to the diagrams describing this state (Figure 10.12a).

We assume the following choice of external four-momenta:

$$\kappa_1 = \kappa + q \quad ; \quad \kappa_1' = -\kappa \quad ; \quad \kappa_2 = \kappa' + q \quad ; \quad \kappa_2' = -\kappa' \tag{1597}$$

so that q is the small sum of (incoming) four-momenta.

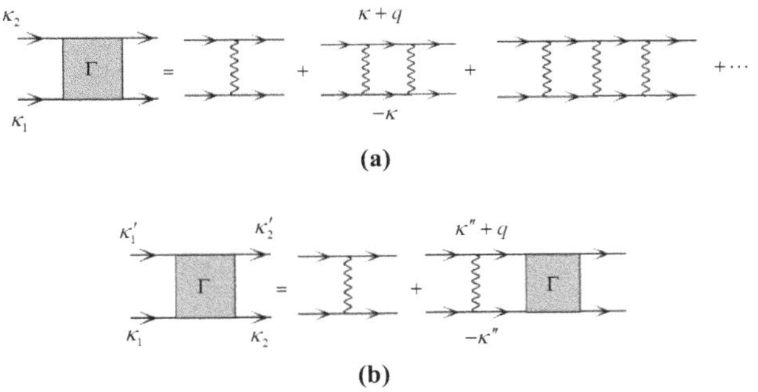

(1598)

(1599)

FIGURE 10.12 "Ladder" in (a) Cooper channel and (b) Integral equation for the appropriate vertex part.

This is a geometric series that also can be summed up diagrammatically as follows:

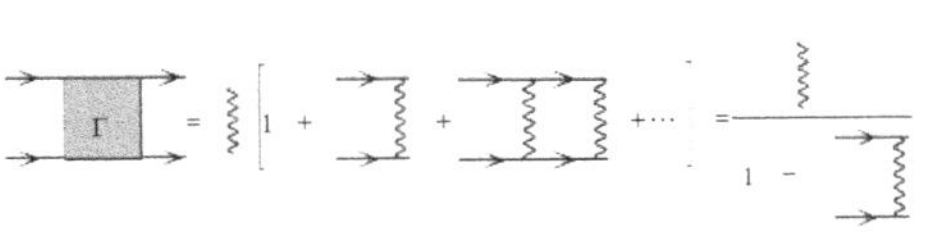

(1600)

FIGURE 10.13 Geometric series for the vertex.

The sum of the given series (without external **tails**) is given by the vertex Γ, which is determined from the integral equation in Figure 10.12b and has the following analytic form:

$$\Gamma(\kappa_2, \kappa_2'; \kappa_1, \kappa_1') = \langle \kappa_2, \kappa_2' | \Gamma | \kappa_1, \kappa_1' \rangle = \langle \kappa' + q - \kappa' | \Gamma | \kappa + q - \kappa \rangle \equiv \Gamma(\kappa, \kappa', q) \tag{1601}$$

or

$$\Gamma(\kappa,\kappa',q)=V(\kappa-\kappa')+i\int\frac{d^4\kappa''}{(2\pi)^4}V(\kappa'-\kappa'')\mathbf{G}_0(\kappa''+q)\mathbf{G}_0(-\kappa'')\Gamma(\kappa'',\kappa',q) \tag{1602}$$

We check the validity of equation (1602) by iterations, and this leads to the **ladder** series. In BCS model interaction, the potential $V(\kappa-\kappa')$ takes the following form:

$$V(\kappa-\kappa')\to V(\kappa,\kappa')=\lambda\theta_{\kappa}\theta_{\kappa'} \tag{1603}$$

Here,

$$\theta_{\kappa}=\begin{cases}1 & , \ |\xi_{\kappa}|<\omega_D\\ 0 & , \ |\xi_{\kappa}|>\omega_D\end{cases} \tag{1604}$$

We consider λ as the electron-phonon coupling strength, and we solve equation (1602) as follows:

$$\Gamma(\kappa,\kappa',q)=\frac{\lambda\theta_{\kappa'+q}\theta_{\kappa+q}}{1-i\lambda\int\frac{d^4\kappa}{(2\pi)^4}\theta^2_{\kappa+q}\mathbf{G}_0(\kappa+q)\mathbf{G}_0(-\kappa)} \tag{1605}$$

This confirms equation (1593):

$$\Gamma=\frac{\lambda}{1-\lambda I} \tag{1606}$$

where,

I = 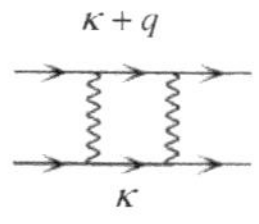

(1607)

and analytically

$$I=i\lambda\int\frac{d^4\kappa}{(2\pi)^4}\theta^2_{\kappa+q}\mathbf{G}_0(\kappa+q)\mathbf{G}_0(-\kappa)=i\lambda\int\frac{d^4\kappa}{(2\pi)^4}\theta^2_{\kappa}\mathbf{G}_0(\kappa)\mathbf{G}_0(q-\kappa) \tag{1608}$$

Here,

$$\mathbf{G}_0(\vec{\kappa})=\frac{1-n(\vec{\kappa})}{\in-\xi_{\vec{\kappa}}+i\delta}+\frac{n(\vec{\kappa})}{\in-\xi_{\vec{\kappa}}-i\delta} \tag{1609}$$

is the bare Green's function where the first term describes the propagation of particles and the second the hole propagation. So, the integral (1608) yields

$$I=i\lambda\int\frac{d\vec{\kappa}}{(2\pi)^3}\frac{d\in}{2\pi}\theta^2_{\kappa}\left(\frac{1-n(\vec{\kappa})}{\in-\xi_{\vec{\kappa}}+i\delta}+\frac{n(\vec{\kappa})}{\in-\xi_{\vec{\kappa}}-i\delta}\right)\left(\frac{1-n(\vec{q}-\vec{\kappa})}{\Omega-\in-\xi_{\vec{q}-\vec{\kappa}}+i\delta}+\frac{n(\vec{q}-\vec{\kappa})}{\Omega-\in-\xi_{\vec{q}-\vec{\kappa}}-i\delta}\right) \tag{1610}$$

or

$$\mathrm{I}=i\lambda\int\frac{d\vec{\kappa}}{(2\pi)^3}\theta_{\kappa}^2\left(\frac{\left(1-n(\vec{\kappa})\right)\left(1-n(\vec{q}-\vec{\kappa})\right)}{\Omega-\xi_{\vec{\kappa}}-\xi_{\vec{q}-\vec{\kappa}}+i\delta}\Bigg|_{\xi_{\vec{\kappa}}>0,\xi_{\vec{q}-\vec{\kappa}}>0}+\frac{n(\vec{\kappa})n(\vec{q}-\vec{\kappa})}{\Omega-\xi_{\vec{\kappa}}-\xi_{\vec{q}-\vec{\kappa}}-i\delta}\Bigg|_{\xi_{\vec{\kappa}}<0,\xi_{\vec{q}-\vec{\kappa}}<0}\right) \tag{1611}$$

We change the integration variable to $\xi=\xi_{\vec{\kappa}}$, with the account of factor ω_{p}^2 cutting off this integration at Debye frequency ω_D. Also

$$\xi_{\vec{\kappa}}+\xi_{\vec{q}-\vec{\kappa}}=\frac{\vec{\kappa}^2}{2m}+\frac{(\vec{q}-\vec{\kappa})^2}{2m}-2\mu\approx\frac{\vec{\kappa}^2}{m}-\frac{\vec{q}\vec{\kappa}}{m}-2\mu\approx2\xi_{\vec{\kappa}}-\frac{\vec{q}\vec{\kappa}}{m}=2\xi_{\vec{\kappa}}-\frac{q\kappa}{m}x\quad,\quad x=\cos\theta \tag{1612}$$

then

$$\int\frac{d\vec{\kappa}}{(2\pi)^3}\{\cdots\}=\int d\in\int\frac{d^3\kappa}{(2\pi)^3}\delta(\in-\xi_{\vec{\kappa}})\{\ldots\}=\int d\in n_{\mathrm{F}}(\in)\{\ldots\} \tag{1613}$$

Here, $n_{\mathrm{F}}(\in)=\frac{m\kappa_{\mathrm{F}}}{2\pi^2}$ is the density of states at the Fermi level.

So,

$$\mathrm{I}=-\frac{\lambda m\kappa_{\mathrm{F}}}{2\pi^2}\left\{\int_{-\omega_D}^{0}d\xi\int_0^1dx\frac{1}{\Omega-2\xi+\frac{q\kappa}{m}x-i\delta}+\int_0^{\omega_D}d\xi\int_0^1dx\frac{1}{2\xi-\Omega-\frac{q\kappa}{m}x-i\delta}\right\} \tag{1614}$$

We change the variable $\xi\to-\xi$ in the first integrand and then

$$\mathrm{I}=-\frac{\lambda m\kappa_{\mathrm{F}}}{2\pi^2}\int_0^{\omega_D}d\xi\int_0^1dx\left\{\frac{1}{\Omega+2\xi+\frac{q\kappa}{m}x-i\delta}+\frac{1}{2\xi-\Omega-\frac{q\kappa}{m}x-i\delta}\right\} \tag{1615}$$

and

$$\mathrm{Re\,I}=-\frac{\lambda m\kappa_{\mathrm{F}}}{2\pi^2}\int_0^{\omega_D}d\xi\frac{m}{q\kappa}\left\{\ln\left|\frac{\Omega+2\xi+\frac{q\kappa}{m}}{\Omega+2\xi}\right|-\ln\left|\frac{2\xi-\Omega-\frac{q\kappa}{m}}{2\xi-\Omega}\right|\right\} \tag{1616}$$

For $\frac{q\kappa}{m}<<1$,

$$\frac{m}{q\kappa}\left(\ln\left|1+\frac{\frac{q\kappa}{m}}{\Omega+2\xi}\right|-\ln\left|1+\frac{\frac{q\kappa}{m}}{\Omega-2\xi}\right|\right)=\frac{1}{\Omega+2\xi}-\frac{1}{\Omega-2\xi} \tag{1617}$$

and

$$\int_0^{\omega_D}d\xi\left(\frac{1}{\Omega+2\xi}-\frac{1}{\Omega-2\xi}\right)=\frac{1}{2}\left(\ln\left|\frac{2\omega_D+\Omega}{\Omega}\right|+\ln\left|\frac{2\omega_D-\Omega}{\Omega}\right|\right)\approx\ln\left|\frac{2\omega_D}{\Omega}\right| \tag{1618}$$

Also,

$$\operatorname{Im} I = -\frac{\lambda m\kappa_F}{2\pi^2}\int_0^{\omega_D} d\xi \frac{m}{q\kappa}\int_0^1 dx i\pi\left(\delta\left(x+\frac{\Omega+2\xi}{q\kappa}m\right)+\delta\left(x+\frac{\Omega-2\xi}{q\kappa}m\right)\right) \tag{1619}$$

For $q = 0$,

$$\operatorname{Im} I = -\frac{\lambda m\kappa_F}{2\pi^2}\int_0^{\omega_D} d\xi\pi\left(\delta(\Omega+2\xi)+\delta(2\xi-\Omega)\right) = -\frac{\pi}{2}\frac{\lambda m\kappa_F}{2\pi^2} \tag{1620}$$

Considering the fact that

$$\ln z = \ln|z| + i\arg z \tag{1621}$$

from the previous solutions, we have

$$\Gamma = \lambda - \frac{\lambda^2 m\kappa_F}{2\pi^2}\left(\ln\left|\frac{2\omega_D}{\Omega}\right| + i\frac{\pi}{2}\right) + \cdots \tag{1622}$$

or

$$\Gamma(\Omega) = \frac{\lambda}{1+\frac{\lambda m\kappa_F}{2\pi^2}\left(\ln\left|\frac{2\omega_D}{\Omega}\right| + i\frac{\pi}{2}\right)} \tag{1623}$$

From here, we read off some essential elements of the transition to the superconducting phase. We observe that the interaction constant λ appears in combination with the density of states $n_F = \frac{m\kappa_F}{2\pi^2}$. This implies that even a weak interaction can lead to sizeable effects if the density of states is large enough. The scaling factor n_F merely measures the number of final states accessible to the scattering mechanism. The effective strength of the Cooper pair correlation grows upon increasing the energetic range ω_D of the attractive force or, equivalently, on lowering Ω.

Now consider $\Gamma(\Omega)$ as a complex valued function of the variable Ω. This defines an analytical continuation of (1623) to the upper halfplane, where $\operatorname{Im}\Omega > 0$. Letting $\Omega = |\Delta|\exp\{i\varphi\}$

$$\Gamma(\Omega) = \frac{\lambda}{1+\frac{\lambda m\kappa_F}{2\pi^2}\left(\ln\left|\frac{2\omega_D}{\Delta}\right| + i\frac{\pi}{2} - i\varphi\right)} \tag{1624}$$

If interaction of electrons is attractive (i.e., $\lambda < 0$), then the vertex (1624) develops a singularity (pole), which is defined by the equation

$$1+\frac{\lambda m\kappa_F}{2\pi^2}\left(\ln\left|\frac{2\omega_D}{\Delta}\right| + i\left(\frac{\pi}{2}-\varphi\right)\right) = 0 \tag{1625}$$

This yields

$$\varphi = \frac{\pi}{2} \quad , \quad 1+\frac{\lambda m\kappa_F}{2\pi^2}\left(\ln\left|\frac{2\omega_D}{\Delta}\right|\right) = 0 \tag{1626}$$

This implies that the pole appears at an **imaginary** frequency, $\Omega = i|\Delta|$, where

$$|\Delta| = 2\omega_D \exp\left\{-\frac{2\pi^2}{m\kappa_F|\lambda|}\right\} \tag{1627}$$

Close to the pole $\Gamma(\Omega)$, we have the following

$$\Gamma(i\Delta+\Omega-i\Delta) = \frac{\lambda}{1+\frac{\lambda m\kappa_F}{2\pi^2}\left(\ln\frac{2\omega_D}{i\Delta}+i\frac{\pi}{2}+\ln\frac{i\Delta}{\Omega}\right)} = \frac{\lambda}{\frac{\lambda m\kappa_F}{2\pi^2}\left(\ln\frac{i\Delta}{\Omega-i\Delta+i\Delta}\right)} \tag{1628}$$

$$\frac{\lambda m\kappa_F}{2\pi^2}\left(\ln\frac{1}{\frac{\Omega-i\Delta}{i\Delta}+1}\right) = -\frac{\lambda m\kappa_F}{2\pi^2}\left(\ln\left[\frac{\Omega-i\Delta}{i\Delta}+1\right]\right) \approx -\frac{\lambda m\kappa_F}{2\pi^2}\frac{\Omega-i\Delta}{i\Delta} \tag{1629}$$

then

$$\Gamma(i\Delta+\Omega-i\Delta) = -\frac{2\pi^2}{m\kappa_F}\frac{i\Delta}{\Omega-i\Delta} \tag{1630}$$

Therefore,

$$\Gamma(\Omega) = -\frac{2\pi^2}{m\kappa_F}\frac{i\Delta}{\Omega-i\Delta} \tag{1631}$$

This corresponds to **Cooper instability**—the pole in the vertex part in the upper half-frequency plane formally signifies the appearance of an unstable collective mode that is exponentially growing in time amplitude:

$$\exp\{-i\Omega t\} \approx \exp\{-i(i\Delta)t\} \approx \exp\{\Delta t\} \tag{1632}$$

This leads to the instability of the system and to reconstruction of its ground state and spectra of excitations. For the case of repulsion $\lambda > 0$, there is no interesting phenomenon, and equation (1624) gives the sum of all **ladder corrections** to the **bare interaction** λ. The large logarithm only leads to the effective suppression of this repulsion, and there is no **pathology** at all.

Consider again the integral

$$\mathrm{I} = -\frac{\lambda m\kappa_F}{2\pi^2}\int_0^{\omega_D} d\xi \int_0^1 dx\left\{\frac{1}{\Omega+2\xi+\frac{q\kappa}{m}x-i\delta}+\frac{1}{2\xi-\Omega-\frac{q\kappa}{m}x-i\delta}\right\} \tag{1633}$$

and

$$\xi_{\vec{\kappa}} + \xi_{\vec{q}-\vec{\kappa}} \approx 2\xi_{\vec{\kappa}} - qv_F x \tag{1634}$$

then for $\Omega > qv_F$ and

$$\ln z = \ln|z| + i\arg z \tag{1635}$$

we have

$$\Gamma(q)=\lambda\left\{1+\lambda\frac{m\kappa_F}{2\pi^2}\left[\ln e\left|\frac{2\omega_D}{\Omega}\right|\right]+\frac{i\pi}{2}-i\varphi-\frac{1}{2}\ln\left(1-\frac{v_F^2q^2}{\Omega^2}\right)+\frac{\Omega}{2v_Fq}\ln\left|\frac{\Omega-v_Fq}{\Omega+v_Fq}\right|\right\}^{-1} \tag{1636}$$

Considering the continuation to the half-plane of $\operatorname{Im}\Omega>0$ and from $|\Delta|$ given in (1627), we have:

$$\Gamma(q,\Delta)=-\frac{2\pi^2}{m\kappa_F}\left\{\ln\frac{\Omega}{i\Delta}-1+\frac{1}{2}\ln\left(1-\frac{v_F^2q^2}{\Omega^2}\right)+\frac{\Omega}{2v_Fq}\ln\left|\frac{\Omega-v_Fq}{\Omega+v_Fq}\right|\right\}^{-1} \tag{1637}$$

For small $qv_F << |\Delta|$, we have:

$$\Gamma(q,\Delta)=-\frac{2\pi^2}{m\kappa_F}\frac{i\Delta}{\Omega-i\Delta+i\dfrac{v_F^2q^2}{6\Delta}} \tag{1638}$$

From here, we find the pole of $\Gamma(q,\Delta)$ as follows:

$$\Omega=i\Delta\left(1-\frac{v_F^2q^2}{6\Delta^2}\right) \tag{1639}$$

The absolute value of Ω decreases with an increase in q. For some v_Fq_{max}, the pole position Ω is zero, and for larger values of v_Fq, the pole in Γ is absent. Because $\vec{q}$ is the sum of the momenta of two electrons, this implies that the tendency of pairing is stronger for electrons with nearly opposite momenta.

The contribution of the **Cooper ladder** to the electron self-energy is shown in Figure 10.14 (a). The existence of the pole in the **ladder** yields the singularity in $\Sigma(\vec{\kappa})$ and for the vertex part as in Figure 10.14 (b).

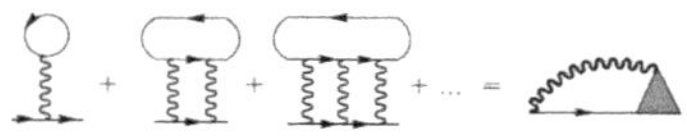

(1640)

(a)

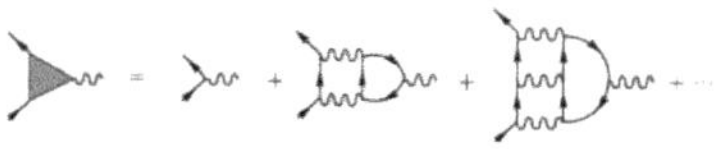

(1641)

(b)

FIGURE 10.14 Corrections to electron self-energy due to scattering in (a) Cooper channel and (b) Diagrams for appropriate vertex part.

Remark

The given results signify the instability of the normal ground state $T=0$ of the Fermi gas due to attractive interaction. Physically, the instability reduces to the ability of particles (with almost zero momentum at their center of inertia) to form bound pairs. These bound pairs are Bose particles that may **condense** in the ground state. The temperature, corresponding to the appearance of the given instability, defines the temperature of superconducting transition. This can be clearly understood within Matsubara formalism.

Consider that we neglect scattering of bound pairs on each other. In this case, the ideal Bose gas of Cooper pairs is formed, and the Matsubara Green's function of this gas can be written in the form:

$$\mathbf{G}(q,\Omega_m) = \frac{1}{i\Omega_m - \frac{q^2}{2m^*} + \mu} \tag{1642}$$

Here, q is the momentum of the bound pair, m^* is its mass and is equal to two masses of an electron and $\Omega_m = 2\pi m T$ the Matsubara frequency. If $\Omega_m = 0$, then equation (1642) becomes

$$\mathbf{G}(q,0) = \frac{1}{\mu - \frac{q^2}{2m^*}} \tag{1643}$$

At the temperature $T = T_0$ for the Bose condensation, the function diverges for $q = 0$, and T_0 is determined from the equation for $\mu = 0$ in accordance with the standard analysis of Bose condensation.

Considering the internal structure of a Cooper pair, the analogue of (1642) is the two-particle fermion Green's function. At the transition point, its analytic behavior is analogous to that of a Bose gas Green's function. This is due to the fact that it has a dependence on $\Omega_m = (\epsilon_1 + \epsilon_2)_m$ and $\vec{q} = \vec{\kappa}_1 + \vec{\kappa}_2$, which corresponds to the center of inertia of a pair. It is instructive to note that single-fermion Green's functions do not have the given singularities. Therefore, we consider the appropriate vertex part $\Gamma(q,\Omega)$ that is given by the same previously mentioned **ladder** diagrams written in Matsubara formalism. The difference is that instead of equation (1608), we consider I for finite temperature T.

10.6.3.1 Finite Temperature Calculation

$$\mathrm{I}_{T\neq 0} = -\lambda T \sum_n \int \frac{d\vec{\kappa}}{(2\pi)^3} \omega_{\vec{\kappa}}^2 \mathbf{G}_0(\vec{\kappa}) \mathbf{G}_0(\vec{q}-\vec{\kappa}) \tag{1644}$$

But

$$\mathbf{G}_0(\vec{\kappa}, i\omega_n) = \frac{1}{i\omega_n - \xi_{\vec{\kappa}}} \tag{1645}$$

then

$$T\sum_{\omega_n} \frac{1}{i\omega_n - \xi_{\vec{\kappa}}} \frac{1}{i\Omega - i\omega_n - \xi_{\vec{q}-\vec{\kappa}}} = T\sum_{\omega_n} \left(\frac{1}{i\omega_n - \xi_{\vec{\kappa}}} + \frac{1}{i\Omega - i\omega_n - \xi_{\vec{q}-\vec{\kappa}}} \right) \frac{1}{i\Omega - \xi_{\vec{q}-\vec{\kappa}} - \xi_{\vec{\kappa}}} \tag{1646}$$

or

$$T\sum_{\omega_n} \frac{1}{i\omega_n - \xi_{\vec{\kappa}}} \frac{1}{i\Omega - i\omega_n - \xi_{\vec{q}-\vec{\kappa}}} = \frac{1}{i\Omega - \xi_{\vec{q}-\vec{\kappa}} - \xi_{\vec{\kappa}}} \left(-\frac{1}{2}\tanh\frac{\xi_{\vec{\kappa}}}{2T} - \frac{1}{2}\tanh\frac{(\xi_{\vec{q}-\vec{\kappa}} - i\Omega_m)}{2T} \right) \tag{1647}$$

Here, $\Omega_m = 2\pi m T$ and therefore,

$$\mathrm{I}_{T\neq 0} = -\lambda \int \frac{d\vec{\kappa}}{(2\pi)^3} \omega_{\vec{\kappa}}^2 \frac{1}{2} \left(\frac{\tanh\frac{\xi_{\vec{\kappa}}}{2T} + \tanh\frac{\xi_{\vec{q}-\vec{\kappa}}}{2T}}{\xi_{\vec{\kappa}} + \xi_{\vec{q}-\vec{\kappa}} - i\Omega_m} \right) \tag{1648}$$

Considering the fact that the pole of $\Gamma(q,\Omega)$ appears for $q=0$, then

$$\Gamma(0)=\frac{\lambda}{1+\frac{\lambda m\kappa_F}{2\pi^2}\int_0^{\omega_D}d\xi\frac{\tanh\frac{\xi}{2T}}{\xi}} \tag{1649}$$

where

$$\int_0^{\omega_D}d\xi\frac{\tanh\frac{\xi}{2T}}{\xi}=\int_0^{\omega_D/2T}dx\frac{\tanh x}{x}=\ln x\tanh x\Big|_0^{\omega_D/2T}-\int_0^{\omega_D/2T}dx\frac{\ln x}{\cosh^2 x} \tag{1650}$$

In the remaining integral, we set the upper limit $x=\frac{\omega_D}{2T}$ to infinity due to fast convergence and $T \ll \omega_D$. Therefore,

$$\int_0^{\infty}dx\frac{\ln x}{\cosh^2 x}=\ln\frac{\pi}{4\gamma} \tag{1651}$$

where $\ln\gamma=C=0.577$ is the Euler constant and $\frac{2\pi}{\gamma}=2\pi\exp\{-0.577\}\approx 3.53$. This is in very good agreement with experimental findings where the ratios typically range between 3 and 4.5. This is just one of the successes of the BCS theory. Consequently,

$$\Gamma(0)=\frac{\lambda}{1+\lambda n_F\ln\frac{2\omega_D\gamma}{\pi T}} \tag{1652}$$

For $\lambda<0$ and the **superconducting transition temperature** T_c as defined by BCS expression [115, 28] be:

$$T_c=\frac{\gamma}{\pi}2\omega_D\exp\left\{-\frac{1}{|\lambda|n_F}\right\} \tag{1653}$$

and

$$\Delta=2\omega_D\exp\left\{-\frac{1}{|\lambda|n_F}\right\}=\frac{\pi}{\gamma}T_c \tag{1654}$$

then

$$\Gamma(0)=\frac{\lambda}{1+\lambda n_F\left(\ln\frac{2\omega_D\gamma}{\pi T_c}+\ln\frac{T_c}{T}\right)}=\frac{\lambda}{\lambda n_F\ln\frac{T_c}{T_c+T-T_c}}=-\frac{1}{n_F}\frac{T_c}{T-T_c} \tag{1655}$$

or

$$\Gamma(0)=-\frac{1}{n_F}\frac{T_c}{T-T_c} \tag{1656}$$

As seen earlier, the temperature T_c marks the transition to the superconducting state. Below this temperature, T_c, a perturbative approach (based on the Fermi sea of the noninteracting system as a reference state) **breaks down**. This **Cooper instability** shows that we have to look for an alternative ground state or **mean-field**. This implies that one accounts for the strong binding of Cooper pairs. It is instructive to note that the density of states n_F enters at the Fermi level for the **single** spin projection. As seen earlier, the value of $|\lambda| n_F$ determines the dimensionless coupling constant of pairing interaction. Both the gap Δ and the critical temperature T_c are reduced by the exponential factor $\exp\left\{-\frac{1}{|\lambda| n_F}\right\}$ as compared to the bare energy scale of the interaction, ω_D. This strong renormalization generates the new scale, $\omega_D \exp\left\{-\frac{1}{|\lambda| n_F}\right\}$. The BCS theory provides an excellent model for the behavior of low-temperature superconductors. However, it is not clear to what extent the theory can be used to explain the superconductivity of high-temperature superconductors and other exotic superconducting materials.

10.6.4 Small Momentum Transfer Vertex Function

We consider a system of fermions at absolute zero temperature $T = 0$ and with arbitrary short-range interaction forces. We examine again some basic relations, such as

$$\in\left(\vec{\kappa}\right) = \frac{\vec{\kappa}^2}{2m} \tag{1657}$$

and

$$\xi_{\vec{\kappa}} = \frac{\vec{\kappa}^2 - \kappa_F^2}{2m} = \frac{1}{2m}\left(\left|\vec{\kappa}\right| - \kappa_F\right)\left(\left|\vec{\kappa}\right| + \kappa_F\right) \approx \frac{\kappa_F}{m}\left(\left|\vec{\kappa}\right| - \kappa_F\right) \tag{1658}$$

We examine the following bare and exact Green's functions:

$$\mathbf{G}_0\left(\in,\vec{\kappa}\right) = \frac{1}{\in - \xi_{\vec{\kappa}} + i\delta\,\mathrm{sgn}\left(\kappa - \kappa_F\right)} = \frac{1}{\in - v_F\left(\left|\vec{\kappa}\right| - \kappa_F\right) + i\delta\,\mathrm{sgn}\left(\kappa - \kappa_F\right)} \tag{1659}$$

and the Green's function in the Dyson form:

$$\mathbf{G}\left(\in,\vec{\kappa}\right) = \frac{1}{\in - \xi_{\vec{\kappa}} - \Sigma\left(\vec{\kappa},\in\right)} \tag{1660}$$

We put the self-energy in series

$$\Sigma\left(\in,\vec{\kappa}\right) \approx \Sigma(0,0) + \in \cdot \left.\frac{\partial \Sigma}{\partial \in}\right|_{\kappa_F,\in\left(\vec{\kappa}\right)} + \left.\frac{\partial \Sigma}{\partial \left|\vec{\kappa}\right|}\right|\left(\left|\vec{\kappa}\right| - \kappa_F\right) + \frac{\partial \Sigma}{\partial \in\left(\vec{\kappa}\right)} \cdot \frac{\partial \in\left(\vec{\kappa}\right)}{\partial \left|\vec{\kappa}\right|} \tag{1661}$$

or

$$\Sigma\left(\in,\vec{\kappa}\right) \approx \Sigma_0 + \in \cdot \left.\frac{\partial \Sigma}{\partial \in}\right|_{\kappa_F,\in\left(\vec{\kappa}\right)} + \left.\frac{\kappa_F}{m}\frac{\partial \Sigma}{\partial \in\left(\vec{\kappa}\right)}\right|_{\kappa_F,\in\left(\vec{\kappa}\right)}\left(\left|\vec{\kappa}\right| - \kappa_F\right) \tag{1662}$$

From here, we write the exact Green's function

$$\mathbf{G}(\in, p) = \frac{1}{\in\left(1 - \frac{\partial\Sigma}{\partial\in(\vec{p})}\right) - \frac{p_{\mathrm{F}}}{m}\left(|\vec{p}| - p_{\mathrm{F}}\right)\left(1 + \frac{\partial\Sigma}{\partial\in(\vec{p})}\right)} \tag{1663}$$

or

$$\mathbf{G}(\in, \vec{\kappa}) = \frac{1 \Big/ \left(1 - \frac{\partial\Sigma}{\partial\in(\vec{\kappa})}\right)}{\in(\vec{\kappa}) - \frac{\kappa_{\mathrm{F}}}{m}\left(|\vec{\kappa}| - \kappa_{\mathrm{F}}\right)\left(1 + \frac{\partial\Sigma}{\partial\in(\vec{\kappa})}\right) \Big/ \left(1 - \frac{\partial\Sigma}{\partial\in(\vec{\kappa})}\right)} \tag{1664}$$

This allows us to have

$$\frac{1}{m^{*}} = \frac{1}{m}\frac{1 + \frac{\partial\Sigma}{\partial\in(\vec{\kappa})}}{1 - \frac{\partial\Sigma}{\partial\in(\vec{\kappa})}} \tag{1665}$$

or

$$m^{*} = m\frac{1 - \frac{\partial\Sigma}{\partial\in(\vec{\kappa})}}{1 + \frac{\partial\Sigma}{\partial\in(\vec{\kappa})}} \tag{1666}$$

From here, we have the following relation

$$0 < Z \equiv \frac{1}{1 - \frac{\partial\Sigma}{\partial\in(\vec{\kappa})}} \leq 1 \tag{1667}$$

The equality is only achieved for the ideal (free) Fermi gas.

We examine the vertex function Γ, which plays an important role in the theory of Fermi liquids. It has almost equal values of the pairs of the 4-momenta κ_1 , κ_3 and κ_2 , κ_4. Considering the 4-momenta conservation law

$$\kappa_1 + \kappa_2 = \kappa_3 + \kappa_4 \tag{1668}$$

and letting

$$\kappa_3 = \kappa_1 + q \quad , \quad \kappa_4 = \kappa_2 - q \tag{1669}$$

this allows us to simplify the vertex function Γ:

$$\Gamma_{\gamma\delta,\alpha\beta}\left(\kappa_1 + q, \kappa_2 - q, \kappa_1, \kappa_2\right) = \Gamma_{\gamma\delta,\alpha\beta}\left(q, \kappa_1, \kappa_2\right) \tag{1670}$$

We consider this vertex function for 4-momentum transfer q small. Considering quasi-particle scattering processes, this implies considering collisions with small transfer 4-momenta that are close to **forward scattering**. We consider the simplest diagrams for Γ as depicted in Figure 10.15:

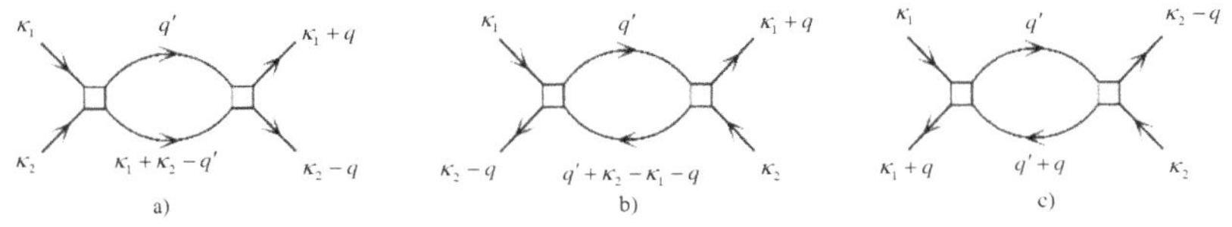

FIGURE 10.15 The simplest diagrams for Γ.

Each diagram in Figure 10.15 contains integrals of two Green's functions. Though for (a) and (b), there is nothing special about the case $q=0$; for (c), the poles of the two Green's functions come together as $q \to 0$. We show that this leads to the appearance of singularities in Γ. We shall be interested in the part of Γ that contains this singularity. Although Figure 10.15 formally relates the 2PI forces, diagram (c) still exhibits a special behavior for the case of arbitrary interacting forces. Diagram (c), as seen earlier, contains a set of two-particle Green's function diagrams that can be cut between the pairs of external lines κ_1, κ_3 and κ_2, κ_4, into two parts linked by two continuous lines. The exact one-particle Green's functions in (c) are $\mathbf{G}(q')$ and $\mathbf{G}(q'+q)$ and relate Γ through an integral relation to be shown later in this book. As $q \to 0$, the arguments of the given Green's functions become closer and, consequently, so do the poles that are the source of the singularity in the vertex function Γ.

Let us denote by

$$\Gamma^{(1)} = \Gamma^{(1)}_{\gamma\delta,\alpha\beta}\left(q=0,\kappa_1,\kappa_2\right) = \quad (1671)$$

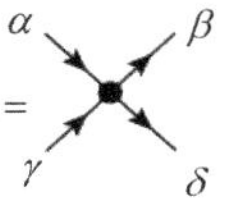

that part of the vertex function (irreducible (2PI) vertex) Γ, which has no singularity at $q=0$. That is, the set of all possible diagrams for Γ that do not contain **anomalous elements**, that is, $\mathbf{G}(q')\mathbf{G}(q'+q)$ lines. Therefore, the exact vertex Γ:

$$\Gamma_{\gamma\delta,\alpha\beta}\left(\kappa_1,\kappa_2,k\right)=\Gamma^{(1)}_{\gamma\delta,\alpha\beta}\left(\kappa_1,\kappa_2\right)-i\int\Gamma^{(1)}_{\gamma\xi,\alpha\eta}(\kappa_1,q)\mathbf{G}(k)\mathbf{G}(q+k)\Gamma_{\eta\delta,\xi\beta}(k,\kappa_2,q)\frac{d^4q}{(2\pi)^4} \quad (1672)$$

This is represented by the following infinite ladder series of diagrams (Bethe-Salpeter equation):

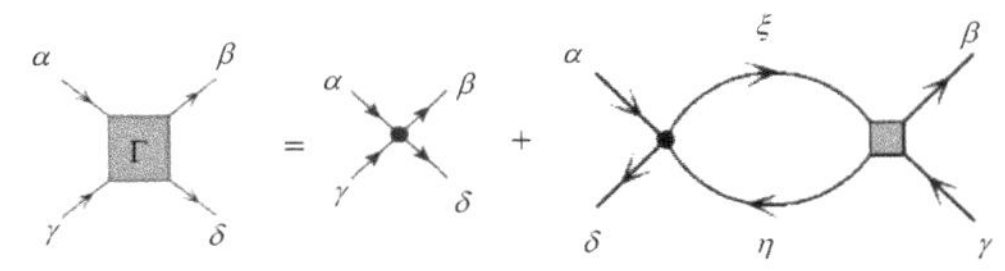

(1673)

FIGURE 10.16 Diagrammatic representation of the Bethe-Salpeter equation relating the 1PI vertex Γ to $\Gamma^{(1)}$ in the Ladder diagrams.

In the previous equation, we have the following four vectors

$$q=\left(\omega,\vec{q}\right) \quad , \quad k=\left(\kappa_F,\vec{k}\right) \quad (1674)$$

To do an integration around p_F in (1672), we replace the contour of integration by an infinite semicircle in either the upper or the lower half-plane. This integral can be calculated by taking the residue of the integrand at the corresponding pole. So, the integral (1672) consists of a contribution from domains far from the $|\vec{\kappa}| = \kappa_F, \in = 0$ and the contribution from around that point. This integral determines the singularities of the entire expression. If q is small, we take around $|\vec{\kappa}| = \kappa_F, \in = 0$ to be small. The only important contribution to the integral will come from the part of the contour around the poles of the Green's functions:

$$\mathbf{G}(\vec{q}) \approx \frac{Z}{\in - \xi_{\vec{q}} + i\delta\,\mathrm{sgn}\left(|\vec{q}| - \kappa_F\right)}, \mathbf{G}\left(\vec{q}+\vec{k}\right) \approx \frac{Z}{\in + \omega - \xi_{\vec{q}+\vec{k}} + i\delta\,\mathrm{sgn}\left(\left|\vec{q}+\vec{k}\right| - \kappa_F\right)} \tag{1675}$$

Here, Z is defined in (1667) and $\delta \to 0$; κ_F is the **Fermi momentum** defined by the relation $\in(\kappa_F) = \mu$. The excitation velocity at the Fermi surface is equal to $\vec{v} = \frac{\vec{q}}{m^*}$ where m^* is the effective mass of the excitations.

The product of the Green's functions $\mathbf{G}(\vec{q})\mathbf{G}\left(\vec{q}+\vec{k}\right)$ determines the poles' position in the upper or lower half-plane of the complex variable κ_F. The singularity in the kernel of the integral equation as well as in the solution of that equation stems from the contour of integration with respect to κ_F (the real axis) between the poles. These poles must be on opposite sides of the contour (opposite half-planes).

As the arguments of these Green's functions are closer together, we assume all other quantities being functions of q in the integrand are slowly varying. Therefore, the poles contribute only when they are located on different sides of the real axis and demands, which for different signs of $\mathrm{sgn}\left(|\vec{q}| - \kappa_F\right)$ and $\mathrm{sgn}\left(\left|\vec{q}+\vec{k}\right| - \kappa_F\right)$, we have

$$|\vec{q}| < \kappa_F \quad , \quad \left|\vec{q}+\vec{k}\right| > \kappa_F \quad \text{or} \quad |\vec{q}| > \kappa_F \quad , \quad \left|\vec{q}+\vec{k}\right| < \kappa_F \tag{1676}$$

This is equivalent to the equation:

$$\kappa_F - k\cos\theta < q < \kappa_F \tag{1677}$$

We evaluate the Cauchy integral

$$\frac{1}{2\pi i}\int d\in 2\pi i\mathbf{G}(\vec{q})\mathbf{G}\left(\vec{q}+\vec{k}\right) = \frac{Z^2 2\pi i}{\xi_{\vec{q}} - \xi_{\vec{q}+\vec{k}} + \omega + i\delta} \tag{1678}$$

But

$$\xi_{\vec{q}} - \xi_{\vec{q}+\vec{k}} = -\frac{\vec{q}^2}{2m} + \frac{\left(\vec{q}+\vec{k}\right)^2}{2m} = -\frac{\vec{q}\cdot\vec{k}}{m} - \frac{\vec{k}^2}{2m} \tag{1679}$$

Then

$$\frac{1}{2\pi i}\int d\in 2\pi i\mathbf{G}(\vec{q})\mathbf{G}\left(\vec{q}+\vec{k}\right) = \frac{Z^2 2\pi i}{-\frac{\vec{q}\cdot\vec{k}}{m} - \frac{\vec{k}^2}{2m} + \omega + i\delta} \tag{1680}$$

Because $\vec{k}^2$ is very small,

$$\frac{1}{2\pi i}\int d\in 2\pi i\mathbf{G}(\vec{q})\mathbf{G}(\vec{q}+\vec{k})\approx\frac{Z^2 2\pi i}{-\frac{\vec{q}\cdot\vec{k}}{m}+\omega+i\delta} \tag{1681}$$

and

$$\frac{1}{2\pi i}\int d\in d\vec{q}\,2\pi i\mathbf{G}(\vec{q})\mathbf{G}(\vec{q}+\vec{k})\approx\int d\vec{q}\frac{Z^2 2\pi i}{-\frac{\vec{q}\cdot\vec{k}}{m}+\omega+i\delta} \tag{1682}$$

We examine the polarization function

$$\Pi_{q,\omega}=-i\int\mathbf{G}(\vec{q})\mathbf{G}(\vec{q}+\vec{k})\frac{d^4q}{(2\pi)^4}=\int\frac{d\vec{q}}{(2\pi)^3}\frac{n(\vec{q}+\vec{k})-n(\vec{q})}{\omega+\xi_{\vec{q}}-\xi_{\vec{q}+\vec{k}}+i\delta} \tag{1683}$$

or

$$\Pi_{q,\omega}=\int\frac{d\vec{q}}{(2\pi)^3}\frac{\partial n}{\partial\xi_{\vec{q}}}\frac{\vec{k}\vec{q}/m}{\omega-\vec{k}\vec{q}/m}=\int\frac{d\vec{q}}{(2\pi)^3}\frac{\partial n}{\partial\xi_{\vec{q}}}\frac{\vec{k}\vec{v}}{\omega-\vec{k}\vec{v}} \tag{1684}$$

Because $\vec{k}$ is small, this implies that $|\vec{q}|\approx\kappa_F$ and $\in\approx 0$. Therefore, $\mathbf{G}(\vec{q})\mathbf{G}(\vec{q}+\vec{k})$ can be replaced by

$$\mathbf{G}(\vec{q})\mathbf{G}(\vec{q}+\vec{k})\approx\mathrm{A}\delta(\in)(|\vec{q}|-\kappa_F) \tag{1685}$$

The factor A is obtained from integrating $\mathbf{G}(\vec{q})\mathbf{G}(\vec{q}+\vec{k})$ with respect to $\in$ and q:

$$\int d\in\mathbf{G}(\vec{q})\mathbf{G}(\vec{q}+\vec{k})=Z^2 2\pi i\int dq\frac{\partial n}{\partial\xi_{\vec{q}}}\frac{\vec{k}\vec{v}}{\omega-\vec{k}\vec{v}}\int\delta(\xi_{\vec{q}}-\xi_{\kappa_F}) \tag{1686}$$

Considering $d\xi_{\vec{q}}=vdq$, then

$$\int d\in\mathbf{G}(\vec{q})\mathbf{G}(\vec{q}+\vec{k})=\frac{Z^2 2\pi i}{|\vec{v}|}\int v\,dq\frac{\partial n}{\partial\xi_{\vec{q}}}\frac{\vec{k}\vec{v}}{\omega-\vec{k}\vec{v}}\int\delta(\xi_{\vec{q}}-\xi_{\kappa_F})=\frac{2\pi iZ^2}{|\vec{v}|}\frac{\vec{k}\vec{v}}{\omega-\vec{k}\vec{v}} \tag{1687}$$

Therefore,

$$\mathrm{A}=\frac{2\pi iZ^2}{|\vec{v}|}\frac{\vec{k}\vec{v}}{\omega-\vec{k}\vec{v}} \tag{1688}$$

From here, we have the product of the Green's functions:

$$\mathbf{G}(\vec{q})\mathbf{G}(\vec{q}+\vec{k})=\frac{2\pi iZ^2}{|\vec{v}|}\frac{\vec{k}\vec{v}}{\omega-\vec{k}\vec{v}}\delta(\in)(|\vec{q}|-\kappa_F)+\varphi(\vec{q})\equiv i\Phi+\varphi(\vec{q}) \tag{1689}$$

Here, $\varphi(\vec{q})$ is the regular part of $\mathbf{G}(\vec{q})\mathbf{G}(\vec{q}+\vec{k})$ that is important only in the integral over the **distant** region. Because the range in (1677) is very narrow when q is small, we let $k=0$ in the quantities Γ, $\Gamma^{(1)}$, and $\varphi(\vec{q})$.

Taking the limit of expression (1689) as $\vec{q},\omega \to 0$ is dependent on the ratio of ω to $|q'|$. This also applies to Γ. We first examine Γ as $\frac{|\vec{q}|}{\omega},\omega \to 0$. Hence, from (1672) and (1689), we have the quasi-particle–quasi-hole vertex Γ^{ω}:

$$\Gamma^{\omega} = \lim_{\frac{|\vec{k}|}{\omega},\omega\to 0} \Gamma(\kappa_1,\kappa_2,k) \quad , \quad \Gamma^{k} = \lim_{\frac{\omega}{|\vec{k}|},|\vec{k}|\to 0} \Gamma(\kappa_1,\kappa_2,k) \tag{1690}$$

and

$$\Gamma^{\omega}_{\gamma\delta,\alpha\beta}(\kappa_1,\kappa_2) = \Gamma^{(1)}_{\gamma\delta,\alpha\beta}(\kappa_1,\kappa_2) - i\int \Gamma^{(1)}_{\gamma\xi,\alpha\eta}(\kappa_1,q)\varphi(q)\Gamma^{\omega}_{\eta\delta,\xi\beta}(q,\kappa_2)\frac{d^4q}{(2\pi)^4} \tag{1691}$$

We write equations (1691) and (1672), respectively, as

$$\Gamma^{\omega} = \lim_{\frac{|\vec{q}|}{\omega},\omega\to 0} \Gamma(\kappa_1,\kappa_2,q') = \Gamma^{(1)} - i\Gamma^{(1)}\phi\Gamma^{\omega} \quad , \quad \Gamma = \Gamma^{(1)} - i\Gamma^{(1)}[i\Phi + \phi]\Gamma \tag{1692}$$

From the first equation of (1692), we have

$$\Gamma^{(\omega)} = \left[1 + i\varphi\Gamma^{(1)}\right]^{-1}\Gamma^{(1)} \tag{1693}$$

From the second equation of (1692), we have

$$\Gamma = \Gamma^{(1)} - i\Gamma^{(1)}(i\Phi + \varphi)\Gamma \tag{1694}$$

But

$$\left[1 + i\varphi\Gamma^{(1)}\right]\Gamma = \Gamma^{(1)} + \Gamma^{(1)}\Phi\Gamma \tag{1695}$$

So

$$\Gamma = \left[1 + i\varphi\Gamma^{(1)}\right]^{-1}\Gamma^{(1)} + \left[1 + i\varphi\Gamma^{(1)}\right]^{-1}\Gamma^{(1)}\Phi\Gamma \tag{1696}$$

Hence,

$$\Gamma = \Gamma^{(\omega)} + \Gamma^{(\omega)}\Phi\Gamma \tag{1697}$$

Using this equation for $\frac{|\vec{q}'|}{\omega},\omega \to 0$ and letting $q' \equiv k$, we write this equation in explicit form:

$$\begin{aligned}\Gamma_{\alpha\beta,\gamma\delta}(\kappa_1,\kappa_2,k) &= \Gamma^{\omega}_{\alpha\beta,\gamma\delta}(\kappa_1,\kappa_2) \\ &+ \frac{2\pi z^2}{|\vec{v}|}\frac{1}{(2\pi)^4}\int d\in dq\,d\Omega q^2\delta(\in)\delta(|\vec{q}| - \kappa_F)\Gamma^{\omega}_{\alpha\xi,\gamma\eta}(\kappa_1,q)\Gamma^{\omega}_{\eta\beta,\xi\delta}(q,\kappa_2,k)\frac{\vec{v}\vec{k}}{\omega - \vec{v}\vec{k}}\end{aligned} \tag{1698}$$

or

$$\Gamma_{\alpha\beta,\gamma\delta}(\kappa_1,\kappa_2,k)=\Gamma^{\omega}_{\alpha\beta,\gamma\delta}(\kappa_1,\kappa_2)+\frac{Z^2\kappa_F^2}{(2\pi)^3|\vec{v}|}\int d\Omega\Gamma^{\omega}_{\alpha\xi,\gamma\eta}(\kappa_1,q)\frac{\vec{v}\vec{k}}{\omega-\vec{v}\vec{k}}\Gamma_{\eta\beta,\xi\delta}(q,\kappa_2) \quad (1699)$$

Here, $d\Omega$ is the element of solid angle.

We now examine Γ in the limiting case $\frac{\omega}{|\vec{k}|},|\vec{k}|\to 0$; this permits us to make the denotation of the limit:

$$\Gamma^{(k)}_{\gamma\delta,\alpha\beta}(\kappa_1,\kappa_2)=\Gamma^{\omega}_{\gamma\delta,\alpha\beta}(\kappa_1,\kappa_2)-\frac{Z^2\kappa_F^2}{(2\pi)^3|\vec{v}|}\int\Gamma^{\omega}_{\alpha\xi,\gamma\eta}(\kappa_1,q)\Gamma^{(k)}_{\eta\beta,\xi\delta}(q,\kappa_2)d\Omega \quad (1700)$$

We investigate the poles of the function $\Gamma(\kappa_1,\kappa_2,k)$ for small $\vec{k}$ and ω.

Because around the pole,

$$\Gamma(\kappa_1,\kappa_2,k)>>\Gamma^{\omega}(\kappa_1,\kappa_2) \quad (1701)$$

we can neglect the term Γ^{ω} at the right had side of equation (1699). In addition, the variable κ_2 and the indices β,δ act as parameters in (1699). Hence, near the pole, the function Γ can be represented:

$$\Gamma=\chi_{\alpha\gamma}(\kappa_1,k)\chi_{\beta\delta}(\kappa_2,k) \quad (1702)$$

From (1699) considering (1701), we have

$$\chi_{\alpha\gamma}(\kappa_1,k)\chi_{\beta\delta}(\kappa_2,k)=\frac{Z^2\kappa_F^2}{(2\pi)^3|\vec{v}|}\int d\Omega\Gamma^{\omega}_{\alpha\xi,\gamma\eta}(\kappa_1,q)\frac{\vec{v}\vec{k}}{\omega-\vec{v}\vec{k}}\chi_{\eta\xi}(\kappa_1,k)\chi_{\beta\delta}(\kappa_2,k) \quad (1703)$$

Cancelling the common factor, we have

$$\chi_{\alpha\gamma}(\kappa_1,k)=\frac{Z^2\kappa_F^2}{(2\pi)^3|\vec{v}|}\int d\Omega\Gamma^{\omega}_{\alpha\xi,\gamma\eta}(\kappa_1,q)\frac{\vec{v}\vec{k}}{\omega-\vec{v}\vec{k}}\chi_{\eta\xi}(\kappa_1,k) \quad (1704)$$

We define the function

$$\nu_{\alpha\gamma}(\vec{n})=\frac{\vec{n}\vec{k}}{\omega-v\vec{n}\vec{k}}\chi_{\alpha\gamma}(\kappa_1,k) \quad (1705)$$

where $\vec{n}$ is a unit vector in the direction of $\vec{\kappa}_1$; therefore, the relation for $\nu_{\alpha\gamma}(\vec{n})$ is:

$$\left(\omega-v\vec{n}\vec{k}\right)\nu_{\alpha\gamma}(\vec{n})=\vec{n}\vec{k}\frac{\kappa_F^2Z^2}{(2\pi)^3}\int\Gamma^{(\omega)}_{\alpha\xi,\gamma\eta}(\vec{n},\vec{n}')\nu_{\eta\zeta}(\vec{n}')d\Omega \quad (1706)$$

This equation imitates the **kinetic equation of phenomenological Landau theory** [58] with $\nu_{\alpha\gamma}(\vec{n})$ being the nonequilibrium part of the distribution function of quasiparticles. If we compare (1706) with the kinetic equation, we may write the correlation between the quasi-particle interaction function (**Landau function**) $f_{\alpha\xi,\gamma\eta}$ and the properties of quasi-particle forward scattering amplitude $\Gamma^{(\omega)}_{\alpha\xi,\gamma\eta}$:

$$f_{\alpha\xi,\gamma\eta}(\vec{n},\vec{n}')=Z^2\Gamma^{(\omega)}_{\alpha\xi,\gamma\eta}(\vec{n},\vec{n}') \quad (1707)$$

Let us use equation (1700) and obtain an explicit relation between $f_{\alpha\xi,\gamma\eta}$ and the **physical forward scattering amplitude** for quasiparticles on the Fermi surface:

$$Z^2\Gamma^{(k)}_{\gamma\delta,\alpha\beta}(\vec{n}_1,\vec{n}_2)=Z^2\Gamma^{\omega}_{\gamma\delta,\alpha\beta}(\vec{n}_1,\vec{n}_2)-\frac{p_F^2}{(2\pi)^3|\vec{v}|}\int Z^2\Gamma^{\omega}_{\alpha\xi,\gamma\eta}(\vec{n}_1,\vec{n}')Z^2\Gamma^{(k)}_{\eta\beta,\xi\delta}(\vec{n}',\vec{n}_2)d\Omega \tag{1708}$$

Denoting

$$A_{\gamma\delta,\alpha\beta}(\vec{n}_1,\vec{n}_2)=Z^2\Gamma^{(k)}_{\gamma\delta,\alpha\beta}(\vec{n}_1,\vec{n}_2)\quad,\quad f_{\alpha\xi,\gamma\eta}(\vec{n}_1,\vec{n}')=Z^2\Gamma^{\omega}_{\alpha\xi,\gamma\eta}(\vec{n}_1,\vec{n}') \tag{1709}$$

equation (1708) then becomes

$$A_{\gamma\delta,\alpha\beta}(\vec{n}_1,\vec{n}_2)=f_{\gamma\delta,\alpha\beta}(\vec{n}_1,\vec{n}_2)-\frac{\kappa_F^2}{(2\pi)^3|\vec{v}|}\int f_{\alpha\xi,\gamma\eta}(\vec{n}_1,\vec{n}')A_{\eta\beta,\xi\delta}(\vec{n}',\vec{n}_2)d\Omega \tag{1710}$$

Assuming the spin-dependent part of the particle interaction is due purely to exchange, we express the spin-dependence of the functions A and f via the Pauli matrices $\boldsymbol{\sigma}$:

$$\frac{\kappa_F^2}{\pi^2|\vec{v}|}f_{\gamma\delta,\alpha\beta}(\vec{n}_1,\vec{n}_2)=F(\theta)\delta_{\alpha\gamma}\delta_{\beta\delta}+G(\theta)\sigma_{\gamma\alpha}\sigma_{\delta\beta}\quad,\quad\frac{\kappa_F^2}{\pi^2|\vec{v}|}A_{\gamma\delta,\alpha\beta}(\vec{n}_1,\vec{n}_2)=B(\theta)\delta_{\alpha\gamma}\delta_{\beta\delta}+C(\theta)\sigma_{\gamma\alpha}\sigma_{\delta\beta} \tag{1711}$$

The coefficients F, G, B, and C in an isotropic liquid are dependent only on the angle θ between $\vec{n}_1$ and $\vec{n}_2$. Suppose we expand these quantities via the Legendre polynomials:

$$B(\theta)=\sum_{\ell=0}^{\infty}(2\ell+1)B_\ell P_\ell(\cos\theta)\quad,\quad\cdots \tag{1712}$$

Substituting (1711) and (1712) into (1710), we have

$$B_\ell=F_\ell(1-B_\ell)\quad,\quad C_\ell=G_\ell(1-C_\ell) \tag{1713}$$

Considering condition (1701), we have

$$\sum_{\ell=0}^{\infty}(2\ell+1)(B_\ell+C_\ell)=0 \tag{1714}$$

Equations (1713) and (1714) show the existence in every stable Fermi liquid of at least one branch (ordinary or spin) of axially symmetric zero sound.

So the determination of the acoustic excitations in a Fermi liquid reduces to the problem of finding the eigenvalues of the integral equation (1704).

For the limiting case $\frac{\omega}{|\vec{k}|},|\vec{k}|\to 0$, then

$$\Gamma^{(k)}_{\gamma\delta,\alpha\beta}(\kappa_1,\kappa_2)=\Gamma^{\omega}_{\gamma\delta,\alpha\beta}(\kappa_1,\kappa_2)-\frac{Z^2\kappa_F^2}{(2\pi)^3|\vec{v}|}\int\Gamma^{\omega}_{\alpha\xi,\gamma\eta}(\kappa_1,q)\Gamma^{(k)}_{\eta\beta,\xi\delta}(q,\kappa_2)d\Omega \tag{1715}$$

10.6.5 Ward Identities: Gauge Invariance

When examining the mathematical formalism of Green's functions, certain identical relations between the derivatives of these functions and quasi-particle scattering amplitude play an important role. We consider the change in the Green's functions caused by some fictitious **external field** $\delta\hat{U}(t)$ that is homogenous in space and slowly varying in time for which the corresponding term in the Hamiltonian determinant:

$$\hat{H}_{\text{int}} = \int \hat{\psi}_\alpha^\dagger(t,\vec{r})\delta\hat{U}(t)\hat{\psi}_\alpha(t,\vec{r})d\vec{r} \tag{1716}$$

With the presence of the external field, the Green's function is dependent on two 4-momenta κ_1 and κ_2. Such a field is represented diagrammatically by

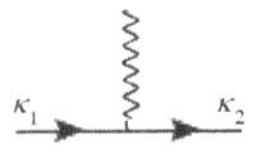

FIGURE 10.17 Green's function dependent on two four-momenta in the presence of the external field.

The wiggly line is the external field

$$-i\delta\hat{U}(\kappa_2,\kappa_1) = -i\int \exp\{i\kappa_2 x\}\delta\hat{U}\exp\{-ix\kappa_1\}d^4x \tag{1717}$$

The first-order correction to the exact Green's function is represented by a sum of two skeleton diagrams:

$i\delta\hat{U}(\kappa_2,\kappa_1) =$ 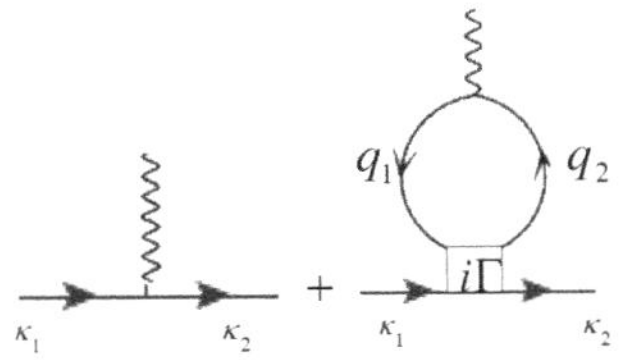 (1718)

FIGURE 10.18 First-order correction to the exact Green's function.

This is represented analytically by the equation

$$\delta\mathbf{G}_{\beta\alpha}(\kappa_2,\kappa_1) = \mathbf{G}_{\beta\gamma}(\kappa_2)\delta U(\kappa_2,\kappa_1)\mathbf{G}_{\gamma\alpha}(\kappa_1) - \\ - i\mathbf{G}_{\beta\gamma}(\kappa_2)\mathbf{G}_{\eta\alpha}(\kappa_1)\int \Gamma_{\gamma\delta,\eta\xi}(\kappa_2,q_1,p_1,q_2)\delta U(q_2,q_1)\mathbf{G}_{\xi\mu}(q_2)\mathbf{G}_{\mu\delta}(q_1)\frac{d^4q_1}{(2\pi)^4} \tag{1719}$$

and

$$\kappa_2 + q_1 = \kappa_1 + q_2 \tag{1720}$$

If we apply the following gauge transformation to a time-dependent Schrödinger equation:

$$\hat{\Psi}_\alpha(x) = \hat{\Psi}'_\alpha(x)\exp\{-i\varphi(x)\} \quad , \quad \hat{\Psi}_\alpha^\dagger(x) = \hat{\Psi}'^\dagger_\alpha(x)\exp\{i\varphi(x)\} \tag{1721}$$

For an infinitesimal change $\delta\varphi$, then

$$\hat{\Psi}'_{\alpha} = \hat{\Psi}_{\alpha}(1+i\delta\varphi) \quad , \quad \hat{\Psi}'^{\dagger}_{\alpha} = \hat{\Psi}^{\dagger}_{\alpha}(1-i\delta\varphi) \tag{1722}$$

Here, $\varphi(x)$ is a real-valued function. From this gauge transformation, then

$$\Delta \to (\nabla - i\nabla\varphi)^2 \quad , \quad \frac{\partial}{\partial t} \to \frac{\partial}{\partial t} - i\frac{\partial\varphi}{\partial t} \tag{1723}$$

and for $\delta\varphi(x)$, we have the **external field**:

$$\delta\hat{U} = -\frac{\partial\delta\varphi}{\partial t} + \frac{i}{2m}(\Delta\delta\varphi + 2\nabla\delta\varphi\cdot\nabla\varphi) \tag{1724}$$

But

$$\delta\varphi(x) = \mathrm{Re}\,\varphi_0 \exp\{-ikx\} \quad , \quad k = \left(\omega, \vec{k}\right) \tag{1725}$$

then

$$\delta U(\kappa_2,\kappa_1) = i(2\pi)^4 \varphi_0 \delta^{(4)}(\kappa_2-\kappa_1-k)\left[\omega - \frac{1}{2m}\vec{k}(\vec{\kappa}_2+\vec{\kappa}_1)\right] \tag{1726}$$

Considering equations (1721) and (1722), the Green's function

$$\delta\mathbf{G}_{\alpha\beta}(x_1,x_2) = i\mathbf{G}_{\alpha\beta}(x_1-x_2)\left[\delta\varphi(x_1)-\delta\varphi(x_2)\right] \tag{1727}$$

The Fourier transform:

$$\delta\mathbf{G}_{\alpha\beta}(\kappa_1,\kappa_2) = \int \delta\mathbf{G}_{\alpha\beta}(x_1-x_2)\exp\{i(\kappa_2 x_1-\kappa_1 x_2)\}d^4x_1 d^4x_2 \tag{1728}$$

or

$$\delta\mathbf{G}_{\alpha\beta}(\kappa_1,\kappa_2) = i\left[\mathbf{G}_{\alpha\beta}(\kappa_1)-\mathbf{G}_{\alpha\beta}(\kappa_2)\right]\delta\varphi(\kappa_2-\kappa_1) \tag{1729}$$

Here,

$$\delta\varphi(\kappa) = \int \delta\varphi(x)\exp\{i\kappa x\}d^4x = (2\pi)^4\varphi_0\delta^{(4)}(\kappa-k) \quad , \quad \kappa = (\kappa_F, \vec{\kappa}) \tag{1730}$$

Substituting (1726) into (1719), equating the results with (1729), and considering

$$\mathbf{G}_{\alpha\beta} = \mathbf{G}\delta_{\alpha\beta} \tag{1731}$$

we have

$$\delta_{\alpha\beta}\left[\mathbf{G}(\kappa+k)-\mathbf{G}(\kappa)\right] = \mathbf{G}(\kappa+k)\mathbf{G}(\kappa)\left[\mathrm{F}\left(-\omega,+\vec{k}\right)\delta_{\alpha\beta} + i\int \Gamma_{\beta\delta,\alpha\delta}(k,\kappa,q)\mathbf{G}(q)\mathbf{G}(q-k)\mathrm{F}\left(+\omega,-\vec{k}\right)\frac{d^4q}{(2\pi)^4}\right] \tag{1732}$$

where

$$F\left(\pm\omega,\vec{k}\right) \equiv \left(\pm\omega + \frac{\mp\vec{k}\left(2\vec{\kappa}\mp\vec{k}\right)}{2m}\right) \tag{1733}$$

We note that

$$\lim_{\omega,\vec{k}\to 0}\left[\mathbf{G}(\kappa+k)-\mathbf{G}(\kappa)\right] \to \omega\frac{\partial\mathbf{G}}{\partial\kappa_F} + \vec{k}\frac{\partial\mathbf{G}}{\partial\vec{\kappa}} \tag{1734}$$

and taking the limit $\frac{k}{\omega} \to 0$, we have

$$\delta_{\alpha\beta}\frac{\partial\mathbf{G}(\kappa)}{\partial\kappa_F} = -\left\{\mathbf{G}^2(\kappa)\right\}_\omega\left[\delta_{\alpha\beta} - i\int\Gamma^{\omega}_{\beta\delta,\alpha\delta}(\kappa,q)\left\{\mathbf{G}^2(q)\right\}_\omega\frac{d^4q}{(2\pi)^4}\right] \tag{1735}$$

Here,

$$\left\{\mathbf{G}^2(\kappa)\right\}_\omega = \lim_{\omega,\vec{k}\to 0}\mathbf{G}(\kappa+k)\mathbf{G}(\kappa) \quad , \quad \frac{k}{\omega}\to 0 \tag{1736}$$

Similarly, for $\frac{\omega}{k} \to 0$, we have

$$\delta_{\alpha\beta}\frac{\partial\mathbf{G}(\kappa)}{\partial\vec{\kappa}} = \left\{\mathbf{G}^2(\kappa)\right\}_k\left[\frac{\vec{\kappa}}{m}\delta_{\alpha\beta} - i\int\Gamma^{k}_{\beta\delta,\alpha\delta}(\kappa,q)\frac{\vec{q}}{m}\left\{\mathbf{G}^2(q)\right\}_k\frac{d^4q}{(2\pi)^4}\right] \tag{1737}$$

10.6.6 Galilean Invariance

We consider the change in the Green's function when the following constant field is applied:

$$\delta\hat{U} = \delta U(\vec{r}) = U_0\exp\left\{i\vec{k}\vec{r}\right\} \tag{1738}$$

If $\vec{k} \to 0$, then this field varies slowly in space. In this case, its influence on the system is treated macroscopically. For a thermodynamic equilibrium condition,

$$\mu + \delta U = \text{constant} \tag{1739}$$

If $\vec{k} \to 0$, then the chemical potential μ changes by $-U_0$. This corresponds to the change in the Green's function:

$$\delta\mathbf{G}_{\alpha\beta}(x_1,x_2) = -U_0\delta_{\alpha\beta}\frac{\partial\mathbf{G}(x_1-x_2)}{\partial\mu} \tag{1740}$$

Consider its Fourier transform in (1729), and then

$$\delta\mathbf{G}_{\alpha\beta}(\kappa_1,\kappa_2) = -(2\pi)^4\delta^{(4)}(\kappa_2-\kappa_1)U_0\delta_{\alpha\beta}\frac{\partial\mathbf{G}(\kappa_1)}{\partial\mu} \tag{1741}$$

Substituting

$$\delta U(\kappa_1,\kappa_2)=(2\pi)^4 U_0\delta^{(4)}(\kappa_2-\kappa_1-k) \quad , \quad k=\left(0,\vec{k}\right) \tag{1742}$$

into (1719) and considering the same for $\omega, k \to 0$ or $\frac{\omega}{k} \to 0$, we have

$$\delta_{\alpha\beta}\frac{\partial \mathbf{G}(\kappa)}{\partial \mu}=-\left\{\mathbf{G}^2(\kappa)\right\}_k\left[\delta_{\alpha\beta}-i\int \Gamma^k_{\beta\delta,\alpha\delta}(\kappa,q)\left\{\mathbf{G}^2(q)\right\}_k\frac{d^4q}{(2\pi)^4}\right] \tag{1743}$$

We examine the Galilean invariance of the system by examining the Fermi liquid in a coordinate system moving with a small velocity:

$$\delta\vec{w}(t)=\vec{w}_0\exp\{-i\omega t\} \tag{1744}$$

This varies slowly with time and changes the external field:

$$\delta\hat{U}=-\delta\vec{w}\hat{\vec{\kappa}}=i\delta\vec{w}\nabla \tag{1745}$$

For the momentum representation,

$$\delta U(\kappa_1,\kappa_2)=-\vec{\kappa}_1\vec{w}_0(2\pi)^4\delta^{(4)}(\kappa_2-\kappa_1-k) \quad , \quad k=(\omega,0) \tag{1746}$$

For $\omega\to 0$, we have the Galilean transformation of the coordinate system from one inertial reference frame to another with a constant velocity $\delta\vec{w}$ and the energy changes by the quantity $-\vec{\kappa}\delta\vec{w}$ referenced to the usual Galilean formula. Therefore, the energy $\in(\kappa)$ of the elementary excitation in the liquid now becomes $\in(\kappa)-\vec{\kappa}\delta\vec{w}$ in the moving frame relative to the liquid with velocity $\delta\vec{w}$. The change in the Green's function is then

$$\delta\mathbf{G}(\kappa_F+\vec{\kappa}\delta\vec{w})=\vec{\kappa}\delta\vec{w}\frac{\partial\mathbf{G}}{\partial\kappa_F} \tag{1747}$$

and

$$\delta_{\alpha\beta}\vec{\kappa}\frac{\partial \mathbf{G}(\kappa)}{\partial \kappa_F}=-\left\{\mathbf{G}^2(\kappa)\right\}_\omega\left[\delta_{\alpha\beta}\vec{\kappa}-i\int \Gamma^k_{\beta\delta,\alpha\delta}(\kappa,q)\vec{q}\left\{\mathbf{G}^2(q)\right\}_\omega\frac{d^4q}{(2\pi)^4}\right] \tag{1748}$$

Around the Fermi surface, the Green's function:

$$\mathbf{G}^{-1}(\kappa)=\frac{\in}{Z}-\frac{\kappa_F}{Zm^*}\left(\left|\vec{\kappa}\right|-\kappa_F\right) \tag{1749}$$

So, on the Fermi surface

$$\frac{\partial\mathbf{G}^{-1}}{\partial\in}=\frac{1}{Z} \quad , \quad \frac{\partial\mathbf{G}^{-1}}{\partial\mu}=\frac{\kappa_F}{Zm^*}\frac{d\kappa_F}{d\mu} \tag{1750}$$

This permits (1735) and (1743) on the Fermi surface to have the form:

$$i\int \Gamma^{\omega}_{\beta\delta,\alpha\delta}(\kappa_F,q)\left\{\mathbf{G}^2(q)\right\}_{\omega}\frac{d^4q}{(2\pi)^4}=\left(1-\frac{1}{Z}\right)\delta_{\alpha\beta} \tag{1751}$$

$$i\int \Gamma^{k}_{\beta\delta,\alpha\delta}(\kappa_F,q)\left\{\mathbf{G}^2(q)\right\}_{k}\frac{d^4q}{(2\pi)^4}=\left(1-\frac{\kappa_F}{Zm^*}\frac{d\kappa_F}{d\mu}\right)\delta_{\alpha\beta} \tag{1752}$$

10.6.7 Response on Vector Potential

Suppose the particles have an infinitely small charge δe and let the system be placed in a magnetic field weakly homogenous in space and constant in time. It follows that we add the term $-\frac{\delta e}{c}\vec{A}$ to the momentum operator in the system's Hamiltonian determinant:

$$\vec{\kappa}\to\vec{\kappa}-\frac{\delta e}{c}\vec{A} \tag{1753}$$

Here, $\vec{A}$ is the vector potential.

The change in the Green's function:

$$\begin{aligned}\delta\mathbf{G}&=-\mathbf{G}(\kappa)\frac{\delta e}{mc}\vec{\kappa}\vec{A}\mathbf{G}(\kappa+k)+\\&+\frac{i}{2}\mathbf{G}(\kappa)\mathbf{G}(\kappa+k)\int\Gamma_{\alpha\beta,\alpha\beta}(\kappa,q,k)\mathbf{G}(q)\frac{\delta e}{mc}\vec{q}\vec{A}\mathbf{G}(q+k)\frac{d^4q}{(2\pi)^4}\quad,\quad k=\left(\vec{k},0\right)\end{aligned} \tag{1754}$$

In diagrammatic form, we have

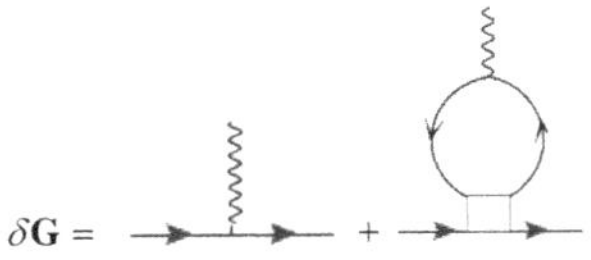

(1755)

FIGURE 10.19 Diagrammatic representation of change in the Green's function.

In the limit $\vec{k}\to 0$, from the gauge invariance, all functions depending on momenta must go over into functions of $\vec{\kappa}-\frac{\delta e}{c}\vec{A}$:

$$\frac{\delta\mathbf{G}}{\frac{\delta e}{c}\vec{A}}=-\frac{\partial\mathbf{G}(\kappa)}{\partial\vec{\kappa}} \tag{1756}$$

So, in the limit

$$\vec{k}\to 0\quad,\quad\delta e\to 0 \tag{1757}$$

we have to find a second relation for $\mathbf{G}(p)$ near the pole:

$$\frac{\partial}{\partial\vec{\kappa}}\mathbf{G}^{-1}=-\frac{\vec{\kappa}}{m^{*}Z}=-\frac{\vec{\kappa}}{m}+\frac{i}{2}\int\Gamma^{k}_{\alpha\beta,\alpha\beta}(\kappa,q)\frac{\vec{q}}{m}\left\{\mathbf{G}^{2}(q)\right\}_{k}\frac{d^{4}q}{(2\pi)^{4}}=-\frac{\vec{\kappa}}{m}\left(1+\frac{\partial\Sigma}{\partial\in(\vec{\kappa})}\right) \tag{1758}$$

or

$$\frac{\partial}{\partial\vec{\kappa}}\mathbf{G}^{-1}=-\frac{\vec{\kappa}}{m}\left(1+\frac{\partial\Sigma}{\partial\in(\vec{\kappa})}\right)=-\vec{\kappa}\frac{\left(1+\frac{\partial\Sigma}{\partial\in(\vec{\kappa})}\right)}{\frac{m}{1-\frac{\partial\Sigma}{\partial\in(\vec{\kappa})}}\left(1-\frac{\partial\Sigma}{\partial\in(\vec{\kappa})}\right)}=-\frac{1}{m^{*}Z}\vec{\kappa} \tag{1759}$$

We examine the change of the Green's function when the entire system moves with a small, slowly varying velocity $\delta\vec{w}(t)$. In this case, the system's Hamiltonian changes by the following additional term:

$$\delta H=-\delta\vec{w}(t)\cdot\hat{\vec{\mathrm{P}}}=-\delta\vec{w}(t)\int\psi^{\dagger}_{\alpha}(\vec{r})\hat{\vec{\kappa}}\psi_{\alpha}(\vec{r})d\vec{r} \tag{1760}$$

Here, $\hat{\vec{\mathrm{P}}}$, is the exact system's momentum operator. The corresponding change in the Green's function:

$$\begin{aligned}\delta\mathbf{G}=&-\mathbf{G}(\kappa)\vec{\kappa}\delta\vec{w}_{\omega}\mathbf{G}(\kappa+k)+\\&+\frac{i}{2}\mathbf{G}(\kappa)\mathbf{G}(\kappa+k)\int\Gamma_{\alpha\beta,\alpha\beta}(\kappa,q,k)\mathbf{G}(q)\vec{q}\delta\vec{w}_{\omega}\mathbf{G}(q+k)\frac{d^{4}q}{(2\pi)^{4}}\quad,\quad k=(\vec{k},0)\end{aligned} \tag{1761}$$

On the one hand, for $\omega=0$, this leads to a transformation to a coordinate system moving with a constant velocity $\delta\vec{w}$. The energy then changes by the quantity $-\vec{\mathrm{P}}\delta\vec{w}$ relative to the usual Galilean formula. Therefore, the frequency $\in$ becomes $\in+\vec{\kappa}\delta\vec{w}$, and the Green's function is changed by the following quantity:

$$\delta\mathbf{G}=\frac{\partial\mathbf{G}}{\partial\in}\vec{\kappa}\delta\vec{w} \tag{1762}$$

This implies that

$$\frac{\delta\mathbf{G}}{\delta\vec{w}}=\frac{\partial\mathbf{G}}{\partial\in}\vec{\kappa} \tag{1763}$$

Therefore, in the limit

$$\omega\to0\quad,\quad\delta\vec{w}\to0 \tag{1764}$$

we have

$$\vec{\kappa}\frac{\partial\mathbf{G}^{-1}}{\partial\in}=\frac{\vec{\mathrm{P}}}{Z}=\vec{\kappa}-\frac{i}{2}\int\Gamma^{\omega}_{\alpha\beta,\alpha\beta}(\kappa,q)\vec{q}\left\{\mathbf{G}^{2}(q)\right\}_{\omega}\frac{d^{4}q}{(2\pi)^{4}} \tag{1765}$$

near the pole.

We consider the change in the Green's function under the influence of a small field $\delta U(\vec{r})$ that is constant in time and weakly inhomogeneous in space:

$$\delta\mathbf{G} = -\mathbf{G}(\kappa)\vec{\kappa}\delta U(\vec{k})\mathbf{G}(\kappa+k)+ \\ +\frac{i}{2}\mathbf{G}(\kappa)\mathbf{G}(\kappa+k)\int\Gamma_{\alpha\beta,\alpha\beta}(\kappa,q,k)\mathbf{G}(q)\delta U(\vec{k})\mathbf{G}(q+k)\frac{d^4q}{(2\pi)^4} \quad , \quad k=(\vec{k},0) \tag{1766}$$

On the one hand, the equilibrium condition must be satisfied in a constant external field:

$$\mu+\delta U(\vec{r}) = \text{const} \tag{1767}$$

In the limit $\vec{k}\to 0$, the chemical potential changes by a small constant $-\delta U$ and therefore for

$$\vec{k}\to 0 \quad , \quad \delta U\to 0 \tag{1768}$$

we have

$$\frac{\delta\mathbf{G}}{\delta U} = -\frac{\partial\mathbf{G}}{\partial\mu} \tag{1769}$$

or

$$\frac{\partial\mathbf{G}^{-1}}{\partial\mu} = 1-\frac{i}{2}\int\Gamma^{k}_{\alpha\beta,\alpha\beta}(\kappa,q)\left\{\mathbf{G}^2(q)\right\}_k\frac{d^4q}{(2\pi)^4} \tag{1770}$$

11

Path Integral Approach to the BCS Theory

Introduction

In this chapter, we study an accurate theory of interacting Fermi mixtures with spin imbalance that shows novel features and a far richer phase diagram compared to the balanced counterpart. Ultracold quantum gases of fermionic atoms are at the crossroads of both experimental and theoretical physicists and present the possibility of experimentally studying various pairing phenomena due to the remarkable experimental control in these gases. This has related to many fundamental discoveries and, in particular, to the realization of the crossover from a **BCS** superfluid of loosely bound Cooper pairs to a **Bose-Einstein condensate (BEC)** of tightly bound molecules, the **so-called the BEC-BCS crossover**.

The **Bose-Einstein condensation** was predicted in 1924 as the outcome of Einstein extending Bose's new calculations on the statistics of a gas of identical bosons. It was not until 1995, that Cornell and Wieman [59] and Ketterle [60] succeeded in cooling a low-density gas of atoms—initially rubidium and sodium atoms—via the Bose-Einstein transition temperature. In the late 1930s, Kapitza [61] first observed the closely related phenomenon of superfluidity in a dense quantum fluid, resulting from a kind of **Bose-Einstein condensation** where interactions among the particles are enhanced.

Recently, **ultracold**, **ultradilute** gases of alkali atoms have been produced using lasers to contain a small quantity of atoms inside a magnetic trap. In laser-cooled atom traps, atoms are localized in a region of space through the Zeeman energy of interaction between the atomic spin and the external field. When the field changes direction, the **up-** and **down-spin** atoms adiabatically evolve to achieve parallel orientations with the magnetic field, and the trapping potential of the **up-spin** atoms is determined by the magnitude of the Zeeman energy. For reduced radius of the trap, the temperature of the gas is reduced while the nano-Kelvin temperature range enhances the production of the Bose-Einstein condensation in the given materials.

The Cooper pairs in the BCS regime, by definition, are formed at zero-momentum and immediately condense, defining the critical temperature T_c. The composite bosons in the BEC regime, on the other hand, are formed at a temperature T. Eventually, as the temperature is lowered, they undergo Bose-Einstein condensation at a different temperature, $T_c < T$. The eventual mechanism for condensation during the entire crossover is Bose-Einstein condensation, which happens in two profoundly different conditions, making the study of the crossover from one limit to the other a very interesting research subject. The BCS theory is one of the greatest theoretical advances in physics in the second half of the twentieth century and was first fully formulated in 1957 by Bardeen, Cooper and Schrieffer [54]. This BCS theory explains a phenomenon observed more than half a century earlier, in 1911, by Onnes [56] and gives a fully microscopic explanation to the phenomenon of pairing leading to **conventional** superconductivity, which refers to a class of superconductive materials whose behavior exhibits a wide range of shared characteristics:

- **The transition temperature T_c is much smaller than the Fermi temperature T_F.**
- **A normal electron gas is a nonsuperconducting state above T_c.**
- **No other kinds of phase transitions exist.**
- **Superconductivity is due to the formation of Cooper pairs mediated by phonon interaction.**

Our interest is only in a temperature range well below the Fermi energy $\in_F$,

$$T << T_F \tag{1771}$$

with all the relevant energy, time, and length scales controlled by the Fermi momentum κ_F of the Fermi sea in the given temperature range and the noninteracting limit. We consider the electron pairs (**quasimolecules**) as local entities in the coordinate space with their size comparable to, or less than, the interpair distance. In this regime, the superconducting state[a] is considered as a **BEC** (**Bose-Einstein Condensate**) state similar to that observed in the quantum liquid where bosonic atoms are partially condensed below the λ-point [62–64]. The characteristic size of the relative coordinate $\vec{r}$ is the lattice spacing for a good metal that is made unstable by some interaction to a many-body state for which $A^{\dagger}_{\sigma,\sigma'}(\vec{r},\vec{r}')$ achieves an expectation value that does not break translational invariance. The Bose-Einstein condensation is an elementary example of a second-order, as well as of a broken symmetry phase, transition. The same phenomenon survives in a more robust form if repulsive interactions among the Bosons are present.

A conduction electron (or hole), together with its self-induced polarization and the phonon cloud in an ionic crystal or a polar semiconductor, forms a quasiparticle that is called a **polaron** [65–68]. A **bipolaron** is a state of two polarons bound through a Coulomb force in a phonon cloud [65, 66, 69]. Here, the phonon-mediated attraction dominates the Coulomb repulsion, which thereby results in a stable state. It is suspected that these coupling mechanisms in the bipolarons might be the basis of pairing in high-temperature superconductivity [65, 66, 70, 71]. The electron-phonon interaction is shared as the underlying mechanism for superconductivity. It is important to note that creating superfluid Fermi gases in the BEC/BCS crossover regime with controlled interactions leads to the improvement of the model for high-temperature superconductivity [72, 73] where $-1 < (\kappa_F a)^{-1} < 1$ and a is the tunable s-wave fermion-fermion scattering length. The tunability of the interaction strength allows us to investigate the crossover regime and may also permit us to find the order parameters for multicomponent gases. If $a < 0$ (attraction between atoms), a cloud of ultracold fermionic atoms undergoes Cooper pairing and thus exhibits superconductivity when cooled below a critical temperature. If $a > 0$, a molecular bound state gets admixed to the scattering state; consequently, a BEC of fermionic dimmers is formed.

Considering the crossover between a BCS state and a Bose-Einstein condensate (BEC) of strongly bound pairs, the path integral approach was first applied in [74] and then recently in [9, 75]. Generally, superconductivity and magnetism do not coexist. However, changing the relative number of spin-up and spin-down electrons upsets the basic mechanism of superconductivity, where **atoms of opposite momentum and spin form Cooper pairs**. However, magnetism can be accommodated by the formation of pairs with finite momentum [76], where a strong exchange field (such as one from ferromagnetically aligned impurities in a metal) will tend to polarize the conduction electron spins. This is only feasible when the spin-exchange field is sufficiently strong compared to the energy gap for a metal superconductor. In [76], when the field is sufficiently strong enough to break many electron pairs, the self-consistent

[a] The paired fermions are not bosons in the sense that they do not obey Bose commutation relations. However, the pairs all have the same wave function. Therefore, the pairs are condensed into the same state—and precisely the **BCS** state, which is a coherent state of fermion pairs.

gap equation is modified and yields a new type of paired superconducting ground state.We idealize a spatially uniform exchange field with no scattering. In this case we observe that the paired state has a spatially dependent complex Gorkov field. This imitates a nonzero pairing momentum in the BCS model. This leads in accordance with Goldstone's theorem to an unusual anisotropic electrodynamics behavior of the superconductor, degenerate ground state as well as the low-lying collective excitations.

In quantum gases, with the help of evaporative cooling and Rabi oscillations, individual "spin-up" and "spin-down" components of the Fermi gas may be tailored independently in contrast to metals. This permits the study of Cooper pairing frustration in the spin-imbalanced Fermi gas. It is important to note that, in this case, superfluidity is ensued. In addition, for the case of the superconducting metal, the magnetic field required to provide a significant **spin-imbalanced Fermi gas** is quenched by the Meisner effect [77].[b] The quenching of electrical resistivity in superconductors is accounted for by a Cooper pair, which is a combination of two fermions that effectively behave as a **bosonic molecule** and, therefore, can undergo Bose-Einstein condensation below some critical temperature. Consequently, the superconducting electrons imitate a highly correlated matter wave and are able to propagate without being scattered by phonons.

Forming a Fermi mixture pairing is always feasible for an equal number of particles in each spin state. Nonetheless, pairing is absent for the noninteracting system with all particles in one spin state. Therefore, as a function of population imbalance, a phase transition must exist at low temperatures. Our focus here will be on spin-imbalanced Fermi gases with strong attractive interactions, for which Cooper-pair formation plays an important role. These two-component mixtures consist of identical fermionic atoms in two different hyperfine states, with each fermion having mass, m. The number of atoms for each component is allowed to be different and leads to a spin imbalance, or spin polarization. Imbalanced Fermi gases have been studied intensively in recent years, following their experimental realization in ultracold atomic Fermi gases [75]. The experimental control in such a system permits a systematic study of the equation of state and the phase diagram as a function of temperature, spin polarization, and interaction strength [78].

To keep the condensate in place, it must be trapped. Magnetic trapping is a widely used technique. The potential energy of an atom with magnetic moment, $\mathcal{M}_B$, within an external magnetic field, $\vec{H}$, is

$$\mathrm{H}_{\mathrm{int}} \equiv -\left(\mathcal{M}_B, \vec{H}\right) \tag{1772}$$

If the magnetic field is not uniform, it will exert a force on the atom:

$$\vec{F} = -grad\,\mathrm{H}_{\mathrm{int}} = grad\left(\mathcal{M}_B, \vec{H}\right) \tag{1773}$$

Generally, magnetic techniques are used to keep the bulk of the condensate in place, whereas finer manipulations are implemented using optical techniques. Sometimes the magnetic and optical trapping techniques are used together in a single device, the so-called **magneto-optical trap** (MOT) needed to realize the BCS-BEC crossover in ultracold Fermi atoms. This gives the possibility of tuning the atomic scattering length with continuity that is possible via a Feshbach resonance, which allows us to study the transition from weakly to strongly coupled systems in a controlled way. In this limit, ultracold Fermi gases are good systems for modeleling properties of strongly interacting many-body systems.

---

[b] We examine superconductivity, which is a property possessed by most metals by which they lose all measurable traces of electrical resistivity when cooled below a well-defined critical temperature T_c. This temperature depends on the external magnetic field in which the sample is placed. Simultaneously, currents are set up in the metal such that the magnetic field vanishes inside the material irrespective of the existence of an applied external field. This leads to the quenching of all magnetic field lines from the bulk of the superconductor and results in a phenomenon known as Meissner effect. It is important to note that the superconducting state exists as long as the temperature is kept below T_c and the external field is sufficiently small.

11.1 Two-Component Fermi Gas Action Functional

We will start with the microscopic partition function for an interacting Fermi mixture that consists of fermions present in two hyperfine states as **path integrals over anticommuting Grassmann variables that allow for an easier and more natural introduction of the beyond-mean-field theory of order parameter fluctuations**:

$$Z = \int d\left[\psi^\dagger\right] d\left[\psi\right] \exp\left\{-S\left[\psi^\dagger, \psi\right]\right\} \tag{1774}$$

The action functional of the fields describing the Fermi gas:

$$S\left[\psi^\dagger, \psi\right] = S_0\left[\psi^\dagger, \psi\right] + S_{\text{int}}\left[\psi^\dagger, \psi\right] \tag{1775}$$

and the interaction action functional

$$S_{\text{int}}\left[\psi^\dagger, \psi\right] = \frac{1}{2}\int d\vec{r}\, d\vec{r}\,' \sum_{\sigma,\sigma'=\uparrow,\downarrow} U_{\text{eff}\,\alpha_1\alpha_2}\left(\vec{r} - \vec{r}\,'\right) \mathrm{A}^\dagger_{\sigma,\sigma'}\left(\vec{r}, \vec{r}\,'\right) \mathrm{A}_{\sigma,\sigma'}\left(\vec{r}, \vec{r}\,'\right) \tag{1776}$$

describe a generic interaction between opposite-spin fermions scattering via a generic potential $U_{\text{eff}\,\alpha\beta}\left(\vec{r} - \vec{r}\,'\right)$ that is an effective attractive potential in some channels and the free particle action functional:

$$S_0\left[\hat{\psi}^\dagger, \hat{\psi}\right] = \int_0^\beta d\tau \int d\vec{r} \sum_\sigma \hat{\psi}^\dagger_\sigma\left(\vec{r}, \tau\right) \mathbf{G}^{-1}_{0\sigma} \hat{\psi}_\sigma\left(\vec{r}, \tau\right) \tag{1777}$$

with the bare fermionic inverse Green's function being

$$\mathbf{G}^{-1}_{0\sigma} = \frac{\partial}{\partial\tau} - \frac{\Delta}{2m} - \mu_\sigma \tag{1778}$$

We consider two different hyperfine spin states trapped so that we include a spin quantum number σ in the description. The quantity μ_σ is a spin-dependent chemical potential fixing the number of atoms of species σ, and the summations in (1776) and (1777) run over all possible indices of the Grassmann variables with each fermion having mass m. Without loss of generality, one can assume that the $\uparrow$ species is the majority component. In the aforementioned information, the field operators $\hat{\psi}^\dagger_\sigma\left(\vec{r}, \tau\right)$ and $\hat{\psi}_\sigma\left(\vec{r}, \tau\right)$ are Grassmann functions that create and annihilate a fermion at position $\vec{r}$ with pseudo-spin σ. This implies that they are anticommuting numbers. Hence, the path integral in (1774) can then be thought of as a sum over every configuration of the field, where each configuration is weighted by a factor $\exp\left\{-S\left[\psi^\dagger, \psi\right]\right\}$.

The spin species may be, for example, alkali atoms easily used (and magnetically trapped) experimentally. For alkali atoms in dilute ultracold Fermi gases, the kinetic energy is low and interactions among atoms are best described by a short-range Van der Waals potential. For dilute gases at low temperatures, spherical s-wave scattering dominates, and scattering into higher-order partial waves is negligible. Due to the short range of the interactions and the dominance of the s-wave scattering, on the lattice scale, the interactions among the fermions are well modeled by a contact interaction with a bare coupling strength λ of the fermion-fermion attractive interaction:

$$U(r) = \lambda\delta(r) \tag{1779}$$

Though the contact potential can be used to sketch a more realistic potential, it has two main drawbacks:

1. **It works only in the dilute limit, that is, only as long as the interatomic distance is greater than the scale length of the potential. We expect this approximation to break down for small distances (i.e., for large momenta).**
2. **It is not obvious how the coupling constant λ relates to the physics of the system. Intuitively, it is interaction strength, relating the intensity of the attractive coupling. But it is not obvious how it relates to observable quantities. This problem may be solved by establishing a relation between the coupling constant λ and some physically observable quantity—with the most natural being the scattering length a that can be easily tuned in with ultracold Fermi gases:**

$$\frac{m}{4\pi\hbar^2 a} = \frac{1}{\lambda} + \frac{1}{V}\sum_{\vec{\kappa}} \frac{1}{2\in_{\vec{\kappa}}} \quad , \quad \in_{\vec{\kappa}} = \frac{\vec{\kappa}^2}{2m} \tag{1780}$$

Because s-wave scattering between fermions of the same (pseudo)spin state is forbidden by the Pauli Exclusion Principle, only the s-wave scattering length a between up and down atoms is relevant in most ultracold Fermi gas experiments. So, the effect of the interactions is entirely described by the dimensionless parameter $\kappa_F a$. In this case, the interaction action functional (1776) takes the form:

$$S_{\text{int}}\left[\psi^\dagger,\psi\right] = \lambda\int d\vec{r} A^\dagger_{\sigma,\sigma'}(\vec{r},\vec{r}) A_{\sigma,\sigma'}(\vec{r},\vec{r}) \tag{1782}$$

where

$$A_{\sigma,\sigma'}(\vec{r},\vec{r}) = \hat{\psi}_\uparrow(\vec{r},\tau)\hat{\psi}_\downarrow(\vec{r},\tau) \tag{1783}$$

We investigate the Fermi gas by applying the path integral approach. We rewrite the partition function in (1774) as:

$$Z = \int d\left[\psi^\dagger\right] d\left[\psi\right] \exp\left\{-S\left[\psi^\dagger,\psi\right]\right\} \tag{1784}$$

with the action functional

$$S\left[\psi^\dagger,\psi\right] = \int_0^\beta d\tau \int d\vec{r} \sum_\sigma \hat{\psi}^\dagger_\sigma(\vec{r},\tau) G^{-1}_{0\sigma} \hat{\psi}_\sigma(\vec{r},\tau) + \lambda\int d\vec{r} A^\dagger_{\sigma,\sigma'}(\vec{r},\vec{r}) A_{\sigma,\sigma'}(\vec{r},\vec{r}) \tag{1785}$$

From the BCS theory, we consider the strength λ to be independent of wave vector (s-wave attraction) and adapt such so that the pseudo potential has the same s-wave scattering length as the true potential. We investigate the thermodynamics of the system at temperature T in a volume V via the partition function, which in path integral formalism is as in (1784) given over all Grassmann field configurations and is weighed by the action functional of the fields as in (1785).

11.2 Hubbard-Stratonovich Fields

The BCS theory can be physically interpreted as the Bose-Einstein condensation of Cooper pairs, and it follows from observation that the order parameter for the BCS transition is proportional to the expectation value of the pairing operators $A(\vec{r},\tau)$ and $A^{\dagger}(\vec{r},\tau)$. This implies that the Cooper pairs are in a coherent state analogous to a condensate of point-like bosons. To make the transition to the paired condensate, we require that the two-body interaction potential be attractive. Otherwise, the formation of pairs would not be energetically favorable. So, we consider an interaction attractive potential.

The right-hand side of (1785) contains a term that is quartic in the fermionic variables; hence, the functional integral appearing in the expression of the partition function (1784) cannot be computed exactly. To deal with this last term in (1785), the condensate of Cooper pairs may be conveniently treated with a Hubbard-Stratonovich transformation, which serves to decouple the interaction. Nonetheless, the price to pay is the introduction of an additional functional integral over auxiliary **Hubbard-Stratonovich bosonic pairing fields** [31, 32] $\Delta(\vec{r},\tau)$ and $\Delta^*(\vec{r},\tau)$, which are coupled to the Fermi fields (Grassmann variables) $\psi^{\dagger}_{\uparrow}(\vec{r},\tau)$, $\psi^{\dagger}_{\downarrow}(\vec{r},\tau)$ and $\psi_{\downarrow}(\vec{r},\tau)$, $\psi_{\uparrow}(\vec{r},\tau)$, respectively. The bosonic **pairing fields**, $\Delta(\vec{r},\tau)$ and $\Delta^*(\vec{r},\tau)$, obey symmetric boundary conditions in imaginary time; $\Delta(\vec{r},\beta)=\Delta(\vec{r},0)$ holds at every point in space and similarly for $\Delta^*(\vec{r},\tau)$. The Hubbard-Stratonovich transformation renders the quartic (Figure 11.1) interaction appearing in the original action functional because the 4-fermion vertex is now decoupled into a quadratic of fermion fields, enabling one to integrate out the fermionic degrees of freedom, as was seen earlier. The Fermi fields are antiperiodic under the shift $\tau \to \tau+\beta$. As previously discussed, the eliminating of the quartic interaction comes at the expense of introducing the new auxiliary fields $\Delta(\vec{r},\tau)$ and $\Delta^*(\vec{r},\tau)$, which are devoid of spin indices and are complex **bosonic fields.**

We resolve the four-product of the fields in the second term of (1785) in equation (1784) into a sum over fields via the **Hubbard-Stratonovich transformation**, which is a Gaussian integral in terms of the pairing operators $A(\vec{r},\tau)$ and $A^{\dagger}(\vec{r},\tau)$ [31, 32, 75] at the expense of introducing an additional functional integration:

$$\exp\left\{-S_{\text{int}}\left[\psi^{\dagger},\psi\right]\right\}=\exp\left\{-\lambda\int d\vec{r}\,A^{\dagger}_{\sigma,\sigma'}(\vec{r},\vec{r})A_{\sigma,\sigma'}(\vec{r},\vec{r})\right\}=\int d\left[\Delta^*\right]d\left[\Delta\right]\exp\left\{\mathbf{F}\left[\Delta^*,\Delta\right]\right\} \tag{1786}$$

where

$$\mathbf{F}\left[\Delta^*,\Delta\right]=\int_0^{\beta}d\tau\int d\vec{r}\left(-\frac{\Delta^*(\vec{r},\tau)\Delta(\vec{r},\tau)}{\lambda}+\left[\begin{array}{cc}\Delta^*(\vec{r},\tau) & \Delta(\vec{r},\tau)\end{array}\right]\left[\begin{array}{c}A(\vec{r},\tau)\\ A^{\dagger}(\vec{r},\tau)\end{array}\right]\right) \tag{1787}$$

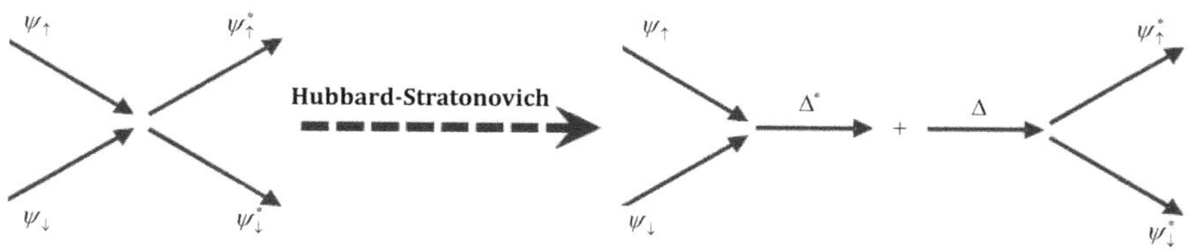

FIGURE 11.1 The Hubbard-Stratonovich transformation illustration with the interaction term originally composed of the product of four-fermion fields decomposed into a term representing two fermions pairing up, a term for the propagation of the pairs, and a term for the pair breaking up into two fermions.

Hence, we rewrite the partition function in (1774) as follows:

$$Z = \int d\left[\psi^{\dagger}\right] d[\psi] d\left[\Delta^{*}\right] d[\Delta] \exp\left\{-\int_{0}^{\beta} d\tau \int d\vec{r} \sum_{\sigma} \hat{\psi}_{\sigma}^{\dagger}(\vec{r},\tau) \mathbf{G}_{0\sigma}^{-1} \hat{\psi}_{\sigma}(\vec{r},\tau) + \mathbf{F}\left[\Delta^{*},\Delta\right]\right\} \tag{1788}$$

The fermions can easily be integrated out in (1788), which will result in a fully bosonic partition function.

Remark

We have just observed that the **Hubbard-Stratonovich transformation** renders the action quadratic in the fermion fields, enabling one to integrate out the fermionic degrees of freedom conveniently. Figure 11.1 shows the **Hubbard-Stratonovich transformation** with the selection of the decoupling of the quartic term. Performing the Hubbard-Stratonovich transformation results in the cost of introducing an additional functional integral over the pairing field (order parameter) $\Delta(\vec{r},\tau)$ in an exact manner into the many-body theory. Integrating out the fermionic fields leads to a **fully bosonic effective action functional** in equation (1788) that is dependent on only the bosonic field $\Delta(\vec{r},\tau)$, which is interpreted as the field describing the fermion pairs. In the given **effective action functional**, the pairing field $\Delta(\vec{r},\tau)$ couples to two fermionic creation fields. This shows that a pair can decay into two fermions of opposite spin, whereas $\Delta^{*}(\vec{r},\tau)$ couples to two fermionic annihilation fields, and this shows that two fermions can form a Cooper pair. The effective action functional physically can be interpreted as an interacting Bose-Fermi mixture.

11.2.1 Nambu-Gorkov Representation

To integrate out the fermionic fields, we diagonalize the action functional in equation (1788) by introducing the **Nambu-Gorkov spinors [79]**, which facilitate integrating out the resulting action functional quadratic in the fermion fields:

$$\psi(\vec{r},\tau) = \begin{bmatrix} \psi_{\uparrow}(\vec{r},\tau) \\ \psi_{\downarrow}^{*}(\vec{r},\tau) \end{bmatrix}, \quad \psi^{\dagger}(\vec{r},\tau) = \begin{bmatrix} \psi_{\downarrow}(\vec{r},\tau) & \psi_{\uparrow}^{\dagger}(\vec{r},\tau) \end{bmatrix} \tag{1789}$$

and the bosonic matrix

$$\Delta = \begin{bmatrix} 0 & \Delta(\vec{r},\tau) \\ \Delta^{*}(\vec{r},\tau) & 0 \end{bmatrix} \tag{1790}$$

This also facilitates conveniently the rewrite of $\mathbf{F}\left[\Delta^{*},\Delta\right]$ in the following matrix form:

$$\mathbf{F}\left[\Delta^{*},\Delta\right] = \int_{0}^{\beta} d\tau \int d\vec{r} \left(-\frac{\Delta^{*}(\vec{r},\tau)\Delta(\vec{r},\tau)}{|\lambda|} + \psi^{\dagger}(\vec{r},\tau)\Delta\psi(\vec{r},\tau)\right) \tag{1791}$$

Integrating out the fermionic fields in (1788) by performing the Grassmann integration:

$$Z = \int d\left[\Delta^{*}\right] d[\Delta] \int d\left[\psi^{\dagger}\right] d[\psi] \exp\left\{-S\left[\Delta^{*},\Delta,\psi^{\dagger},\psi\right]\right\} \tag{1792}$$

where the action functional

$$S\left[\Delta^*,\Delta,\psi^\dagger,\psi\right]=\int_0^\beta d\tau\int d\vec{r}\left[-\frac{\Delta^*(\vec{r},\tau)\Delta(\vec{r},\tau)}{|\lambda|}+\psi^\dagger(\vec{r},\tau)\left[-\mathbf{G}^{-1}\right]\psi(\vec{r},\tau)\right] \tag{1793}$$

with the physics of the system being encoded in the **inverse Nambu-Gorkov** 2×2 **Green's function** for the interacting fermions in spin space representation:

$$\left[-\mathbf{G}^{-1}\right]=\begin{bmatrix} \mathbf{G}_{0\uparrow}^{-1} & -\Delta(\vec{r},\tau) \\ -\Delta^*(\vec{r},\tau) & \mathbf{G}_{0\downarrow}^{-1} \end{bmatrix} \tag{1794}$$

and

$$\mathbf{G}_{0\uparrow}^{-1}=\frac{\partial}{\partial\tau}-\frac{\nabla_{\vec{r}}^2}{2m}-\mu_\uparrow \quad , \quad \mathbf{G}_{0\downarrow}^{-1}=\frac{\partial}{\partial\tau}+\frac{\nabla_{\vec{r}}^2}{2m}+\mu_\downarrow \tag{1795}$$

with m being mass of the atom. As the action in (1793) is quadratic in the fermionic fields, the corresponding functional integral is Gaussian and can be performed exactly:

$$\int d\left[\psi^\dagger\right]d[\psi]\exp\left\{-\int_0^\beta d\tau\int d\vec{r}\,\psi^\dagger(\vec{r},\tau)\left[-\mathbf{G}^{-1}\right]\psi(\vec{r},\tau)\right\}=\prod_{\vec{r},\tau}(-1)\det_\sigma\left[-\mathbf{G}^{-1}\right] \tag{1796}$$

This can be rewritten as

$$\prod_{\vec{r},\tau}(-1)\det_\sigma\left[-\mathbf{G}^{-1}\right]=\prod_{\vec{r},\tau}\exp\left\{\ln\left[-\det_\sigma\left[-\mathbf{G}^{-1}\right]\right]\right\}=\exp\left\{\sum_{\vec{r},\tau}\ln\left[-\det_\sigma\left[-\mathbf{G}^{-1}\right]\right]\right\} \tag{1797}$$

or

$$\exp\left\{\sum_{\vec{r},\tau}\ln\left[-\det_\sigma\left[-\mathbf{G}^{-1}\right]\right]\right\}=\exp\left\{\mathrm{Tr}\left[\ln\left[-\det_\sigma\left[-\mathbf{G}^{-1}\right]\right]\right]\right\}\equiv\exp\left\{\int_0^\beta d\tau\int d\vec{r}\left[\ln\left[-\det_\sigma\left[-\mathbf{G}^{-1}\right]\right]\right]\right\} \tag{1798}$$

The grand-canonical partition function of the system can be written fully in terms of the bosonic field Δ:

$$Z=\int d\left[\Delta^*\right]d[\Delta]\exp\left\{-S^{\mathrm{eff}}\left[\Delta^*,\Delta\right]\right\} \tag{1799}$$

where the effective action functional

$$S^{\mathrm{eff}}\left[\Delta^*,\Delta\right]=\int_0^\beta d\tau\int d\vec{r}\left[-\frac{\Delta^*(\vec{r},\tau)\Delta(\vec{r},\tau)}{|\lambda|}-\ln\left[-\det_\sigma\left[-\mathbf{G}^{-1}(\vec{r},\tau)\right]\right]\right] \tag{1800}$$

Generally, for interactions other than the delta function, the inverse Green's matrix $\left[-\mathbf{G}^{-1}\right]$ is not diagonal in $(\vec{r},\tau)$. We should treat $\left[-\mathbf{G}^{-1}\right]$ as an operator in, for example, space representation:

$$\langle \vec{r}',\tau' | \left[-\mathbf{G}^{-1}\right] | \vec{r},\tau\rangle = \langle \vec{r}',\tau' | \vec{r},\tau\rangle \left[-\mathbf{G}^{-1}\right] = \delta(\vec{r}-\vec{r}')\delta(\tau-\tau')\left[-\mathbf{G}^{-1}\right] \tag{1801}$$

where we have the orthogonality relation

$$\langle \vec{r}',\tau' | \vec{r},\tau\rangle = \delta(\vec{r}-\vec{r}')\delta(\tau-\tau') \tag{1802}$$

from where

$$\hat{1} = \int_0^\beta d\tau \int d\vec{r} \, |\vec{r},\tau\rangle\langle \vec{r},\tau| \tag{1803}$$

It is instructive to note that for a **noninteracting system**, there will be no pairs, and we arrive at space representation of the noninteracting (inverse) Green's function describing a bare fermion

$$\left[-\mathbf{G}_0^{-1}\right] = \begin{bmatrix} \mathbf{G}_{0\uparrow}^{-1} & 0 \\ 0 & \mathbf{G}_{0\downarrow}^{-1} \end{bmatrix} \tag{1804}$$

and

$$\langle \vec{r}',\tau' | \left[-\mathbf{G}_0^{-1}\right] | \vec{r},\tau\rangle = \langle \vec{r}',\tau' | \vec{r},\tau\rangle \left[-\mathbf{G}_0^{-1}\right] = \delta(\vec{r}-\vec{r}')\delta(\tau-\tau')\left[-\mathbf{G}_0^{-1}\right] \tag{1805}$$

We therefore observe that the fermionic part has exactly the same matrix structure as the condensed Bose gas. The **self-energy** that is a function of the pairing field Δ has the form

$$\Sigma = \delta(\vec{r}-\vec{r}')\delta(\tau-\tau')\Delta \tag{1806}$$

where

$$\Sigma = \begin{bmatrix} \Sigma_{11} & \Sigma_{12} \\ \Sigma_{21} & \Sigma_{22} \end{bmatrix} \tag{1807}$$

and

$$\mathbf{G}_0^{-1} = \mathbf{G}^{-1} + \Sigma \;\rightarrow\; \mathbf{G} = \left(\mathbf{G}_0^{-1} - \Sigma\right)^{-1} \tag{1808}$$

From here

$$\mathbf{G}_0\mathbf{G}_0^{-1}\mathbf{G} = \mathbf{G}_0\left(\mathbf{G}^{-1} + \Sigma\right)\mathbf{G} \tag{1809}$$

or

$$\mathbf{G} = \mathbf{G}_0 + \mathbf{G}_0\Sigma\mathbf{G} \tag{1810}$$

and diagrammatically represented as

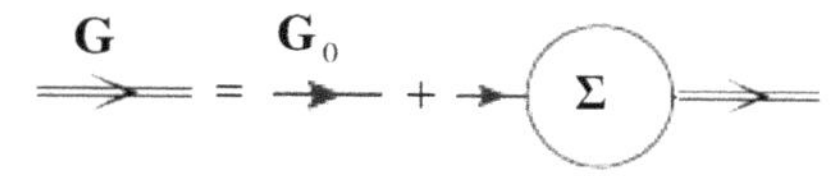

(1811)

FIGURE 11.2 Dyson equation for the full Green's function.

11.2.2 Pairing-Order Parameter Effective Action

It is important to note that $\left[-\mathbf{G}^{-1}\right]$ is dependent on $\Delta(\vec{r},\tau)$ and space time derivatives; therefore, there is difficulty in calculating the logarithm in equation (1800). It is useful to note that the partition function (1799) achieves its maximum value for configurations and minimize the action $S^{\text{eff}}\left[\Delta^*,\Delta\right]$. This gives us the so-called saddle point and permits us to expand the action about its minimum value, which results in the BCS grand potential.

11.2.3 Reciprocal Space

In order to evaluate the effective action functional $S^{\text{eff}}\left[\Delta^*,\Delta\right]$ in (1800), we evaluate first the quadratic action functional for the order parameter:

$$S_{\text{ord}}^{\text{eff}}\left[\Delta^*,\Delta\right]=-\int_0^{\beta}d\tau\int d\vec{r}\,\frac{\Delta^*(\vec{r},\tau)\Delta(\vec{r},\tau)}{|\lambda|} \tag{1812}$$

This can be done by going to the reciprocal space via the Matsubara expansion for the fermionic field:

$$\psi_\sigma(\vec{r},\tau)=\frac{1}{\sqrt{\beta V}}\sum_{n\vec{\kappa}}\exp\left\{-i\left(\omega_n\tau-\vec{\kappa}\vec{r}\right)\right\}\psi_{n\vec{\kappa}\sigma}\quad,\quad \psi_\sigma^\dagger(\vec{r},\tau)=\frac{1}{\sqrt{\beta V}}\sum_{n\vec{\kappa}}\exp\left\{i\left(\omega_n\tau-\vec{\kappa}\vec{r}\right)\right\}\psi_{n\vec{\kappa}\sigma}^\dagger \tag{1813}$$

and

$$\left\langle\vec{r},\tau\middle|\vec{\kappa},n\right\rangle=\frac{\exp\left\{i\vec{\kappa}\vec{r}\right\}}{\sqrt{V}}\frac{\exp\left\{-i\omega_n\tau\right\}}{\sqrt{\beta}} \tag{1814}$$

Here, $\omega_n=\frac{2n+1}{\beta}\pi$ is the fermionic Matsubara frequency. We require that the reciprocal space kets obey the following completeness relation:

$$\hat{1}=\sum_{m\vec{\kappa}}\left|\vec{\kappa},n\right\rangle\left\langle\vec{\kappa},n\right| \tag{1815}$$

This yields the orthonormality relation:

$$\left\langle\vec{\kappa},n\middle|\vec{\kappa}',n'\right\rangle=\delta\left(\vec{\kappa}-\vec{\kappa}'\right)\delta_{nn'} \tag{1816}$$

For the bosonic field, $\Delta(\vec{r},\tau)$ has the following Matsubara expansion

$$\Delta(\vec{r},\tau)\equiv\left\langle\vec{r},\tau\middle|\Delta\right\rangle=\sum_{\vec{q}n}\frac{\exp\left\{-i\left(\Omega_n\tau-\vec{q}\vec{r}\right)\right\}}{\sqrt{\beta V}}\Delta_{\vec{q}n}\equiv\sum_{\vec{q}n}\frac{\exp\left\{-i\left(\Omega_n\tau-\vec{q}\vec{r}\right)\right\}}{\sqrt{\beta V}}\left\langle\vec{q},n\middle|\Delta\right\rangle \tag{1817}$$

and its Fourier transform:

$$\Delta_{\vec{q}n} = \int_0^\beta d\tau \int d\vec{r} \frac{\exp\{-i(\Omega_n\tau - \vec{q}\vec{r})\}}{\sqrt{\beta V}} \Delta(\vec{r},\tau) \tag{1818}$$

with the bosonic Matsubara frequency being $\Omega_n = \frac{2n}{\beta}\pi$. From here, we have

$$\Delta^*(\vec{r},\tau) = \sum_{\vec{q}n} \frac{\exp\{i(\Omega_n\tau - \vec{q}\vec{r})\}}{\sqrt{\beta V}} \Delta^*_{\vec{q}n} \tag{1819}$$

and its Fourier transform:

$$\Delta^*_{\vec{q}n} = \int_0^\beta d\tau \int d\vec{r} \frac{\exp\{i(\Omega_n\tau - \vec{q}\vec{r})\}}{\sqrt{\beta V}} \Delta^*(\vec{r},\tau) \tag{1820}$$

We therefore evaluate (1812) via reciprocal space:

$$\int_0^\beta d\tau \int d\vec{r} \Delta^*(\vec{r},\tau)\Delta(\vec{r},\tau) = \int_0^\beta d\tau \int d\vec{r} \left(\sum_{\vec{q}n} \frac{\exp\{i(\Omega_n\tau - \vec{q}\vec{r})\}}{\sqrt{\beta V}} \Delta^*_{\vec{q}n} \right) \left(\sum_{\vec{q}'n'} \frac{\exp\{-i(\Omega_{n'}\tau - \vec{q}'\vec{r})\}}{\sqrt{\beta V}} \Delta_{\vec{q}'n'} \right) \tag{1821}$$

or

$$\int_0^\beta d\tau \int d\vec{r} \Delta^*(\vec{r},\tau)\Delta(\vec{r},\tau) = \sum_{\vec{q}\vec{q}'nn'} \frac{1}{\beta V} \left[\int_0^\beta d\tau \int d\vec{r} \exp\{i(\Omega_n - \Omega_{n'})\tau - i(\vec{q} - \vec{q}')\vec{r}\} \right] \Delta^*_{\vec{q}n}\Delta_{\vec{q}'n'} = \sum_{\vec{q}n} \Delta^*_{\vec{q}n}\Delta_{\vec{q}n} \tag{1822}$$

Now, we will evaluate the second integral in (1800), which is a quadratic action for the Grassmann variables taking into consideration relations (1817) and (1819) for the Grassmann fields:

$$\begin{aligned} &\int_0^\beta d\tau \int d\vec{r} \psi^\dagger_\uparrow(\vec{r},\tau) \mathbf{G}^{-1}_{0\uparrow} \psi_\uparrow(\vec{r},\tau) \\ &= \sum_{\vec{\kappa}n\vec{\kappa}'n'} \frac{1}{\beta V} \int_0^\beta d\tau \int d\vec{r} \exp\{i(\omega_n\tau - \vec{\kappa}\vec{r})\} \psi^\dagger_{\vec{\kappa}n\uparrow} \left[\mathbf{G}^{-1}_{0\uparrow}\right]_{\vec{\kappa}n\vec{\kappa}'n'} \exp\{-i(\omega_{n'}\tau - \vec{\kappa}'\vec{r})\} \psi_{\vec{\kappa}'n'\uparrow} \end{aligned} \tag{1823}$$

or

$$\int_0^\beta d\tau \int d\vec{r} \psi^\dagger_\uparrow(\vec{r},\tau) \mathbf{G}^{-1}_{0\uparrow} \psi_\uparrow(\vec{r},\tau) = \sum_{\vec{\kappa}n} \left[\mathbf{G}^{-1}_{0\uparrow}\right] \psi^\dagger_{\vec{\kappa}n\uparrow} \psi_{\vec{\kappa}n\uparrow} \tag{1824}$$

where

$$\left[\mathbf{G}^{-1}_{0\uparrow}\right]_{\vec{\kappa}n\vec{\kappa}'n'} = \left[-i\omega_{n'} + \frac{\vec{\kappa}'^2}{2m} - \mu_\uparrow \right] \delta_{\vec{\kappa}\vec{\kappa}'}\delta_{nn'} \quad , \quad \left[\mathbf{G}^{-1}_{0\uparrow}\right] = \left[-i\omega_n + \frac{\vec{\kappa}^2}{2m} - \mu_\uparrow \right] \tag{1825}$$

Similarly,

$$\int_0^\beta d\tau \int d\vec{r} \psi_{\downarrow}(\vec{r},\tau) \mathbf{G}_{0\downarrow}^{-1} \psi_{\downarrow}^{\dagger}(\vec{r},\tau) = \sum_{\vec{\kappa}n} \left[\mathbf{G}_{0\downarrow}^{-1}\right] \psi_{\vec{\kappa}n\downarrow} \psi_{\vec{\kappa}n\downarrow}^{\dagger} \tag{1826}$$

where

$$\left[\mathbf{G}_{0\downarrow}^{-1}\right] = \left[i\omega_n - \frac{\vec{\kappa}^2}{2m} + \mu_{\downarrow}\right] \tag{1827}$$

We consider the terms with a coupling between the order parameter and bilinear in the Grassmann variables:

$$\begin{aligned} &\int_0^\beta d\tau \int d\vec{r} \Delta(\vec{r},\tau) \psi_{\uparrow}^{\dagger}(\vec{r},\tau) \psi_{\downarrow}^{\dagger}(\vec{r},\tau) \\ &= \frac{1}{\sqrt{\beta V}} \sum_{\vec{\kappa}n\vec{\kappa}'n'} \left[\frac{1}{\sqrt{\beta V}} \int_0^\beta d\tau \int d\vec{r} \exp\left\{i(\omega_n + \omega_{n'})\tau - i(\vec{\kappa} + \vec{\kappa}')\vec{r}\right\} \Delta(\vec{r},\tau)\right] \psi_{\vec{\kappa}n\uparrow}^{\dagger} \psi_{\vec{\kappa}'n'\downarrow}^{\dagger} \end{aligned} \tag{1828}$$

or

$$\int_0^\beta d\tau \int d\vec{r} \Delta(\vec{r},\tau) \psi_{\uparrow}^{\dagger}(\vec{r},\tau) \psi_{\downarrow}^{\dagger}(\vec{r},\tau) = \frac{1}{\sqrt{\beta V}} \sum_{\vec{\kappa}n\vec{\kappa}'n'} \Delta_{\vec{\kappa}+\vec{\kappa}',n+n'} \psi_{\vec{\kappa}n\uparrow}^{\dagger} \psi_{\vec{\kappa}'n'\downarrow}^{\dagger} \tag{1829}$$

Similarly,

$$\int_0^\beta d\tau \int d\vec{r} \Delta^*(\vec{r},\tau) \psi_{\downarrow}(\vec{r},\tau) \psi_{\uparrow}(\vec{r},\tau) = \frac{1}{\sqrt{\beta V}} \sum_{\vec{\kappa}n\vec{\kappa}'n'} \Delta_{\vec{\kappa}+\vec{\kappa}',n+n'}^{*} \psi_{\vec{\kappa}n\downarrow} \psi_{\vec{\kappa}'n'\uparrow} \tag{1830}$$

From equations (1824) to (1830) and (1822), the full partition function (1792) has the Grassmann-path integral representation:

$$Z = \int d\left[\psi^{\dagger}\right] d\left[\psi\right] \int d\left[\Delta_{\vec{q}m}^{*}\right] d\left[\Delta_{\vec{q}m}\right] \exp\left\{-S\left[\Delta^*, \Delta, \psi^{\dagger}, \psi\right]\right\} \tag{1831}$$

or

$$Z = \int d\left[\psi_{\vec{\kappa}'n'}^{\dagger}\right] d\left[\psi_{\vec{\kappa}''n''}\right] \int d\left[\Delta_{\vec{q}m}^{*}\right] d\left[\Delta_{\vec{q}m}\right] \exp\left\{-S\left[\Delta^*, \Delta, \psi^{\dagger}, \psi\right]\right\} \tag{1832}$$

with the action functional in momentum space being

$$S\left[\Delta^*, \Delta, \psi^{\dagger}, \psi\right] = -\sum_{\vec{q}m} \frac{\Delta_{\vec{q}m}^{*} \Delta_{\vec{q}m}}{|\lambda|} + \sum_{\vec{\kappa}'n'\vec{\kappa}''n''} \psi_{\vec{\kappa}'n'}^{\dagger} \left[-\mathbf{G}^{-1}\right] \psi_{\vec{\kappa}''n''} \tag{1833}$$

and the Nambu-Gorkov spinors:

$$\psi_{n,\vec{\kappa}} = \begin{bmatrix} \psi_{n,\vec{\kappa},\uparrow} \\ \psi^{*}_{n,\vec{\kappa},\downarrow} \end{bmatrix}, \quad \psi^{\dagger}_{n,\vec{\kappa}} = \begin{bmatrix} \psi_{n,\vec{\kappa},\downarrow}(\vec{r},\tau) & \psi^{\dagger}_{n,\vec{\kappa},\uparrow} \end{bmatrix} \tag{1834}$$

and

$$\left[-\mathbf{G}^{-1}\right] \equiv \left\langle \vec{\kappa}'n' \right| \left[-\mathbf{G}^{-1}\right] \left| \vec{\kappa}''n'' \right\rangle \tag{1835}$$

Hence, the integration over the $\psi_{\vec{\kappa}''n''}$ and $\psi^{\dagger}_{\vec{\kappa}'n'}$ fields is Gaussian and can be carried out exactly using the relation

$$\int d\left[\psi^{\dagger}_{\vec{\kappa}'n'}\right] d\left[\psi_{\vec{\kappa}''n''}\right] \exp\left\{-\sum_{\vec{\kappa}'n'\vec{\kappa}''n''} \psi^{\dagger}_{\vec{\kappa}'n'}\left[-\mathbf{G}^{-1}\right]\psi_{\vec{\kappa}''n''}\right\} = \det\left[-\mathbf{G}^{-1}\right] \tag{1836}$$

Where, as shown earlier,

$$\det\left[-\mathbf{G}^{-1}\right] = \exp\left\{\ln\left[-\det\left[-\mathbf{G}^{-1}\right]\right]\right\} = \exp\left\{\mathrm{Tr}\left[\ln\left[-\det\left[-\mathbf{G}^{-1}\right]\right]\right]\right\} \tag{1837}$$

The det and Tr are taken on the spinor degrees of freedom as well on the momentum and frequency; then, the partition function is written in an exact form as a path integral over the $\Delta_{\vec{q}m}$ and $\Delta^{*}_{\vec{q}m}$ fields:

$$Z = \int d\left[\Delta^{*}_{\vec{q}m}\right] d\left[\Delta_{\vec{q}m}\right] \exp\left\{-S^{\mathrm{eff}}\left[\Delta^{*},\Delta\right]\right\} \tag{1838}$$

Here, the effective action functional is

$$S^{\mathrm{eff}}\left[\Delta^{*},\Delta\right] = -\sum_{\vec{q}m} \frac{\Delta^{*}_{\vec{q}m}\Delta_{\vec{q}m}}{|\lambda|} + \mathrm{Tr}\left[\ln\left[-\det\left[-\mathbf{G}^{-1}\right]\right]\right] \tag{1839}$$

Considering (1816), the reciprocal space representation of the inverse Green's function $\left[-\mathbf{G}^{-1}\right]$ can be rewritten

$$-\mathbf{G}^{-1} = -\mathbf{G}_0^{-1} + \mathbf{F} \tag{1840}$$

The inverse bare Green's function

$$-\mathbf{G}_0^{-1} = \delta(\vec{\kappa} - \vec{\kappa}')\delta_{nn'} \begin{bmatrix} \left[\mathbf{G}^{-1}_{0\uparrow}\right] & 0 \\ 0 & \left[\mathbf{G}^{-1}_{0\downarrow}\right] \end{bmatrix} \tag{1841}$$

is the noninteracting part that is diagonal, while the pairing term

$$\mathbf{F}=\frac{1}{\sqrt{\beta V}}\begin{bmatrix} 0 & -\Delta_{\vec{\kappa}+\vec{\kappa}',n+n'} \\ -\Delta^*_{\vec{\kappa}+\vec{\kappa}',n+n'} & 0 \end{bmatrix} \tag{1842}$$

relating the fermion pair field in reciprocal space is nondiagonal. It is worthy to note that the quantity $\left[-\mathbf{G}^{-1}\right]$ encodes the full physics of the system, and different approximation schemes will correspond to its different choices.

From the effective action functional in (1839) and (1840), we have

$$\begin{aligned}\operatorname{Tr}\left[\ln\left[-\det\left[-\mathbf{G}_0^{-1}+\mathbf{F}\right]\right]\right]&=\operatorname{Tr}\left[\ln\left[-\det\left[-\mathbf{G}_0^{-1}\left(1+\mathbf{G}_0\mathbf{F}\right)\right]\right]\right]\\&=\operatorname{Tr}\ln\left[-\det\left[-\mathbf{G}_0^{-1}\right]\right]+\operatorname{Tr}\left[\ln\left[-\det\left(1+\mathbf{G}_0\mathbf{F}\right)\right]\right]\end{aligned} \tag{1843}$$

With the help of the expansion for the logarithmic function

$$\ln(1-x)=-\sum_{n=1}^{\infty}\frac{x^n}{n} \tag{1844}$$

then

$$\operatorname{Tr}\left[\ln\left[-\det\left[-\mathbf{G}_0^{-1}+\mathbf{F}\right]\right]\right]=\operatorname{Tr}\ln\left[-\det\left[-\mathbf{G}_0^{-1}\right]\right]+\sum_{n=1}^{\infty}\frac{1}{n}\operatorname{Tr}\left[-\det\left(\mathbf{G}_0\mathbf{F}\right)\right]^n \tag{1845}$$

Thus, the action in terms of an expansion in powers of the pairing field is

$$S^{\text{eff}}\left[\Delta^*,\Delta\right]=-\sum_{\vec{q}m}\frac{\Delta^*_{\vec{q}m}\Delta_{\vec{q}m}}{|\lambda|}+\operatorname{Tr}\ln\left[-\det\left[-\mathbf{G}_0^{-1}\right]\right]+\sum_{n=1}^{\infty}\frac{1}{n}\operatorname{Tr}\left[-\det\left(\mathbf{G}_0\mathbf{F}\right)\right]^n \tag{1846}$$

This is the pivot in many different approaches beyond the mean field treatment of ultracold Fermi gases. An example is found in the Ginzburg-Landau theory. Note that our result still contains a path integral over the bosonic pair fields $\Delta_{n,\vec{\kappa}}$ and $\Delta_{n,\vec{\kappa}}$ that cannot be done analytically. The saddle point, which is a mean field approximation and valid only if fluctuation effects are small or if the coherence length is long [48, 75, 80], will be very appropriate.

11.3 Saddle-Point Approximation

The saddle-point approximation is a loop expansion in the language of Feynman diagrams and is equivalent to minimizing the free energy of a spatially homogeneous system. Because the true system is not homogenous, this is only an approximation. We investigate the superfluid state, where we assume the bosonic pair fields $\Delta^*_{n,\vec{\kappa}}$ and $\Delta_{n,\vec{\kappa}}$ to be Bose condensed with the contribution only from the $\vec{\kappa}$, $n=0$ (and requiring that pairing must happen in a zero-momentum state). This term is dominant and imitates the assumption of Bogoliubov for the case of superfluid helium [48, 75]:

$$\Delta_{n,\vec{\kappa}}=\sqrt{\beta V}\delta(\vec{\kappa})\delta_{n,0}\times\Delta \quad , \quad \Delta^*_{n,\vec{\kappa}}=\sqrt{\beta V}\delta(\vec{\kappa})\delta_{n,0}\times\Delta^* \tag{1847}$$

This approximation will greatly simplify the evaluation of the Grassmann integration over the fermionic fields. The factor $\sqrt{\beta V}$ is introduced for ease of calculation and to give Δ the energy unit. We use the property of inversion symmetry by reindexing the indices in the spin-down, two-particle term as $\vec{\kappa} \to -\vec{\kappa}, n \to -n$ and observe that none of the terms $\Delta_{\vec{\kappa}+\vec{\kappa}',n+n'}$ and $\Delta^*_{\vec{\kappa}+\vec{\kappa}',n+n'}$ with $\vec{\kappa}+\vec{\kappa}' \neq 0$ or $n+n' \neq 0$ survives due to the delta functions in equations (1847). Consequently, the saddle-point partition function:

$$Z_{sp} = \int d\left[\psi^{\dagger}_{\vec{\kappa}m\sigma}\right] d\left[\psi_{\vec{\kappa}m\sigma}\right] \exp\left\{-S_{sp}\left[\Delta^*, \Delta, \psi^{\dagger}, \psi\right]\right\} \tag{1848}$$

where

$$S_{sp}\left[\Delta^*, \Delta, \psi^{\dagger}, \psi\right] = -\frac{\beta V}{|\lambda|}|\Delta|^2 + \sum_{\vec{\kappa}n} \boldsymbol{\psi}^{\dagger}_{\vec{\kappa}n} \cdot \left[-\mathbf{G}^{-1}_{sp}\right]_{\vec{\kappa}n} \cdot \boldsymbol{\psi}_{\vec{\kappa}n} \tag{1849}$$

with

$$\left[-\mathbf{G}^{-1}_{sp}\right]_{\vec{\kappa}n} = \begin{bmatrix} \left[\mathbf{G}^{-1}_{0\uparrow}\right] & -\Delta \\ -\Delta^* & \left[\mathbf{G}^{-1}_{0\downarrow}\right] \end{bmatrix}_{\vec{\kappa}n} \tag{1850}$$

and we introduce the Nambu-Gorkov spinors

$$\boldsymbol{\psi}_{\vec{\kappa}n} = \begin{bmatrix} \psi_{\vec{\kappa}n\uparrow} \\ \psi^{\dagger}_{-\vec{\kappa},-n,\downarrow} \end{bmatrix}, \quad \boldsymbol{\psi}^{\dagger}_{\vec{\kappa}n} = \begin{bmatrix} \psi^{\dagger}_{\vec{\kappa}n\uparrow} & \psi_{-\vec{\kappa},-n,\downarrow} \end{bmatrix} \tag{1851}$$

Integration (1848) occurs over the $\boldsymbol{\psi}_{\vec{\kappa}n}$ and $\boldsymbol{\psi}^{\dagger}_{\vec{\kappa}n}$ fields and then the partition function is written as a path integral over the Δ and Δ^* fields:

$$Z = \int d\left[\Delta^*_{\vec{q}m}\right] d\left[\Delta_{\vec{q}m}\right] Z_{sp}\left[\Delta^*, \Delta\right] \tag{1852}$$

where

$$Z_{sp}\left[\Delta^*, \Delta\right] = \exp\left\{-S^{\mathrm{eff}}\left[\Delta^*, \Delta\right]\right\} \tag{1853}$$

and then the effective action can be written

$$S^{\mathrm{eff}}\left[\Delta^*, \Delta\right] = -\frac{\beta V}{|\lambda|}|\Delta|^2 - \sum_{\vec{\kappa}n} \ln\left[-\det{}_{\sigma}\left[-\mathbf{G}^{-1}_{sp}\right]_{\vec{\kappa}n}\right] \tag{1854}$$

and

$$-\det\left[-\mathbf{G}^{-1}_{sp}\right]_{\vec{\kappa}n} = -\det \begin{bmatrix} \left[\mathbf{G}^{-1}_{0\uparrow}\right] & -\Delta \\ -\Delta^* & \left[\mathbf{G}^{-1}_{0\downarrow}\right] \end{bmatrix} = \mathbf{G}^{-1}_{0\uparrow}\mathbf{G}^{-1}_{0\downarrow} + |\Delta|^2 \tag{1855}$$

where

$$\mathbf{G}_{0\uparrow}^{-1} = i\omega_n - \varsigma_{\vec{\kappa}\uparrow} \quad , \quad \mathbf{G}_{0\downarrow}^{-1} = i\omega_n - \varsigma_{\vec{\kappa}\downarrow} \quad , \quad \varsigma_{\vec{\kappa}\sigma} = \frac{\vec{\kappa}^2}{2m} - \mu_\sigma \tag{1856}$$

Setting the averaged chemical potential as the half-sum

$$\mu = \frac{\mu_\uparrow + \mu_\downarrow}{2} \tag{1857}$$

will determine the total number of fermions and the chemical potential imbalance as half-difference

$$\varsigma = \frac{\mu_\uparrow - \mu_\downarrow}{2} \tag{1858}$$

Without loss of generality, we assume that the $\uparrow$ species is the majority component and that the fermion energy measured from the chemical potential μ as

$$\varsigma_{\vec{\kappa}} = \frac{\vec{\kappa}^2}{2m} - \mu \equiv \epsilon_{\vec{\kappa}} - \mu \tag{1859}$$

and

$$E_{\vec{\kappa}} = \sqrt{\varsigma_{\vec{\kappa}}^2 + |\Delta|^2} \tag{1860}$$

The Bogoliubov dispersion is the quasi-particle energy and Δ plays the role of the order parameter or the gap at the Fermi surface. So, Δ should be called the BCS **gap parameter** and

$$-\det\left[-\mathbf{G}_{\text{sp}}^{-1}\right]_{\vec{\kappa}n} = -\left(-i\omega_n + \varsigma_{\vec{\kappa}} - \varsigma\right)\left(i\omega_n - \varsigma_{\vec{\kappa}} + \varsigma\right) - |\Delta|^2 \equiv \left(i\omega_n + \varsigma - E_{\vec{\kappa}}\right)\left(-i\omega_n - \varsigma - E_{\vec{\kappa}}\right) \tag{1861}$$

Therefore, the saddle-point action function in equation (1854) becomes

$$S^{\text{eff}}\left[\Delta^*, \Delta\right] = -\frac{\beta V}{|\lambda|}|\Delta|^2 - \sum_{\vec{\kappa}n} \ln\left[\left(i\omega_n + \varsigma - E_{\vec{\kappa}}\right)\left(-i\omega_n - \varsigma - E_{\vec{\kappa}}\right)\right] \tag{1862}$$

or

$$S^{\text{eff}}\left[\Delta^*, \Delta\right] = -\frac{\beta V}{|\lambda|}|\Delta|^2 - \sum_{\vec{\kappa}n} \ln\left[\left(i\nu_n - E_{\vec{\kappa}}\right)\left(-i\nu_n - E_{\vec{\kappa}}\right)\right] \tag{1863}$$

Here, $i\nu_n = i\omega_n + \varsigma$ is the Matsubara Fermi frequency shift.

Finally, in terms of the inverse Green's function, the approximation is then achieved as:

$$\left[-\mathbf{G}^{-1}\right] \to \left[-\mathbf{G}_{\text{sp}}^{-1}\right] \tag{1864}$$

This completely neglects the fluctuation contribution coming from $\mathbf{F}$. Hence, the action functional (1854) is greatly simplified because the $\Delta_{\vec{q}m}$ field loses the dynamics, and it is replaced by its saddle-point value Δ. That is why it is no longer necessary to integrate over the bosonic fields $\Delta_{\vec{q}m}$ in (1852). Calculating the mean-field thermodynamic potential from the effective action via the formula is straightforward:

$$Z_{\text{sp}} = \exp\left\{\beta V \Omega_{\text{sp}}(T, V, \varsigma, \mu)\right\} \tag{1865}$$

Here, the effective or saddle-point grand thermodynamic potential per unit volume $\Omega_{\text{sp}}(T, V, \varsigma, \mu)$:

$$\Omega_{\text{sp}}(T, V, \varsigma, \mu) = -\frac{|\Delta|^2}{|\lambda|} - \frac{1}{\beta V} \sum_{\vec{\kappa}n} \ln\left[(i\nu_n - \text{E}_{\vec{\kappa}})(-i\nu_n - \text{E}_{\vec{\kappa}})\right] \tag{1866}$$

The gap parameter Δ for the imbalanced Fermi gas is determined via the minimization of the saddle-point thermodynamic potential $\Omega_{\text{sp}}(T, V, \varsigma, \mu)$ as a function of Δ at given T, ς, μ. However, for a complete determination of thermodynamic parameters at a given temperature, the minimum condition for the thermodynamic potential should be solved jointly with the number equations. The gap equation is obtained by imposing the saddle-point condition on the grand potential:

$$\frac{\partial \Omega_{\text{sp}}\left[\Delta^*, \Delta\right]}{\partial \Delta} = \frac{\partial \Omega_{\text{sp}}\left[\Delta^*, \Delta\right]}{\partial \Delta^*} = 0 \tag{1867}$$

or

$$\frac{\partial \Omega_{\text{sp}}\left[\Delta^*, \Delta\right]}{\partial \Delta} = -\frac{2\Delta \operatorname{sgn} \Delta}{|\lambda|} - \Pi \Delta \tag{1868}$$

where

$$\Pi = \frac{2}{\beta V} \sum_{\vec{\kappa}n} \frac{1}{(i\nu_n - \text{E}_{\vec{\kappa}})(-i\nu_n - \text{E}_{\vec{\kappa}})} \tag{1869}$$

is the polarization function for the Cooper pair, which consists of the product of two fermionic propagators. The factor of two in the polarization operator is due to the spin. Relation (1869) is also the **(causal) density-density propagator** for a noninteracting Fermi gas and contains information about the actual excitation spectrum $\text{E}_{\vec{\kappa}}$ of the Fermi gas that has a fixed number of particles.

From (1868), the order parameter Δ may relate the coupling strength λ:

$$\frac{1}{|\lambda|} = -\frac{1}{2} \Pi \operatorname{sgn} \Delta \tag{1870}$$

This coupling strength is observed to be energy dependent. It is instructive to note that the coupling strength relates to the contact potential in real space via

$$U(\vec{r} - \vec{r}') = \lambda \delta(\vec{r} - \vec{r}') \tag{1871}$$

As mentioned, earlier, this relation leads to an ultraviolet divergence in three dimensions, and we have to renormalize the bare interaction λ to remove the divergence at high momentum. This renormalization is equivalent to introducing the zero-range pseudo potential [81]:

$$U(\vec{r}) = \lambda \delta(\vec{r}) \frac{\partial}{\partial \vec{r}} \vec{r} \tag{1872}$$

So, the regularized form of the coupling constant

$$\frac{m}{4\pi\hbar^2 a} = \frac{1}{\lambda} + \frac{1}{V}\sum_{\vec{\kappa}} \frac{1}{2\in_{\vec{\kappa}}} \tag{1873}$$

To obtain the thermodynamic potential of the sum over the Matsubara frequencies in (1866), we consider the following

$$v_n^2 = (\omega_0 n + \alpha)^2 \quad , \quad \alpha = \frac{a}{\beta} \quad , \quad a = \frac{1-\chi}{2}\pi - i\beta\varsigma \quad , \quad \omega_0 = \frac{2\pi}{\beta} \tag{1874}$$

where $\chi = -1$ for fermions and $\chi = +1$ for bosons. The parameter χ defines the system exhibiting a fermionic or bosonic behavior. So, the polarization function for the Cooper pair

$$\Pi = \frac{2}{\beta V}\sum_{\vec{\kappa}n} \frac{1}{(i\mathrm{v}_n - \mathrm{E}_{\vec{\kappa}})(-i\mathrm{v}_n - \mathrm{E}_{\vec{\kappa}})} = \frac{2}{\beta V}\sum_{\vec{\kappa}n} \frac{1}{\mathrm{v}_n^2 + (\mathrm{E}_{\vec{\kappa}})^2} = \frac{2}{\beta V}\sum_{\vec{\kappa}n} \frac{1}{(\omega_0 n + \alpha)^2 + (\mathrm{E}_{\vec{\kappa}})^2} \tag{1875}$$

or

$$\Pi = \frac{2}{\beta V}\sum_{\vec{\kappa}n} \mathbf{G}_0(\vec{\kappa}, i\omega_n)\mathbf{G}_0(\vec{\kappa}, -i\omega_n) \tag{1876}$$

It is important to note that the Matsubara summations in (1866) are convergent and that the summation can be done by transforming an infinite series to a contour integration in the complex plane and deforming it.

We apply the sum over Matsubara frequencies considering that we sum up terms with all n that are complex. The terms in this sum can be reordered and, in particular, we set

$$n' = -1 - n \tag{1877}$$

and sum over all n'. Consequently,

$$i\omega_{n'} = i\omega_{-n-1} = -i\omega_n \tag{1878}$$

which holds the general Matsubara summation:

$$\frac{1}{\beta}\sum_{n=-\infty}^{\infty} \mathrm{f}(i\omega_n) = \frac{1}{\beta}\sum_{n=-\infty}^{\infty} \mathrm{f}(-i\omega_n) \tag{1879}$$

Therefore, representing the polarization function in (1875) also as

$$\Pi = \frac{2}{\beta V}\sum_{\vec{\kappa},n} \frac{1}{(\omega_0 n + \alpha)^2 + (\mathrm{E}_{\vec{\kappa}})^2} = \frac{1}{\beta V}\sum_{\vec{\kappa}} \frac{1}{\mathrm{E}_{\vec{\kappa}}}\sum_{n=-\infty}^{\infty}\left[\frac{1}{i\omega_0 n + i\alpha - \mathrm{E}_{\vec{\kappa}}} - \frac{1}{i\omega_0 n + i\alpha + \mathrm{E}_{\vec{\kappa}}}\right] \tag{1880}$$

with the help of the infinite series expansion [82] of $\coth(\pi y)$, then:

$$\frac{1}{\mathrm{E}_{\vec{\kappa}}}\sum_{n=-\infty}^{\infty}\left[\frac{1}{i\omega_0 n + i\alpha - \mathrm{E}_{\vec{\kappa}}} - \frac{1}{i\omega_0 n + i\alpha + \mathrm{E}_{\vec{\kappa}}}\right] = \frac{1}{\omega_0 \mathrm{E}_{\vec{\kappa}}}\sum_{n=-\infty}^{\infty}\left[\frac{1}{in + \dfrac{i\alpha - \mathrm{E}_{\vec{\kappa}}}{\omega_0}} - \frac{1}{in + \dfrac{i\alpha + \mathrm{E}_{\vec{\kappa}}}{\omega_0}}\right] \tag{1881}$$

or

$$\frac{1}{E_{\vec{\kappa}}}\sum_{n=-\infty}^{\infty}\left[\frac{1}{i\omega_0 n+i\alpha-E_{\vec{\kappa}}}-\frac{1}{i\omega_0 n+i\alpha+E_{\vec{\kappa}}}\right]=\frac{\pi}{\omega_0 E_{\vec{\kappa}}}\left[\coth\frac{\pi}{\omega_0}(i\alpha-E_{\vec{\kappa}})-\coth\frac{\pi}{\omega_0}(i\alpha+E_{\vec{\kappa}})\right] \quad (1882)$$

Therefore,

$$\Pi=\frac{1}{\beta V}\sum_{\vec{\kappa}}\frac{\pi}{\omega_0 E_{\vec{\kappa}}}\left[\coth\frac{\pi}{\omega_0}(i\alpha-E_{\vec{\kappa}})-\coth\frac{\pi}{\omega_0}(i\alpha+E_{\vec{\kappa}})\right] \quad (1883)$$

or

$$\Pi=\frac{1}{V}\sum_{\vec{\kappa}}\frac{1}{2E_{\vec{\kappa}}}\left[\coth\frac{\beta}{2}(i\alpha-E_{\vec{\kappa}})-\coth\frac{\beta}{2}(i\alpha+E_{\vec{\kappa}})\right] \quad (1884)$$

For the case of fermions, $\chi=-1$, and then

$$\Pi=\frac{1}{V}\sum_{\vec{\kappa}}\frac{1}{2E_{\vec{\kappa}}}\left[\coth\frac{\beta}{2}(\varsigma-E_{\vec{\kappa}})-\coth\frac{\beta}{2}(\varsigma+E_{\vec{\kappa}})\right] \quad (1885)$$

But,

$$\coth\frac{\beta}{2}(\varsigma-E_{\vec{\kappa}})-\coth\frac{\beta}{2}(\varsigma+E_{\vec{\kappa}})=\frac{1}{\tanh\frac{\beta}{2}(\varsigma-E_{\vec{\kappa}})}-\frac{1}{\tanh\frac{\beta}{2}(\varsigma+E_{\vec{\kappa}})}=\frac{\sinh\beta E_{\vec{\kappa}}}{\cosh\frac{\beta}{2}(\varsigma+E_{\vec{\kappa}})\cosh\frac{\beta}{2}(\varsigma-E_{\vec{\kappa}})} \quad (1886)$$

and

$$2\cosh\frac{\beta}{2}(\varsigma+E_{\vec{\kappa}})\cosh\frac{\beta}{2}(\varsigma-E_{\vec{\kappa}})=\cosh\beta\varsigma+\cosh\beta E_{\vec{\kappa}} \quad (1887)$$

So,

$$\coth\frac{\beta}{2}(\varsigma-E_{\vec{\kappa}})-\coth\frac{\beta}{2}(\varsigma+E_{\vec{\kappa}})=\frac{2\sinh\beta E_{\vec{\kappa}}}{\cosh\beta\varsigma+\cosh\beta E} \quad (1888)$$

and

$$\Pi=\frac{2}{V}\sum_{\vec{\kappa}}\frac{1}{2E_{\vec{\kappa}}}\frac{\sinh\beta E_{\vec{\kappa}}}{\cosh\beta\varsigma+\cosh\beta E} \quad (1889)$$

Therefore, it follows that

$$\frac{1}{\beta}\sum_{n}\frac{1}{(i\nu_n-E_{\vec{\kappa}})(-i\nu_n-E_{\vec{\kappa}})}=\frac{1}{\beta}\sum_{n}\frac{1}{\nu_n^2+E_{\vec{\kappa}}^2}=\frac{1}{2E_{\vec{\kappa}}}\frac{\sinh\beta E_{\vec{\kappa}}}{\cosh\beta\varsigma+\cosh\beta E_{\vec{\kappa}}}\equiv f_1(\beta,E_{\vec{\kappa}},\varsigma) \quad (1890)$$

and

$$\Pi=\frac{2}{V}\sum_{\vec{\kappa}}f_1(\beta,E_{\vec{\kappa}},\varsigma) \quad (1891)$$

We can also show that

$$\frac{1}{\beta}\sum_{n}\frac{1}{\left(\nu_n^2+E_{\vec{\kappa}}^2\right)^s}=f_s\left(\beta,E_{\vec{\kappa}},\varsigma\right) \tag{1892}$$

as well as

$$f_{s+1}\left(\beta,E_{\vec{\kappa}},\varsigma\right)=\frac{1}{2sE_{\vec{\kappa}}}\frac{\partial f_s\left(\beta,E_{\vec{\kappa}},\varsigma\right)}{\partial E_{\vec{\kappa}}} \tag{1893}$$

Therefore, for fermions, the coupling strength as well as the gap equation can now be obtained from

$$\frac{1}{|\lambda|}=-\frac{1}{V}\sum_{\vec{\kappa}}f_1\left(\beta,E_{\vec{\kappa}},\varsigma\right)\operatorname{sgn}\Delta \tag{1894}$$

For the case of bosons $\chi=+1$, we have the same equation

$$\frac{1}{|\lambda|}=-\frac{1}{V}\sum_{\vec{\kappa}}f_1\left(\beta,E_{\vec{\kappa}},\varsigma\right)\operatorname{sgn}\Delta \tag{1895}$$

From (1866), we have the thermodynamic potential

$$\Omega_{sp}\left[\Delta^*,\Delta\right]=-\frac{|\Delta|^2}{|\lambda|}-\frac{1}{\beta V}\sum_{\vec{\kappa}}\ln\left[-\sinh\beta\left(i\alpha+E_{\vec{\kappa}}\right)\sinh\beta\left(i\alpha-E_{\vec{\kappa}}\right)\right]\times\text{Const} \tag{1896}$$

or

$$\Omega_{sp}\left[\Delta^*,\Delta\right]=-\frac{|\Delta|^2}{|\lambda|}-\frac{1}{\beta V}\sum_{\vec{\kappa}}\ln\left[2\cosh\beta E_{\vec{\kappa}}-2\cos\beta\alpha\right]+\ln C \tag{1897}$$

But,

$$\beta\alpha=\frac{1-\chi}{2}\pi-i\beta\varsigma \tag{1898}$$

and

$$\cos\beta\alpha=\cos\frac{1-\chi}{2}\pi\cosh\beta\varsigma+i\sin\frac{1-\chi}{2}\pi\sinh\beta\varsigma \tag{1899}$$

For the case of fermions, $\chi=-1$, and then

$$\Omega_{sp}\left[\Delta^*,\Delta\right]=-\frac{|\Delta|^2}{|\lambda|}-\frac{1}{\beta V}\sum_{\vec{\kappa}}\ln\left[2\cosh\beta E_{\vec{\kappa}}+2\cosh\beta\varsigma\right]+\ln C_1 \tag{1900}$$

For the case of bosons, $\chi=+1$, and then

$$\Omega_{sp}\left[\Delta^*,\Delta\right]=-\frac{|\Delta|^2}{|\lambda|}-\frac{1}{\beta V}\sum_{\vec{\kappa}}\ln\left[2\cosh\beta E_{\vec{\kappa}}-2\cosh\beta\varsigma\right]+\ln C_2 \tag{1901}$$

Therefore, the compact formula for both fermions and bosons:

$$\Omega_{sp}\left[\Delta^*,\Delta\right]=-\frac{|\Delta|^2}{|\lambda|}-\frac{1}{\beta V}\sum_{\vec{\kappa}}\ln\left[2\cosh\beta E_{\vec{\kappa}}-2\chi\cosh\beta\varsigma\right]+\ln C \tag{1902}$$

The constants $\ln C_1$ and $\ln C_2$ can be found by investigating the convergence of $\Omega_{sp}\left[\Delta^*,\Delta\right]$ via the following:

$$\lim_{\kappa\to\infty}\left[2\cosh\beta E_{\vec{\kappa}}-2\chi\cosh\beta\varsigma\right]\cong\lim_{\kappa\to\infty}\exp\left\{\beta\varsigma_{\vec{\kappa}}\right\} \tag{1903}$$

Therefore, for the convergence of $\Omega_{sp}\left[\Delta^*,\Delta\right]$, we have

$$\ln C_1=\ln C_2=\ln C=\frac{1}{\beta V}\sum_{\vec{\kappa}}\ln\exp\left\{\beta\varsigma_{\vec{\kappa}}\right\} \tag{1904}$$

So,

$$\Omega_{sp}\left[\Delta^*,\Delta\right]=-\frac{|\Delta|^2}{|\lambda|}-\frac{1}{\beta V}\sum_{\vec{\kappa}}\left(\ln\left[2\cosh\beta E_{\vec{\kappa}}-2\chi\cosh\beta\varsigma\right]-\beta\varsigma_{\vec{\kappa}}\right) \tag{1905}$$

We can relate the saddle-point thermodynamic potential $\Omega_{sp}(T,\varsigma,\mu)$ to the time-independent and space-independent **free energy** $F_{sp}(T,V,N,\delta N)$:

$$\Omega_{sp}(T,\varsigma,\mu)=\frac{F_{sp}(T,V,N,\delta N)}{V}-\mu n-\varsigma\delta n \tag{1906}$$

Here, $n=\frac{N}{V}$ is the total density, $\delta n=\frac{\delta N}{V}$ is the excess particle density, which accounts for a majority of particles in one spin component in the presence of imbalance, $N=N_\uparrow+N_\downarrow$, the total population, and $\delta N=\delta N_\uparrow-\delta N_\downarrow$ its population imbalance:

$$n_{sp}=\sum_{\vec{\kappa}}\left(1-\frac{\xi_{\vec{\kappa}}}{E_{\vec{\kappa}}}\frac{\sinh\beta E_{\vec{\kappa}}}{\cosh\beta E_{\vec{\kappa}}-\chi\cosh\beta\varsigma}\right) \tag{1907}$$

$$\delta n_{sp}=-\frac{\partial\Omega_{sp}(T,\mu,\varsigma)}{\partial\varsigma}=-\sum_{\vec{\kappa}}\left(\frac{\chi\sinh\beta\varsigma}{\cosh\beta E_{\vec{\kappa}}-\chi\cosh\beta\varsigma}\right) \tag{1908}$$

For fermions,

$$n_{sp}=\sum_{\vec{\kappa}}\left(1-2\xi_{\vec{\kappa}}\,f_1\left(\beta,E_{\vec{\kappa}},\varsigma\right)\right) \tag{1909}$$

$$\delta n_{sp}=2\varsigma\sum_{\vec{\kappa}}F_1\left(\beta,E_{\vec{\kappa}},\varsigma\right) \tag{1910}$$

where

$$\frac{1}{2\varsigma}\frac{\sinh\beta\varsigma}{\cosh\beta\varsigma+\cosh\beta E_{\vec{\kappa}}}\equiv F_1\left(\beta,E_{\vec{\kappa}},\varsigma\right) \tag{1911}$$

For bosons,

$$n_{sp} = \sum_{\vec{\kappa}} \left(1 + 2\xi_{\vec{\kappa}} \Phi_1(\beta, E_{\vec{\kappa}}, \varsigma)\right) \tag{1912}$$

$$\delta n_{sp} = 2\varsigma \sum_{\vec{\kappa}} \phi_1(\beta, E_{\vec{\kappa}}, \varsigma) \tag{1913}$$

where

$$\frac{1}{2E_{\vec{\kappa}}} \frac{\sinh \beta E_{\vec{\kappa}}}{\cosh \beta\varsigma - \cosh \beta E_{\vec{\kappa}}} \equiv \Phi_1(\beta, E_{\vec{\kappa}}, \varsigma) \quad , \quad \frac{1}{2\varsigma} \frac{\sinh \beta\varsigma}{\cosh \beta\varsigma + \cosh \beta E_{\vec{\kappa}}} \equiv \phi_1(\beta, E_{\vec{\kappa}}, \varsigma) \tag{1914}$$

But because

$$\frac{1}{3\pi^2} = n_{sp} \tag{1915}$$

then for fermions,

$$\frac{1}{3\pi^2} \frac{\delta n_{sp}}{n_{sp}} = 2\varsigma \sum_{\vec{\kappa}} F_1(\beta, E_{\vec{\kappa}}, \varsigma) \tag{1916}$$

and for bosons,

$$\frac{1}{3\pi^2} \frac{\delta n_{sp}}{n_{sp}} = 2\varsigma \sum_{\vec{\kappa}} \phi_1(\beta, E_{\vec{\kappa}}, \varsigma) \tag{1917}$$

The gap equation can be regularized to eliminate λ, yielding

$$\frac{m}{4\pi\hbar^2 a} = \frac{1}{2V} \sum_{\vec{\kappa}} \left(\frac{1}{\epsilon_{\vec{\kappa}}} - \frac{1}{E_{\vec{\kappa}}} \frac{\sinh \beta E_{\vec{\kappa}}}{\cosh \beta\varsigma + \cosh \beta E_{\vec{\kappa}}} \right) \tag{1918}$$

Equation (1905) is the thermodynamic potential of Cooper pairs as a function of energy gap Δ, average chemical potential μ, imbalance chemical potential ς, and temperature T. The gap Δ may be treated as a separate input through (1894) or (1895). For each T, μ, ς, this equation allows us to have $\Delta_{sp}(T, \mu, \varsigma)$, which minimizes $\Omega_{sp}(T, \mu, \varsigma)$.

11.4 Generalized Correlation Functions

In this section, we examine two quantities that tailor the fermionic superfluid, such as the condensate fraction, which gives a measure of the number of particles participating in the pairing mechanism, and the pair correlation length, which is an estimate of the characteristic size of the Cooper pairs forming the fermionic condensate. We calculate these quantities via the correlation functions that are easily obtained via the generating functionals, which are the cornerstone of the functional techniques sufficient to derive all correlation functions. Generally, these generating functionals are achieved by coupling the field operators to one or more external fields, called the source fields, which are c-numbers for bosons and Grassmann variables for fermions. These source fields may be thought of as our probes inside the quantum system. The propagators are probed by varying the source fields. For the BCS-BEC theory, we can achieve the correlation functions by expressing the generating functional $Z\left[\eta,\eta^*\right]$ via source variables $\eta_\alpha(\vec{r},\tau)$ and $\eta^*_\alpha(\vec{r},\tau)$ coupled linearly to the fields $\psi^\dagger$ and $\hat{\psi}$:

$$Z\left[\eta,\eta^*\right]=\int d\left[\hat{\psi}^\dagger\right]d\left[\hat{\psi}\right]\exp\left\{-\mathbf{S}\left[\hat{\psi}^\dagger,\hat{\psi},\eta,\eta^*\right]\right\} \tag{1919}$$

The extended action functional $\mathbf{S}\left[\hat{\psi}^\dagger,\hat{\psi},\eta,\eta^*\right]$ is defined as

$$\mathbf{S}\left[\hat{\psi}^\dagger,\hat{\psi},\eta,\eta^*\right]=S_0\left[\psi^\dagger,\psi\right]+S_{\text{int}}\left[\psi^\dagger,\psi\right]+S\left[\eta,\eta^*\right] \tag{1920}$$

with the interaction and free particle action functionals being

$$S_{\text{int}}\left[\psi^\dagger,\psi\right]=\lambda\int_0^\beta d\tau\int d\vec{r}A^\dagger_{\sigma\sigma'}(\vec{r},\tau)A_{\sigma\sigma'}(\vec{r},\tau) \tag{1921}$$

and

$$S_0\left[\hat{\psi}^\dagger,\hat{\psi}\right]=\int_0^\beta d\tau\int d\vec{r}\sum_\sigma\hat{\psi}^\dagger_\sigma(\vec{r},\tau)\mathbf{G}^{-1}_{0\sigma}\hat{\psi}_\sigma(\vec{r},\tau) \tag{1922}$$

respectively.
The contribution to the action functional from the source fields is:

$$S\left[\eta,\eta^*\right]=-\sum_{\alpha=\uparrow,\downarrow}\int_0^\beta d\tau\int d\vec{r}\left(\hat{\psi}^\dagger_\alpha(\vec{r},\tau)\eta_\alpha(\vec{r},\tau)+\eta^*_\alpha(\vec{r},\tau)\hat{\psi}_\alpha(\vec{r},\tau)\right) \tag{1923}$$

The bare fermionic inverse Green's function is

$$\mathbf{G}^{-1}_{0\sigma}=\frac{\partial}{\partial\tau}-\frac{\Delta}{2m}-\mu_\sigma \tag{1924}$$

and the pairing operators are

$$A^\dagger_{\sigma\sigma'}(\vec{r},\tau)=\hat{\psi}^\dagger_\uparrow(\vec{r},\tau)\hat{\psi}^\dagger_\downarrow(\vec{r},\tau)\quad,\quad A_{\sigma\sigma'}(\vec{r},\tau)=\hat{\psi}_\downarrow(\vec{r},\tau)\hat{\psi}_\uparrow(\vec{r},\tau) \tag{1925}$$

The extra terms $\hat{\psi}^\dagger_\alpha(\vec{r},\tau)\eta_\alpha(\vec{r},\tau)$ and $\eta^*_\alpha(\vec{r},\tau)\hat{\psi}_\alpha(\vec{r},\tau)$ in (1920) are considered as the effects of external sources.

We define the generating functional via the normal and extended partition functions as the normalized expression

$$\mathbf{G}\left[\eta,\eta^*\right]\equiv\frac{Z\left[\eta,\eta^*\right]}{Z[0,0]} \tag{1926}$$

This achieves the value unity when all the sources are switched off. That is when all η,η^* are equal to zero. The explicit form of $\mathbf{G}\left[\eta,\eta^*\right]$:

$$\mathbf{G}\left[\eta,\eta^*\right]=\left\langle\exp\left\{\sum_{\alpha=\uparrow,\downarrow}\int_0^\beta d\tau\int d\vec{r}\left(\hat{\psi}^\dagger_\alpha(\vec{r},\tau)\eta_\alpha(\vec{r},\tau)+\eta^*_\alpha(\vec{r},\tau)\hat{\psi}_\alpha(\vec{r},\tau)\right)\right\}\right\rangle \tag{1927}$$

Here, the average $\langle \cdots \rangle$ is taken in terms of the normal fermionic action functional S. The quantity $Z[\eta,\eta^*]$ imitates the partition function of a noninteracting system, while $Z[0,0]$ imitates the noninteracting partition function for vanishing sources, as seen earlier.

We perform the Hubbard-Stratonovich transformation on $\exp\{-S_{\text{int}}[\psi^\dagger,\psi]\}$, such as in relation (1786), and this yields the following extended action functional:

$$\mathbf{S}[\hat{\psi},\eta]=\int_0^\beta d\tau\int d\vec{r}\left(\hat{\psi}^\dagger(\vec{r},\tau)[-\mathbf{G}^{-1}]\hat{\psi}(\vec{r},\tau)+\left(\hat{\psi}^\dagger(\vec{r},\tau)\eta+\eta^*\hat{\psi}(\vec{r},\tau)\right)+\frac{\Delta^*(\vec{r},\tau)\Delta(\vec{r},\tau)}{\lambda}\right) \quad (1930)$$

Here, we have reintroduced the fermionic **Nambu-Gorkov spinor [79]** in (1930) together with that of the respective source fields as

$$\psi(\vec{r},\tau)=\begin{bmatrix}\psi_\uparrow(\vec{r},\tau)\\ \psi_\downarrow^\dagger(\vec{r},\tau)\end{bmatrix},\quad \psi^\dagger(\vec{r},\tau)=\begin{bmatrix}\psi_\downarrow(\vec{r},\tau) & \psi_\uparrow^\dagger(\vec{r},\tau)\end{bmatrix} \quad (1931)$$

$$\eta(\vec{r},\tau)=\begin{bmatrix}\eta_\uparrow(\vec{r},\tau)\\ \eta_\downarrow^*(\vec{r},\tau)\end{bmatrix},\quad \eta^*(\vec{r},\tau)=\begin{bmatrix}\eta_\downarrow(\vec{r},\tau) & \eta_\uparrow^*(\vec{r},\tau)\end{bmatrix} \quad (1932)$$

and

$$[-\mathbf{G}^{-1}]=\begin{bmatrix}\mathbf{G}_{0\uparrow}^{-1} & -\Delta(\vec{r},\tau)\\ -\Delta^*(\vec{r},\tau) & \mathbf{G}_{0\downarrow}^{-1}\end{bmatrix} \quad (1933)$$

is the **inverse Nambu-Gorkov** 2×2 **Green's function** for the interacting fermions in spin space with

$$\mathbf{G}_{0\uparrow}^{-1}=\frac{\partial}{\partial\tau}-\frac{\nabla_{\vec{r}}^2}{2m}-\mu_\uparrow,\quad \mathbf{G}_{0\downarrow}^{-1}=\frac{\partial}{\partial\tau}+\frac{\nabla_{\vec{r}}^2}{2m}+\mu_\downarrow \quad (1934)$$

After considering the Hubbard-Stratonovich transformation, for the generating functional to be easily solvable, we introduce new shifted fermionic Nambu-Gorkov spinors $\hat{\psi}'_\alpha$ and $\hat{\psi}^\dagger_\alpha$:

$$\hat{\psi}=\hat{\psi}'+\mathbf{G}\eta,\quad \hat{\psi}^\dagger=\hat{\psi}'^\dagger+\eta^*\mathbf{G} \quad (1935)$$

So, the action functional in (1930) becomes

$$\mathbf{S}[\hat{\psi},\eta,\Delta]=\int_0^\beta d\tau\int d\vec{r}\left(\hat{\psi}'^\dagger[-\mathbf{G}^{-1}]\hat{\psi}'+\eta^*\mathbf{G}\eta+\frac{\Delta^*(\vec{r},\tau)\Delta(\vec{r},\tau)}{\lambda}\right) \quad (1936)$$

Because the shift of the fermionic variables does not affect the result of the integration in the partition function, so we may safely drop the primes and then

$$\mathbf{S}[\hat{\psi},\eta,\Delta]=\mathbf{S}[\hat{\psi},\Delta]+\mathbf{S}[\eta] \quad (1937)$$

where the action functional

$$\mathbf{S}\left[\hat{\psi},\Delta\right]=\int_0^\beta d\tau\int d\vec{r}\left(\hat{\psi}^\dagger\left[-\mathbf{G}^{-1}\right]\hat{\psi}+\frac{\Delta^*(\vec{r},\tau)\Delta(\vec{r},\tau)}{\lambda}\right) \tag{1938}$$

is dependent on the original fermionic fields $\hat{\psi}^\dagger$ and $\hat{\psi}$ as well as on the pair fields Δ^* and Δ, while the action functional

$$\mathbf{S}[\eta]=\int_0^\beta d\tau d\tau'\int d\vec{r}\,d\vec{r}'\eta^*(\vec{r}',\tau')\left[-\mathbf{G}(\vec{r}',\tau'|\vec{r},\tau)\right]\eta(\vec{r},\tau) \tag{1939}$$

is devoid of the original fermionic fields $\hat{\psi}^\dagger$ and $\hat{\psi}$.

We observe that the sources are therefore successfully decoupled from the fermionic part of the action functional. We then express the generating functional $\mathbf{G}\left[\eta,\eta^*\right]$ as an average via the effective bosonic action rather than the fermionic one:

$$\mathbf{G}\left[\eta,\eta^*\right]=\left\langle\exp\{-\mathbf{S}[\eta]\}\right\rangle=\left\langle\exp\left\{\int_0^\beta d\tau\int d\vec{r}\eta^*\mathbf{G}\eta\right\}\right\rangle \tag{1940}$$

or

$$\mathbf{G}\left[\eta,\eta^*\right]=\left\langle\exp\left\{\int_0^\beta d\tau d\tau'\int d\vec{r}d\vec{r}'\eta^*(\vec{r}',\tau')\mathbf{G}(\vec{r}',\tau',\vec{r},\tau)\eta(\vec{r},\tau)\right\}\right\rangle \tag{1941}$$

This generating functional permits us to determine any fermionic correlation function of any order. This can be done via functional derivatives of the generating functional with respect to the sources η,η^*. We then set $\eta=\eta^*=0$:

$$\mathbf{G}^{(2n)}(\alpha_1\tau_1,\cdots,\alpha_n\tau_n,\alpha_n'\tau_n',\cdots,\alpha_1'\tau_1')=\left.\frac{\partial^{(2n)}\mathbf{G}\left[\eta,\eta^*\right]}{\partial\eta^*_{\alpha_1}(\tau_1)\cdots\partial\eta^*_{\alpha_n}(\tau_n)\partial\eta_{\alpha_n'}(\tau_n')\cdots\partial\eta_{\alpha_1'}(\tau_1')}\right|_{\eta,\eta^*=0} \tag{1942}$$

11.5 Condensate Fraction

The condensate fraction ν_c can be calculated via (1942) and (1941):

$$\left\langle\hat{\psi}_\uparrow(\vec{r},\tau)\hat{\psi}_\downarrow(\vec{r}',\tau)\right\rangle=-\left\langle\mathbf{G}_{12}(\vec{r}',\tau',\vec{r},\tau)\right\rangle \tag{1943}$$

Here, $\mathbf{G}_{12}$ is the matrix element of the matrix in (1794), and the condensate density is defined:

$$n_c=\frac{1}{V}\int d\vec{r}d\vec{r}'\left|\left\langle\mathbf{G}_{12}(\vec{r}',\tau',\vec{r},\tau)\right\rangle\right|^2 \tag{1944}$$

This equation can be conveniently solved via the Fourier transform of $\mathbf{G}_{12}(\vec{r}',\tau',\vec{r},\tau)$:

$$\mathbf{G}_{12}(\vec{r}',\tau',\vec{r},\tau)=\frac{1}{\beta V}\sum_{\vec{\kappa}\vec{\kappa}'nn'}\mathbf{G}_{12}(\vec{\kappa},n,\vec{\kappa}',n')\exp\left\{-i\left(\vec{\kappa}'\vec{r}'-\omega_{n'}\tau'\right)\right\}\exp\left\{i\left(\vec{\kappa}\vec{r}-\omega_n\tau\right)\right\} \tag{1945}$$

Substituting this into equation (1944), we then have

$$n_c = \frac{1}{V}\int d\vec{r}d\vec{r}' \frac{1}{\beta V}\sum_{\vec{\kappa}\vec{\kappa}'nn'} \left\langle \mathbf{G}_{12}\left(\vec{\kappa},n,\vec{\kappa}',n'\right)\right\rangle \exp\left\{-i\left(\vec{\kappa}'\vec{r}' - \omega_{n'}\tau'\right)\right\}\exp\left\{i\left(\vec{\kappa}\vec{r} - \omega_n\tau\right)\right\}\times$$
$$\times\frac{1}{\beta V}\sum_{\vec{q}\vec{q}'mm'} \left\langle \mathbf{G}^*_{12}\left(\vec{q},m,\vec{q}',m'\right)\right\rangle \exp\left\{i\left(\vec{q}'\vec{r}' - \omega_{m'}\tau'\right)\right\}\exp\left\{-i\left(\vec{q}\vec{r} - \omega_m\tau\right)\right\} \tag{1946}$$

or

$$n_c = \frac{1}{\beta^2 V}\sum_{\vec{\kappa}\vec{\kappa}'nn'\vec{q}\vec{q}'mm'} \left\langle \mathbf{G}_{12}\left(\vec{\kappa},n,\vec{\kappa}',n'\right)\right\rangle\left\langle \mathbf{G}^*_{12}\left(\vec{q},m,\vec{q}',m'\right)\right\rangle \exp\left\{-i\left(\omega_n + \omega_{m'} - \omega_m - \omega_{n'}\right)\tau\right\} \tag{1947}$$

Because the system under consideration is stationary, the condensate fraction n_c must be time independent. Hence, we let

$$n + m' - m - n' = 0 \tag{1948}$$

From here, we have

$$n - n' = m - m' \equiv s \tag{1949}$$

and expression (1947) is reduced to

$$n_c = \frac{1}{\beta^2 V}\sum_{\vec{\kappa}\vec{q}nn's} \left\langle \mathbf{G}_{12}\left(\vec{\kappa},n,\vec{\kappa}+\vec{q},n+s\right)\right\rangle\left\langle \mathbf{G}^*_{12}\left(\vec{\kappa},n',\vec{\kappa}+\vec{q},n'+s\right)\right\rangle \tag{1950}$$

Let $s \to m$, then

$$n_c = \frac{1}{\beta^2 V}\sum_{\vec{\kappa}\vec{q}nn'm} \left\langle \mathbf{G}_{12}\left(\vec{\kappa},n,\vec{\kappa}+\vec{q},n+m\right)\right\rangle\left\langle \mathbf{G}^*_{12}\left(\vec{\kappa},n',\vec{\kappa}+\vec{q},n'+m\right)\right\rangle \tag{1951}$$

We calculate the saddle-point expression for the condensate density by rewriting the saddle-point inverse fermionic propagator:

$$-\mathbf{G}^{-1}_{\mathrm{sp}}\left(\vec{\kappa},n\middle|\vec{\kappa}',n'\right) \equiv \left[-\mathbf{G}^{-1}_{\mathrm{sp}}\right]_{\vec{\kappa}n\vec{\kappa}'n'} = \delta_{\vec{\kappa}\vec{\kappa}'}\delta_{nn'}\left[-\mathbf{G}^{-1}_{\mathrm{sp}}\right] \tag{1952}$$

where

$$\left[-\mathbf{G}^{-1}_{\mathrm{sp}}\right] = \begin{bmatrix} \mathbf{G}^{-1}_{0\uparrow} & -\Delta \\ -\Delta^* & \mathbf{G}^{-1}_{0\downarrow} \end{bmatrix} \tag{1953}$$

is the **inverse Nambu-Gorkov** 2×2 **Green's function** for the interacting fermions and

$$\mathbf{G}^{-1}_{0\uparrow} = i\omega_n - \varsigma_{\vec{\kappa}\uparrow} \quad , \quad \mathbf{G}^{-1}_{0\downarrow} = i\omega_n - \varsigma_{\vec{\kappa}\downarrow} \quad , \quad \varsigma_{\vec{\kappa}\sigma} = \frac{\vec{\kappa}^2}{2m} - \mu_\sigma \tag{1954}$$

By inverting (1952), we obtained the fermionic Green's function

$$-\mathbf{G}_{\rm sp}\left(\vec{\kappa},n\middle|\vec{\kappa}',n'\right)=\delta_{\vec{\kappa}\vec{\kappa}'}\delta_{nn'}\frac{1}{-\det\left[-\mathbf{G}_{\rm sp}^{-1}\right]_{\vec{\kappa}n}}\begin{bmatrix}-i\omega_n+\xi_{\vec{\kappa}}-\varsigma & \Delta\\ \Delta^* & i\omega_n-\xi_{\vec{\kappa}}+\varsigma\end{bmatrix} \tag{1955}$$

where

$$-\det\left[-\mathbf{G}_{\rm sp}^{-1}\right]_{\vec{\kappa}n}=-(-i\omega_n+\xi_{\vec{\kappa}}-\varsigma)(i\omega_n-\xi_{\vec{\kappa}}+\varsigma)-|\Delta|^2\equiv(i\omega_n+\varsigma-\mathrm{E}_{\vec{\kappa}})(-i\omega_n-\varsigma-\mathrm{E}_{\vec{\kappa}}) \tag{1956}$$

or

$$-\det\left[-\mathbf{G}_{\rm sp}^{-1}\right]_{\vec{\kappa}n}=(i\omega_n+\varsigma-\mathrm{E}_{\vec{\kappa}})(-i\omega_n-\varsigma-\mathrm{E}_{\vec{\kappa}})\equiv(i\nu_n-\mathrm{E}_{\vec{\kappa}})(-i\nu_n-\mathrm{E}_{\vec{\kappa}}) \tag{1957}$$

Then, from equation (1955),

$$-\mathbf{G}_{12}\left(\vec{\kappa},n\middle|\vec{\kappa}',n'\right)=\delta_{\vec{\kappa}\vec{\kappa}'}\delta_{nn'}\frac{\Delta}{-\det\left[-\mathbf{G}_{\rm sp}^{-1}\right]_{\vec{\kappa}n}} \tag{1958}$$

So, the condensate density at saddle point via the Matsubara frequency ω_n is written:

$$n_{\rm c}^{(\rm sp)}=|\Delta|^2\int\frac{d\vec{\kappa}}{(2\pi)^3}\left(\frac{1}{\beta}\sum_n\frac{1}{(i\nu_n-\mathrm{E}_{\vec{\kappa}})(-i\nu_n-\mathrm{E}_{\vec{\kappa}})}\right)^2 \tag{1959}$$

But, as seen earlier,

$$\frac{1}{\beta}\sum_n\frac{1}{(i\nu_n-\mathrm{E}_{\vec{\kappa}})(-i\nu_n-\mathrm{E}_{\vec{\kappa}})}=\frac{1}{4\mathrm{E}_{\vec{\kappa}}}\left[\coth\frac{\beta}{2}(\varsigma-\mathrm{E}_{\vec{\kappa}})-\coth\frac{\beta}{2}(\varsigma+\mathrm{E}_{\vec{\kappa}})\right] \tag{1960}$$

then

$$n_{\rm c}^{(\rm sp)}=|\Delta|^2\int\frac{d\vec{\kappa}}{(2\pi)^3}\left(\frac{1}{4\mathrm{E}_{\vec{\kappa}}}\left[\coth\frac{\beta}{2}(\varsigma-\mathrm{E}_{\vec{\kappa}})-\coth\frac{\beta}{2}(\varsigma+\mathrm{E}_{\vec{\kappa}})\right]\right)^2 \tag{1961}$$

or

$$n_{\rm c}^{(\rm sp)}=|\Delta|^2\int\frac{d\vec{\kappa}}{(2\pi)^3}\left(\frac{1}{2\mathrm{E}_{\vec{\kappa}}}\frac{\sinh\beta\mathrm{E}_{\vec{\kappa}}}{\cosh\beta\varsigma+\cosh\beta\mathrm{E}_{\vec{\kappa}}}\right)^2=|\Delta|^2\int\frac{d\vec{\kappa}}{(2\pi)^3}\left(\mathrm{f}_1(\beta,\mathrm{E}_{\vec{\kappa}},\varsigma)\right)^2 \tag{1962}$$

This corresponds to the result found in [83].

The condensate fraction $\nu_{\rm c}$ is defined as the ratio of twice the condensate density to the total density:

$$\nu_{\rm c}=\frac{2n_{\rm c}}{n}\quad,\quad n=\frac{1}{3\pi^2} \tag{1963}$$

Here, the factor 2 is introduced to normalize the condensate fraction to 1. So,

$$\nu_{\rm c}^{(\rm sp)}=\frac{3}{2}|\Delta|^2\int_0^\infty\kappa^2\,d\kappa\left(\mathrm{f}_1(\beta\mathrm{E}_{\vec{\kappa}},\varsigma)\right)^2 \tag{1964}$$

11.6 Pair Correlation Length

The pair correlation length gives an estimate of the typical size of a Cooper pair. We calculate it via the generating functional method seen earlier:

$$\xi_{\text{pair}} = \left(\frac{\int d\vec{r} r^2 \lambda_{\uparrow\downarrow}(\vec{r})}{\int d\vec{r} \lambda_{\uparrow\downarrow}(\vec{r})} \right)^{\frac{1}{2}} \tag{1965}$$

where the correlation function [83]:

$$\lambda_{\uparrow\downarrow}(\vec{r}) = \left\langle \psi_{\uparrow}^{\dagger}\left(\vec{R}+\frac{\vec{r}}{2}\right)\psi_{\downarrow}^{\dagger}\left(\vec{R}-\frac{\vec{r}}{2}\right)\psi_{\downarrow}\left(\vec{R}-\frac{\vec{r}}{2}\right)\psi_{\uparrow}\left(\vec{R}+\frac{\vec{r}}{2}\right)\right\rangle - \left(\frac{n}{2}\right)^2 \tag{1966}$$

Here, n is the fermionic density, $\vec{R}$ is the center of mass coordinate, and $\vec{r}$ is the relative coordinate. At the mean-field level, the pair correlation function $\lambda_{\uparrow\downarrow}(\vec{r})$ can be factorized as a product of single particle averages [83]:

$$\lambda_{\uparrow\downarrow}^{(\text{sp})}(\vec{r}) = \left\langle \psi_{\downarrow}\left(\vec{R}-\frac{\vec{r}}{2}\right)\psi_{\uparrow}\left(\vec{R}+\frac{\vec{r}}{2}\right)\right\rangle \left\langle \psi_{\uparrow}^{\dagger}\left(\vec{R}+\frac{\vec{r}}{2}\right)\psi_{\downarrow}^{\dagger}\left(\vec{R}-\frac{\vec{r}}{2}\right)\right\rangle = \left| \left\langle \psi_{\downarrow}\left(\vec{R}-\frac{\vec{r}}{2}\right)\psi_{\uparrow}\left(\vec{R}+\frac{\vec{r}}{2}\right)\right\rangle \right|^2 \tag{1967}$$

Because our system is uniform, this quantity must be independent of the center of mass coordinate $\vec{R}$. Therefore,

$$\lambda_{\uparrow\downarrow}^{(\text{sp})}(\vec{r}) = \left| \left\langle \psi_{\downarrow}\left(-\frac{\vec{r}}{2}\right)\psi_{\uparrow}\left(\frac{\vec{r}}{2}\right)\right\rangle \right|^2 \tag{1968}$$

For convenience, we write the expressions at the numerator and denominator of (1965) in reciprocal space. Hence,

$$\int d\vec{r} r^2 \lambda_{\uparrow\downarrow}(\vec{r}) = |\Delta|^2 \int \frac{d\vec{\kappa}}{(2\pi)^3} \left(\nabla_{\vec{\kappa}} \frac{1}{2E_{\vec{\kappa}}} \frac{\sinh \beta E_{\vec{\kappa}}}{\cosh \beta\varsigma + \cosh \beta E_{\vec{\kappa}}} \right)^2 = |\Delta|^2 \int \frac{d\vec{\kappa}}{(2\pi)^3} \left(\nabla_{\vec{\kappa}} f_1(\beta, E_{\vec{\kappa}}, \varsigma)\right)^2 \tag{1969}$$

or

$$\int d\vec{r} r^2 \lambda_{\uparrow\downarrow}(\vec{r}) = |\Delta|^2 \int \frac{d\vec{\kappa}}{(2\pi)^3} \left(\nabla_{\vec{\kappa}} f_1(\beta, E_{\vec{\kappa}}, \varsigma)\right)^2 = |\Delta|^2 \int \frac{d\vec{\kappa}}{(2\pi)^3} \left(-4\vec{\kappa}\xi_{\vec{\kappa}} f_2(\beta, E_{\vec{\kappa}}, \varsigma)\right)^2 \tag{1970}$$

while

$$\int d\vec{r} \lambda_{\uparrow\downarrow}(\vec{r}) = |\Delta|^2 \int \frac{d\vec{\kappa}}{(2\pi)^3} \left(\frac{1}{2E_{\vec{\kappa}}} \frac{\sinh \beta E_{\vec{\kappa}}}{\cosh \beta\varsigma + \cosh \beta E_{\vec{\kappa}}} \right)^2 = |\Delta|^2 \int \frac{d\vec{\kappa}}{(2\pi)^3} \left(f_1(\beta, E_{\vec{\kappa}}, \varsigma)\right)^2 \tag{1971}$$

Therefore,

$$\xi_{\text{pair}} = \left(\frac{\int \frac{d\vec{\kappa}}{(2\pi)^3} \left(-4\vec{\kappa}\xi_{\vec{\kappa}} f_2(\beta, E_{\vec{\kappa}}, \varsigma)\right)^2}{\int \frac{d\vec{\kappa}}{(2\pi)^3} \left(f_1(\beta, E_{\vec{\kappa}}, \varsigma)\right)^2} \right)^{\frac{1}{2}} \tag{1972}$$

11.7 Improvement of the Saddle-Point Solution

It is important to note that the BCS-BEC crossover theory extended to a finite temperature must include the effects of fluctuations of pairs. The problems of the BCS theory can already be overcome significantly by considering the Gaussian fluctuations of the order parameter around its mean-field value. Because Gaussian functional integrals can be performed analytically, the contribution of these pair fluctuations can be studied exactly. Diagrammatically, this corresponds to summing the so-called ladder diagram contributions to the grand potential.

So far, the treatment has been kept exact, and equation (1952), albeit impossible to treat analytically, is an exact description of the system. To proceed with the analytical treatment, it is necessary to decompose the pairing field Δ as the sum of a uniform and constant mean-field value (uniform in coordinate space and constant with respect to the imaginary time τ), which will be self-consistently determined later as the saddle point of the action, and a fluctuation bosonic fields $\xi_{n,\vec{\kappa}}$ (Bogoliubov shift):

$$\Delta_{n,\vec{\kappa}} = \sqrt{\beta V}\delta(\vec{\kappa})\delta_{n,0}\times\Delta+\xi_{n,\vec{\kappa}} \quad , \quad \Delta^*_{n,\vec{\kappa}} = \sqrt{\beta V}\delta(\vec{\kappa})\delta_{n,0}\times\Delta^*+\xi^*_{n,\vec{\kappa}} \tag{1973}$$

The complex fields $\xi_{n,\vec{\kappa}}$ describe pairing fluctuations. The saddle-point value is independent of space coordinates because it is interpreted as the order parameter of the condensate of pairs, which condense in the zero momentum (i.e., uniform density state). The saddle point serves to factorize out the dominant contribution. We consider up to second order in the fluctuations $\xi_{n,\vec{\kappa}}$ and $\xi^*_{n,\vec{\kappa}}$, which captures the essence of the thermodynamics of the noncondensed pairs, and leads to a quadratic (bosonic) path integral that we perform exactly.

We substitute (1973) into expression (1840) for the inverse Green's function. We then decompose the inverse Green's function as the sum of a saddle-point contribution $-\mathbf{G}_{sp}^{-1}$ and a fluctuation contribution (fluctuation propagator) $\mathbf{F}$:

$$\psi^\dagger(\vec{r},\tau)\cdot\left[-\mathbf{G}^{-1}\right]\cdot\psi(\vec{r},\tau)=\psi^\dagger(\vec{r},\tau)\cdot\left[-\mathbf{G}_{sp}^{-1}+\mathbf{F}\right]\cdot\psi(\vec{r},\tau)=\sum_{n\kappa n'\kappa'}\psi^\dagger_{n\kappa}\left[-\mathbf{G}_{sp}^{-1}+\mathbf{F}\right]_{n\kappa n'\kappa'}\psi_{n'\kappa'} \tag{1974}$$

with the saddle-point part that is diagonal in momentum space

$$\left(-\mathbf{G}_{sp}^{-1}\right)_{n,\vec{\kappa},n',\vec{\kappa}'}=\begin{bmatrix} -i\omega_n+\varsigma_{\vec{\kappa}\uparrow} & -\Delta \\ -\Delta^* & -i\omega_n-\varsigma_{\vec{\kappa}\downarrow}\end{bmatrix}\delta_{n',n}\delta(\vec{\kappa}-\vec{\kappa}') \tag{1975}$$

and the correction matrix (fluctuation propagator):

$$\left(\mathbf{F}\right)_{n,\vec{\kappa},n',\vec{\kappa}'}=-\frac{1}{\sqrt{\beta V}}\begin{bmatrix} 0 & \xi_{\vec{\kappa}+\vec{\kappa}',n+n'} \\ \xi^*_{\vec{\kappa}+\vec{\kappa}',n+n'} & 0\end{bmatrix} \tag{1976}$$

The notation of the inverse Green's function $-\mathbf{G}^{-1}$ is particularly convenient as its inverse encodes the full physics of the system. Different approximation schemes correspond to different choices for $-\mathbf{G}^{-1}$. As mentioned earlier, the simplest approximation scheme consists of completely ignoring the fluctuations via the following replacement for the coordinate-space pairing field:

$$\Delta(\vec{r},\tau)\to\Delta \tag{1977}$$

In momentum space, this approximation requires that the pairing should occur in a zero-momentum state, which is exactly the pairing Ansatz analyzed previously:

$$\Delta_{n,\vec{\kappa}}\to\sqrt{\beta V}\delta(\vec{\kappa})\delta_{n,0}\times\Delta \quad , \quad \Delta^*_{n,\vec{\kappa}}\to\sqrt{\beta V}\delta(\vec{\kappa})\delta_{n,0}\times\Delta^* \tag{1978}$$

For the inverse Green's function, the approximation yields:

$$-\mathbf{G}^{-1} \to -\mathbf{G}_{\text{sp}}^{-1} \tag{1979}$$

This completely neglects the fluctuation contribution coming from **F**. It is obvious that the action will be greatly simplified because the Δ field loses the dynamics and is replaced by its saddle-point value. So, it is no longer necessary to integrate over the Δ and Δ^* fields, and the mean-field partition function is obtained via (1853).

Hence, in this section, we have derived a more accurate approach that takes into account the fluctuations encoded in **F** in an approximate manner, rather than completely ignoring them as is done when deriving the mean field theory. So,

$$Z_{\text{Fluc}} = \int d\left[\xi^*\right] d\left[\xi\right] \exp\left\{-S^{\text{eff}}\left[\Delta^*, \Delta\right]\right\} \tag{1980}$$

where

$$S^{\text{eff}}\left[\Delta^*, \Delta\right] = -\sum_{n,\vec{\kappa}} \frac{\left(\sqrt{\beta V}\delta(\vec{\kappa})\delta_{n,0} \times \Delta^* + \xi^*_{n,\vec{\kappa}}\right)\left(\sqrt{\beta V}\delta(\vec{\kappa})\delta_{n,0} \times \Delta + \xi_{n,\vec{\kappa}}\right)}{|\lambda|} - \text{Tr}\left[\ln\left(-\mathbf{G}^{-1}\right)\right] \tag{1981}$$

The next step consists of expanding $\ln\left(-\mathbf{G}^{-1}\right)$ in expression (1980), considering that the matrix elements of **F** are small in comparison to the elements of $\mathbf{G}_{\text{sp}}^{-1}$:

$$\ln\left(-\mathbf{G}^{-1}\right) = \ln\left(-\mathbf{G}_{\text{sp}}^{-1} + \mathbf{F}\right) = \ln\left(-\mathbf{G}_{\text{sp}}^{-1} + \mathbf{G}_{\text{sp}}^{-1}\mathbf{G}_{\text{sp}}\mathbf{F}\right) = \ln\left(-\mathbf{G}_{\text{sp}}^{-1}\left(\hat{1} - \mathbf{G}_{\text{sp}}\mathbf{F}\right)\right) = \ln\left(-\mathbf{G}_{\text{sp}}^{-1}\right) + \ln\left(\left(\hat{1} - \mathbf{G}_{\text{sp}}\mathbf{F}\right)\right) \tag{1982}$$

The first term $\ln\left(-\mathbf{G}_{\text{sp}}^{-1}\right)$ gives the mean-field contribution exactly, where $\mathbf{G}_{\text{sp}}$ is the inverse of $\mathbf{G}_{\text{sp}}^{-1}$:

$$\left(\mathbf{G}_{\text{sp}}\right)_{n,\vec{\kappa},n',\vec{\kappa}'} = \delta(\vec{\kappa} - \vec{\kappa}')\delta_{nn'} \begin{bmatrix} i\omega_n - \varsigma_{\vec{\kappa}\uparrow} & \Delta \\ \Delta^* & i\omega_n + \varsigma_{\vec{\kappa}\downarrow} \end{bmatrix}^{-1} = \frac{\delta(\vec{\kappa} - \vec{\kappa}')\delta_{nn'}}{\det\left[-\mathbf{G}_{\text{sp}}^{-1}\right]_{\vec{\kappa}n}} \begin{bmatrix} i\omega_n + \varsigma_{\vec{\kappa}\downarrow} & -\Delta \\ -\Delta^* & i\omega_n - \varsigma_{\vec{\kappa}\uparrow} \end{bmatrix} \tag{1983}$$

where

$$\det\left[-\mathbf{G}_{\text{sp}}^{-1}\right]_{\vec{\kappa}n} = (i\nu_n - \text{E}_{\vec{\kappa}})(i\nu_n + \text{E}_{\vec{\kappa}}) \tag{1984}$$

Therefore, the effective action functional

$$S^{\text{eff}}\left[\Delta^*, \Delta\right] = -\sum_{n,\vec{\kappa}} \frac{\left(\sqrt{\beta V}\delta(\vec{\kappa})\delta_{n,0} \times \Delta^* + \xi^*_{n,\vec{\kappa}}\right)\left(\sqrt{\beta V}\delta(\vec{\kappa})\delta_{n,0} \times \Delta + \xi_{n,\vec{\kappa}}\right)}{|\lambda|} - \text{Tr}\ln\left(-\mathbf{G}_{\text{sp}}^{-1}\right) - \text{Tr}\ln\left(\left(\hat{1} - \mathbf{G}_{\text{sp}}\mathbf{F}\right)\right) \tag{1985}$$

Let us make a change in sign in the index of the ξ terms:

$$\xi^*_{\vec{\kappa}'+\vec{\kappa},n'+n} \to \xi^*_{\vec{\kappa}'-\vec{\kappa},n'-n} \quad , \quad \xi_{\vec{\kappa}+\vec{\kappa}',n+n'} \to \xi_{\vec{\kappa}-\vec{\kappa}',n-n'} \tag{1986}$$

Then,

$$(\mathbf{F})_{n,\vec{\kappa},n',\vec{\kappa}'} = -\frac{1}{\sqrt{\beta V}} \begin{bmatrix} 0 & \xi_{\vec{\kappa}-\vec{\kappa}',n-n'} \\ \xi^*_{\vec{\kappa}'-\vec{\kappa},n'-n} & 0 \end{bmatrix} \tag{1987}$$

and

$$\left(\mathbf{G}_{\mathrm{sp}}\mathbf{F}\right)_{n,\vec{\kappa},n',\vec{\kappa}'} = -\frac{1}{\sqrt{\beta V}}\frac{1}{\det\left[-\mathbf{G}_{\mathrm{sp}}^{-1}\right]_{\vec{\kappa}n}}\begin{bmatrix} i\omega_n+\varsigma_{\vec{\kappa}\downarrow} & -\Delta \\ -\Delta^* & i\omega_n-\varsigma_{\vec{\kappa}\uparrow}\end{bmatrix}\begin{bmatrix} 0 & \xi_{\vec{\kappa}'-\vec{\kappa},n'-n} \\ \xi^*_{\vec{\kappa}-\vec{\kappa}',n-n'} & 0\end{bmatrix} \tag{1988}$$

After substituting (1982) in equation (1980), we have

$$Z_{\mathrm{Fluc}} = \int d\left[\xi^*\right]d\left[\xi\right]\exp\left\{-S^{\mathrm{eff}}\left[\Delta^*,\Delta\right]\right\} \tag{1989}$$

where $S^{\mathrm{eff}}\left[\Delta^*,\Delta\right]$ is the effective action functional in bosonic fields:

$$S^{\mathrm{eff}}\left[\Delta^*,\Delta\right]$$

$$= -\sum_{n,\vec{\kappa}}\frac{1}{|\lambda|}\left(\sqrt{\beta V}\delta(\vec{\kappa})\delta_{n,0}\times\Delta^*+\xi^*_{n,\vec{\kappa}}\right)\left(\sqrt{\beta V}\delta(\vec{\kappa})\delta_{n,0}\times\Delta+\xi_{n,\vec{\kappa}}\right)-\mathrm{Tr}\left[\ln\left(-\mathbf{G}_{\mathrm{sp}}^{-1}\right)-\mathrm{Tr}\left[\ln\left(\hat{1}-\mathbf{G}_{\mathrm{sp}}\mathbf{F}\right)\right]\right] \tag{1990}$$

or

$$S^{\mathrm{eff}}\left[\Delta^*,\Delta\right] = -\frac{\beta V|\Delta|^2}{|\lambda|}-\mathrm{Tr}\left[\ln\left(-\mathbf{G}_{\mathrm{sp}}^{-1}\right)\right]-\mathrm{Tr}\left[\ln\left(\hat{1}-\mathbf{G}_{\mathrm{sp}}\mathbf{F}\right)\right]-\frac{\sqrt{\beta V}}{|\lambda|}\left(\Delta^*\xi_{0,0}+\Delta\xi^*_{0,0}\right)-\sum_{n,\vec{\kappa}}\frac{\xi^*_{n,\vec{\kappa}}\xi_{n,\vec{\kappa}}}{|\lambda|} \tag{1991}$$

From here, the partition function for the system is the product of the **old** mean-field contribution Z_{sp} times the fluctuation Z_{Fluc} contribution:

$$Z_{\mathrm{F}} = Z_{\mathrm{sp}}Z_{\mathrm{Fluc}} \tag{1992}$$

which is up to quadratic order, where the saddle-point partition function is as follows

$$Z_{\mathrm{sp}} = \exp\left\{\frac{\beta V|\Delta|^2}{|\lambda|}+\mathrm{Tr}\left[\ln\left(-\mathbf{G}_{\mathrm{sp}}^{-1}\right)\right]\right\} \equiv \exp\left\{\beta V\Omega_{\mathrm{sp}}\left(T,V,\varsigma,\mu\right)\right\} \tag{1993}$$

with

$$\Omega_{\mathrm{sp}}\left(T,V,\varsigma,\mu\right) = -\frac{|\Delta|^2}{|\lambda|}-\frac{1}{\beta V}\mathrm{Tr}\left[\ln\left(-\mathbf{G}_{\mathrm{sp}}^{-1}\right)\right] \tag{1994}$$

The partition function with the action function from the Gaussian pairing fluctuations are, respectively, as follows

$$Z_{\mathrm{Fluc}} = \int d\left[\xi^*\right]d\left[\xi\right]\exp\left\{-S^{\mathrm{eff}}_{\mathrm{Fluc}}\left[\Delta^*,\Delta\right]\right\} \tag{1995}$$

and

$$S^{\mathrm{eff}}_{\mathrm{Fluc}}\left[\Delta^*,\Delta\right] = -\mathrm{Tr}\left[\ln\left(\hat{1}-\mathbf{G}_{\mathrm{sp}}\mathbf{F}\right)\right]-\frac{\sqrt{\beta V}}{|\lambda|}\left(\Delta^*\xi_{0,0}+\Delta\xi^*_{0,0}\right)-\sum_{n,\vec{\kappa}}\frac{\xi^*_{n,\vec{\kappa}}\xi_{n,\vec{\kappa}}}{|\lambda|} \tag{1996}$$

Because the saddle-point value Δ for the pairing field is calculated self-consistently by minimizing the action, we expect that the fluctuations $\xi_{n,\vec{\kappa}}$ and $\xi^*_{n,\vec{\kappa}}$ around the saddle point should be small. This justifies the expansion of the last summand of (1996) in powers of **F** because each nonzero entry of **F** is proportional to $\xi_{n,\vec{\kappa}}$ or $\xi^*_{n,\vec{\kappa}}$. Because we have neglected terms above Gaussian order in the fluctuations, we have

$$\ln\left(\hat{1}-\mathbf{G}_{\text{sp}}\mathbf{F}\right)=-\sum_{n=1}^{\infty}\frac{\left(\mathbf{G}_{\text{sp}}\mathbf{F}\right)^n}{n}\approx-\mathbf{G}_{\text{sp}}\mathbf{F}-\frac{1}{2}\mathbf{G}_{\text{sp}}\mathbf{F}\mathbf{G}_{\text{sp}}\mathbf{F}+O\left(\left(\mathbf{G}_{\text{sp}}\mathbf{F}\right)^3\right) \tag{1997}$$

Therefore, the Gaussian approximation scheme can be written in terms of the inverse Green's function:

$$-\mathbf{G}^{-1}\to-\mathbf{G}_{\text{sp}}^{-1}-\mathbf{G}_{\text{sp}}\mathbf{F}-\frac{1}{2}\mathbf{G}_{\text{sp}}\mathbf{F}\mathbf{G}_{\text{sp}}\mathbf{F} \tag{1998}$$

This confirms the partition function for the system as

$$Z_{\text{F}}=Z_{\text{sp}}Z_{\text{Fluc}} \tag{1999}$$

and

$$S^{\text{eff}}_{\text{Fluc}}\left[\Delta^*,\Delta\right]=\text{Tr}\left[\mathbf{G}_{\text{sp}}\mathbf{F}\right]+\frac{1}{2}\text{Tr}\left[\mathbf{G}_{\text{sp}}\mathbf{F}\mathbf{G}_{\text{sp}}\right]-\frac{\sqrt{\beta V}}{|\lambda|}\left(\Delta^*\xi_{0,0}+\Delta\xi^*_{0,0}\right)-\sum_{n,\vec{\kappa}}\frac{\xi^*_{n,\vec{\kappa}}\xi_{n,\vec{\kappa}}}{|\lambda|} \tag{2000}$$

It is instructive to note that for the saddle-point approximation, linear fluctuation terms vanish due to the fact that first-order derivatives vanish. Therefore, the Gaussian partition function follows as

$$Z_{\text{Fluc}}=\int d\left[\xi^*_{n,\vec{\kappa}}\right]d\left[\xi_{n,\vec{\kappa}}\right]\exp\left\{-\tilde{S}^{\text{eff}}_{\text{Fluc}}\left[\Delta^*,\Delta\right]\right\} \tag{2001}$$

and is characterized by the Gaussian action

$$\tilde{S}^{\text{eff}}_{\text{Fluc}}\left[\Delta^*,\Delta\right]=-\sum_{n,\vec{\kappa}}\frac{\xi^*_{n,\vec{\kappa}}\xi_{n,\vec{\kappa}}}{|\lambda|}+\frac{1}{2}\text{Tr}\left[\mathbf{G}_{\text{sp}}\mathbf{F}\mathbf{G}_{\text{sp}}\mathbf{F}\right] \tag{2002}$$

Calculating $\text{Tr}\left[\mathbf{G}_{\text{sp}}\mathbf{F}\mathbf{G}_{\text{sp}}\mathbf{F}\right]$ allows us to calculate (2001).

In (1988), letting $\vec{q}=\vec{\kappa}-\vec{\kappa}'\quad m=n-n'$, then

$$\text{Tr}\left(\mathbf{G}_{\text{sp}}\mathbf{F}\mathbf{G}_{\text{sp}}\mathbf{F}\right)_{n,\vec{\kappa},n,\vec{\kappa}}=\frac{1}{\beta V}\sum_{m,\vec{q},n,\vec{\kappa}}\frac{1}{\det\left[-\mathbf{G}_{\text{sp}}^{-1}\right]_{\vec{q}+\vec{\kappa},m+n}\det\left[-\mathbf{G}_{\text{sp}}^{-1}\right]_{\vec{\kappa}n}}\times$$
$$\times\begin{bmatrix}\xi^*_{\vec{q},m} & \xi_{-\vec{q},-m}\end{bmatrix}\begin{bmatrix}\left(i\nu_{m+n}+\varsigma_{\vec{q}+\vec{\kappa}}\right)\left(i\nu_{m+n}-\varsigma_{\vec{\kappa}}\right) & \Delta^2\\ \left(\Delta^*\right)^2 & \left(i\nu_{m+n}-\varsigma_{\vec{q}+\vec{\kappa}}\right)\left(i\nu_n+\varsigma_{\vec{\kappa}}\right)\end{bmatrix}\begin{bmatrix}\xi_{\vec{q},m}\\ \xi^*_{-\vec{q},-m}\end{bmatrix} \tag{2003}$$

where

$$i\nu_{m+n}=i\omega_m+i\omega_n+\varsigma\quad,\quad i\nu_n=i\omega_n+\varsigma \tag{2004}$$

with

$$i\nu_n=i\omega_n+\varsigma=i\frac{(2n+1)\pi}{\beta}+\varsigma \tag{2005}$$

and

$$\omega_m = \frac{2m\pi}{\beta} \tag{2006}$$

are the shifted fermionic and bosonic Matsubara frequencies, respectively.

11.8 Fluctuation Partition Function

We now calculate the fluctuation Gaussian partition function:

$$Z_{\text{Fluc}} = \int d\left[\xi^*_{m,\vec{q}}\right] d\left[\xi_{m,\vec{q}}\right] \exp\left\{-\tilde{S}^{\text{eff}}_{\text{Fluc}}\left[\Delta^*, \Delta\right]\right\} \tag{2007}$$

with the fluctuation Gaussian action being

$$\tilde{S}^{\text{eff}}_{\text{Fluc}}\left[\Delta^*, \Delta\right] = -\sum_{m,\vec{q}} \frac{\xi^*_{m,\vec{q}}\xi_{m,\vec{q}}}{|\lambda|} + \frac{1}{2}\text{Tr}\left[\mathbf{G}_{\text{sp}}\mathbf{F}\mathbf{G}_{\text{sp}}\mathbf{F}\right] \tag{2008}$$

First, we calculate $\sum_{m,\vec{q}} \frac{\xi^*_{m,\vec{q}}\xi_{m,\vec{q}}}{|\lambda|}$ and consider that $\vec{q}$ takes positive and negative values:

$$\sum_{m,\vec{q}} \frac{\xi^*_{m,\vec{q}}\xi_{m,\vec{q}}}{|\lambda|} = \frac{1}{2}\sum_{m,\vec{q}} \frac{1}{|\lambda|}\left(\xi^*_{-m,-\vec{q}}\xi_{-m,-\vec{q}} + \xi^*_{m,\vec{q}}\xi_{m,\vec{q}}\right) = \frac{1}{2}\sum_{m,\vec{q}} \xi^*_{m,\vec{q}} \frac{1}{|\lambda|}\begin{bmatrix} 1 & 0 \\ 0 & 1 \end{bmatrix}\xi_{m,\vec{q}} \tag{2009}$$

Therefore, the fluctuation correction to the effective action is given by

$$\tilde{S}^{\text{eff}}_{\text{Fluc}}\left[\Delta^*, \Delta\right] = -\frac{1}{2}\sum_{m,\vec{q}} \xi^*_{m,\vec{q}} \mathrm{A}\left(\vec{q}, i\Omega_m\right)\xi_{m,\vec{q}} \tag{2010}$$

where the Nambu-Gorkov spinors are

$$\xi^*_{m,\vec{q}} = \begin{bmatrix} \xi^*_{\vec{q},m} & \xi_{-\vec{q},-m} \end{bmatrix}, \quad \xi_{m,\vec{q}} = \begin{bmatrix} \xi_{\vec{q},m} \\ \xi^*_{-\vec{q},-m} \end{bmatrix} \tag{2011}$$

and the 2×2 fluctuation matrix

$$\mathrm{A}\left(\vec{q}, i\omega_n\right) = \begin{bmatrix} \mathrm{A}_{11}\left(\vec{q}, i\Omega_n\right) & \mathrm{A}_{12}\left(\vec{q}, i\Omega\omega_n\right) \\ \mathrm{A}_{21}\left(\vec{q}, i\Omega_n\right) & \mathrm{A}_{22}\left(\vec{q}, i\Omega_n\right) \end{bmatrix} \tag{2012}$$

is the inverse pair fluctuation propagator, which describes the dynamics of the bosonic collective excitations of the theory, with its matrix elements given by the following fluctuation matrix elements

$$\mathrm{A}_{11}\left(\vec{q}, i\Omega_n\right) = \frac{1}{\beta V}\sum_{n,\vec{\kappa}} \frac{\left(i\nu_{m+n} + \varsigma_{\vec{q}+\vec{\kappa}}\right)\left(i\nu_n - \varsigma_{\vec{\kappa}}\right)}{\det\left[-\mathbf{G}^{-1}_{\text{sp}}\right]_{\vec{q}+\vec{\kappa},m+n} \det\left[-\mathbf{G}^{-1}_{\text{sp}}\right]_{\vec{\kappa}n}} - \frac{1}{|\lambda|} \tag{2013}$$

$$A_{22}(\vec{q},i\Omega_n)=\frac{1}{\beta V}\sum_{n,\vec{\kappa}}\frac{(i\nu_{m+n}-\varsigma_{\vec{q}+\vec{\kappa}})(i\nu_n+\varsigma_{\vec{\kappa}})}{\det\left[-\mathbf{G}_{\rm sp}^{-1}\right]_{\vec{q}+\vec{\kappa},m+n}\det\left[-\mathbf{G}_{\rm sp}^{-1}\right]_{\vec{\kappa}n}}-\frac{1}{|\lambda|} \tag{2014}$$

$$A_{12}(\vec{q},i\Omega_n)=\frac{1}{\beta V}\sum_{n,\vec{\kappa}}\frac{1}{\det\left[-\mathbf{G}_{\rm sp}^{-1}\right]_{\vec{q}+\vec{\kappa},m+n}\det\left[-\mathbf{G}_{\rm sp}^{-1}\right]_{\vec{\kappa}n}}\Delta^2 \tag{2015}$$

$$A_{12}(\vec{q},i\Omega_n)=\frac{1}{\beta V}\sum_{n,\vec{\kappa}}\frac{1}{\det\left[-\mathbf{G}_{\rm sp}^{-1}\right]_{\vec{q}+\vec{\kappa},m+n}\det\left[-\mathbf{G}_{\rm sp}^{-1}\right]_{\vec{\kappa}n}}\left(\Delta^*\right)^2 \tag{2016}$$

These matrix elements are dependent on the energy gap Δ, average chemical potential μ, imbalance chemical potential ς, and temperature T. It is important to note from the aforementioned that at low temperatures, the unique single-particle excitation is the Bogoliubov dispersion $E_{\vec{\kappa}}$, which is the quasi-particle energy (energy spectrum for breaking a Cooper pair), where Δ plays the role of the order parameter.

The evaluation of $A_{22}(\vec{q},i\Omega_n)$ is obtained from $A_{11}(\vec{q},i\Omega_n)$ by setting $\varsigma_{\vec{q}+\vec{\kappa}}\to-\varsigma_{\vec{q}+\vec{\kappa}}$ and $\varsigma_{\vec{\kappa}}\to-\varsigma_{\vec{\kappa}}$:

$$A_{11}(\vec{q},i\Omega_n)=A_{22}(-\vec{q},-i\Omega_n) \tag{2017}$$

This is because the saddle-point result shows that $|\Delta|$ is fixed by the gap equation and we can independently choose the phase and set Δ real:

$$A_{12}(\vec{q},i\Omega_n)=A_{21}(\vec{q},i\Omega_n)=A_{21}(-\vec{q},-i\Omega_n)=A_{12}(-\vec{q},-i\Omega_n) \tag{2018}$$

11.9 Fluctuation Bosonic Partition Function

Next, we calculate the fluctuation partition function:

$$Z_{\rm Fluc}=\int d\left[\xi^*_{m,\vec{q}}\right]d\left[\xi_{m,\vec{q}}\right]\exp\left\{-\frac{1}{2}\sum_{m,\vec{q}}\xi^\dagger_{m,\vec{q}}A(\vec{q},\Omega_m)\xi_{m,\vec{q}}\right\} \tag{2019}$$

by integrating out the fluctuations. It should be noted that

$$i\nu_{m+n}=i\Omega_m+i\omega_n+\varsigma\ ,\quad i\nu_n=i\omega_n+\varsigma\ ,\quad i\Omega_m=i\frac{2\pi m}{\beta} \tag{2020}$$

Considering symmetry properties of A, each term appears twice. Hence, we restrict summation in (2010) to half the $(\vec{q},m)$ space so that from

$$\frac{1}{2}\sum_{m,\vec{q}}\xi^\dagger_{m,\vec{q}}A(\vec{q},\Omega_m)\xi_{m,\vec{q}}=\sum_{\substack{m,\vec{q}\\ q_z\ge0}}\xi^\dagger_{m,\vec{q}}A(\vec{q},\Omega_m)\xi_{m,\vec{q}} \tag{2021}$$

and

$$\begin{aligned}Z_{\rm Fluc}&=\int d\left[\xi^*_{m,\vec{q}}\right]d\left[\xi_{m,\vec{q}}\right]\exp\left\{-\frac{1}{2}\sum_{m,\vec{q}}\xi^*_{m,\vec{q}}A(\vec{q},\Omega_m)\xi_{m,\vec{q}}\right\}\\&=\prod_{m,\vec{q}}\int d\xi^*_{m,\vec{q}}\,d\xi_{m,\vec{q}}\exp\left\{-\frac{1}{2}\sum_{m,\vec{q}}\xi^*_{m,\vec{q}}A(\vec{q},\Omega_m)\xi_{m,\vec{q}}\right\}\end{aligned} \tag{2022}$$

or

$$Z_{\text{Fluc}} = \prod_{\substack{m,\vec{q} \\ q_z \geq 0}} \int d\xi^*_{m,\vec{q}}\, d\xi_{m,\vec{q}} \int d\xi^*_{-m,-\vec{q}}\, d\xi_{-m,-\vec{q}} \exp\left\{ -\frac{1}{2} \sum_{m,\vec{q}} \xi^*_{m,\vec{q}} \mathrm{A}(\vec{q},\Omega_m) \xi_{m,\vec{q}} \right\} = \prod_{\substack{m,\vec{q} \\ q_z \geq 0}} \frac{\pi^4}{\left\| \mathrm{A}(\vec{q},\Omega_m) \right\|} \tag{2023}$$

where

$$\left\| \mathrm{A}(\vec{q},i\Omega_n) \right\| = \mathrm{A}_{11}(\vec{q},i\Omega_n)\mathrm{A}_{22}(\vec{q},i\Omega_n) - \mathrm{A}_{12}(\vec{q},i\Omega_n)\mathrm{A}_{21}(\vec{q},i\Omega_n) = \Gamma(\vec{q},i\Omega_n) \tag{2024}$$

We can therefore rewrite (2023) as follows

$$Z_{\text{Fluc}} = \exp\left\{ \sum_{\substack{m,\vec{q} \\ q_z \geq 0}} \ln\left[\frac{\pi^4}{\Gamma(\vec{q},\Omega_m)} \right] \right\} = \exp\left\{ \frac{1}{2} \sum_{m,\vec{q}} \ln\left[\frac{\pi^4}{\Gamma(\vec{q},\Omega_m)} \right] \right\} \tag{2025}$$

By shifting the overall free energy by a constant amount, the correction to the grand canonical partition function of the system is:

$$Z_{\text{Fluc}} = \exp\left\{ \frac{1}{2} \sum_{m,\vec{q}} \ln\left[\pi^4 \right] \right\} \exp\left\{ -\frac{1}{2} \sum_{m,\vec{q}} \ln\left[\Gamma(\vec{q},\Omega_m) \right] \right\} = \text{const} \exp\left\{ -\frac{1}{2} \sum_{m,\vec{q}} \ln\left[\Gamma(\vec{q},\Omega_m) \right] \right\} \tag{2026}$$

Considering that

$$Z_{\text{sp}} = \exp\left\{ \beta V \Omega_{\text{sp}}(T,V,\varsigma,\mu) \right\} \tag{2027}$$

then

$$Z_{\text{Fluc}} = \exp\left\{ \beta V \Omega_{\text{Fluc}}(T,V,\varsigma,\mu) \right\} \tag{2028}$$

This leads to the expression for the collective mode Gaussian contribution to the finite-temperature grand thermodynamic potential

$$\Omega_{\text{Fluc}} = -\frac{1}{2\beta V} \sum_{m,\vec{q}} \ln\left[\Gamma(\vec{q},i\Omega_m) \right] \tag{2029}$$

where

$$\Gamma(\vec{q},i\Omega_m) = \mathrm{A}_{11}(\vec{q},i\Omega_m)\mathrm{A}_{11}(\vec{q},-i\Omega_m) - \mathrm{A}^2_{12}(\vec{q},i\Omega_m) \tag{2030}$$

So, the thermodynamic will be the sum of the saddle-point contribution and fluctuations:

$$\Omega(T,\mu,\varsigma;\Delta) = \Omega_{\text{sp}}(T,\mu,\varsigma;\Delta) + \Omega_{\text{Fluc}}(T,\mu,\varsigma;\Delta) \tag{2031}$$

which is the sum of a saddle-point potential $\Omega_{\text{sp}}(T,\mu,\varsigma;\Delta)$ (for the condensed pairs) and a fluctuation potential $\Omega_{\text{Fluc}}(T,\mu,\varsigma;\Delta)$, which contributes to the total potential coming from the noncondensed pairs.

11.10 Number Equation Fluctuation Contributions

We now find the fluctuation contribution to the number equations, which is determined on the basis of the fluctuation contribution to the thermodynamic potential:

$$n_{\text{Fluc}} = -\frac{1}{\beta}\int \frac{d\vec{q}}{(2\pi)^3}\sum_{m=-\infty}^{\infty} \mathrm{M}(\vec{q}, i\Omega_m) \quad , \quad \delta n_{\text{Fluc}} = -\frac{1}{\beta}\int \frac{d\vec{q}}{(2\pi)^3}\sum_{m=-\infty}^{\infty} \mathrm{N}(\vec{q}, i\Omega_m) \tag{2032}$$

Here, for a complex augment z, the functions $\mathrm{M}(\vec{q},z)$ and $\mathrm{N}(\vec{q},z)$ are as follows

$$\mathrm{M}(\vec{q},z) = \frac{1}{\Gamma(\vec{q},z)}\left[\mathrm{A}_{11}(\vec{q},-z)\frac{\partial \mathrm{A}_{11}(\vec{q},z)}{\partial \mu} - \mathrm{A}_{12}(\vec{q},-z)\frac{\partial \mathrm{A}_{12}(\vec{q},z)}{\partial \mu}\right] \tag{2033}$$

$$\mathrm{N}(\vec{q},z) = \frac{1}{\Gamma(\vec{q},z)}\left[\mathrm{A}_{11}(\vec{q},-z)\frac{\partial \mathrm{A}_{11}(\vec{q},z)}{\partial \varsigma} - \mathrm{A}_{12}(\vec{q},-z)\frac{\partial \mathrm{A}_{12}(\vec{q},z)}{\partial \varsigma}\right] \tag{2034}$$

The functions $\mathrm{A}_{11}(\vec{q},z)$ and $\mathrm{A}_{12}(\vec{q},z)$ with the complex argument z are analytical in the complex z-plane except in the branching line, which lies at the real axis $z = \omega$. The summations over the boson Matsubara frequencies in (2032) can be converted to the contour integrals in the complex plane, as shown in Figure 11.3.

We consider the contour integral on the contour C shown in Figure 11.3 via formula

$$\mathrm{I}(\beta) = \frac{1}{2\pi i}\oint_{\mathrm{C}} \frac{f(z)}{\exp\{\beta z\}-1}dz \tag{2035}$$

Here, the points $z = i\Omega_m$ with $|m| > m_0$ lie inside the contour while the other points $z = i\Omega_m$ are outside the contour. The function $f(z)$:

I. Is analytic in the entire complex z-plane except, probably, in the branching line on the real axis.

II. Decreases at $\mathrm{Re}\, z \to -\infty$ faster than z^{-1} so that the integral

$$\int_{-\infty}^{0} f(\omega \pm i\in)\, d\omega \tag{2036}$$

converges, with ω and $\in$ being real variables.

These conditions are applied to the functions $\mathrm{M}(\vec{q},z)$ and $\mathrm{N}(\vec{q},z)$ defined in (2032). The function $\frac{1}{\exp\{\beta z\}-1}$ has the poles at the points $z = i\Omega_m$, $m = 0, \pm 1, \pm 2, \cdots$, where its residues at the given points are

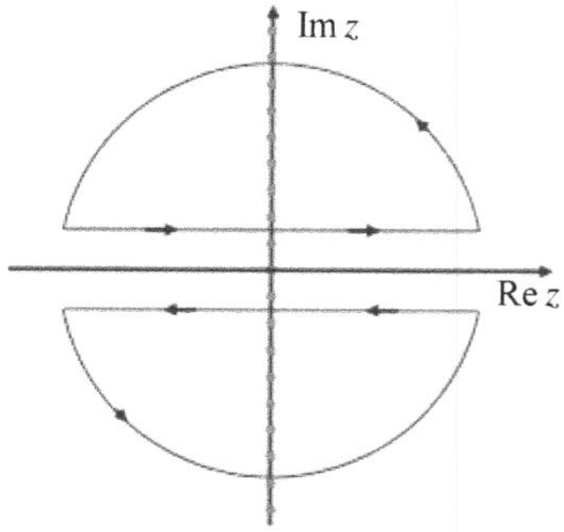

FIGURE 11.3 Integration contour in the complex z-plane. The full dots indicate the poles $z = i\Omega_m$, $m = 0, \pm 1, \pm 2, \cdots$.

equal to $\frac{1}{\beta}$. Hence, the integral $I(\beta)$ is equal to the sum of the residues of the function $\frac{f(z)}{\exp\{\beta z\}-1}$ at the points $z = i\Omega_m$ inside the contour C:

$$I(\beta) = \frac{1}{\beta} \sum_{|m|>m_0} f(i\Omega_m) \tag{2037}$$

On the one hand, this integral also can be represented as

$$I(\beta) = \frac{1}{2\pi i} \int_{-\infty}^{\infty} \frac{f(\omega+i\in)}{\exp\{\beta(\omega+i\in)\}-1} d\omega - \frac{1}{2\pi i} \int_{-\infty}^{\infty} \frac{f(\omega-i\in)}{\exp\{\beta(\omega-i\in)\}-1} d\omega \tag{2038}$$

Here, the parameter $\in$ satisfies the inequality

$$\Omega_{m_0} < \in < \Omega_{m_0+1} \tag{2039}$$

From the previous formulae, we have

$$\sum_m f(i\Omega_m) = \frac{\beta}{\pi} \int_{-\infty}^{\infty} \operatorname{Im}\left[\frac{f(\omega+i\in)}{\exp\{\beta(\omega+i\in)\}-1}\right] d\omega + \sum_{m=-m_0}^{m_0} f(i\Omega_m) \tag{2040}$$

Consequently, the fluctuation contributions to n and δn are:

$$n_{\text{Fluc}} = -\int \frac{d\vec{q}}{(2\pi)^3} \left(\frac{1}{\pi} \int_{-\infty}^{\infty} \operatorname{Im}\left[\frac{N(\vec{q},\omega+i\in)}{\exp\{\beta(\omega+i\in)\}-1}\right] d\omega + \frac{1}{\beta} \sum_{m=-m_0}^{m_0} N(\vec{q},i\Omega_m) \right) \tag{2041}$$

$$\delta n_{\text{Fluc}} = -\int \frac{d\vec{q}}{(2\pi)^3} \left(\frac{1}{\pi} \int_{-\infty}^{\infty} \operatorname{Im}\left[\frac{M(\vec{q},\omega+i\in)}{\exp\{\beta(\omega+i\in)\}-1}\right] d\omega + \frac{1}{\beta} \sum_{m=-m_0}^{m_0} M(\vec{q},i\Omega_m) \right) \tag{2042}$$

11.11 Collective Mode Excitations

For the Matsubara summation (2013)–(2016), the fermionic Matsubara summation in $A_{11}(\vec{q},i\Omega_n)$ as well as $A_{12}(\vec{q},i\Omega_n)$ yields, respectively,

$$A_{11}(\vec{q},i\Omega_m) = \frac{1}{V} \sum_{\vec{\kappa}} f_1(\beta, E_{\vec{\kappa}}, \varsigma) \left[\frac{(i\Omega_m + \varsigma_{\vec{q}+\vec{\kappa}} - E_{\vec{\kappa}})(E_{\vec{\kappa}} - \varsigma_{\vec{\kappa}})}{\det\left[-\mathbf{G}_{sp}^{-1}(-i\Omega_m)\right]_{\vec{q}+\vec{\kappa},m}} - \frac{(i\Omega_m + E_{\vec{\kappa}} + \varsigma_{\vec{q}+\vec{\kappa}})(E_{\vec{\kappa}} + \varsigma_{\vec{\kappa}})}{\det\left[-\mathbf{G}_{sp}^{-1}(i\Omega_m)\right]_{\vec{q}+\vec{\kappa},m}} \right] - \frac{1}{|\lambda|} \tag{2043}$$

$$A_{12}(\vec{q},i\Omega_m) = -\frac{\Delta^2}{V} \sum_{\vec{\kappa}} f_1(\beta, E_{\vec{\kappa}}, \varsigma) \left[\frac{1}{\det\left[-\mathbf{G}_{sp}^{-1}(-i\Omega_m)\right]_{\vec{q}+\vec{\kappa},m}} + \frac{1}{\det\left[-\mathbf{G}_{sp}^{-1}(i\Omega_m)\right]_{\vec{q}+\vec{\kappa},m}} \right] \tag{2044}$$

where

$$\det\left[-\mathbf{G}_{sp}^{-1}(i\Omega_m)\right]_{\vec{q}+\vec{\kappa},m} = (i\Omega_m + E_{\vec{\kappa}} - E_{\vec{q}+\vec{\kappa}})(i\Omega_m + E_{\vec{\kappa}} + E_{\vec{q}+\vec{\kappa}}) \tag{2045}$$

It is important to note that the thermodynamic potential $\Omega(T,\mu,\varsigma,\Delta)$ depends (via A_{11} and A_{21}) on the choice of the saddle point $|\Delta|$. This implies that the excitation spectrum is also dependent on the choice of the saddle point. In addition to nonzero temperatures, these excitations will be populated through Bose statistics. The excitation spectrum is temperature dependent, and single-particle (fermionic) and collective (bosonic) modes are temperature dependent and associated with fluctuations of Δ. However, at finite temperatures, there are excitations that give the atoms of a broken Cooper pair an extra energy $E_{\vec{\kappa}+\vec{q}} - E_{\vec{\kappa}}$. In addition, single-particle excitations may have collective modes when

$$\Gamma\left(\vec{q}, i\Omega_m \to \omega\right) = 0 \tag{2046}$$

These excitations correspond to plasmons when the Fermi gas is in the normal state, and the solution of the given equation is expanded in powers of $\vec{q}$ up to the fourth order yielding

$$\omega_q^2 = c_s^2 q^2 + \left(\frac{1}{2m}\right)^2 \Lambda q^4 + O\left(q^6\right) \tag{2047}$$

where Λ is a correction to the mass of the bosonic excitation. The Bogoliubov excitation is linear at small momenta from the Goldstone theorem. This implies

$$\omega_q = c_s q + o\left(q^2\right) \tag{2048}$$

and introduces the sound velocity c_s, which can also be obtained via equation (2047) by taking the square root of the coefficient at q^2. The second term in (2047) also allows us to find the effective mass of the particles

$$\frac{1}{m} = \frac{2}{\hbar q^2}\sqrt{\left(\omega_q^2 - c_s^2 q^2\right)\Lambda^{-1}} \tag{2049}$$

The dispersion relation of the pair excitation (the Bogoliubov-Anderson mode) can therefore be written as

$$\hbar\omega_q = \hbar q\sqrt{c_s^2 + \left(\frac{\hbar q}{2m}\right)^2 \Lambda} \tag{2050}$$

We have introduced $\hbar$ to get a better picture of the units. We then expand the square root in powers of q up to second order. This yields the following energy of the collective excitations:

$$\hbar\omega_q = \hbar q c_s\left(1 + \left(\frac{\hbar q}{c_s m}\right)^2 \frac{\Lambda}{8}\right) + \cdots \tag{2051}$$

This relation shows that the first correction to the linear dispersion is due to terms in q^3.

The BEC $c_s \to \frac{v_F}{2}\sqrt{\frac{\mu}{m}}$ as well as the BCS limits $c_s \to \frac{v_F}{\sqrt{3}}$ are all reproduced and in good agreement in the entire interaction domain—in particular, at unitarity and on the BCS side of the resonance.

12

Green's Function Averages over Impurities

Introduction

In this section, we use the Green's function formalism in describing the electron gas in the presence of an impurity with the main aim being introduction of the diagrammatic perturbation theory. The diagrammatic perturbation theory can be a useful technique in solving the problem of electrons propagating in a disordered system. An electron travelling from one point to another while being scattered at impurities on its way describes a one-particle-system. For complete knowledge of the system, it is important to find the impurity-averaged Green's function for this process where over all possible impurity configurations are averaged.

12.1 Scattering Potential and Disordered System

We apply the Green's function approach to another realistic problem concerning an electron in an impure metal. For convenience, we suppose the regular array of lattice ions is nonexistent so that we now have only a set of N randomly distributed impurity ions assumed to be identical in a volume Ω. For a **random ensemble**, the coordinate for the jth impurity, $\vec{R}_j$, is equally likely to be found anywhere in the volume Ω. **Random impurities in disordered metals are examples of elastic scattering by external potentials.** The role of impurities on kinetic (transport) properties in normal metals is very essential where the familiar transport equations for free quasiparticles may provide adequate mathematical description from the Landau Fermi liquid theory. Phase excitations have a more complex character in superconductivity; where the Cooper channel pairing introduces a new energy scale, T_c-, which is the phase transition temperature. To account for these new scales, the diagrammatic technique is the most suitable tool for addressing the problem of scattering by impurities [84]. The technique reproduces all well-known results obtained by the standard kinetic equation approach. Scattering from lattice vibrations and other electrons at low temperatures is greatly influenced by defects or impurities.

We will consider some diagrammatic applications via the propagator for an electron in a disordered potential. In addition, we will introduce the concept of **impurity average** where we study disorder in an otherwise free spinless electron gas. This is described by the following Hamiltonian determinant in the second-quantized form:

$$\hat{H} = \sum_{\vec{K}} \xi_{\vec{K}} \hat{\psi}^{\dagger}_{\vec{K}} \hat{\psi}_{\vec{K}} + \hat{H}_{sc} \tag{2052}$$

where the scattering Hamiltonian is

$$\hat{H}_{sc} = \int d\vec{r}\, d\tau \hat{\psi}_{\alpha} \hat{\psi}^{\dagger}_{\alpha} U(\vec{r}) \tag{2053}$$

The kinetic energy $\in(\vec{\kappa})$ referenced to the chemical potential μ is:

$$\xi_{\vec{\kappa}} = \in(\vec{\kappa}) - \mu \tag{2054}$$

and a point impurity creates a local potential

$$U(\vec{r}) = \lambda\delta^{(d)}(\vec{r}) \tag{2055}$$

with $\alpha \equiv (\vec{r},\tau)$. Performing a Fourier transform on the scattering Hamiltonian $\hat{H}_{sc}$, then

$$\hat{H}_{sc} = \sum_{\vec{\kappa}} U(\vec{\kappa}-\vec{\kappa}')\hat{\psi}^{\dagger}_{\vec{\kappa}}\hat{\psi}_{\vec{\kappa}'} \quad , \quad U(\vec{\kappa}-\vec{\kappa}') = \int d\vec{r}\exp\{i(\vec{\kappa}-\vec{\kappa}')\vec{r}\}U(\vec{r}) \tag{2056}$$

The full Hamiltonian $\hat{H}$ is quadratic and not easily solvable. Therefore, we use the Feynman diagrammatic technique.

12.2 Disorder Diagrams

It is not clear a priori that the perturbation $\hat{H}_{sc}$ is small. However in the diagrammatic perturbation theory technique, it is assumed an expansion in $\hat{H}_{sc}$ is likely, and resummation of that expansion gives the correct results. Since we expand in $\hat{H}_{sc}$ then we require the temperature Green's function for the unperturbed Hamiltonian that is the first term in (2052). From the higher-order Green's function in (474), we write the following two-point Green's function:

$$\mathbf{G}(\vec{r},\vec{r}',\tau-\tau') = \frac{1}{Z_0}\mathrm{Tr}\left[\hat{T}\exp\left\{-\int_0^{\beta} d\tau'\hat{H}(\tau')\right\}\hat{\psi}(\vec{r},\tau)\hat{\psi}^{\dagger}(\vec{r}',\tau')\right] \tag{2057}$$

The perturbation expansion to the first-order term can be represented in terms of the bare Green's function $\mathbf{G}_0$:

$$\mathbf{G}^{(1)}(\vec{r},\vec{r}',\tau-\tau') = \int d\vec{r}_1\, d\tau_1 U(\vec{r}_1)\mathbf{G}_0(\vec{r},\vec{r}_1,\tau-\tau_1)\mathbf{G}_0(\vec{r}',\vec{r}_1,\tau_1-\tau') \tag{2058}$$

and diagrammatically represented as showing a particle traveling from α' to α via an intermediate scattering event at $\alpha_1' \equiv (\vec{r}_1',\tau_1')$. We solve our problem in momentum space with the Green's function being

$$\mathbf{G}(\vec{\kappa},\vec{\kappa}',\tau-\tau') = \int d\vec{r}\, d\vec{r}'\exp\{i\vec{\kappa}\vec{r}\}\exp\{-i\vec{\kappa}'\vec{r}'\}\mathbf{G}(\vec{r},\vec{r}',\tau-\tau') \tag{2059}$$

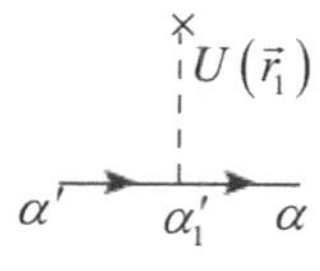

FIGURE 12.1 First-order correction to **G** showing particle traveling from α' to α via an intermediate scattering event at $\alpha_1' \equiv (\vec{r}_1',\tau_1')$. The dashed lines with the cross correspond to the impurity potential U.

and in diagrammatic form in the $(\vec{\kappa},\omega)$ space:

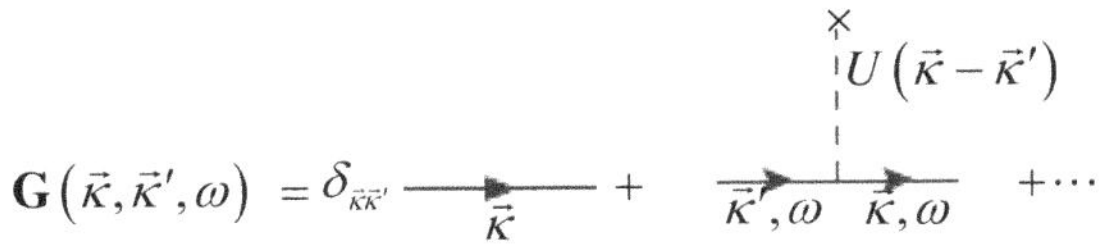

(2060)

In equation (2060), the momentum argument of the impurity potential is the difference in electron momenta at the scattering vertex. Lack of translational invariance implies momentum is not conserved and the static nature of the disorder potential implies

- **Energy is conserved at the scattering vertex.**
- **The vertex is energy independent.**

12.3 Perturbation Series T-Matrix Expansion

In this section, we use the Green's function formalism to describe the electron gas in the presence of an impurity with the main goal of introducing the formal diagrammatic perturbation theory technique. We sum over all diagrams leading to the formal expansion of the Green's function in powers of the impurity potential in Figure 12.2:

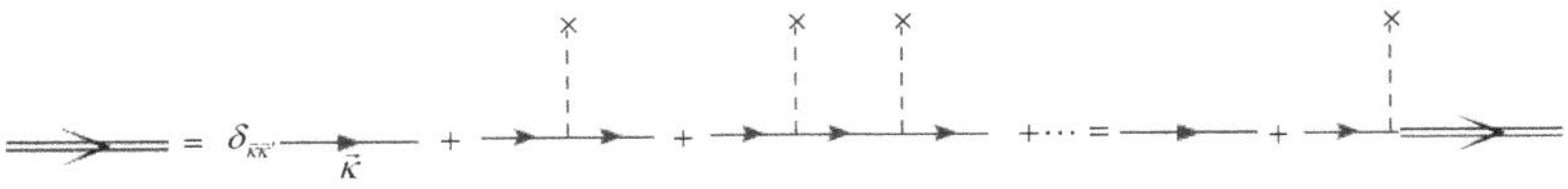

(2061)

FIGURE 12.2 Diagrammatic series expansion of the impurity averaged Green's function. The double directed line, corresponds to the full Green's function while the single directed line correspond to the Green's functions of the free electron Hamiltonian. First diagram represents amplitude transmission without scattering, and subsequent diagrams multiple scattering processes involving one, two, three, and more scattering events. The last equality is the diagrammatic representation of the Dyson equation for the impurity scattering.

The first diagram represents the amplitude transmission without scattering, and then subsequent diagrams show multiple scattering processes involving one, two, three, and more scattering events. Note that the Matsubara frequency ω is conserved at every scattering event. The diagrammatic expansion in Figure 12.2 can be represented by the following self-consistent equation for $\mathbf{G}$:

$$\mathbf{G}(\vec{\kappa},\vec{\kappa}',\omega)=\delta_{\vec{\kappa}\vec{\kappa}'}\mathbf{G}_0(\vec{\kappa},\omega)+\sum_{\vec{\kappa}''}\mathbf{G}_0(\vec{\kappa},\omega)U(\vec{\kappa}-\vec{\kappa}'')\mathbf{G}(\vec{\kappa}',\vec{\kappa}'',\omega) \tag{2062}$$

Writing an ansatz for the solution to equation (2062) as an infinite series, then

$$\mathbf{G}(\vec{\kappa},\vec{\kappa}',\omega)=\sum_{n=0}^{\infty}\mathbf{G}^{(n)}(\vec{\kappa},\vec{\kappa}',\omega) \tag{2063}$$

where $\mathbf{G}^{(n)}$ has n powers of the nondiagonal matrix elements $U(\vec{\kappa}_i-\vec{\kappa}_j)$. If the nondiagonal part achieves the value of zero, then (2062) is obviously the unperturbed Green's function:

$$\mathbf{G}_0(\vec{\kappa},\vec{\kappa}',\omega)=\mathbf{G}_0(\vec{\kappa},\omega)\delta_{\vec{\kappa}\vec{\kappa}'} \quad , \quad \mathbf{G}(\vec{\kappa},\vec{\kappa}',\omega)\equiv\mathbf{G}^0(\vec{\kappa},\vec{\kappa}',\omega) \tag{2064}$$

Substituting equation (2063) into (2062) cancels the zero-order terms resulting in the following equation:

$$\mathbf{G}^{(n)}(\vec{\kappa},\vec{\kappa}',\omega)=\mathbf{G}^{(0)}(\vec{\kappa},\omega)\sum_{\vec{\kappa}''}U(\vec{\kappa}-\vec{\kappa}'')\mathbf{G}^{(n-1)}(\vec{\kappa}',\vec{\kappa}'',\omega) \quad , \quad n\geq 1 \tag{2065}$$

12.4 T-Matrix Expansion

Writing equation (2062) via the **T-matrix** $\mathrm{T}(\vec{\kappa},\vec{\kappa}',\omega)$, which is denoted by a bold dot in Figure 12.3, represents a self-consistent set of Feynman diagrams that in potential scattering diagrams plays the same role as the self-energy in two-body interaction diagrams:

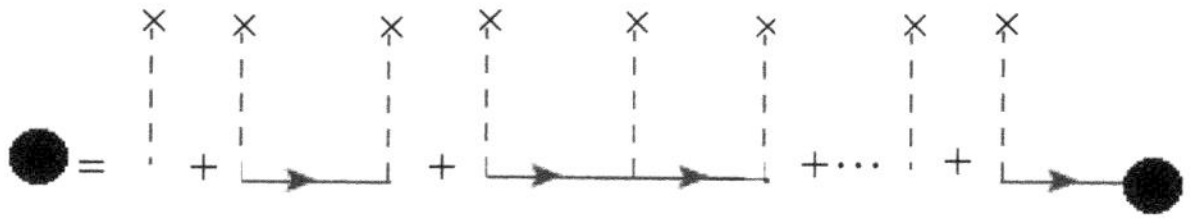

(2066)

FIGURE 12.3 Self-Consistent Set of Feynman Diagrams in Potential Scattering Diagrams.

The Green's function is represented as

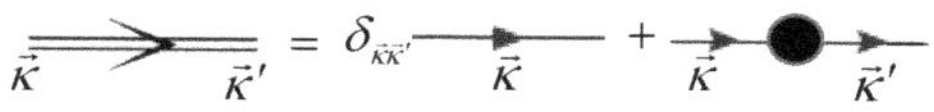

(2067)

FIGURE 12.4 Two-Body Interaction Green's Function Diagrammatic Representation in the Dyson-like form.

Because the T-matrix is a scattering matrix, for consistency, we perform the following denotation to permit proper understanding in other sections of this book:

$$\mathrm{T}(\vec{\kappa},\vec{\kappa}',\omega)\equiv\Gamma(\vec{\kappa},\vec{\kappa}',\omega) \tag{2068}$$

Therefore, the Green's function can be represented by the equation:

$$\mathbf{G}(\vec{\kappa},\vec{\kappa}',\omega)=\delta_{\vec{\kappa}\vec{\kappa}'}\mathbf{G}_0(\vec{\kappa},\omega)+\mathbf{G}_0(\vec{\kappa},\omega)\Gamma(\vec{\kappa},\vec{\kappa}',\omega)\mathbf{G}_0(\vec{\kappa}',\omega) \tag{2069}$$

The T-matrix imitates the renormalized scattering and takes into account not just a single scattering event but the sum of all possible multiple scattering events as well. The T-matrix from equation (2066) can be rewritten by the self-consistent Dyson equation

$$\Gamma(\vec{\kappa},\vec{\kappa}',\omega)=U(\vec{\kappa}-\vec{\kappa}')+\sum_{\vec{\kappa}''}U(\vec{\kappa}-\vec{\kappa}'')\mathbf{G}_0(\vec{\kappa}'',\omega)\Gamma(\vec{\kappa}'',\vec{\kappa}',\omega) \tag{2070}$$

that fully describes the scattering off the impurity.

Feynman Diagrammatic (T-Matrix) Technique

To consider an electron scattering off an attractive central scattering potential, one example we can use is s-wave scattering off a point-like scattering center:

$$U(\vec{r})=\lambda\delta^{(d)}(\vec{r}) \tag{2071}$$

Resumming the Feynman diagrams (T-matrix) shows that in $d \leq 2$ dimensions, an arbitrarily weak attractive potential yields bound states with $U(\vec{\kappa}-\vec{\kappa}')=\lambda$ being independent of momentum transfer. From observation, the T-matrix is momentum independent, and the Dyson equation in (2070) becomes

$$\Gamma(\vec{\kappa},\vec{\kappa}',\omega)=U(\vec{\kappa}-\vec{\kappa}')+\lambda\sum_{\vec{\kappa}''}G_0(\vec{\kappa}'',\omega)\Gamma(\vec{\kappa}'',\vec{\kappa}',\omega) \tag{2072}$$

Because T is independent of $\vec{\kappa}$ or $\vec{\kappa}'$, the equation then becomes

$$\Gamma(\omega)=\lambda+\lambda F(\omega)\Gamma(\omega) \quad , \quad F(\omega)=\sum_{\vec{\kappa}''}G_0(\vec{\kappa}'',\omega) \tag{2073}$$

So,

$$\Gamma(\omega)=\frac{\lambda}{1-\lambda F(\omega)} \tag{2074}$$

where the d-dimensional case

$$F(\omega)=\int\frac{d^d\vec{\kappa}}{(2\pi)^d}\frac{1}{\omega-\xi_{\vec{\kappa}}+i\,\mathrm{sgn}\,\xi_{\vec{\kappa}}} \tag{2075}$$

and for any spherically symmetric spectrum $\in(\vec{\kappa})=\in(|\vec{\kappa}|)$, then

$$F(\omega)=\int d\in\frac{\nu(\in)}{\omega-\in+i\delta\,\mathrm{sgn}\in} \tag{2076}$$

where $\nu(\in)$ is the density of states. For the case of single-particle behavior (i.e., the chemical potential $\mu=0$), then

$$F(\omega)=\int_0^{\Lambda}d\in\frac{\nu(\in)}{\omega-\in+i\delta} \tag{2077}$$

Here, the high-energy cut-off Λ is introduced to guarantee the convergence of the given integral where, physically, such a cutoff matches the energy scale beyond which the scattering potential behavior is no longer point-like. At low energies, $F(\omega)<0$, when $\lambda<0$, there is the possibility that poles in the T-matrix will correspond to bound states. **Integrals that do not converge at high energy** in the field theory are known as **ultraviolet divergence**, whereas **infrared divergence occurs at low energy**.

The scattering T-matrix describes scattering in the presence of a Fermi sea. To recover free-particle behavior, we imagine the Fermi sea to be empty, so that the chemical potential is zero:

$$\in=\frac{\kappa^2}{2m} \tag{2078}$$

In d-dimensions, the density of states:

$$\nu(\in)\approx\kappa^{d-1}\frac{d\kappa}{d\in}\approx\in^{\frac{d}{2}-1} \tag{2079}$$

The low energy behavior for $F(\omega)$ is then

$$F(\omega) \approx -\omega^{\frac{d}{2}-1} \tag{2080}$$

and diverges in dimensions $d \leq 2$. So, there are bound states for arbitrarily small attractive potentials. In $d = 2$ dimension, the density of states is

$$\nu(\omega) = \nu(0) \tag{2081}$$

and

$$F(\omega) = -\nu(\omega)\ln\frac{\Lambda}{-\omega} \tag{2082}$$

So, for arbitrarily small attraction (negative) $\lambda = -|\lambda|$, then

$$\Gamma(\omega) = -\frac{|\lambda|}{1-|\lambda|\nu(0)\ln\frac{\Lambda}{-\omega}} = \frac{1}{\nu(0)\ln\left(\frac{\omega_0}{-\omega}\right)} \tag{2083}$$

where

$$\omega_0 = \Lambda \exp\left\{-\frac{1}{|\lambda|\nu(0)}\right\} \tag{2084}$$

This gives rise to a bound state at energies $\omega = -\omega_0$. So, a delta-function potential will have at least one bound state for arbitrarily small potential in $d < 2$ dimensions; in $d = 3$, a finite critical λ is needed to form a bound state. It is worth nothing that

- **The energy scale ω_0 cannot be written as a power series in λ because it is an example of a nonperturbative result. The bound state appears due to resummation of an infinite class of Feynman diagrams.**
- **The presence of a bound state for electrons scattering off an arbitrarily weak attractive potential imitates Cooper instability.**

12.5 Disorder Averaging

The previous analysis makes sense for a single impurity. We now consider a finite density of impurities where the perturbation series as well as the Green's function depend on the exact position of each of the impurities, which implies that the exact microscopic properties are dependent on the exact realization of disorder. Though typical macroscopic properties are independent of this detail, we find a method to average over disorder realizations to achieve a typical impure material. For concreteness,

$$\hat{H} = \sum_{\vec{\kappa}} \xi_{\vec{\kappa}} \hat{\psi}^{\dagger}_{\vec{\kappa}} \hat{\psi}_{\vec{\kappa}} + \hat{H}_{\text{disorder}} \tag{2085}$$

This contains no interactions among electrons. Therefore, the energy of each individual electron is conserved (all interactions are elastic). Here, $\hat{H}_{\text{disorder}}$ is of the same form as the scattering potential seen earlier:

$$\hat{H}_{\text{disorder}} = \int d\vec{r}U(\vec{r})\hat{\psi}^{\dagger}(\vec{r})\hat{\psi}(\vec{r}) \tag{2086}$$

and $U(\vec{r})$ represents the scattering potential generated by a random array of N_i impurities located at positions $\vec{R}_j$, each with an atomic potential $V\left(\vec{r}-\vec{R}_j\right)$ in the neighborhood of each impurity:

$$U(\vec{r}) = \sum_j V\left(\vec{r}-\vec{R}_j\right) \tag{2087}$$

So, the Green's function **G** seen previously should be for a particular set of $\vec{R}_j$s (i.e., for a particular array of impurities in the system), and for each different set of $\vec{R}_j$s, we obtain a different value of the Green's function **G**. We will now consider an ensemble consisting of all possible arrays of impurities where we suppose the ensemble is **random**. This implies that the coordinate $\vec{R}_j$ for the impurity j is equally likely to be located anywhere in the volume Ω. Suppose we compute $\langle \mathbf{G}\rangle \equiv \overline{\mathbf{G}}$ for the average value of **G** for the ensemble. It is obvious that $\mathbf{G} \neq \overline{\mathbf{G}}$. Commonly, for large systems in the limit $N \to \infty$ and $\frac{N}{\Omega} = \text{constant}$ while $\frac{\langle \mathbf{G}^2\rangle - \langle \mathbf{G}\rangle^2}{\langle \mathbf{G}\rangle^2} \to 0$ and so $\mathbf{G} = \overline{\mathbf{G}}$, there is all but a negligible number of arrays of impurities. Therefore, it is necessary to calculate $\overline{\mathbf{G}}$.

Representing the Hamiltonian $\hat{H}_{\text{disorder}}$ in the first quantized form, we state:

$$\hat{H}_{\text{disorder}} = \sum_{i=1}^{N} U(\vec{r}_i) \tag{2088}$$

where $\vec{r}_i$ are the electron coordinates and N the total number of electrons. The single-particle operator, $\hat{H}_{\text{disorder}}$, is not diagonal in the momentum basis. So, from equation (2087) and the basis functions, we can write

$$\varphi_{\vec{\kappa}}(\vec{r}) = \frac{1}{\sqrt{\Omega}}\exp\left\{i\vec{\kappa}\vec{r}\right\} \tag{2089}$$

then

$$\hat{H}_{\text{disorder}} = \sum_{\vec{\kappa}\vec{\kappa}'} \left\langle \vec{\kappa}' \middle| U(\vec{r}) \middle| \vec{\kappa} \right\rangle \hat{\psi}^{\dagger}_{\vec{\kappa}'}\hat{\psi}_{\vec{\kappa}} \tag{2090}$$

where

$$\left\langle \vec{\kappa}' \middle| U(\vec{r}) \middle| \vec{\kappa} \right\rangle = \int d\vec{r}\varphi^{*}_{\vec{\kappa}'}(\vec{r})V(\vec{r})\varphi_{\vec{\kappa}}(\vec{r}) = \frac{1}{\Omega}\int d\vec{r}\exp\left\{-i\left(\vec{\kappa}'-\vec{\kappa}\right)\vec{r}\right\}U(\vec{r}) \tag{2091}$$

or

$$\left\langle \vec{\kappa}' \middle| U(\vec{r}) \middle| \vec{\kappa} \right\rangle = \frac{1}{\Omega}\sum_{j=1}^{N}\int d\vec{r}\exp\left\{-i\left(\vec{\kappa}'-\vec{\kappa}\right)\vec{r}\right\}U\left(\vec{r}-\vec{R}_j\right) = \frac{1}{\Omega}\sum_{j=1}^{N}\int d\vec{r}\exp\left\{-i\left(\vec{\kappa}'-\vec{\kappa}\right)\left(\vec{r}'+\vec{R}_j\right)\right\}U(\vec{r}') \tag{2092}$$

or

$$\langle \vec{\kappa}' | U(\vec{r}) | \vec{\kappa} \rangle = U(\vec{\kappa}' - \vec{\kappa}) \rho(\vec{\kappa}' - \vec{\kappa}) \tag{2093}$$

Here,

$$U(\vec{\kappa}) = \frac{1}{\Omega} \int d\vec{r} \exp\{-i\vec{\kappa}\vec{r}\} U(\vec{r}) \quad , \quad \rho(\vec{\kappa}) = \sum_{j=1}^{N} \exp\{-i\vec{\kappa}\vec{R}_j\} \tag{2094}$$

The volume Ω of the system serves as a normalization factor, and the impurity density $\rho(\vec{\kappa})$ has all the information about the position of the impurities.

12.6 Green's Function Perturbation Series

We determine from the section and with the random potential (2087) that the Green's function has the form

$$\mathbf{G}(\vec{\kappa}, \vec{\kappa}', \omega) = \sum_{n=0}^{\infty} \mathbf{G}^{(n)}(\vec{\kappa}, \vec{\kappa}', \omega) \tag{2095}$$

where the n order term

$$\begin{aligned} \mathbf{G}^{(n)}(\vec{\kappa}, \vec{\kappa}', \omega) = \sum_{\vec{\kappa}_1 \cdots \vec{\kappa}_{n-1}} & \mathbf{G}^{(0)}(\vec{\kappa}, \omega) U(\vec{\kappa} - \vec{\kappa}_1) \rho(\vec{\kappa} - \vec{\kappa}_1) \mathbf{G}^{(0)}(\vec{\kappa}_1, \omega) \times \cdots \times \\ & \times \mathbf{G}^{(0)}(\vec{\kappa}_{n-1}, \omega) U(\vec{\kappa}_{n-1} - \vec{\kappa}') \rho(\vec{\kappa}_{n-1} - \vec{\kappa}') \mathbf{G}^{(0)}(\vec{\kappa}', \omega) \end{aligned} \tag{2096}$$

However, $\mathbf{G}(\vec{\kappa}, \vec{\kappa}', \omega)$ is not diagonal in $\vec{\kappa}$ because the impurities render the system not translationally invariant. This n-order contribution (2096) is interpreted as the sum over all processes involving n scattering events in all possible combinations of impurities. Indeed, this problem may not be solved exactly. For all practical purposes, it is impossible to know where all the impurities in a given metallic sample actually are situated. Perhaps it is possible that no simple solution for the Green's function can be found.

12.7 Quenched Average and White Noise Potential

Suppose the electron wave functions are completely coherent throughout an entire disordered metal. Then each true electronic eigenfunction exhibits an extremely complex diffraction pattern spawned by the randomly positioned scatterers. If we change some external parameter—for example, the average electron density or an external magnetic field—then each individual diffraction pattern significantly changes due to the sensitivity of the scattering phases of the wave functions. So, substantial quantum fluctuations must occur in any observable at sufficiently low temperatures.

Our interest is focused on **impurity-averaged properties** obtained by averaging over all possible impurity configurations. This is a valid procedure for describing any real macroscopic system of interest for experimentally realizable temperatures due to **self-averaging**. We carry out such an **impurity average** for the Green's function with knowledge of the fluctuations of the impurity scattering potential about its average because these fluctuations are responsible for scattering the electrons.

Locations of various impurities are assumed to be independent of one another. Hence, the probability distribution for the impurity configuration is simply a product of probability distributions for individual impurities' locations taken to be uniform in space. So, obtaining the impurity average merely consists of averaging the positions of N impurities over all space and considering the expectation value of some operator. Then

$$\left\langle \hat{F} \right\rangle = F\left(\left\{\vec{R}_j\right\}\right) \tag{2097}$$

which is dependent on all impurity position $\left\{\vec{R}_j\right\}$. We then calculate the so-called **quenched average** of $\left\langle \hat{F} \right\rangle$:

$$\overline{\left\langle \hat{F} \right\rangle} = \int \prod_j \frac{d\vec{R}_j}{\Omega} \left\langle F\left(\left\{\vec{R}_j\right\}\right) \right\rangle \tag{2098}$$

with the **impurity average** taken after the thermodynamic average. Generally, electrons scatter off the fluctuations in the potential μ, and the average impurity potential $U(\vec{r})$ imitates a shifted chemical potential where, if the chemical potential μ has a shift of $\Delta\mu$, then the scattering potential becomes $U(\vec{r}) - \Delta\mu$. Therefore, we choose $\Delta\mu$ such that $U(\vec{r}) - \Delta\mu = 0$, with the most important quantity being the fluctuations about the average potential:

$$\delta U(\vec{r}) = U(\vec{r}) - \overline{U}(\vec{r}) \tag{2099}$$

If we shift the origin to $\overline{U}(\vec{r})$, then $\delta U(\vec{r}) \to U(\vec{r})$. So, considering first averages of the potential itself $\overline{U}(\vec{r})$, we then assume $\overline{\delta U}(\vec{r}) \to \overline{U}(\vec{r}) = 0$. We consider fluctuations around the given average and then

$$\overline{\delta U(\vec{r})\delta U(\vec{r}\,')} \to \overline{U(\vec{r})U(\vec{r}\,')} = \int \prod_j \frac{d\vec{R}_j}{\Omega} \sum_i V\left(\vec{r} - \vec{R}_i\right) \sum_l V\left(\vec{r}\,' - \vec{R}_l\right) \tag{2100}$$

Recall that U is defined in such a way that $\overline{U}(\vec{r}) = 0$ and then the sum over l gives zero except $l = i$. Also, in the last term in (2100), because each of the N terms is identical in form, we can simply integrate over one of them and multiply by N:

$$\overline{U(\vec{r})U(\vec{r}\,')} = \sum_i \int \prod_j \frac{d\vec{R}_j}{\Omega} V\left(\vec{r} - \vec{R}_j\right) V\left(\vec{r}\,' - \vec{R}_j\right) = \frac{N}{\Omega} \int d\vec{R} V\left(\vec{r} - \vec{R}\right) V\left(\vec{r}\,' - \vec{R}\right) \tag{2101}$$

If $\frac{N}{\Omega} = n_i$ is the density of the impurities and if point impurities exist then, neglecting any possible higher moments of the average disorder potential,

$$V\left(\vec{r} - \vec{R}\right) = \lambda\delta\left(\vec{r} - \vec{R}\right) \tag{2102}$$

Hence, the fluctuations imitate white noise:

$$\overline{U(\vec{r})U(\vec{r}\,')} = n_i \lambda^2 \delta(\vec{r} - \vec{r}\,') \tag{2103}$$

We assume equation (2103) along with $\overline{U}(\vec{r})=0$ will define the disorder potential. For the transfer momentum $\vec{q}=\vec{\kappa}'-\vec{\kappa}$, and considering the Fourier transform, we have the disordered potential

$$U(\vec{q})=\int d\vec{r}\exp\{-i\vec{q}\vec{r}\}U(\vec{r})=\sum_j \exp\{-i\vec{q}\vec{R}_j\}\int d\vec{r}\exp\{-i\vec{q}(\vec{r}-\vec{R}_j)\}V(\vec{r}-\vec{R}_j) \tag{2104}$$

or

$$U(\vec{q})=\sum_j \exp\{-i\vec{q}\vec{R}\}V(\vec{q})=V(\vec{q})\sum_j \exp\{-i\vec{q}\vec{R}_j\} \tag{2105}$$

Certainly, $\overline{U}(\vec{q})=0$, and we find the correlation:

$$\overline{U(\vec{q})U(\vec{q}')}=n_i\lambda^2\int d\vec{r}\int d\vec{r}'\exp\{-i\vec{q}\vec{r}\}\exp\{-i\vec{q}'\vec{r}'\}\delta(\vec{r}-\vec{r}')=n_i\lambda^2\int d\vec{r}\exp\{-i(\vec{q}+\vec{q}')\vec{r}\}=n_i\lambda^2\delta_{\vec{q}+\vec{q}'} \tag{2106}$$

12.8 Average over Impurities' Locations

The impurity-averaged Green's function $\overline{\mathbf{G}}(\vec{\kappa},\vec{\kappa}')$ is evaluated from

$$\overline{\mathbf{G}}(\vec{\kappa},\vec{\kappa}')=\prod_{i=1}^{N}\frac{1}{\Omega}\int d\vec{R}_i\mathbf{G}(\vec{\kappa},\vec{\kappa}') \tag{2107}$$

which simply is an integral over all impurity coordinates. Because the impurity average is a linear operation, it can be carried out separately for each term in the perturbation series:

$$\overline{\mathbf{G}}(\vec{\kappa},\vec{\kappa}')=\sum_{n=0}^{\infty}\overline{\mathbf{G}^{(n)}}(\vec{\kappa},\vec{\kappa}') \tag{2108}$$

In equation (2096), the only factors dependent on impurity positions are functions of $\rho(\vec{\kappa})$. To find $\overline{\mathbf{G}^{(n)}}(\vec{\kappa},\vec{\kappa}')$, we calculate the quantity

$$\overline{\rho(\vec{\kappa}-\vec{\kappa}_1)\rho(\vec{\kappa}_1-\vec{\kappa}_2)\cdots\rho(\vec{\kappa}_{n-1}-\vec{\kappa}')} \tag{2109}$$

We consider the lowest n order, and $n=1$ in particular, where for a random ensemble, the probability of finding the jth impurity within the volume $d\vec{R}_i$ surrounding the point $\vec{R}_i$ is independent of $\vec{R}_i$ and equal to $\frac{d\vec{R}_i}{\Omega}$. Therefore,

$$\overline{\rho}(\vec{\kappa}-\vec{\kappa}')=\prod_{i=1}^{N}\frac{1}{\Omega}\int d\vec{R}_i\rho(\vec{\kappa}-\vec{\kappa}')=\prod_{i=1}^{N}\frac{1}{\Omega}\int d\vec{R}_i\sum_{j=1}^{N}\exp\{-i(\vec{\kappa}-\vec{\kappa}')\vec{R}_j\}=\sum_{j=1}^{N}\left(\frac{1}{\Omega}\times\Omega\delta_{\vec{\kappa}\vec{\kappa}'}\right)\prod_{i\neq j}^{N}1=N\delta_{\vec{\kappa}\vec{\kappa}'} \tag{2110}$$

For $n = 2$,

$$\overline{\rho(\vec{\kappa}-\vec{\kappa}_1)\rho(\vec{\kappa}_1-\vec{\kappa}')} = \prod_{i=1}^{N}\frac{1}{\Omega}\int d\vec{R}_i \sum_{j_1=1}^{N}\exp\left\{-i(\vec{\kappa}-\vec{\kappa}_1)\vec{R}_{j_1}\right\}\sum_{j_2=1}^{N}\exp\left\{-i(\vec{\kappa}_1-\vec{\kappa}')\vec{R}_{j_2}\right\} \tag{2111}$$

If $j_1 \neq j_2$, then $N-2$ integrals achieve the value unity, whereas the integrals over j_1 and j_2 yield

$$\frac{1}{\Omega}\int d\vec{R}_{j_1}\exp\left\{-i(\vec{\kappa}-\vec{\kappa}_1)\vec{R}_{j_1}\right\}\frac{1}{\Omega}\int d\vec{R}_{j_2}\exp\left\{-i(\vec{\kappa}_1-\vec{\kappa}')\vec{R}_{j_2}\right\} = \delta_{\vec{\kappa}\vec{\kappa}_1}\delta_{\vec{\kappa}_1\vec{\kappa}'} \tag{2112}$$

If $j_1 = j_2$, then $N-1$ integrals achieve the value unity, whereas the integral over $j_1 = j_2$ yields

$$\frac{1}{\Omega}\int d\vec{R}_{j_1}\exp\left\{-i(\vec{\kappa}-\vec{\kappa}_1)\vec{R}_{j_1}\right\}\exp\left\{-i(\vec{\kappa}_1-\vec{\kappa}')\vec{R}_{j_1}\right\} = \frac{1}{\Omega}\int d\vec{R}_{j_1}\exp\left\{-i(\vec{\kappa}-\vec{\kappa}')\vec{R}_{j_1}\right\} = \delta_{\vec{\kappa}\vec{\kappa}'} \tag{2113}$$

and so

$$\overline{\rho(\vec{\kappa}-\vec{\kappa}_1)\rho(\vec{\kappa}_1-\vec{\kappa}')} = \sum_{j_1,j_2=1}^{N}\left[\left(1-\delta_{j_1j_2}\right)\delta_{\vec{\kappa}\vec{\kappa}_1}\delta_{\vec{\kappa}_1\vec{\kappa}'} + \delta_{j_1j_2}\delta_{\vec{\kappa}\vec{\kappa}'}\right] = \left(N^2-N\right)\delta_{\vec{\kappa}\vec{\kappa}_1}\delta_{\vec{\kappa}_1\vec{\kappa}'} + N\delta_{\vec{\kappa}\vec{\kappa}'} \tag{2114}$$

Consider

$$N^2 - N = N(N-1) \cong N^2 \tag{2115}$$

because the error is of order N^{-1} for large numbers of impurities. In addition, the product of Kronecker deltas can be rewritten as $\delta_{\vec{\kappa}\vec{\kappa}'}\delta_{\vec{\kappa}\vec{\kappa}_1}$. So,

$$\overline{\rho(\vec{\kappa}-\vec{\kappa}_1)\rho(\vec{\kappa}_1-\vec{\kappa}')} = \left(N^2\delta_{\vec{\kappa}\vec{\kappa}_1} + N\right)\delta_{\vec{\kappa}\vec{\kappa}'} \tag{2116}$$

For $n = 3$,

$$\overline{\rho(\vec{\kappa}-\vec{\kappa}_1)\rho(\vec{\kappa}_1-\vec{\kappa}_2)\rho(\vec{\kappa}_2-\vec{\kappa}')} = \sum_{j_1,j_2,j_3=1}^{N}\prod_{i=1}^{N}\frac{1}{\Omega}\int d\vec{R}_i\exp\left\{-i(\vec{\kappa}-\vec{\kappa}_1)\vec{R}_{j_1}\right\}\exp\left\{-i(\vec{\kappa}_1-\vec{\kappa}_2)\vec{R}_{j_2}\right\} \exp\left\{-i(\vec{\kappa}_2-\vec{\kappa}')\vec{R}_{j_3}\right\} \tag{2117}$$

and so we determine the following cases and their corresponding Kronecker deltas (**Table 12.1**).

TABLE 12.1 All possible cases of $n = 3$ with their corresponding kronecker deltas.

Case	Kronecker Delta
$j_1 \neq j_2 \neq j_3$	$\delta_{\vec{\kappa}\vec{\kappa}_1}\delta_{\vec{\kappa}_1\vec{\kappa}_2}\delta_{\vec{\kappa}_2\vec{\kappa}'}$
$j_1 = j_2 \neq j_3$	$\delta_{\vec{\kappa}\vec{\kappa}_2}\delta_{\vec{\kappa}_2\vec{\kappa}'}$
$j_1 \neq j_2 = j_3$	$\delta_{\vec{\kappa}\vec{\kappa}_1}\delta_{\vec{\kappa}_1\vec{\kappa}'}$
$j_1 = j_3 \neq j_2$	$\delta_{\vec{\kappa}+\vec{\kappa}_2,\vec{\kappa}_1+\vec{\kappa}'}\delta_{\vec{\kappa}_1\vec{\kappa}_2}$
$j_1 = j_2 = j_3$	$\delta_{\vec{\kappa}\vec{\kappa}'}$

Considering the sums in (2117),

$$\overline{\rho(\vec{\kappa}-\vec{\kappa}_1)\rho(\vec{\kappa}_1-\vec{\kappa}_2)\rho(\vec{\kappa}_2-\vec{\kappa}')} = \left(N^3\delta_{\vec{\kappa}\vec{\kappa}_1}\delta_{\vec{\kappa}_1\vec{\kappa}_2}+N^2\delta_{\vec{\kappa}\vec{\kappa}_2}+N^2\delta_{\vec{\kappa}_1\vec{\kappa}_2}+N\right)\delta_{\vec{\kappa}\vec{\kappa}'} \tag{2118}$$

where we consider the approximation

$$N(N-1)\approx N^2 \tag{2119}$$

and rewrite the Kronecker delta products in all terms to contain a $\delta_{\vec{\kappa}\vec{\kappa}'}$.

12.9 Disorder Average Green's Function

Each term in the impurity average gives rise to a term in the perturbation expansion for the Green's function. But not all impurity averages contain the factor $\delta_{\vec{\kappa}\vec{\kappa}'}$. This also is valid for higher n-orders. In contrast to the original Green's function $\mathbf{G}(\vec{\kappa},\vec{\kappa}',\omega)$, the impurity-averaged Green's function is diagonal in $\vec{\kappa}$:

$$\overline{\mathbf{G}}(\vec{\kappa},\vec{\kappa}',\omega)=\overline{\mathbf{G}}(\vec{\kappa},\omega)\delta_{\vec{\kappa}\vec{\kappa}'} \tag{2120}$$

Hence, the impurity average tailors the system to be translationally invariant and implies that the electrons see the same **average** environment everywhere in the system.

The $\vec{\kappa}$-diagonalization is an important simplification resulting from the impurity average. Each term is represented by a **Feynman diagram** with specific **Feynman rules** that translate the diagrammatic form of each term into its corresponding mathematical expression and vice versa. The perturbation expansion for this problem is of the same form as previously indicated:

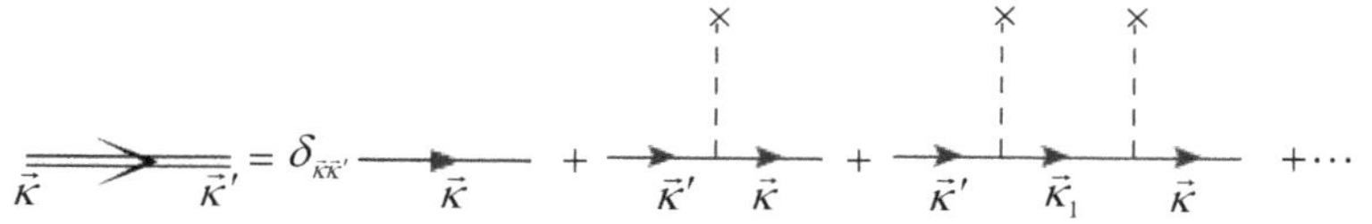

(2121)

FIGURE 12.5 Diagrammatic representation of the impurity-averaged Green's function.

This can be rewritten in the form

$$\begin{aligned}\mathbf{G}(\vec{\kappa},\vec{\kappa}',\omega)&=\delta_{\vec{\kappa}\vec{\kappa}'}\mathbf{G}_0(\vec{\kappa},\omega)+\mathbf{G}_0(\vec{\kappa}',\omega)U(\vec{\kappa}-\vec{\kappa}')\mathbf{G}_0(\vec{\kappa},\omega)+\\&+\sum_{\vec{\kappa}_1}\mathbf{G}_0(\vec{\kappa}',\omega)U(\vec{\kappa}-\vec{\kappa}_1)\mathbf{G}_0(\vec{\kappa}_1,\omega)U(\vec{\kappa}_1-\vec{\kappa})\mathbf{G}_0(\vec{\kappa},\omega)+\cdots\end{aligned} \tag{2122}$$

So, we can calculate $\mathbf{G}$ by the perturbation series for any particular realization of the disorder where interest is on the **quenched average** $\overline{\mathbf{G}}(\vec{\kappa},\vec{\kappa}',\omega)$. The zero-order Green's function is unaffected by the disorder average procedure:

$$\overline{\mathbf{G}_0}(\vec{\kappa},\omega)=\mathbf{G}_0(\vec{\kappa},\omega) \tag{2123}$$

However, for the single scattering event:

$$\overline{\mathbf{G}_0(\vec{\kappa}',\omega)U(\vec{\kappa}-\vec{\kappa}')\mathbf{G}_0(\vec{\kappa},\omega)} = \mathbf{G}_0(\vec{\kappa}',\omega)\overline{U}(\vec{\kappa}-\vec{\kappa}')\mathbf{G}_0(\vec{\kappa},\omega) = 0 \tag{2124}$$

This implies that the first-order contribution to **G** is zero after performing the disorder average.

We now examine the second-order term, which is a double scattering event:

$$\overline{\mathbf{G}_0(\vec{\kappa}',\omega)U(\vec{\kappa}-\vec{\kappa}_1)\mathbf{G}_0(\vec{\kappa}_1,\omega)U(\vec{\kappa}_1-\vec{\kappa})\mathbf{G}_0(\vec{\kappa},\omega)} = \mathbf{G}_0(\vec{\kappa}',\omega)\mathbf{G}_0(\vec{\kappa}_1,\omega)\mathbf{G}_0(\vec{\kappa},\omega)\overline{U(\vec{\kappa}-\vec{\kappa}_1)U(\vec{\kappa}_1-\vec{\kappa})} \tag{2125}$$

From the definition of the disorder average in equation (2106), we see that

$$\overline{U(\vec{\kappa}-\vec{\kappa}_1)U(\vec{\kappa}_1-\vec{\kappa})} = n_i\lambda^2\delta(\vec{\kappa}-\vec{\kappa}_1+\vec{\kappa}_1-\vec{\kappa}') = n_i\lambda^2\delta(\vec{\kappa}-\vec{\kappa}') \tag{2126}$$

The delta function is the most important consequence of the given expression and defines momentum conservation. It implies that, while any particular realization of the disorder breaks **translational invariance**, the average over all realizations restores **translational invariance**. This restoration of momentum conservation is represented diagrammatically by a line above the first diagram (2127) indicating a disorder average:

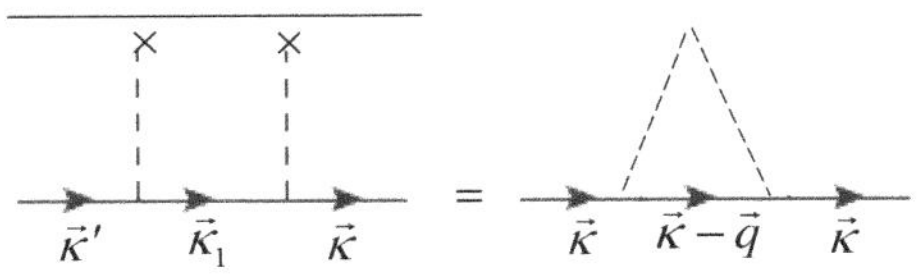

(2127)

FIGURE 12.6 Diagram indicating a disorder average.

It is the same for a higher-order diagram and, therefore, for a nonzero result. We select pairs of disorder crosses to merge together with the strength $n_i\lambda^2$, and this yields momentum conservation for the resultant dashed-line, which then links with fermions at both ends, imitating an effective interaction:

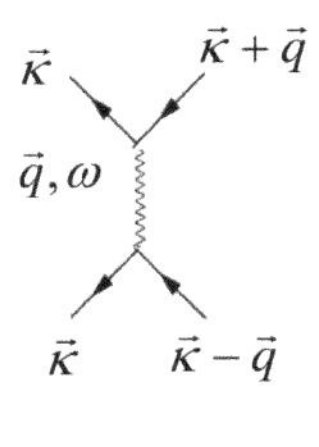

FIGURE 12.7 Diagram indicating momentum conservation.

or

$$V_{\text{eff}}(\vec{q},\omega) = -in_i\lambda^2\delta(\omega=0) \tag{2128}$$

The energy is conserved at each individual scattering cross.

From time representation, we consider two scattering events happening simultaneously. The independence of the scattering on the relative times yields the delta function in energy space, while the interaction independent of the relative times is an **infinitely retarded (effective) interaction**. The physical sense of this **effective attractive infinitely retarded interaction** can be seen by considering a small region of the disordered potential as depicted in Figure 12.8.

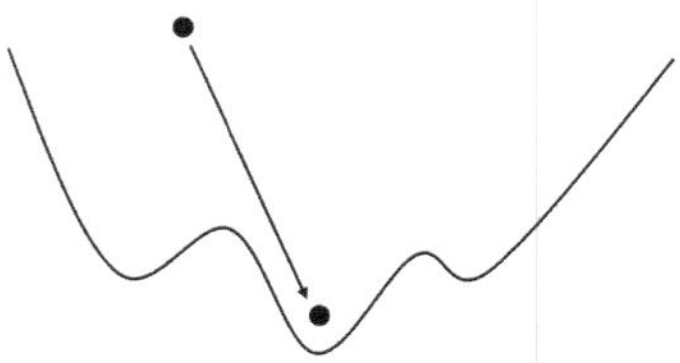

FIGURE 12.8 When a second electron is added, it also has a tendency to go into the lowest energy well, which could be viewed as an attraction to the first electron.

Adding one electron, it has the tendency to go into the lowest energy state in the neighborhood of the region where the disorder potential is lowest. Adding a second electron, it has the tendency to go into the remaining lowest energy state that is in the neighborhood of the same region as the first electron, thereby implying attraction to the first.

12.10 Disorder Diagrams

We derive a disordered system when the disorder averaging is equivalent to an effective interaction and use the same diagrammatic expansions for the average electronic Green's function $\overline{G_0}(\vec{\kappa},\omega)$ (Figure 12.9):

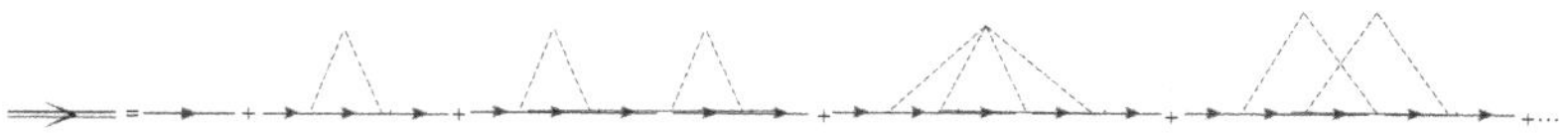

(2129)

FIGURE 12.9 Diagrammatic expansion of the disorder averaged single-particle Green's function $\overline{G}(\vec{\kappa},\omega)$.

We introduce the concept of the average electronic **self-energy** to enable us to understand the interacting environment feedback on a propagating particle. The average electronic Green's function of a fermion (2129) in an interacting environment can be achieved by grouping all scattering processes into the average electronic **self-energy** $\overline{\Sigma}(\vec{\kappa},\omega)$:

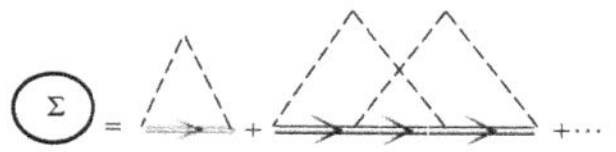

(2130)

FIGURE 12.10 Diagrammatic representation of all scattering processes by the average electronic self-energy $\overline{\Sigma}(\vec{\kappa},\omega)$.

This self-energy $\overline{\Sigma}(\vec{\kappa},\omega)$ physically describes the cloud of particle-hole excitations forming the wake that accompanies the propagating electron. We represent the average electronic Green's function in equation (2129) in terms of the **self-energy** via the Dyson series:

(2131)

or

(2132)

or

$$\overline{G}(\vec{\kappa},\omega)=\overline{G_0}(\vec{\kappa},\omega)+\overline{G_0}(\vec{\kappa},\omega)\overline{\Sigma}(\vec{\kappa},\omega)\overline{G}(\vec{\kappa},\omega) \tag{2133}$$

from where we have

$$\overline{\mathbf{G}}(\vec{\kappa},\omega)=\frac{1}{\omega-\xi_{\vec{\kappa}}-\overline{\Sigma}(\vec{\kappa},\omega)+i\delta\,\mathrm{sgn}\,\xi_{\vec{\kappa}}} \tag{2134}$$

with the self-energy to lowest order:

$$\overline{\Sigma}(\omega)=n_i\lambda^2\int\frac{d\vec{q}}{(2\pi)^3}\overline{\mathbf{G}}_0(\vec{q},\omega)=n_i\lambda^2\int\frac{d\vec{q}}{(2\pi)^3}\frac{1}{\omega-\xi_{\vec{q}}+i\delta\,\mathrm{sgn}\,\xi_{\vec{q}}}=n_i\lambda^2\int d\in\frac{\nu(\in)}{\omega-\in+i\delta\,\mathrm{sgn}\in} \tag{2135}$$

This integral is easily solvable if:

1. We assume weak disorder; therefore, only states near the Fermi surface scatter and $\nu(\in)$ achieves its value N_F at the Fermi surface. We stretch the limits of integration from $\in$ to $\pm\infty$ because processes of physical significance are at the Fermi surface and not at exceedingly far removed energies.
2. Because there are no energy integrals in our expression, we make the analytic continuation $\mathbf{G}_R(\vec{\kappa},\omega)=\mathbf{G}(\vec{\kappa},\omega-i\delta)$ before evaluating the integral. Hence, ω is treated as a complex variable and implies ignoring $i\delta\,\mathrm{sgn}\in$:

$$\overline{\Sigma}(\omega)=n_i\lambda^2N_F\int_{-\infty}^{\infty}\frac{d\in}{\omega-\in}=-n_i\lambda^2N_F\left[i\pi\,\mathrm{sgn}(\mathrm{Im}\,\omega)\right]=-\frac{i}{2\tau}\mathrm{sgn}(\mathrm{Im}\,\omega)\quad,\quad\frac{1}{\tau}=2\pi n_i\lambda^2N_F \tag{2136}$$

Here, τ is the elastic scattering rate.
So, the Green's function is

$$\overline{\mathbf{G}}(\vec{\kappa},\omega)=\frac{1}{\omega-\xi_{\vec{\kappa}}+\frac{i}{2\tau}\mathrm{sgn}(\mathrm{Im}\,\omega)} \tag{2137}$$

and the spectral function for the weakly disordered free fermionic gas is

$$\mathrm{A}(\vec{\kappa},\omega)=-\frac{1}{\pi}\mathrm{Im}\,\overline{\mathbf{G}}(\vec{\kappa},\omega-i\delta)=\frac{1}{\pi}\frac{(2\tau)^{-1}}{(\omega-\xi_{\vec{\kappa}})^2+(2\tau)^{-2}} \tag{2138}$$

We observe that the original delta function is broadened to a Lorentzian of width $\frac{1}{\tau}$, which implies that the electron now has a lifetime τ due to the disorder and so it is called **elastic scattering time**. We observe that the free electrons are turned into **quasiparticles** by impurity scattering and that the quasiparticle has a finite lifetime τ.

12.11 Gorkov Equation with Impurities

We now extend the diagrammatic cross technique to superconductors where we again examine the Nambu-Gorkov propagators [84] this time with impurities and, in particular, the **normal Green's function** $\mathscr{G}$ and the **anomalous Green's function** $\mathscr{F}^\dagger$ **and** $\mathscr{F}$ (Figure 12.11a):

$$\mathscr{G}_{\sigma\sigma'}(\vec{r},\tau;\vec{r}',\tau')=-\langle N|\mathrm{T}_\tau\hat{\rho}_{\sigma\sigma'}(\vec{r},\tau;\vec{r}',\tau')|N\rangle \tag{2139}$$

$$\langle N|\mathrm{T}_\tau\mathrm{A}|N+2\rangle=\mathscr{F}\quad,\quad\langle N+2|\mathrm{T}_\tau\mathrm{A}^\dagger|N\rangle=\mathscr{F}^\dagger \tag{2140}$$

$$\mathscr{F}_{\sigma\sigma'}(\vec{r},\tau;\vec{r}',\tau') = -\langle N|\mathrm{T}_\tau \mathrm{A}_{\alpha\beta}(\vec{r},\tau;\vec{r}',\tau')|N+2\rangle \tag{2141}$$

$$\mathscr{F}^\dagger_{\sigma\sigma'}(\vec{r},\tau;\vec{r}',\tau') = -\langle N+2|\mathrm{T}_\tau \mathrm{A}^\dagger_{\sigma\sigma'}(\vec{r},\tau;\vec{r}',\tau')|N\rangle \tag{2142}$$

which satisfies the following symmetry properties:

$$\mathscr{F}_{\sigma\sigma'}(\vec{r},\tau;\vec{r}',\tau') = -\mathscr{F}_{\sigma'\sigma}(\vec{r}',\tau';\vec{r},\tau) \quad , \quad \mathscr{F}^\dagger_{\sigma\sigma'}(\vec{r},\tau;\vec{r}',\tau') = -\mathscr{F}^\dagger_{\sigma'\sigma}(\vec{r}',\tau';\vec{r},\tau) \tag{2143}$$

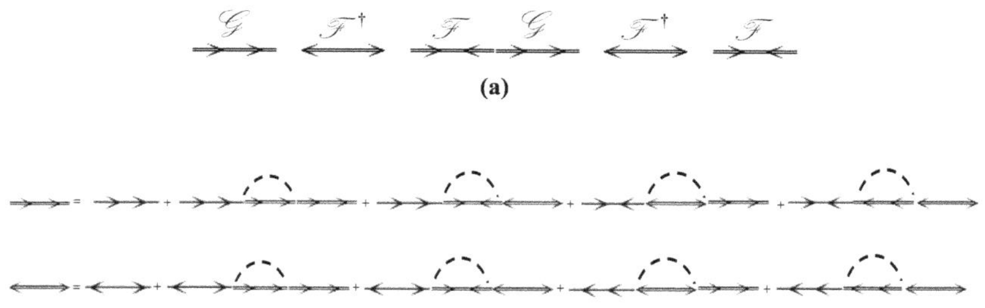

(2144)

(b)

FIGURE 12.11 **(a)** Directions of arrows identify the ordinary and anomalous Green's functions. **(b)** Equation for all three functions generalizing the Dyson equation for the superconducting phase.

In the absence of impurities (and any external fields), we assume all the functions have translational symmetry and can be expressed in terms of the momentum and frequency representation in the standard manner. Scattering by static defects only breaks the spatial homogeneity. In the absence of an external magnetic field, both anomalous Green's functions can be expressed in terms of one another by using their definitions and their symmetry properties. To proceed with discussions on the average of the Green's functions $\mathscr{G}(\vec{r},\vec{r}',\omega_n)$, $\mathscr{F}(\vec{r},\vec{r}',\omega_n)$, and $\mathscr{F}^\dagger(\vec{r},\vec{r}',\omega_n)$, we examine how the averaged values $\overline{\Delta}$ and $\overline{\Delta^*}$ relate to Δ and Δ^*. When introducing an impurity into a superconductor, $\Delta(\vec{r})$ is expected to experience local changes dependent on the strength of the impurity potential. But local variations of $\Delta(\vec{r})$ rapidly decrease away from an impurity center. So, the order parameter, $\Delta(\vec{r})$, is a self-averaging quantity, at least at low enough impurity concentrations. So, we assume that $\Delta(\vec{r}) \simeq \overline{\Delta}$.

Considering again the impurity Hamiltonian $\hat{\mathrm{H}}_{sc}$, we then obtain an expansion for each of the two Green's functions that formally imitates Figure 12.2. The main difference is that, for the Green's functions $\mathscr{G}_0(\vec{r}-\vec{r}',\omega_n)$, connecting the crosses in Figure 12.2 now alternate with the bare functions $\mathscr{F}_0(\vec{r}-\vec{r}',\omega_n)$ and $\mathscr{F}_0^\dagger(\vec{r}-\vec{r}',\omega_n)$ of the pure superconductor. This reasoning is repeated for the equations for the averaged $\mathscr{G}(\vec{\kappa},\omega_n)$, $\mathscr{F}(\vec{\kappa},\omega_n)$, and $\mathscr{F}^\dagger(\vec{\kappa},\omega_n)$. Hence, the two equations shown in Figure 12.11 (b) follow. Here, double lines stand for the averaged (exact) functions $\mathscr{G}(\vec{\kappa},\omega_n)$, $\mathscr{F}(\vec{\kappa},\omega_n)$, and $\mathscr{F}^\dagger(\vec{\kappa},\omega_n)$ (Figure 12.11a). We write them in a more compact form via Gorkov equations in the absence of impurities, such as in Figure 12.11b.

From the previous sections, we rewrite the following equations

$$(i\omega_n - \xi_{\vec{\kappa}})\mathscr{G}_0 + \Delta\mathscr{F}_0^\dagger = 1 \quad , \quad \Delta^*\mathscr{G}_0 + (i\omega_n + \xi_{\vec{\kappa}})\mathscr{F}_0^\dagger = 0 \tag{2145}$$

with solutions

$$\mathscr{F}_0^\dagger = \frac{\Delta^*}{\omega_n^2 + \mathrm{E}_{\vec{\kappa}}^2} \quad , \quad \mathscr{G}_0 = -\frac{i\omega_n + \xi_{\vec{\kappa}}}{\omega_n^2 + \mathrm{E}_{\vec{\kappa}}^2} \quad , \quad \mathrm{E}_{\vec{\kappa}}^2 = \xi_{\vec{\kappa}}^2 + \Delta^2 \tag{2146}$$

Considering impurities, we then have the following Dyson equations:

$$\mathbf{G} = \mathcal{G}_0 + \mathcal{G}_0\overline{\mathbf{G}_\omega}\mathbf{G} + \mathcal{G}_0\overline{\mathcal{F}_\omega}\mathcal{F}^\dagger + \mathcal{F}_0\overline{\mathcal{F}_\omega}^\dagger\mathbf{G} + \mathcal{F}_0\overline{\mathbf{G}_{-\omega}}\mathcal{F}^\dagger \tag{2147}$$

$$\mathcal{F}^\dagger = \mathcal{F}_0 + \mathcal{F}_0\overline{\mathbf{G}_\omega}\mathbf{G} + \mathcal{F}_0\overline{\mathcal{F}_\omega}\mathcal{F}^\dagger + \mathcal{G}_{-\omega}\overline{\mathcal{F}_\omega}^\dagger\mathbf{G} + \mathcal{G}_{-\omega}\overline{\mathbf{G}_{-\omega}}\mathcal{F}^\dagger \tag{2148}$$

where the quenched averages are

$$\overline{\mathcal{F}_\omega}^\dagger = \frac{N_{\text{imp}}}{(2\pi)^3\Omega}\int\left|u(\vec{\kappa}-\vec{\kappa}')\right|^2\mathcal{F}^\dagger(\vec{\kappa}')d\vec{\kappa}' \quad , \quad \overline{\mathbf{G}_\omega} = \frac{N_{\text{imp}}}{(2\pi)^3\Omega}\int\left|u(\vec{\kappa}-\vec{\kappa}')\right|^2\mathbf{G}(\vec{\kappa}')d\vec{\kappa}' \tag{2149}$$

and N_{imp} the impurity concentration, $u(\vec{\kappa}-\vec{\kappa}')$, is the Fourier component of the interaction potential between electron and impurity atom. From (2147) and (2148), we have, respectively,

$$\mathbf{G} = \mathcal{G}_0\left(1+\overline{\mathbf{G}_\omega}\mathbf{G}+\overline{\mathcal{F}_\omega}\mathcal{F}^\dagger\right)+\mathcal{F}_0\left(\overline{\mathcal{F}_\omega}^\dagger\mathbf{G}+\overline{\mathbf{G}_{-\omega}}\mathcal{F}^\dagger\right) \tag{2150}$$

$$\mathcal{F}^\dagger = \mathcal{F}_0\left(1+\overline{\mathbf{G}_\omega}\mathbf{G}+\overline{\mathcal{F}_\omega}\mathcal{F}^\dagger\right)+\mathcal{G}_{-\omega}\left(\overline{\mathcal{F}_\omega}^\dagger\mathbf{G}+\overline{\mathbf{G}_{-\omega}}\mathcal{F}^\dagger\right) \tag{2151}$$

and from where

$$\frac{\mathbf{G}}{\mathcal{G}_0} = \left(1+\overline{\mathbf{G}_\omega}\mathbf{G}+\overline{\mathcal{F}_\omega}\mathcal{F}^\dagger\right)+\frac{\mathcal{F}_0}{\mathcal{G}_0}\left(\overline{\mathcal{F}_\omega}^\dagger\mathbf{G}+\overline{\mathbf{G}_{-\omega}}\mathcal{F}^\dagger\right) \tag{2152}$$

$$\frac{\mathcal{F}^\dagger}{\mathcal{F}_0} = \left(1+\overline{\mathbf{G}_\omega}\mathbf{G}+_0\overline{\mathcal{F}_\omega}\mathcal{F}^\dagger\right)+\frac{\mathcal{G}_{-\omega}}{\mathcal{F}_0}\left(\overline{\mathcal{F}_\omega}^\dagger\mathbf{G}+\overline{\mathbf{G}_{-\omega}}\mathcal{F}^\dagger\right) \tag{2153}$$

From (2152) and (2153), we have

$$\frac{\mathbf{G}}{\mathcal{G}_0}-\frac{\mathcal{F}_0}{\mathcal{G}_0}\left(\overline{\mathcal{F}_\omega}^\dagger\mathbf{G}+\overline{\mathbf{G}_{-\omega}}\mathcal{F}^\dagger\right) = \frac{\mathcal{F}^\dagger}{\mathcal{F}_0}-\frac{\mathcal{G}_{-\omega}}{\mathcal{F}_0}\left(\overline{\mathcal{F}_\omega}^\dagger\mathbf{G}+\overline{\mathbf{G}_{-\omega}}\mathcal{F}^\dagger\right) \tag{2154}$$

or

$$\left(\frac{\mathcal{G}_{-\omega}}{\mathcal{F}_0}-\frac{\mathcal{F}_0}{\mathcal{G}_0}\right)\left(\overline{\mathcal{F}_\omega}^\dagger\mathbf{G}+\overline{\mathbf{G}_{-\omega}}\mathcal{F}^\dagger\right) = \frac{\mathcal{F}^\dagger}{\mathcal{F}_0}-\frac{\mathbf{G}}{\mathcal{G}_0} \tag{2155}$$

Also from (2152) and (2153) then

$$\frac{\mathbf{G}}{\mathcal{F}_0} = \frac{\mathcal{G}_0}{\mathcal{F}_0}\left(1+\overline{\mathbf{G}_\omega}\mathbf{G}+\overline{\mathcal{F}_\omega}\mathcal{F}^\dagger\right)+\left(\overline{\mathcal{F}_\omega}^\dagger\mathbf{G}+\overline{\mathbf{G}_{-\omega}}\mathcal{F}^\dagger\right) \tag{2156}$$

$$\frac{\mathcal{F}^\dagger}{\mathcal{G}_{-\omega}} = \frac{\mathcal{F}_0}{\mathcal{G}_{-\omega}}\left(1+\overline{\mathbf{G}_\omega}\mathbf{G}+_0\overline{\mathcal{F}_\omega}\mathcal{F}^\dagger\right)+\left(\overline{\mathcal{F}_\omega}^\dagger\mathbf{G}+\overline{\mathbf{G}_{-\omega}}\mathcal{F}^\dagger\right) \tag{2157}$$

from where

$$\frac{\mathbf{G}}{\mathcal{F}_0}-\frac{\mathcal{G}_0}{\mathcal{F}_0}\left(1+\overline{\mathbf{G}_\omega}\mathbf{G}+\overline{\mathcal{F}_\omega}\mathcal{F}^\dagger\right) = \frac{\mathcal{F}^\dagger}{\mathcal{G}_{-\omega}}-\frac{\mathcal{F}_0}{\mathcal{G}_{-\omega}}\left(1+\overline{\mathbf{G}_\omega}\mathbf{G}+_0\overline{\mathcal{F}_\omega}\mathcal{F}^\dagger\right) \tag{2158}$$

or

$$\left(\frac{\mathscr{F}_0}{\mathscr{G}_{-\omega}}-\frac{\mathscr{G}_0}{\mathscr{F}_0}\right)\left(1+\overline{\mathbf{G}_\omega\mathbf{G}}+\overline{\mathscr{F}_\omega\mathscr{F}^\dagger}\right)=\frac{\mathscr{F}^\dagger}{\mathscr{G}_{-\omega}}-\frac{\mathbf{G}}{\mathscr{F}_0} \tag{2159}$$

Rearranging equations (2155) and (2159) and considering (2146), we then have the following more general equations for (2145):

$$\left(i\omega_n-\xi_{\vec{\kappa}}-\overline{\mathbf{G}_\omega}\right)\mathbf{G}+\left(\Delta+\overline{\mathscr{F}_\omega^\dagger}\right)\mathscr{F}^\dagger=1 \quad , \quad \left(i\omega_n+\xi_{\vec{\kappa}}+\overline{\mathbf{G}_{-\omega}}\right)\mathscr{F}^\dagger+\left(\Delta+\overline{\mathscr{F}_\omega^\dagger}\right)\mathbf{G}=0 \tag{2160}$$

where the solution of this set of equations is:

$$\mathbf{G}(\vec{\kappa})=-\frac{i\omega_n+\xi_{\vec{\kappa}}-\overline{\mathbf{G}_\omega}}{\left(\omega_n+i\overline{\mathbf{G}_\omega}\right)^2+\xi_{\vec{\kappa}}^2+\left(\Delta+\overline{\mathscr{F}_\omega^\dagger}\right)^2} \quad , \quad \mathscr{F}^\dagger(\vec{\kappa})=-\frac{\Delta+\overline{\mathscr{F}_\omega}}{\left(\omega_n+i\overline{\mathbf{G}_\omega}\right)^2+\xi_{\vec{\kappa}}^2+\left(\Delta+\overline{\mathscr{F}_\omega^\dagger}\right)^2} \tag{2161}$$

12.11.1 Properties of Homogeneous Superconductors

In the absence of external electromagnetic fields and currents, the order parameter Δ is chosen to be real $\mathscr{F}(\vec{\kappa},\omega_n)=\mathscr{F}^\dagger(\vec{\kappa},\omega_n)$, and the averages for the isotropic model are independent of $\vec{\kappa}$. From equations (2160) and (2161), we introduce the notation

$$\tilde{\Delta}=\Delta+\overline{\mathscr{F}_{\omega_n}^\dagger}=\Delta\eta_{\omega_n} \quad , \quad i\tilde{\omega}_n=i\omega_n-\overline{\mathbf{G}_{\omega_n}}=i\omega_n\eta_{\omega_n} \tag{2162}$$

and use the property

$$\overline{\mathbf{G}}(\vec{\kappa},-\omega_n)=-\overline{\mathbf{G}}(\vec{\kappa},\omega_n) \tag{2163}$$

The contributions in $\overline{\mathbf{G}}(\vec{\kappa},\omega_n)$ are due to integration over ξ far from the Fermi surface that is defined by $\xi=0$. These contributions are independent of the system being in the normal or superconducting state. These are safely included as corrections to the chemical potential. Therefore, we integrate equation (2149) over ξ in the symmetric interval $\xi\in[-\Lambda,\Lambda]$ about $\xi=0$, and this yields

$$\frac{\overline{\mathbf{G}_\omega}}{-i\omega}=\frac{\overline{\mathscr{F}^\dagger}}{\Delta} \quad , \quad \overline{\mathscr{F}^\dagger}=\Delta(\eta_\omega-1) \tag{2164}$$

where

$$\eta_\omega=1+\frac{\eta_\omega}{2\pi\tau}\int\frac{d\xi}{\xi_{\vec{\kappa}}^2+\left(\omega^2+\Delta^2\right)\eta_\omega^2}=1+\frac{\eta_\omega}{2\pi\tau\sqrt{(\omega^2+\Delta^2)}}\int\frac{d\xi}{\xi^2+1}=1+\frac{1}{2\tau\sqrt{\omega^2+\Delta^2}} \tag{2165}$$

We rewrite the expression for the function $\mathscr{F}^\dagger(\vec{r}-\vec{r}',\omega_n)$ in the coordinate representation $|\vec{r}-\vec{r}'|=R$ as follows

$$\mathscr{F}^\dagger(R,\omega_n)=-\frac{i\Delta}{(2\pi)^2R}\int\kappa\,d\kappa\frac{\left(\exp\{i\kappa R\}-\exp\{-i\kappa R\}\right)\eta_{\omega_n}^2}{\xi^2+\left(\omega_n^2+|\Delta|^2\right)\eta_{\omega_n}^2} \tag{2166}$$

Setting κ as $\kappa \equiv \kappa_F + \frac{\xi}{v_F}$ and integrating over ξ, we have

$$\mathscr{F}^{\dagger}(R,\omega_n) = \frac{\Delta}{2\pi R}\frac{\cos(\kappa_F R)}{\sqrt{\omega_n^2 + |\Delta|^2}}\exp\left\{-\frac{R}{v_F}\eta_{\omega_n}\sqrt{\omega_n^2 + |\Delta|^2}\right\} \tag{2167}$$

From the expression for η_{ω_n}, the only change in the spatial variation of $\mathscr{F}^{\dagger}(R,\omega_n)$ as well as in the Green's function $\mathscr{G}(R,\omega_n)$ is the introduction of the factor

$$\exp\left\{-\frac{R}{2\ell}\right\} \tag{2168}$$

(where $\ell = v_F\tau$ is the mean free path). From the isotropic model for the average gap, the following quantity remains invariant:

$$\Delta = |\lambda|\mathscr{F}^{\dagger} \tag{2169}$$

This implies the gap invariance for a superconductor without defects as well as with impurities. So, for exceedingly low defect concentrations, the thermodynamics of homogeneous superconductors together with the superconducting transition temperature T_c are invariant for a superconductor. This is because the thermodynamics of a superconductor depend only on the value of the gap [84] (**Anderson theorem**) [85]. This only applies to the isotropic model. For an anisotropic metal, there are variations of T_c upon alloying. So, the range of applicability of the previous results is for exceedingly low concentrations and implies we neglect any terms of order $\frac{1}{\epsilon_F \tau} \ll 1$ or higher.

13

Classical and Quantum Theory of Magnetism

Introduction

Magnetism is a key driver in modern technology and stimulates research in numerous branches of condensed matter physics. The study of spin systems is an extensive part of many-body theory because numerous solids display magnetic properties among their electrons. From the famous Stern-Gerlach experiment, the magnetic moment of an electron corresponds to the magnetic moment of one Bohr magneton, which is a natural quantum theoretical measure for the magnetic moment of atomic systems. The passage of a beam of electrons through an inhomogeneous magnetic field results in the splitting of that beam of electrons into two distinct partial beams (quantization of the result of measurement) or quantization of the angular momentum (directional quantization). This confirms the fact that the magnetic moment of one Bohr magneton may take two values and assumes an intrinsic magnetic moment carried by the electrons, which is then attributed to an intrinsic angular momentum, the spin. This leads us to Quantum magnetism that is a bit different from classical magnetism, because individual atoms have a quality called spin, which is quantized, or in discrete states (usually called up or down). Spins play a vital role in many impurity problems. These phenomena may be conveniently explained by different types of spin models. Among these are localized spins that interact among themselves and other localized spins that interact with free electrons. This subject matter has few exactly solvable models that are nontrivial. Not many models have solutions that are well understood, even though many of them have been intensively studied. The devolution of models results in real solids that are strongly interacting systems. But this has yet to be very effective.

13.1 Classical Theory of Magnetism

We know, however, that electrons carry spin $S = \pm\frac{1}{2}$ (e.g., from Stern-Gerlach-type experiments) and that they carry a magnetic moment of one Bohr magneton $\mu_B = \frac{e\hbar}{2mc}$, where, for the electron number N in a given volume V,

$$N = 2V \int \frac{d\vec{p}}{(2\pi)^3} \theta(\mu - \epsilon_k) \tag{2170}$$

Here, the factor 2 is due to the electron spin: $\theta(\mu - \epsilon_k)$ is the step function; μ is the chemical potential; and ϵ_k is the kinetic energy of the electron. At the Fermi surface, the Fermi energy is

$$\mu = \epsilon_k = \frac{\hbar^2 \kappa_F^2}{2m} \equiv \frac{p_F^2}{2m} \tag{2171}$$

and κ_F and p_F are the Fermi wave vector and momentum, respectively:

$$N = 2V\int_0^{p_F} \frac{p^2 dp 4\pi}{(8\pi\hbar)^3} = \frac{V}{\pi^2\hbar^3}\frac{p_F^3}{3} \tag{2172}$$

From here, the number density is found to be

$$\frac{N}{V} = \frac{p_F^3}{3\pi^2\hbar^3} \tag{2173}$$

and the Fermi momentum

$$p_F = \left(3\pi^2\hbar^3\frac{N}{V}\right)^{\frac{1}{3}} \tag{2174}$$

So, the Fermi energy

$$\epsilon_F = \frac{1}{2m}\left(3\pi^2\hbar^3\frac{N}{V}\right)^{\frac{2}{3}} \tag{2175}$$

The Stern-Gerlach-type experiment confirms the fact that the quantity μ_B may take two values, and that the angular momentum associated with it:

$$\vec{\mathcal{M}} = \mu_B\frac{\vec{S}}{S} \tag{2176}$$

is not only affected by a torque but is also affected by a force $\vec{F}$ in the inhomogeneous magnetic field $\vec{H}$:

$$\vec{F} = -grad\mathrm{H}_{\mathrm{int}} \tag{2177}$$

where the interaction energy:

$$\mathrm{H}_{\mathrm{int}} \equiv -\left(\vec{\mathcal{M}}, \vec{H}\right) \tag{2178}$$

The splitting of the beam of electrons has its origin in the force in (2177) that acts on the magnetic moment $\vec{\mathcal{M}}$ in the field $\vec{H}$.

Suppose the spins do not interact with the magnetic field $\vec{H}$. Then we have the scenario in Figure 13.1 where the energies of the spin species are lifted in the same manner.

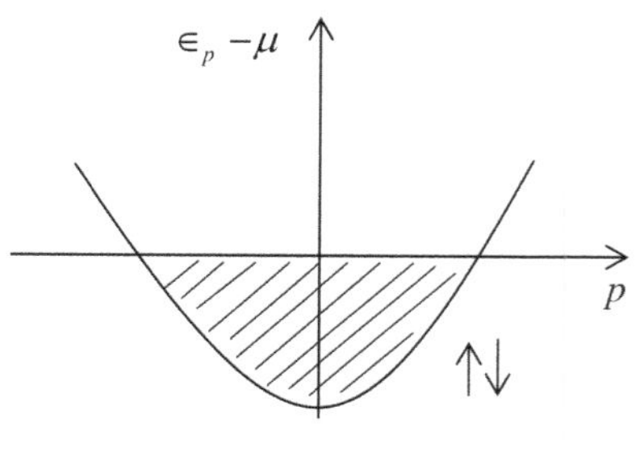

FIGURE 13.1 Electron with spin species in the absence of a magnetic field.

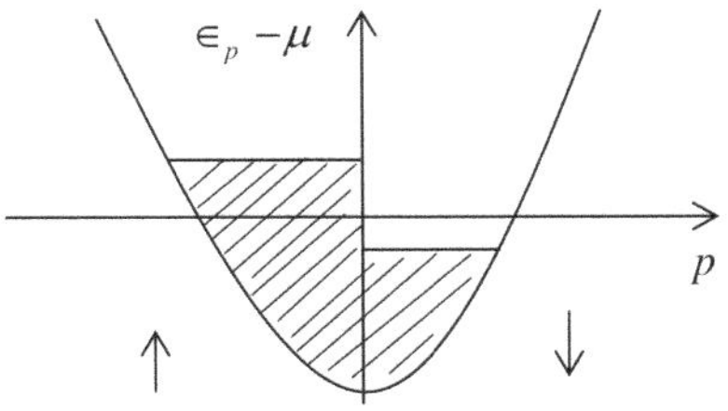

FIGURE 13.2 Electron with spin species in the presence of a magnetic field.

If the spins do interact with the magnetic field, then we have the scenario in Figure 13.2 where the energies for the spin species are lifted differently as

$$\in_{\downarrow} = \in + \mu_B H \quad , \quad \in_{\uparrow} = \in - \mu_B H \tag{2179}$$

This corresponds to the magnetic moments for the given spin species:

$$\mathcal{M}_{\uparrow} = \mu + \mu_B H \quad , \quad \mathcal{M}_{\downarrow} = \mu - \mu_B H \tag{2180}$$

The corresponding number densities are $n_{\uparrow}$ and $n_{\downarrow}$; hence, the magnetic moment per unit volume is computed as follows

$$\mathcal{M} = \frac{\mathcal{M}_{\text{tot}}}{V} = \mu_B \left(n_{\uparrow} - n_{\downarrow}\right) \tag{2181}$$

Its projection on the oz-axis:

$$\mathcal{M}^z = \frac{1}{2}\mu_B \left(n_{\uparrow} - n_{\downarrow}\right) = \frac{\mu_B}{2} \int_{\in_F - \mu_B H}^{\in_F + \mu_B H} \theta\left(p_F - \left|\vec{p}\right|\right) \frac{d\vec{p}}{(2\pi\hbar)^3} \tag{2182}$$

The susceptibility may be computed as follows

$$\chi = \left.\frac{\partial \mathcal{M}}{\partial H}\right|_{H \to 0} = \text{const}\, \mu_B^2 > 0 \tag{2183}$$

This effect is called **Pauli paramagnetism** and exists only in those atoms with a **permanent** magnetic moment. The applied external field H aligns the magnetic moments against thermal fluctuations (along the field).

In this section, we examine atomic or molecular systems with a nonzero magnetic moment at the ground state. Examples include some atomic systems with a permanent magnetic moment in the ground state:

1. Atoms, molecules, ions, or free radicals with an odd number of electrons—such as hydrogen, free sodium atoms, gaseous nitric oxide, and organic free radicals such as triphenymethyl and F centers in alkali halides;
2. Some molecules with an even number of electrons—such as oxygen molecules and some organic compounds;
3. Atoms or ions with unfilled electronic shells—such as:
   - Transition elements (3d shell incomplete);
   - Rare earth (series of the lanthanides) (4f shell incomplete);
   - Actinides (5f shell incomplete)

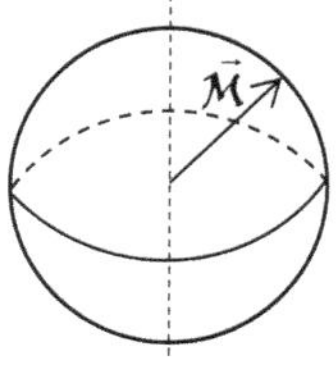

FIGURE 13.3 Sphere described by magnetic moment vector.

From the classical point of view, it is the moment $\vec{\mathcal{M}}$ that can be arbitrarily directed relative to the field $\vec{H}$. We will examine the classical picture of paramagnetism by considering the magnetic moment that describes a sphere of radius vector $\vec{\mathcal{M}}$, such as in Figure 13.3.

This magnetic moment has the following components:

$$\mathcal{M}_x = \mathcal{M}_0 \sin\theta\cos\varphi \quad , \quad \mathcal{M}_y = \mathcal{M}_0 \sin\theta\sin\varphi \quad , \quad \mathcal{M}_z = \mathcal{M}_0 \cos\theta \tag{2184}$$

Here, $\mathcal{M}_0$ is the total magnetic moment. Hence, the interaction energy in the direction of the field can be written as:

$$\mathrm{H}_{\mathrm{int}} = -\vec{\mathcal{M}}\vec{H} = -\mathcal{M}_0 H\cos\theta \tag{2185}$$

The mean value of the projection of the moment $\vec{\mathcal{M}}$ on the direction of the magnetic field yields the mean of its z-component [36, 13]:

$$\langle \mathcal{M}_z \rangle = \frac{\int \sin\theta d\theta d\varphi \mathcal{M}_z \exp\{-\beta \mathrm{H}_{\mathrm{int}}\}}{\int \sin\theta d\theta d\varphi \exp\{-\beta \mathrm{H}_{\mathrm{int}}\}} = \frac{\int_0^\pi m\cos\theta \exp\{\beta\mathcal{M}_0 H\cos\theta\}\sin\theta d\theta}{\int_0^\pi \exp\{\beta\mathcal{M}_0 H\cos\theta\}\sin\theta d\theta} = \mathcal{M}_0 \frac{d}{d\alpha}\ln \mathrm{I}(\lambda) \tag{2186}$$

where

$$\mathrm{I}(\lambda) = \int_0^\pi \exp\{\alpha\cos\theta\}\sin\theta d\theta = \frac{\exp\{\lambda\}-\exp\{-\lambda\}}{\lambda} \quad , \quad \beta\mathcal{M}_0 H = \lambda \quad , \quad \beta = \frac{1}{T} \tag{2187}$$

So,

$$\langle \mathcal{M}_z \rangle = \mathcal{M}_0 \left(\coth\lambda - \frac{1}{\lambda} \right) = \mathcal{M}_0 L(\lambda) \tag{2188}$$

where $L(\lambda)$ is the **Langevin function** and λ measures the ratio of a typical magnetic energy to the typical thermal energy. For weak fields $\lambda << 1$, then

$$\langle \mathcal{M}_z \rangle = \mathcal{M}_0 L(\lambda) \cong \mathcal{M}_0 \frac{\lambda}{3} = \frac{\mathcal{M}_0^2 H}{3T} \tag{2189}$$

and the magnetization:

$$\mathcal{M} = n\langle \mathcal{M}_z \rangle = n\frac{\mathcal{M}_0^2 H}{3T} \tag{2190}$$

which is proportional to n-atom concentration as well as to the magnetic field $\vec{H}$ (**paramagnetism**) [36]. So, the **magnetic susceptibility**:

$$\chi = \left.\frac{\partial \mathcal{M}}{\partial H}\right|_{H\to 0} = n\frac{\mathcal{M}_0^2}{3T} = \frac{C}{T} > 0 \tag{2191}$$

This is the **Curie law of paramagnetism**, and C is the **Curie constant** for the system. This was first determined experimentally by **Curie**, and **Langevin** later derived it classically. The Curie constant C assumes different values depending on the materials, with temperature T measured from the absolute zero. In diamagnetic substances, $\chi < 0$.

If we have a strong field where $\lambda >> 1$, then

$$\langle \mathcal{M}_z \rangle = \mathcal{M}_0 L(\lambda) = \mathcal{M}_0 \times 1 = \mathcal{M}_0 \tag{2192}$$

The magnetic moment of the gas achieves the saturation value, which implies that all the dipoles tend to become aligned with the field.

13.1.1 Molecular Field (Weiss Field)

Some materials possess a spontaneous magnetic moment, that is, even in the absence of an applied magnetic field, they have the **spontaneous magnetization** $\mathcal{M}_s(T)$ that vanishes at the so-called **Curie temperature** T_c. The simplest way to account for the spontaneous alignment of magnetic moments is by postulating the existence of an internal field, $H_{\text{Weiss}} = \Gamma \cdot \langle \mathcal{M}_z \rangle$. This is called the **Weiss field**, and it causes the magnetic moments of the atoms to line up. The value of H_{Weiss} is determined by the Curie temperature T_c. Typically, H_{Weiss} has a value of about 500 tesla. We shall see that the effective field is not magnetic in origin.

The greatest contributions of the theory of magnetism to general physics are in the fields of quantum statistical mechanics and thermodynamics. In 1907, Pierre Weiss (1865–1940) gave us the first modern theory of magnetism [86]. Weiss assumed that the interactions among magnetic molecules could be described empirically by a so-called "**molecular**" or "**internal**" field (**Weiss field**):

$$H_{\text{eff}} = H + \Gamma \cdot \langle \mathcal{M}_z \rangle \tag{2193}$$

where $\Gamma \cdot \langle \mathcal{M}_z \rangle$ is the molecular field (**Weiss field**) of other atoms, and Γ is a constant physical property of the material. Here, each atom is subject not only to the external field H but also to an internally generated molecular field (Weiss field) $\Gamma \cdot \langle \mathcal{M}_z \rangle$. This field simulates the physical interactions with all other atoms. Weiss's modification of the Langevin formula is given by [36]:

$$\langle \mathcal{M}_z \rangle = n\mathcal{M}_0 L\left(\frac{\mathcal{M}_0}{T}\left(H + \Gamma\langle \mathcal{M}_z \rangle\right)\right) \tag{2194}$$

The effect of neighboring atoms has thus been replaced by the field $\Gamma \cdot \langle \mathcal{M}_z \rangle$. If the molecular field results from the demagnetizing field caused by the free north and south poles on the surface of a spherical ferromagnetic, the Weiss constant would be $\Gamma \approx \frac{4\pi}{3}$ in some appropriate units.

Suppose

1. $H = 0$, then the molecular field vanishes. At times, we may have a spontaneous magnetization in the absence of the external field. This is a distinguishing characteristic of ferromagnetism.
2. $H = 0$ and $\langle \mathcal{M}_z \rangle \to 0$ as $T \to T_c$ then

$$\langle \mathcal{M}_z \rangle = n\mathcal{M}_0 L\left(\frac{\mathcal{M}_0 \Gamma \langle \mathcal{M}_z \rangle}{T_c}\right) = n\frac{\mathcal{M}_0^2 \Gamma \langle \mathcal{M}_z \rangle}{3T_c} \tag{2195}$$

Hence, this yields the critical temperature T_c:

$$T_c = \frac{n\mathcal{M}_0^2\Gamma}{3} \tag{2196}$$

We next analyze equation (2194) similarly as in reference [36] and we suppose $\vec{\mathcal{M}}_0$ to be an atom's intrinsic magnetic moment and n to be its atom concentration. Therefore, the absolute saturation magnetization:

$$\vec{\mathcal{M}}_\infty = n\vec{\mathcal{M}}_0 \tag{2197}$$

Considering that $\vec{\mathcal{M}}$ and $\vec{H}$ are parallel, then

$$\mathcal{M} = \mathcal{M}_\infty L\left(\frac{\left|\vec{\mathcal{M}}_0\right|\mathrm{B}'}{T}\right) = \mathcal{M}_\infty L\left(\frac{\mathcal{M}_0}{T}\left(H+\Gamma\langle\mathcal{M}_z\rangle\right)\right) \tag{2198}$$

which is transcendental relative to $\mathcal{M}$. To properly analyze relation (2198), we parametrize it:

$$x \equiv \frac{\mathcal{M}_0}{T}\left(H+\Gamma\langle\mathcal{M}_z\rangle\right) \quad , \quad y \equiv \frac{\mathcal{M}}{\mathcal{M}_\infty} \tag{2199}$$

where the relative magnetization

$$y = ax - b = L(x) \tag{2200}$$

is the Langevin function with argument x and

$$a = \frac{T}{\Gamma\mathcal{M}_0\mathcal{M}_\infty} = \frac{T}{\Gamma\mathcal{M}_0^2 n} \quad , \quad b = \frac{H}{\Gamma\mathcal{M}_\infty} = \frac{H}{\Gamma\mathcal{M}_0 n} \tag{2201}$$

For large x, the Langevin function achieves the value of 1, and the relative magnetization achieves its saturation value (Figure 13.4).

We may determine the solution of equation (2200) graphically by finding the intersection point of Langevin curve $y = L(x)$ and line $y = ax - b$ as in Figure 13.5. From

$$L'(x) \le \frac{1}{3} \tag{2202}$$

we consider two domains of definition for parameter a:

$$1)\quad a > \frac{1}{3} \quad , \quad (\beta > \alpha) \tag{2203}$$

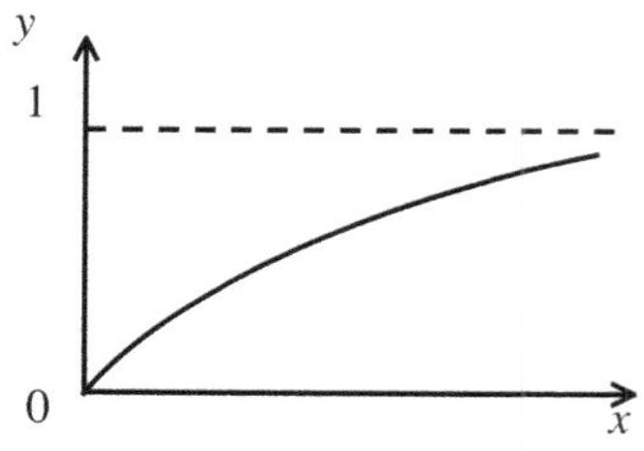

FIGURE 13.4 Plot of the relative magnetization y versus the ratio of a typical magnetic energy to the typical thermal energy x where y achieves the value 1 for large x and the relative magnetization achieves its saturation value.

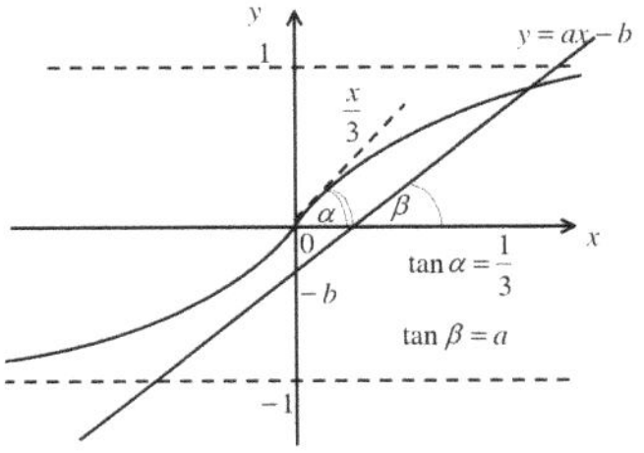

FIGURE 13.5 Shows the variation of the relative magnetization y versus the ratio of a typical magnetic energy to the typical thermal energy x. It depicts the graphical solution of equation of the relative magnetization (2200) by finding the intersection point of Langevin curve $y = L(x)$ and line $y = ax - b$.

For this case, there exists a unique intersection point of the Langevin curve $y = L(x)$ and line $y = ax - b$ that corresponds to a paramagnetic susceptibility. This implies the absence of spontaneous magnetization: If $H = 0$ then $\mathcal{M} = 0$. Therefore, the temperature defined from condition $a = \frac{1}{3}$ is the Curie temperature:

$$T_c = \frac{\Gamma \mathcal{M}_0^2 n}{3} \tag{2204}$$

For high temperatures:

$$T >> T_c \tag{2205}$$

and so $x << 1$, then $L(x) \approx 3$ and

$$y = 3b(3a-1)^{-1} \tag{2206}$$

This yields the magnetization $\mathcal{M}$:

$$\mathcal{M} = \mathcal{M}_\infty y = \frac{3HT_c}{\Gamma(T - T_c)} \tag{2207}$$

The magnetization $\mathcal{M}$ varies with temperature T according to the Curie-Weiss law (Figure 13.6). In this case, the ferromagnetic and paramagnetic Curie points coincide and indicate the approximate nature of the Weiss theorem.

We have seen that once the molecular field is determined, the total magnetization is known. It has been observed that there exists the possibility of ferromagnetism below a certain critical temperature called the Curie temperature. This ferromagnetic state, where all spins are preferentially aligned to one another, has lower free energy than the state with $\Gamma \cdot \langle \mathcal{M}_z \rangle = 0$. The ferromagnetic state with the temperature below T_c ($T < T_c$) is therefore a stable one. So, near T_c, the ferromagnet is close to a **phase transition** and a **spontaneous magnetization**, even when $H = 0$.

In all ferromagnetic materials, there is a relatively rapid decrease of the **spontaneous magnetization** $\langle \mathcal{M}_z \rangle$ as the temperature is raised to a critical value T_c. **Above the critical temperature** T_c ($T > T_c$),

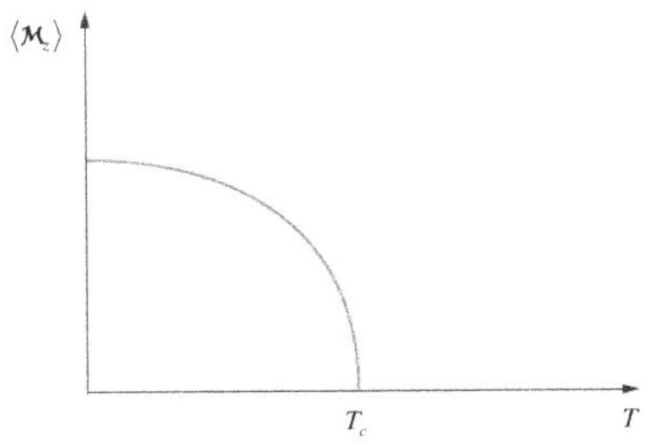

FIGURE 13.6 Magnetization $\mathcal{M}$ varies with temperature T according to Curie-Weiss law.

the ferromagnet behaves much like an ordinary paramagnetic substance. This is a **phase transition.** From there, the susceptibility for paramagnetic substances may be obtained from the following relation that considers equation (2207):

$$\chi = \left.\frac{\partial \mathcal{M}}{\partial H}\right|_{H\to 0} = \frac{3T_c}{\Gamma(T-T_c)} = \frac{C}{T-T_c} \tag{2208}$$

This is the famous **Curie-Weiss law** that is nearly, if not perfectly, obeyed by all ferromagnets. The **magnetic susceptibility** χ becomes infinite **when** $T \to T_c$**, that is, at the Curie temperature, where the substance becomes ferromagnetic.**

$$2)\quad a < \frac{1}{3} \quad , \quad (\beta < \alpha) \tag{2209}$$

For a weak magnetic field, $b << 1$ where there is a possibility of three intersection points (see Figure 13.7) with only one being thermodynamically stable. So, condition $H \geq 0$ corresponds to stable point maps as well as condition $\mathcal{M} > 0$, which is the domain of spontaneous magnetization.

For $H = 0$ and $T << T_c$ that implies $b = 0$,$a << 1$ and $x >> 1$. Therefore,

$$ax = L(x) \approx 1 - x^{-1} \tag{2210}$$

So, $x \approx a^{-1} - 1$ and $y = ax \approx 1 - a$ correspond to a linear dependence of spontaneous magnetization on temperature:

$$\mathcal{M} = \mathcal{M}_\infty \left[1 - \frac{T}{3T_c}\right] \tag{2211}$$

Hence, from the Weiss theorem, it is not possible to derive the **Bloch** $T^{\frac{3}{2}}$ **law**:

$$\mathcal{M} = \mathcal{M}_s(0)\left[1 - AT^{\frac{3}{2}}\right] \tag{2212}$$

where $A \approx 10^{-6}$ is the absolute saturation magnetization.

If field direction is changed (i.e., if $H < 0$), then the thermodynamic stability point corresponds to a minimum value for $\mathcal{M} < 0$ and corresponds to the remagnetization of the Weiss region. For $H > 0$, the solution for $\mathcal{M} > 0$ is absent. Thus, there is a jump in the dependence $\mathcal{M}(H)$ (Figure 13.7). Therefore, for the Weiss elementary region, there is a hysteresis phenomenon. Suppose a macroscopic specimen constitutes many Weiss regions. In that case, remagnetization results in a smooth hysteresis curve and mean values are taken relative to many Weiss regions.

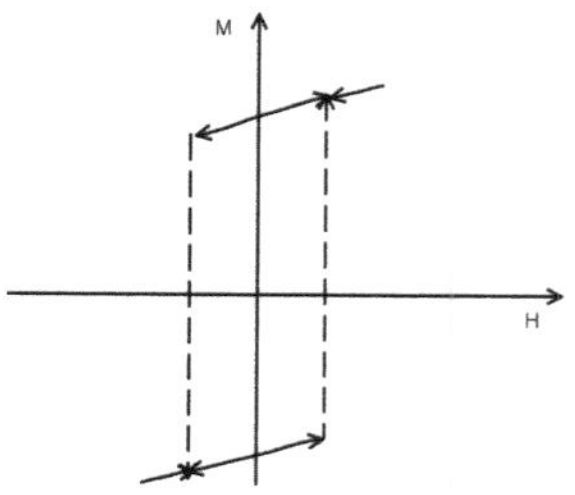

FIGURE 13.7 Variation of the magnetization $\mathcal{M}$ versus the magnetic field H.

13.2 Quantum Theory of Magnetism

We next examine paramagnetism as a quantum mechanical phenomenon. In quantum mechanics, the magnetic moment $\vec{\mathcal{M}}$ of an atom is quantized. We also consider paramagnetic substances, which are ionic crystals with a nonzero permanent magnetic moment in the ground state. Examples of substances with non-vanishing magnetic moments even when $H = 0$ are atoms and ions with

- an odd number of electrons, such as Na;
- partially filled inner shells, like Mn^{2+}, Gd^{3+}, or U^{4+} (transition elements, ions which are isoelectronic with transition elements, rare-earth and actinide elements).

From quantum mechanics, the interaction energy

$$\mathrm{H}_{\text{int}} = -\vec{\mathcal{M}}\vec{H} \tag{2213}$$

where

$$\vec{\mathcal{M}} = g_J \mu_B \left(\vec{J} + \vec{S}\right) \tag{2214}$$

and $\vec{J}$ is the total angular momentum vector:

$$\vec{J} = \vec{L} + \vec{S} \tag{2215}$$

with g_J being the Landé factor. Hence, from (2214), the mean value of the projection of the moment $\vec{\mathcal{M}}$ on the direction of the magnetic field yields the mean of its z-component:

$$\langle \mathcal{M}_z \rangle = g_J \mu_B \langle J^z \rangle \tag{2216}$$

where

$$\langle J^z \rangle = \frac{\sum_{\mathcal{M}_0=-J}^{J} \mathcal{M}_0 \exp\{-\beta \mathrm{H}_{\text{int}}\}}{\sum_{\mathcal{M}_0=-J}^{J} \exp\{-\beta \mathrm{H}_{\text{int}}\}} = \frac{\sum_{\mathcal{M}_0=-J}^{J} \mathcal{M}_0 \exp\{\beta g_J \mu_B \mathcal{M}_0 H\}}{\sum_{\mathcal{M}_0=-J}^{J} \exp\{\beta g_J \mu_B \mathcal{M}_0 H\}} = J\mathrm{B}_J\left(\beta g_J \mu_B J H\right) \tag{2217}$$

So,

$$\langle \mathcal{M}_z \rangle = g_J \mu_B J \mathrm{B}_J\left(\beta g_J \mu_B J H\right) \tag{2218}$$

where $\mathrm{B}_J(z)$ is called a **Brillouin function** of order J:

$$\mathrm{B}_J(z) = \left(1 + \frac{1}{2J}\right)\coth\left(\left(1 + \frac{1}{2J}\right)z\right) - \frac{1}{2J}\coth\frac{z}{2J} \tag{2219}$$

From here, the mean magnetic moment per unit volume (magnetization) is as follows:

$$\langle \mathcal{M} \rangle = n g_J \mu_B J \mathrm{B}_J\left(\beta g_J \mu_B J H\right) \tag{2220}$$

This is known as the **Curie-Brillouin law**, which is a quantum mechanical result. The magnetization is oriented parallel to the magnetic field H. The **magnetic susceptibility**:

$$\chi_T = \frac{\partial \mathcal{M}}{\partial H} = n\beta\left(Jg_J\mu_B\right)^2 B_J'\left(\beta g_J\mu_B JH\right) \tag{2221}$$

1. Let $J = \frac{1}{2}$ (smallest value), which stems from the single electron spins and allows only two orientations:

$$\left\langle J^z \right\rangle = \frac{\sum_{\mathcal{M}_0=-\frac{1}{2}}^{\frac{1}{2}} \mathcal{M}_0 \exp\left\{\beta g_J\mu_B\mathcal{M}_0 H\right\}}{\sum_{\mathcal{M}_0=-\frac{1}{2}}^{\frac{1}{2}} \exp\left\{\beta g_J\mu_B\mathcal{M}_0 H\right\}} = \frac{\sum_{\mathcal{M}_0=-\frac{1}{2}}^{\frac{1}{2}} \mathcal{M}_0 \exp\left\{\beta g_J\mu_B\mathcal{M}_0 H\right\}}{\sum_{\mathcal{M}_0=-\frac{1}{2}}^{\frac{1}{2}} \exp\left\{\beta g_J\mu_B\mathcal{M}_0 H\right\}} = \frac{1}{2}\mathrm{B}_{\frac{1}{2}}\left(\frac{\beta g_J\mu_B H}{2}\right) \tag{2222}$$

where

$$\mathrm{B}_{\frac{1}{2}}\left(\frac{\beta g_J\mu_B H}{2}\right) = \tanh\left(\frac{\beta g_J\mu_B H}{2}\right) = \tanh\left(\beta\mu_B H\right) \tag{2223}$$

This is because for the electron spin, $g_J = 2$, and this implies that the average magnetic moment per unit volume:

$$\langle \mathcal{M} \rangle = n\mu_B \mathrm{B}_{\frac{1}{2}}\left(\beta\mu_B H\right) = n\mu_B \tanh\left(\beta\mu_B H\right) \tag{2224}$$

with the magnetic susceptibility χ_T:

$$\chi_T = n\beta\mu_B^2 \operatorname{sec}h^2\left(\beta\mu_B H\right) \tag{2225}$$

2. If $J \to \infty$ (large J, classical limit), then

$$\mathrm{B}_\infty\left(\beta g_J\mu_B JH\right) \equiv L\left(\beta g_J\mu_B JH\right) \tag{2226}$$

This is the case if dipoles are in any arbitrary direction. For this limiting case, we arrive at classical mechanics. For classical moments, B_∞ is identical to the Langevin function. A classical magnetic moment $\vec{\mathcal{M}}$ can be oriented in any direction in space. Its energy is

$$-\left(\vec{\mathcal{M}}, \vec{H}\right) = -\mathcal{M}H\cos\theta \tag{2227}$$

where θ is the angle between the dipole and the applied magnetic field.

Next, we will look at the limiting cases for the magnetic field, $\vec{H}$, or for the temperature, T.

a. Consider weak magnetic fields $\vec{H}$ or high temperatures T, that is,

$$\beta g_J\mu_B JH << 1 \tag{2228}$$

then

$$\langle \mathcal{M} \rangle = n(g_J \mu_B)^2 \frac{J(J+1)H}{3T} \quad , \quad \chi_T = \frac{\partial \langle \mathcal{M} \rangle}{\partial H} = n(g_J \mu_B)^2 \frac{J(J+1)}{3T} \tag{2229}$$

The magnetic behavior characterized by (2229) is termed **paramagnetism.** From (2229), we have

$$\langle \mathcal{M} \rangle = \chi_T H \tag{2230}$$

and

$$\chi_T = \frac{C_J}{T} \tag{2231}$$

This variation of the susceptibility is inversely proportional to the temperature in equations (2229) or (2231) and known as **Curie's law** and the **Curie constant** C_J:

$$C_J = n \frac{(g_J \mu_B)^2 J(J+1)}{3} \equiv \frac{n \mathcal{M}_{\text{eff}}^2}{3} \tag{2232}$$

This variation of susceptibility (2229) or (2231) characterizes paramagnetic systems with **permanent moments**. The quantity $\mathcal{M}_{\text{eff}}$ is the effective magnetic moment. Though the condition

$$\beta g_J \mu_B J H << 1 \tag{2233}$$

holds true for the validity of Curie law and is achieved for a large range of fields and temperatures, the **permanent moments'** alignment is favored by the field and opposed by thermal disorder.

As $J \to \infty$, then

$$C_J \to C_\infty = \frac{n \mathcal{M}^2}{3} \tag{2234}$$

which is the **classical result of the Curie constant.** If $J = \frac{1}{2}$, then

$$C_{\frac{1}{2}} = n \mathcal{M}^2 = 3 C_\infty \tag{2235}$$

If the magnetic field $\vec{H}$ is exceedingly high or the temperature T exceedingly low,

$$\beta g_J \mu_B J H >> 1 \tag{2236}$$

then

$$\mathrm{B}_J(\beta g_J \mu_B J H) \cong 1 \tag{2237}$$

for all values of J. This implies saturation magnetization (complete alignment of the magnetic moments) at exceedingly strong magnetic fields or low temperatures.

Suppose we consider a molecular field (Weiss field), then

$$H_{\text{eff}} = H + \Gamma \cdot \langle \mathcal{M}_z \rangle \tag{2238}$$

and

$$\langle \mathcal{M}_z \rangle = ng_J \mu_B J \mathrm{B}_J \left(\beta g_J \mu_B J \left(H + \Gamma \langle \mathcal{M}_z \rangle \right) \right) \tag{2239}$$

The Curie temperature, T_c, is the temperature above which the spontaneous magnetization vanishes. The temperature T_c separates the disordered paramagnetic phase at $T > T_c$ from the ordered ferromagnetic phase at $T < T_c$. For spontaneous symmetry breaking, we have $H = 0$ and

$$\langle \mathcal{M}_s \rangle \equiv \langle \mathcal{M}_z \rangle = ng_J \mu_B J \mathrm{B}_J \left(\beta g_J \mu_B J \Gamma \langle \mathcal{M}_z \rangle \right) \tag{2240}$$

If $T \to T_c$ then $\langle \mathcal{M}_z \rangle \to 0$, so

$$\langle \mathcal{M}_z \rangle = ng_J \mu_B J \mathrm{B}_J \left(\beta g_J \mu_B J \Gamma \langle \mathcal{M}_z \rangle \right) = ng_J \mu_B J \frac{(J+1) \beta_c g_J \mu_B J \Gamma \langle \mathcal{M}_z \rangle}{3J} \tag{2241}$$

and

$$T_c = \frac{n\Gamma \left(g_J \mu_B \right)^2 J(J+1)}{3} \tag{2242}$$

So the molecular field parameter can be written

$$\Gamma^{-1} = \frac{n \left(g_J \mu_B \right)^2 J(J+1)}{3T_c} \tag{2243}$$

For iron (Fe), we have $\Gamma = 5000$.

The **magnetic susceptibility**:

$$\chi_T = \left. \frac{\partial \langle \mathcal{M} \rangle}{\partial H} \right|_{H \to 0} = \frac{C}{T - T_c} \tag{2244}$$

13.2.1 Spin Wave: Model of Localized Magnetism

It is instructive to note that exchange **forces** are not forces in the usual sense. Therefore, the energy splitting between quantized parallel and antiparallel spin configurations has to be evaluated by some independent means. This should be parametrized by what is known as a scalar. This is the **exchange constant** J that considers the two-body electrostatic repulsion. For the simplest case, where the given splitting can be calculated explicitly, this is a result of a concentration of two effects due to the Pauli Exclusion Principle. This affects only parallel spin electrons.

13.2.2 Heisenberg Hamiltonian

We examine the case of the **Heisenberg Hamiltonian** determinant that puts one spin on each lattice site; the spins then interact via a vector interaction. So, for nearest neighbors' spin interaction, the Hamiltonian of the system may be written

$$\mathrm{H}_{\mathrm{int}} = -\sum_{n \neq m} J_{nm} \vec{S}_n \vec{S}_m = -\sum_{n \neq m} J_{nm} \left(S_n^x S_m^x + S_n^y S_m^y + S_n^z S_m^z \right) \tag{2245}$$

where J_{nm} is the exchange integral. For the ferromagnetic ordering, $J_{nm} > 0$. Here, we take the usual practice that the symbol $\vec{S}_n$ represents the total angular momentum of the nth ion. The origin of the Weiss effective field is found in the **exchange field** between two interacting electrons on different atoms. For simplicity, assume that atoms n and m are neighbors and that each atom has one electron. The term $-J_{nm}\vec{S}_n\vec{S}_m$ denotes the contribution to the energy from a pair of atoms (or ions) located at sites n and m. Normally, one assumes that J_{nm} is nonzero only for nearest neighbors and perhaps next-nearest neighbors. The introduction of the interaction term $-J_{nm}\vec{S}_n\vec{S}_m$ is the source of the Weiss internal field that produces ferromagnetism.

Due to interaction terms in the Hamiltonian determinant, including noncommutative spin operators, the Heisenberg Hamiltonian is often **solved approximately** for spin greater than one-half or for coupling between spins that may be further neighbors.

Anisotropic Heisenberg Model

For the **anisotropic Heisenberg model**, the coupling constant in one direction, for example, z, usually is different from that in other directions:

$$\mathrm{H}_{\text{int}} = -\sum_{n\neq m} J_{||nm} S_n^z S_m^z - \sum_{n\neq m} J_{\perp nm}\left(S_n^x S_m^x + S_n^y S_m^y\right) \tag{2246}$$

or

$$\mathrm{H}_{\text{int}} = \mathrm{H}_{||} + \mathrm{H}_{\perp} \tag{2247}$$

where

$$\mathrm{H}_{||} = -\sum_{n\neq m} J_{||nm} S_n^z S_m^z \quad , \quad \mathrm{H}_{\perp} = -\frac{1}{2}\sum_{n\neq m} J_{\perp nm}\left(S_n^+ S_m^- + S_n^- S_m^+\right) \tag{2248}$$

We realize from these equations that crystals are not spherically symmetric but have finite point group symmetry. In real crystals, certain directions are easy to magnetize while others are **hard** to magnetize. One example would be Co, which is a hexagonal crystal where magnetization is **easy** along the hexagonal axis and **hard** along any axis perpendicular to the hexagonal axis. If we consider Hamiltonian (2248), then we have two types of magnetization axes:

- If $J_{\perp nm} = 0$ then the **Ising model** and

$$\mathrm{H}_{\text{int}} = -\sum_{n\neq m} J_{nm} S_n^z S_m^z \tag{2249}$$

  This may be solved exactly in one dimension even if a magnetic field H is added to the Hamiltonian:

$$\mathrm{H}_{\text{int}} = -\sum_{n\neq m} J_{nm} S_n^z S_m^z + \mathrm{H}_H \quad , \quad \mathrm{H}_H = -g\mu_B H \sum_n S_n^z \tag{2250}$$

- In two dimensions, this may be solved exactly without the magnetic field H [87]. This **X-Y model** has only $J_{\perp nm}$:

$$\mathrm{H}_{\text{int}} = -\sum_{n\neq m} J_{nm}\left(S_n^x S_m^x + S_n^y S_m^y\right) \tag{2251}$$

  and therefore may be solved exactly only in one dimension.

13.2.3 X-Y Model

To solve (2251) in one dimension and for any n, we introduce, respectively, spin raising and lowering operators:

$$S_n^+ = S_n^x + iS_n^y \quad , \quad S_n^- = S_n^x - iS_n^y \tag{2252}$$

The terms raising and lowering are applied to these operators because they raise or lower the magnetic quantum number of the spin state. So,

$$S_n^x = \frac{S_n^+ + S_n^-}{2} \quad , \quad S_n^y = \frac{S_n^+ - S_n^-}{2i} \tag{2253}$$

with

$$S_n^x S_m^x + S_n^y S_m^y = \frac{1}{2}\left(S_n^- S_m^+ + S_n^+ S_m^-\right) \tag{2254}$$

So,

$$\mathrm{H}_{\text{int}} = -\frac{1}{2}\sum_{n \neq m} J_{nm}\left(S_n^+ S_m^- + S_n^- S_m^+\right) \tag{2255}$$

We assume the static magnetic field H to be in the direction of the oz axis. So, the Heisenberg Hamiltonian for a ferromagnet (with unit volume) consisting of N spins with the nearest neighbor interactions can be written

$$\mathrm{H} = -\sum_{n \neq m} J_{nm}\vec{S}_n\vec{S}_m - g\mu_B H \sum_n S_n^z \tag{2256}$$

From the Fourier transformation:

$$\vec{S}_{\vec{q}} = \int d\vec{r} \exp\left\{-i\vec{q}\vec{r}\right\}\vec{S}(\vec{r}) = \sum_n \int d\vec{r} \exp\left\{-i\vec{q}\vec{r}\right\}\vec{S}_n\delta\left(\vec{r} - \vec{R}_n\right) = \sum_n \vec{S}_n \exp\left\{-i\vec{q}\vec{R}_n\right\} \tag{2257}$$

Then the interaction Hamiltonian

$$\mathrm{H}_{\text{int}} = -\sum_{n \neq m} J_{nm}\vec{S}_n\vec{S}_m = -\sum_{n \neq m} J_{nm}\frac{1}{N}\sum_{\vec{q}'}\frac{1}{N}\sum_{\vec{q}''}\vec{S}_{\vec{q}'}\vec{S}_{\vec{q}''}\exp\left\{-i\vec{q}'\vec{R}_n\right\}\exp\left\{i\vec{q}''\vec{R}_m\right\} \tag{2258}$$

Letting

$$\vec{R}_n = \vec{R}_m + \vec{R} \tag{2259}$$

then

$$\mathrm{H}_{\text{int}} = -\sum_{n \neq m\vec{R}\vec{q}'\vec{q}''} J_{\vec{R}}\frac{1}{N^2}\vec{S}_{\vec{q}'}\vec{S}_{\vec{q}''}\exp\left\{i\left(\vec{q}' + \vec{q}''\right)\vec{R}_m\right\}\exp\left\{i\vec{q}''\vec{R}\right\} \tag{2260}$$

Because

$$\sum_m \exp\left\{i\left(\vec{q}' + \vec{q}''\right)\vec{R}_m\right\} = N\delta_{\vec{q}',-\vec{q}''} \tag{2261}$$

then

$$\mathrm{H}_{\mathrm{int}} = -\sum_{\vec{q}} \vec{S}_{\vec{q}} \vec{S}_{-\vec{q}} \frac{1}{N} \sum_{\vec{R}} J_{\vec{R}} \exp\left\{i\vec{q}\vec{R}\right\} \tag{2262}$$

Considering that

$$J_{\vec{q}} = \frac{1}{N} \sum_{\vec{R}} J_{\vec{R}} \exp\left\{i\vec{q}\vec{R}\right\} \tag{2263}$$

So,

$$\mathrm{H}_{\mathrm{int}} = -\sum_{\vec{q}} J_{\vec{q}} \vec{S}_{\vec{q}} \vec{S}_{-\vec{q}} = -\sum_{n \neq m} J_{nm} \vec{S}_n \vec{S}_m \tag{2264}$$

Next, we consider a simple cubic crystal with a lattice constant a. For the interaction between nearest neighbors, the translation invariance of the Hamiltonian determinant is manifest in the fact that

$$J_{nm} = \begin{cases} J & , \left|\vec{R}_n - \vec{R}_m\right| = a \\ 0 & , \left|\vec{R}_n - \vec{R}_m\right| \neq a \end{cases} \tag{2265}$$

and

$$J_{\vec{q}} = \frac{2J}{N}\left(\cos q_x a + \cos q_y a + \cos q_z a\right) \tag{2266}$$

For a square lattice, we have

$$J_{\vec{q}} = \frac{2J}{N}\left(\cos q_x a + \cos q_y a\right) \tag{2267}$$

From here, we observe that

$$J_{\vec{q}} = J_{-\vec{q}} \tag{2268}$$

Consider the case where $q \to 0$. For the cubic crystal, we have

$$J_{\vec{q}\to 0} = \frac{6J}{N} \tag{2269}$$

and for the square lattice,

$$J_{\vec{q}\to 0} = \frac{4J}{N} \tag{2270}$$

So, for the d-dimensional lattice, we have

$$J_{\vec{q}\to 0} = \frac{2dJ}{N} = \frac{ZJ}{N} \tag{2271}$$

where Z is the **coordination number or number of nearest neighbors**.

We introduce the contribution of the magnetic field:

$$\mathrm{H} = -g\mu_B H \frac{1}{2}\sum_n S_n^z\left(\exp\left\{i\vec{q}_0\vec{R}_n\right\}+\exp\left\{-i\vec{q}_0\vec{R}_n\right\}\right) = -\frac{1}{2}g\mu_B H\left(\sum_n S_n^z\exp\left\{i\vec{q}_0\vec{R}_n\right\}+\sum_n S_n^z\exp\left\{-i\vec{q}_0\vec{R}_n\right\}\right) \tag{2272}$$

or

$$\mathrm{H} = -\frac{1}{2}g\mu_B H\left(S_{\vec{q}_0}^z + S_{-\vec{q}_0}^z\right) \tag{2273}$$

To connect our model to molecular field treatment, we consider the effective magnetic field applying random phase approximation (RPA):

$$\vec{S}_{-\vec{q}} \to \left\langle \vec{S}_{-\vec{q}} \right\rangle \tag{2274}$$

and so

$$\mathrm{H}_{\mathrm{RPA}} = -\frac{1}{2}g\mu_B \sum_{\sigma=+,-} S_{\sigma q_0}^z \left\{\frac{1}{2}H - \frac{J_{q_0}V}{g\mu_B}\left\langle S_{-\sigma q_0}^z\right\rangle\right\} \tag{2275}$$

From here, the effective field:

$$H_{\mathrm{eff}}\left(-q_0\right) = \frac{H}{2}V + \frac{J_{q_0}}{g\mu_B}\left\langle S_{-q_0}^z\right\rangle \tag{2276}$$

Therefore, the magnetization

$$\left\langle \mathcal{M}_z\left(q_0\right)\right\rangle = -g\mu_B\left\langle S_{q_0}^z\right\rangle \tag{2277}$$

and the susceptibility

$$\chi_0\left(q_0\right) = \lim_{H\to 0}\frac{\left\langle \mathcal{M}_z\left(q_0\right)\right\rangle}{H_{\mathrm{eff}}} \tag{2278}$$

From the previous relation, we have

$$\left\langle \mathcal{M}_z\left(q_0\right)\right\rangle = -g\mu_B\left\langle S_{q_0}^z\right\rangle = \chi_0\left(q_0\right)H_{\mathrm{eff}}\left(q_0\right) = \chi_0\left(q_0\right)\frac{HV}{2} - \frac{J_{q_0}\chi_0\left(q_0\right)V}{g\mu_B}\left\langle S_{q_0}^z\right\rangle \tag{2279}$$

Here, the Weiss field is

$$H_{\mathrm{Weiss}}\left(q_0\right) = \frac{J_{q_0}\chi_0\left(q_0\right)V}{g\mu_B}\left\langle S_{q_0}^z\right\rangle \tag{2280}$$

and

$$-g\mu_B\left\langle S_{q_0}^z\right\rangle = \frac{\chi_0\left(q_0\right)\frac{HV}{2}}{1-\frac{J_{q_0}\chi_0\left(q_0\right)V}{\left(g\mu_B\right)^2}} \tag{2281}$$

Because

$$J_{q_0} = J_{-q_0} \tag{2282}$$

then

$$g\mu_B\left(\left\langle S_{q_0}^z\right\rangle+\left\langle S_{-q_0}^z\right\rangle\right)=\frac{\chi_0(q_0)H}{1-\dfrac{J_{q_0}\chi_0(q_0)}{(g\mu_B)^2}} \tag{2283}$$

The **susceptibility of interacting system** is

$$\chi_q=\lim_{H\to 0}\frac{\langle \mathcal{M}_z\rangle}{H}=\frac{\chi_0}{1-\dfrac{J_q\chi_0 V}{(g\mu_B)^2}} \tag{2284}$$

and from

$$\chi_0=\frac{N}{V}(g\mu_B)^2\frac{S(S+1)}{3T} \tag{2285}$$

then

$$\chi_q=\lim_{H\to 0}\frac{\langle \mathcal{M}_z\rangle}{H}=\frac{N(g\mu_B)^2\dfrac{S(S+1)}{3T}}{1-J_qN\dfrac{VS(S+1)}{3T}} \tag{2286}$$

Considering that only nearest neighbors interact, for a simple cubic crystal we have

$$\chi_q=\frac{C}{T}\frac{1}{1-2J\left(\cos q_xa+\cos q_ya+\cos q_za\right)\dfrac{S(S+1)}{3T}} \tag{2287}$$

where

$$C=N(g\mu_B)^2\frac{S(S+1)}{3} \tag{2288}$$

is the Curie constant. For $q\to 0$, then

$$\cos qa\cong 1-\frac{(qa)^2}{2} \tag{2289}$$

and

$$\chi_q=\frac{C}{T-\dfrac{6JS(S+1)}{3}-\dfrac{JS(S+1)}{3}(\vec{q}\vec{a})^2} \tag{2290}$$

In this case, because the coordination number is $Z=6$, then

$$\chi_q=\frac{C}{T-\dfrac{ZJS(S+1)}{3}-\dfrac{JS(S+1)}{3}(\vec{q}\vec{a})^2} \tag{2291}$$

Then, for $q \to 0$, we have the **Curie-Weiss law**:

$$\chi_{q\to 0} = \frac{C}{T - T_c} \tag{2292}$$

where

$$T_c = \frac{ZJS(S+1)}{3} > 0 \tag{2293}$$

In this case, the fact that the exchange constant J is positive makes S_i and S_j totally align parallel to one another so that the energy is minimized. It is not uncommon to have spin systems where J is negative.

Considering again that $q \to 0$, then

$$\chi_q = \frac{C}{T_c} \frac{1}{\frac{T-T_c}{T_c} + \frac{(qa)^2}{6}} \tag{2294}$$

Denoting

$$\tau = \frac{T-T_c}{T_c} \quad , \quad \lambda^2 = \frac{6(T-T_c)}{T_c a^2} = \frac{6\tau}{a^2} \quad , \quad \lambda^{-1} = a\sqrt{\frac{T_c}{6(T-T_c)}} \tag{2295}$$

and substituting in (2294), then

$$\chi_q = \frac{\text{const}}{\lambda^2 + q^2} \tag{2296}$$

Here, λ^{-1} has the dimension of length that maps the correlation length:

$$\xi = \lambda^{-1} \approx \tau^{-\nu} \quad , \quad \nu = \frac{1}{2} \tag{2297}$$

and

$$\chi_{q\to 0} = \frac{\text{const}}{\tau} \approx \tau^{-\gamma} \quad , \quad \gamma = 1 \tag{2298}$$

So,

$$\chi_R = \int \chi_q \exp\left\{i\vec{q}\vec{R}\right\} d\vec{q} \approx \frac{1}{R} \exp\left\{-\frac{R}{\xi}\right\} \tag{2299}$$

When, $T = T_c$ then $\chi_R \approx \frac{1}{R}$ as $\xi \to \infty$ and long-range order takes place.

For $S = \frac{1}{2}$,

$$\left\langle S^z \right\rangle = \frac{1}{2} \tanh \frac{ZJ\left\langle S^z \right\rangle}{2T} \cong \frac{1}{2}\left(\frac{ZJ\left\langle S^z \right\rangle}{2T} - \frac{1}{3}\left(\frac{ZJ\left\langle S^z \right\rangle}{2T} \right)^3 \right) \tag{2300}$$

from where

$$\left\langle S^z \right\rangle = \sqrt{\frac{3}{4}\frac{T - T_c}{T_c}} \quad (2301)$$

with the Curie temperature, T_c, at which point the long-range order (or magnetization) vanishes:

$$T_c = \frac{ZJ}{4} \quad (2302)$$

So,

$$\left\langle S^z \right\rangle \approx \tau^{\beta} \quad , \quad \beta = \frac{1}{2} \quad (2303)$$

In 1932, Néel [88–90] put forth the idea of **antiferromagnetism** to explain the temperature-independent paramagnetic susceptibility of such materials as chromium and manganese that were too large to be explained by Pauli theory. In our case, $J < 0$ will attempt to align the neighboring spins anti-parallel with materials called **antiferromagnets** and **Néel temperature** T_N and the susceptibility $\chi_{q\to 0}$:

$$\chi_{q\to 0} = \frac{C}{T - T_N} \quad , \quad T_N = \frac{6JS(S+1)}{3} \quad (2304)$$

For the antiferromagnet, the magnetic susceptibility increases as the temperature increases up to a **transition** or **Néel temperature** T_N. Above T_N, the antiferromagnetic crystal is in a standard paramagnetic state.

Also, the magnetic susceptibility:

$$\chi_q = \frac{C}{T_N}\frac{1}{\frac{T - T_N}{T_N} + \frac{(\vec{q}\vec{a})^2}{6}} = \frac{\tilde{C}}{\lambda^2 + q^2} \quad (2305)$$

In an antiferromagnetic state, we may think of two different **sublattices (Figure 13.8)** where, if the two sublattices happen to have different spins from one another, we have a **ferrimagnet** instead of an antiferromagnet.

13.2.4 Spin Waves in Ferromagnets

We have just seen how ferromagnetism displays spontaneous magnetization or polarization. Suppose electrons are confined to their respective atoms. We can then study the distribution of the electrons over the energy levels of each atom. At low temperatures, the electrons completely fill the lowest energy shells while the upper occupied shell may remain partially filled. To avoid double occupation of the same spatial quantum state where the Coulomb repulsion is strongest, the electrons in the partially filled shell tend to align their spin states in an effect called **Hund-rule coupling**. This gives rise to a local dipole moment of each atom. For the case of the Heisenberg model, a local dipole moment is related to each atom localized in a lattice. These local dipole moments interact with one another with the nearest neighbor interactions being dominant and the other interactions neglected. In such a case, the Heisenberg model corresponds to the Ising model.

FIGURE 13.8 Sublattice structure of spins in a ferromagnet.

It is instructive to note that the Heisenberg model is very realistic for electrically insulating materials where the electrons are assumed to be confined to their respective atoms. However, in metals, not all of the electrons are confined to their atoms. So, the use of local moments may not always be realistic. This example may be found in d-band transition metals (such as iron, nickel, and cobalt) where the Heisenberg model cannot predict the noninteger magnetic moment per atom and the large specific heat capacity. Elsewhere in this book, we examine itinerant ferromagnetism, which requires moving electrons. Hence, this can only occur in conductive materials.

We study spin-wave excitations (magnons) on top of a ferromagnetically and antiferromagnetically ordered state of quantum spins arranged on a lattice. The Heisenberg model Hamiltonian of the system (2245) for (anti-)ferromagnetism in a homogenous magnetic field is as follows:

$$\mathrm{H}=-\frac{1}{2}\sum_{n\neq m}J_{nm}\vec{S}_n\vec{S}_m-g\mu_B H\sum_n S_n^z \tag{2306}$$

For $J_{nm}>0$, the interactions favor spin alignment that is a **ferromagnetic arrangement**. For $J_{nm}<0$, the ordering has alternate spins up and down, when the lattice permits, and this is called an **antiferromagnetic arrangement**. But

$$\vec{S}_n\cdot\vec{S}_m=S_n^xS_m^x+S_n^yS_m^y+S_n^zS_m^z \tag{2307}$$

and from (2253) then

$$\vec{S}_n\cdot\vec{S}_m=\frac{1}{2}\left(S_n^+S_m^-+S_n^-S_m^+\right)+S_n^zS_m^z \tag{2308}$$

Taking the summation over all n,m, then the Hamiltonian of the system:

$$\mathrm{H}=-\frac{1}{2}\sum_{n\neq m}J_{nm}\left(S_n^-S_m^++S_n^zS_m^z\right) \tag{2309}$$

It is instructive to note that the difficulty with solving spin problems is shown by defining collective operators. The operators are transformed into wave vector space via the Fourier (plane-wave) transformation:

$$S_n^\pm=\frac{1}{\sqrt{N}}\sum_q S_q^\pm\exp\left\{\pm i\vec{q}\vec{R}_n\right\}\quad S_q^\pm=\frac{1}{\sqrt{N}}\sum_{qn}S_n^\pm\exp\left\{\mp i\vec{q}\vec{R}_n\right\} \tag{2310}$$

This transformation is appropriate for solving fermionic and bosonic problems and helps in transforming our Hamiltonian as follows:

$$\mathrm{H}=\mathrm{H}_\perp+\mathrm{H}_{||}+\mathrm{H}_H \tag{2311}$$

with

$$\mathrm{H}_\perp=-\frac{1}{2}\sum_{n\neq m}J_{nm}\left(S_n^+S_m^-+S_n^-S_m^+\right) \tag{2312}$$

If we do the change $n \leftrightarrow m$, then

$$\mathrm{H}_{\perp} = -\sum_{n \neq m} J_{nm} S_n^- S_m^+ \tag{2313}$$

From the transformation (2310), we then have

$$\mathrm{H}_{\perp} = -\sum_{n \neq m} J_{nm} S_n^- S_m^+ = -\sum_{n \neq m} J_{nm} \frac{1}{N} \sum_{\vec{q}'\vec{q}''} S_{\vec{q}'}^- S_{\vec{q}''}^+ \exp\left\{-i\vec{q}'\vec{R}_n + i\vec{q}''\vec{R}_m\right\} \tag{2314}$$

Changing $\vec{R}_n = \vec{R}_m + \vec{R}$, then

$$\mathrm{H}_{\perp} = -\sum_{m\vec{q}'\vec{q}''} \frac{1}{N} \exp\left\{i\left(\vec{q}'' - \vec{q}'\right)\vec{R}_m\right\} \sum_{\vec{R}} J_{\vec{R}} S_{\vec{q}'}^- S_{\vec{q}''}^+ \exp\left\{-i\vec{q}'\vec{R}\right\} \tag{2315}$$

From

$$\sum_m \exp\left\{i\left(\vec{q}'' - \vec{q}'\right)\vec{R}_m\right\} = N\delta_{\vec{q}''\vec{q}'} \tag{2316}$$

then

$$\mathrm{H}_{\perp} = -\sum_{\vec{q}} S_{\vec{q}}^- S_{\vec{q}}^+ \sum_{\vec{R}} J_{\vec{R}} \exp\left\{-i\vec{q}\vec{R}\right\} = -\sum_{\vec{q}} J_{\vec{q}} S_{\vec{q}}^- S_{\vec{q}}^+ \tag{2317}$$

where

$$\sum_{\vec{R}} J_{\vec{R}} \exp\left\{-i\vec{q}\vec{R}\right\} = J_{\vec{q}} \tag{2318}$$

Let us examine the commutation relations for the operators (2310):

$$\left[S_m^+, S_n^-\right] = 2S_n^z \delta_{nm} \tag{2319}$$

The inverse Fourier transformation:

$$\left[S_{\vec{q}'}^+, S_{\vec{q}''}^-\right] = \frac{1}{N} \sum_{n \neq m} \left[S_m^+, S_n^-\right] \exp\left\{-i\vec{q}'\vec{R}_n + i\vec{q}''\vec{R}_m\right\} = \frac{2}{N} \sum_n S_n^z \exp\left\{-i\vec{R}_n\left(\vec{q}'' - \vec{q}'\right)\right\} \tag{2320}$$

Here, we consider commutation rest (2319). We observe from (2320) that the operators S_m^+ and S_n^- commute except on the same site, and their commutation is (2319). Because the right-hand side of (2320) is not simple, it is preferable to find something that imitates oscillators, such as:

$$\left[S_{\vec{q}'}^+, S_{\vec{q}''}^-\right] = \mathrm{const}\,\delta_{\vec{q}''\vec{q}'} \tag{2321}$$

Here, we show that $S_{\vec{q}'}^+$ and $S_{\vec{q}''}^-$ are independent operators, except at $\vec{q}'' = \vec{q}'$. Their behavior is similar to that of bosons and Bose statistics would be applicable. Unfortunately, relation (2321) does not exhibit this property. Therefore, we base our approximation on Callen [91] in relation (2321):

- If the temperature $T \neq 0$, then **all spins do not totally line up** and

$$\left[S_{\vec{q}'}^+, S_{\vec{q}''}^-\right] = \frac{2}{N} \sum_n S_n^z \exp\left\{-i\vec{R}_n\left(\vec{q}'' - \vec{q}'\right)\right\} = \left\langle S_n^z \right\rangle \frac{2}{N} \sum_n \exp\left\{-i\vec{R}_n\left(\vec{q}'' - \vec{q}'\right)\right\} = 2\left\langle S_n^z \right\rangle \delta_{\vec{q}''\vec{q}'} \tag{2322}$$

- If the temperature $T=0$, then **all spins do totally line up** and

$$\left[S^{+}_{\vec{q}'},S^{-}_{\vec{q}''}\right]=\frac{2}{N}\sum_{n}S^{z}_{n}\exp\left\{-i\vec{R}_n\left(\vec{q}''-\vec{q}'\right)\right\}=S\frac{2}{N}\sum_{n}\exp\left\{-i\vec{R}_n\left(\vec{q}''-\vec{q}'\right)\right\}=2S\delta_{\vec{q}''\vec{q}'} \quad (2323)$$

So,

$$\left[S^{+}_{\vec{q}'},S^{-}_{\vec{q}''}\right]=\frac{2}{N}\sum_{n}S^{z}_{n}\exp\left\{-i\vec{R}_n\left(\vec{q}''-\vec{q}'\right)\right\}=\begin{cases}2S\delta_{\vec{q}''\vec{q}'} & , \quad T=0\\ 2\left\langle S^{z}\right\rangle\delta_{\vec{q}''\vec{q}'} & , \quad T\neq 0\end{cases} \quad (2324)$$

We then try to find the self-consistent equation for the average magnetization $\left\langle S^{z}\right\rangle$ that is an approximate method. The operators $S^{+}_{\vec{q}'}$ and $S^{-}_{\vec{q}''}$ do not describe independent eigenstates of the system except in special cases. The absence of collective eigenstates is the difficulty faced when solving spin systems. This results in the difficulty in solving Hamiltonians determinants such as in (2251). The Hamiltonian with bilinear boson or fermion operators may be solved exactly in a way that is contrary to solving that of bilinear spin operators.

13.2.5 Bosonization of Operators

We now introduce bosonic operators that map back to physical states in the spin variables

$$S^{-}_{\vec{q}}=\sqrt{2S}b^{\dagger}_{\vec{q}} \quad , \quad S^{+}_{\vec{q}}=\sqrt{2S}b_{\vec{q}} \quad (2325)$$

where the condition $S>>1$ is necessary and the boson commutation relation

$$\left[b_{\vec{q}},b^{\dagger}_{\vec{q}}\right]=1 \quad (2326)$$

It is instructive to note that the operator $b^{\dagger}_{\vec{q}}$ creates a magnon with the wave vector $\vec{q}$ and the operator $b_{\vec{q}}$ annihilates it. We use the symbols S and S^{z} for quantum mechanical operators and for numbers associated with eigenvalues. Where confusion might arise, we write $\hat{S}$ and $\hat{S}^{z}$ for the quantum mechanical operators. From quantum mechanics, we know that $\hat{S}^{2}$ and $\hat{S}^{z}$ can be diagonalized in the same representation because they commute. We usually write

$$\hat{S}^{2}\left|S,S^{z}\right\rangle=S(S+1)\left|S,S^{z}\right\rangle \quad , \quad \hat{S}^{z}\left|S,S^{z}\right\rangle=S^{z}\left|S,S^{z}\right\rangle \quad (2327)$$

Here, S is either a positive half odd integer or a positive integer. So, for our case,

$$\vec{S}^{2}_{n}=\left(S^{z}_{n}\right)^{2}+\left(S^{x}_{n}\right)^{2}+\left(S^{y}_{n}\right)^{2}=S(S+1)=\frac{1}{2}\left(S^{-}_{n}S^{+}_{n}+S^{+}_{n}S^{-}_{n}\right)+S^{z}_{n}S^{z}_{n} \quad (2328)$$

$$\vec{S}^{2}_{n}=\frac{1}{2}\left(S^{-}_{n}S^{+}_{n}+S^{+}_{n}S^{-}_{n}\right)=\left(S^{x}_{n}\right)^{2}+\left(S^{y}_{n}\right)^{2}=\frac{1}{2N}\sum_{\vec{q}'\vec{q}''}\left(S^{+}_{\vec{q}'}S^{-}_{\vec{q}''}+S^{-}_{\vec{q}''}S^{+}_{\vec{q}'}\right)\exp\left\{i\left(\vec{q}'-\vec{q}''\right)\vec{R}_n\right\} \quad (2329)$$

Making the change $\vec{q}'\leftrightarrow\vec{q}''$ in the factor of the exponential function and also considering (2325), then

$$\vec{S}^{2}_{n}=\frac{1}{2N}2S\sum_{\vec{q}'\vec{q}''}\left(b_{\vec{q}'}b^{\dagger}_{\vec{q}''}+b^{\dagger}_{\vec{q}''}b_{\vec{q}'}\right)\exp\left\{i\left(\vec{q}'-\vec{q}''\right)\vec{R}_n\right\} \quad (2330)$$

From the commutation relation

$$b_{\vec{q}'}b_{\vec{q}''}^{\dagger} = b_{\vec{q}''}^{\dagger}b_{\vec{q}'} + \delta_{\vec{q}'\vec{q}''} \tag{2331}$$

then

$$\vec{S}_n^2 = \frac{S}{N}\sum_{\vec{q}'\vec{q}''}\delta_{\vec{q}'\vec{q}''}\exp\left\{i\left(\vec{q}'-\vec{q}''\right)\vec{R}_n\right\} + \frac{2S}{N}\sum_{\vec{q}'\vec{q}''}b_{\vec{q}''}^{\dagger}b_{\vec{q}'}\exp\left\{i\left(\vec{q}'-\vec{q}''\right)\vec{R}_n\right\} \tag{2332}$$

and

$$\vec{S}_n^2 = S + \frac{2S}{N}\sum_{\vec{q}'\vec{q}''}b_{\vec{q}''}^{\dagger}b_{\vec{q}'}\exp\left\{i\left(\vec{q}'-\vec{q}''\right)\vec{R}_n\right\} \tag{2333}$$

From here, and considering equation (2328), we have

$$\vec{S}_n^2 = \left(S_n^z\right)^2 + \left(S_n^x\right)^2 + \left(S_n^y\right)^2 = \left(S_n^z\right)^2 + S + \frac{2S}{N}\sum_{\vec{q}'\vec{q}''}b_{\vec{q}''}^{\dagger}b_{\vec{q}'}\exp\left\{i\left(\vec{q}'-\vec{q}''\right)\vec{R}_n\right\} = S(S+1) \tag{2334}$$

or

$$\left(S_n^z\right)^2 = S(S+1) - S - \frac{2S}{N}\sum_{\vec{q}'\vec{q}''}b_{\vec{q}''}^{\dagger}b_{\vec{q}'}\exp\left\{i\left(\vec{q}'-\vec{q}''\right)\vec{R}_n\right\} \tag{2335}$$

and

$$S_n^z = \sqrt{S^2 - \frac{2S}{N}\sum_{\vec{q}'\vec{q}''}b_{\vec{q}''}^{\dagger}b_{\vec{q}'}\exp\left\{i\left(\vec{q}'-\vec{q}''\right)\vec{R}_n\right\}} = S\sqrt{1 - \frac{2}{SN}\sum_{\vec{q}'\vec{q}''}b_{\vec{q}''}^{\dagger}b_{\vec{q}'}\exp\left\{i\left(\vec{q}'-\vec{q}''\right)\vec{R}_n\right\}} \tag{2336}$$

For $S >> 1$ and considering

$$\left.\sqrt{1-x}\right|_{x<<1} \approx 1 - \frac{x}{2} \tag{2337}$$

then

$$S_n^z = S - \frac{1}{N}\sum_{\vec{q}'\vec{q}''}b_{\vec{q}''}^{\dagger}b_{\vec{q}'}\exp\left\{i\left(\vec{q}'-\vec{q}''\right)\vec{R}_n\right\} \tag{2338}$$

Finally, the Hamiltonian is rewritten in terms of oscillators as follows

$$\mathrm{H}_{||} = -\sum_{n\neq m}J_{nm}S_n^zS_m^z = -\sum_{n\neq m}J_{nm}\left(S - \frac{1}{N}\sum_{\vec{q}'\vec{q}''}b_{\vec{q}''}^{\dagger}b_{\vec{q}'}\exp\left\{i\left(\vec{q}'-\vec{q}''\right)\vec{R}_n\right\}\right)\left(S - \frac{1}{N}\sum_{\vec{\bar{q}}''\vec{\bar{q}}'}b_{\vec{\bar{q}}''}^{\dagger}b_{\vec{\bar{q}}'}\exp\left\{i\left(\vec{\bar{q}}'-\vec{\bar{q}}''\right)\vec{R}_m\right\}\right) \tag{2339}$$

or

$$\mathrm{H}_{||} = -\sum_{n\neq m}J_{nm}\left(S^2 - \frac{2S}{N}\sum_{\vec{q}'\vec{q}''}b_{\vec{q}''}^{\dagger}b_{\vec{q}'}\exp\left\{i\left(\vec{q}'-\vec{q}''\right)\vec{R}_n\right\} + \frac{1}{N^2}\sum_{\vec{q}'\vec{q}''}\sum_{\vec{\bar{q}}''\vec{\bar{q}}'}b_{\vec{q}''}^{\dagger}b_{\vec{q}'}b_{\vec{\bar{q}}''}^{\dagger}b_{\vec{\bar{q}}'}\exp\left\{i\left(\vec{q}'-\vec{q}''\right)\vec{R}_n\right\}\exp\left\{i\left(\vec{\bar{q}}'-\vec{\bar{q}}''\right)\vec{R}_m\right\}\right) \tag{2340}$$

Ignoring the third summand, we have

$$\mathrm{H}_{||} = -\sum_{m,n\neq m} J_{nm}\left(S^2 - \frac{2S}{N}\sum_{\vec{q}'\vec{q}''} b^{\dagger}_{\vec{q}''} b_{\vec{q}'} \exp\left\{ i\left(\vec{q}' - \vec{q}''\right)\vec{R}_n \right\} \right) \tag{2341}$$

Doing the change of variable $\vec{R}_n = \vec{R}_m + \vec{R}$, then

$$\mathrm{H}_{||} = -\sum_{m\vec{R}} J_{\vec{R}}\left(S^2 - \frac{2S}{N}\sum_{\vec{q}'\vec{q}''} b^{\dagger}_{\vec{q}''} b_{\vec{q}'} \exp\left\{ i\left(\vec{q}' - \vec{q}''\right)\vec{R}_m \right\}\exp\left\{ i\left(\vec{q}' - \vec{q}''\right)\vec{R} \right\} \right) \tag{2342}$$

as

$$\sum_{\vec{R}} J_{\vec{R}} \exp\left\{ -i\vec{q}\vec{R} \right\} = J_{\vec{q}} \tag{2343}$$

From

$$\sum_{m} \exp\left\{ i\left(\vec{q}'' - \vec{q}'\right)\vec{R}_m \right\} = N\delta_{\vec{q}''\vec{q}'} \tag{2344}$$

and letting

$$\sum_{\vec{R}} J_{\vec{R}} S^2 = JNZS^2 \tag{2345}$$

then

$$\mathrm{H}_{||} = -JNZS^2 + \frac{2S}{N}\sum_{\vec{R}} J_{\vec{R}} \sum_{\vec{q}'\vec{q}''} b^{\dagger}_{\vec{q}''} b_{\vec{q}'} N\delta_{\vec{q}''\vec{q}'} \exp\left\{ i\left(\vec{q}' - \vec{q}''\right)\vec{R} \right\} = -JNZS^2 + 2S\sum_{\vec{R}} J_{\vec{R}} \sum_{\vec{q}} b^{\dagger}_{\vec{q}} b_{\vec{q}} \tag{2346}$$

or

$$\mathrm{H}_{||} = -JNZS^2 + 2S\sum_{\vec{R}} J_{\vec{R}} \sum_{\vec{q}} b^{\dagger}_{\vec{q}} b_{\vec{q}} = -JNZS^2 + 2SZJ\sum_{\vec{q}} b^{\dagger}_{\vec{q}} b_{\vec{q}} \tag{2347}$$

Here N is the number of sites, and Z is the number of next-nearest neighbors.
We now examine the transversal component:

$$\mathrm{H}_{\perp} = -\sum_{n\neq m} J_{nm} S^{-}_{n} S^{+}_{m} = -\frac{1}{2}\sum_{n\neq m} J_{nm}\frac{1}{N}\sum_{\vec{q}'\vec{q}''}\left(S^{-}_{\vec{q}'} S^{+}_{\vec{q}''} + S^{+}_{\vec{q}''} S^{-}_{\vec{q}'} \right)\exp\left\{ -i\vec{q}'\vec{R}_n + i\vec{q}''\vec{R}_m \right\} \tag{2348}$$

From the bosonization of operators in (2325), then

$$\mathrm{H}_{\perp} = -\frac{1}{2}\sum_{n\neq m} J_{nm}\frac{2S}{N}\sum_{\vec{q}'\vec{q}''}\left(b^{\dagger}_{\vec{q}'} b_{\vec{q}''} + b_{\vec{q}''} b^{\dagger}_{\vec{q}'} \right)\exp\left\{ -i\vec{q}'\vec{R}_n + i\vec{q}''\vec{R}_m \right\} \tag{2349}$$

Consider

$$b_{\vec{q}'} b^{\dagger}_{\vec{q}''} = b^{\dagger}_{\vec{q}''} b_{\vec{q}'} + \delta_{\vec{q}'\vec{q}''} \tag{2350}$$

then

$$H_\perp = -\frac{1}{2}\sum_{n\neq m} J_{nm}\frac{2S}{N}\sum_{\vec{q}'\vec{q}''}\left(2b^\dagger_{\vec{q}'}b_{\vec{q}''}+\delta_{\vec{q}'\vec{q}''}\right)\exp\left\{-i\vec{q}'\vec{R}_n+i\vec{q}''\vec{R}_m\right\} \tag{2351}$$

and

$$H_\perp = -\sum_{n\neq m} J_{nm}\frac{2S}{N}\sum_{\vec{q}'\vec{q}''} b^\dagger_{\vec{q}'}b_{\vec{q}''}\exp\left\{-i\vec{q}'\vec{R}_n+i\vec{q}''\vec{R}_m\right\}-\sum_{n\neq m} J_{nm}\frac{S}{N}\sum_{\vec{q}'\vec{q}''}\delta_{\vec{q}'\vec{q}''}\exp\left\{-i\vec{q}'\vec{R}_n+i\vec{q}''\vec{R}_m\right\} \tag{2352}$$

Doing the change of variable $\vec{R}_n = \vec{R}_m + \vec{R}$, then

$$H_\perp = -\sum_{m\vec{R}} J_{\vec{R}}\frac{2S}{N}\sum_{\vec{q}'\vec{q}''} b^\dagger_{\vec{q}'}b_{\vec{q}''}\exp\left\{-i\vec{q}'\vec{R}+i\left(\vec{q}''-\vec{q}'\right)\vec{R}_m\right\}-\sum_{m\vec{R}} J_{\vec{R}}\frac{S}{N}\sum_{\vec{q}'\vec{q}''}\delta_{\vec{q}'\vec{q}''}\exp\left\{-i\vec{q}'\vec{R}+i\left(\vec{q}''-\vec{q}'\right)\vec{R}_m\right\} \tag{2353}$$

and from the change of variable $\vec{q} = \vec{q}' - \vec{q}''$, then

$$\sum_m \exp\left\{i\left(\vec{q}''-\vec{q}'\right)\vec{R}_m\right\} = \sum_m \exp\left\{-i\vec{q}\vec{R}_m\right\} = N\delta_{\vec{q}0} \tag{2354}$$

and

$$H_\perp = -\sum_{m\vec{R}} J_{\vec{R}}\frac{2S}{N}\sum_{\vec{q}'\vec{q}} b^\dagger_{\vec{q}'}b_{\vec{q}'-\vec{q}}\exp\left\{-i\vec{q}'\vec{R}-i\vec{q}\vec{R}_m\right\}-\sum_{m\vec{R}} J_{\vec{R}}\frac{S}{N}\sum_{\vec{q}'\vec{q}}\delta_{\vec{q}',\vec{q}'-\vec{q}}\exp\left\{-i\vec{q}'\vec{R}-i\vec{q}\vec{R}_m\right\} \tag{2355}$$

or

$$H_\perp = -\sum_{\vec{R}} J_{\vec{R}}\frac{2S}{N}\sum_{\vec{q}'\vec{q}} b^\dagger_{\vec{q}'}b_{\vec{q}'-\vec{q}}N\delta_{\vec{q}0}\exp\left\{-i\vec{q}'\vec{R}\right\}-\sum_{\vec{R}} J_{\vec{R}}\frac{S}{N}\sum_{\vec{q}'\vec{q}}\delta_{\vec{q}',\vec{q}'-\vec{q}}N\delta_{\vec{q}0}\exp\left\{-i\vec{q}'\vec{R}\right\} \tag{2356}$$

or

$$H_\perp = -2S\sum_{\vec{R}} J_{\vec{R}}\sum_{\vec{q}} b^\dagger_{\vec{q}}b_{\vec{q}}\exp\left\{-i\vec{q}\vec{R}\right\}-S\sum_{\vec{R}} J_{\vec{R}}\sum_{\vec{q}}\delta_{\vec{q}\vec{q}}\exp\left\{-i\vec{q}\vec{R}\right\} \tag{2357}$$

Because no two spins occupy the same site, we then ignore the last term:

$$H_\perp = 2S\sum_{\vec{q}} b^\dagger_{\vec{q}}b_{\vec{q}}ZJ\left(1-\sum_{\vec{R}} J_{\vec{R}}\frac{\exp\left\{-i\vec{q}\vec{R}\right\}}{ZJ}-1\right) = 2S\sum_{\vec{q}} b^\dagger_{\vec{q}}b_{\vec{q}}ZJ\left(1-\gamma_{\vec{q}}\right)-2SZJ\sum_{\vec{q}} b^\dagger_{\vec{q}}b_{\vec{q}} \tag{2358}$$

where the factor

$$\gamma_{\vec{q}} = \frac{1}{ZJ}\sum_{\vec{R}} J_{\vec{R}}\exp\left\{-i\vec{q}\vec{R}\right\} \tag{2359}$$

includes a sum over the Z vectors $\vec{R}$, which are the directed connections to nearest-neighbor sites. The total Hamiltonian of a state containing magnons is then

$$H = H_\perp + H_{||} = \epsilon_0 + \sum_{\vec{q}} \epsilon_{\vec{q}} b^\dagger_{\vec{q}}b_{\vec{q}} \tag{2360}$$

where

$$\in_{\vec{q}} = 2JZS\left(1-\gamma_{\vec{q}}\right) \tag{2361}$$

is the dispersion relation for spin waves and $n_{\vec{q}} = b_{\vec{q}}^{\dagger}b_{\vec{q}}$ is interpreted as the number operator of spin waves with wave vector $\vec{q}$. This shows that the elementary excitations are waves of energy $\in_{\vec{q}}$. The quantity

$$\in_0 = -g\mu_B HNS - JNZS^2 \tag{2362}$$

is the energy of the completely saturated state where all spins are parallel to the applied field. We assume that $H > 0$. Hence, the magnetic moments should be aligned in the positive direction of the oz axis when the system is found in its ground state.

For a simple cubic crystal where $q \to 0$ (long wave lengths), we then have the following Taylor expansion

$$\gamma_q = \frac{1}{3}\left(\cos q_x a + \cos q_y a + \cos q_z a\right) \cong 1 - \frac{q_x^2 + q_y^2 + q_z^2}{6} a^2 = 1 - \frac{q^2 a^2}{6} \tag{2363}$$

Here, a is the lattice constant and now

$$\in_{\vec{q}} = 2JZS\left(1-\gamma_{\vec{q}}\right) = 2JZS\frac{q^2a^2}{6} = JZS\frac{q^2a^2}{3} \equiv \lambda q^2 \quad , \quad \lambda = \frac{JZS}{3}a^2 \tag{2364}$$

This is a **characteristic of spin waves in ferromagnets:** the energy $\in_{\vec{q}}$ is quadratic in the wave number q for long wavelengths. This is depicted in Figure 13.9. This dispersion relation is appropriate for a Bravais lattice. In a reciprocal space, the allowed q values are restricted to the first Brillouin zone. In a simple lattice, the magnon energy is lifted in the same manner as the energy of a free particle in a constant potential

$$\in_{\vec{q}} = g\mu_B H + JZS\frac{q^2a^2}{3} \equiv U_0 + \frac{\hbar^2 q^2}{2m^*} \tag{2365}$$

where

$$\frac{1}{m^*} = \frac{4JSa^2}{\hbar^2} \quad , \quad U_0 = g\mu_B H \tag{2366}$$

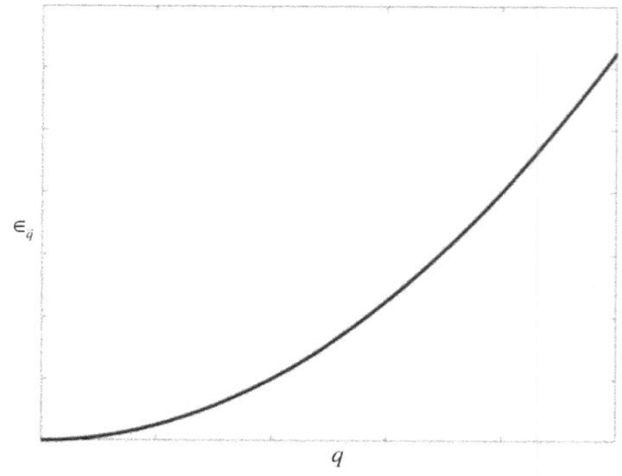

FIGURE 13.9 The figure depicts the energy versus the wave vector.

The magnons (bosons) have gapless Goldstone modes with a quadratic dispersion. The internal energy associated with these excitations is given by

$$E = \sum_{\vec{q}} \in_{\vec{q}} \left\langle b_{\vec{q}}^{\dagger} b_{\vec{q}} \right\rangle = \sum_{\vec{q}} \in_{\vec{q}} \left\langle n_{\vec{q}} \right\rangle \quad , \quad \in_{\vec{q}} = \lambda q^2 \tag{2367}$$

where

$$\left\langle n_{\vec{q}} \right\rangle = \frac{1}{\exp\left\{\frac{\in_{\vec{q}}}{T}\right\} - 1} \tag{2368}$$

is the Bose-Einstein distribution function because magnons are bosons. The energy in (2367) is depicted in Figure 13.9. Converting the sum in (2367) to an integral over q gives the internal energy

$$E = \frac{1}{(2\pi)^3} \int d\vec{q} \frac{\lambda q^2}{\exp\left\{\frac{\lambda q^2}{T}\right\} - 1} \tag{2369}$$

Letting

$$\in_{\vec{q}} = \lambda q^2 \equiv T\xi \quad , \quad q \equiv \left(\frac{T\xi}{\lambda}\right)^{\frac{1}{2}} \quad , \quad 2\lambda q dq \equiv T d\xi \tag{2370}$$

then

$$d\vec{q} = 4\pi q^2 dq = \frac{2\pi}{\lambda}\left(\frac{T\xi}{\lambda}\right)^{\frac{1}{2}} T d\xi = 2\pi\lambda^{-\frac{3}{2}} T^{\frac{3}{2}} \xi^{\frac{1}{2}} d\xi \tag{2371}$$

So

$$\sum_q \bar{n}_q = \int \frac{d\vec{q}}{(2\pi)^3} \frac{1}{\exp\left\{\frac{\lambda q^2}{T}\right\} - 1} = T^{\frac{3}{2}} 2\pi\lambda^{-\frac{3}{2}} \int \frac{\xi^{\frac{1}{2}} d\xi}{(2\pi)^3} \frac{1}{\exp\{\xi\} - 1} \equiv B T^{\frac{3}{2}} \tag{2372}$$

and

$$E = T^{\frac{5}{2}} 2\pi\lambda^{-\frac{3}{2}} \frac{1}{(2\pi)^3} \int \frac{\xi^{\frac{3}{2}} d\xi}{\exp\{\xi\} - 1} \equiv A T^{\frac{5}{2}} \tag{2373}$$

The specific heat due to magnons

$$C = \frac{\partial E}{\partial T} = \frac{5}{2} A T^{\frac{3}{2}} \tag{2374}$$

Consider the case

$$T << \mu_B H \tag{2375}$$

then

$$\sum_q \bar{n}_q = \int \frac{d^d q}{(2\pi)^d} \frac{1}{\exp\left\{\frac{\lambda q^2}{T}\right\} - 1} \cong \int \frac{d^d q}{(2\pi)^d} \frac{T}{\lambda q^2} \tag{2376}$$

From

$$\lambda q^2 \equiv T\xi \quad , \quad q \equiv \left(\frac{T\xi}{\lambda}\right)^{\frac{1}{2}} \tag{2377}$$

then

$$d^d q \equiv \text{const}\, q^{d-1} dq = \text{const}\left(\frac{T\xi}{\lambda}\right)^{\frac{d-1}{2}} \left(\frac{T\xi}{\lambda}\right)^{-\frac{1}{2}} \frac{T}{\lambda} d\xi = \text{const}\frac{1}{2} T^{\frac{d}{2}} \xi^{\frac{d-2}{2}} \lambda^{\frac{1-d}{2}} d\xi \tag{2378}$$

and

$$\sum_q \bar{n}_q = \int \frac{d^d q}{(2\pi)^d} \frac{T}{\lambda q^2} = T^{\frac{d}{2}} \int \frac{1}{(2\pi)^d} \text{const}\frac{1}{2} \xi^{\frac{d-4}{2}} \lambda^{\frac{1-d}{2}} d\xi \equiv \Lambda T^{\frac{d}{2}} \tag{2379}$$

where

$$\Lambda \equiv \Lambda_0 \int \xi^{\frac{d-4}{2}} d\xi \quad , \quad \Lambda_0 \equiv \frac{\text{const}}{2(2\pi)^d} \lambda^{\frac{1-d}{2}} \tag{2380}$$

If $d = 2$, then $\Lambda \approx \ln q$ and there is no long-range order at any $T \neq 0$. If $d = 1$, then $\Lambda \approx \frac{1}{q}$ and there is no long-range order at any $T \neq 0$ as well as $T = 0$.

Mermin-Wagner Theorem

There exists no spontaneous breaking of a continuous symmetry group in $d = 1$ and $d = 2$ dimensions of space at nonvanishing temperature T if the Hamiltonian has only short-ranged interactions [92].

In contrast, discrete symmetries can be broken spontaneously in $d = 2$ dimensions for a nonvanishing temperature T by short-range interactions; a prominent example is the Ising model.

13.2.6 Magnetization

We show how the magnetization varies with temperature T in a ferromagnetic state in which the low-lying excitations are spin waves. We consider that the excitations are bosons. The thermal average of the magnetization at a temperature T is referred to as the spontaneous magnetization at temperature T. It is given by

$$\langle \mathcal{M}_z \rangle = g\mu_B \langle S^z \rangle \tag{2381}$$

But,

$$S^z = \sum_m S_m^z = \sum_m S - \frac{1}{N} \sum_{m\vec{q}'\vec{q}''} b_{\vec{q}''}^\dagger b_{\vec{q}'} \exp\left\{i(\vec{q}' - \vec{q}'')\vec{R}_m\right\} \tag{2382}$$

and

$$\sum_m \exp\left\{i\left(\vec{q}'' - \vec{q}'\right)\vec{R}_m\right\} = N\delta_{\vec{q}''\vec{q}'} \tag{2383}$$

then

$$S^z = NS - \frac{1}{N}\sum_{\vec{q}'\vec{q}''} b^\dagger_{\vec{q}''} b_{\vec{q}'} N\delta_{\vec{q}''\vec{q}'} = NS - \sum_{\vec{q}} b^\dagger_{\vec{q}} b_{\vec{q}} \tag{2384}$$

Therefore

$$\left\langle \mathcal{M}_z \right\rangle = g\mu_B \left\langle S^z \right\rangle = g\mu_B \left(NS - \sum_{\vec{q}} \left\langle b^\dagger_{\vec{q}} b_{\vec{q}} \right\rangle \right) \tag{2385}$$

We define

$$\Delta\mathcal{M} = \mathcal{M}_s(0) - \mathcal{M}_s(T) = g\mu_B \sum_{\vec{q}} \left\langle \hat{n}_{\vec{q}} \right\rangle \tag{2386}$$

and restrict ourselves to low temperatures where the initial assumption of interacting spin waves is valid when the occupation numbers of the lowest excited states are significant. Here, we consider the energy $\epsilon_{\vec{q}}$ to be quadratic in the wave number q for long wavelengths. So,

$$\Delta\mathcal{M} = \mathcal{M}_s(0) - \mathcal{M}_s(T) = g\mu_B \mathrm{B} T^{\frac{3}{2}} \equiv \tilde{\mathrm{B}} T^{\frac{3}{2}} \tag{2387}$$

Therefore, the **three-halves power law** or the **Bloch law**

$$\mathcal{M}_s(T) = \mathcal{M}_s(0) - \tilde{\mathrm{B}} T^{\frac{3}{2}} \tag{2388}$$

or

$$\mathcal{M}_s(T) = g\mu_B \left(NS - \mathrm{B} T^{\frac{3}{2}} \right) \tag{2389}$$

The temperatrure-**three-halves power law** $T^{\frac{3}{2}}$ magnetization dependence is a well-known result associated with the presence of noninteracting spin waves. Higher-order terms are obtained if the full expression for $\gamma_{\vec{q}}$ is used instead of just the long wavelength expansion (correct up to the q^2 term) and the q-integral is performed over the first Brillouin zone and not integrated to infinity.

13.2.7 Experiments Revealing Magnons

The most important factors demonstrating the existence of magnons are:

1. The existence of side bands in ferromagnetic resonances: A uniform precession mode in a ferromagnetic resonance experiment excites a $q = 0$ spin wave. In a ferromagnetic film, it is possible to couple to modes with wave length λ satisfying $\frac{1}{2}\lambda = \frac{d}{n}$, where the thickness of the film is d. This yields resonance at magnon wave numbers $q_n = \frac{n\pi}{d}$.
2. The existence of inelastic neutron scattering peaks associated with magnons.
3. The coupling of magnons to phonons in ferromagnetic crystals.

13.2.8 Spin Waves in Antiferromagnets

In order to study spin waves in an antiferromagnetic ground state, we make the following three modifications:

- **We assume the lattice, Λ, to be bipartite. This implies that it can be divided into two interpenetrating sublattices, A and B, such that all nearest neighbors of any lattice site in A are lattice sites in B and vice versa. So,**

$$\Lambda = \mathrm{A} \cup \mathrm{B} \quad , \quad \mathrm{A} \cap \mathrm{B} = 0 \tag{2390}$$

  For example, in two dimensions, a square lattice is bipartite while the triangular lattice is not (the triangular lattice is tripartite).
- **For an antiferromagnetic coupling, the sign of the interaction parameter, J_{nm}, entering the Hamiltonian has to be changed (later we implement this sign change as $J_{nm} < 0$).**
- **The ferromagnetic source field, $H\vec{e}_z$, must be replaced by a staggered source field, $\pm H\vec{e}_z$, where one sign is assigned to sublattice A and the other sign is assigned to sublattice B.**

The state in which all N spins on sublattice A are $\uparrow$ and all N spins on sublattice B are $\downarrow$ is highly degenerate because the direction for $\uparrow$ (or $\downarrow$) is completely arbitrary. This degeneracy is not removed by introducing an external field H. When H is not very large, the spins align themselves antiferromagnetically in the plane perpendicular to H. However, the direction of a given sublattice magnetization is still arbitrary in that plane.

The limit to this case is when the lattice Λ is assumed to be bipartite and made of $N >> 1$ sites, which implies that the lattice is a macroscopically large subset of the hyper cubic lattice with lattice spacing a. We assume the couplings J_{nm} are nonvanishing if n belongs to one sublattice. However, if m belongs to the other sublattice, they are taken as negative $J_{nm} < 0$. The condition $J_{nm} \leq 0$ defines a quantum spin-S Heisenberg antiferromagnet whereas the former condition ensures the absence of geometric frustration. The interaction $|J_{nm}|\vec{S}_n\vec{S}_m$ favors the singlet state for the two-site problem. If the degrees of freedom $\vec{S}_n$ and $\vec{S}_m$ are not operator-valued vectors but classical vectors in real space of a fixed magnitude, then the classical interaction $|J_{nm}|\vec{S}_n\vec{S}_m$ would favor an antiparallel alignment of these classical vectors. Hence, the classical configuration minimizes the classical energy when $H = 0$ in the classical counterpart to equation (2306) has all spins pointing along one direction on one sublattice and all spins pointing in the opposite direction on the other sublattice for any $d \geq 1$. This is known as the **Néel collinear antiferromagnetic state.** The fate of the classical long-range order in the quantum ground state as a result of quantum fluctuations should be the fundamental question to be addressed at the quantum level when $H = 0$.

So, the Heisenberg Hamiltonian of an antiferromagnet has $J_{nm} < 0$. Therefore,

$$\mathrm{H} = -\frac{1}{2}\sum_{n\neq m} J_{nm}\vec{S}_n\vec{S}_m - g\mu_B H\sum_{n} \vec{S}_n^z \quad , \quad n \in \mathrm{A} \quad , \quad m \in \mathrm{B} \tag{2391}$$

Here, the sum is over all possible distinct nearest neighbor pairs. Note that the Heisenberg exchange interaction $J_{nm}\vec{S}_n\vec{S}_m$ is the simplest interaction between two quantum spins that is invariant under a global SU(2) rotation of all the quantum spins. Also, the Zeeman term breaks the local SU(2) symmetry in spin space down to the subgroup U(1) of local rotations around the direction in spin space corresponding to H.

From (2391), we have

$$\vec{S}_n \cdot \vec{S}_m = S_n^x S_m^x + S_n^y S_m^y + S_n^z S_m^z \tag{2392}$$

and for any n and m, we introduce spin raising and lowering operators (2252) and (2254), respectively, which helps in transforming the Hamiltonian as follows:

$$\mathrm{H} = \mathrm{H}_{\perp} + \mathrm{H}_{||} + \mathrm{H}_{H} \tag{2393}$$

with

$$\mathrm{H}_{\perp} = -\frac{1}{2}\sum_{n\neq m} J_{nm}\left(S_n^+ S_m^- + S_n^- S_m^+\right) \tag{2394}$$

We introduce collective coordinates into the given Hamiltonian:

$$\vec{S}_n^{\pm} = \frac{1}{\sqrt{N}}\sum_{\vec{q}} \vec{S}_{\vec{q}}^{\pm} \exp\left\{\pm i\vec{q}\vec{R}_n\right\} \quad , \quad \vec{S}_{\vec{q}}^{\pm} = \frac{1}{\sqrt{N}}\sum_{n} \vec{S}_n^{\pm} \exp\left\{\mp i\vec{q}\vec{R}_n\right\} \tag{2395}$$

It is easy to show that, for the sublattice A, we have

$$S_{\vec{q}}^- = \sqrt{2S}a_{\vec{q}}^\dagger \quad , \quad S_{\vec{q}}^+ = \sqrt{2S}a_{\vec{q}} \tag{2396}$$

and for the sublattice B, we have

$$S_{\vec{q}}^- = \sqrt{2S}b_{\vec{q}}^\dagger \quad , \quad S_{\vec{q}}^+ = \sqrt{2S}b_{\vec{q}} \tag{2397}$$

From here, for the sublattice A, we have

$$\vec{S}_n^- = \sqrt{\frac{2S}{N}}\sum_{\vec{q}} a_{\vec{q}}^\dagger \exp\left\{-i\vec{q}\vec{R}_n\right\}, \quad \vec{S}_n^+ = \sqrt{\frac{2S}{N}}\sum_{\vec{q}} a_{\vec{q}} \exp\left\{i\vec{q}\vec{R}_n\right\} \tag{2398}$$

and for the sublattice B, we have

$$\vec{S}_m^- = \sqrt{\frac{2S}{N}}\sum_{\vec{q}} b_{\vec{q}}^\dagger \exp\left\{-i\vec{q}\vec{R}_m\right\}, \quad \vec{S}_m^+ = \sqrt{\frac{2S}{N}}\sum_{\vec{q}} b_{\vec{q}} \exp\left\{i\vec{q}\vec{R}_m\right\} \tag{2399}$$

From the transformation (2398) and (2399) and the change of variable $\vec{R}_n = \vec{R}_m + \vec{R}$, we then have

$$\mathrm{H}_{\perp} = -\frac{1}{2}\sum_{n\neq m} J_{nm}\left(S_n^+ S_m^- + S_n^- S_m^+\right) = -\frac{1}{2}\sum_{m\vec{R}} J_{\vec{R}} \frac{2S}{N}\sum_{\vec{q}'\vec{q}''}\left(a_{\vec{q}'}^\dagger b_{\vec{q}''} \exp\left\{-i\vec{q}'\vec{R}_n + i\vec{q}''\vec{R}_m\right\} + a_{\vec{q}'} b_{\vec{q}''}^\dagger \exp\left\{i\vec{q}'\vec{R}_n - i\vec{q}''\vec{R}_m\right\}\right) \tag{2400}$$

or

$$\mathrm{H}_{\perp} = -S\sum_{\vec{R}} J_{\vec{R}} \sum_{\vec{q}} \exp\left\{-i\vec{q}\vec{R}\right\}\left(a_{\vec{q}}^\dagger b_{\vec{q}} + b_{\vec{q}}^\dagger a_{\vec{q}}\right) \tag{2401}$$

So, if only nearest neighbors interact, then we have

$$\mathrm{H}_{\perp} = -SJZ\sum_{\vec{q}} \gamma_{\vec{q}}\left(a_{\vec{q}}^\dagger b_{\vec{q}} + b_{\vec{q}}^\dagger a_{\vec{q}}\right) \tag{2402}$$

where

$$\gamma_{\vec{q}} = \frac{1}{ZJ}\sum_{\vec{R}} J_{\vec{R}} \exp\left\{-i\vec{q}\vec{R}\right\} \tag{2403}$$

The longitudinal part of the Hamiltonian is rewritten as follows

$$\mathrm{H}_{||} = -\sum_{n\neq m} J_{nm} S_n^z S_m^z = -\sum_{n\neq m} J_{nm}\left(S - \frac{1}{N}\sum_{\vec{q}'\vec{q}''} a_{\vec{q}''}^{\dagger} a_{\vec{q}'} \exp\left\{i\left(\vec{q}'-\vec{q}''\right)\vec{R}_n\right\}\right)\left(S - \frac{1}{N}\sum_{\vec{\bar{q}}''\vec{\bar{q}}'} b_{\vec{\bar{q}}''}^{\dagger} b_{\vec{\bar{q}}'} \exp\left\{i\left(\vec{\bar{q}}'-\vec{\bar{q}}''\right)\vec{R}_m\right\}\right) \tag{2404}$$

or

$$\begin{aligned}\mathrm{H}_{||} = &-\sum_{n\neq m} J_{nm}\left(S^2 - \frac{S}{N}\sum_{\vec{q}'\vec{q}''} a_{\vec{q}''}^{\dagger} a_{\vec{q}'} \exp\left\{i\left(\vec{q}'-\vec{q}''\right)\vec{R}_n\right\} - \frac{S}{N}\sum_{\vec{q}'\vec{q}''} b_{\vec{q}''}^{\dagger} b_{\vec{q}'} \exp\left\{i\left(\vec{q}'-\vec{q}''\right)\vec{R}_m\right\}\right) - \\ &-\sum_{n\neq m} J_{nm}\frac{1}{N^2}\sum_{\vec{q}'\vec{q}''}\sum_{\vec{\bar{q}}''\vec{\bar{q}}'} a_{\vec{q}''}^{\dagger} a_{\vec{q}'} b_{\vec{\bar{q}}''}^{\dagger} b_{\vec{\bar{q}}'} \exp\left\{i\left(\vec{q}'-\vec{q}''\right)\vec{R}_n\right\}\exp\left\{i\left(\vec{\bar{q}}'-\vec{\bar{q}}''\right)\vec{R}_m\right\}\end{aligned} \tag{2405}$$

Ignoring the fourth summand, we have

$$\mathrm{H}_{||} = -\sum_{n\neq m} J_{nm}\left(S^2 - \frac{S}{N}\sum_{\vec{q}'\vec{q}''} a_{\vec{q}''}^{\dagger} a_{\vec{q}'} \exp\left\{i\left(\vec{q}'-\vec{q}''\right)\vec{R}_n\right\} - \frac{S}{N}\sum_{\vec{q}'\vec{q}''} b_{\vec{q}''}^{\dagger} b_{\vec{q}'} \exp\left\{i\left(\vec{q}'-\vec{q}''\right)\vec{R}_m\right\}\right) \tag{2406}$$

Doing the change of variable $\vec{R}_n = \vec{R}_m + \vec{R}$, then

$$\begin{aligned}\mathrm{H}_{||} = -\sum_{m\vec{R}} J_{\vec{R}}\Bigg(&S^2 - \frac{S}{N}\sum_{\vec{q}'\vec{q}''} a_{\vec{q}''}^{\dagger} a_{\vec{q}'} \exp\left\{i\left(\vec{q}'-\vec{q}''\right)\vec{R}_m\right\}\exp\left\{i\left(\vec{q}'-\vec{q}''\right)\vec{R}\right\} \\ &-\frac{S}{N}\sum_{\vec{q}'\vec{q}''} b_{\vec{q}''}^{\dagger} b_{\vec{q}'} \exp\left\{i\left(\vec{q}'-\vec{q}''\right)\vec{R}_m\right\}\exp\left\{i\left(\vec{q}'-\vec{q}''\right)\vec{R}\right\}\Bigg)\end{aligned} \tag{2407}$$

as

$$\sum_{\vec{R}} J_{\vec{R}} \exp\left\{-i\vec{q}\vec{R}\right\} = J_{\vec{q}} \tag{2408}$$

From

$$\sum_{m} \exp\left\{i\left(\vec{q}''-\vec{q}'\right)\vec{R}_m\right\} = N\delta_{\vec{q}''\vec{q}'} \tag{2409}$$

and letting

$$\sum_{\vec{R}} J_{\vec{R}} S^2 = JNZS^2 \tag{2410}$$

then

$$\mathrm{H}_{||} = -JNZS^2 + SZJ\sum_{\vec{q}}\left(a_{\vec{q}}^{\dagger}a_{\vec{q}} + b_{\vec{q}}^{\dagger}b_{\vec{q}}\right) \tag{2411}$$

We assume that $H > 0$. Therefore, the magnetic moments should be aligned in the positive direction of the oz axis when the system is found in its ground state, and an anisotropy field H' can be rewritten

$$\mathrm{H} = -\frac{1}{2}\sum_{n\neq m} J_{nm}\vec{S}_n \cdot \vec{S}_m + \mathrm{H}_2 \tag{2412}$$

where

$$\mathrm{H}_2 = -g\mu_B(H'+H)\sum_n S_n^z + g\mu_B(H'-H)\sum_m S_m^z \tag{2413}$$

The anisotropy field H' is a mathematical convenience that accounts for anisotropic interaction in real crystals. It is not that essential in ferromagnets but is essential in antiferromagnets. But if

$$S_n^z = S - \frac{1}{N}\sum_{\vec{q}'\vec{q}''} b_{\vec{q}''}^{\dagger}b_{\vec{q}'}\exp\left\{i\left(\vec{q}'-\vec{q}''\right)\vec{R}_n\right\} \tag{2414}$$

then

$$\mathrm{H}_2 = \mathrm{E}_{02} + g\mu_B(H'+H)\frac{1}{N}\sum_{n\vec{q}'\vec{q}''} a_{\vec{q}''}^{\dagger}a_{\vec{q}'}\exp\left\{i\left(\vec{q}'-\vec{q}''\right)\vec{R}_n\right\} - \frac{1}{N}g\mu_B(H'-H)\sum_{m\vec{q}'\vec{q}''} b_{\vec{q}''}^{\dagger}b_{\vec{q}'}\exp\left\{i\left(\vec{q}'-\vec{q}''\right)\vec{R}_m\right\} \tag{2415}$$

where

$$\mathrm{E}_{02} = -2g\mu_B HNS \tag{2416}$$

We then do the change of variable $\vec{q} = \vec{q}' - \vec{q}''$,

$$\sum_m \exp\left\{i\left(\vec{q}''-\vec{q}'\right)\vec{R}_m\right\} = \sum_m \exp\left\{-i\vec{q}\vec{R}_m\right\} = N\delta_{\vec{q}0} \tag{2417}$$

and

$$\mathrm{H}_2 = \mathrm{E}_{02} + g\mathcal{M}_B(H'+H)\frac{1}{N}\sum_{\vec{q}\vec{q}''} a_{\vec{q}+\vec{q}''}^{\dagger}a_{\vec{q}'}N\delta_{\vec{q}0} - \frac{1}{N}g\mathcal{M}_B(H'-H)\sum_{\vec{q}\vec{q}''} b_{\vec{q}+\vec{q}''}^{\dagger}b_{\vec{q}'}N\delta_{\vec{q}0} \tag{2418}$$

or

$$\mathrm{H}_2 = \mathrm{E}_{02} + g\mu_B(H'+H)\sum_{\vec{q}} a_{\vec{q}}^{\dagger}a_{\vec{q}} - g\mu_B(H'-H)\sum_{\vec{q}} b_{\vec{q}}^{\dagger}b_{\vec{q}} \tag{2419}$$

The total Hamiltonian is then

$$\mathrm{H} = \mathrm{H}_2 - JNZS^2 + \omega\sum_{\vec{q}}\left(a_{\vec{q}}^{\dagger}a_{\vec{q}} + b_{\vec{q}}^{\dagger}b_{\vec{q}}\right) - \omega\sum_{\vec{q}}\gamma_{\vec{q}}\left(a_{\vec{q}}^{\dagger}b_{\vec{q}} + b_{\vec{q}}^{\dagger}a_{\vec{q}}\right) \quad , \quad \omega \equiv SJZ \tag{2420}$$

Unlike the ferromagnetic case, this cannot be trivially solved. We have to diagonalize (2420) by a suitable transformation.

13.2.9 Bogoliubov Transformation

Equation (2420) is the Hamiltonian quadratic in the creation and annihilation operators. It can then be diagonalized by means of a canonical transformation in order to find the energy levels. The diagonalization is done by introducing the canonical transformation known as a **Bogoliubov transformation [47, 48]**:

$$\begin{bmatrix} a_{\vec{q}} \\ b_{\vec{q}} \end{bmatrix} = \begin{bmatrix} u_{\vec{q}} & -v_{\vec{q}} \\ -v_{\vec{q}} & u_{\vec{q}} \end{bmatrix} \begin{bmatrix} \alpha_{\vec{q}} \\ \beta_{\vec{q}} \end{bmatrix} \tag{2421}$$

From here, for the diagonal elements, we have

$$a_{\vec{q}}^{\dagger} a_{\vec{q}} = \left(u_{\vec{q}} \alpha_{\vec{q}}^{\dagger} - v_{\vec{q}} \beta_{\vec{q}}^{\dagger}\right)\left(u_{\vec{q}} \alpha_{\vec{q}} - v_{\vec{q}} \beta_{\vec{q}}\right) = u_{\vec{q}}^{2} \alpha_{\vec{q}}^{\dagger} \alpha_{\vec{q}} + v_{\vec{q}}^{2} \beta_{\vec{q}}^{\dagger} \beta_{\vec{q}} - u_{\vec{q}} v_{\vec{q}}\left(\alpha_{\vec{q}}^{\dagger} \beta_{\vec{q}} + \beta_{\vec{q}}^{\dagger} \alpha_{\vec{q}}\right) \tag{2422}$$

and

$$b_{\vec{q}}^{\dagger} b_{\vec{q}} = \left(-v_{\vec{q}} \alpha_{\vec{q}}^{\dagger} + u_{\vec{q}} \beta_{\vec{q}}^{\dagger}\right)\left(-v_{\vec{q}} \alpha_{\vec{q}} + u_{\vec{q}} \beta_{\vec{q}}\right) = v_{\vec{q}}^{2} \alpha_{\vec{q}}^{\dagger} \alpha_{\vec{q}} + u_{\vec{q}}^{2} \beta_{\vec{q}}^{\dagger} \beta_{\vec{q}} - u_{\vec{q}} v_{\vec{q}}\left(\alpha_{\vec{q}}^{\dagger} \beta_{\vec{q}} + \beta_{\vec{q}}^{\dagger} \alpha_{\vec{q}}\right) \tag{2423}$$

For the off-diagonal elements, we have

$$a_{\vec{q}}^{\dagger} b_{\vec{q}} = \left(u_{\vec{q}} \alpha_{\vec{q}}^{\dagger} - v_{\vec{q}} \beta_{\vec{q}}^{\dagger}\right)\left(-v_{\vec{q}} \alpha_{\vec{q}} + u_{\vec{q}} \beta_{\vec{q}}\right) = u_{\vec{q}}^{2} \alpha_{\vec{q}}^{\dagger} \beta_{\vec{q}} + v_{\vec{q}}^{2} \beta_{\vec{q}}^{\dagger} \alpha_{\vec{q}} - u_{\vec{q}} v_{\vec{q}}\left(\alpha_{\vec{q}}^{\dagger} \alpha_{\vec{q}} + \beta_{\vec{q}}^{\dagger} \beta_{\vec{q}}\right) \tag{2424}$$

$$b_{\vec{q}}^{\dagger} a_{\vec{q}} = \left(-v_{\vec{q}} \alpha_{\vec{q}}^{\dagger} + u_{\vec{q}} \beta_{\vec{q}}^{\dagger}\right)\left(u_{\vec{q}} \alpha_{\vec{q}} - v_{\vec{q}} \beta_{\vec{q}}\right) = u_{\vec{q}}^{2} \beta_{\vec{q}}^{\dagger} \alpha_{\vec{q}} + v_{\vec{q}}^{2} \alpha_{\vec{q}}^{\dagger} \beta_{\vec{q}} - u_{\vec{q}} v_{\vec{q}}\left(\alpha_{\vec{q}}^{\dagger} \alpha_{\vec{q}} + \beta_{\vec{q}}^{\dagger} \beta_{\vec{q}}\right) \tag{2425}$$

From here, the Hamiltonian (2420) becomes

$$\mathrm{H} = \mathrm{E}_{01} + \mathrm{H}_{1} \quad , \quad \mathrm{E}_{01} = -JNZS^{2} \tag{2426}$$

where

$$\mathrm{H}_{1} = \omega \sum_{\vec{q}} \left\{ \left(\left(u_{\vec{q}}^{2} + v_{\vec{q}}^{2}\right) + 2 u_{\vec{q}} v_{\vec{q}} \gamma_{\vec{q}} \right)\left(\alpha_{\vec{q}}^{\dagger} \alpha_{\vec{q}} + \beta_{\vec{q}}^{\dagger} \beta_{\vec{q}}\right) - \left(2 u_{\vec{q}} v_{\vec{q}} + \gamma_{\vec{q}}\left(u_{\vec{q}}^{2} + v_{\vec{q}}^{2}\right)\right)\left(\alpha_{\vec{q}}^{\dagger} \beta_{\vec{q}} + \beta_{\vec{q}}^{\dagger} \alpha_{\vec{q}}\right) \right\} \tag{2427}$$

Also

$$\mathrm{H}_{2} = \mathrm{E}_{02} + \omega_{\mathrm{A}} \sum_{\vec{q}} \left(u_{\vec{q}}^{2} - v_{\vec{q}}^{2}\right)\left(\alpha_{\vec{q}}^{\dagger} \alpha_{\vec{q}} - \beta_{\vec{q}}^{\dagger} \beta_{\vec{q}}\right) + \omega_{\mathrm{B}} \sum_{\vec{q}} \left(u_{\vec{q}}^{2} + v_{\vec{q}}^{2}\right)\left(\alpha_{\vec{q}}^{\dagger} \alpha_{\vec{q}} + \beta_{\vec{q}}^{\dagger} \beta_{\vec{q}}\right) + 2\omega_{\mathrm{A}} \sum_{\vec{q}} u_{\vec{q}} v_{\vec{q}}\left(\alpha_{\vec{q}}^{\dagger} \beta_{\vec{q}} + \beta_{\vec{q}}^{\dagger} \alpha_{\vec{q}}\right) \tag{2428}$$

where

$$\omega = 2JZS \quad , \quad \omega_{\mathrm{A}} = g\mu_{B} H' \quad , \quad \omega_{\mathrm{B}} = g\mu_{B} H \tag{2429}$$

Considering that

$$u_{\vec{q}}^{2} - v_{\vec{q}}^{2} = 1 \tag{2430}$$

then

$$\mathrm{H}_{2} = \mathrm{E}_{02} + \omega_{\mathrm{A}} \sum_{\vec{q}} \left(\alpha_{\vec{q}}^{\dagger} \alpha_{\vec{q}} - \beta_{\vec{q}}^{\dagger} \beta_{\vec{q}}\right) + \omega_{\mathrm{B}} \sum_{\vec{q}} \left(u_{\vec{q}}^{2} + v_{\vec{q}}^{2}\right)\left(\alpha_{\vec{q}}^{\dagger} \alpha_{\vec{q}} + \beta_{\vec{q}}^{\dagger} \beta_{\vec{q}}\right) + 2\omega_{\mathrm{A}} \sum_{\vec{q}} u_{\vec{q}} v_{\vec{q}}\left(\alpha_{\vec{q}}^{\dagger} \beta_{\vec{q}} + \beta_{\vec{q}}^{\dagger} \alpha_{\vec{q}}\right) \tag{2431}$$

For the Hamiltonian to be diagonalized, we request that in (2427),

$$2u_{\vec{q}}v_{\vec{q}} + \gamma_{\vec{q}}\left(u_{\vec{q}}^2 + v_{\vec{q}}^2\right) = 0 \tag{2432}$$

It is instructive to note that

$$u_{\vec{q}} = u_{-\vec{q}} \quad , \quad v_{\vec{q}} = v_{-\vec{q}} \tag{2433}$$

It follows from (2432) that

$$u_{\vec{q}}v_{\vec{q}} < 0 \tag{2434}$$

From the first term of (2432), we have

$$\left[\gamma_{\vec{q}}\left(u_{\vec{q}}^2 + v_{\vec{q}}^2\right)\right]^2 = \left[2u_{\vec{q}}v_{\vec{q}}\right]^2 \tag{2435}$$

then considering the second term of (2432), we have

$$v_{\vec{q}}^4 + v_{\vec{q}}^2 - \frac{\gamma_{\vec{q}}^2}{4\left(1-\gamma_{\vec{q}}^2\right)} = 0 \tag{2436}$$

From here, we have

$$v_{\vec{q},1,2}^2 = -\frac{1}{2}\left(1 \pm \frac{2}{\lambda_{\vec{q}}}\right) \quad , \quad \lambda_{\vec{q}}^2 = 4\left(1-\gamma_{\vec{q}}^2\right) \tag{2437}$$

Considering that $v_{\vec{q}}^2 > 0$ and using the second term of (2432), then

$$v_{\vec{q}}^2 = \frac{1}{2}\left(\frac{2}{\lambda_{\vec{q}}} - 1\right) \quad , \quad u_{\vec{q}}^2 = \frac{1}{2}\left(\frac{2}{\lambda_{\vec{q}}} + 1\right) \tag{2438}$$

From here,

$$u_{\vec{q}}^2 v_{\vec{q}}^2 = \frac{1}{4}\left(\frac{4}{\lambda_{\vec{q}}^2} - 1\right) = \frac{\gamma_{\vec{q}}^2}{\lambda_{\vec{q}}^2} \tag{2439}$$

From the inequality (2434), we have

$$u_{\vec{q}}v_{\vec{q}} = -\frac{\gamma_{\vec{q}}}{\lambda_{\vec{q}}} < 0 \tag{2440}$$

and

$$H_2 = E_{02} + \omega_A \sum_{\vec{q}}\left(\alpha_{\vec{q}}^\dagger \alpha_{\vec{q}} - \beta_{\vec{q}}^\dagger \beta_{\vec{q}}\right) + \omega_B \sum_{\vec{q}} \frac{2}{\lambda_{\vec{q}}}\left(\alpha_{\vec{q}}^\dagger \alpha_{\vec{q}} + \beta_{\vec{q}}^\dagger \beta_{\vec{q}}\right) - 2\omega_A \sum_{\vec{q}} \frac{\gamma_{\vec{q}}}{\lambda_{\vec{q}}}\left(\alpha_{\vec{q}}^\dagger \beta_{\vec{q}} + \beta_{\vec{q}}^\dagger \alpha_{\vec{q}}\right) \tag{2441}$$

We can now rewrite the total Hamiltonian of our system as

$$H = E_0 + \sum_{\vec{q}} \left[E_{\vec{q}} \alpha_{\vec{q}}^{\dagger} \alpha_{\vec{q}} + \left(E_{\vec{q}} - 2\omega_A \right) \beta_{\vec{q}}^{\dagger} \beta_{\vec{q}} \right] - 2\omega_A \sum_{\vec{q}} \frac{\gamma_{\vec{q}}}{\lambda_{\vec{q}}} \left(\alpha_{\vec{q}}^{\dagger} \beta_{\vec{q}} + \beta_{\vec{q}}^{\dagger} \alpha_{\vec{q}} \right) \tag{2442}$$

where the ground state energy:

$$E_0 = -JNZS^2 - 2g\mu_B HNS \tag{2443}$$

and

$$E_{\vec{q}} = \omega\sqrt{1-\gamma_{\vec{q}}^2} + \omega_A + \frac{2\omega_B}{\lambda_{\vec{q}}} \tag{2444}$$

is the dispersion relation.

Consider the case

$$H = H' = 0 \tag{2445}$$

Thus $\omega_A, \omega_B \to 0$ and

$$\omega_{\vec{q}} \to \omega\sqrt{1-\gamma_{\vec{q}}^2} \tag{2446}$$

Hence, the ground state energy is given by

$$E_{GS} = E_0 + \sum_{\vec{q}} E_{\vec{q}} \left(\alpha_{\vec{q}}^{\dagger} \alpha_{\vec{q}} + \beta_{\vec{q}}^{\dagger} \beta_{\vec{q}} \right) \quad , \quad E_{\vec{q}} = \omega\sqrt{1-\gamma_{\vec{q}}^2} \tag{2447}$$

For a simple cubic crystal where $q \to 0$, then

$$\gamma_q = 1 - \frac{q^2 a^2}{6} \tag{2448}$$

and

$$\gamma_{\vec{q}}^2 = 1 - \frac{q^2 a^2}{3} \tag{2449}$$

So,

$$E_{\vec{q}} = \frac{\omega}{\sqrt{3}} qa = cq \tag{2450}$$

This is a sound-like magnon, and the energy is linear in the wave vector q. It is a basic characteristic relation of an antiferromagnetic state. Knowledge of these elementary excitations permit us to perform calculations of the temperature dependence of the **sublattice magnetization** as well as of the specific heat.

We define the deviation of the sublattice magnetization from its ground state value:

$$\Delta\mathcal{M} = \mathcal{M}_s(0) - \mathcal{M}_s(T) = g\mu_B \sum_{\vec{q}} \langle \hat{n}_{\vec{q}} \rangle \tag{2451}$$

where the mean number of bosons is given by

$$\langle n_{\vec{q}} \rangle = \frac{1}{\exp\left\{\frac{cq}{T}\right\} - 1} \quad (2452)$$

and so

$$\Delta\mathcal{M} = \mathcal{M}_s(0) - \mathcal{M}_s(T) = g\mu_B \sum_{\vec{q}} \frac{1}{\exp\left\{\frac{cq}{T}\right\} - 1} \quad (2453)$$

$$E_{\vec{q}} = cq \equiv T\xi \quad , \quad q \equiv \frac{T\xi}{c} \quad , \quad dq \equiv \frac{T}{c} d\xi \quad (2454)$$

then

$$d\vec{q} = 4\pi q^2 dq = 4\pi \left(\frac{T\xi}{c}\right)^2 \frac{T}{c} d\xi = \frac{2\pi}{c^2} T^3 \xi^2 d\xi \quad (2455)$$

and

$$\Delta\mathcal{M} = T^3 g\mu_B \frac{1}{(2\pi)^3} \frac{2\pi}{c^2} \int d\xi \frac{\xi^2}{\exp\{\xi\} - 1} = AT^3 \quad (2456)$$

Therefore, the **Bloch law**

$$\mathcal{M}_s(T) = \mathcal{M}_s(0) - AT^3 \quad (2457)$$

The internal energy

$$E = \frac{1}{(2\pi)^3} \int d\vec{q} \frac{cq}{\exp\left\{\frac{cq}{T}\right\} - 1} = T^4 \frac{1}{(2\pi)^3} \frac{2\pi}{c^2} \int d\xi \frac{\xi^3}{\exp\{\xi\} - 1} = BT^4 \quad (2458)$$

The specific heat due to magnons

$$C_V = \frac{\partial E}{\partial T} = 4BT^3 \quad (2459)$$

13.2.10 Stability

In the ferromagnetic ground state, spins are aligned. However, the direction of the resulting magnetization is arbitrary because the Hamiltonian H has a complete rotational symmetry. This results in a degenerate ground state. The system is unstable if we select a certain direction for $\mathcal{M}$ as the starting point of magnon theory. An infinitesimal amount of thermal energy excites a huge number of spin waves. We recall that when $H = 0$, the $q = 0$ spin waves have zero energy. The difficulty of having an unstable ground state with $\mathcal{M}$ in a particular direction is removed by removing the degeneracy caused by the spherical symmetry of the Hamiltonian H. This is realized either by

1. Applying a field H in a particular direction or
2. Introducing an effective anisotropy field H'.

Comparing the antiferromagnetic and ferromagnetic cases, we can observe that both cases can be understood as an instance of spontaneous symmetry breaking if the magnitude $|H|$ of the source field is taken to zero at the end of the calculation. So, the operator multiplying the source field should be the order parameter. We can show that the Hamiltonian $\sum_{n\neq m} J_{nm}\vec{S}_n \cdot \vec{S}_m$ commutes with the symmetry-breaking field for the case of ferromagnetism and does not commute with the symmetry-breaking field in the case of antiferromagnetism.

13.2.11 Spin Dynamics, Dynamical Response Function

Two ways of describing magnetic systems are localized moments and itinerant moments, with the choice between these two ways dependent on the nature of the material. In many cases, the choice may be a difficult one to make. In certain cases, the relevant current distributions may be localized within a lattice cell where the ionic magnetic moment is relatively unambiguous. For this, the interaction with external charge and current distributions is then expressed via the given moment. This approach yields the **spin Hamiltonian** and has proven very suitable. In some cases, we assume that the current distributions are linked to free electrons. Because these electrons may extend throughout the lattice, they may be approximated as an **electron gas**, which is a suitable simplification. These two types of current distributions correspond to very localized electrons as well as to itinerant electrons. This permits us to consider the average moment that relies on statistical mechanics techniques for which we will provide a somewhat basic introduction.

13.2.11.1 Spin Dynamics

To proceed with the quantum-mechanical solution of our spin system, we examine some classical spin systems. For this, we consider the spin particle subjected to a static magnetic field $\vec{H}_0$:

$$\mathrm{H} = -\vec{H}_0\vec{\mathcal{M}} \quad , \quad \vec{\mathcal{M}} = g\mu_B \sum_n \vec{S}_n \tag{2460}$$

The Heisenberg equations of motion for the spatial components of the spin operators S_m^α on the m site then are:

$$\dot{S}_m^\alpha = \frac{i}{\hbar}\left[\mathrm{H}, S_m^\alpha\right] = -\frac{i}{\hbar} g\mu_B H_0^\beta \sum_n \left[S_n^\beta, S_m^\alpha\right] \tag{2461}$$

Here, H_0^β are the components of the magnetic field. The components of the spin operator at each site obey

$$\left[S_n^\alpha, S_m^\beta\right] = i \,\epsilon_{\alpha\beta\gamma}\, S_m^\gamma \delta_{nm} \tag{2462}$$

where $\epsilon_{\alpha\beta\gamma}$ represents a Levi-Civita tensor. So, from (2461) and (2462), then

$$\dot{S}_m^\alpha = -\frac{g\mu_B}{\hbar} \epsilon_{\alpha\beta\gamma}\, H_0^\beta S_m^\gamma \tag{2463}$$

Considering the second equation of (2460), we have

$$\dot{\mathcal{M}}^{\alpha} = g\mu_B \sum_m \dot{S}_m^{\alpha} = -\frac{g\mu_B}{\hbar} \epsilon_{\alpha\beta\gamma} H_0^{\beta} g\mu_B \sum_m S_m^{\gamma} = -\frac{g\mu_B}{\hbar} \epsilon_{\alpha\beta\gamma} H_0^{\beta} \mathcal{M}^{\gamma} \quad (2464)$$

and consequently,

$$\dot{\vec{\mathcal{M}}} = -\frac{g\mu_B}{\hbar}\left[\vec{\mathcal{M}}, \vec{H}_0\right] \equiv -\gamma\left[\vec{\mathcal{M}}, \vec{H}_0\right] \quad , \quad \gamma = \frac{g|\mu_B|}{\hbar} \quad (2465)$$

This equation describes the dynamics of the magnetization $\vec{\mathcal{M}} = \vec{\mathcal{M}}(t)$ due to a particle placed in a magnetic field $\vec{H}_0$ that exerts a torque $\left[\vec{\mathcal{M}}, \vec{H}_0\right]$ on the system. If we perform a scalar multiplication of both sides of equation (2465) by either $\vec{\mathcal{M}}$ or $\vec{H}_0$, this yields:

$$\frac{d\vec{\mathcal{M}}^2}{dt} = 0 \quad , \quad \frac{d}{dt}\left(\vec{\mathcal{M}}, \vec{H}_0\right) = 0 \quad (2466)$$

From here, $\vec{\mathcal{M}}$ evolves with a constant magnitude, maintaining a constant angle with $\vec{H}_0$.

Let us project equation (2465) onto the plane perpendicular to $\vec{H}_0$. We observe that $\vec{\mathcal{M}}$ rotates about $\vec{H}_0$ (Larmor precession) with an angular velocity of $\omega_0 = -\gamma H_0$ (the rotation is counterclockwise). Consider now that we add $\vec{H}_0$ a field $\vec{H}_1(t)$, perpendicular to $\vec{H}_0$ to the static field. This field should be of constant magnitude and should rotate about $\vec{H}_0$ with angular velocity ω.

For convenience, we set

$$\omega_0 = -\gamma H_0 \quad , \quad \omega_1 = -\gamma H_1 \quad (2467)$$

and consider an absolute reference frame $Oxyz$ with the field $\vec{H}_0$ in the direction of the z-axis. We also consider a rotating reference frame $OXYZ$, with the axes obtained from $Oxyz$ by rotation through the angle ωt about oz where OX is the direction of the rotating field $\vec{H}_1(t)$. In this case, the equation of motion for $\vec{\mathcal{M}} = \vec{\mathcal{M}}(t)$ in the presence of the total field $H(t) = H_0 + H_1(t)$ becomes:

$$\frac{d\vec{\mathcal{M}}(t)}{dt} = -\gamma\vec{\mathcal{M}}(t) \times \vec{H}(t) \quad (2468)$$

We show that the magnetization is the average magnetic moment. To calculate this average, it is necessary to know the probabilities of the system being in its various configurations. This information is inherent in the distribution function associated with the system. For the case of a time-dependent field, the distribution function must be obtained from its equation of motion, and for the case of localized moments, this consists of solving the **Bloch equations**.

Consider our system possesses translational invariance. If so, the statistical average over numerous unit cells of the crystal is equivalent to the time average over one cell. Hereafter $\vec{\mathcal{M}}$ will be understood to be an operator $\hat{\mathcal{M}}$. In order to find the magnetization, we must take the expectation value of the magnetic moment operator:

$$\left\langle \hat{\mathcal{M}}^{(n)}(t) \right\rangle = \int \phi^n(t)^* \hat{\mathcal{M}} \phi^n(t) d\vec{R} \quad , \quad d\vec{R} = d\vec{r}_1 d\vec{r}_2 \cdots d\vec{r}_n \quad (2469)$$

The calculation is straightforward if the wave function $\phi^n(t)$ is known. Because we describe a system at a temperature T, this implies that the system is in equilibrium with some temperature bath. If we describe the system in terms of its eigenfunctions, φ_k, the effect of the bath temperature causes the system to move through different accessible states k in the same manner as a classical system moves through phase space. Therefore, the wave function $\phi^n(t)$ can be written as a superposition of states:

$$\phi^n(t) = \sum_k C_k^n(t)\varphi_k \tag{2470}$$

So, the ensemble average over numerous unit cells of the crystal:

$$\langle \hat{\mathcal{M}} \rangle = \frac{1}{N}\sum_n \hat{\mathcal{M}}^{(n)} \tag{2471}$$

or

$$\langle \hat{\mathcal{M}} \rangle = \frac{1}{N}\sum_n \hat{\mathcal{M}}^{(n)} = \frac{1}{N}\sum_n \int d\vec{R}\left(\phi_t^n\right)^* \hat{\mathcal{M}}\phi_t^n = \sum_{nk,k'} \int d\vec{R}\left(C_t^n\right)^* C_{k'}^n(t)\varphi_n^* \hat{\mathcal{M}}\varphi_{k'} = \sum_{nk,k'} C_k^{n*}(t)C_{k'}^n(t)\int d\vec{R}\varphi_n^*\hat{\mathcal{M}}\varphi_{k'} \tag{2472}$$

where the matrix elements

$$\mathcal{M}_{kk'} = \int d\vec{R}\varphi_n^*\hat{\mathcal{M}}\varphi_{k'} \tag{2473}$$

and so

$$\langle \hat{\mathcal{M}} \rangle = \sum_{nk,k'} C_k^{n*}(t)C_{k'}^n(t)\mathcal{M}_{kk'} \tag{2474}$$

Measuring the magnetization, we actually do a chronological average. Hence, the mean value of the magnetization

$$\langle \overline{\hat{\mathcal{M}}} \rangle = \sum_{k,k'} \overline{\sum_n C_k^{n*}(t)C_{k'}^n(t)}\mathcal{M}_{kk'} = \sum_{k,k'} \rho_{k'k}\mathcal{M}_{kk'} = \mathrm{Tr}\left\{\rho\hat{\mathcal{M}}\right\} \tag{2475}$$

This gives the average of the magnetic moment over the entire system. The quantity

$$\rho_{kk'} = \overline{\sum_n C_k^{n*}(t)C_{k'}^n(t)} \tag{2476}$$

is defined as the **statistical density matrix**.

Suppose the system is isolated from the temperature bath. Then, the coefficients C_k^n are independent of time:

$$\rho_{kk'} = \sum_n \left|C_k^n\right|^2 \delta_{kk'} \tag{2477}$$

These states involve the microcanonical ensemble. Considering time-dependent fields, we then solve for $\rho_{kk'}$ from its equation of motion:

$$\frac{d\rho}{dt} = \frac{\partial\rho}{\partial t} + \frac{i}{\hbar}\left[\mathrm{H},\rho\right] = 0 \tag{2478}$$

Therefore,

$$\frac{\partial \rho}{\partial t} + \frac{i}{\hbar}[\mathrm{H}, \rho] = 0 \tag{2479}$$

This is a more convenient approach to the density matrix. By applying perturbation theory, we solve this iteratively.

Considering the ensemble average of $\hat{\mathcal{M}}$, then

$$\frac{d\hat{\mathcal{M}}}{dt} = \mathrm{Tr}\left\{\frac{\partial \rho}{\partial t}\hat{\mathcal{M}}\right\} = -\mathrm{Tr}\left\{[\mathrm{H}, \rho]\hat{\mathcal{M}}\right\} \tag{2480}$$

As

$$[\mathrm{H}, \rho]\hat{\mathcal{M}} = \mathrm{H}\rho\hat{\mathcal{M}} - \rho\mathrm{H}\hat{\mathcal{M}} \tag{2481}$$

and

$$\mathrm{Tr}\frac{\partial \rho}{\partial t}\hat{\mathcal{M}} = \sum_{kk'}\frac{\partial \rho_{k'k}}{\partial t}\hat{\mathcal{M}}_{kk'} = \frac{1}{\hbar}\sum_{kk'}\left[\rho_{k'k}\mathrm{H}_{k'k}\hat{\mathcal{M}}_{kk'} - \mathrm{H}_{k'k}\rho_{k'k}\hat{\mathcal{M}}_{kk'}\right] = -\frac{1}{\hbar}\mathrm{Tr}\left\{\rho\left[\hat{\mathcal{M}}, \mathrm{H}\right]\right\} \tag{2482}$$

so,

$$\frac{d\hat{\mathcal{M}}}{dt} = -\frac{i}{\hbar}\mathrm{Tr}\left\{\rho\left[\hat{\mathcal{M}}, \mathrm{H}\right]\right\} \tag{2483}$$

Let the field

$$H_1(t) = H_1 \cos \omega t \tag{2484}$$

be parallel to the x-axis while rotating about H_0 with angular velocity ω. If the particle is placed in the field H_0, then in thermal equilibrium at a temperature T, the magnetization will be along the oz axis:

$$m_x = 0 \quad , \quad m_y = 0 \quad , \quad m_z = \mathcal{M}_0 \tag{2485}$$

When the magnetization $\vec{\mathcal{M}}$ is not in thermal equilibrium, we suppose that it approaches equilibrium at a rate proportional to the departure from the equilibrium value $\mathcal{M}_0$:

$$\vec{\mathcal{M}} = m_x\vec{e}_x + m_y\vec{e}_y + (\mathcal{M}_0 - m_z)\vec{e}_z \tag{2486}$$

The last term in parentheses is the departure from the equilibrium value of the magnetization $\mathcal{M}_0$. We consider the case where $H_1(t)$ and H_0 are parallel to the ox -and oz-axes, respectively. From (2486), we have:

$$\vec{\mathcal{M}} \times \vec{H} = m_y H_0 \vec{e}_x + \left((\mathcal{M}_0 - m_z)H_1 - m_x H_0\right)\vec{e}_y - m_y H_1 \vec{e}_z \tag{2487}$$

and letting $\gamma H_0 = \omega_0$, then

$$\frac{dm_x}{dt} = -\omega_0 m_y \quad , \quad \frac{dm_y}{dt} = -\gamma H_1(t)(\mathcal{M}_0 - m_z) + \omega_0 m_x \quad , \quad \frac{dm_z}{dt} = \gamma H_1(t) m_y \tag{2488}$$

For

$$H_1 << H_0 \quad , \quad m_x, m_y, m_z << \mathcal{M}_0 \tag{2489}$$

we then arrive at the equation of motion that allows us to find the **response function** for our system:

$$\frac{dm_z}{dt} = 0 \quad , \quad \frac{dm_x}{dt} = -\omega_0 m_y \quad , \quad \frac{dm_y}{dt} = \omega_0 m_x - \gamma H_1(t)\mathcal{M}_0 \tag{2490}$$

Differentiating the second equation and substituting it in the expression of the last equation then

$$\frac{d^2 m_x}{dt^2} = -\omega_0 \frac{dm_y}{dt} = -\omega_0^2 m_x + \gamma\omega_0 H_1(t)\mathcal{M}_0 \tag{2491}$$

Considering (2484), then

$$\frac{d^2 m_x}{dt^2} + \omega_0^2 m_x = \gamma\omega_0 H_1 \mathcal{M}_0 \cos\omega t \tag{2492}$$

Solution by Green's Function Method

Equation (2492) is conveniently solved by the Green's function method. First, we find the homogenous solution

$$m_{0x} = \alpha\cos\omega_0 t + \beta\sin\omega_0 t \tag{2493}$$

Considering the Green's function method, the nonhomogenous equation is conveniently solved by first defining the linear differential operator $\hat{\mathrm{A}}$:

$$\hat{\mathrm{A}} m_x(t) = \mathrm{F}(t) \equiv \gamma\omega_0 H_1 \mathcal{M}_0 \cos\omega t \tag{2494}$$

Here, the quantity $\mathrm{F}(t)$ imitates external source fields and is analogous to driving forces that can be tailored to probe the system.

We investigate how our system reacts to the presence of the source or the driving force. Our goal is to investigate how the correlation functions of the system change when we turn on a source $\mathrm{F}(t)$. Generally, investigating how the system is tailored by the sources is difficult. However, this problem can be simplified by the assumption that the source is a small perturbation of the original system. From equation (2494), this implies the magnetization is linear in the perturbing source. Hence, we write the solution of (2494) as

$$m_x(t) = \hat{\mathrm{A}}^{-1}\mathrm{F}(t) = \int_{-\infty}^{\infty} \hat{\mathrm{A}}^{-1}\mathrm{F}(t')\delta(t-t')dt' = \int_{-\infty}^{\infty} \mathbf{G}(t-t')\mathrm{F}(t')dt' \tag{2495}$$

Here,

$$\mathbf{G}(t-t') = \hat{\mathrm{A}}^{-1}\delta(t-t') \tag{2496}$$

is the **Green's function** (also known as the **response function**), which effectively solves the dynamics of the system. The equation for the Green's function:

$$\hat{A}\mathbf{G}(t-t') = \delta(t-t') \tag{2497}$$

We observe from equation (2496) that our system is invariant under time translations:

$$\mathbf{G}(t,t') = \mathbf{G}(t-t') \tag{2498}$$

From here, and setting $t-t' \equiv \tau$, it is useful to perform a Fourier transform to work in frequency space that will permit us to find the Green's function $\mathbf{G}(\tau)$:

$$\mathbf{G}(\omega) = \int_{-\infty}^{\infty} d\tau \exp\{i\omega\tau\}\mathbf{G}(\tau) \quad , \quad \mathbf{G}(\tau) = \int_{-\infty}^{\infty} \frac{d\omega}{2\pi} \exp\{-i\omega\tau\}\mathbf{G}(\omega) \tag{2499}$$

and

$$\hat{A}\mathbf{G}(\tau) = \left(\frac{d^2}{d\tau^2} + \omega_0^2\right) \int_{-\infty}^{\infty} \frac{d\omega}{2\pi} \exp\{-i\omega\tau\}\mathbf{G}(\omega) = \int_{-\infty}^{\infty} \frac{d\omega}{2\pi}\left(\omega_0^2 - \omega^2\right)\mathbf{G}(\omega)\exp\{-i\omega\tau\} \equiv \delta(\tau) = \int_{-\infty}^{\infty} \frac{d\omega}{2\pi} \exp\{-i\omega\tau\} \tag{2500}$$

So

$$\left(\omega_0^2 - \omega^2\right)\mathbf{G}(\omega) = 1 \tag{2501}$$

or

$$\mathbf{G}(\omega) = \frac{1}{\omega_0^2 - \omega^2} \tag{2502}$$

We observe the response to be **local** in frequency space. Using the Fourier transform of (2502), we have

$$\mathbf{G}(\tau) = \int_{-\infty}^{\infty} \frac{d\omega}{2\pi} \frac{\exp\{-i\omega\tau\}}{\omega_0^2 - \omega^2} \tag{2503}$$

This is an integral taken within the interval $\omega \in [-\infty, \infty]$ of a real line. Nevertheless, if ω is within this interval, then the integral diverges at $\omega = \omega_0$. To avoid this, we deform the contour of integration into the complex plane, either running just above the singularity along $\omega + i\delta$ or just below the singularity along $\omega - i\delta$. For that, we consider

$$\frac{1}{\omega^2 - \omega_0^2} = \frac{1}{(\omega - \omega_0)(\omega + \omega_0)} = \left(\frac{1}{\omega - \omega_0} - \frac{1}{\omega + \omega_0}\right)\frac{1}{2\omega_0} \tag{2504}$$

and let

$$\omega = \omega' + i\omega'' \tag{2505}$$

then

$$-i\omega\tau = -i\omega'\tau + \omega''\tau \tag{2506}$$

This provides the $\exp\{\omega''\tau\}$ in the Green's function. So, if $\tau < 0$, then the factor $\exp\{\omega''\tau\}$ decays and $\mathbf{G}(\tau < 0) = 0$, which implies that there are no poles in the upper-half-plane. For $\tau > 0$, the factor $\exp\{\omega''\tau\}$ does not decay and $\mathbf{G}(\tau > 0) \neq 0$, which implies that all poles are in the lower-half-plane:

$$\mathbf{G}(\tau > 0) = -\int_{-\infty}^{\infty} \frac{d\omega}{2\pi} \frac{\exp\{-i\omega\tau\}}{(\omega - \omega_0 + i\delta)(\omega + \omega_0 + i\delta)} \tag{2507}$$

where δ is an infinitesimally small positive number. From the Cauchy integral and residue theory, we have

$$\mathbf{G}(\tau) = 2\pi i \frac{1}{2\pi} \frac{1}{2\omega_0} \left(\exp\{-i\omega_0\tau - \tau\delta\} - \exp\{i\omega_0\tau - \tau\delta\} \right) \tag{2508}$$

or

$$\mathbf{G}(\tau) = \frac{\sin\omega_0\tau}{\omega_0} \theta(\tau) \tag{2509}$$

Because we have found the **unknown (Green's) function** (2495), then the nonhomogeneous solution now becomes

$$m_{1x}(t) = \int_{-\infty}^{t} \mathbf{G}(t - t')\mathrm{F}(t')dt' = \int_{-\infty}^{t} \frac{\sin\omega_0(t - t')}{\omega_0} \theta(t - t')\gamma\omega_0 H_1 \mathcal{M}_0 \cos\omega t' dt' \tag{2510}$$

Since $t - t' > 0$, then

$$m_{1x}(t) = \gamma H_1 \mathcal{M}_0 \int_{-\infty}^{t} \sin\omega_0(t - t')\cos\omega t' dt' \tag{2511}$$

or

$$m_{1x}(t) = \gamma H_1 \mathcal{M}_0 \omega_0 \frac{\cos\omega t}{\omega_0^2 - \omega^2} \tag{2512}$$

Susceptibility: Dynamical Response Function

Equation (2512) permits us to have the susceptibility

$$\chi_{xx}(\omega) = \gamma \mathcal{M}_0 \omega_0 \mathbf{G}(\omega) \tag{2513}$$

where

$$\mathbf{G}(\omega) = \frac{1}{\omega_0^2 - (\omega + i\delta)^2} = \left(\frac{1}{\omega - \omega_0 + i\delta} - \frac{1}{\omega + \omega_0 + i\delta} \right) \left(-\frac{1}{2\omega_0} \right) \tag{2514}$$

By definition,

$$\lim_{\delta \to +0} \int_0^{\infty} \exp\{is\tau - \delta\tau\} d\tau = i \frac{1}{s + i\delta} \tag{2515}$$

where

$$\frac{1}{s+i\delta} = \frac{\mathrm{P}}{s} - i\pi\delta(s) \tag{2516}$$

Here, the symbol P denotes the principal value and implies that when it is substituted into an integral, the integral routine must be performed in a symmetric fashion about the singularity at $s = 0$, so that it can be well defined. Hence, the susceptibility

$$\chi_{xx}(\omega) = \gamma \mathcal{M}_0\omega_0 \frac{\mathrm{P}}{\omega_0^2 - \omega^2} + \frac{i\pi\gamma\mathcal{M}_0}{2}\left(\delta(\omega-\omega_0) - \delta(\omega+\omega_0)\right) \equiv \chi'_{xx}(\omega) + i\chi''_{xx}(\omega) \tag{2517}$$

This function is of particular interest because its singularities determine the magnetic-excitation spectrum of the system. The function is depicted in Figure 13.10:

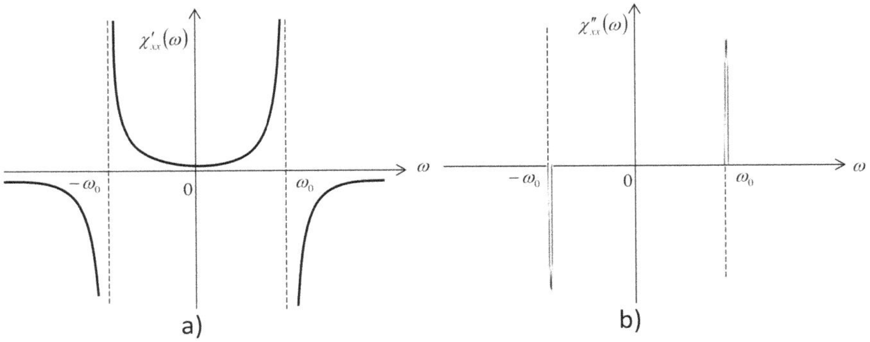

FIGURE 13.10 Representation of the (a) Real and (b) Imaginary parts of the susceptibility.

The solution of the differential equation (2492) can now be written as:

$$m_x(t) = \alpha\cos\omega_0 t + \beta\sin\omega_0 t + H_1\left(\chi'_{xx}\cos\omega t + \chi''_{xx}\sin\omega t\right) \tag{2518}$$

Remark

We observe from the previous example that the real and imaginary parts of the response function $\chi(\omega)$ have different interpretations.

Imaginary Part

$$\chi''(\omega) = -\frac{i}{2}\left(\chi(\omega) - \chi^*(\omega)\right) = -\frac{i}{2}\int_{-\infty}^{\infty} d\tau\chi(\tau)\left(\exp\{i\omega\tau\} - \exp\{-i\omega\tau\}\right) = -\frac{i}{2}\int_{-\infty}^{\infty} d\tau \exp\{i\omega\tau\}\left(\chi(\tau) - \chi(-\tau)\right) \tag{2519}$$

In this case, the imaginary part of $\chi(\omega)$ is due to the part of the response function that is **not** invariant under time reversal $\tau \to -\tau$. This implies $\chi''(\omega)$ knows about the arrow of time. Because microscopic systems typically are invariant under time reversal, the imaginary part $\chi''(\omega)$ must stem from dissipative processes. So $\chi''(\omega)$ should be called the **dissipative** or **absorptive part of the response function.**

It is also known as the **spectral function** and has information about the density of states in the system that take part in absorptive processes. It is also an odd function:

$$\chi''(-\omega) = -\chi''(\omega) \tag{2520}$$

Real Part

Similarly, we have

$$\chi'(\omega) = \frac{1}{2}\int_{-\infty}^{\infty} d\tau \exp\{i\omega\tau\}\big(\chi(\tau) + \chi(-\tau)\big) \tag{2521}$$

The real part does not consider the arrow of time and is called the **reactive part of the response function.** It is an even function,

$$\chi'(-\omega) = +\chi'(\omega) \tag{2522}$$

Causality

Causality implies that any response function must satisfy

$$\chi(\tau) = 0 \quad , \quad \tau < 0 \tag{2523}$$

For this reason, we often refer to χ as the causal **Green's or retarded Green's function** that sometimes is denoted as $\mathbf{G}^R(\tau)$. Consider the Fourier expansion of χ:

$$\chi(\tau) = \int_{-\infty}^{\infty} \frac{d\omega}{2\pi} \exp\{-i\omega\tau\}\chi(\omega) \tag{2524}$$

If $\tau < 0$, then we perform the integral by completing the contour in the upper-half-plane. This permits the exponent to become $-i\omega \times (-i|\tau|) \to -\infty$ and then $\chi(\tau) = 0$. It should be noted that the integral is given by the sum of the residues inside the contour. So, if the response function has to vanish for all $\tau < 0$, then $\chi(\omega)$ should have no poles in the upper-half-plane. This implies the causality requires that $\chi(\omega)$ is analytic for $\operatorname{Im}\omega > 0$.

Because χ is analytic in the upper-half-plane, this implies a correlation between the real and imaginary parts, χ' and χ''. This is called the **Kramers-Kronig relation**, and this concept will be examined in further detail later in this chapter.

13.2.12 Response Function and Relaxation Time

If we consider the **relaxation time**, then the equation of motion can written as

$$\frac{d\mathcal{m}_z}{dt} = \gamma\left(\vec{\mathcal{M}} \times \vec{H}\right) - \frac{\mathcal{m}_z}{T_1} \tag{2525}$$

and

$$\frac{d\mathcal{m}_{x,y}}{dt} = \gamma\left(\vec{\mathcal{M}} \times \vec{H}\right)_{x,y} - \frac{\mathcal{m}_{x,y}}{T_2} \tag{2526}$$

In (2525), the notation T_1 is called the longitudinal relaxation time or the spin-lattice relaxation time. If at time $t = 0$ we place an unmagnetized specimen in a static magnetic field H_0, the magnetization will increase from the initial value $m_z = 0$ to a final value $m_z = M_0$. What happens before and just after the specimen is placed in the field? The population n_1 will be equal to n_2, as is appropriate to thermal equilibrium in zero magnetic fields. In (2526), the notation T_2 (the measure of the times during which the individual moments contribute to $m_{x,y}$ remain in phase with one another) is **called the transverse relaxation time**, and $\frac{1}{T_1}$ with $\frac{1}{T_2}$ are relaxation rates. It is instructive to note that the magnetization energy $-\vec{M} \cdot \vec{H}$ does not change as $m_{x,y}$ changes provided that H_0 is along the oz-axis. Depending on local conditions, the two times may be nearly equal and sometimes $T_1 \gg T_2$. Different local magnetic fields at different spins will cause $m_{x,y}$ to precess at different frequencies.

Equations (2525) and (2526) are called the **Bloch equations**. The last term in (2525) arises from the spin-lattice interactions. Besides precessing about the magnetic field, $\vec{M}$ will relax to equilibrium value M_0. Suppose initially the spins have a common phase, then the phases will become random with time and the values of $m_{x,y}$ will achieve the value zero. Then T_2 can be understood as a dephasing time.

Let us find the susceptibility for the spin system in the total magnetic field. We begin by linearizing the following equations:

$$\frac{dm_z}{dt} = -\frac{m_z}{T_2} \quad , \qquad \frac{dm_x}{dt} = -\omega_0 m_y - \frac{m_x}{T_2} \quad , \qquad \frac{dm_y}{dt} = \omega_0 m_x - \gamma H_1(t) M_0 - \frac{m_y}{T_2} \tag{2527}$$

The solution of the first equation of (2527) is:

$$m_z = M_0 \left(1 - \exp\left\{ -\frac{t}{T_1} \right\} \right) \tag{2528}$$

We differentiate the second equation (2527) with respect to time and then

$$\frac{d^2 m_x}{dt^2} = -\omega_0 \frac{dm_y}{dt} - \frac{1}{T_2} \frac{dm_x}{dt} \tag{2529}$$

Next, we substitute the third equation of (2527) in (2529):

$$\frac{d^2 m_x}{dt^2} = -\omega_0 \left(\omega_0 m_x - \gamma H_1(t) M_0 - \frac{m_y}{T_2} \right) - \frac{1}{T_2} \frac{dm_x}{dt} \tag{2530}$$

From (2530) and considering the second equation of (2527), then

$$\frac{d^2 m_x}{dt^2} + \frac{2}{T_2} \frac{dm_x}{dt} + \left(\omega_0^2 + \frac{1}{T_2^2} \right) m_x = \gamma H_1(t) \omega_0 M_0 \equiv F(t) \tag{2531}$$

Solution by Green's Function Method

We solve equation (2531) via the Green's function method:

$$m_x(t) = \int_{-\infty}^{\infty} G(t - t') F(t') dt' \tag{2532}$$

Therefore, we have to find the **unknown (Green's or response) function** $\mathbf{G}(t-t')$ described in (2496) and (2497). We consider our system is invariant under time translations as also is indicated in (2498). Setting $t-t' \equiv \tau$ and considering the Fourier transforms in (2499), from (2497) we have

$$\hat{A}\mathbf{G}(\tau)=\left(\frac{d^2}{d\tau^2}+\frac{2}{T_2}\frac{d}{d\tau}+\omega_0^2+\frac{1}{T_2^2}\right)\int_{-\infty}^{\infty}\frac{d\omega}{2\pi}\exp\{-i\omega\tau\}\mathbf{G}(\omega)=\int_{-\infty}^{\infty}\frac{d\omega}{2\pi}\left(\omega_0^2+\frac{1}{T_2^2}-\omega^2-\frac{2i\omega}{T_2}\right)\mathbf{G}(\omega)\exp\{-i\omega\tau\} \tag{2533}$$

or

$$\hat{A}\mathbf{G}(\tau)=\int_{-\infty}^{\infty}\frac{d\omega}{2\pi}\left(\omega_0^2+\frac{1}{T_2^2}-\omega^2-\frac{2i\omega}{T_2}\right)\mathbf{G}(\omega)\exp\{-i\omega\tau\}\equiv\delta(\tau)=\int_{-\infty}^{\infty}\frac{d\omega}{2\pi}\exp\{-i\omega\tau\} \tag{2534}$$

From here

$$\left(\omega_0^2+\frac{1}{T_2^2}-\omega^2-\frac{2i\omega}{T_2}\right)\mathbf{G}(\omega)=1 \tag{2535}$$

or

$$\mathbf{G}(\omega)=\frac{1}{\omega_0^2+\frac{1}{T_2^2}-\omega^2-\frac{2i\omega}{T_2}}\equiv\frac{1}{\omega_0^2-(\omega+i\delta)^2}\quad,\quad\delta=\frac{1}{T_2} \tag{2536}$$

or

$$\mathbf{G}(\omega)=\left(\frac{1}{\omega-\omega_0+i\delta}-\frac{1}{\omega+\omega_0+i\delta}\right)\left(-\frac{1}{2\omega_0}\right) \tag{2537}$$

So,

$$\mathbf{G}(\tau>0)=\int_{-\infty}^{\infty}\frac{d\omega}{2\pi}\frac{\exp\{-i\omega\tau\}}{\omega_0^2-(\omega+i\delta)^2}=\frac{1}{4\pi\omega_0}\int_{-\infty}^{\infty}d\omega\exp\{-i\omega\tau\}\left(\frac{1}{\omega+\omega_0+i\delta}-\frac{1}{\omega-\omega_0+i\delta}\right) \tag{2538}$$

or

$$\mathbf{G}(\tau>0)=\frac{1}{4\pi\omega_0}(-2\pi i)\left(\exp\{i\omega_0\tau-\tau\delta\}-\exp\{-i\omega_0\tau-\tau\delta\}\right)=\frac{\sin\omega_0\tau}{\omega_0}\exp\{-\tau\delta\} \tag{2539}$$

or

$$\mathbf{G}(t-t')=\frac{\sin\omega_o(t-t')}{\omega_0}\theta(t-t')\exp\{-(t-t')\delta\} \tag{2540}$$

Susceptibility: Dynamical Response Function

From the expression of the Green's function (2537), we write the susceptibility:

$$\chi_{xx}(\omega)=\gamma\mathcal{M}_0\omega_0\mathbf{G}(\omega) \tag{2541}$$

or

$$\chi_{xx}(\omega)=\gamma\mathcal{M}_0\omega_0\frac{1}{2\omega_0}\left(\frac{1}{\omega+\omega_0+i\delta}-\frac{1}{\omega-\omega_0+i\delta}\right)=\frac{\gamma\mathcal{M}_0}{2}\left(\frac{\omega+\omega_0-i\delta}{(\omega+\omega_0)^2+\delta^2}-\frac{\omega-\omega_0-i\delta}{(\omega-\omega_0)^2+\delta^2}\right) \quad (2542)$$

From here, we may write the susceptibility in its real and imaginary parts:

$$\chi_{xx}(\omega)=\chi'_{xx}(\omega)+i\chi''_{xx}(\omega) \quad (2543)$$

where

$$\chi'_{xx}(\omega)=\frac{\gamma\mathcal{M}_0}{2}\left(\frac{\omega_0-\omega}{(\omega_0-\omega)^2+\delta^2}+\frac{\omega_0+\omega}{(\omega_0+\omega)^2+\delta^2}\right) \quad (2544)$$

is the reactive part and

$$\chi''_{xx}(\omega)=\frac{\gamma\mathcal{M}_0\delta}{2}\left(\frac{1}{(\omega_0-\omega)^2+\delta^2}-\frac{1}{(\omega_0+\omega)^2+\delta^2}\right) \quad (2545)$$

is the dissipative part of the response function.

Note that $\chi'_{xx}(\omega)$ is an even function and $\chi''_{xx}(\omega)$ is an odd function as expected. For $\chi''_{xx}(\omega)$, the function peaks around $\pm\omega_0$ at frequencies where the system naturally vibrates. This is where the system is able to absorb energy. However, as $\delta\to 0$, the imaginary part does not achieve the value zero and instead tends toward two delta functions located at $\pm\omega_0$.

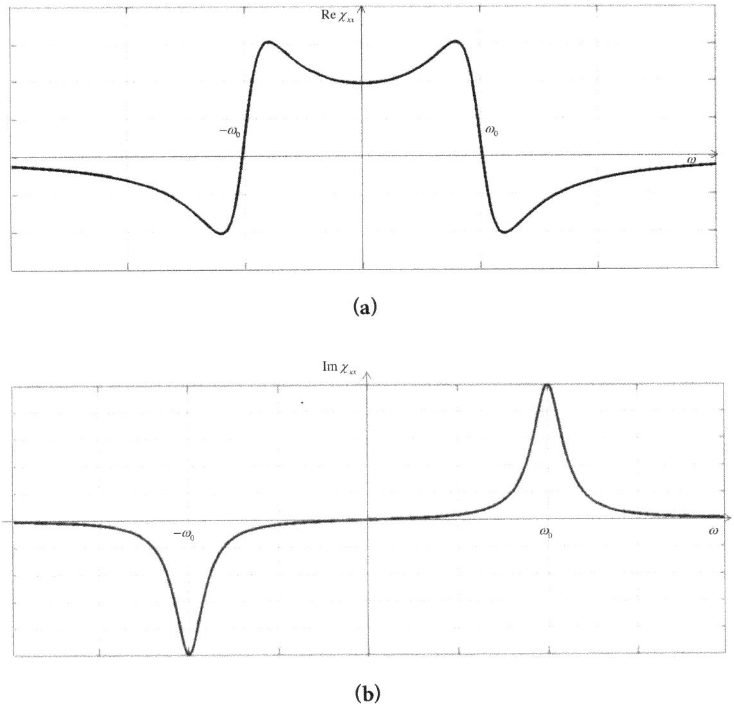

FIGURE 13.11 **(a)** Plot of the real part of the susceptibility function $\mathrm{Re}\,\chi_{xx}(\omega)$ versus the frequency ω. **(b)** Plot of the imaginary part of the susceptibility function $\mathrm{Im}\,\chi_{xx}(\omega)$ versus frequency ω.

13.2.12.1 Linear Response Function

We devote this section to a brief presentation of linear response theory. This provides a general framework for analyzing the dynamical properties of a condensed-matter system close to thermal equilibrium. Dynamical processes may either be due to spontaneous fluctuations or external perturbations. These two kinds of phenomena are correlated. If an ordinary system is left alone, sooner or later it will achieve an equilibrium state. This equilibrium state is dependent on the temperature of the environment and on external parameters. If the temperature or the external parameters change slowly enough, the system can achieve the new equilibrium state practically instantaneously, and we call this a **reversible process**. If, on the one hand, the external parameters vary so rapidly that the system has no chance to adapt, it remains away from equilibrium, and we call this **irreversibility**. With the application of an external field to a system at equilibrium, properties of the system that couple to the external field change accordingly. For a low enough field, the change is proportional to the external field. The proportionality constant is called the **linear response function**, and it provides valuable information about the system. There is a strong correlation between the time-dependent response functions and dynamical properties of the system at equilibrium.

Therefore, it is obvious that any macroscopic system may be characterized by its response to external field $f(t)$:

$$\hat{H}(t)=\hat{H}_0+\hat{H}_s(t) \quad , \quad \hat{H}_s(t)=-\hat{A}(t)f(t) \tag{2546}$$

This response function relates the change of an ensemble-averaged physical observable, say $\langle \hat{A}(t)\rangle$, to the external field $f(t)$. The quantity $\hat{A}(t)$ could be the angular momentum of an ion or the magnetization, and $f(t)$ is a time-dependent applied magnetic field. The applicability of linear response theory is limited to the regime where $\langle \hat{A}(t)\rangle$ changes linearly with the external stimulus. So, the external stimulus $f(t)$ is considered sufficiently weak to ensure that the response is linear.

Speaking of susceptibility, we refer to a medium in which the response is proportional in some sense to the excitation. If the medium is linear, then the response is directly proportional to the excitation; if the medium is nonlinear, then the proportionality involves higher powers of the excitation. If the excitation is very small, then the response can be approximated fairly well by the linear susceptibility. Because time- and space-varying magnetic fields generally are quite small, a linear response theory is adequate. When dealing with hysteresis phenomena or high-power absorption in magnetic materials, nonlinear effects become essential.

Our concern here is mainly with the response of such a system to a magnetic field where the **output** is the magnetization and the response function is the **magnetic susceptibility** χ. As seen previously, determining the susceptibility requires the evaluation of the magnetization produced by an applied magnetic field that generally may depend on space and time. Consequently, the resulting magnetization will vary similarly in space and time. Suppose the spatial and temporal dependence of the applied field may be characterized by wave vector $\vec{q}$ and frequency ω, respectively. Hence, the **magnetization** $\vec{\mathcal{M}}(\vec{q},\omega)$ as well as the **susceptibility** $\chi(\vec{q},\omega)$ would depend on the wave vector $\vec{q}$ as well as the frequency ω.

Next, we define the dynamical susceptibility that plays a vital role in modern theories of magnetism. We consider the magnetization $\vec{\mathcal{M}}(\vec{r},t)$ linked to a weak magnetic field $\vec{H}(\vec{r},t)$ that has the following Fourier transforms

$$\vec{\mathcal{M}}(\vec{r},t)=\frac{1}{2\pi V}\sum_{\vec{q}}\int d\omega\vec{\mathcal{M}}(\vec{q},\omega)\exp\{i(\vec{q}\vec{r}-\omega t)\} \tag{2547}$$

$$\vec{H}(\vec{r},t)=\frac{1}{2\pi V}\sum_{\vec{q}}\int d\omega\vec{H}(\vec{q},\omega)\exp\{i(\vec{q}\vec{r}-\omega t)\} \tag{2548}$$

These expansions can be inverted via the following relations:

$$\int dt \exp\{i(\omega-\omega')t\} = 2\pi\delta(\omega-\omega') \tag{2549}$$

$$\int d\vec{r} \exp\{i(\vec{q}-\vec{q}\,')\vec{r}\} = V\delta_{(\vec{q}\vec{q}\,')} \tag{2550}$$

$$\sum_{\vec{q}} \exp\{i\vec{q}(\vec{r}-\vec{r}\,')\} = \frac{V}{(2\pi)^3}\int d\vec{q} \exp\{i\vec{q}(\vec{r}-\vec{r}\,')\} = V\delta(\vec{r}-\vec{r}\,') \tag{2551}$$

We define the generalized wave-vector-dependent frequency-dependent susceptibility by

$$\vec{\mathcal{m}}_\beta(\vec{q},\omega) = \sum_{\vec{q}\,'}\int d\omega' \sum_{\alpha} \chi_{\alpha\beta}(\vec{q},\vec{q}\,';\omega,\omega')\vec{H}_\alpha(\vec{q},\omega') \tag{2552}$$

Here α and β are Cartesian labels. The defined linear response function is called **generalized susceptibility**. We may write equation (2552) in a more convenient dyadic form as

$$\vec{\mathcal{M}}(\vec{q},\omega) = \sum_{\vec{q}\,'}\int d\omega' \chi(\vec{q},\vec{q}\,';\omega,\omega')\vec{H}(\vec{q},\omega') \tag{2553}$$

Generally, $\chi(\vec{q},\vec{q}\,';\omega,\omega')$ is dependent on the form of $\vec{H}(\vec{r},t)$, or equivalently, $\vec{H}(\vec{q},\omega')$, which implies that the susceptibility is a functional of the field. We observe from (2552) that the susceptibility is a tensor as well. Because the magnetization may be out of phase with the exciting field, the susceptibility is also complex. If we substitute (2553) into (2547), we have

$$\vec{\mathcal{M}}(\vec{r},t) = \frac{1}{2\pi V}\sum_{\vec{q}}\int d\omega \sum_{\vec{q}\,'}\int d\omega' \chi(\vec{q},\vec{q}\,';\omega,\omega')\vec{H}(\vec{q},\omega')\exp\{i(\vec{q}\vec{r}-\omega t)\} \tag{2554}$$

This may also be written as

$$\vec{\mathcal{M}}(\vec{r},t) = \int d\vec{r}\,' dt' \chi(\vec{r},\vec{r}\,',t,t')\vec{H}(\vec{r}\,',t') \tag{2555}$$

where the generalized spatial-temporal susceptibility density is defined as

$$\chi(\vec{r},\vec{r}\,',t,t') = \frac{1}{2\pi V}\sum_{\vec{q}}\int d\omega \sum_{\vec{q}\,'}\int d\omega' \chi(\vec{q},\vec{q}\,';\omega)\exp\{i\vec{q}(\vec{r}-\vec{r}\,')\}\exp\{-i\omega(t-t')\}\times$$
$$\times\exp\{i(\vec{q}-\vec{q}\,')\vec{r}\,'\}\exp\{-i(\omega-\omega')t'\} \tag{2556}$$

Suppose the magnetic medium possesses translational invariance. Therefore, the susceptibility $\chi(\vec{r},\vec{r}\,',t,t')$ should only be a function of the relative coordinate $(\vec{r}-\vec{r}\,')$. This implies that in the wave-vector-dependent susceptibility, $\vec{q}\,'$ is equal to $\vec{q}$. If the medium is stationary, then the temporal dependence is $(t-t')$. This implies a monochromatic response to a monochromatic excitation with the same frequency, that is $\omega=\omega'$. From the aforementioned conditions, the susceptibility takes the form

$$\chi(\vec{q},\vec{q}';\omega,\omega')=\chi(\vec{q},\omega')\delta_{\vec{q}\vec{q}'}\delta(\omega-\omega') \tag{2557}$$

Therefore,

$$\vec{\mathcal{M}}(\vec{r},t)=\int d\vec{r}'\,dt'\chi(\vec{r}-\vec{r}',t-t')\vec{H}(\vec{r}',t') \tag{2558}$$

and letting

$$\tau=t-t' \quad , \quad \vec{R}=\vec{r}-\vec{r}' \tag{2559}$$

we have

$$\chi(\vec{R},\tau)=\frac{1}{2\pi V}\sum_{\vec{q}}\int d\omega\chi(\vec{q},\omega)\exp\{i\vec{q}\vec{R}\}\exp\{-i\omega\tau\} \tag{2560}$$

where the Fourier transform is as follows

$$\chi(\vec{q},\omega)=\int d\vec{R}\,d\tau\chi(\vec{R},\tau)\exp\{-i\vec{q}\vec{R}\}\exp\{i\omega\tau\} \tag{2561}$$

We consider a perfect lattice with the lattice vector $\vec{R}$. Generally, the applied fields macroscopically vary so that $\vec{q}$ is small. In this case, $\chi_{\alpha\beta}(\vec{q},\vec{q}',\tau)$ gives the macroscopic response of the system to the field, whereas $\chi_{\alpha\beta}(\vec{q},\vec{q}'+\vec{q}'',\tau)$ for $\vec{q}''\neq 0$ is dependent on the microscopic variation of the response within the individual cell. However, for spins completely localized at the lattice sites (such as in the case of the Heisenberg model), we have

$$\chi_{\alpha\beta}(\vec{q}+\vec{q}'',\vec{q},\tau)=\chi_{\alpha\beta}(\vec{q},\vec{q},\tau)=\chi_{\alpha\beta}(\vec{q}+\vec{q}'',\vec{q}+\vec{q}'',\tau) \tag{2562}$$

This is due to the fact that the only material values of $\vec{H}(\vec{r})$ are those at the lattice sites:

$$\vec{H}(\vec{R})=\vec{H}\exp\{i\vec{q}\vec{R}\}=\vec{H}\exp\{i(\vec{q}+\vec{q}'')\vec{R}\} \tag{2563}$$

When impurities destroy the translational invariance, we require a more general susceptibility. A typical example is the response of a paramagnet. Because it has such a general nature, susceptibility is involved in various important theorems. Three such theorems are:

1. **Kramers–Kronig relations** involve the real and imaginary parts of the susceptibility.
2. **The fluctuation-dissipation theorem** involves the susceptibility to thermal fluctuations in the magnetization.
3. **The Onsager relation** describes the symmetry of the susceptibility tensor.

The Kramers–Kronig Relations

We consider equation (2561) again:

$$\chi(\vec{q},\omega)=\int d\vec{R}\,d\tau\chi(\vec{R},\tau)\exp\{-i\vec{q}\vec{R}\}\exp\{i\omega\tau\} \tag{2564}$$

Then from

$$\chi(\vec{q},\tau)=\int d\vec{R}\chi(\vec{R},\tau)\exp\{-i\vec{q}\vec{R}\} \tag{2565}$$

we have

$$\chi(\vec{q},\omega)=\int d\tau\chi(\vec{q},\tau)\exp\{i\omega\tau\} \tag{2566}$$

It is obvious from (2566) that the dynamical susceptibility is complex:

$$\chi(\vec{q},\omega)=\chi'(\vec{q},\omega)+i\chi''(\vec{q},\omega) \tag{2567}$$

Considering some rather general properties of (2567), the real part $\chi'(\vec{q},\omega)$ and the imaginary part $\chi''(\vec{q},\omega)$ are connected on the real axis ω by integral relations known as **Kramers–Kronig relations** (or **dispersion relations**). We consider a medium that is **linear** and **stationary** as well as **translationally invariant**. So, $\chi(\vec{q},\omega)$ relates $\chi(\vec{R},\tau)$ via (2561). In (2561), the response is independent of any future perturbations. This **causal** behavior may be incorporated in the response function by the requirement (**principle of causality**):

$$\chi(\vec{R},\tau)=0 \quad , \quad \tau<0 \tag{2568}$$

So, the time integral in (2561) runs only from 0 to ∞. This implies that

$$\chi(\vec{q},\omega)=\int_0^\infty d\tau\chi(\vec{q},\tau)\exp\{i\omega\tau\} \tag{2569}$$

So, as seen earlier, $\chi(\vec{q},\omega)$ is a complex valued function of ω and has no singularities at the ends of the real axis, provided that for $\omega\to 0$, we have a finite integral and, consequently, the full response function

$$\chi(\vec{q},\omega\to 0)=\int d\tau\chi(\vec{q},\tau) \tag{2570}$$

This is the **static susceptibility**. This implies that the response on the finite excitation is also finite. Finite values of $\chi(\vec{q},\omega)$ at the ends of the real axis are identified with the real part of the susceptibility $\chi'(\vec{q},\infty)$. We give the physical explanation why $\chi''(\vec{q},\omega)$ vanishes as $\omega\to\infty$. Further, we see that the rate of energy absorption by a magnetic system is proportional to $\chi''(\vec{q},\omega)$. For this to be finite as $\omega\to\infty$, then $\chi''(\vec{q},\omega)$ achieves the value 0 as $\omega\to\infty$. This result is also derived from the finite-response assumption.

We take advantage of the causality condition in (2568) and consider the Laplace transform of $F(\vec{q},\tau)$:

$$\chi(\vec{q},z)=\int_0^\infty d\tau F(\vec{q},\tau)\exp\{iz\tau\} \tag{2571}$$

Here, $z=z_1+iz_2$ is a complex variable. If

$$\chi(\vec{q},z)=\int_0^\infty d\tau\left|F(\vec{q},\tau)\right|\exp\{-\in\tau\} \tag{2572}$$

is assumed to be finite in the limit $\in \to 0^+$, the converse relation is

$$F(\vec{q},\tau) = \int_{-\infty+i\in}^{\infty+i\in} d\tau \chi(\vec{q},z)\exp\{-iz\tau\} \quad , \quad \in > 0 \tag{2573}$$

If $F(\vec{q},\tau)$ satisfies the previous condition as well as condition (2568), then it can readily be shown that $\chi(\vec{q},z)$ is an analytic function in the upper part of the complex z-plane $(z_2 > 0)$.

The conditions (2568) and (2571) through (2573) on $F(\vec{q},\tau)$ have direct physical significance that the system is **causal** and **stable** against small perturbations. These two conditions ensure that $\chi(\vec{q},z)$ has no poles in the upper half-plane. If this were not the case, then the response $\langle \mathcal{M}(\vec{q},\tau) \rangle$ to small disturbances would diverge exponentially as a function of time. There is an absence of poles in $\chi(\vec{q},z)$, when $z_2 > 0$ results in a relation between the real and imaginary part of $\chi(\vec{q},\omega)$ (**Kramers–Kronig dispersion relation**), as seen earlier. If $\chi(\vec{q},z)$ has no poles within the contour C, then we may express it via the Cauchy integral along C by the identity

$$\chi(\vec{q},z) = \frac{1}{2\pi i} \int_C \frac{\chi(\vec{q},z')}{z'-z} dz' \tag{2574}$$

We chose the contour C to be the half-circle, in the upper half-plane, centered at the origin, and bounded below by the line parallel to the z_1-axis through $z_2 = \in'$, with z being a point lying within this contour. Because $F(\vec{q},\tau)$ is a bounded function in the domain $\in' > 0$, $\chi(\vec{q},z')$ must go to zero as $|z'| \to \infty$, when $z_2' > 0$. Therefore, it follows that the part of the contour integral along the half-circle vanishes when its radius goes to infinity. So,

$$\chi(\vec{q},z) = \lim_{\in' \to 0^+} \frac{1}{2\pi i} \int_{-\infty+i\in'}^{\infty+i\in'} dz' \frac{\chi(\vec{q},z')}{z'-z} \quad , \quad z' = \omega' + i\in' \tag{2575}$$

Letting $z = \omega + i\in$ and with the help of the **Dirac** formula

$$\lim_{\in' \to 0^+} \frac{1}{\omega' - \omega - i\in} = P\frac{1}{\omega'-\omega} + i\pi\delta(\omega'-\omega) \tag{2576}$$

Considering (2575) and (2576), we obtain the **Kramers–Kronig relation**:

$$\chi(\vec{q},\omega) = \frac{1}{i\pi} P\int_{-\infty}^{\infty} d\omega' \frac{\chi(\vec{q},\omega')}{\omega'-\omega} \tag{2577}$$

Here, P is the **principal value** of the integral (2575).

We define

$$\lim_{\omega \to \infty} \chi'(\vec{q},\omega) = \chi'(\vec{q},\infty) \tag{2578}$$

and

$$F(\vec{q},\omega) = \chi(\vec{q},\omega) - \chi(\vec{q},\infty) \tag{2579}$$

The complex valued function $F(\vec{q},\omega)$ vanishes at the ends of the real axis. The function $F(\vec{q},z)$, where z is a complex variable, is analytic in the upper half-plane. We apply (2579) to equation (2577):

$$F(\vec{q},\omega) = \frac{1}{i\pi} P\int_{-\infty}^{\infty} d\omega' \frac{F(\vec{q},\omega')}{\omega'-\omega} \tag{2580}$$

If we equate the real and imaginary parts separately, then

$$\chi'(\vec{q},\omega)-\chi(\vec{q},\infty)=\frac{1}{\pi}\mathrm{P}\int_{-\infty}^{\infty}\frac{\chi''(\vec{q},\omega')}{\omega'-\omega}d\omega' \quad , \quad \chi''(\vec{q},\omega)=-\frac{1}{\pi}\mathrm{P}\int_{-\infty}^{\infty}\frac{\chi'(\vec{q},\omega')-\chi(\vec{q},\infty)}{\omega'-\omega}d\omega' \tag{2581}$$

These are the **Kramers-Kronig relations** that follow from **causality** alone and show that the dissipative, imaginary part of the response function $\chi''(\vec{q},\omega)$ is determined in terms of the **reactive**, real part, $\chi'(\vec{q},\omega)$ and vice versa. Nevertheless, **the relationship is not local in frequency space**: Knowledge of $\chi'(\vec{q},\omega)$ for all frequencies is needed in order to reconstruct $\chi''(\vec{q},\omega)$ for any single frequency.

The relation (2581) is useful in that $\chi''(\vec{q},\omega)$ is proportional to the absorption spectrum of the medium. The first equation shows that the static susceptibility may be obtained by integrating over the absorption spectrum. This is an experimental technique used in obtaining the static susceptibility of certain systems.

Considering the fact that the response $\vec{\mathcal{M}}(\vec{r},t)$ is a real quantity, then $\chi'(\vec{q},\omega)$ is an even function, while $\chi''(\vec{q},\omega)$ is an odd function of ω. **Kramers–Kronig relations** can be expressed in terms of integrals over positive frequencies and, in particular, the following integral

$$\chi''(\vec{q},\omega)=-\frac{2\omega}{\pi}\mathrm{P}\int_{0}^{\infty}\frac{\chi'(\vec{q},\omega')}{\omega'^2-\omega^2}d\omega' \tag{2582}$$

The term with $\chi(\vec{q},\infty)$ vanishes because the principal value of the integral of $\frac{1}{\omega'^2-\omega^2}$ is zero.

13.2.12.2 The Fluctuation-Dissipation Theorem

We examine the fluctuation-dissipation theorem [93] and suppose we have charged particles accelerated by an external electric field. Due to impacts with molecules, these particles experience a resistive force proportional to their velocity. So, the mechanism producing the random fluctuations in the position of the particle may be responsible for its response to an external excitation. The relationship between the response to an external perturbation of a system and its spontaneous thermal fluctuation spectrum is the so-called **fluctuation-dissipation theorem**. Though this relationship is general, our concern here will be only on its specific application to a magnetic medium.

Consider a linearly polarized magnetic field

$$H_\alpha = H_1\cos(\vec{q}\vec{r})\cos\omega t \tag{2583}$$

oscillating at a frequency ω in the α direction. The principle of superposition applies due to the linear system. So, it is possible to construct the response to an arbitrary field if we know the response to the given field. Relation (2551) gives the response in the β direction to such an excitation. From

$$H_\alpha(\vec{q}',\omega')=\frac{\pi H_1 V}{2}\left[\delta(\vec{q}'-\vec{q})\delta(\omega'+\omega)+\delta(\vec{q}'-\vec{q})\delta(\omega'-\omega)+\delta(\vec{q}'+\vec{q})\delta(\omega'+\omega)+\delta(\vec{q}'+\vec{q})\delta(\omega'-\omega)\right] \tag{2584}$$

then

$$\mathcal{m}_\beta(\vec{q}'',\omega'')=\frac{\pi H_1 V}{2}\left[\chi_{\alpha\beta}(\vec{q}'',\vec{q},\omega'',-\omega)+\chi_{\alpha\beta}(\vec{q}'',\vec{q},\omega'',\omega)+\chi_{\alpha\beta}(\vec{q}'',-\vec{q},\omega'',-\omega)+\chi_{\alpha\beta}(\vec{q}'',-\vec{q},\omega'',\omega)\right] \tag{2585}$$

We calculate $\mathcal{m}_\beta(\vec{q}'',\omega'')$. The magnetization is

$$\mathcal{m}_\beta(\vec{r},\tau)=\mathrm{Tr}\{\rho\,\mathcal{m}_\beta(\vec{r})\} \tag{2586}$$

In this case, the density matrix ρ is time-dependent. Because the time-varying field upsets the thermodynamic equilibrium, we must solve for ρ. The total Hamiltonian determinant is written as

$$\mathrm{H} = \mathrm{H}_0 + \mathrm{H}_1 \tag{2587}$$

Here,

$$\mathrm{H}_1 = -\int d\vec{r}\mathcal{M}(\vec{r})H(\vec{r},t) \tag{2588}$$

For our particular case, we have

$$\mathrm{H}_1 = -H_1\int d\vec{r}\mathcal{m}_\alpha(\vec{r})\cos(\vec{q}\vec{r})\cos\omega t = -\frac{H_1}{2}\left(\mathcal{m}_\alpha(\vec{q}) + \mathcal{m}_\alpha(-\vec{q})\right)\cos\omega t \tag{2589}$$

We write the equation of motion for the density matrix

$$\frac{\partial\rho}{\partial t} = \frac{i}{\hbar}[\rho, \mathrm{H}_0 + \mathrm{H}_1] \tag{2590}$$

It is convenient to introduce the interaction picture for the density matrix

$$\rho(t) = \exp\left\{\frac{i\mathrm{H}_0 t}{\hbar}\right\}\rho\exp\left\{-\frac{i\mathrm{H}_0 t}{\hbar}\right\} \tag{2591}$$

We differentiate (2591) and then, with the help of (2590), we have

$$\frac{d\rho(t)}{dt} = \frac{i}{\hbar}\left[\rho(t), \mathrm{H}_1(t)\right] \tag{2592}$$

where

$$\mathrm{H}_1(t) = \exp\left\{\frac{i\mathrm{H}_0 t}{\hbar}\right\}\mathrm{H}_1\exp\left\{-\frac{i\mathrm{H}_0 t}{\hbar}\right\} \tag{2593}$$

The solution of equation (2592) is as follows

$$\rho(t) = \rho(-\infty) + \frac{i}{\hbar}\int_{-\infty}^{t} dt'\left[\rho(t'), \mathrm{H}_1(t')\right] \tag{2594}$$

Suppose the interaction is turned on adiabatically, then

$$\rho(-\infty) = \rho_0 \equiv \frac{\exp\{-\beta\mathrm{H}_0\}}{\mathrm{Tr}\left[\exp\{-\beta\mathrm{H}_0\}\right]} \tag{2595}$$

is the equilibrium density matrix. Here, the inverse temperature is $\beta = \frac{1}{T}$. We invert (2594) with the help of (2589) and replace ρ within the commutator by ρ_0:

$$\rho \cong \rho_0 - \frac{iH_1}{2\hbar}\int_0^{\infty} dt'\left\{\rho_0, \exp\left\{-\frac{i\mathrm{H}_0 t'}{\hbar}\right\}\left(\mathcal{m}_\alpha(\vec{q}) + \mathcal{m}_\alpha(-\vec{q})\right)\exp\left\{\frac{i\mathrm{H}_0 t'}{\hbar}\right\}\right\} \tag{2596}$$

Suppose the system is ordered in the absence of the applied field, then

$$\mathrm{Tr}\{\rho_0 m_\beta\} = m_\beta(-\infty) \neq 0 \tag{2597}$$

The response of such a system then is defined by

$$\mathcal{M}(\vec{r},t) = m_\beta(\vec{r},t) - m_\beta(-\infty) \tag{2598}$$

This is a result of the applied field. We show $\mathcal{M}_\beta(\vec{r},t)$ being the response to the applied field and so,

$$m_\beta(\vec{r},t) = -\frac{iH_1}{2\hbar}\mathrm{Tr}\left\{\int_0^\infty dt'\left\{\rho_0, \exp\left\{-\frac{i\mathrm{H}_0 t'}{\hbar}\right\}\left(m_\alpha(\vec{q}) + m_\alpha(-\vec{q})\right)\exp\left\{\frac{i\mathrm{H}_0 t'}{\hbar}\right\}\right\} m_\beta(\vec{r})\right\}\cos\omega(t-t') \tag{2599}$$

We take the Fourier transform of this equation, and we have

$$m_\beta(\vec{q}'',\omega'') = -\frac{\pi H_1}{2\hbar}\mathrm{Tr}\left\{\int_0^\infty dt' \exp\{i\omega t'\}\left[\rho_0, m_\alpha(\vec{q},-t')\right] m_\beta(\vec{q}'')\right\}\delta(\omega''+\omega) + (\text{terms involving } -\vec{q} \text{ and } -\omega) \tag{2600}$$

The delta function in (2600) stems from the linearization of the expression for ρ by replacing ρ by ρ_0 within the commutator.

We commute the integral with the trace in (2600) and apply the cyclic invariance of the trace, and we have

$$\mathrm{Tr}\left\{\int_0^\infty dt' \exp\{i\omega t'\}\left[\rho_0, m_\alpha(\vec{q},-t')\right] m_\beta(\vec{q}'')\right\} = \int dt' \exp\{i\omega t'\}\left\langle\left[m_\alpha(\vec{q},-t'), m_\beta(\vec{q}'')\right]\right\rangle \tag{2601}$$

We compare the resulting expression for $m_\beta(\vec{q}',\omega')$ with (2585), and then we have

$$\chi_{\alpha\beta}(\vec{q}',\vec{q},\omega',\omega) = \frac{i}{\hbar V}\int_0^\infty dt \exp\{i\omega t\}\left[m_\beta(\vec{q}',t), m_\alpha(-\vec{q})\right]\delta(\omega'-\omega) \tag{2602}$$

As the $\vec{q}$ component of the applied field couples to the $-\vec{q}$ component of the magnetization, we consider $\chi_{\alpha\beta}(\vec{q},\vec{q},\omega)$, which we write as $\chi_{\alpha\beta}(\vec{q},\omega)$. So,

$$\chi_{\alpha\beta}(\vec{q},\omega) = \frac{i}{\hbar V}\int_0^\infty dt \exp\{i\omega t\}\left\langle\left[m_\beta(\vec{q},t), m_\alpha(-\vec{q})\right]\right\rangle \tag{2603}$$

Here,

$$m_\beta(\vec{q},t) = \exp\left\{\frac{i\mathrm{H}_0 t}{\hbar}\right\} m_\beta(\vec{q}) \exp\left\{-\frac{i\mathrm{H}_0 t}{\hbar}\right\} \tag{2604}$$

In the literature, the function

$$\left\langle\left[m_\beta(\vec{q},t), m_\alpha(-\vec{q})\right]\right\rangle = \mathrm{Tr}\left\{\left[m_\alpha(-\vec{q}), \rho_0\right] m_\beta(\vec{q},t)\right\} \tag{2605}$$

is referred to as the **response function** of the system that may as well be multiplied by the theta function $\theta(t)$, which yields the so-called **double-time-retarded Green's function** and is represented by double angular brackets:

$$\left\langle\left\langle\left[m_\beta(\vec{q},t),m_\alpha(-\vec{q})\right]\right\rangle\right\rangle \equiv -i\left\langle\left[m_\beta(\vec{q},t),m_\alpha(-\vec{q})\right]\right\rangle\theta(t) \tag{2606}$$

The response function has no classical analog and is not a well-defined observable. In this case, it is convenient to relate the susceptibility to the **correlation function** $\left\langle\left\{m_\beta(\vec{q},t),m_\alpha(-\vec{q})\right\}\right\rangle$. Here, $\{\cdots\}$ is the symmetrized product and is defined:

$$\left\{m_\beta(\vec{q},t),m_\alpha(-\vec{q})\right\} \equiv \frac{1}{2}\left[m_\beta(\vec{q},t)m_\alpha(-\vec{q})+m_\alpha(-\vec{q})m_\beta(\vec{q},t)\right] \tag{2607}$$

To relate the response function to the correlation function, we consider their Fourier transforms:

$$f_{\alpha\beta}(\vec{q},\omega)=\frac{i}{\hbar}\int_{-\infty}^{\infty}dt\exp\{i\omega t\}\left\langle\left[m_\beta(\vec{q},t),m_\alpha(-\vec{q})\right]\right\rangle,\quad g_{\alpha\beta}(\vec{q},\omega)=\frac{i}{\hbar}\int_{-\infty}^{\infty}dt\exp\{i\omega t\}\left\langle\left\{m_\beta(\vec{q},t),m_\alpha(-\vec{q})\right\}\right\rangle \tag{2608}$$

But,

$$\left\langle\left[m_\beta(\vec{q},t),m_\alpha(-\vec{q})\right]\right\rangle = \mathrm{Tr}\left\{\left[m_\alpha(-\vec{q}),\rho_0\right]m_\beta(\vec{q},t)\right\} \tag{2609}$$

and the density operator of the thermal-equilibrium state:

$$\rho_0=\frac{\exp\{-\beta H_0\}}{\mathrm{Tr}\left[\exp\{-\beta H_0\}\right]} \tag{2610}$$

Also

$$\left[m_\beta(\vec{q},t),m_\alpha(-\vec{q})\right]=m_\beta(\vec{q},t)m_\alpha(-\vec{q})-m_\alpha(-\vec{q})m_\beta(\vec{q},t) \tag{2611}$$

then

$$\begin{aligned}
&\int_{-\infty}^{\infty}dt\exp\{i\omega t\}\left\langle\left[m_\alpha(-\vec{q}),m_\beta(\vec{q},t)\right]\right\rangle=\int_{-\infty}^{\infty}dt\exp\{i\omega t\}\mathrm{Tr}\left[m_\alpha(-\vec{q}),\rho_0\right]m_\beta(\vec{q},t)=\\
&=\mathrm{Tr}\left[\int_{-\infty}^{\infty}dt\exp\{i\omega t\}m_\alpha(-\vec{q})\rho_0 m_\beta(\vec{q},t)-\int_{-\infty}^{\infty}dt\exp\{i\omega t\}\rho_0 m_\alpha(-\vec{q})m_\beta(\vec{q},t)\right]=\\
&=\mathrm{Tr}\left[\int_{-\infty}^{\infty}dt\exp\{i\omega t\}m_\alpha(-\vec{q})\rho_0 m_\beta(\vec{q},t)-\int_{-\infty}^{\infty}dt\exp\{i\omega t\}\rho_0 m_\alpha(-\vec{q})\exp\left\{\frac{iH_0t}{\hbar}\right\}m_\beta(\vec{q})\exp\left\{-\frac{iH_0t}{\hbar}\right\}\right]=\\
&=\mathrm{Tr}\left[\int_{-\infty}^{\infty}dt\exp\{i\omega t\}\rho_0 m_\alpha(-\vec{q})m_\beta(\vec{q},t)-\right.\\
&\qquad\left.\int_{-\infty}^{\infty}dt\exp\{i\omega t\}\rho_0\exp\left\{\frac{iH_0}{\hbar}(t-i\hbar\beta)\right\}m_\alpha(-\vec{q})\exp\left\{-\frac{iH_0}{\hbar}(t-i\hbar\beta)\right\}m_\beta(\vec{q})\right]=\\
&\underset{t-i\hbar\beta=\tau}{=}\mathrm{Tr}\left[\int_{-\infty}^{\infty}dt\exp\{i\omega t\}\rho_0 m_\alpha(-\vec{q})m_\beta(\vec{q},t)-\exp\{-\beta\hbar\omega\}\int_{-\infty}^{\infty}d\tau\exp\{i\omega\tau\}\rho_0 m_\beta(\vec{q},\tau)m_\alpha(-\vec{q})\right]=\\
&=\left(1-\exp\{-\beta\hbar\omega\}\right)\int_{-\infty}^{\infty}dt\exp\{i\omega t\}\mathrm{Tr}\left[m_\alpha(-\vec{q}),\rho_0\right]m_\beta(\vec{q},t)
\end{aligned} \tag{2612}$$

So,

$$f_{\alpha\beta}(\vec{q},\omega)=\frac{i}{\hbar}\left(1-\exp\{-\beta\hbar\omega\}\right)\int_{-\infty}^{\infty}dt\exp\{i\omega t\}\left\langle\left[m_\beta(\vec{q},t),m_\alpha(-\vec{q})\right]\right\rangle \tag{2613}$$

We find how $g_{\alpha\beta}$ relates $f_{\alpha\beta}$ from its definition:

$$g_{\alpha\beta}(\vec{q},\omega)=\frac{1}{2}\left(1+\exp\{-\beta\hbar\omega\}\right)\int_{-\infty}^{\infty}dt\exp\{i\omega t\}\left\langle\left[m_\beta(\vec{q},t),m_\alpha(-\vec{q})\right]\right\rangle \tag{2614}$$

But,

$$\int_{-\infty}^{\infty}dt\exp\{i\omega t\}\left\langle\left[m_\beta(\vec{q},t),m_\alpha(-\vec{q})\right]\right\rangle=\frac{\hbar}{i}\frac{f_{\alpha\beta}(\vec{q},\omega)}{1-\exp\{-\beta\hbar\omega\}} \tag{2615}$$

then

$$g_{\alpha\beta}(\vec{q},\omega)=\frac{\hbar}{2i}\coth\frac{\hbar\omega}{2T}f_{\alpha\beta}(\vec{q},\omega) \tag{2616}$$

We also relate $f_{\alpha\beta}$ to the susceptibility. For this, we separate the time integral into two parts:

$$f_{\alpha\beta}(\vec{q},\omega)=\frac{i}{\hbar}\int_{-\infty}^{0}dt\exp\{i\omega t\}\left\langle\left[m_\beta(\vec{q},t),m_\alpha(-\vec{q})\right]\right\rangle+\frac{i}{\hbar}\int_{0}^{\infty}dt\exp\{i\omega t\}\left\langle\left[m_\beta(\vec{q},t),m_\alpha(-\vec{q})\right]\right\rangle \tag{2617}$$

We do the change of variable $t\rightarrow -t$ in the first integral:

$$f_{\alpha\beta}(\vec{q},\omega)=\frac{i}{\hbar}\int_{0}^{\infty}dt\exp\{-i\omega t\}\left\langle\left[m_\beta(\vec{q},-t),m_\alpha(-\vec{q})\right]\right\rangle+\frac{i}{\hbar}\int_{0}^{\infty}dt\exp\{i\omega t\}\left\langle\left[m_\beta(\vec{q},t),m_\alpha(-\vec{q})\right]\right\rangle \tag{2618}$$

Consider again

$$\chi_{\alpha\beta}(\vec{q},\omega)=\frac{i}{\hbar V}\int_{0}^{\infty}dt\exp\{i\omega t\}\left\langle\left[m_\beta(\vec{q},t),m_\alpha(-\vec{q})\right]\right\rangle \tag{2619}$$

then, from here,

$$\chi_{\alpha\beta}(-\vec{q},-\omega)=\chi^*_{\alpha\beta}(\vec{q},\omega) \tag{2620}$$

Consequently,

$$f_{\alpha\beta}(\vec{q},\omega)=\left[\chi_{\alpha\beta}(\vec{q},\omega)-\chi^*_{\alpha\beta}(\vec{q},\omega)\right]V \tag{2621}$$

and so

$$g_{\alpha\beta}(\vec{q},\omega)=\frac{\hbar V}{2i}\coth\frac{\hbar\omega}{2T}\left[\chi_{\alpha\beta}(\vec{q},\omega)-\chi^*_{\alpha\beta}(\vec{q},\omega)\right] \tag{2622}$$

Considering

$$\chi_{\alpha\beta}(\vec{q},\omega)=\chi'_{\alpha\beta}(\vec{q},\omega)+i\chi''_{\alpha\beta}(\vec{q},\omega) \tag{2623}$$

then

$$\chi_{\alpha\beta}(\vec{q},\omega)-\chi^*_{\alpha\beta}(\vec{q},\omega)=2i\chi''_{\alpha\beta}(\vec{q},\omega) \tag{2624}$$

and

$$f_{\alpha\beta}(\vec{q},\omega)=2i\chi''_{\alpha\beta}(\vec{q},\omega)V \tag{2625}$$

with

$$g_{\alpha\beta}(\vec{q},\omega) = \hbar V \coth\frac{\hbar\omega}{2T}\chi''_{\alpha\beta}(\vec{q},\omega) \tag{2626}$$

So, the Fourier transform of the correlation function is proportional to the imaginary part of the susceptibility that describes the absorptive, or loss, response of the magnetic system. Therefore, the fluctuation-dissipation theorem relates the fluctuations in the magnetization to energy loss.

13.2.12.3 Onsager Relation

Generally, when we probe a magnetic system, it is in the presence of a constant field H. Therefore, H_0 (and, consequently, the response function) is a function of the given field. In 1931, Onsager showed that microscopic reversibility requires the simultaneous reversal of both the magnetic field and time. This can be shown via the response function for the susceptibility:

$$\left\langle\left[m_\beta(\vec{q},t), m_\alpha(-\vec{q})\right]\right\rangle = \sum_n \langle n|\rho_0(H_0)\left[m_\beta(\vec{q},t), m_\alpha(-\vec{q})\right]|n\rangle \tag{2627}$$

Let us consider the operator $\overline{\mathrm{A}}$ under the time reversal procedure:

$$\langle n|\overline{\mathrm{A}}|m\rangle = \langle \mathrm{T}n|\mathrm{A}|\mathrm{T}m\rangle^* \tag{2628}$$

From here,

$$\overline{\mathrm{A}} = \mathrm{T}^{-1}\mathrm{A}\mathrm{T} \tag{2629}$$

and T is the time reversal operator. So,

$$\overline{\mathrm{A}(t)} = \overline{\mathrm{A}}(-t) \quad , \quad \overline{\mathrm{A}(t,H_0)} = \overline{\mathrm{A}}(-t,H_0) \tag{2630}$$

and

$$\left\langle\left[m_\beta(\vec{q},t), m_\alpha(-\vec{q})\right]\right\rangle = \sum_n \langle \mathrm{T}n|\rho_0(H_0)\left[m_\beta(\vec{q},t), m_\alpha(-\vec{q})\right]|\mathrm{T}n\rangle^* \tag{2631}$$

We insert $\mathrm{T}^{-1}\mathrm{T}$ between all the factors, and then

$$\begin{aligned}\left\langle\left[m_\beta(\vec{q},t), m_\alpha(-\vec{q})\right]\right\rangle &= \sum_n \langle \mathrm{T}n|\rho_0(H_0)\mathrm{T}\left[\overline{m_\beta}(\vec{q},t), \overline{m_\alpha}(-\vec{q})\right]\mathrm{T}^{-1}|\mathrm{T}n\rangle^* \\ &= \sum_n \langle n|\mathrm{T}^{-1}\rho_0(H_0)\mathrm{T}\left[\overline{m_\beta}(\vec{q},t), \overline{m_\alpha}(-\vec{q})\right]|n\rangle^*\end{aligned} \tag{2632}$$

or

$$\left\langle\left[m_\beta(\vec{q},t), m_\alpha(-\vec{q})\right]\right\rangle = \sum_n \langle n|\rho_0\left(\overline{H_0}\right)\left[\overline{m_\beta}(\vec{q},-t), \overline{m_\alpha}(-\vec{q})\right]|n\rangle \tag{2633}$$

So, the **Onsager relation**

$$\chi_{\alpha\beta}\left(\vec{q},\omega,\vec{H}\right) = \chi_{\beta\alpha}\left(-\vec{q},\omega,-\vec{H}\right) \tag{2634}$$

This implies that the diagonal components of the susceptibility tensor must be even functions of the field:

$$\chi_{\alpha\alpha}\left(\vec{H}\right) = \chi_{\alpha\alpha}\left(-\vec{H}\right) \tag{2635}$$

13.2.13 Itinerant Ferromagnetism

Now, we examine metallic systems in which correlation plays a much larger role in the dynamics of the electron. Because strong correlation causes itinerant magnetism, the theory of magnetism in metals continues to be among the more challenging subjects of modern physics. In this section, we examine itinerant ferromagnetism, which requires moving electrons. Hence, it can only occur in conductive materials such as d- or f-band transition metals like iron, nickel, and cobalt.

Itinerant models permit us to explain the noninteger magnetic moment per atom and large specific heat capacity. Nonetheless, the temperature dependence of their magnetization and magnetic susceptibilities are best described by the Heisenberg model of local moments. For pure itinerant models, the ferromagnetic-to-paramagnetic phase transition occurs via a uniform shrinking of the material's magnetic moments. This phase transition for the Heisenberg model is achieved by a directional disorder of the local moment caused by thermal fluctuations. For d-band transition metals, electron correlations and electron spin density fluctuations are responsible for the ferromagnetic-to-paramagnetic phase transition. These effects are best described by the interaction of local moments.

Notwithstanding the major progress in d-band transition metals, the realization of hybrid models that capture their full ferromagnetic behavior remains an open topic in solid state physics. In condensed matter systems, itinerant ferromagnetism is known to coexist together with localized ferromagnetism as pure itinerant ferromagnetism has not yet been observed. Because the strong interactions and correlations involved are very challenging theoretically, it is very important to study itinerant models. It is useful to note that nearly all theoretical models resolve to approximations though their validity is yet to be verified.

It is yet to be clarified whether a free electron gas with a uniform positive background can achieve itinerant ferromagnetism without auxiliary conditions such as coupling to a periodic potential, coupling to lattice vibrations, the presence of Heisenberg ferromagnetism, and so on. We have a few lattice models with specific band fillings where itinerant ferromagnetism has rigorously been proven to exist. Experimentally, itinerant ferromagnetism is predicted in liquid He-3, at high pressure (for low temperatures). The bad news is that liquid He-3 solidifies long before it achieves the itinerant ferromagnetic phase transition. Therefore, ultracold atomic gases should be more promising for the study of pure itinerant ferromagnetic models.

13.2.13.1 Quantum Impurities and the Kondo Effect

A quantum impurity basically is a collection of discrete quantum states coupled to a continuum of noninteracting degrees of freedom that can either be fermions or bosons, or both. The Kondo effect (named, in 1964, for the Japanese theoretician Jun Kondo) [94] is the simplest example of a **phenomenon driven by strongly correlated ("bad metals") electron systems such as heavy fermion compounds** in artificial nanosized structures. This effect has great significance in the development of strongly correlated quantum systems. Most heavy fermion compounds are noted for their anomalous electronic and magnetic properties as well as anomalous superconductivity and a dramatically sharper scattering resonance near the Fermi level. This narrow resonance appears due to the many-body effects. Within the framework of Landau's Fermi-liquid theory, this leads to strongly renormalized electronic quasiparticles with very heavy masses. The formation of these heavy quasiparticles is feasible at low temperatures.

Though the Kondo effect is a widely studied phenomenon in condensed-matter physics, it continues to attract the interest of experimentalists and theoreticians due to recent technological developments in both sample materials and experimental techniques that have prompted the possibility of reaching control over the dynamics of individual electrons in nanoscaled devices, such as single-electron transistors. Particular attention is devoted to the **strong correlation effects** in the transport properties of nanoscaled devices where electrons can now be confined and tailored, allowing for a myriad of single-particle and many-body effects to be probed in detail. Prominent among these is the **Kondo effect**, arising from the interactions between a single magnetic atom, such as cobalt, and the many electrons in an otherwise nonmagnetic metal. Such an impurity typically has an intrinsic angular momentum or spin that interacts with the electrons, resulting in a many-body problem that at the moment seems difficult.

There is recent interest in the Kondo effect due to its provision of clues in understanding the electronic properties of a wide variety of materials where the interactions among electrons are particularly strong, like heavy-fermion materials and high-temperature superconductors, as well as to new advances in experimental techniques from the rapidly developing field of nanotechnology, which gives unprecedented control over Kondo systems. There is a drop in electrical resistance of a pure metal as its temperature is lowered such that below a critical temperature value, the **resistance saturates** due to static defects in the material. This results from electrons that move more easily through a metallic crystal when the vibrations of the atoms are small. Some metals like lead, niobium, and aluminum suddenly lose all their resistance to electrical current and become superconducting. This phase transition from a conducting to a superconducting state occurs at a critical temperature below which the electrons behave as a single entity. Superconductivity is a prime example of a many-electron phenomenon. The value of the low-temperature resistance is dependent on the number of defects in the material. **Introducing defects increases the value of the saturation resistance** while leaving the character of the temperature dependence invariant. There is a dramatic change in this behavior when magnetic atoms, such as cobalt, are added. Apart from saturating, the electrical resistance increases as the temperature is further lowered. Though this behavior does not involve a phase transition, there is the so-called **Kondo temperature** T_{K} at which the resistance starts to increase again. This temperature completely determines the low-temperature electronic properties of the material.

The electrical resistance is associated with the amount of backscattering from defects that hampers the motion of the electrons through the crystal. In 1964, Kondo considered the scattering from a magnetic ion that interacts with the spins of the conducting electrons and observed that the second term in the calculation could be much larger than the first. The outcome of this result shows that the resistance of a metal increases logarithmically when the temperature is lowered. The Kondo theory correctly describes the observed upturn of the resistance at low temperatures with the validity of the results only above a certain temperature, the **Kondo temperature** T_{K}. In dilute alloys, the characteristic energy scale $\in \cong T_{\mathrm{K}}$ related to the resonance at $\in_{\mathrm{F}}$ usually is related with the Kondo effect—the resonance scattering of an electron on a magnetic impurity with a simultaneous change of spin projection.

The theoretical framework for understanding the physics below T_{K} began in the late 1960s from Phil Anderson's idea of **scaling** in the Kondo problem, where scaling assumes the low-temperature properties of a real system to be adequately represented by a coarse-grained model. As the temperature is lowered, the model becomes coarser, and its number of degrees of freedom is reduced. Later, in 1974, Kenneth Wilson [95, 83] devised a **numerical renormalization method** that overcame the shortcomings of conventional perturbation theory and confirmed the scaling hypothesis as well as proving that at temperatures well below T_{K}, the magnetic moment of the impurity ion is screened entirely by the spins of the electrons in the metal. This spin-screening imitates the screening of an electric charge inside a metal, though with very different microscopic processes. Because it is impossible to form a bound state from the impurity spin and single conduction electron, interaction leads to a complicated many-body scattering state—a screening cloud made of conduction electrons. For the impurity, spin equal to one-half ground state of the system becomes a singlet. Its energy is proportional to the Kondo temperature and depends nonanalytically on the strength of the coupling, which demonstrates a breakdown of the conventional perturbation theory.

We therefore observe that the Kondo effect only arises when the defects are magnetic in nature. This implies that the total spin of all the electrons in the impurity atom is nonzero. The given electrons coexist with the mobile electrons in the host metal and behave like a Fermi sea—with all the states with energies below the Fermi level occupied, while the higher-energy states are empty.

In this chapter, we develop the low-energy theory of the **Kondo impurity system** and describe the interaction of a local impurity with an itinerant band of carriers. We determine an effective Hamiltonian for the coupled system, describing the spin exchange interaction acting between the local moment of the impurity state and the itinerant band. We apply perturbation theory methods to explore the impact of magnetic fluctuations on transport while explaining the mechanism responsible for the observed minimum of electrical resistance found in **magnetic quantum impurity systems**.

A **quantum impurity** is a collection of discrete quantum states coupled to a continuum of noninteracting degrees of freedom, with the latter being either fermions or bosons or both. A very simple (and important) example would be to consider one quantum spin $\mathbf{S} = \frac{1}{2}$ coupled to fermions described by the following Hamiltonian

$$\mathrm{H} = \sum_{\vec{\kappa}\sigma} \xi(\vec{\kappa}) \hat{\psi}^{\dagger}_{\vec{\kappa}\sigma} \hat{\psi}_{\vec{\kappa}\sigma} + \mathrm{H}^{J} \quad , \quad \sigma = \uparrow, \downarrow \tag{2636}$$

where the first term is the kinetic energy of free fermions in a band width Λ measured relative to the Fermi surface, and

$$\mathrm{H}^{J} = -J \vec{\mathbf{S}}_i \vec{\mathbf{s}}(0) \tag{2637}$$

describes the coupling of an impurity (localized) spin $\mathbf{S}$ where the local spin density of the fermions (itinerant electrons):

$$\vec{\mathbf{s}}(\vec{r}) = \frac{1}{2} \sum_{\vec{\kappa}\vec{\kappa}'\sigma\sigma'} \vec{\sigma}_{\sigma\sigma'} \hat{\psi}^{\dagger}_{\vec{\kappa}'\sigma'} \hat{\psi}_{\vec{\kappa}\sigma} \tag{2638}$$

Here, σ and σ' denote the spin of the electron before and after, respectively, the scattering event. We can rewrite (2636) as follows

$$\mathrm{H} = \sum_{\vec{\kappa}\sigma} \xi(\vec{\kappa}) \hat{\psi}^{\dagger}_{\vec{\kappa}\sigma} \hat{\psi}_{\vec{\kappa}\sigma} - \frac{J}{2} \sum_{\vec{\kappa}\vec{\kappa}'\sigma\sigma'} \hat{\psi}^{\dagger}_{\vec{\kappa}'\sigma'}\left(\vec{R}_i\right) \vec{\sigma}_{\sigma\sigma'} \hat{\psi}_{\vec{\kappa}\sigma}\left(\vec{R}_i\right) \vec{\mathbf{S}}_i \quad , \quad \sigma = \uparrow, \downarrow \tag{2639}$$

Here, $\vec{\sigma} = \left(\sigma^x, \sigma^y, \sigma^z\right)^{\mathrm{T}}$ denotes the vector formed by three Pauli matrices that stem from spin polarization of the conduction-electron spin density:

$$\sigma^x = \begin{pmatrix} 0 & 1 \\ 1 & 0 \end{pmatrix} \quad \sigma^y = \begin{pmatrix} 0 & -i \\ i & 0 \end{pmatrix} \quad \sigma^z = \begin{pmatrix} 1 & 0 \\ 0 & -1 \end{pmatrix} \tag{2640}$$

The spin of the impurity, $\vec{\mathbf{S}}_i$, is regarded as a fixed c-number, which sometimes is termed f-spin because they typically are realized in heavy fermion systems by the rare-earth 4f electrons; $\vec{R}_i$, is the impurity position, while J is the strength of the electron-impurity interaction. Here, values with $J > 0$ are called **ferromagnetic** because the local spin tends to line up parallel with the conduction band spins, and values with $J < 0$ are called **antiferromagnetic** because the local spin tends to line up antiparallel with the conduction band spins; $\hat{\psi}^{\dagger}_{\vec{\kappa}\sigma}$ creates a conduction electron of energy $\xi(\vec{\kappa})$ and momentum $\vec{\kappa}$ with

$$\vec{\sigma}\vec{\mathbf{S}} = \sigma^x S^x + \sigma^y S^y + \sigma^z S^z \tag{2641}$$

We examine the interaction between magnetic impurity embedded into a metal and the conduction band Fermi sea described by the following effective interaction Hamiltonian:

$$\mathrm{H} = -\frac{J}{2} \sum_{i} \vec{\sigma} \vec{S}_i \delta\left(\vec{r} - \vec{R}_i\right) \tag{2642}$$

The sum in (2642) runs over all impurities in the system ($\vec{R}_i$ are the impurity positions). It is instructive to note that the best-known kind of **pair-breaking effect** (for a simple s-wave-paired superconductor) is that of random magnetic impurities. Such impurities may be regarded as coupling to the conduction-electron system via an interaction of the form (2642). The presence of the term (2642) renders the single-particle Hamiltonian no longer invariant under time reversal. Unlike the case of a uniform **Zeeman field**, the time reversal of an energy eigenstate generally is not an eigenstate itself. So, magnetic impurities are indeed **pair-breaking**.

13.2.13.1.1 Kondo Model

Considering the fact that perturbative manipulations on quantum spins are difficult to formulate in a field integral language, for the perturbation theory of the Kondo effect, the traditional formalism of second quantized operators is superior to the field integral. Precisely, the method of choice would be second-quantized perturbation theory formulated in the language of the interaction picture. Kondo suggested this in his study on the simplest model considering the local exchange interaction J between the magnetic impurity and itinerant electrons at the impurity site. So, the scattering of electrons by a single impurity is described by the Hamiltonian (2642) model bearing Kondo's name.

The electron scattering from a nonmagnetic impurity contributes to the resistivity that is independent of temperature. However, as mentioned earlier, a magnetic impurity causes a resistance minimum at a nonzero temperature as a result of the spin-flip scattering between the conduction electrons and the localized spin.

We consider the scattering of a single fermion above the Fermi sea, described by the state

$$\left|\vec{\kappa}\sigma\mathbf{S}\right\rangle = \hat{\psi}^{\dagger}_{\vec{\kappa}\sigma}\left|\mathrm{FS}\right\rangle \otimes \left|\mathbf{S}\right\rangle_{\mathrm{imp}} \tag{2643}$$

where, $\left|\mathrm{FS}\right\rangle$ is the Fermi sea ground state of the Fermi gas while $\left|\mathrm{S}\right\rangle_{\mathrm{imp}}$ describes the impurity spin. We use the formalism of second-quantized operators in the perturbation theory of the Kondo effect because perturbative manipulations on quantum spins are difficult to manipulate in a field integral language. So, our formulation should be on second-quantized perturbation theory formulated in the interaction picture. We then consider a time-dependent perturbation theory in the coupling between the impurity and the fermions, and then we compute the scattering amplitude by expanding the interaction picture evolution operator

$$\mathbf{U}_{\mathrm{I}}(t) = \hat{\mathrm{T}}\exp\left\{-i\int_0^t \mathrm{H}^J(t')dt'\right\} \tag{2644}$$

with $\hat{\mathrm{T}}$ being the time-ordering operator, while

$$\mathrm{H}^J(t) = -\frac{J}{2}\sum_{\vec{\kappa}\vec{\kappa}'\sigma\sigma'} \vec{\mathbf{S}}_i(t)\vec{\sigma}_{\sigma\sigma'}\hat{\psi}^{\dagger}_{\vec{\kappa}'\sigma'}(t)\hat{\psi}_{\vec{\kappa}\sigma}(t) \tag{2645}$$

where

$$\hat{\psi}_{\vec{\kappa}\sigma}(t) = \exp\left\{-it\xi(\vec{\kappa})\right\}\hat{\psi}_{\vec{\kappa}\sigma} \quad , \quad \hat{\psi}^{\dagger}_{\vec{\kappa}\sigma}(t) = \exp\left\{it\xi(\vec{\kappa})\right\}\hat{\psi}^{\dagger}_{\vec{\kappa}\sigma} \tag{2646}$$

Because the impurity spin does not appear in the unperturbed Hamiltonian, $\vec{\mathbf{S}}_i(t)$ has no time dependence. However, we label it to keep track of the order in which it appears, coupled to operators with true time dependence, when we expand the evolution operator in equation (2644).

13.2.13.2 Localized and Itinerant Spins Interaction

Let us examine the interaction between localized and itinerant spins described by the following effective interaction Hamiltonian that considers **scattering on the impurity beyond the Born approximation**

$$\mathrm{H} = \sum_{\vec{\kappa}\sigma} \xi(\vec{\kappa}) \hat{\psi}^{\dagger}_{\vec{\kappa}\sigma} \hat{\psi}_{\vec{\kappa}\sigma} + \mathrm{H}^{J} \quad , \quad \sigma = \uparrow, \downarrow \tag{2647}$$

where the first term is the kinetic energy of free fermions and the effective interaction due to scattering from local spin at the site $\vec{R}_m$:

$$\mathrm{H}^{J} = -\frac{J}{2} \sum_{mm'} \hat{\psi}^{\dagger}_{\sigma'}\left(\vec{R}_m\right) \vec{\sigma}_{\sigma\sigma'} \hat{\psi}_{\sigma}\left(\vec{R}_m\right) \vec{\mathbf{S}}_{mm'} \delta\left(\vec{r} - \vec{R}_m\right) \equiv -\frac{J}{2} \sum_{m} \hat{\psi}^{\dagger}_{\sigma'}\left(\vec{R}_m\right) \vec{\sigma}_{\sigma\sigma'} \hat{\psi}_{\sigma}\left(\vec{R}_m\right) \vec{\mathbf{S}}_{m} \tag{2648}$$

It is obvious from here that spin-flip processes, which change the spin state of the impurity and that of the scattered electron, are enabled. Here, $\vec{\mathbf{S}}_m$ is the local spin operator for the magnetic impurity (atomic shells) at impurity position $\vec{R}_m$:

$$\vec{\mathbf{S}}_m = \sum_{\alpha\alpha'} \hat{f}^{\dagger}_{\alpha}\left(\vec{R}_m\right) \vec{\mathbf{S}}_{\alpha\alpha'} \hat{f}_{\alpha'}\left(\vec{R}_m\right) \tag{2649}$$

The pseudo-fermion operators $\hat{f}^{\dagger}_{\alpha}$ and $\hat{f}_{\alpha'}$ each create or destroy, respectively, one **pseudo-fermion**. This implies, writing $|0\rangle$ for the pseudo-fermion **vacuum** state (a nonphysical state), with

$$|\alpha\rangle = \hat{f}^{\dagger}_{\alpha} |0\rangle \tag{2650}$$

representing the impurity in the spin state labeled α. Here, $\vec{\mathbf{S}}_{\alpha\alpha'}$ is the matrix element of the spin $\vec{\mathbf{S}}$. The quasiparticle spin

$$\vec{\sigma} = \left(\sigma^x, \sigma^y, \sigma^z\right)^{\mathrm{T}} \tag{2651}$$

is the Pauli spin matrices and

$$\vec{\sigma}\vec{\mathbf{S}} = \sigma^x S^x + \sigma^y S^y + \sigma^z S^z \quad , \quad \left(\sigma^x\right)^2 + \left(\sigma^y\right)^2 + \left(\sigma^z\right)^2 = 1 \tag{2652}$$

If a local spin is from the d-shell then, usually, the conduction band is formed from atomic orbitals having s- and p-symmetry; if a local spin is from an f-orbital, then the conduction band could be from s-, p-, or d-electrons.

We develop a perturbative expansion in $\frac{J}{2}$ to explore the scattering properties of the Kondo model where we assume that the exchange constant $\frac{J}{2}$ is characterized by a single parameter that is positive in sign (i.e., antiferromagnetic). For the calculation of the resistivity, we need the scattering amplitude of the Kondo Hamiltonian (2647) that, in first-order approximation, is independent of momentum $\vec{\kappa}$ and energy $\in$ (Figure 13.12):

$$\Gamma^{(1)} = -\frac{J}{2}\left(\vec{\sigma}_{\sigma\sigma'} \vec{\mathbf{S}}_{mm'}\right) \tag{2653}$$

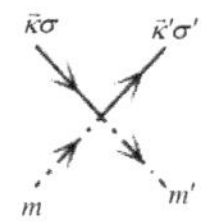

FIGURE 13.12 First-order scattering amplitude. The conduction electron propagators are denoted by solid, pseudo-fermion propagators (dashed lines).

Let us now consider scattering of an electron from an initial state $|\vec{\kappa}\sigma\rangle$ (where $\vec{\kappa}$ is its momentum and σ its spin) to a final state $|\vec{\kappa}'\sigma'\rangle$ through an intermediate state $|\vec{\kappa}_1\sigma_1\rangle$ (13.3). For this to happen, we have two possibilities:

- The electron first gets scattered into the intermediate state $|\vec{\kappa}\sigma\rangle \to |\vec{\kappa}_1\sigma_1\rangle$ and then to the final state $|\vec{\kappa}_1\sigma_1\rangle \to |\vec{\kappa}'\sigma'\rangle$ with $\vec{\kappa}_1$ being above the Fermi surface. In order to calculate the scattering amplitude for this process, we have to bear in mind that the intermediate state has to be unoccupied by a factor $1-f(\vec{\kappa}_1)$, where $f(\vec{\kappa})$ is the Fermi distribution function. Taking the sum over all intermediate states, we have for the scattering amplitude

$$\Gamma_1^{(2)} = \left(\frac{J}{2}\right)^2 \sum_{\vec{\kappa}_1\sigma_1} \frac{\left(\vec{\sigma}\vec{S}\right)_{\sigma'\sigma_1}\left(\vec{\sigma}\vec{S}\right)_{\sigma_1\sigma}\left(1-f(\vec{\kappa}_1)\right)}{\xi(\vec{\kappa})-\xi(\vec{\kappa}_1)} \tag{2654}$$

- An electron from the already occupied intermediate state gets scattered into the final state $|\vec{\kappa}_1\sigma_1\rangle \to |\vec{\kappa}'\sigma'\rangle$ and the initial electron fills up the now free intermediate state $|\vec{\kappa}\sigma\rangle \to |\vec{\kappa}_1\sigma_1\rangle$.

We observe that the indistinguishability (and hence fermionic statistics) play a fundamental role in the second process. For this process, the scattering amplitude is written as

$$\Gamma_2^{(2)} = -\left(\frac{J}{2}\right)^2 \sum_{\vec{\kappa}_1\sigma_1} \frac{\left(\vec{\sigma}\vec{S}\right)_{\sigma_1\sigma}\left(\vec{\sigma}\vec{S}\right)_{\sigma'\sigma_1} f(\vec{\kappa}_1)}{\xi(\vec{\kappa}_1)-\xi(\vec{\kappa}')} \tag{2655}$$

The minus sign takes the asymmetry of the electronic wave function into account as the particles are permutated considering (2654).

Assuming elastic scattering, then we have

$$\xi(\vec{\kappa}) = \xi(\vec{\kappa}') \tag{2656}$$

Employing the commutator and eigenvalue relations for the spin operators, we have

$$\left(\vec{\sigma}\vec{S}\right)\left(\vec{\sigma}\vec{S}\right) = S(S+1) - \vec{\sigma}\vec{S} \tag{2657}$$

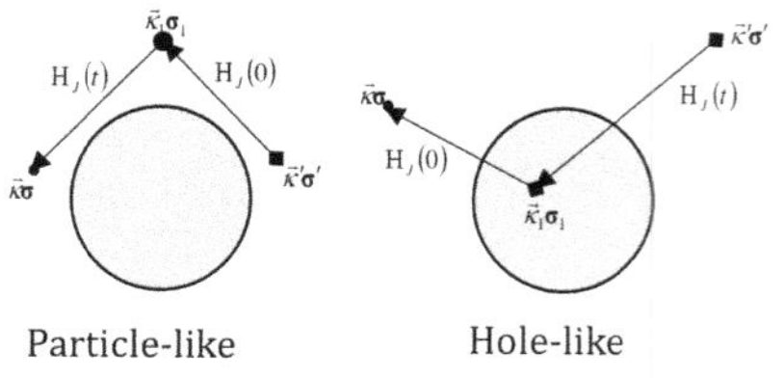

FIGURE 13.13 Scattering of an electron from an initial state $|\vec{\kappa}\sigma\rangle$ (where $\vec{\kappa}$ is its momentum and σ its spin) to a fnal state $|\vec{\kappa}'\sigma'\rangle$ through an intermediate state $|\vec{\kappa}_1\sigma_1\rangle$.

From (2656) and (2657), at finite temperature, the scattering amplitudes (2654) and (2655) can be written as follows

$$\Gamma^{(2)} = \Gamma_1^{(2)} + \Gamma_2^{(2)} = \left(\frac{J}{2}\right)^2 \sum_{\vec{\kappa}_1} \frac{1}{\xi(\vec{\kappa}) - \xi(\vec{\kappa}_1)} \left(S(S+1)\delta_{\sigma'\sigma} + \left(2f(\vec{\kappa}_1) - 1\right)\left(\vec{\sigma}\vec{S}\right)_{\sigma'\sigma} \right) \quad (2658)$$

Considering that $f(\vec{\kappa}_1)$ only depends on $\xi(\vec{\kappa}_1) \equiv \xi_1$ then,

$$\Gamma^{(2)} = \left(\frac{J}{2}\right)^2 \int \frac{d\xi_1}{\xi - \xi_1} \left(S(S+1)\delta_{\sigma'\sigma} + \left(2f(\xi_1) - 1\right)\left(\vec{\sigma}\vec{S}\right)_{\sigma'\sigma} \right) \frac{\nu(\in)}{2} \quad (2659)$$

Here, $\nu(\in)$ is the density of states that is considered constant ($\nu(\mu)$) at the vicinity of the Fermi energy $\in \approx \mu$. Upon integration over ξ_1, the first term in the integrand of (2659) yields a value of order $\frac{\xi}{\mu}$. This can be neglected when we consider only electrons at the vicinity of the Fermi energy. The second term is antisymmetric with respect to ξ_i and easily shows that

$$2f(\xi_1) - 1 = \frac{2}{\exp\{\beta\xi_1\} + 1} - 1 = \frac{2\exp\left\{-\frac{\beta\xi_1}{2}\right\}}{\exp\left\{\frac{\beta\xi_1}{2}\right\} + \exp\left\{-\frac{\beta\xi_1}{2}\right\}} - 1 = \frac{-\sinh\frac{\beta\xi_1}{2}}{\cosh\frac{\beta\xi_1}{2}} = -\tanh\frac{\beta\xi_1}{2} \quad (2660)$$

We consider the asymmetry of the integrand in the integral (2659) and take the limits of the integral to be $\pm\mu$. So, considering (2659), then

$$\int_{-\mu}^{\mu} d\xi_1 \frac{2f(\xi_1) - 1}{\xi - \xi_1} = -\int_{-\mu}^{\mu} d\xi_1 \tanh\left(\frac{\beta\xi_1}{2}\right) \frac{1}{\xi - \xi_1} = -\int_0^{\mu} d\xi_1 \tanh\left(\frac{\beta\xi_1}{2}\right)\left(\frac{1}{\xi - \xi_1} - \frac{1}{\xi + \xi_1}\right) \quad (2661)$$

or

$$\int_{-\mu}^{\mu} d\xi_1 \frac{2f(\xi_1) - 1}{\xi - \xi_1} = -\int_0^{\mu} d\xi_1 \tanh\left(\frac{\beta\xi_1}{2}\right) \frac{2\xi_1}{\xi^2 - \xi_1^2} \quad (2662)$$

For $|\xi_1| >> \xi$, the ξ^2 in the denominator of (2662) may be safely ignored. Also, for $\xi_1 >> T$, the Fermi distribution $f(\xi_1)$ vanishes and the integral becomes a logarithmic function justifying the choice of the limits $\pm\mu$ because for logarithmic integrals, it is sufficient to know only the order of their limits. So, the integral (2662) becomes

$$\int_{-\mu}^{\mu} d\xi_1 \frac{2f(\xi_1) - 1}{\xi - \xi_1} = 2\ln\left(\frac{\mu}{\max(|\xi|, T)}\right) \quad (2663)$$

and depends sensitively on the bandwidth 2μ of the itinerant electrons and on the energy $|\xi| = \in$ of the reference state. So, the contribution of the Born approximation to the scattering amplitude:

$$\Gamma = \Gamma^{(1)} + \Gamma^{(2)} + \cdots \quad (2664)$$

with

$$\Gamma^{(1)} = -\frac{J}{2}\left(\vec{\sigma}\vec{S}\right)_{\sigma'\sigma} \quad , \quad \Gamma^{(2)} = 2\left(\frac{J}{2}\right)^2 \nu(\mu)\ln\left(\frac{\mu}{\max\left(|\xi|,T\right)}\right)\left(\vec{\sigma}\vec{S}\right)_{\sigma'\sigma} \tag{2665}$$

So,

$$\Gamma = -\frac{J}{2}\left(\vec{\sigma}\vec{S}\right)_{\sigma'\sigma}\left[1 - J\nu(\mu)\ln\left(\frac{\mu}{\max\left(|\xi|,T\right)}\right) + \cdots\right] \tag{2666}$$

In normal metals, the behavior of conduction electrons is well described by Landau's theory of Fermi liquids, which predicts that when the temperature, T, decreases, the electrical resistivity, ρ, of the metal drops for small T according to

$$\rho(T) = \rho_0 + aT^2 + bT^5 \tag{2667}$$

The resistivity, ρ_0, is due to impurities and defects in the metal; the aT^2 term describes scattering of electrons against other electrons, while the bT^5 term describes electron-phonon scattering where b and a are constants. But, when some metals are doped with impurities having magnetic moments, there is an anomalous increase in electrical resistance for decreasing T. Apart from converging to a constant value as in (2667), the resistance again increases below some specific temperature, $T_{\min}$, and stops at a finite value at $T = 0$. Hence, a pronounced resistivity minimum arises. The increase in resistivity below $T_{\min}$ is due to electron spin-flip scattering events against the inserted magnetic impurities:

$$\rho(T) = \rho_0 + na\ln\frac{\epsilon_F}{T} + bT^5 \tag{2668}$$

The last summand is the phonon contribution (Bloch T^5 law), the second is spin-dependent contribution to the resistivity, and the first, ρ_0 is the nonmagnetic contribution due to the particle-particle interaction that is temperature independent.

From (2668), we have

$$\rho'(T) = -\frac{na}{T} + 5bT^4 = 0 \tag{2669}$$

and

$$T_{\min} = \left(\frac{na}{5b}\right)^{\frac{1}{5}} \tag{2670}$$

This temperature, at which the electrical resistivity achieves a minimum, varies as one-fifth power of the concentration of the magnetic impurities, in agreement with experiments (at least for Cu diluted with Fe). The aforementioned behavior of the resistivity is called the **Kondo effect** and is used to describe many-body scattering processes from impurities or ions that have low-energy quantum mechanical degrees of freedom.

The Kondo effect has become a key concept in condensed matter physics for understanding the behavior of metallic systems with strongly interacting electrons. From this Kondo effect, the resistance ρ is the square of the scattering amplitude Γ:

$$\rho = \Gamma^2 = \left[\frac{J}{2}\left(\vec{\sigma}\vec{S}\right)_{\sigma'\sigma}\right]^2\left[1 - J\nu(\mu)\ln\left(\frac{\mu}{\max\left(|\xi|,T\right)}\right) + \cdots\right]^2 \tag{2671}$$

or

$$\rho = \Gamma^2 = \left[\frac{J}{2}\left(\vec{\sigma}\vec{S}\right)_{\sigma'\sigma}\right]^2 \left[1 - 2J\nu(\mu)\ln\left(\frac{\mu}{\max(|\xi|, T)}\right) + \cdots\right] \tag{2672}$$

or

$$\rho = \rho_J \frac{1}{1 + 2J\nu(\mu)\ln\left(\frac{\mu}{\max(|\xi|, T)}\right)} \tag{2673}$$

Here,

$$\rho_J = \left[\frac{J}{2}\left(\vec{\sigma}\vec{S}\right)_{\sigma'\sigma}\right]^2 \tag{2674}$$

is due to scattering in first-order approximation. From (2671), we find that the resistivity diverges logarithmically with temperature. Therefore, considering that the spin exchange coupling J between the localized moments and the itinerant conduction electrons is antiferromagnetic, **breaking down of phase transition** yields

$$1 = 2|J|\nu(\epsilon_F)\ln\left(\frac{\epsilon_F}{T}\right) \tag{2675}$$

and consequently, the **Kondo temperature**

$$T_K = \epsilon_F \exp\left\{-\frac{1}{2|J|\nu(\epsilon_F)}\right\} \tag{2676}$$

This Kondo temperature is

- A crossover temperature below which the coupling between the conduction electrons and the dynamical magnetic impurity grows nonperturbatively; the resistance saturates experimentally for $T << T_K$.
- The temperature of an electron in the itinerant band combines with the electron on the impurity site to form a singlet bound state effectively screening the magnetic impurity. These states modify the density of a state's spectrum where the low-energy physics is reconciled by the so-called Kondo resonance, which is manifested by a temperature-dependent peak (sharpness) at the Fermi level in the spectrum.

The expression of the Kondo temperature, T_K, imitates that for the change in ground state energy. The difference is the factor of 2/3 in the exponent that is now 1/2. The general definition is the change in ground state energy that defines an energy scale that also defines a characteristic temperature. Though the perturbation theory predicts divergence of the resistivity with temperature, the results are valid only up to the characteristic **Kondo temperature scale** T_K where the logarithmic correction is of the order of the first term.

Consider various localized magnetic moments in a metal that are arranged on a lattice. The same spin-exchange coupling J inducing the Kondo effect also induces a magnetic interaction among the localized spins. Hence, local moments can exchange their spins mediated by two conduction electrons scattering from and traveling among the impurity sites. This yields an effective, long-range spin-exchange coupling involving two elementary scattering events between electron and impurity spins. This can be ferromagnetic as well as antiferromagnetic due to the long-range, spatial oscillations of the conduction electron density correlations. The conduction-electron-mediated spin interaction was first studied by Ruderman, Kittel, Kasuya and Yosida (RKKY) [37, 39, 96]. It is therefore called **RKKY** interaction. This interaction sometimes dominates the magnetic dipole-dipole coupling as well as the direct exchange coupling between neighboring local moments. This is due to the short spatial extent of these couplings or to the exponentially small overlap of the local moment wave functions on neighboring lattice sites.

13.2.13.3 Ruderman-Kittel-Kasuya-Yosida (RKKY) Interaction

We have seen earlier that there are impurity atoms inducing so-called resonant scattering. If this resonance occurs close to the Fermi energy, then the scattering rate is strongly energy dependent. This induces a more pronounced temperature dependence of the resistivity. An example is the scattering of magnetic impurities with a spin degree of freedom, which yields a dramatic energy dependence of the scattering rate. We consider again the following Hamiltonian describing the **Kondo lattice model**

$$\mathrm{H} = \sum_{\vec{\kappa}\sigma} \epsilon_{\vec{\kappa}} \hat{\psi}^{\dagger}_{\vec{\kappa}\sigma} \hat{\psi}_{\vec{\kappa}\sigma} + \mathrm{H}^{J} \tag{2677}$$

where

$$\mathrm{H}^{J} = -\frac{J}{2} \sum_{\vec{\kappa}\vec{\kappa}'\sigma\sigma' k} \exp\left\{ i\left(\vec{\kappa}' - \vec{\kappa}\right)\vec{r}_i \right\} \vec{\sigma}_{\sigma\sigma'} \vec{\mathbf{S}}_k \hat{\psi}^{\dagger}_{\vec{\kappa}'\sigma} \hat{\psi}_{\vec{\kappa}\sigma'} \tag{2678}$$

Here, we have added phase factors associated with the location of each different impurity. We assume $T > T_{\mathrm{K}}$, and ignore singlet formation, and then start from a **ground state** of a filled Fermi sea and localized spin impurities. We find an effective interaction among the localized spins due to creation and annihilation of electron-hole spin-flip pairs by eliminating excitations of the filled Fermi sea. From the perturbation theory, we find only those terms such as $\mathbf{S}_i^{+}\mathbf{S}_k^{-}$ and keep only terms that return the Fermi sea to its original state. This term will be indicative of the full spin-spin interaction. So, for randomly located impurities, we have the interaction term

$$\mathrm{H}_{SS} = -2\left(\frac{J}{2}\right)^2 \sum_{i \neq k} J_{ik} \vec{\mathbf{S}}_i \vec{\mathbf{S}}_k \tag{2679}$$

where the effective interaction coupling strength:

$$J_{ik} = \sum_{\vec{\kappa}\vec{\kappa}'} \frac{f(\vec{\kappa})\left(1 - f(\vec{\kappa}')\right)}{\epsilon_{\vec{\kappa}} - \epsilon_{\vec{\kappa}'}} \exp\left\{ i\left(\vec{\kappa} - \vec{\kappa}'\right)\left(\vec{r}_i - \vec{r}_k\right) \right\} \tag{2680}$$

Consider the coordinates of relative motion

$$\vec{q} = \vec{\kappa} - \vec{\kappa}' \quad , \quad \vec{R}_{ik} = \vec{r}_i - \vec{r}_k \tag{2681}$$

Here, $\vec{R}_{ik}$ is the distance between the moments $\vec{\mathbf{S}}_i$ and $\vec{\mathbf{S}}_k$. For the center of mass

$$\vec{q}' = \frac{\vec{\kappa} + \vec{\kappa}'}{2} \tag{2682}$$

So,

$$\vec{\kappa} = \vec{q}' + \frac{\vec{q}}{2} \quad , \quad \vec{\kappa}' = \vec{q}' - \frac{\vec{q}}{2} \tag{2683}$$

and

$$J_{ik} = \sum_{\vec{q}} h(\vec{q}) \exp\{i\vec{q}\vec{R}_{ik}\} \tag{2684}$$

is the interaction strength that oscillates with the distance $\vec{R}_{ik}$ and its sign depends on $\vec{q}\vec{R}_{ik}$. The RKKY interaction alone can lead to ferro-, antiferro-, or helimagnetism and, in heavy fermions, the magnetic order is often antiferromagnetic. The bare Lindhard function is

$$h(\vec{q}) = \sum_{\vec{q}'} \frac{f\left(\vec{q}' + \frac{\vec{q}}{2}\right)\left(1 - f\left(\vec{q}' - \frac{\vec{q}}{2}\right)\right)}{\epsilon_{\vec{q}' + \frac{\vec{q}}{2}} - \epsilon_{\vec{q}' - \frac{\vec{q}}{2}}} \tag{2685}$$

and the **polarization function** is

$$\Pi^0_{q,\omega=0} = -N_F h(q) \tag{2686}$$

where

$$N_F = \frac{m\kappa_F}{2\pi^2} \tag{2687}$$

is the noninteracting density of states per spin at the Fermi surface $\epsilon_{\vec{q}} = 0$. As seen earlier,

$$\epsilon_{\vec{q}' + \frac{\vec{q}}{2}} \epsilon_{\vec{q}' - \frac{\vec{q}}{2}} < 0 \tag{2688}$$

and implies the difference of the Fermi-Dirac distributions at two single-particle energies are nonvanishing at zero temperature only when one of the two single-particle states is above the Fermi energy and the other single-particle state is below the Fermi energy.

At low temperatures, the effective interaction coupling strength (2684) is thus tailored by the geometrical properties of the Fermi sea, the unperturbed ground state of the Fermi gas.

Introducing the following dimensionless variable

$$q' = \frac{|\vec{q}|}{2\kappa_F} \tag{2689}$$

then

$$\Pi^0_{q,\omega=0} = -N_F h\left(\frac{|\vec{q}|}{2\kappa_F}\right) \tag{2690}$$

and the bare **Lindhard function** is

$$h(z) = 1 + \frac{1 - z^2}{2z} \ln\left|\frac{1+z}{1-z}\right| \tag{2691}$$

Hence, in (2684), we find the presence of logarithmic singularities when $|\vec{q}| = 2\kappa_F$, where there are spin density waves for the low-dimensional cases. These are responsible for the so-called Ruderman-Kittel-Kasuya-Yosida (RKKY) oscillations of the static spin susceptibility induced by a magnetic impurity in a free electron gas. The lower the dimension, the stronger the singularity at $q = 2\kappa_F$.

It appears that

$$h(z \to 0) = 1 \quad , \quad h(z \to \infty) = \frac{1}{2z^2} \tag{2692}$$

So, for the three-dimensional (3D) RKKY interaction, then

$$J(z) = -\left(\frac{\cos x}{z^3} - \frac{\sin x}{z^4}\right) \quad , \quad z = \frac{2q_F R_{ik}}{\hbar} \tag{2693}$$

and

$$\lim_{z \to 0} J(z) = \frac{1}{3z} \quad , \quad \lim_{z \to \infty} J(z) = -\frac{\cos z}{z^3} \tag{2694}$$

We observe that the sign of the RKKY interaction shows oscillatory change that depends on the separation distance between two impurities. Any magnetic order or glassing behavior quenches the Kondo effect. Hence, any magnetic transition quenches the Kondo effect. From (2684), we deduce that $h(q')$ characterizes correlations and maps the susceptibility of free electron gas $\chi_0(q')$:

$$\chi(q') = \frac{\chi_0(q')}{1 - \lambda\chi_0(q')} \tag{2695}$$

Therefore, we observe from the aforementioned that the interaction term has random sign and magnitude. This disordered spin-spin coupling can lead to glassy behavior, with very many nearly degenerate

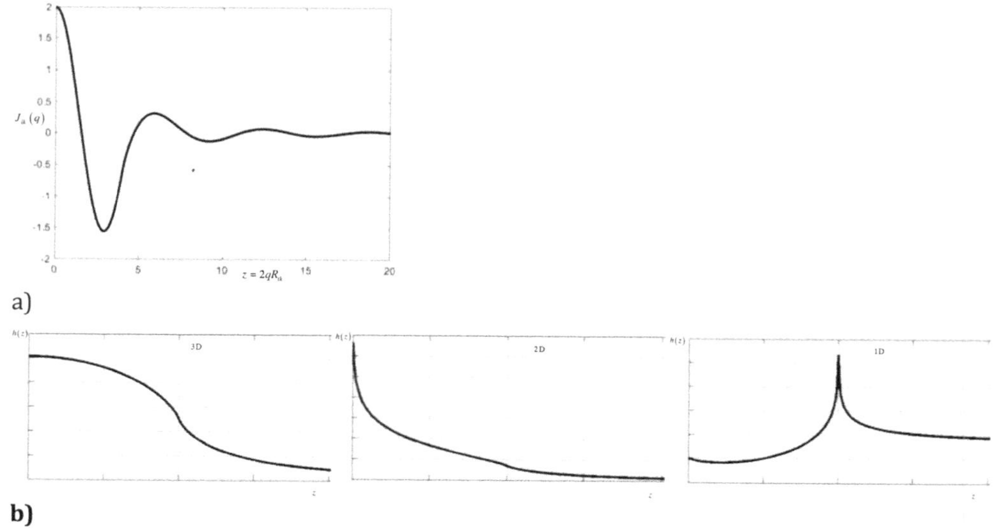

FIGURE 13.14 **(a)** Effective interaction coupling strength versus separation distance between two impurities and **(b)** Bare lindhard functions for different dimensions.

ground states. The Kondo coupling yields a paramagnetic Fermi liquid state without local moments. For the given state, the local orbitals (whose spectrum achieves the Kondo resonance at the Fermi energy) hybridize with each other and eventually achieve lattice coherence at low temperatures, forming Bloch-like quasiparticle states. Consequently, a narrow band crossing the Fermi energy is formed where the bandwidth is controlled by the Kondo resonance width T_K. This yields an exponentially strong effective mass enhancement (**heavy Fermi liquid**).

By contrast, the RKKY interaction tends to induce magnetic order of the local moments. So, the Kondo spin screening of the local moments should eventually break down, giving rise to magnetic order, when the RKKY coupling energy becomes larger than the characteristic energy scale for Kondo singlet formation, the Kondo temperature T_K. Hence, for $T = 0$, the quantum phase transition (QPT) occurs with the local spin exchange coupling J serving as the control parameter.

Though the reason for the occurrence of the Kondo breakdown at a magnetic QPT is controversial, several QPT scenarios in heavy-fermion systems are achievable.

1. The instability occurs at $q = 2\kappa_F$ where there are spin density waves (SDW) leading to critical fluctuations of the bosonic magnetic order parameter but leaving the fermionic, heavy quasiparticles intact.
2. The magnetization local fluctuations, coupling to the nearly localized, heavy quasiparticles, may become critical (divergent) and thereby destroy the heavy Fermi liquid (local quantum criticality) [40].
3. At phase transition, the Kondo effect and, hence, the heavy-fermion band, vanish. This yields an abrupt change in the Fermi surface (**Fermi volume collapse**). The Fermi surface fluctuations associated with this change self-consistently destroy the Kondo singlet state [41].
4. Recently, a scenario of critical quasiparticles characterized by a diverging effective mass and a singular quasiparticle interaction has been self-consistently generated by the nonlocal order-parameter fluctuations of an impending SDW instability [42–45].

Though scenario (1) describes a critical field theory of the bosonic, magnetic order parameter alone, a complete understanding of breakdown scenarios (2), (3), and (4) requires fermionic degrees of freedom field theory forming the Kondo effect and heavy quasiparticles coupled with the bosonic-order parameter field. In the absence of such a complete theory, these scenarios presume specific fluctuations: (2) local fluctuations, (3) Fermi surface fluctuations, or (4) antiferromagnetic fluctuations become **soft** for certain values of the system parameters and, thus, dominate the QPT. So, conditions for realization of these scenarios are controversial.

13.2.13.4 Abrikosov Technique

Pseudo-Fermion Representation of Spin

We use the Green's function techniques to calculate the self-energy and lifetime of electrons scattered by magnetic impurities. We consider small enough temperatures that quench electron-phonon scattering but not so small that perturbation theory breaks down. To apply a field theoretical treatment imitating the standard functional integral or Wick theorem and many-body perturbation theory, it is necessary for the corresponding field operators to obey canonical commutation rules. This implies that their (anti) commutators should be proportional to the unit operator. Nevertheless, the spin operators $\vec{\mathbf{S}}$ obey the $SU(2)$ algebra that leads to absence of a Wick theorem for the generators. This difficulty can be overcome by using the spin fermionic representation introduced for the first time by Abrikosov [46, 97]. We introduce the pseudo-fermion operators $\hat{f}_\sigma^\dagger$ and $\hat{f}_{\sigma'}$ for each of the basis states spanning the impurity spin Hilbert space, $|\sigma\rangle, \sigma = \uparrow, \downarrow$, which create or destroy, respectively, one **pseudo-fermion** (as seen earlier) and imply writing $|0\rangle$ for the pseudo-fermion **vacuum** state (a nonphysical state),

$$|\sigma\rangle = \hat{f}_\sigma^\dagger |0\rangle \tag{2696}$$

representing the impurity in the spin state labeled by σ. Here,

$$\sigma = -S, \cdots, S \tag{2697}$$

labels the eigenstates of S^z. So, the combination $\hat{f}_\sigma^\dagger \hat{f}_{\sigma'}$ changes the impurity from spin-state σ' to σ:

$$\hat{f}_\sigma^\dagger \hat{f}_{\sigma'} |\sigma'\rangle = |\sigma\rangle \tag{2698}$$

The impurity spin operator $\vec{\mathbf{S}}$ in a compact form is written via **Abrikosov pseudo-fermion representation**:

$$\vec{\mathbf{S}} = S \sum_{\sigma\sigma'} \hat{f}_\sigma^\dagger \vec{\sigma}_{\sigma\sigma'} \hat{f}_{\sigma'} \tag{2699}$$

For $S = \frac{1}{2}$, then

$$\vec{\mathbf{S}} = \frac{1}{2} \sum_{\sigma\sigma'} \hat{f}_\sigma^\dagger \vec{\sigma}_{\sigma\sigma'} \hat{f}_{\sigma'} \tag{2700}$$

This implies, the components of the spin operator $\vec{\mathbf{S}}$ expressed in the pseudo-fermion representation:

$$\mathbf{S}^+ = \hat{f}_\uparrow^\dagger \hat{f}_\downarrow \quad , \quad \mathbf{S}^- = \hat{f}_\downarrow^\dagger \hat{f}_\uparrow \quad , \quad \mathbf{S}^z = \frac{1}{2}\left(\hat{f}_\uparrow^\dagger \hat{f}_\uparrow - \hat{f}_\downarrow^\dagger \hat{f}_\downarrow\right) = \frac{1}{2}\left(n_\uparrow - n_\downarrow\right) \quad , \quad \left[\mathbf{S}^+, \mathbf{S}^-\right] = 2\mathbf{S}^z \tag{2701}$$

So, $\mathbf{S}^x$ and $\mathbf{S}^y$ are given by

$$\mathbf{S}^x = \frac{1}{2}\left(\mathbf{S}^- + \mathbf{S}^+\right) = \frac{1}{2}\left(\hat{f}_\downarrow^\dagger \hat{f}_\uparrow + \hat{f}_\uparrow^\dagger \hat{f}_\downarrow\right) \quad , \quad \mathbf{S}^y = \frac{1}{2}\left(\mathbf{S}^- - \mathbf{S}^+\right) = \frac{i}{2}\left(\hat{f}_\downarrow^\dagger \hat{f}_\uparrow - \hat{f}_\uparrow^\dagger \hat{f}_\downarrow\right) \tag{2702}$$

The operators $\hat{f}_\sigma^\dagger$ and $\hat{f}_{\sigma'}$ obey the usual fermionic anticommutation relations

$$\left\{\hat{f}_\sigma, \hat{f}_{\sigma'}^\dagger\right\} = \delta_{\sigma\sigma'} \tag{2703}$$

Operators on the left-hand side and the right-hand side of equation (2703) have identical matrix elements in the physical spin Hilbert space. Nevertheless, repeated action of the fermionic operators would lead to unphysical double occupancy or no occupancy of the spin states $|\uparrow\rangle, |\downarrow\rangle$. So, for $S = \frac{1}{2}$, the **spurious states** are:

$$|00\rangle \quad , \quad |11\rangle \tag{2704}$$

and the **physical states** are:

$$|10\rangle = |\uparrow\rangle, |01\rangle = |\downarrow\rangle \tag{2705}$$

So, **only the singly occupied fermion states have any physical relevance.**

To apply the perturbation theory and Wick theorem for expressions involving spin operators, we **ensure that unphysical pseudo-fermion (spurious) states do not contribute to thermal averages.**

So, dynamics are restricted to the physical spin space by imposing the local constraint (**Abrikosov technique**) [97] considering only states with exactly one pseudo-fermion are physical:

$$\hat{Q} = \sum_{\sigma} \hat{f}^{\dagger}_{m\sigma} \hat{f}_{m\sigma} = \hat{1} \tag{2706}$$

This constraint of the pseudo-fermion number operator in (2706) originates from the fact that only the singly occupied fermion states have any physical relevance. So, this representation must be supplemented by a projection onto the physical subspace, where double and empty occupied states are excluded. The state

$$\begin{bmatrix} 0 \\ 0 \end{bmatrix} = |00\rangle \tag{2707}$$

has eigenvalue 0 and the state

$$\begin{bmatrix} 1 \\ 1 \end{bmatrix} = |11\rangle \tag{2708}$$

is eliminated by introducing λ, Abrikosov pseudo-fermions **associated** fictitious **chemical potential (Lagrange multiplier)** [97]. Introducing the **chemical potential** λ, whereupon the physically relevant expectation value of an observable $\langle \hat{A}(\lambda) \rangle$ is obtained as the limiting value for which we set λ to the limit $\lambda \to \infty$, **quenches** all unphysical states. This implies the existence of an additional $U(1)$ gauge field that quenches charge fluctuations associated with this representation. This technique applies to dilute spin subsystems where all the spins are considered independently. Projectors spontaneously **quench** a state with an opposite projection in creating a fermion with a given spin projection. This guarantees that the creation operator acts only on a state from the physical subspace (2705). Equations (2703) and (2706) constitute the exact pseudo-fermion representation for spin $S = \frac{1}{2}$. The impurity-spin operator and, consequently, the equation of motion with Hamiltonian determinant (2677) are symmetric under the local $U(1)$ gauge transformation:

$$\hat{f}_{\sigma} \to \exp\{-i\phi(t)\} \hat{f}_{\sigma} \quad , \quad i\frac{d}{dt} \to i\frac{d}{dt} - \frac{\partial \phi(t)}{\partial t} \tag{2709}$$

Here, $\phi(t)$ is an arbitrary, time-dependent phase. This is closely related to the conservation of the pseudo-fermion number $\hat{Q}$.

Projection onto the Physical Hilbert Space

To apply the perturbation theory and Wick theorem for expressions involving spin operators, we ensure unphysical pseudo-fermion states do not contribute to thermal averages. We note that only states with exactly **one** pseudo-fermion are physical and so, as seen previously, $|0\rangle$ and $\hat{f}^{\dagger}_{\sigma} \hat{f}^{\dagger}_{\sigma'} |0\rangle$ are unphysical states. The problem is resolved by using the **Abrikosov projection technique**, which entails adding the term:

$$\hat{H}^{\lambda} = \lambda \hat{Q} \equiv \sum_{m} \lambda_m \left(\sum_{\sigma} \hat{f}^{\dagger}_{\sigma m} \hat{f}_{\sigma m} - 1 \right) \tag{2710}$$

to the Hamiltonian determinant.

The aim is to give each pseudo-fermion a very large energy. Hence, considering (2710), we evaluate the grand canonical ensemble defined as the statistical operator

$$\hat{\rho} = \frac{\exp\left\{-\beta\left(\hat{H}+\lambda\hat{Q}\right)\right\}}{Z} \tag{2711}$$

where

$$Z = \mathrm{Tr}\left[\exp\left\{-\beta\left(\hat{H}+\lambda\hat{Q}\right)\right\}\right] \tag{2712}$$

is the grand canonical partition function. The trace extends over the complete Fock space, including summation over $\mathbf{Q} = 0,1,2$. The grand canonical expectation value of the observable $\hat{A}$ acting on the impurity spin space is defined:

$$\left\langle \hat{A}(\lambda)\right\rangle = \mathrm{Tr}\left[\hat{\rho}\hat{A}\right] \tag{2713}$$

The physical expectation value $\left\langle \hat{A}\right\rangle$ of the operator $\hat{A}$ in the canonical ensemble with fixed $\mathbf{Q} = 1$ [46] and is evaluated:

$$\left\langle \hat{A}(\lambda)\right\rangle = \frac{\mathrm{Tr}_{Q=1}\left[\hat{A}\exp\left\{-\beta\hat{H}\right\}\right]}{\mathrm{Tr}_{Q=1}\left[\exp\left\{-\beta\hat{H}\right\}\right]} = \lim_{\lambda\to\infty}\frac{\mathrm{Tr}_{Q=1}\left[\hat{A}\exp\left\{-\beta\left(\hat{H}+\lambda\hat{Q}\right)\right\}\right]}{\mathrm{Tr}_{Q=1}\left[\hat{Q}\exp\left\{-\beta\left(\hat{H}+\lambda\hat{Q}\right)\right\}\right]} = \lim_{\lambda\to\infty}\frac{\left\langle \hat{A}(\lambda)\right\rangle}{\left\langle \hat{Q}(\lambda)\right\rangle} \tag{2714}$$

Because λ imitates the chemical potential, thermal averages taken with finite λ contain various powers of $\exp\{-\beta\lambda\}$ and, in particular, $\left\langle \hat{Q}(\lambda)\right\rangle \sim \exp\{-\beta\lambda\}$. So, the limit $\lambda \to \infty$ in equation (2714) quenches terms in $\left\langle \hat{A}(\lambda)\right\rangle$ proportional to $\exp\{-\beta\lambda\}$. Hence, when calculating any thermal average at finite λ, one is allowed to retain only terms of lowest order in $\exp\{-\beta\lambda\}$. Thus, in (2714), all terms of the grand canonical traces in the numerator and in the denominator for $\mathbf{Q} > 1$ are projected away by the limit $\lambda \to \infty$, while in the denominator, the operator $\hat{\mathbf{Q}}$ tailors all terms with $\mathbf{Q} = 0$ so they vanish. In the numerator, the observable $\hat{A}$ acts on the impurity-spin space and so is a power of $\vec{\mathbf{S}}$. This observable vanishes in the $\mathbf{Q} = 0$ subspace. Therefore, in the numerator and in the denominator, the canonical traces over the physical sector $\mathbf{Q} = 1$ remain, as expected. Consequently, any impurity-spin correlation function can be evaluated as a pseudo-fermion correlation function in the unrestricted Fock space. Here, Wick theorem and the decomposition in terms of Feynman diagrams with pseudo-fermion propagators are valid when taking the limit $\lambda \to \infty$ at the end of the calculation. It is important to note that for the c-electron spin, equation (2638), the $\mathbf{Q} = 1$ projection is irrelevant because doubly occupied or empty states are allowed for the noninteracting c-electrons.

Finally, the full Hamiltonian for our system:

$$\mathrm{H} = \sum_{\vec{\kappa}\sigma} \epsilon_{\vec{\kappa}}\hat{\psi}^{\dagger}_{\vec{\kappa}\sigma}\hat{\psi}_{\vec{\kappa}\sigma} + \mathrm{H}^{J} + \hat{\mathrm{H}}^{\lambda} \tag{2715}$$

where the interaction energy,

$$\mathrm{H}^{J} = \sum_{\vec{\kappa}\vec{\kappa}'\sigma\sigma'\alpha\alpha' lm} J_{lm}\hat{\psi}^{\dagger}_{\vec{\kappa}'\sigma}\hat{\psi}_{\vec{\kappa}\sigma'}\hat{f}^{\dagger}_{\alpha m}\hat{f}_{\alpha' m} \tag{2716}$$

and the effective interaction coupling strength,

$$J_{lm} = -\frac{J}{2}\exp\left\{i\left(\vec{\kappa}' - \vec{\kappa}\right)\vec{r}_l\right\}\vec{\sigma}_{\sigma\sigma'}\vec{\mathbf{S}}_m \tag{2717}$$

Electrons are assumed to lie in a band of width 2Λ that is symmetric about $\in_{\vec{\kappa}} = \in_F$. This is with a constant density of states $\nu(\in_F)$ per spin up or spin down (Λ is of order $\in_F$).

Diagrammatic Rules

We write down the Feynman diagrammatic rules in momentum-Matsubara space after applying impurity averaging neglecting interference among impurities because the impurity concentration is small. We show that the limit $\lambda \to \infty$ translates into the diagrammatic rules for the evaluation of impurity Green's and correlation functions. We denote the **local c-electron Green's function (c-Green's function)** at the impurity site by

$$\mathscr{G}_{\alpha\alpha'} = -\left\langle \hat{T}_\tau \hat{\psi}_\alpha \hat{\psi}^\dagger_{\alpha'} \right\rangle = \sum_{\vec{\kappa}} \frac{\delta_{\alpha\alpha'}}{i\omega - \in_{\vec{\kappa}}} \tag{2718}$$

as a **solid line** ─►─ and the bare grand canonical **pseudo-fermion Green's function (f-Green's function)** by

$$\mathscr{F}_{\sigma\sigma'} = -\left\langle \hat{T}_\tau \hat{f}_\sigma \hat{f}^\dagger_{\sigma'} \right\rangle = \frac{\delta_{\sigma\sigma'}}{i\omega_n - \lambda} \tag{2719}$$

as a **broken line** --►-- with the fermionic Matsubara frequencies $\omega_n = \frac{2\pi}{\beta}\left(n + \frac{1}{2}\right)$.

Because only states with exactly one pseudo-fermion are physical and otherwise unphysical, we resolve the problem by accomplishing the Abrikosov projection technique:

$$\hat{\mathbf{H}} = \hat{\mathrm{H}} - \mu\hat{N} + \hat{\mathrm{H}}^\lambda = -\int d\vec{r} \sum_\sigma \hat{\psi}^\dagger_\sigma(\vec{r})\hat{\mathbf{M}}_0\hat{\psi}_\sigma(\vec{r}) + \hat{\mathrm{H}}^J + \hat{\mathrm{H}}^\lambda \quad , \quad \hat{\mathbf{M}}_0 = \frac{\Delta}{2m} + \mu \tag{2720}$$

where

$$\mathrm{H}^J = -\frac{J}{2}\sum_{m\sigma\sigma'\alpha\alpha'} \int d\vec{r}\hat{\psi}^\dagger_\sigma(\vec{r})\hat{\psi}_{\sigma'}(\vec{r})\vec{\sigma}_{\sigma\sigma'}\vec{\sigma}_{\alpha\alpha'}\hat{f}^\dagger_\alpha(\vec{r}_m)\hat{f}_{\alpha'}(\vec{r}_m)\delta(\vec{r} - \vec{r}_m) \tag{2721}$$

and $\hat{N}$ is the electron number operator. With the help of (2720) and the imaginary time $\tau = it$ within the interval $[0, \beta]$ where $\beta = \frac{1}{T}$, we write the following equations of motion:

$$\frac{\partial\hat{\psi}_\sigma(\vec{r},\tau)}{\partial\tau} = \hat{\mathbf{M}}_0\hat{\psi}_\sigma(\vec{r},t) - \left[\hat{\psi}_\sigma(\vec{r},\tau), \mathrm{H}^J\right] \tag{2722}$$

$$\frac{\partial\hat{f}_\alpha(\vec{r}_m,\tau)}{\partial\tau} = -\lambda_m\hat{f}_\alpha(\vec{r}_m,\tau) - \left[\hat{f}_\alpha(\vec{r}_m,\tau), \mathrm{H}^J\right] \tag{2723}$$

From the following **fermion**

$$\mathscr{G}_{\sigma\sigma'}(\vec{r},\tau;\vec{r}',\tau') = -\left\langle \hat{T}_\tau \hat{\psi}_\sigma(\vec{r},\tau)\hat{\psi}^\dagger_{\sigma'}(\vec{r}',\tau')\right\rangle \tag{2724}$$

and **pseudo-fermion Green's functions**

$$\mathscr{F}_{\sigma\sigma'}(\vec{r}_m,\tau;\vec{r}_{m'},\tau')=-\left\langle \hat{T}_\tau \hat{f}_\sigma(\vec{r}_m,\tau)\hat{f}^\dagger_{\sigma'}(\vec{r}_{m'},\tau')\right\rangle \tag{2725}$$

and considering (2722) and (2723), then

$$\mathscr{G}_0^{-1}\mathscr{G}_{\sigma\sigma'}(\vec{r},\tau;\vec{r}',\tau')=\hat{1}_{\vec{r}-\vec{r}'}-\left\langle \hat{T}_\tau\left(\left[\hat{\psi}_\sigma(\vec{r},\tau),H^J\right],\hat{\psi}^\dagger_{\sigma'}(\vec{r}',\tau')\right)\right\rangle \tag{2726}$$

and

$$\mathscr{F}_0^{-1}\mathscr{F}_{\sigma\sigma'}(\vec{r}_m,\tau;\vec{r}_{m'},\tau')=\hat{1}_{\vec{r}_m,-\vec{r}_{m'}}+\frac{J}{2}\sum_{\sigma\sigma'\alpha\alpha'}\left\langle \hat{T}_\tau\left(\hat{\psi}_\sigma(\vec{r}_m,\tau)\hat{\psi}^\dagger_{\sigma'}(\vec{r}_m,\tau)\vec{\sigma}_{\sigma\sigma'}\vec{\sigma}_{\alpha\alpha'}\hat{f}_\alpha(\vec{r}_m,\tau)\hat{f}^\dagger_{\alpha'}(\vec{r}_{m'},\tau')\right)\right\rangle \tag{2727}$$

where

$$\mathscr{G}_0^{-1}=-\frac{\partial}{\partial\tau}+\frac{\Delta}{2m}+\mu \quad , \quad \mathscr{F}_0^{-1}=-\frac{\partial}{\partial\tau}-\lambda_m \tag{2728}$$

with

$$\hat{1}_{\vec{r}-\vec{r}'}=\delta_{\sigma\sigma'}\delta(\tau-\tau')\delta(\vec{r}-\vec{r}') \quad , \quad \hat{1}_{\vec{r}_m,-\vec{r}_{m'}}=\delta_{\alpha\alpha'}\delta(\tau-\tau')\delta(\vec{r}_m-\vec{r}_{m'}) \tag{2729}$$

The **fermion and pseudo-fermion Green's functions** are diagrammatically represented, respectively, as follows:

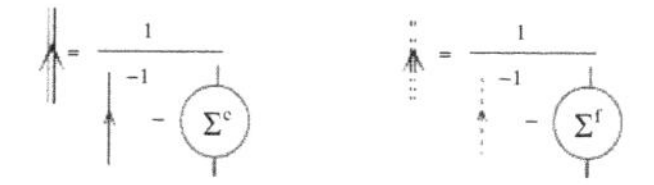

(2730)

or

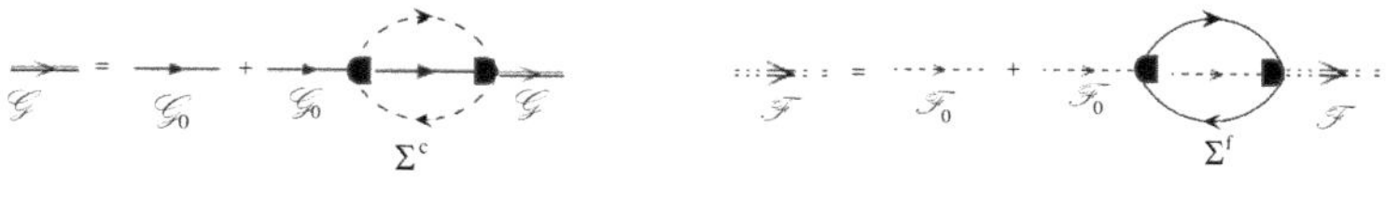

(2731)

In the absence of an external magnetic field, the Green's functions are diagonalized:

$$\mathscr{G}_{\sigma\sigma'}=\mathscr{G}\,\delta_{\sigma\sigma'} \quad , \quad \mathscr{F}_{\sigma\sigma'}=\mathscr{F}\,\delta_{\sigma\sigma'} \tag{2732}$$

From equation (2714) and (2719), we consider first $\lim_{\lambda\to\infty}\left\langle \hat{Q}(\lambda)\right\rangle$ via the Cauchy integral:

$$\left\langle \hat{Q}(\lambda)\right\rangle=\sum_\sigma\frac{1}{\beta}\sum_n\mathscr{F}_{\sigma\sigma}(i\omega_n)=-\sum_\sigma\oint\frac{dz}{2\pi i}f(z)\mathscr{F}_{\sigma\sigma}(z) \tag{2733}$$

or

$$\left\langle \hat{Q}(\lambda)\right\rangle=-\sum_\sigma\int_{-\infty}^{+\infty}\frac{d\in}{2\pi i}f(\in)\left[\mathscr{F}_{\sigma\sigma}(\in+i0)-\mathscr{F}_{\sigma\sigma}(\in-i0)\right] \tag{2734}$$

where the Fermi function is

$$f(\in)=\frac{1}{\exp\{\beta\in\}+1} \tag{2735}$$

and the $\in$-integral extends along the branch cut off $\mathcal{F}_{\sigma\sigma}(z)$ at the real frequency axis, $\mathrm{Im}\,z=0$. We do a gauge transformation of the operators

$$\hat{f}_\sigma \to \exp\{-i\lambda\tau\}\hat{f}_\sigma \tag{2736}$$

that shifts all pseudo-fermion energies in a diagram by

$$\in\to\in+\lambda \tag{2737}$$

This procedure eliminates λ from the pseudo-fermion propagator and translates it into the argument of the Fermi function:

$$\left\langle\hat{Q}(\lambda)\right\rangle=-\sum_\sigma\int_{-\infty}^{+\infty}\frac{d\in}{\pi}f(\in+\lambda)\mathrm{Im}\,\mathcal{F}_{\sigma\sigma}(\in+i0) \tag{2738}$$

or

$$\left\langle\hat{Q}(\lambda)\right\rangle\overset{\lambda\to\infty}{=}\exp\{-\beta\lambda\}\sum_\sigma\int_{-\infty}^{+\infty}\frac{d\in}{\pi}\exp\{-\beta\in\}\mathrm{Im}\,\mathcal{F}_{\sigma\sigma}(\in+i0) \tag{2739}$$

Here,

$$\mathcal{F}_{\sigma\sigma}(\in+i0)\equiv\mathcal{F}_{\sigma\sigma}(\in+\lambda+i0)=\frac{1}{\in+i0} \tag{2740}$$

Equation (2739) can be generalized to arbitrary Feynman diagrams involving f- and c-Green's functions:

a. Each complex contour integral has one distribution function, $f(z)$. The integral is written as a sum of integrals along the branch cuts at the real energy axis of all propagators appearing in the diagram.
b. Taking one term of this sum, the argument of the distribution function $f(\in)$ in that term is real and equal to the argument $\in$ of that propagator along whose branch cut the integration extends.
c. The energy-shift gauge transformation is applicable to all pseudo-fermion energies ω in the diagram and, hence, eliminates the parameter λ in all pseudo-fermion propagators.
d. If the given term in the integral is along a pseudo-fermion branch cut, then the gauge transformation also shifts the argument of the distribution function,

$$f(\in)\to f(\in+\lambda) \tag{2741}$$

This is by virtue of (c) and implies that the pseudo-fermion branch cut integral vanishes $\approx\exp\{-\beta\lambda\}$, as in equation (2739). If the integral is along a c-electron branch cut, the argument of $f(\in)$ is unaffected by the gauge transformation, and the integral does not vanish.

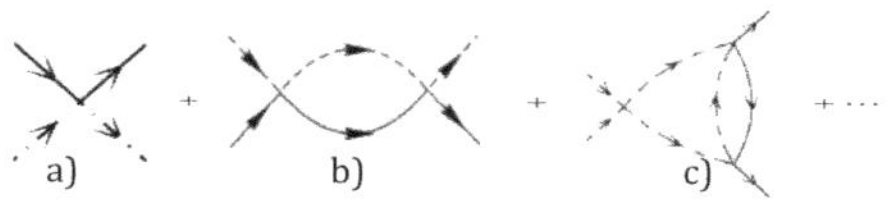

FIGURE 13.15 Electron-impurity Feynman diagrams spin vertex of a single-impurity Kondo model up to second-order in the spin exchange coupling J. The conduction electron propagators are denoted by solid, pseudo-fermion propagators (dashed lines). Figures (a) first-order scattering amplitude; (b) and (c) represent the second-order processes in the Kondo interaction vertex. The external lines, though drawn for clarity, are not part of the vertex.

We summarize this derivation in the following diagrammatic rules for $\left(\mathbf{Q}=1\right)$, projected expectation values:

1. In a diagrammatic part consisting of a product of c- and f-Green's functions, only integrals along the c–electron branch cuts contribute.
2. A closed pseudo-fermion loop contains only pseudo-fermion branch cut integrals and so carries a factor $\exp\{-\beta\lambda\}$.
3. Each diagram contributing to the projected expectation value of an impurity spin observable, $\langle \hat{A} \rangle$, contains exactly one closed pseudo-fermion loop per impurity site because the factor $\exp\{-\beta\lambda\}$ cancels in the numerator and denominator of equation (2714), and higher-order loops vanish by virtue of rule 2.

It is worth noting that the pseudo-fermion representation can be generalized to higher local spins than $S=\frac{1}{2}$ by choosing a respective higher-dimensional representation of the spin matrices in equation (2703) and defining the constraint $\hat{Q}=\hat{1}$ with a summation over all possible spin orientations σ.

Perturbation Theory

Next, we again analyze the scattering of a conduction electron from a spin impurity via perturbation theory because this will show the physical origin of its singular behavior. The first- and second-order terms of the scattering vertex are depicted in Figure 13.15. Considering the vector of Pauli matrices $\vec{\sigma}$ acting in c-electron spin space and the vector of Pauli matrices $\vec{S}$ in f–spin space, then the scattering vertex Γ of the Kondo Hamiltonian (2647) in first order of $\frac{J}{2}$:

$$\Gamma^{(1)}=-\frac{J}{2}\vec{\sigma}_{\sigma\sigma'}\vec{S}_{mm'} \tag{2742}$$

The first-order term of the scattering vertex (Figure 13.16), $\Gamma^{(1)}$, for the scattering of conduction electron:

$$|\vec{\kappa}\sigma\rangle \to |\vec{\kappa}'\sigma'\rangle \tag{2743}$$

The first-order perturbation theory is not found to contribute to the self-energy considering that the overlap between a one-particle and a two-particle-one-hole state vanishes. So, we consider higher orders of the perturbation theory that may be obtained from equation (2727).

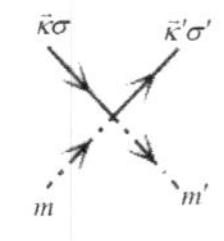

FIGURE 13.16 First-order term of the scattering vertex. The conduction electron propagators are denoted by solid, pseudo-fermion propagators (dashed lines).

13.2.13.5 Self-Energy of the Pseudo-Fermion

Feynman Diagrammatic Technique and Dyson Equation

Self-Energy of the Pseudo-Fermion

We evaluate the spin-dependent part of equation (2727) via the $SU(2)$ spin algebra

$$\sigma^{\mu}_{\sigma'\sigma}\sigma^{\nu}_{\sigma\sigma'} = \sigma^{\nu}_{\sigma\sigma'}\sigma^{\mu}_{\sigma'\sigma} = \delta^{\mu\nu}\delta_{\sigma'\sigma} + i\in_{\eta\nu\mu}\sigma^{\eta}_{\sigma'\sigma} \tag{2744}$$

So,

$$\sum_{\sigma'}\sigma^{\mu}_{\sigma'\sigma}\sigma^{\nu}_{\sigma\sigma'} = \sum_{\sigma'}\sigma^{\nu}_{\sigma\sigma'}\sigma^{\mu}_{\sigma'\sigma} = 2\delta^{\mu\nu} \quad , \quad \mathrm{Tr}\left(\vec{\sigma}\right)^2 = 3 \tag{2745}$$

Also

$$\sum_{mm'} \mathbf{S}^{\mu}_{m'm}\mathbf{S}^{\mu}_{mm'} = \sum_{m'}\left(\vec{\mathbf{S}}^2\right)_{m'm'} = S(S+1) \tag{2746}$$

Moving to the momentum representation of equation (2726) while setting $\lambda_m \equiv \lambda$ and $\omega_i \equiv \omega_{n_i}, i = 1,2$ as well as neglecting the frequency dependence of the vertex, then

$$(i\omega_n - \lambda)\mathscr{G}(\vec{\kappa},\omega_n) = 1 + \Sigma^{(2)}\mathscr{G}(\vec{\kappa},\omega_n) \tag{2747}$$

from where

$$\mathscr{G}(\vec{\kappa},\omega_n) = \frac{1}{i\omega_n - \lambda - \Sigma^{(2)}} \tag{2748}$$

and the second-order self-energy

$$\Sigma^{(2)} = -\frac{2S(S+1)J^2}{\beta^2}\sum_{\omega_1,\omega_2}\sum_{\vec{\kappa}_1\vec{\kappa}_2}\mathscr{F}_0(\vec{\kappa}_1,\omega_1)\mathscr{F}_0(\vec{\kappa}_2,\omega_2)\mathscr{G}(\vec{\kappa}+\vec{\kappa}_2-\vec{\kappa}_1,\omega_n+\omega_2-\omega_1) \tag{2749}$$

Also, this expression and, in particular, the coupling constant for temperatures of the order of the Kondo temperature may be renormalized. For further calculations, we introduce the polarization function

$$\Pi_{\vec{\kappa},\omega_m} = \frac{1}{\beta}\sum_{\vec{q}n}\mathscr{F}_0(\vec{\kappa}+\vec{q},\omega_n+\omega_m)\mathscr{F}_0(\vec{\kappa},\omega_n) \quad , \quad \omega_m = \frac{2m\pi}{\beta} \tag{2750}$$

This permits us to rewrite the self-energy in (2749):

$$\Sigma^{(2)}(\vec{\kappa},\omega_n) = -\frac{2S(S+1)J^2}{\beta}\sum_{\vec{q}m}\Pi_{\vec{\kappa}-\vec{q},\omega_n-\omega_m}\mathscr{G}(\vec{q},\omega_m) \tag{2751}$$

Note that a closed pseudo-fermion loop contains only pseudo-fermion branch cut integrals as indicated earlier and so carries a factor $\exp\{-\beta\lambda\}$. This condition is absent in our case for the self-energy, and the factor of $\exp\{-\beta\lambda\}$ can be eliminated by setting $\lambda = 0$, which corresponds to a half-filled pseudo-fermion band. This defines the Green's function via the complete nonreducible part of $\Sigma^{(2)}(\vec{\kappa},\omega_n)$:

$$\mathscr{G}(\vec{\kappa},\omega_n) = \frac{1}{i\omega_n - \Sigma^{(2)}(\vec{\kappa},\omega_n)} \tag{2752}$$

If we consider interaction between next-nearest neighbors in a simple cubic crystal and consider the shift of the lattice radius vector $\vec{r}$ from $\vec{r}_m$ to $\vec{R}$, then the Fourier transform of the polarization function can be rewritten as

$$\Pi_{\vec{q},\omega_m} = \sum_{\vec{r}} \Pi_{\vec{r},\omega_m} \exp\{-i\vec{q}\vec{r}\} = \Pi_{\vec{r}_m,\omega_m} + \Pi_{R,\omega_m}\varphi(\vec{q}) \tag{2753}$$

where the form factor is

$$\varphi(\vec{q}) = \sum_{\vec{R}} \exp\{-i\vec{q}\vec{R}\} = 2\left[\cos(q_x R) + \cos(q_y R) + \cos(q_z R)\right] \tag{2754}$$

Then, considering

$$\mathscr{F}_0(0,\omega_{n_i}) = \mathscr{F}_0(r_m, r_m, \omega_{n_i}) \quad , \quad i = 1,2 \tag{2755}$$

and

$$\Sigma^{(2)}(\omega_n) = -\frac{2S(S+1)J^2}{\beta^2} \sum_{\omega_1,\omega_2} \mathscr{F}_0(0,\omega_1)\mathscr{F}_0(0,\omega_2)\mathscr{G}(0,\omega_n+\omega_2-\omega_1) \tag{2756}$$

equation (2747) can be rewritten

$$\left[i\omega_n - \Sigma^{(2)}(\omega_n)\right]\mathscr{G}(\vec{\kappa},\omega_n) = 1 + \mathcal{M}(\vec{\kappa},\omega_n)\mathscr{G}(\vec{\kappa},\omega_n) \tag{2757}$$

with the magnetization of the system:

$$\mathcal{M}(\vec{\kappa},\omega_n) = -\frac{2S(S+1)J^2}{\beta} \sum_{\vec{q}m} \varphi(\vec{\kappa}-\vec{q})\Pi_{R,\omega_n-\omega_m}\mathscr{G}(\vec{q},\omega_m) \tag{2758}$$

then the Green's function

$$\mathscr{G}(\vec{\kappa},\omega_n) = \frac{1}{i\omega_n - \Sigma(\vec{\kappa},\omega_n)} \tag{2759}$$

and total self-energy

$$\Sigma(\vec{\kappa},\omega_n) = \Sigma^{(2)}(\omega_n) + \mathcal{M}(\vec{\kappa},\omega_n) \tag{2760}$$

From equation (2758) and with the help of the infinite series expansion of [82] $\tanh(\pi y)$ and $\coth(\pi y)$, the magnetization may take the form

$$\mathcal{M}(\vec{\kappa}) = S(S+1)J^2 \sum_{\vec{q}} \varphi(\vec{\kappa}-\vec{q})\Pi_{R,0} \tanh\frac{\beta\mathcal{M}(\vec{q})}{2} \tag{2761}$$

13.2.13.6 Effective Spin Screening, Spin Susceptibility

In this section we will study the properties of spin liquids at high temperatures $T >> T_K$ and investigate the effective magnetic moment as well as the effective spin susceptibility. At finite temperatures, the magnetic moment may be evaluated via the formula

$$\mathcal{M}(H,T) = g\mu_B \frac{\sum_{\mathcal{M}_z=-\frac{1}{2}}^{\frac{1}{2}} \mathcal{M}_z T \sum_{\omega} \mathcal{G}_{\mathcal{M}_z\mathcal{M}_z}(\omega)\exp\{i\omega\tau\}}{\sum_{\mathcal{M}_z=-\frac{1}{2}}^{\frac{1}{2}} T \sum_{\omega} \mathcal{G}_{\mathcal{M}_z\mathcal{M}_z}(\omega)\exp\{i\omega\tau\}} \tag{2762}$$

where H is an external magnetic field, $\mathcal{M}_z$; the projection of magnetic moment is μ_B (the Bohr magneton); g is the Lande g-factor; and the Green's function is

$$\mathcal{G}_{\mathcal{M}_z\mathcal{M}_z'} = \mathcal{G}\delta_{\mathcal{M}_z\mathcal{M}_z'} = \delta_{\mathcal{M}_z\mathcal{M}_z'} \frac{1}{i\omega_n - \lambda + H\delta\,\mathrm{sgn}(\mathcal{M}_z - S)} \quad , \quad \delta = \frac{1}{2g\mu_B} \tag{2763}$$

Because the operator average (defining the physical properties of the system) of the nonphysical state appears to be linked with the pseudo-fermion representation, the first-order magnetic moment from (2762) is found to be

$$\mathcal{M}^1(H,T) = \frac{1}{2} g\mu_B H \left[\tanh\frac{\beta g\mu_B H}{2} + N^1(H,T)\coth\frac{\beta g\mu_B H}{2} \right] \tag{2764}$$

where the magnetization of local impurity in the presence of Kondo effect

$$\mathrm{N}^1(H,T) = \frac{1}{2}\left(\frac{J}{\beta}\right)^2 \sum_{\omega_1\omega_2\vec{\kappa}_1\vec{\kappa}_2} \mathcal{F}(\vec{\kappa}_1,\omega_1)\mathcal{F}(\vec{\kappa}_2,\omega_2)\mathrm{F}(\omega_1,\omega_2) \tag{2765}$$

and

$$\mathrm{F}(\omega_1,\omega_2) = \frac{1}{2\beta} \sum_{\omega,k=0,1} \exp\{-ik\pi\}\mathcal{G}^0_{\sigma k}(i\omega)\mathcal{G}^0_{\sigma k}(i\omega)\left(\mathcal{G}^0_{\sigma k}(i\omega + i\omega_1 - i\omega_2) + 2\mathcal{G}^0_{\hat{f}^\dagger_{\sigma k}|0\rangle}(i\omega + i\omega_1 - i\omega_2)\right) \tag{2766}$$

with

$$\hat{f}^\dagger_{\sigma k}|0\rangle \mapsto \sigma_k' \quad , \quad \sigma_0 \mapsto \uparrow \quad , \quad \sigma_1 \mapsto \downarrow \quad , \quad \sigma_0' \mapsto \downarrow \quad , \quad \sigma_1' \mapsto \uparrow \tag{2767}$$

We calculate the sum over ω for the limit $T >> \mu_B H$ and make the transition from sum over the momenta to integration in phase volume and then obtain this result in the magnetic moment:

$$\mathcal{M}^1(H,T) = \frac{S(S+1)(g\mu_B)^2 H}{3T}\left[1 - \left(\frac{2J}{\beta}\right)^2 \sum_{\omega_1\omega_2} \int \frac{d\vec{\kappa}_1 d\vec{\kappa}_2}{(2\pi)^{2d}} \frac{1}{i\omega_1 - \xi_{\vec{\kappa}_1}} \frac{1}{i\omega_2 - \xi_{\vec{\kappa}_2}} \frac{1}{(i\omega_1 - i\omega_2)^2}\right] \tag{2768}$$

Calculating (2768) in the logarithmic approximation and considering a constant density of states $\nu(0)$, we arrive at the expression

$$\mathcal{M}^1(H,T)=\frac{S(S+1)(g\mu_B)^2 H}{3T}\left(1-(2\nu(0)J)^2\ln\left(\frac{\epsilon_F}{T}\right)\right) \tag{2769}$$

From the perturbation theory, the magnetic moment now takes the form

$$\mathcal{M}(H,T)=\frac{S(S+1)(g\mu_B)^2 H}{3T}\left(1-\frac{(2\nu(0)J)^2\ln\left(\frac{\epsilon_F}{T}\right)}{1+2\nu(0)J\ln\left(\frac{\epsilon_F}{T}\right)}+\cdots\right) \tag{2770}$$

Because

$$\frac{1}{1+2\nu(0)J\ln\left(\frac{\epsilon_F}{T}\right)}=\frac{1}{2\nu(0)Jn\left(\frac{T}{T_K}\right)} \tag{2771}$$

then

$$\mathcal{M}(H,T)=\frac{S(S+1)(g\mu_B)^2 H}{3T}\left(1-\frac{1}{\ln\left(\frac{T}{T_K}\right)}+\cdots\right) \tag{2772}$$

This equation expresses the screening tendency of the conduction electron impurity spins when the temperature is reduced and the parameter of the perturbation theory is $\ln\left(\frac{T}{T_K}\right)\ll 1$. The competition between the antiferromagnetic fluctuations and the spin liquid occurs when the magnetic (RKKY) and the Kondo interactions are of the same order of magnitude. So, the critical antiferromagnetic fluctuations are observed to be essential in the link between local and nonlocal correlations.

From the magnetic moment, we calculate the magnetic susceptibility via

$$\chi_0=\lim_{H\to 0}\frac{\mathcal{M}(H,T)}{H} \tag{2773}$$

Therefore, for our case,

$$\chi_0(T)=\frac{S(S+1)(g\mu_B)^2}{3T}\left(1-\frac{1}{\ln\left(\frac{T}{T_K}\right)}+\cdots\right) \tag{2774}$$

The effective susceptibility also screens the electron conductivity. The approximation of noncrossing diagrams in the Kondo lattice for the complete statistical spin susceptibility takes the following form:

$$\chi(\vec{q},T)=\frac{\chi_0(T)}{1+\tilde{J}_{\text{RKKY}}(\vec{q},T)\chi_0(T)} \tag{2775}$$

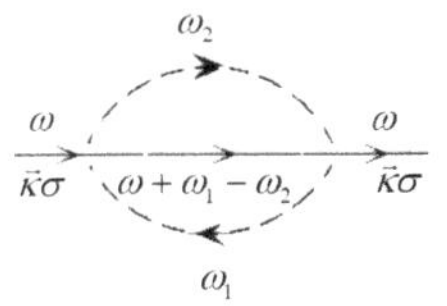

FIGURE 13.17 Second-order diagram contributing to the self-energy part of pseudo-fermions.

The pole of this expression defines the critical temperature for the magnetic ordering. The wave vector for which the susceptibility is divergent depends on the form of the function $\tilde{J}_{\mathrm{RKKY}}$.

13.2.13.7 Second-Order Self-Energy Diagrams

Examining the closed loop for $\lambda \neq 0$, the diagrammatic representation for the second-order self-energy $\Sigma^{(2)}$ is shown in Figure 13.17.

Summing over the frequencies ω_1 and ω_2 for the closed loop, the second-order self-energy is analytically evaluated as follows:

$$\Sigma^{(2)} = -J_\lambda^2 \frac{1}{\beta^2} \sum_{\omega_1,\omega_2} \sum_{\vec{\kappa}'} \mathscr{F}_0(-\lambda,\omega_1)\mathscr{F}_0(-\lambda,\omega_2)\mathscr{G}_0(-\xi_{\vec{\kappa}'},\omega+\omega_1-\omega_2) \tag{2776}$$

or

$$\Sigma^{(2)} = -J_\lambda^2 \frac{1}{\beta^2} \sum_{\omega_1,\omega_2} \sum_{\vec{\kappa}'} \frac{1}{i\omega_1-\lambda} \frac{1}{i\omega_2-\lambda} \frac{1}{i(\omega+\omega_1-\omega_2)-\xi_{\vec{\kappa}'}} \tag{2777}$$

where

$$J_\lambda^2 = \left(\frac{J}{2}\right)^2 \frac{\exp\{\beta\lambda\}}{2S+1} \mathrm{Tr}\left(\vec{\sigma}_{\sigma'\sigma}\vec{\mathbf{S}}_{m'm}\right)\left(\vec{\sigma}_{\sigma\sigma'}\vec{\mathbf{S}}_{mm'}\right) \tag{2778}$$

and the normalization factor for the closed loops is:

$$\frac{\exp\{\beta\lambda\}}{2S+1} \tag{2779}$$

From partial fractions, we have

$$\frac{1}{i\omega_1-\lambda} \frac{1}{i(\omega+\omega_1-\omega_2)-\xi_{\vec{\kappa}'}} = \frac{a}{i\omega_1-\lambda} + \frac{b}{i(\omega+\omega_1-\omega_2)-\xi_{\vec{\kappa}'}} \tag{2780}$$

from where

$$1 = a\left[i(\omega+\omega_1-\omega_2)-\xi_{\vec{\kappa}'}\right]+b\left[i\omega_1-\lambda\right] \tag{2781}$$

or

$$1 = a\left[i(\omega-\omega_2)-\xi_{\vec{\kappa}'}\right]+b\left[-\lambda\right] \quad , \quad a=-b \tag{2782}$$

and

$$a = \frac{1}{i(\omega-\omega_2)-\xi_{\vec{\kappa}'}+\lambda} \tag{2783}$$

So,

$$\frac{1}{i\omega_1-\lambda}\frac{1}{i(\omega+\omega_1-\omega_2)-\xi_{\vec{\kappa}'}}=\frac{1}{i(\omega-\omega_2)-\xi_{\vec{\kappa}'}+\lambda}\left(\frac{1}{i\omega_1-\lambda}-\frac{1}{i(\omega+\omega_1-\omega_2)-\xi_{\vec{\kappa}'}}\right) \tag{2784}$$

and

$$\Sigma^{(2)}=-J_\lambda^2\frac{1}{\beta^2}\sum_{\omega_1,\omega_2}\sum_{\vec{\kappa}'}\frac{1}{i\omega_2-\lambda}\frac{1}{i(\omega-\omega_2)-\xi_{\vec{\kappa}'}+\lambda}\left(\frac{1}{i\omega_1-\lambda}-\frac{1}{i(\omega+\omega_1-\omega_2)-\xi_{\vec{\kappa}'}}\right) \tag{2785}$$

Also, from partial fractions,

$$\frac{1}{i\omega_2-\lambda}\frac{1}{i(\omega-\omega_2)-\xi_{\vec{\kappa}'}+\lambda}=\frac{a}{i\omega_2-\lambda}+\frac{b}{i(\omega-\omega_2)-\xi_{\vec{\kappa}'}+\lambda} \tag{2786}$$

and

$$1=a\left[i(\omega-\omega_2)-\xi_{\vec{\kappa}'}+\lambda\right]+b\left[i\omega_2-\lambda\right] \tag{2787}$$

or

$$1=a\left[i\omega-\xi_{\vec{\kappa}'}+\lambda\right]-b\lambda \quad , \quad a=b \tag{2788}$$

then

$$a=\frac{1}{i\omega-\xi_{\vec{\kappa}'}} \tag{2789}$$

So,

$$\frac{1}{i\omega_2-\lambda}\frac{1}{i(\omega-\omega_2)-\xi_{\vec{\kappa}'}+\lambda}=\frac{1}{i\omega-\xi_{\vec{\kappa}'}}\left(\frac{1}{i\omega_2-\lambda}+\frac{1}{i(\omega-\omega_2)-\xi_{\vec{\kappa}'}+\lambda}\right) \tag{2790}$$

We, therefore, have

$$\Sigma^{(2)}=-J_\lambda^2\frac{1}{\beta^2}\sum_{\omega_1,\omega_2}\sum_{\vec{\kappa}'}\mathrm{I}_{\vec{\kappa}'}(\omega,\omega_1,\omega_2,\lambda) \tag{2791}$$

Where

$$\mathrm{I}_{\vec{\kappa}'}(\omega,\omega_1,\omega_2,\lambda)=\frac{1}{i\omega-\xi_{\vec{\kappa}'}}\left(\frac{1}{i\omega_2-\lambda}+\frac{1}{i(\omega-\omega_2)-\xi_{\vec{\kappa}'}+\lambda}\right)\left(\frac{1}{i\omega_1-\lambda}-\frac{1}{i(\omega+\omega_1-\omega_2)-\xi_{\vec{\kappa}'}}\right) \tag{2792}$$

Considering (2757), then

$$\mathrm{I}_{\vec{\kappa}'}(\omega,\omega_1,\omega_2,\lambda)=\mathrm{I}^{(1)}_{\vec{\kappa}'}(\omega,\omega_1,\omega_2,\lambda)+\mathrm{I}^{(2)}_{\vec{\kappa}'}(\omega,\omega_1,\omega_2,\lambda) \tag{2793}$$

with

$$\mathrm{I}_{\vec{\kappa}'}^{(1)}(\omega,\omega_1,\omega_2,\lambda)=\frac{1}{i\omega-\xi_{\vec{\kappa}'}}\left(\frac{1}{i\omega_2-\lambda}\frac{1}{i\omega_1-\lambda}+\frac{1}{i(\omega-\omega_2)-\xi_{\vec{\kappa}'}+\lambda}\frac{1}{i\omega_1-\lambda}\right) \tag{2794}$$

$$\mathrm{I}_{\vec{\kappa}'}^{(2)}(\omega,\omega_1,\omega_2,\lambda)=-\frac{1}{i\omega-\xi_{\vec{\kappa}'}}\left(\frac{1}{i\omega_2-\lambda}\frac{1}{i(\omega+\omega_1-\omega_2)-\xi_{\vec{\kappa}'}}+\frac{1}{i(\omega-\omega_2)-\xi_{\vec{\kappa}'}+\lambda}\frac{1}{i(\omega+\omega_1-\omega_2)-\xi_{\vec{\kappa}'}}\right) \tag{2795}$$

So, with the help of the infinite series expansion [82] of $\tanh(\pi y)$ and $\coth(\pi y)$, then

$$\frac{1}{\beta^2}\sum_{\omega_1,\omega_2}\mathrm{I}_{\vec{\kappa}'}(\omega,\omega_1,\omega_2,\lambda)=\frac{1}{4}\left(\tanh\frac{\beta\lambda}{2}-\coth\frac{\beta(\lambda-\xi_{\vec{\kappa}'})}{2}\right)\left(\tanh\frac{\beta\lambda}{2}-\tanh\frac{\beta\xi_{\vec{\kappa}'}}{2}\right) \tag{2796}$$

But

$$\tanh\frac{\beta\lambda}{2}-\coth\frac{\beta(\lambda-\xi_{\vec{\kappa}'})}{2}=\tanh\frac{\beta\lambda}{2}-\frac{1-\tanh\frac{\beta\lambda}{2}\tanh\frac{\beta\xi_{\vec{\kappa}'}}{2}}{\tanh\frac{\beta\lambda}{2}-\tanh\frac{\beta\xi_{\vec{\kappa}'}}{2}}=\frac{\tanh^2\frac{\beta\lambda}{2}-1}{\tanh\frac{\beta\lambda}{2}-\tanh\frac{\beta\xi_{\vec{\kappa}'}}{2}} \tag{2797}$$

then

$$\Sigma^{(2)}=-J_\lambda^2\frac{1}{4}\sum_{\vec{\kappa}'}\frac{1}{i\omega_{\vec{\kappa}'}-\xi_{\vec{\kappa}'}}\left(\tanh^2\frac{\beta\lambda}{2}-1\right) \tag{2798}$$

From

$$\tanh^2\frac{\beta\lambda}{2}-1=\left(\frac{\exp\left\{\frac{\beta\lambda}{2}\right\}-\exp\left\{-\frac{\beta\lambda}{2}\right\}}{\exp\left\{\frac{\beta\lambda}{2}\right\}+\exp\left\{-\frac{\beta\lambda}{2}\right\}}\right)^2-1=\frac{-4}{\left(\exp\left\{\frac{\beta\lambda}{2}\right\}+\exp\left\{-\frac{\beta\lambda}{2}\right\}\right)^2} \tag{2799}$$

then

$$\Sigma^{(2)}=\left(\frac{J}{2}\right)^2\frac{1}{2S+1}\mathrm{Tr}\left(\vec{\sigma}_{\sigma'\sigma}\vec{\mathrm{S}}_{m'm}\right)\left(\vec{\sigma}_{\sigma\sigma'}\vec{\mathrm{S}}_{mm'}\right)\sum_{\vec{\kappa}'}\frac{1}{i\omega_{\vec{\kappa}'}-\xi_{\vec{\kappa}'}}\frac{\exp\{\beta\lambda\}}{\left(\exp\left\{\frac{\beta\lambda}{2}\right\}+\exp\left\{-\frac{\beta\lambda}{2}\right\}\right)^2} \tag{2800}$$

Also from

$$\frac{\exp\{\beta\lambda\}}{\left(\exp\left\{\frac{\beta\lambda}{2}\right\}+\exp\left\{-\frac{\beta\lambda}{2}\right\}\right)^2}=\frac{1}{\left(1+\exp\{-\beta\lambda\}\right)^2} \tag{2801}$$

then

$$\Sigma^{(2)} = \frac{1}{2S+1}\left(\frac{J}{2}\right)^2 \mathrm{Tr}\left(\vec{\sigma}_{\sigma'\sigma}\vec{\mathbf{S}}_{m'm}\right)\left(\vec{\sigma}_{\sigma\sigma'}\vec{\mathbf{S}}_{mm'}\right)\sum_{\vec{\kappa}'}\frac{1}{i\omega_{\vec{\kappa}'}-\xi_{\vec{\kappa}'}}\frac{1}{\left(1+\exp\{-\beta\lambda\}\right)^2} \tag{2802}$$

So, after executing the Matsubara sum and the limit $T \to 0\ (\beta \to \infty)$, $\Sigma^{(2)}$ then has the form:

$$\Sigma^{(2)} = \frac{1}{2S+1}\left(\frac{J}{2}\right)^2 \mathrm{Tr}\left(\vec{\sigma}_{\sigma'\sigma}\vec{\mathbf{S}}_{m'm}\right)\left(\vec{\sigma}_{\sigma\sigma'}\vec{\mathbf{S}}_{mm'}\right)\int\frac{d\vec{\kappa}'}{(2\pi)^3}\frac{1}{i\omega_{\vec{\kappa}'}-\xi_{\vec{\kappa}'}} \tag{2803}$$

Considering that

$$\mathrm{Tr}\left(\vec{\sigma}\vec{\mathbf{S}}\right)\left(\vec{\sigma}\vec{\mathbf{S}}\right) = 2S(S+1) \tag{2804}$$

we then have the second-order to the self-energy equivalent to the Born approximation:

$$\Sigma^{(2)} = \left(\frac{J}{2}\right)^2 \frac{2S(S+1)}{2S+1}\int\frac{d\vec{\kappa}'}{(2\pi)^3}\frac{1}{i\omega_{\vec{\kappa}'}-\xi_{\vec{\kappa}'}} \tag{2805}$$

For $S = \frac{1}{2}$, then

$$2S+1 = 2\frac{1}{2}+1 = 2 \tag{2806}$$

and

$$\Sigma^{(2)} = \left(\frac{J}{2}\right)^2 S(S+1)\int\frac{d\vec{\kappa}'}{(2\pi)^3}\frac{1}{i\omega_{\vec{\kappa}'}-\xi_{\vec{\kappa}'}} \tag{2807}$$

The spin enters the self-energy only as an effective interaction of $\left(\frac{J}{2}\right)^2 S(S+1)$. Determining the momentum integral in (2807), then the self-energy $\Sigma^{(2)}$ relates the lifetime τ_0:

$$\Sigma^{(2)}\left(\vec{\kappa}, i\omega_{\vec{\kappa}'}\right) = -\frac{i}{2\tau_0}\delta_{\sigma\sigma'}\,\mathrm{sgn}\,\omega_{\vec{\kappa}'} \tag{2808}$$

where

$$\frac{1}{\tau_0} = \frac{mp_{\mathrm{F}}}{\pi^2} S(S+1)\left(\frac{J}{2}\right)^2 \int_{-\infty}^{\infty} d\epsilon \frac{\omega_{\vec{\kappa}'}}{\omega_{\vec{\kappa}'}^2+\epsilon^2} \tag{2809}$$

or

$$\frac{1}{\tau_0} = \frac{mp_{\mathrm{F}}}{\pi} S(S+1)\left(\frac{J}{2}\right)^2 \tag{2810}$$

So,

$$\operatorname{Im}\Sigma = \frac{mp_F}{\pi} S(S+1)\Gamma^2 \tag{2811}$$

as well as the resistivity, which is the Γ^2 and is independent of temperature. So, the scattering of magnetic impurities clearly does not explain the existence of a resistance minimum. From our result, we can infer that a resistance

$$\rho \approx \Gamma^2 = \frac{\pi}{mp_F S(S+1)} \operatorname{Im}\Sigma \tag{2812}$$

is independent of temperature. So, the fermionic self-energy with the anomalous behavior (existence of a resistance minimum) reveals the Kondo effect is not evident.

13.2.13.7.1 Popov-Fedotov Self-Energy, Scattering Amplitudes

Popov-Fedotov Self-Energy

Many variants of diagrammatic techniques based on different spin representations exist. However, the dimensionality of the space in which the spin operators act is always greater than the dimensionality of the spin matrices. This presents a serious problem in the elimination of superfluous states as well as a serious complication for diagrammatic rules and analytical expressions for the diagrams. The Popov-Fedotov technique [98, 99] addresses this problem. We consider equations (2745) and (2746):

$$\Upsilon = \left(\vec{\sigma}_{\sigma'\sigma} \cdot \vec{\mathbf{S}}_{m'm}\right)\left(\vec{\sigma}_{\sigma\sigma'} \cdot \vec{\mathbf{S}}_{mm'}\right) = \left(\vec{\sigma}^{\mu}_{\sigma'\sigma} \cdot \vec{\mathbf{S}}^{\mu}_{m'm}\right)\left(\vec{\sigma}^{\nu}_{\sigma\sigma'} \cdot \vec{\mathbf{S}}^{\nu}_{mm'}\right) = \sum_{\sigma\sigma'} \sigma^{\nu}_{\sigma'\sigma}\sigma^{\mu}_{\sigma\sigma'} \sum_{m'} \mathbf{S}^{\nu}_{mm'}\mathbf{S}^{\mu}_{m'm} = 2\left(\vec{\mathbf{S}}^2\right)_{mm'} = 2S(S+1)\delta_{mn} \tag{2813}$$

So,

$$\Sigma = -\Upsilon\left(\frac{J}{\beta}\right)^2 \sum_{\omega_1,\omega_2} \sum_{\vec{\kappa}_1'\vec{\kappa}_2'} \mathscr{G}_0(-\xi_1,\omega_1)\mathscr{G}_0(-\xi_2,\omega_2)\mathscr{F}_0(-\lambda,\omega+\omega_1-\omega_2) \tag{2814}$$

or

$$\Sigma = -\Upsilon\left(\frac{J}{\beta}\right)^2 \sum_{\omega_1,\omega_2} \sum_{\vec{\kappa}_1'\vec{\kappa}_2'} \frac{1}{i\omega_2-\xi_2} \frac{1}{i(\omega-\omega_2)+\xi_1-\lambda}\left(\frac{1}{i\omega_1-\xi_1} - \frac{1}{i(\omega+\omega_1-\omega_2)-\lambda}\right) \tag{2815}$$

But,

$$\frac{1}{i\omega_2-\xi_2} \frac{1}{i(\omega-\omega_2)+\xi_1-\lambda} = \frac{1}{i\omega+\xi_1-\xi_2-\lambda}\left(\frac{1}{i\omega_2-\xi_2} - \frac{1}{i(\omega-\omega_2)+\xi_1-\lambda}\right) \tag{2816}$$

then

$$\Sigma = \frac{\Upsilon J^2}{4} \sum_{\vec{\kappa}_1\vec{\kappa}_2} \frac{1}{i\omega+\xi_1-\xi_2-\lambda}\left(\tanh\frac{\beta\lambda}{2} - \tanh\frac{\beta\xi_1}{2}\right)\left(\tanh\frac{\beta\xi_2}{2} - \coth\frac{\beta(\xi_1-\lambda)}{2}\right) \tag{2817}$$

We investigate the behaviors near the Kondo temperatur T_K qualitatively and suppose

$$\lambda = \lambda_T = \mu_s \to \infty \tag{2818}$$

then

$$\Sigma = \Upsilon(J\nu)^2 \int_{-\Lambda}^{\Lambda}\int_{-\Lambda}^{\Lambda} d\xi_1 d\xi_2 \frac{n(\xi_1)(1-n(\xi_2))}{i\omega+\xi_1-\xi_2-\mu_s} \tag{2819}$$

Here, $\nu(\in)$ is the density of states that is considered constant $\nu(\mu)$ at the vicinity of the Fermi surface $\in_F \approx \mu$. Also, applying an analytic continuation, we have

$$\mathrm{Im}\,\Sigma \underset{\substack{\omega\to i\delta \\ \mu_s \to i\mu_s}}{=} -\Upsilon(J\nu)^2 \int d\xi_1 d\xi_2 n(\xi_1)(1-n(\xi_2))\delta(\omega+\xi_1-\xi_2) = -\Upsilon(J\nu)^2 \int_{-\Lambda}^{0}\int_{0}^{\Lambda} d\xi_1 d\xi_2 \delta(\omega+\xi_1-\xi_2) \tag{2820}$$

$$0<\omega+\xi_1=\xi_2<\Lambda \quad , \quad -\omega<\xi_1<0 \tag{2821}$$

Here, Λ is an upper cutoff arising from the microscopic scales of the problem. For,

$$\mathrm{Im}\,\Sigma \underset{\substack{\omega\to i\delta \\ \mu_s \to i\mu_s}}{=} -\Upsilon(J\nu)^2 \int_{-\Lambda}^{0} d\xi_1 \int_{0}^{\Lambda} d\xi_2 \delta(\omega+\xi_1-\xi_2) \tag{2822}$$

then

$$-\Lambda<\xi_1=\xi_2-\omega<\xi_1<0 \tag{2823}$$

and

$$\mathrm{Im}\,\Sigma \underset{\omega\to i\delta}{=} -2S(S+1)(J\nu)^2\omega \tag{2824}$$

But,

$$\mathrm{Re}\,\Sigma = \Upsilon(J\nu)^2 \mathrm{P}\int_{0}^{\Lambda} d\xi_2 \int_{-\Lambda}^{0} d\xi_1 \frac{1}{\omega+\xi_1-\xi_2} = \Upsilon(J\nu)^2 \mathrm{P}\int_{0}^{\Lambda} d\xi_2 \ln\left|\frac{\xi_2-\omega}{\xi_2-\omega+\Lambda}\right| \tag{2825}$$

or

$$\mathrm{Re}\,\Sigma = \Upsilon(J\nu)^2 \mathrm{P}\int_{0}^{\Lambda} d\xi_2 \ln\left|\frac{\xi_2-\omega}{\xi_2-\omega+\Lambda}\right| \approx \Upsilon(J\nu)^2 \mathrm{P}\int_{0}^{\Lambda} d\xi_2 \ln\left|\frac{\xi_2-\omega}{\Lambda}\right| \tag{2826}$$

where P is the principal value of the integral. From,

$$\mathrm{P}\int_{-a}^{a} \ln|y|dy = \lim_{\delta\to 0}\left(\int_{\delta}^{a} \ln y\,dy + \int_{-a}^{-\delta} \ln(-y)dy\right) = 2\lim_{\delta\to 0}(y\ln y - y)\Big|_{\delta}^{a} = 2(a\ln a - a) \tag{2827}$$

then

$$\mathrm{Re}\,\Sigma = -\Upsilon(Jv)^2\,\omega\ln\frac{\omega}{\Lambda} \tag{2828}$$

The interesting feature of this expression is that the ω dependence of the self-energy scales with the cutoff temperature Λ imitates the Kondo temperature. So, we are able to describe all of the self-energy effects by this single parameter. For $\Lambda \to 0$, the self-energy diverges. This is known as the **infrared divergence** because it comes from the low-energy range.

Considering the imaginary and real parts of the self-energy for a shift of

$$-\lambda \to -\lambda + 2S(S+1)(Jv)^2\,\omega\ln\frac{\omega}{\Lambda} \tag{2829}$$

then the Green's function

$$\mathscr{G} = \frac{1}{\omega - \lambda + 2S(S+1)(Jv)^2\,\omega\ln\frac{\omega}{\Lambda} + 2iS(S+1)(Jv)^2\,\omega} = \frac{1}{\omega + i\delta}\,\frac{1}{1 + 2S(S+1)(Jv)^2\ln\frac{\omega}{\Lambda}} \tag{2830}$$

The denominator of (2830) has a pole when

$$1 + 2S(S+1)(Jv)^2\ln\frac{\omega}{\Lambda} = 0 \tag{2831}$$

or

$$\omega = \Lambda\exp\left\{-\frac{1}{2S(S+1)(Jv)^2}\right\} \tag{2832}$$

This shows that there is collective behavior for all J.

13.2.13.8 Scattering Amplitudes

In this section, we examine the antiferromagnetic interaction of fermions with impurities. Here, the interaction with conduction electrons yields coherent effects connected with resonance scattering due to spin-flip enhanced by a small concentration of the magnetic impurities. The resonance effect is due to the minimum in the dependence of the electric resistance in metals such as pure gold, silver, and copper hosting impurity atoms with uncompleted internal shells that have spins different from zero. As seen earlier, spin density waves for the low-dimensional cases are responsible for the so-called RKKY oscillations of the static spin susceptibility, which is induced by a magnetic impurity in a free electron gas. The sign of the RKKY interaction shows oscillatory change tailored by the separation distance between impurities. Any magnetic ordering or glassing behavior quenches the Kondo effect. This also implies that any magnetic transition quenches the Kondo effect. As seen earlier, due to random sign and magnitude of the interaction constant, the disordered spin-spin coupling yields a glassy behavior with an effective temperature proportional to the concentration of the impurities. In this case, the impurities of interwoven spins that hinder spin-flip during scattering are excluded.

Because the Kondo coupling yields a paramagnetic Fermi liquid state without local moments, the local orbitals hybridize with each other. This subsequently achieves a coherent lattice at low temperatures forming Bloch-like quasiparticle states. We find that the RKKY interaction tends to induce magnetic order of the local moments resulting in the breakdown of the Kondo spin screening of the local moments. This yields the magnetic order, which is when the RKKY coupling energy becomes larger than the characteristic energy scale for Kondo singlet formation, the Kondo temperature T_K. So, for $T = 0$, a quantum phase transition exists with the local spin exchange coupling J serving as the control parameter.

For the scattering amplitudes, we first evaluate summation topologically with respect to the spin indices. We examine when the fermion comes in, when it interacts once with the impurity spin, and when it is itself flipped. In the subsequent time, the flipped spin interacts back on the fermion. Note that the spin-flip makes the Kondo problem a many-body problem in spite of there being no direct interaction between two electrons in the Kondo Hamiltonian determinant. Consequently, this alters the fermion's energy by scattering it and giving it a lifetime. So, fermion m flips the spin, and the flipped spin interacts with fermion m''. Hence, two fermions indirectly interact via the impurity spin as follows:

1. 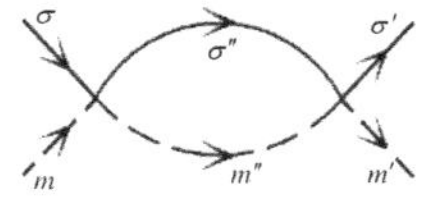

or

$$\Upsilon_1 = \left(\vec{\sigma}^{\mu}_{\sigma''\sigma}\cdot\vec{S}^{\mu}_{m''m}\right)\left(\vec{\sigma}^{\nu}_{\sigma'\sigma''}\cdot\vec{S}^{\nu}_{m'm''}\right) = \sigma^{\nu}_{\sigma'\sigma''}\sigma^{\mu}_{\sigma''\sigma}S^{\nu}_{m'm''}S^{\mu}_{m''m} = S(S+1)\delta_{m'm}\delta_{\sigma'\sigma} - 2\vec{\sigma}_{\sigma'\sigma}\vec{S}_{mm'} \quad (2833)$$

Also,

2. 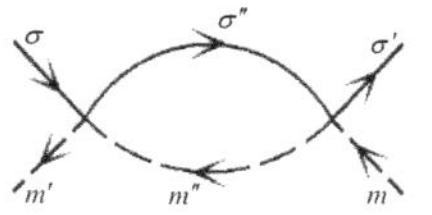

or

$$\Upsilon_2 = \left(\vec{\sigma}^{\mu}_{\sigma''\sigma}\cdot\vec{S}^{\mu}_{m'm''}\right)\left(\vec{\sigma}^{\nu}_{\sigma'\sigma''}\cdot\vec{S}^{\nu}_{m''m}\right) = \sigma^{\mu}_{\sigma''\sigma}\sigma^{\nu}_{\sigma'\sigma''}S^{\mu}_{m'm''}S^{\nu}_{m''m} = S(S+1)\delta_{mm'}\delta_{\sigma\sigma'} + 2\vec{\sigma}_{\sigma'\sigma}\vec{S}_{mm'} \quad (2834)$$

We now calculate the following integrals for

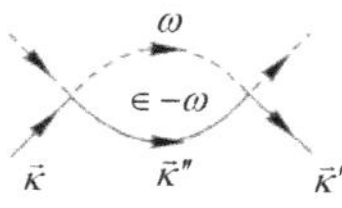

or

$$\Gamma_1 = -\Upsilon_1\left(\frac{J}{2}\right)^2\frac{1}{\beta}\sum_{\omega}\sum_{\vec{\kappa}'}\mathscr{F}_0\left(-\lambda,\omega_{\vec{\kappa}'}\right)\mathscr{G}_0\left(-\xi_{\vec{\kappa}'},\in-\omega_{\vec{\kappa}'}\right) \quad (2835)$$

But,

$$\frac{1}{i\omega_{\vec{\kappa}'}-\lambda}\frac{1}{i(\in-\omega_{\vec{\kappa}'})-\xi_{\vec{\kappa}'}} = \frac{1}{i\in-\xi_{\vec{\kappa}'}}\left(\frac{1}{i\omega_{\vec{\kappa}'}-\lambda}+\frac{1}{i(\in-\omega_{\vec{\kappa}'})-\xi_{\vec{\kappa}'}}\right) \quad (2836)$$

then

$$\Gamma_1 = -\Upsilon_1\left(\frac{J}{2}\right)^2\frac{1}{\beta}\sum_{\omega}\sum_{\vec{\kappa}'}\frac{1}{i\in-\xi_{\vec{\kappa}'}}\left(\frac{1}{i\omega_{\vec{\kappa}'}-\lambda}+\frac{1}{i(\in-\omega_{\vec{\kappa}'})-\xi_{\vec{\kappa}'}}\right) \quad (2837)$$

or

$$\Gamma_1 = \frac{\Upsilon_1}{2}\left(\frac{J}{2}\right)^2 \sum_{\vec{\kappa}'} \frac{\tanh\frac{\beta\lambda}{2} + \tanh\frac{\beta\xi_{\vec{\kappa}'}}{2}}{i\in - \xi_{\vec{\kappa}'}} \underset{\lambda\to\infty}{=} \frac{\Upsilon_1}{2}\left(\frac{J}{2}\right)^2 \sum_{\vec{\kappa}'} \frac{1+\tanh\frac{\beta\xi_{\vec{\kappa}'}}{2}}{i\in - \xi_{\vec{\kappa}'}} \tag{2838}$$

Also for

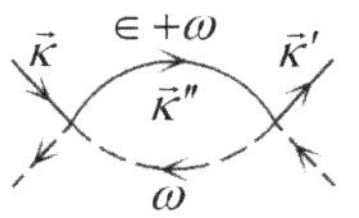

we have

$$\Gamma_2 = -\Upsilon_2\left(\frac{J}{2}\right)^2 \frac{1}{\beta}\sum_{\omega}\sum_{\vec{\kappa}'} \mathscr{F}_0(-\lambda,\omega)\mathscr{G}_0(-\xi_{\vec{\kappa}'}, \in+\omega) = -\Upsilon_2\left(\frac{J}{2}\right)^2 \frac{1}{\beta}\sum_{\omega}\sum_{\vec{\kappa}'} \frac{1}{i\in-\xi_{\vec{\kappa}'}}\left(\frac{1}{i\omega-\lambda} - \frac{1}{i(\in+\omega)-\xi_{\vec{\kappa}'}}\right) \tag{2839}$$

or

$$\Gamma_2 = -\frac{\Upsilon_2}{2}\left(\frac{J}{2}\right)^2 \sum_{\vec{\kappa}'} \frac{\tanh\frac{\beta\xi_{\vec{\kappa}'}}{2} - \tanh\frac{\beta\lambda}{2}}{i\in - \xi_{\vec{\kappa}'}} \underset{\lambda\to\infty}{=} \frac{-\Upsilon_2}{2}\left(\frac{J}{2}\right)^2 \sum_{\vec{\kappa}'} \frac{\tanh\frac{\beta\xi_{\vec{\kappa}'}}{2} - 1}{i\in - \xi_{\vec{\kappa}'}} \tag{2840}$$

So, the total scattering amplitudes

$$\Gamma = \Gamma_1 + \Gamma_2 = \frac{\Upsilon_1}{2}\left(\frac{J}{2}\right)^2 \sum_{\vec{\kappa}'} \frac{1+\tanh\frac{\beta\xi_{\vec{\kappa}'}}{2}}{i\in - \xi_{\vec{\kappa}'}} + \frac{-\Upsilon_2}{2}\left(\frac{J}{2}\right)^2 \sum_{\vec{\kappa}'} \frac{\tanh\frac{\beta\xi_{\vec{\kappa}'}}{2} - 1}{i\in - \xi_{\vec{\kappa}'}} \tag{2841}$$

or

$$\Gamma = (\Upsilon_1 - \Upsilon_2)\frac{1}{2}\left(\frac{J}{2}\right)^2 \sum_{\vec{\kappa}'} \frac{\tanh\frac{\beta\xi_{\vec{\kappa}'}}{2}}{i\in - \xi_{\vec{\kappa}'}} + (\Upsilon_1 + \Upsilon_2)\frac{1}{2}\left(\frac{J}{2}\right)^2 \sum_{\vec{\kappa}'} \frac{1}{i\in - \xi_{\vec{\kappa}'}} \tag{2842}$$

Because

$$\left(\frac{J}{2}\right)^2 \sum_{\vec{\kappa}'} \frac{\tanh\frac{\beta\xi_{\vec{\kappa}'}}{2}}{i\in - \xi_{\vec{\kappa}'}} = \nu\left(\frac{J}{2}\right)^2 \int d\xi_{\vec{\kappa}'} \frac{\tanh\frac{\beta\xi_{\vec{\kappa}'}}{2}}{i\in - \xi_{\vec{\kappa}'}} \underset{i\in\to\in+i\delta}{=} -\nu\left(\frac{J}{2}\right)^2 \ln\left(\frac{\in_F}{\max\{T,\in\}}\right) \tag{2843}$$

$$\int \frac{d\vec{\kappa}'}{(2\pi)^3}\frac{1}{i\in - \xi_{\vec{\kappa}'}} = \mathrm{P}\int \frac{d\xi_{\vec{\kappa}'}}{i\in - \xi_{\vec{\kappa}'}} = 0 \tag{2844}$$

Hence,

$$\Gamma = (\Upsilon_1 - \Upsilon_2)\frac{1}{2}\left(\frac{J}{2}\right)^2 \sum_{\vec{\kappa}'} \frac{\tanh\frac{\beta\xi_{\vec{\kappa}'}}{2}}{i\in - \xi_{\vec{\kappa}'}} = -2\vec{\boldsymbol{\sigma}}_{\sigma'\sigma}\vec{\mathbf{S}}_{mm'}\left(\frac{J}{2}\right)^2 \sum_{\vec{\kappa}'} \frac{\tanh\frac{\beta\xi_{\vec{\kappa}'}}{2}}{i\in - \xi_{\vec{\kappa}'}} \tag{2845}$$

or

$$\Gamma = 2\vec{\sigma}_{\sigma'\sigma}\vec{S}_{mm'}\nu\left(\frac{J}{2}\right)^2 \ln\left(\frac{\epsilon_F}{\max(|\epsilon|, T)}\right) \tag{2846}$$

Because electrons with energies $|\epsilon| \approx T$ are most interesting for the physical description of the system, we neglect the case discrimination $\max(|\epsilon|, T)$ in the argument of the logarithm.

From the aforementioned, the total scattering amplitude:

$$\Gamma = -\frac{J}{2}\vec{\sigma}_{\sigma'\sigma}\vec{S}_{mm'}\nu\left[1 - 2J\nu\ln\left(\frac{\mu}{\max(|\xi|, T)}\right) + \cdots\right] = -\frac{J}{2}\vec{\sigma}_{\sigma'\sigma}\vec{S}_{mm'}\nu\frac{1}{1 + 2J\nu\ln\left(\frac{\mu}{\max(|\xi|, T)}\right)} \tag{2847}$$

or

$$\Gamma = -\frac{J}{2}\vec{\sigma}_{\sigma'\sigma}\vec{S}_{mm'}\nu\frac{1}{1 + 2J\nu\ln\left(\frac{\mu}{\max(|\xi|, T)}\right)} \equiv -\frac{\vec{\sigma}_{\sigma'\sigma}\vec{S}_{mm'}}{2}\frac{1}{2\ln\left(\frac{T}{T_K}\right)} \tag{2848}$$

This converges for $T > T_K$ and exhibits a logarithmic divergence for low temperatures T and signals a breakdown of perturbation theory. This occurs at a characteristic temperature scale that can be read from (2848), the Kondo temperature T_K. However, the logarithmic behavior of the perturbation expansion paves the way for the development of the **renormalization group method**. This is particularly useful for analytically studying the interplay of Kondo screening and RKKY interaction. In (2848), as a consequence of the logarithmic behavior, the parameters J, ϵ_F, and ν indeed conspire to form the Kondo temperature T_K as the only scale in the problem.

Considering that the **logarithm is a scale invariant function**, it is possible that the resummation of a logarithmic perturbation expansion yields a universal behavior. This is in the sense that some variables (such as energy, temperature, and so forth) can be expressed in units of a single scale, T_K, such that all physical quantities are functions of dimensionless variables. For the Kondo model, this property can be visualized via the scattering amplitude Γ. The renormalized coupling constant may be obtained from

$$\Gamma = -\frac{1}{2}\vec{\sigma}_{\sigma'\sigma}\vec{S}_{mm'}\frac{J\nu}{1 + 2J\nu\ln\left(\frac{\mu}{\max(|\xi|, T)}\right)} = -\frac{J\nu}{2}\vec{\sigma}_{\sigma'\sigma}\vec{S}_{mm'}\sum_n (-2J\nu)^n \ln^n\left(\frac{\mu}{\max(|\xi|, T)}\right) \tag{2849}$$

and

$$J_T = \frac{J\nu}{1 + 2J\nu\ln\left(\frac{\mu}{\max(|\xi|, T)}\right)} \equiv \frac{1}{2\ln\left(\frac{T}{T_K}\right)} \tag{2850}$$

In this procedure, the Kondo interaction J is renormalized and is equivalent to the renormalization group (RG) or poor man's scaling. This scaling law implies the attractive force becomes stronger for lower energies. From (2850), we conclude that

$$T = T_K = \epsilon_F \exp\left\{-\frac{1}{2\nu J}\right\} \tag{2851}$$

and J_T diverges, which implies the occurrence of some kind of instability. Because (2850) is derived under the assumption that J is small, our theory should be qualitative. Our result shows the perturbative method to be invalid because T achieves the value T_K from the aforementioned, signaling a breakdown of the given scaling approach. Nevertheless, the solution J_T is obtained in the zero temperature limit. For a finite temperature regime, $T >> T_K$, the renormalization group flow is cut off at this temperature yielding a renormalized coupling J_T. So, in the regime $T >> T_K$, this is the leading term; physical quantities may be calculated by replacing the bare interaction vertex by this renormalized coupling.

Considering equations (2850) and (2848), we rewrite the antiferromagnetic scattering amplitude as the sum of the most divergent terms in perturbation theory expansion—the **leading logarithmic contributions** of each order summed up in a geometric series [100]:

$$\Gamma = \frac{J\nu}{2}\vec{\sigma}_{\sigma'\sigma}\vec{\mathbf{S}}_{mm'}\sum_n (2J\nu)^n \ln^n\left(\frac{\mu}{\max(|\xi|, T)}\right) \tag{2852}$$

The diagrammatic representations are shown in Figure 13.18.

$$\Gamma^{(1)} = \frac{1}{2}\vec{\boldsymbol{\sigma}}_{\sigma'\sigma}\vec{\mathbf{S}}_{mm'}(2\nu J)^2 \ln\frac{\epsilon_F}{T} \quad + \quad \Gamma^{(2)} = \frac{1}{2}\vec{\boldsymbol{\sigma}}_{\sigma'\sigma}\vec{\mathbf{S}}_{mm'}(2\nu J)^3 \ln^2\frac{\epsilon_F}{T} \quad + \quad \Gamma^{(3)} = \frac{1}{2}\vec{\boldsymbol{\sigma}}_{\sigma'\sigma}\vec{\mathbf{S}}_{mm'}(2\nu J)^4 \ln^3\frac{\epsilon_F}{T} \quad +\cdots \tag{2853}$$

FIGURE 13.18 Diagrammatic representation of scattering amplitude in a geometric series.

Therefore, from equation (2852), all higher orders of perturbation theory contain powers of that logarithmic dependency and summing up the terms for scattering amplitude results in a geometric series. Summation of the leading logarithms relates an accumulation of spin-flip processes that leads to the Kondo resonance, which is a narrow peak centered at the Fermi level ϵ_F and broadened by T_K in the spectral function of the single energy level.

If $J < 0$, then $\vec{\sigma}$ and $\vec{S}$ are antiparallel. In this case, we have an **antiferromagnetic state** and equation (2852) has a logarithmic divergence (singularity) when T approaches the Kondo temperature T_K that indicates the breakdown of perturbation theory. The behavior of the scattering amplitude (whose square is the resistivity) in (2852) is a consequence of a large **soft** cloud of conduction electrons' spin polarization resonating about the impurity's local moment. In this case, the role of temperature is to break up this cloud and reduce the strong scattering that such resonance produces. Hence, it is possible to predict the concentration of magnetic impurities at which the overlap of neighboring clouds becomes more enhanced than the energy binding each to its local moment. For such a value of the concentration, the **spin glass** phase is enhanced. For higher concentrations, the **ordered magnetic alloys** are relevant.

From the many body character of the Kondo effect, the coupling of the localized spin with the electrons in the reservoir is no longer a small perturbation. If $J > 0$, then $\vec{\sigma}$ and $\vec{S}$ are parallel. In this case, we have a **ferromagnetic state**. Here, there is no divergence (no singularity). We observe that the Kondo effect appears only if $J < 0$. We find a correction due to the scattering in first and second Born approximation.

The magnitudes of the Kondo temperature T_K estimated from the resistivity measurements range from less than 1K for MgMn, CuMn, CdMn, and so on to greater than 1000K for AuTi and CuNi. This leads to the fact that even at higher Kondo temperatures T_K, we have the **weak coupling Kondo regime**. Notwithstanding, the ground state might still exist.

The Kondo model shows that, unlike nonmagnetic impurities that produce temperature-independent scattering as well as residual resistance of metals, scattering from a magnetic impurity is enhanced at low energies or temperatures. From the relaxation time, we have

$$\frac{1}{\tau_0} = \frac{mp_F}{\pi} S(S+1)(J_T\nu)^2 \tag{2854}$$

13.2.13.9 Scaling and Parquet Equation

When finite-order perturbation theory breaks down, we have to use partial summation such as for the case of the electron gas in RPA. Abrikosov [100] proved that at high temperatures, T, the dominant diagrams were **parquets**. These are diagrams such as in Figure 13.18 that result from **creating** more and more interaction vertices and inserting in them a fermionic-impurity pair bubble. So, from these discussions, the renormalization of Γ can be iterated repeatedly via a differential equation:

$$\frac{d\Gamma(x)}{dx} = -\frac{mp_F}{\pi^2}\Gamma^2(x) \tag{2855}$$

Here,

$$\Gamma(x) = J - \frac{mp_F}{\pi^2}\int_0^x dy\left[\Gamma(y)\right]^2 \tag{2856}$$

and

$$x = \ln\left(\frac{\Lambda}{\Lambda'}\right) \tag{2857}$$

represents the scale factor between the original scale Λ and the current scale Λ'. This is the result of the renormalization group and can be considered via the Parquet equation in Figure 13.19:

(2858)

FIGURE 13.19 Result of the renormalization group via the parquet equation.

Equation (2855) is an example of an **RG flow equation**. There are similar equations that outline the scaling of parameters in theories describing continuous phase transitions. In this case, there are fluctuations at all length scales. We also have massless quantum field theories where zero mass implies the absence of any length or time scale.

13.2.13.10 Kondo Effect and Numerical Renormalization Group

The simplest model of a magnetic impurity is the **Anderson model**, which was introduced in 1961 [101]. This model describes the physics of single impurities hosted in a metal and has only one electron level with energy ϵ_0. Here, the electron can quantum mechanically tunnel from the impurity and escape provided its energy lies above the Fermi level; otherwise, it remains trapped. For this case, the defect has a spin of $\frac{1}{2}$ and its z-components are fixed either as spin up or spin down. Nonetheless, exchange processes can take place that effectively flip the spin of the impurity from spin up to spin down, or vice versa, while simultaneously creating a spin excitation in the Fermi sea.

In the leading logarithmic approximation, the Kondo interaction J is renormalized as in (2850). This is equivalent to the renormalization group (RG), or poor man's scaling. The summation of leading logarithms corresponds to an accumulation of spin-flip processes that leads to the **Kondo resonance**, which is a narrow peak centered at the Fermi level $\in_F$ and broadened by T_K in the spectral function of the single energy level. The Kondo resonance is due to the formation of a quasi–bounded state of conduction electrons at the vicinity of the impurity site when the density of states become enhanced at the Fermi level. This, in turn, provides additional contributions to the thermodynamic and transport properties (such as specific heat, magnetic susceptibility, conductivity, and so on), with the weight proportional to the impurity concentration.

The concept of universality is the starting point for the renormalization group (RG) method. The renormalization begins with the transformation of the Hamiltonian determinant describing the high-energy physics. Because our interest is on low-energy physics, we then find a transformation for the Hamiltonian that removes the high-energy state and absorbs it into a Hamiltonian with the same form but different parameters. This is because in the Kondo effect, the kinetic energy of free conduction electrons typically is on the order of several electron-volts, whereas the energy scale of the electron-impurity interaction is one of few meV. Therefore, in order to appreciate the physics in the low-temperature regime $T < T_K$, the Kondo model at higher energies should be scaled down to lower energies by lowering an energy cutoff scale, Λ. The ultraviolet cutoff leads to the maximal allowed energy of the excitations being taken into account in the model while ignoring physical quantities and excitations with energies above the cutoff. As only low-energy (infrared limit) properties of the system are of primary interest, successive lowering of the cutoff tailors the behavior of the system to lower energies and thereby eliminates any higher energy contributions. Once the system is renormalized, it is rescaled again to a new energy cutoff. Integrating out the degrees of freedom outside the newly rescaled energy range yields an effective Hamiltonian that depends on the running energy scale Λ after rescaling of the cutoff energy. So, it is the cutoff dependence of physical quantities that is the basis of the renormalization group, which should be a set of transformations of the quantities in the Dyson equation. These transformations are simple multiplications by a factor that leaves the Dyson equation invariant. The transformed quantities, however, obey the Lie equations, which turn out be simpler to solve than the Dyson equation when we examine this topic further. Not only does this prove the strength of this method, it may lead to a new series expansion that should be an improvement over the original expansion.

For the poor man's scaling approach, the high-energy states are integrated out step by step by reducing the half-bandwidth Λ by $\delta\Lambda$ and absorbing it into the Kondo coupling J. The introduction of the bandwidth Λ as a cutoff for the energy integration yielded a logarithmic dependency on the bandwidth in (2850). The terms do not vanish for $\Lambda \to \infty$, and this shows the importance of high-energy excitations in the Kondo problem that cannot be neglected—for example, those with energies close to the bandwidth. The scaling approach provides an excellent solution for taking them into account. The basic principle of a scaling theory maps a system onto a reduced version of itself by integrating out high-energy contributions beginning from a cutoff energy that may be given, for example, by the bandwidth of the system. It is obvious that the new, energetically reduced system will have different and even newly generated couplings. Hence, it is necessary to rewrite the Hamiltonian determinant in an invariant form for scaling purposes. This is to obtain the relation between the old and the new coupling constants with the relations called **flow equations**. Analyzing these relations for fix points, invariants, or divergences will reveal the importance of the physical properties of the system under consideration.

13.2.13.10.1 Scaling

If we integrate out high-energy shells from $-\Lambda$ to $-\Lambda+\delta\Lambda$ and $\Lambda-\delta\Lambda$ to Λ (Figure 13.20), this results in the Hamiltonian dependent on the running energy scale Λ:

$$\mathrm{H}_\Lambda = \sum_{|\in_{\vec{\kappa}}|<\Lambda} \in_{\vec{\kappa}} \hat{\psi}^\dagger_{\vec{\kappa}\sigma}\hat{\psi}_{\vec{\kappa}\sigma} + J_{\Lambda z}\sum \hat{\psi}^\dagger_{\vec{\kappa}'\sigma'}\sigma^z_{\sigma'\sigma}\hat{\psi}_{\vec{\kappa}\sigma}\mathbf{S}_z + \frac{J_{\Lambda\pm}}{2}\sum\left(\hat{\psi}^\dagger_{\vec{\kappa}'\uparrow}\hat{\psi}_{\vec{\kappa}\downarrow}\mathbf{S}_- + \hat{\psi}^\dagger_{\vec{\kappa}'\downarrow}\hat{\psi}_{\vec{\kappa}\uparrow}\mathbf{S}_+\right) \tag{2859}$$

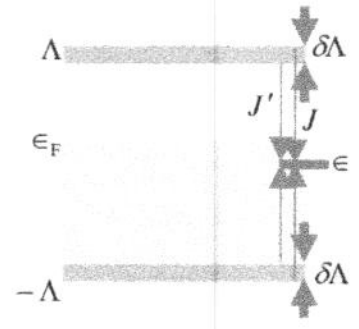

FIGURE 13.20 Principle of scaling: integrating out high-energy shells from $-\Lambda$ to $-\Lambda+\delta\Lambda$ and $\Lambda-\delta\Lambda$ to Λ. This is reflected in renormalizing the coupling constants $J \to J'$.

The last two terms correspond to the original Kondo Hamiltonian with the renormalized coupling constant J_Λ. We perform the renormalization of the Kondo Hamiltonian H_Λ explicitly in a perturbative way, following Anderson [102]. Therefore, it is convenient to introduce new variables (dimensionless) such as

$$J = J_\Lambda \nu \quad , \quad J^{\pm} = J^{z} = J \tag{2860}$$

To enforce the invariance of physical quantities under the RG flow, it is sufficient to keep the total conduction pseudo-fermion electron vertex Γ invariant because all physical quantities are derived from it within the Kondo model. So, from the renormalization group, we define the following set of transformations:

$$\mathbf{G} \to z_1 \mathbf{G} \quad , \quad \mathcal{G} \to z_2 \mathcal{G} \tag{2861}$$

$$\Gamma^{J} \to z_3^{-1}\Gamma^{J} \quad , \quad \bar{J} = \nu J \quad , \quad \bar{J} \to z_1^{-1} z_2^{-1} z_3 \bar{J} \tag{2862}$$

where the multipliers z_1, z_2, and z_3 are real numbers. Because z_1, z_2, and z_3 can vary continuously, the renormalization group is a **continuous** or **Lie group.** We note that $\mathbf{G}$, $\mathcal{G}$, Γ^{J}, and $\bar{J}$ are true **physical** quantities, while $z_1\mathbf{G}$, $z_2\mathcal{G}$, $z_3^{-1}\Gamma^{J}$, and $z_1^{-1}z_2^{-1}z_3\bar{J}$ are **unphysical**. So, the following Dyson equation is invariant under the renormalization group transformation:

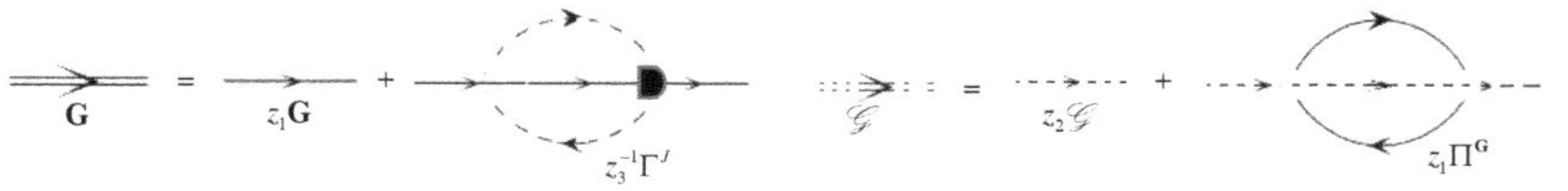

(2863)

$$\Sigma \to \bar{J}.\Pi.\Gamma^{J}\mathcal{G} \to \bar{J}.z_1\Pi z_3^{-1}\Gamma^{J} z_2\mathcal{G} = \bar{J}.z_1 z_2 z_3^{-1}\Pi\Gamma^{J}\mathcal{G} \tag{2864}$$

and so,

$$\bar{J} \to \bar{J}.z_1 z_2 z_3^{-1} \tag{2865}$$

From these equations, we infer that the **unphysical** or **renormalized** quantities obey the same Dyson equation as the physical ones. It is worthy to note that the resulting equations are exact for high-density electron gas.

We consider all physical quantities of the system

$$A \equiv F\left(\frac{\omega}{\lambda}, \frac{\Lambda}{\lambda}, J\right) \tag{2866}$$

dependent on energy ω and temperature Λ in a universal way, with universal functions F and some (**yet unknown**) characteristic scale λ that depends on the microscopic parameters of the Hamiltonian, $J(\epsilon_F)$, ϵ_F, and $\nu(\epsilon_F)$. The dependence of F on the given parameters is only implicit via λ. This implies that different values of this set of parameters realize the same physical system (defined by its observables, F), and that is only if the different sets of parameter values lead to the same scale λ. Hence, as seen previously, calculations are done at each step so that the coupling constant J_Λ tailors the transformation of the Hamiltonian such that the physical observable F is invariant. Generally, the cutoff reduction may generate new types of interaction operators in the Hamiltonian thereby guaranteeing that physical observables are invariant.

For **scaling**, the physical observable F, which may be some function d (that may be a **screening** or **shielding** factor), or Γ^J obeys the so-called **normalization condition**:

$$\mathrm{F}\left(\frac{\omega}{\lambda},\frac{\Lambda}{\lambda},J\right)=1 \tag{2867}$$

Here, λ is the normalizing energy. If d is the **screening** factor, then the aforementioned condition corresponds to short distances where the charge is unscreened. Through a scaling factor, the physical observable F can be expressed as follows:

1.
$$\mathrm{F}\left(\frac{\omega}{\lambda'},\frac{\Lambda}{\lambda'},J'\right)=z_2 d\left(\frac{\omega}{\lambda},\frac{\Lambda}{\lambda},J\right) \tag{2868}$$

Here λ and λ' are two different normalizing energies and

$$z_2=d\left(\frac{\lambda}{\lambda'},\frac{\Lambda}{\lambda'},J'\right)\quad,\quad z_2^{-1}=d\left(\frac{\lambda'}{\lambda},\frac{\Lambda}{\lambda},J\right) \tag{2869}$$

Similarly,

$$z_3=\Gamma^J\left(\frac{\lambda'}{\lambda},\frac{\Lambda}{\lambda},J\right)\quad,\quad J'=\psi^J\left(\frac{\lambda'}{\lambda},\frac{\Lambda}{\lambda},J\right)\cdot J \tag{2870}$$

And, introducing the ψ^J -function, then

$$\psi^J\left(\frac{\lambda'}{\lambda},\frac{\Lambda}{\lambda},J\right)=\Gamma^J\left(\frac{\lambda}{\lambda'},\frac{\Lambda}{\lambda'},J'\right)\cdot d\left(\frac{\lambda'}{\lambda},\frac{\Lambda}{\lambda},J\right) \tag{2871}$$

So,

$$d\left(\frac{\omega}{\lambda'},\frac{\Lambda}{\lambda'},J'\right)=z_2 d\left(\frac{\omega}{\lambda},\frac{\Lambda}{\lambda},J\right)\quad,\quad z_2=d\left(\frac{\lambda}{\lambda'},\frac{\Lambda}{\lambda'},J'\right)\quad,\quad z_2^{-1}=d\left(\frac{\lambda'}{\lambda},\frac{\Lambda}{\lambda},J\right) \tag{2872}$$

and

$$d\left(\frac{\omega}{\lambda'},\frac{\Lambda}{\lambda'},J'\right)\cdot d\left(\frac{\lambda'}{\lambda},\frac{\Lambda}{\lambda},J\right)=d\left(\frac{\omega}{\lambda},\frac{\Lambda}{\lambda},J\right) \tag{2873}$$

It is convenient to relabel the quantities found in the last equation as follows

$$x=\frac{\omega}{\lambda}\quad,\quad y=\frac{\Lambda}{\lambda}\quad,\quad t=\frac{\lambda'}{\lambda}\quad,\quad \frac{\omega}{\lambda'}=\frac{x}{t}\quad,\quad \frac{\Lambda}{\lambda'}=\frac{y}{t} \tag{2874}$$

Therefore, we have the functional equation (universal) of the renormalization group for the given case:

$$d(x,y,J)=d\left(\frac{x}{t},\frac{y}{t},J\psi^{J}\right)\cdot d(t,y,J) \tag{2875}$$

13.2.13.10.2 Lie Equation for the Renormalization Group

We now transform equation (2875) into differential form via

$$\ln d(x,y,J)=\ln d\left(\frac{x}{t},\frac{y}{t},J\psi^{J}\right)+\ln d(t,y,J) \tag{2876}$$

This is done by differentiating both sides of (2876) with respect to x and letting $\xi=\frac{x}{t}$ on one side:

$$\frac{\partial}{\partial x}\ln d(x,y,J)=\frac{1}{t}\frac{\partial}{\partial \xi}\ln d\left(\xi,\frac{y}{t},J\psi^{J}\right) \tag{2877}$$

Now, we set $t=x(\xi=1)$ so as to obtain an equation with a single variable x, yielding the **Lie equation**:

$$\frac{\partial}{\partial x}\ln d(x,y,J)=\frac{1}{x}\left[\frac{\partial}{\partial \xi}\ln d\left(\xi,\frac{y}{x},J\psi^{J}\right)\right]\Bigg|_{\xi=1} \tag{2878}$$

where we have used the so-called **normalization condition** (i.e., $d=1$ when $\xi=1$):

$$d\left(\xi=1,\frac{y}{x},J\psi^{J}\right)=1 \tag{2879}$$

Solution of the Lie Equation

Considering (2878) and the normalization condition (2879), we find the following first-order perturbation expansion:

$$\Gamma^{J}=1+2J\ln\frac{\Lambda}{\omega} \tag{2880}$$

For convenience, we define an invariant charge under the renormalization group

$$J_{\text{inv}}=\bar{J}\psi^{J} \tag{2881}$$

So, from

$$\frac{\partial}{\partial x}\ln\psi^{J}(x,y,J)=\frac{1}{t}\frac{\partial}{\partial x}\ln\psi^{J}\left(\frac{x}{t},\frac{y}{x},J\psi^{J}\right) \tag{2882}$$

and letting $\xi=\frac{x}{t}$, then

$$\frac{\partial}{\partial x}\ln\psi^{J}(x,y,J)=\frac{1}{x}\left[\frac{\partial}{\partial \xi}\ln\psi^{J}\left(\xi,\frac{y}{x},J\psi^{J}\right)\right]\Bigg|_{\xi=1} \tag{2883}$$

Considering

$$J_{\text{inv}} = \bar{J}\psi^J \equiv \bar{J}\Gamma^J = \bar{J}\Gamma d = z_3 z_1^{-1} z_2^{-1} z_3^{-1} z_2 \bar{J}\Gamma d \tag{2884}$$

then

$$J_{\text{inv}}\left(\frac{\omega}{\lambda}, \frac{\Lambda}{\lambda}, J\right) = J_{\text{inv}}\left(\frac{\omega}{\lambda'}, \frac{\Lambda}{\lambda'}, J'\right) \tag{2885}$$

$$\psi^J\left(\frac{\omega}{\lambda'}, \frac{\Lambda}{\lambda'}, J'\right)\psi^J\left(\frac{\lambda'}{\lambda}, \frac{\Lambda}{\lambda}, J\right) = \psi^J\left(\frac{\omega}{\lambda}, \frac{\Lambda}{\lambda}, J\right) \tag{2886}$$

$$\ln\psi^J\left(\frac{\omega}{\lambda'}, \frac{\Lambda}{\lambda'}, J'\right) + \ln\psi^J\left(\frac{\lambda'}{\lambda}, \frac{\Lambda}{\lambda}, J\right) = \ln\psi^J\left(\frac{\omega}{\lambda}, \frac{\Lambda}{\lambda}, J\right) \tag{2887}$$

So,

$$\frac{\partial}{\partial x}\ln\psi^J\left(x, y, \bar{J}\right) = \frac{1}{x}\left[\frac{\partial}{\partial \xi}\ln\psi^J\left(\xi, \frac{y}{x}, \bar{J}\psi\right)\right]\Bigg|_{\xi=1} \tag{2888}$$

$$J_{\text{inv}}\left(x, y, \bar{J}\right) = J_{\text{inv}}\left(\frac{x}{t}, \frac{y}{t}, J_{inv}\left(t, y, \bar{J}\right)\right) \tag{2889}$$

where

$$J_{\text{inv}}\left(1, y, \bar{J}\right) = \bar{J} \tag{2890}$$

and

$$\frac{\partial}{\partial x}\left[J_{\text{inv}}\left(x, y, \bar{J}\right)\right] = \frac{1}{x}\left[\frac{\partial}{\partial \xi}J_{\text{inv}}\left(\xi, \frac{y}{x}, \bar{J}\right)\right]\Bigg|_{\xi=1} = \frac{1}{x}\varphi\left(y, J\right) \tag{2891}$$

Here, we introduce the **Gell-Mann-Low function** [103]:

$$\varphi\left(y, J\right) = \left[\frac{\partial}{\partial \xi}J_{\text{inv}}\left(\xi, \frac{y}{x}, \bar{J}\right)\right]\Bigg|_{\xi=1} \tag{2892}$$

13.2.13.10.3 Calculation of the Gell-Mann-Low Function

One-Loop Approximation

Gell-Mann and Low [103] originally developed the renormalization group to improve on perturbation theory in particles:

$$J_{\text{inv}} = \bar{J}\Gamma d \quad , \quad \xi = \frac{\omega}{\lambda} \quad , \quad \lambda \to \Lambda \tag{2893}$$

where

$$\Gamma = 1 + 2\bar{J}\ln\frac{\Lambda}{\omega} + 2\bar{J}^2\left[S(S+1)-1\right]\ln\frac{\Lambda}{\omega} + \cdots \quad , \quad d = 1 - 2\bar{J}^2 S(S+1)\ln\frac{\Lambda}{\omega} + \cdots \tag{2894}$$

In the one-loop approximation, we have

$$J_{inv} = \bar{J}\Gamma d = \bar{J}\left(1 + 2\bar{J}\ln\frac{\Lambda}{\omega}\right) = \bar{J} + 2\bar{J}^2\ln\frac{\Lambda}{\omega} \tag{2895}$$

$$\varphi(y, J) = \left[\frac{\partial}{\partial \xi} J_{inv}\left(\xi, \frac{y}{x}, \bar{J}\right)\right]\Bigg|_{\xi=1} = -2\bar{J}^2 + 2\bar{J}^3 + \cdots \tag{2896}$$

Letting $x = \frac{\Lambda}{\omega}$, then the **differential renormalization group equation (of one-loop approximation)**:

$$\frac{d\bar{J}}{d\ln x} = -2\bar{J}^2 \tag{2897}$$

This is the Gell-Mann-Low equation, and the **Gell-Mann-Low function**

$$\beta(\bar{J}) = -2\bar{J}^2 \tag{2898}$$

controls the running coupling constant renormalization and is called the β-function or **relevant** of the RG. The RG equation can be integrated:

$$\int_{\bar{J}}^{J_{inv}} \frac{d\bar{J}}{2\bar{J}^2} = -\int_x^1 d\ln x \tag{2899}$$

or

$$\ln x = \frac{1}{2J_{inv}} - \frac{1}{2\bar{J}} \tag{2900}$$

This gives the renormalized coupling constant

$$J_{inv} = \frac{\bar{J}}{1 + 2\bar{J}\ln x} = \frac{\nu J}{1 + 2\nu J\ln\frac{\omega}{\Lambda}} \equiv J'(\Lambda) \equiv \frac{1}{2\ln\left(\frac{T}{T_K}\right)} \tag{2901}$$

with boundary condition

$$J'(\omega) = \nu J \tag{2902}$$

From here, in the scaling calculation, the Kondo temperature marks the end of the validity of the results because the coupling constant again diverges for antiferromagnetic $J > 0$ when T achieves the value

$$T_K = \Lambda\exp\left\{-\frac{1}{2J\nu}\right\} \equiv \Lambda\exp\left\{-\frac{1}{2J'(\Lambda)}\right\} \tag{2903}$$

This is a consequence of the previous perturbative RG treatment. Nevertheless, this divergence permits us to conclude that the ground state of the single-impurity Kondo model is a spin-singlet state between the impurity spin and the spin cloud of the surrounding conduction electron spins. Also, it permits a more general definition of the Kondo spin screening scale T_K (i.e., the value of the running cutoff Λ where the coupling constant diverges and the singlet starts to be formed). So, the renormalization procedure should be halted when Λ becomes of the order of the temperature T of the system.

The expression of the Kondo temperature T_K is a constant during the scaling where J_Λ is given in (2901). The values with this property are called scaling invariants and underline the role of the Kondo temperature as the important energy scale for the Kondo effect. Systems that are characterized by the same $T_K(J_\Lambda, \nu)$ are on the same trajectories. Hence, they show the same low-energy behavior, where the attractive force becomes stronger.

Therefore, the scaling method provides a means of including contributions of high-energy processes in first-order calculations. This is done by summation, which leads to new coupling constants J_Λ for a system with reduced bandwidth that is characterized by the energy scale T_K. Higher-order contributions could be considered. However, they are irrelevant because they behave like $\frac{1}{T}$ rather than $\ln T$. Therefore, they are safely neglected for the high-energy region.

The disturbing issue is the divergence of the renormalized Kondo coupling J_Λ when the energy achieves the range defined by the Kondo temperature T_K. Because perturbation theory works only for small interactions J_Λ, the divergence first indicates the breakdown of the perturbative approach and, second defines a range where the perturbation theory is more justifiable. This is the so-called **weak-coupling regime**. So, T_K might be thought of as a characteristic energy scale indicating a crossover between the weakly and strongly interacting regimes rather than a phase transition.

It is instructive to note that an important feature of equation (2897) is that it represents an example of **asymptotic freedom**. This is because the effective coupling J_{inv} is weak at large energy scales (i.e., at the beginning of the poor man's scaling procedure) and grows beyond unity at low-energy scales.

For $T >> T_K$, the impurity is effectively weakly coupled to the electron gas, and the spin $S = 1/2$ is not screened out, although the properties of the system can be computed within a perturbative approach provided the renormalized interaction J_{inv} is used. Here, the scaling behavior of J is certain and reflected in physical quantities, that is, they can be written as universal functions of $\frac{T}{T_K}$. This concerns, for example, the magnetic susceptibility or the resistance ρ.

$$\rho \equiv \Gamma^2 = \left(\frac{J}{2}\vec{\sigma}_{\sigma'\sigma}\vec{\mathbf{S}}_{mm'}\nu\right)^2 \frac{1}{\left(1+2J\nu\ln\left(\frac{\mu}{T}\right)\right)^2} \tag{2904}$$

If the temperature is lowered $T \to T_K$, the spin scattering of electrons at the magnetic impurity increases. This results in the Kondo resistance minimum that is the famous signature of the Kondo effect.

When the effective coupling J_{inv} is strong, the impurity is strongly bound to the conduction electrons by antiferromagnetic exchange, while a paramagnetic **Kondo singlet** is established and decouples from the system. The remaining (low-energy) conduction electrons for $T << T_K$ would then represent a **special kind of Fermi liquid** with a resonance because moving to the Kondo cloud sites is accompanied by a large energy cost. Here, the impurity spin is almost completely screened by the electron gas. So, perturbative renormalization analysis and, in particular (2901), are not applicable in this regime.

13.2.13.10.4 Two-Loop-Beyond Parquet Approximation

Similarly, we consider again equation (2897) so that

$$J_{inv} = \bar{J}\Gamma d = \bar{J}\left(1+2\bar{J}\ln\frac{\Lambda}{\omega}+2\bar{J}^2\left[S(S+1)-1\right]\ln\frac{\Lambda}{\omega}+\cdots\right)\left(1-2\bar{J}^2 S(S+1)\ln\frac{\Lambda}{\omega}+\cdots\right) \tag{2905}$$

and

$$\varphi(y,J)=\frac{\partial}{\partial\xi}J_{inv}\left(\xi,\frac{y}{x},\bar{J}\right)\Bigg|_{\xi=1}=-2\bar{J}^2+2\bar{J}^3+\cdots \tag{2906}$$

Therefore, it follows that

$$\int_{\bar{J}}^{J_{inv}}\frac{d\xi}{-2\xi^2+2\xi^3}=\ln x \tag{2907}$$

or

$$\frac{1}{J_{inv}}-\frac{1}{\bar{J}}-\ln\frac{J_{inv}}{\bar{J}}=2\ln x \tag{2908}$$

and

$$\frac{1}{J_{inv}}=\frac{1+2\bar{J}\ln x}{\bar{J}}+\ln\frac{\bar{J}_{inv}}{J} \tag{2909}$$

$$\varphi(J_{inv})-\varphi(J)=\ln\frac{\omega}{\Lambda} \tag{2910}$$

$$-\varphi(J)=\ln\frac{\omega}{\Lambda} \tag{2911}$$

Hence,

$$\varphi(\bar{J})=-\ln\frac{T_K}{\Lambda} \tag{2912}$$

and

$$\varphi(J_{inv})=\ln\frac{\omega}{T_K} \tag{2913}$$

$$\frac{1}{2\bar{J}}-\frac{1}{2}\ln\bar{J}=-\ln\frac{T_K}{\Lambda} \tag{2914}$$

$$\frac{1}{2\bar{J}}=\ln\sqrt{\bar{J}}-\ln\frac{T_K}{\Lambda}=\ln\frac{\sqrt{\bar{J}}\Lambda}{T_K} \tag{2915}$$

and

$$T_K=\Lambda\sqrt{\bar{J}}\exp\left\{-\frac{1}{2J}\right\} \tag{2916}$$

13.2.13.11 Anisotropic Kondo Model

We solve equation (2855) by introducing anisotropy into the original Kondo coupling. This equation is now replaced by

$$\frac{dJ_z}{d\Lambda} = 2J_\pm^2 \quad , \quad \frac{dJ_\pm}{d\Lambda} = 2J_\pm J_z \tag{2917}$$

Also,

$$\frac{dJ_{\text{inv}}^{\pm}}{d\ln x} = -2J_{\text{inv}}^{z} J_{\text{inv}}^{\pm} \tag{2918}$$

with

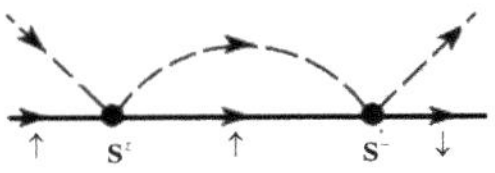

and

$$\frac{dJ_{\text{inv}}^{z}}{d\ln x} = -2\left(J_{\text{inv}}^{\pm}\right)^2 \tag{2919}$$

with

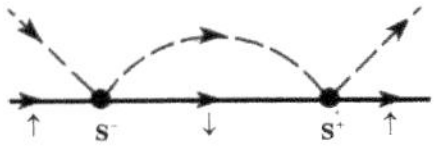

Dividing the second equation in (2917) by the first, we then have

$$\frac{dJ^\pm}{dJ^z} = \frac{J^z}{J^\pm} \tag{2920}$$

From here, we have

$$J^\pm dJ^\pm - J^z dJ^z = 0 \tag{2921}$$

The character of the solutions can be readily understood by observing that

$$\left(J^\pm\right)^2 - \left(J^z\right)^2 = \text{const} \tag{2922}$$

which is invariant along the flow, leading to a series of hyperbolae that are the required scaling trajectories. The next step is to figure out the direction of the flow (Figure 13.21):

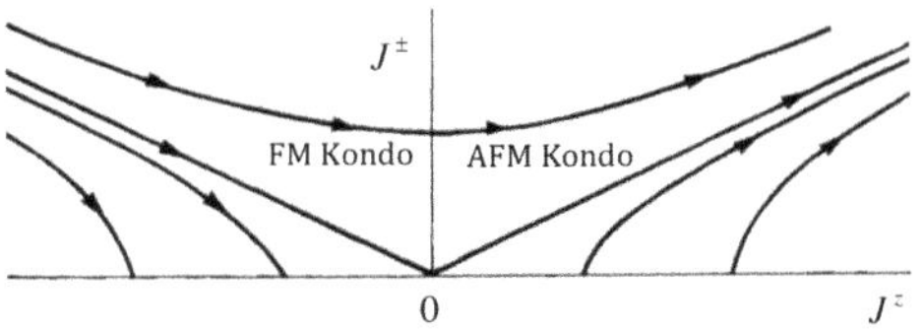

FIGURE 13.21 The RG flow for the anisotropic Kondo problem, a poor man's derivation of scaling laws for the Kondo problem [115].

Summarily,

1. **For the ferromagnetic couplings, $J_z < 0$ and $|J_z| > |J^{\pm}|$. The flow is to $J_{\perp} = 0$. This implies that at low energies the amplitude for spin-flips vanishes, leaving only a $S^z s^z(0)$ coupling.**
2. **For the antiferromagnetic coupling, $J_z > 0$, or for ferromagnetic coupling, $|J_z| < |J^{\pm}|$. The flow is to strong coupling. This implies that our analysis based on the smallness of J breaks down.**
3. **Along the line $|J_z| = |J^{\pm}|$, we have the usual isotropic Kondo effect.**

13.2.14 Schwinger-Wigner Representation

Originally, this method was used to transform spins into bosons and was also useful for large values of spin or high dimensionality. It imitates the results of spin wave theory. We study the Schwinger-Wigner representation for both bosons and fermions:

$$S^{\alpha} = f_{\alpha}^{\dagger}\left(\frac{\sigma_{\alpha\beta}^{\alpha}}{2}\right)f_{\beta} \tag{2923}$$

and explicitly represented:

$$S^x = \frac{1}{2}\left(f_1^{\dagger}f_2 + f_2^{\dagger}f_1\right) \quad , \quad S^y = \frac{i}{2}\left(f_2^{\dagger}f_1 - f_1^{\dagger}f_2\right) \quad , \quad S^z = \frac{1}{2}\left(f_1^{\dagger}f_1 - f_2^{\dagger}f_2\right) \tag{2924}$$

Here, σ are Pauli matrices, and f_1 or f_2 can either be fermionic or bosonic operators where commutation rules for spins are satisfied for the fermions as well as bosons. We now verify:

$$\left[S^{\alpha}, S^{\beta}\right] = i\,\epsilon_{\alpha\beta\gamma}\,S^{\gamma} \tag{2925}$$

For the following commutators, the upper sign is for bosons and the lower for fermions:

$$\left[f_{\alpha}^{\dagger}f_{\beta}, f_{\beta}^{\dagger}f_{\alpha}\right] = f_{\alpha}^{\dagger}f_{\beta}f_{\beta}^{\dagger}f_{\alpha} - f_{\beta}^{\dagger}f_{\alpha}f_{\alpha}^{\dagger}f_{\beta} = f_{\alpha}^{\dagger}f_{\alpha} \pm f_{\alpha}^{\dagger}f_{\alpha}f_{\beta}^{\dagger}f_{\beta} - f_{\beta}^{\dagger}f_{\beta} \mp f_{\alpha}^{\dagger}f_{\alpha}f_{\beta}^{\dagger}f_{\beta} = f_{\alpha}^{\dagger}f_{\alpha} - f_{\beta}^{\dagger}f_{\beta} \tag{2926}$$

$$\left[f_{\alpha}^{\dagger}f_{\beta}, f_{\alpha}^{\dagger}f_{\alpha}\right] = f_{\alpha}^{\dagger}f_{\beta}f_{\alpha}^{\dagger}f_{\alpha} - f_{\alpha}^{\dagger}f_{\alpha}f_{\alpha}^{\dagger}f_{\beta} = f_{\alpha}^{\dagger}f_{\alpha}^{\dagger}f_{\alpha}f_{\beta} - f_{\alpha}^{\dagger}f_{\beta} \mp f_{\alpha}^{\dagger}f_{\alpha}^{\dagger}f_{\alpha}f_{\beta} = -f_{\alpha}^{\dagger}f_{\beta} \tag{2927}$$

$$\left[f_{\alpha}^{\dagger}f_{\beta}, f_{\beta}^{\dagger}f_{\beta}\right] = f_{\alpha}^{\dagger}f_{\beta} \tag{2928}$$

So, considering (2923) or (2924), from here we verify the commutator (2925):

$$\left[S^x, S^y\right] = \frac{i}{2}\left[f_1^{\dagger}f_2, f_2^{\dagger}f_1\right] = \frac{i}{2}\left(f_1^{\dagger}f_1 - f_2^{\dagger}f_2\right) = iS^z \tag{2929}$$

$$\left[S^y, S^z\right] = \frac{i}{4}\left[f_2^{\dagger}f_1 - f_1^{\dagger}f_2, f_1^{\dagger}f_1 - f_2^{\dagger}f_2\right] = \frac{i}{2}\left(f_2^{\dagger}f_1 + f_1^{\dagger}f_2\right) = iS^x \tag{2930}$$

$$\left[S^z, S^x\right] = \frac{1}{4}\left[f_1^{\dagger}f_1 - f_2^{\dagger}f_2, f_1^{\dagger}f_2 + f_2^{\dagger}f_1\right] = \frac{1}{2}\left(f_1^{\dagger}f_2 - f_2^{\dagger}f_1\right) = iS^y \tag{2931}$$

The spin algebra spawns a $(2S+1)$-dimensional Hilbert space. However, in the representation (2923), the dimensionality is infinity for bosons, whereas for fermions, it is four. The constraint

$$\vec{S}^2 = S(2S+1) \tag{2932}$$

is introduced to shrink the space. So, considering

$$n_\alpha = f_\alpha^\dagger f_\alpha \tag{2933}$$

Hence, for fermions, we have

$$n_\alpha^2 = n_\alpha \tag{2934}$$

then

$$\left(S^x\right)^2 + \left(S^y\right)^2 = \frac{1}{2}\left(f_1^\dagger f_2 f_2^\dagger f_1 + f_2^\dagger f_1 f_1^\dagger f_2\right) = \frac{1}{2}\left(f_1^\dagger f_1 + f_2^\dagger f_2 \pm 2 f_1^\dagger f_1 f_2^\dagger f_2\right) = \frac{1}{2}\left(n_1 + n_2 \pm 2n_1n_2\right) \tag{2935}$$

$$\left(S^z\right)^2 = \frac{1}{4}\left(n_1^2 + n_2^2 - 2n_1n_2\right) \tag{2936}$$

and

$$\vec{S}^2 = \left(S^x\right)^2 + \left(S^y\right)^2 + \left(S^z\right)^2 = \frac{1}{4}\left(2\left(n_1 \pm n_2\right)^2 + \left(n_1 - n_2\right)^2\right) \tag{2937}$$

Note that for fermions we can only represent spin $\frac{1}{2}$, which is a special case. Therefore, to properly represent spin $\frac{1}{2}$ we must impose the following constraint, which is valid for fermions as well as bosons:

$$f_1^\dagger f_1 + f_2^\dagger f_2 = 1 \tag{2938}$$

This is equivalent to forcing an infinite repulsion among particles at the same site. The constraint seems not as strong for high dimensionality. However, the probability for two particles to occupy the same site decreases with the dimensionality. Indeed the constraint is significant for one or two dimensions.

13.2.15 Jordan-Wigner

In previous headings, we studied the bosonization technique within the Holstein-Primakoff approach to ferromagnetic and antiferromagnetic spin systems. In this section, we investigate a similar approach known as a **fermionization method** and, in particular, the application of the Wigner-Jordan transformation to the isotropic quantum XY model. This entails transforming the Hamiltonian that describes a spin system by the use of new operators that obey the fermion anticommutation rules. The Bethe's approach [54] to the spectrum of the XY model—though exact—is rather abstract and, hence, makes it difficult to understand even such basic properties as long-range order. However, a much more natural approach to the problem of interacting spin $\frac{1}{2}$ systems was originally introduced by Jordan and Wigner [58], who developed simple mathematical transformations converting spin $\frac{1}{2}$ systems into problems of interacting spinless fermions (and even noninteracting ones in some cases). Indeed, a special case of the Heisenberg Hamiltonian determinant, the XY model, reduces to a free theory of spinless fermions via the Jordan-Wigner transformations.

Now, we will study a linear chain of N spin-$\frac{1}{2}$ atoms interacting antiferromagnetically with its nearest neighbors and described by the Hamiltonian

$$\mathrm{H} = -\frac{1}{2}\sum_{n \neq m} J_{nm}\vec{S}_n\vec{S}_m \tag{2939}$$

where for $J_{nm} > 0$, the interactions favor spin alignment that is a **ferromagnetic arrangement**. But for $J_{nm} < 0$, the ordering has alternate spins up and down, when the lattice permits, and is called **antiferromagnetic**. According to Jordan and Wigner, the up and down state of a single spin is assumed to be an empty or singly occupied fermion state:

$$\left|\uparrow\right\rangle = \hat{f}_m^\dagger|0\rangle \quad , \quad \left|\downarrow\right\rangle = |0\rangle \tag{2940}$$

The explicit representation of the spin raising and lowering operators is then

$$S_m^+ \equiv \hat{f}_m^\dagger = \begin{bmatrix} 0 & 1 \\ 0 & 0 \end{bmatrix} \quad , \quad S_m^- \equiv \hat{f}_m = \begin{bmatrix} 0 & 0 \\ 1 & 0 \end{bmatrix} \tag{2941}$$

The transverse spin operators:

$$S_m^x = \frac{1}{2}\left(S_m^+ + S_m^-\right) \equiv \frac{1}{2}\left(\hat{f}_m^\dagger + \hat{f}_m\right) \quad , \quad S_m^y = \frac{1}{2i}\left(S_m^+ - S_m^-\right) \equiv \frac{1}{2}\left(\hat{f}_m^\dagger - \hat{f}_m\right) \tag{2942}$$

while the z component of the spin operator:

$$S_m^z = \frac{1}{2}\left(\left|\uparrow\right\rangle\left\langle\uparrow\right| - \left|\downarrow\right\rangle\left\langle\downarrow\right|\right) = \hat{f}_m^\dagger \hat{f}_m - \frac{1}{2} \tag{2943}$$

These operators satisfy the algebra

$$\left[\vec{S}_m, \vec{S}_n\right] = i \in_{mnk} \vec{S}_k \tag{2944}$$

From supersymmetry, they also satisfy an anticommuting algebra

$$\left\{\vec{S}_m, \vec{S}_n\right\} = \frac{1}{4}\left\{\vec{\sigma}_m, \vec{\sigma}_n\right\} = \frac{1}{2}\delta_{mn} \tag{2945}$$

Therefore, the Pauli spin operators provided Jordan and Wigner with an elementary model of a fermion. Nevertheless, the representation needs modification if there is more than one spin because independent spin operators commute while independent fermions anticommute. This difficulty is addressed by Jordan and Wigner in one dimension by attaching a phase factor called a **string** to the fermions. For a spin chain in one dimension, the Jordan-Wigner representation of the spin operator at site m is defined as:

$$S_m^+ = \hat{f}_m^\dagger \exp\{i\phi_m\} \quad , \quad \phi_m = \pi \sum_{n>m} \hat{n}_m \tag{2946}$$

Here, the phase operator ϕ_m contains the sum over all fermion occupancies at sites to the left of m. We introduce phase factors so that spins on different sites commute.

So, for the complete Jordan-Wigner transformation:

$$S_m^z = \hat{f}_m^\dagger \hat{f}_m - \frac{1}{2} \quad , \quad S_m^+ = \hat{f}_m^\dagger \exp\{i\phi_m\} \quad , \quad S_m^- = \hat{f}_m \exp\{-i\phi_m\} \tag{2947}$$

It is instructive to note that the overall sign of the phase factors can be reversed without changing the spin operator. Also, in this representation, the operator $\exp\{i\pi\hat{n}_m\}$ anticommutes with the fermion operators at the same site:

$$\left\{\exp\{i\pi\hat{n}_m\},\hat{f}_m^\dagger\right\}=\exp\{i\pi\hat{n}_m\}\hat{f}_m^\dagger+\hat{f}_m^\dagger\exp\{i\pi\hat{n}_m\}=\exp\{i\pi\hat{n}_m\}\left[\hat{f}_m^\dagger-\hat{f}_m^\dagger\right]=0 \tag{2948}$$

Therefore, multiplying a fermion by the string operator transforms it into a boson. We verify that the transverse spin operators now satisfy the correct commutation algebra. If we consider $m<k$, then $\exp\{i\phi_m\}$ commutes with both $\hat{f}_m$ and $\hat{f}_k$. However, $\exp\{i\phi_m\}$ commutes with $\hat{f}_k$ but contains $\exp\{i\pi\hat{n}_m\}$, which does not commute with $\hat{f}_m$ or $\hat{f}_m^\dagger$. So,

$$\left[S_m^\pm,S_k^\pm\right]=\left[\hat{f}_m^\dagger\exp\{i\phi_m\},\hat{f}_k^\dagger\exp\{i\phi_k\}\right]=\left[\hat{f}_m^\dagger,\hat{f}_k^\dagger\exp\{i\phi_m\}\right]=\left\{\hat{f}_m^\dagger,\hat{f}_k^\dagger\right\}\exp\{i\phi_m\}-\hat{f}_k^\dagger\left\{\hat{f}_m^\dagger,\exp\{i\phi_m\}\right\}=0 \tag{2949}$$

We apply this to the one-dimensional Heisenberg model

$$\mathrm{H}=-\frac{1}{2}\sum_{n\neq m}J_{\perp nm}\left[S_m^xS_n^x+S_m^yS_n^y\right]-\frac{1}{2}\sum_{n\neq m}J_{\|nm}S_m^zS_n^z \tag{2950}$$

Local moments can interact via ferromagnetic as well as antiferromagnetic interactions in real magnetic systems. Ferromagnetic interactions generally arise from **direct exchange**, whereby the Coulomb repulsion energy is lowered when electrons are in a triplet state. This is the case when the wave function is spatially antisymmetric. Usually, antiferromagnetic interactions are due to the mechanism of **double exchange**. Here, electrons in antiparallel spin states lower their energy by undergoing virtual fluctuations into high-energy states where two electrons occupy the same orbital. The model in this case is written as if the interactions are ferromagnetic.

Here, we rewrite the model for convenience in terms of spin raising and lowering operators:

$$\mathrm{H}=-\frac{1}{2}\sum_{n\neq m}J_{\perp nm}\left[S_n^+S_m^-+\mathrm{h.c}\right]-\frac{1}{2}\sum_{n\neq m}J_{\|nm}S_n^zS_m^z \tag{2951}$$

For fermionization of the first term, all terms in the strings cancel, except the redundant term $\exp\{i\pi\hat{n}_m\}$:

$$\sum_{n\neq m}J_{\perp nm}S_n^+S_m^-=\sum_{n\neq m}J_{\perp nm}\hat{f}_n^\dagger\exp\{i\pi\hat{n}_m\}\hat{f}_m=\sum_{n\neq m}J_{\perp nm}\hat{f}_n^\dagger\hat{f}_m \tag{2952}$$

In this way, a **hopping** term in the fermionized Hamiltonian is induced by the transverse component of the interaction. The string terms would enter if the spin interaction involved next-nearest neighbors. We rewrite the z -component of the Hamiltonian:

$$\sum_{n\neq m}J_{\|nm}S_n^zS_m^z=\sum_{n\neq m}J_{\|nm}\left(\hat{n}_n-\frac{1}{2}\right)\left(\hat{n}_m-\frac{1}{2}\right) \tag{2953}$$

The Ferromagnetic interaction shows that spin fermions attract one another and the transformed Hamiltonian:

$$\hat{\mathrm{H}}=-\frac{1}{2}\sum_{n\neq m}J_{\perp nm}\left(\hat{f}_n^\dagger\hat{f}_m+\hat{f}_m^\dagger\hat{f}_n\right)+\sum_{n\neq m}J_{\|nm}\hat{n}_m-\sum_{n\neq m}J_{\|nm}\hat{n}_m\hat{n}_n \tag{2954}$$

Because the pure X-Y model has no interaction term in it, it can be solved as a noninteracting fermion problem. We write out the fermionized Hamiltonian in its most compact form by passing to momentum space (by Fourier transforming) because the Hamiltonian is translational invariant, and we obtain:

$$\hat{f}_m = \frac{1}{\sqrt{N}} \sum_{\vec{\kappa}} \hat{C}_{\vec{\kappa}} \exp\left\{ i\vec{\kappa}\vec{R}_m \right\} \tag{2955}$$

Here, N is the number of spin sites of the chain and $\hat{C}^{\dagger}_{\vec{\kappa}}$, is the creation operator of the spin excitation in momentum space numbered by the momentum $\vec{\kappa}$. So, for the one-particle terms, we have

$$\sum_{n\neq m} J_{||nm}\hat{n}_m = \sum_{m} J_{||nm} \frac{1}{N} \sum_{\vec{\kappa}\vec{\kappa}'} \hat{C}^{\dagger}_{\vec{\kappa}}\hat{C}_{\vec{\kappa}'} \exp\left\{ -i\vec{\kappa}\vec{R}_n \right\} \exp\left\{ i\vec{\kappa}'\vec{R}_m \right\} \tag{2956}$$

Letting

$$\vec{R}_n = \vec{R}_m + \vec{R} \tag{2957}$$

then

$$\sum_{n\neq m} J_{||nm}\hat{n}_m = \frac{1}{N} \sum_{m\vec{\kappa}\vec{\kappa}'\vec{R}} J_{||R}\hat{C}^{\dagger}_{\vec{\kappa}}\hat{C}_{\vec{\kappa}'} \exp\left\{ -i\left(\vec{\kappa}-\vec{\kappa}'\right)\vec{R}_m \right\} \exp\left\{ -i\vec{\kappa}\vec{R} \right\} \tag{2958}$$

But,

$$\sum_{m} \exp\left\{ -i\left(\vec{\kappa}-\vec{\kappa}'\right)\vec{R}_m \right\} = N\delta_{\vec{\kappa},-\vec{\kappa}'} \tag{2959}$$

then

$$\sum_{n\neq m} J_{||nm}\hat{n}_m = \frac{1}{N} \sum_{\vec{\kappa}\vec{\kappa}'\vec{R}} J_{||R}\hat{C}^{\dagger}_{\vec{\kappa}}\hat{C}_{\vec{\kappa}'} N\delta_{\vec{\kappa},-\vec{\kappa}'} \exp\left\{ -i\vec{\kappa}\vec{R} \right\} = \sum_{\vec{\kappa}'\vec{R}} J_{||R}\hat{C}^{\dagger}_{\vec{\kappa}'}\hat{C}_{\vec{\kappa}'} \exp\left\{ i\vec{\kappa}'\vec{R} \right\} \tag{2960}$$

or

$$\sum_{n\neq m} J_{||nm}\hat{n}_m = \sum_{\vec{\kappa}} J_{||\vec{\kappa}}\hat{C}^{\dagger}_{\vec{\kappa}}\hat{C}_{\vec{\kappa}} \tag{2961}$$

where

$$J_{||\vec{\kappa}} = \sum_{\vec{R}} J_{||R} \exp\left\{ i\vec{\kappa}\vec{R} \right\} \tag{2962}$$

Also,

$$-\frac{1}{2}\sum_{n\neq m} J_{\perp nm}\left(\hat{f}^{\dagger}_n\hat{f}_m + \hat{f}^{\dagger}_m\hat{f}_n \right) = -\frac{1}{2N}\sum_{\vec{\kappa}\vec{\kappa}'\vec{R}} J_{\perp R}\left(\exp\left\{ -i\vec{\kappa}\vec{a} \right\} + \exp\left\{ -i\vec{\kappa}\vec{a} \right\} \right)\hat{C}^{\dagger}_{\vec{\kappa}}\hat{C}_{\vec{\kappa}'} N\delta_{\vec{\kappa},-\vec{\kappa}'} \exp\left\{ -i\vec{\kappa}\vec{R} \right\} \tag{2963}$$

or

$$-\frac{1}{2}\sum_{n\neq m} J_{\perp nm}\left(\hat{f}^{\dagger}_n\hat{f}_m + \hat{f}^{\dagger}_m\hat{f}_n \right) = -\sum_{\vec{\kappa}} J_{\perp\vec{\kappa}}\hat{C}^{\dagger}_{\vec{\kappa}}\hat{C}_{\vec{\kappa}'} \cos\left(\vec{\kappa}\vec{a} \right) \tag{2964}$$

where

$$J_{\perp\vec{\kappa}} = \sum_{\vec{R}} J_{\perp R} \exp\left\{ i\vec{\kappa}\vec{R} \right\} \tag{2965}$$

Considering (2954), (2961), and (2964), we therefore rewrite the Heisenberg Hamiltonian:

$$\hat{\mathrm{H}} = \sum_{\vec{\kappa}} \omega_{\vec{\kappa}} \hat{C}^{\dagger}_{\vec{\kappa}} \hat{C}_{\vec{\kappa}} - \sum_{n \neq m} J_{|\,|nm} \hat{n}_m \hat{n}_n \tag{2966}$$

where magnon excitation energy:

$$\omega_{\vec{\kappa}} = J_{|\,|\vec{\kappa}} - J_{\perp\vec{\kappa}} \cos\left(\vec{\kappa}\vec{a}\right) \tag{2967}$$

The second interaction term in (2966) is quartic and can be seen as an interaction term written in the position basis. It can easily be cast in momentum space considering (2955) and supposing two spins within the system at positions $\vec{R}'_m$ and $\vec{R}_n$ are translated by the same vector, $\vec{R}'$:

$$\vec{R} = \frac{\vec{R}'_m + \vec{R}_n}{2} \quad , \quad \vec{R}' = \vec{R}'_m - \vec{R}_n \tag{2968}$$

Hence, their interaction energy does not change:

$$\sum_{n \neq m} J_{|\,|nm} \hat{n}_m \hat{n}_n = \frac{1}{N_C} \sum_{\vec{\kappa}\vec{\kappa}'\vec{\kappa}''\vec{R}'} \exp\left\{ -i\left(\vec{\kappa}' - \vec{\kappa}''\right)\vec{R}' \right\} \hat{C}^{\dagger}_{\vec{\kappa}} \hat{C}^{\dagger}_{\vec{\kappa}'} J_{|\,|(\vec{\kappa}'-\vec{\kappa}'')} \cos\left(\left(\vec{\kappa}' - \vec{\kappa}''\right)\vec{a} \right) \hat{C}_{\vec{\kappa}''} \hat{C}_{\vec{\kappa}+\vec{\kappa}'-\vec{\kappa}''} \tag{2970}$$

Substituting new momentum variables

$$\vec{\kappa}_1 = \vec{\kappa} + \vec{\kappa}' - \vec{\kappa}'' \quad , \quad \vec{\kappa}_2 = \vec{\kappa}'' \quad , \quad \vec{q} = \vec{\kappa}' - \vec{\kappa}'' \tag{2971}$$

we then obtain the interaction term in second quantization

$$\sum_{n \neq m} J_{|\,|nm} \hat{n}_m \hat{n}_n = \frac{1}{N} \sum_{\vec{\kappa}_1\vec{\kappa}_2\vec{q}} J_{|\,|\vec{q}} \cos\left(\vec{q}\vec{a}\right) \hat{C}^{\dagger}_{\vec{\kappa}_1-\vec{q}} \hat{C}^{\dagger}_{\vec{\kappa}_2+\vec{q}} \hat{C}_{\vec{\kappa}_2} \hat{C}_{\vec{\kappa}_1} \tag{2972}$$

or

$$\sum_{n \neq m} J_{|\,|nm} \hat{n}_m \hat{n}_n = \frac{1}{N} \sum_{\vec{\kappa}\vec{\kappa}'\vec{q}} J_{|\,|\vec{q}} \cos\left(\vec{q}\vec{a}\right) \hat{C}^{\dagger}_{\vec{\kappa}-\vec{q}} \hat{C}^{\dagger}_{\vec{\kappa}'+\vec{q}} \hat{C}_{\vec{\kappa}'} \hat{C}_{\vec{\kappa}} \tag{2973}$$

where

$$J_{|\,|\vec{q}} = \sum_{\vec{R}'} J_{|\,|R'} \exp\left\{ i\vec{q}\vec{R}' \right\} \tag{2974}$$

and

$$\hat{\mathrm{H}} = \sum_{\vec{\kappa}} \omega_{\vec{\kappa}} \hat{C}^{\dagger}_{\vec{\kappa}} \hat{C}_{\vec{\kappa}} - \frac{1}{N} \sum_{\vec{\kappa}\vec{\kappa}'\vec{q}} J_{|\,|\vec{q}} \cos\left(\vec{q}\vec{a}\right) \hat{C}^{\dagger}_{\vec{\kappa}-\vec{q}} \hat{C}^{\dagger}_{\vec{\kappa}'+\vec{q}} \hat{C}_{\vec{\kappa}'} \hat{C}_{\vec{\kappa}} \tag{2975}$$

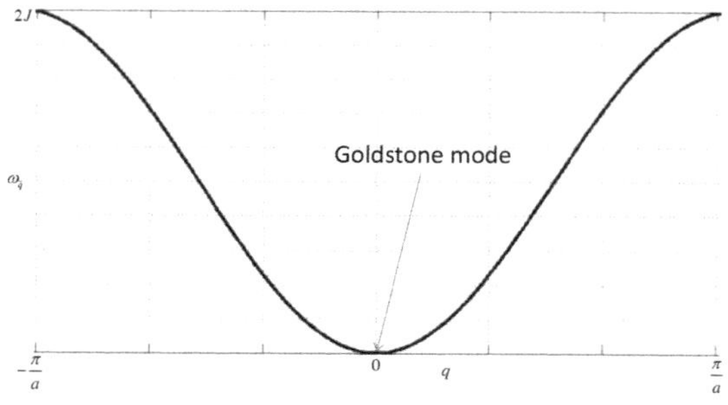

FIGURE 13.22 Excitation spectrum of the one-dimensional Heisenberg ferromagnet.

This transformation is applicable for both the ferromagnet (where the fermionic spin excitations correspond to the magnons of the ferromagnet) and the antiferromagnet (where the fermionic spin excitations often are called **spinons**).

We examine the physical significance of Hamiltonian (2975) by first neglecting the interactions. This is reasonable for the limiting cases of:

- The Heisenberg ferromagnet (Figure 13.22), $J_{||\vec{q}} = J_{\vec{q}}$
- The x-y model, $J_{||\vec{q}} = 0$
- Heisenberg ferromagnet.

$$J_{||\vec{q}} = J_{\vec{q}} \tag{2976}$$

For this case, the spectrum

$$\omega_{\vec{q}} = 2J_{\vec{q}} \sin^2\left(\frac{\vec{q}\vec{a}}{2}\right) \tag{2977}$$

This is always positive, and there are no magnons present in the ground state, which can be written

$$|0\rangle = |\downarrow\downarrow\downarrow\cdots\rangle \tag{2978}$$

This corresponds to a state with a spontaneous magnetization $M = -\frac{N}{2}$. Because $\omega_{\vec{q}=0} = 0$, no energy is needed to add a magnon of arbitrarily long wavelength. This is an example of a Goldstone mode, which arises due to spontaneous magnetization that could point in any direction.

Suppose we rotate the magnetization infinitesimally upward, then

$$S^{+}_{\text{total}}|\downarrow\downarrow\downarrow\cdots\rangle = \sum_m \hat{f}^{\dagger}_m \exp\{i\phi_m\}|0\rangle = \sum_m \hat{f}^{\dagger}_m|0\rangle = \sqrt{N}C^{\dagger}_{\vec{\kappa}=0}|0\rangle \tag{2979}$$

This implies adding a single magnon at $\vec{q} = 0$ and rotates the magnetization infinitesimally upward. Rotating the magnetization demands no energy and is the reason why the $\vec{\kappa} = 0$ magnon is a zero energy excitation.

For the x-y ferromagnetic, as $J_{||\vec{q}}$ is reduced from $J_{\perp\vec{q}}$, the spectrum develops a negative part (Figure 13.23), and magnon states with negative energy become occupied. For the pure X–Y model, where $J_{||\vec{q}} = 0$, the interaction term now identically vanishes, yielding a simple **hopping** Hamiltonian. The excitation spectrum is that of magnons (**Figure 13.23**):

$$\omega_{\vec{q}} = -J_{\perp\vec{q}} \cos(\vec{q}\vec{a}) \tag{2980}$$

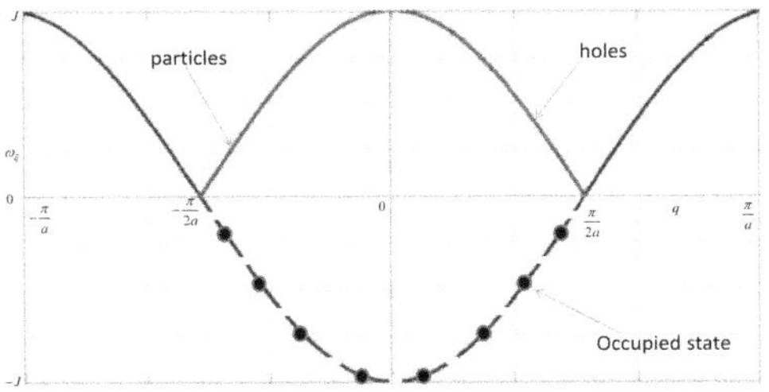

FIGURE 13.23 Excitation spectrum of the one-dimensional x-y ferromagnet, showing how the negative energy states are filled and the negative energy dispersion curve is folded over to describe the positive hole excitation energy.

This spectrum has all the negative fermion energy states that are occupied for

$$|\vec{q}| < \frac{\pi}{2a} \tag{2981}$$

We define the ground state for this system as the state with all filled negative states:

$$|\Psi_g\rangle = \prod_{|\vec{q}|<\frac{\pi}{2a}} C_{\vec{q}}^{\dagger}|0\rangle \tag{2982}$$

where $|0\rangle$ is the vacuum state for the fermions. Hence, we have a half-filled band, as can be verified by explicit calculation

$$\langle S^z\rangle = \left\langle \hat{n}_f - \frac{1}{2}\right\rangle = 0 \tag{2983}$$

So, there is no ground state magnetization. This loss of ground state magnetization may be interpreted as a consequence of the growth of quantum spin fluctuations when going from the Heisenberg to the X-Y ferromagnet.

Excitations of the ground state are feasible, either by adding a magnon at wave vectors

$$|\vec{q}| > \frac{\pi}{2a} \tag{2984}$$

or by annihilating a magnon at wave vectors

$$|\vec{q}| < \frac{\pi}{2a} \tag{2985}$$

in order to form a **hole** with the energy to form a hole being $-\omega_{\vec{q}}$.

In order to represent the hole excitations, we make a **particle-hole** transformation for the occupied states via **physical** excitation operators:

$$\hat{\tilde{C}}_{\vec{q}} = \begin{cases} \hat{C}_{\vec{q}} & , \ |\vec{q}| > \frac{\pi}{2a} \\ \hat{C}_{-\vec{q}}^{\dagger} & , \ |\vec{q}| < \frac{\pi}{2a} \end{cases} \tag{2986}$$

Because

$$\hat{C}_{\vec{q}}^{\dagger}\hat{C}_{\vec{q}} = 1 - \hat{C}_{\vec{q}}\hat{C}_{\vec{q}}^{\dagger} \tag{2987}$$

the Hamiltonian of the pure X-Y ferromagnet is written

$$\hat{H}_{xy} = \sum_{\vec{q}} J_{\perp\vec{q}} \left|\cos(\vec{q}\vec{a})\right| \left(\hat{\tilde{C}}_{\vec{q}}^{\dagger} \hat{\tilde{C}}_{\vec{q}} - \frac{1}{2} \right) \tag{2988}$$

Notice that unlike the pure ferromagnet, the magnon excitation spectrum is now linear. Evidently, the ground state energy is

$$E_g = -\frac{1}{2} \sum_{\vec{q}} J_{\perp\vec{q}} \left|\cos(\vec{q}\vec{a})\right| \tag{2989}$$

Some characteristic of the spectrum are worth noting:

- First, though there is no true long-range order, the spin correlations in the X-Y model display power-law correlations, with an infinite spin correlation length generated by the gapless magnons at the vicinity of $q = \pm\frac{\pi}{2a}$. So, there are **Goldstone modes** that mark the presence of long-range correlation.
- Second, the spectrum for low-energy excitations is linear. Low-energy excitations are responsible for long-range correlations. So, the only important states used to study long-range correlations are those close to the Fermi points, that is, states with:

$$q = \pm q_F = \pm\frac{\pi}{2a} \tag{2990}$$

So, we linearize the single-particle spectrum

$$\omega_{\vec{q}} = -J_{\perp\vec{q}} \cos(\vec{q}\vec{a}) \tag{2991}$$

at the vicinity of these points:

$$\omega_{\vec{q}} \approx \begin{cases} -J_{\perp\vec{q}}a(q+q_F) & , \quad |q+q_F| \le \Lambda \\ J_{\perp\vec{q}}a(q-q_F) & , \quad |q-q_F| \le \Lambda \end{cases} \tag{2992}$$

where Λ is a cutoff:

$$\Lambda << \frac{\pi}{2a} \tag{2993}$$

13.2.16 Semi-Fermionic Representation: Hubbard Model

We employ a semi-fermionic (SF) technique that imitates the fundamental Pauli spin nature proposed by Popov and Fedotov [98, 99] as well as by M. Kiselev, K. Kikoin and R. Oppermann [104]. The dimensionality of the space in which the spin operators act is always greater than the dimensionality of the spin matrices and presents a serious problem in the elimination of superfluous states and must be considered when treating spin systems.

13.2.16.1 Semi-Fermionic Representation

Let us consider the Hamiltonian for the spin system to be, for example

$$\hat{\mathrm{H}}_{\mathrm{S}} = \sum_{i\neq j} J_{ij}\left(\frac{1}{4}\sigma_i^z\sigma_j^z + \lambda^2\sigma_i^+\sigma_j^-\right) - \frac{h}{2}\sum_i \sigma_i^z \tag{2994}$$

Here,

$$\sigma_i^+ = \begin{bmatrix} 0 & 1 \\ 0 & 0 \end{bmatrix} \quad , \quad \sigma_i^- = \begin{bmatrix} 0 & 0 \\ 1 & 0 \end{bmatrix} \quad , \quad \sigma_i^z = \begin{bmatrix} 1 & 0 \\ 0 & -1 \end{bmatrix} \tag{2995}$$

and $h = g_{\mathrm{L}}\mu_{\mathrm{B}}H$ is some external magnetic field where the Bohr magneton is μ_{B} and g_{L} is the Landé factor. The parameter λ for $\lambda = 0$ leads to the Ising model, and for $\lambda = 1$ leads to a spherical symmetric Heisenberg model. For spin $S = \frac{1}{2}$ systems, we define an auxiliary full fermionic model in terms of fermionic raising and lowering operators $a_i^\dagger$, a_i, $b_i^\dagger$, and b_i, which satisfy the following anticommutator relations:

$$a_i a_j^\dagger + a_j^\dagger a_i = \delta_{ij} \quad , \quad b_i b_j^\dagger + b_j^\dagger b_i = \delta_{ij} \tag{2996}$$

This permits us to define the following bilinear combination of fermionic operators at the ith site:

$$a_i^\dagger a_i - b_i^\dagger b_i \quad , \quad a_i^\dagger b_i \quad , \quad b_i^\dagger a_i \tag{2997}$$

These obey the same commutation relations as the pseudo-spin operators σ_i^z, σ_i^+, and σ_i^- where the correspondence is given by

$$\sigma_i^z \to a_i^\dagger a_i - b_i^\dagger b_i \quad , \quad \sigma_i^+ \to a_i^\dagger b_i \quad , \quad \sigma_i^- \to b_i^\dagger a \tag{2998}$$

Substituting these relations in the Hamiltonian determinant (2994), we obtain the Hamiltonian $\hat{\mathrm{H}}_{\mathrm{F}}$ of the auxiliary full fermionic model. The Hamiltonians $\hat{\mathrm{H}}_{\mathrm{S}}$ and $\hat{\mathrm{H}}_{\mathrm{F}}$ are defined in different Hilbert spaces. This is because each operator σ_i^α in $\hat{\mathrm{H}}_{\mathrm{S}}$ acts on a two-dimensional Hilbert space, whereas each of the fermionic operators $a_i^\dagger$, a_i, $b_i^\dagger$, and b_i in $\hat{\mathrm{H}}_{\mathrm{F}}$ acts on a four-dimensional Fock space. The number of particle operators for the given system is

$$\hat{N} = \sum_{i=1}^{N}\left(a_i^\dagger a_i + b_i^\dagger b_i\right) \tag{2999}$$

and N is the number of sites. We observe from (2995) that the dimensionality of the space in which the spin operators act is two, whereas the fermion space corresponding to the ith site is four-dimensional and is generated by the vectors:

$$a_i^\dagger\Phi_0 = |10\rangle_i \quad , \quad b_i^\dagger\Phi_0 = |01\rangle_i \quad , \quad \Phi_0 = |00\rangle_i \quad , \quad a_i^\dagger b_i^\dagger\Phi_0 = |11\rangle_i \tag{3000}$$

For $S = \frac{1}{2}$, the **spurious (unphysical) states** are:

$$|00\rangle \quad , \quad |11\rangle \tag{3001}$$

and the **physical states** are:

$$|10\rangle = |\uparrow\rangle, |01\rangle = |\downarrow\rangle \tag{3002}$$

The physical vectors generate a two-dimensional physical subspace characterized by the following condition:

$$\hat{N}_i \Phi = \Phi \tag{3003}$$

with the number operator of fermions at the ith site being

$$\hat{N}_i = a_i^\dagger a_i + b_i^\dagger b_i \tag{3004}$$

So, the Hamiltonians $\hat{\mathrm{H}}_{\mathrm{F}}$ and $\hat{\mathrm{H}}_{\mathrm{S}}$ are defined in the direct product of the physical subspaces of all the sites. The operator $\hat{N}_i$ acts on the following unphysical states

$$|00\rangle \quad , \quad |11\rangle \tag{3005}$$

and gives, respectively, the following eigenvalues

$$0 \quad , \quad 2 \tag{3006}$$

We now consider the case of $S = 1$ where the Hamiltonian $\hat{\mathrm{H}}_S$ is represented via three-component spin operators:

$$S^+ = \begin{bmatrix} 0 & 1 & 0 \\ 0 & 0 & 1 \\ 0 & 0 & 0 \end{bmatrix} \quad , \quad S^- = \begin{bmatrix} 0 & 0 & 0 \\ 1 & 0 & 0 \\ 0 & 1 & 0 \end{bmatrix} \quad , \quad S^z = \begin{bmatrix} 1 & 0 & 0 \\ 0 & 0 & 0 \\ 0 & 0 & -1 \end{bmatrix} \tag{3007}$$

and $\hat{\mathrm{H}}_F$ is obtained from $\hat{\mathrm{H}}_S$ via

$$S_i^z \to a_i^\dagger a_i - b_i^\dagger b_i \quad , \quad S_i^+ \to a_i^\dagger c_i + c_i^\dagger b_i \quad , \quad S_i^- \to b_i^\dagger c_i + c_i^\dagger a_i \tag{3008}$$

with

$$\hat{N} = \sum_{i=1}^{N} \hat{n}_i \quad , \quad \hat{n}_i = a_i^\dagger a_i + b_i^\dagger b_i + c_i^\dagger c_i \tag{3009}$$

It is obvious that the states with occupation numbers 0 and 3 cancel each other (unphysical states):

$$\hat{n}_i \Phi = 0 \quad , \quad \hat{n}_i \Phi = 3\Phi \tag{3010}$$

Due to particle-hole symmetry, the states with occupation 1 and 2 are equivalent (physical states):

$$\hat{n}_i \Phi = \Phi \quad , \quad \hat{n}_i \Phi = 2\Phi \tag{3011}$$

13.2.16.2 Kondo Lattice: Effective Action

We consider the following Heisenberg Hamiltonian for spin $\frac{1}{2}$ with antiferromagnetic interaction with local $SU(2)$ symmetry [30]:

$$\hat{H} = \hat{H}_{int} + h\sum_i \mathbf{S}_i^z \tag{3012}$$

where

$$\hat{H}_{int} = J\sum_{i \neq j}^{n} \hat{H}_{ij} \tag{3013}$$

and

$$\hat{H}_{ij} = \mathbf{S}_i\mathbf{S}_j - \frac{1}{4} \tag{3014}$$

$$\mathbf{S}_i\mathbf{S}_j = \frac{1}{2}\left(\mathbf{S}_i^+\mathbf{S}_j^- + \mathbf{S}_i^-\mathbf{S}_j^+\right) + \mathbf{S}_i^z\mathbf{S}_j^z \tag{3015}$$

The second term in (3012) describes Zeeman splitting in an infinitesimal magnetic field $h = g\mu_B H$, and we express the spin operator components $\vec{S}$ via the Abrikosov pseudo-fermion representation:

$$\mathbf{S}^+ = \hat{f}_\sigma^\dagger \hat{f}_{\sigma'} \quad , \quad \mathbf{S}^- = \hat{f}_{\sigma'}^\dagger \hat{f}_\sigma \quad , \quad \mathbf{S}^z = \frac{1}{2}\left(\hat{f}_\sigma^\dagger \hat{f}_\sigma - \hat{f}_{\sigma'}^\dagger \hat{f}_{\sigma'}\right) \tag{3016}$$

$$\mathbf{S}^x = \frac{1}{2}\left(\mathbf{S}^- + \mathbf{S}^+\right) = \frac{1}{2}\left(\hat{f}_{\sigma'}^\dagger \hat{f}_\sigma + \hat{f}_\sigma^\dagger \hat{f}_{\sigma'}\right) \quad , \quad \mathbf{S}^y = \frac{1}{2}\left(\mathbf{S}^- - \mathbf{S}^+\right) = \frac{i}{2}\left(\hat{f}_{\sigma'}^\dagger \hat{f}_\sigma - \hat{f}_\sigma^\dagger \hat{f}_{\sigma'}\right) \tag{3017}$$

Here, the Grassmann operators $\hat{f}_\sigma^\dagger$ and $\hat{f}_{\sigma'}$ obey the local constraint condition at each site:

$$n = \hat{f}_\sigma^\dagger \hat{f}_\sigma + \hat{f}_{\sigma'}^\dagger \hat{f}_{\sigma'} = 1 \tag{3018}$$

For the $SU(2)$ invariance, we express the spin operators $\left\{\mathbf{S}^+, \mathbf{S}^-, \mathbf{S}^z\right\}$ as arbitrary linear combinations of spin fermions $\left\{\hat{f}_{\sigma'}, \hat{f}_\sigma, \hat{f}_{\sigma'}^\dagger, \hat{f}_\sigma^\dagger\right\}$:

$$\mathbf{S}_i^+ = \left(\cos\theta \hat{f}_{i\sigma}^\dagger + \sin\theta \hat{f}_{i\sigma'}\right)\left(\cos\theta \hat{f}_{i\sigma'} - \sin\theta \hat{f}_{i\sigma}^\dagger\right), \mathbf{S}_i^- = \left(\cos\theta \hat{f}_{i\sigma'}^\dagger - \sin\theta \hat{f}_{i\sigma}\right)\left(\cos\theta \hat{f}_{i\sigma} + \sin\theta \hat{f}_{i\sigma'}^\dagger\right) \tag{3019}$$

$$\mathbf{S}^z = \frac{1}{2}\left(\hat{f}_{i\sigma}^\dagger \hat{f}_{i\sigma} - \hat{f}_{i\sigma'}^\dagger \hat{f}_{i\sigma'}\right) \tag{3020}$$

From equation (3019) and $\theta = 0$, $\pi/2$, follows the relation for the complete particle–hole symmetry

$$\hat{f}_{i\sigma}^\dagger \hat{f}_{i\sigma} = \hat{f}_{i-\sigma} \hat{f}_{i-\sigma}^\dagger \tag{3021}$$

So, from the Hamiltonian (3012), we have the following pseudo-fermion representation:

$$\begin{aligned} \hat{H}_{ij} = {} & \frac{1}{2}\hat{f}_{i\sigma}^\dagger \hat{f}_{j\sigma'} \hat{f}_{j\sigma'}^\dagger \hat{f}_{i\sigma} + \frac{1}{2}\hat{f}_{i\sigma'}^\dagger \hat{f}_{j\sigma} \hat{f}_{j\sigma}^\dagger \hat{f}_{i\sigma'} + \frac{1}{4}\hat{f}_{i\sigma}^\dagger \hat{f}_{j\sigma} \hat{f}_{j\sigma}^\dagger \hat{f}_{i\sigma} + \\ & + \frac{1}{4}\hat{f}_{i\sigma'}^\dagger \hat{f}_{j\sigma'} \hat{f}_{j\sigma'}^\dagger \hat{f}_{i\sigma'} - \frac{1}{4}\hat{f}_{i\sigma}^\dagger \hat{f}_{j\sigma} \hat{f}_{j\sigma'}^\dagger \hat{f}_{i\sigma'} - \frac{1}{4}\hat{f}_{i\sigma'}^\dagger \hat{f}_{j\sigma'} \hat{f}_{j\sigma}^\dagger \hat{f}_{i\sigma} - \frac{1}{4} \end{aligned} \tag{3022}$$

We write the Hamiltonian of the spin or strongly correlated electrons in terms of the Hubbard's pseudo-fermion projection operators, which imitate the Hubbard real electron operator:

$$X_{ij}^{\sigma} = \hat{f}_{i\sigma}^{\dagger}\hat{f}_{j\sigma} \quad , \quad \left(X_{ij}^{\sigma}\right)^{\dagger} = -\hat{f}_{i\sigma}\hat{f}_{j\sigma}^{\dagger} \quad , \quad \sigma = \uparrow,\downarrow \tag{3023}$$

with

$$\left(X_{ij}^{\sigma}\right)^{\dagger} = X_{ji}^{\sigma} \tag{3024}$$

This may permit us to express (3022) via (3023). As in (3022), the indices of the first and second terms are different and thus can be expressed via (3023), considering an even permutation. We apply the fermionic anticommutation relation (3018) to the third and fourth terms of (3022), and then we apply an odd permutation. We apply relation (3021) to the fifth and sixth terms. So, considering $\sigma = \uparrow$ and $\sigma' = \downarrow$, then (3022) now becomes

$$\begin{aligned}\hat{H}_{ij} &= \frac{1}{2}X_{ij}^{\sigma}\left(X_{ji}^{\sigma'}\right)^{\dagger} + \frac{1}{2}X_{ij}^{\sigma'}\left(X_{ji}^{\sigma}\right)^{\dagger} + \frac{1}{4}X_{ij}^{\sigma}\left(X_{ji}^{\sigma}\right)^{\dagger} + \\ &+ \frac{1}{4}X_{ij}^{\sigma'}\left(X_{ji}^{\sigma'}\right)^{\dagger} + \frac{1}{4}X_{ij}^{\sigma}\left(X_{ji}^{\sigma}\right)^{\dagger} + \frac{1}{4}X_{ij}^{\sigma'}\left(X_{ji}^{\sigma'}\right)^{\dagger} + \frac{1}{2}\left[\hat{f}_{j\sigma}^{\dagger}\hat{f}_{j\sigma} + \hat{f}_{j\sigma'}^{\dagger}\hat{f}_{j\sigma'} - 1\right]\end{aligned} \tag{3025}$$

Because the local constraint

$$\hat{f}_{j\sigma}^{\dagger}\hat{f}_{j\sigma} + \hat{f}_{j\sigma'}^{\dagger}\hat{f}_{j\sigma'} = 1 \tag{3026}$$

then

$$\hat{H}_{ij} = \frac{1}{2}X_{ij}^{\sigma}\left(X_{ji}^{\sigma'}\right)^{\dagger} + \frac{1}{2}X_{ij}^{\sigma'}\left(X_{ji}^{\sigma}\right)^{\dagger} + \frac{1}{4}X_{ij}^{\sigma}\left(X_{ji}^{\sigma}\right)^{\dagger} + \frac{1}{4}X_{ij}^{\sigma'}\left(X_{ji}^{\sigma'}\right)^{\dagger} + \frac{1}{4}X_{ij}^{\sigma}\left(X_{ji}^{\sigma}\right)^{\dagger} + \frac{1}{4}X_{ij}^{\sigma'}\left(X_{ji}^{\sigma'}\right)^{\dagger} \tag{3027}$$

So, from (3024), we arrive at

$$\hat{H}_{int} = \frac{J}{2}\sum_{ij\sigma\sigma'} \hat{f}_{i\sigma}^{\dagger}\hat{f}_{j\sigma}\hat{f}_{j\sigma'}^{\dagger}\hat{f}_{i\sigma'} \tag{3028}$$

and

$$\hat{H} = \hat{H}_0 + \hat{H}_{int} \tag{3029}$$

where

$$\hat{H}_0 = -\frac{h}{2}\sum_{ij\sigma} \sigma\hat{f}_{i\sigma}^{\dagger}\hat{f}_{i\sigma} \quad , \quad \hat{H}_{int} = \frac{J}{2}\sum_{ij\sigma\sigma'} \hat{f}_{i\sigma}^{\dagger}\hat{f}_{j\sigma}\hat{f}_{j\sigma'}^{\dagger}\hat{f}_{i\sigma'} \tag{3030}$$

We examine the partition function of the Heisenberg model via Feynman path integration

$$Z = \int d\left[\hat{f}^{\dagger}\right]d\left[\hat{f}\right]\exp\left\{-\int_0^{\beta} d\tau\left(\sum_{i\sigma} \hat{f}_{i\sigma}^{\dagger}\partial_{\tau}\hat{f}_{i\sigma} + \hat{H}(\tau)\right)\right\}\prod_i \delta\left(\sum_{\sigma} \hat{f}_{i\sigma}^{\dagger}\hat{f}_{i\sigma} - 1\right) \tag{3031}$$

Here, the Feynman path integration measures

$$d\left[\hat{f}^{\dagger}\right]=\prod_{i} d\hat{f}_{i}^{\dagger} \quad , \quad d\left[\hat{f}\right]=\prod_{i} d\hat{f}_{i} \tag{3032}$$

and the Dirac delta function describes the local constraint in (3018). The Hamiltonian $\hat{H}(\tau)$ has the form of (3029). To conveniently solve the integral in (3031) with Grassmann variables, we have to decouple the variables. We do this by first expressing the Dirac delta function in (3031) via the new field λ_i:

$$\delta\left(\sum_{\sigma} \hat{f}_{i\sigma}^{\dagger} \hat{f}_{i\sigma}-1\right)=\int d\lambda_{i} \exp\left\{i\lambda_{i}\left(\sum_{\sigma} \hat{f}_{i\sigma}^{\dagger} \hat{f}_{i\sigma}-1\right)\right\} \tag{3033}$$

In addition, we consider the local distribution, keeping in mind the antiferromagnetic ordering of the spin. For further decoupling of variables, we again introduce the new variable Δ in the partition function (3031) via the following identity for the Dirac delta functions:

$$1=\int d\Delta_{ij}\, d\Delta_{ij}^{*} \delta\left(\Delta_{ij}-\sum_{\sigma} \hat{f}_{i\sigma}^{\dagger} \hat{f}_{j\sigma}\right) \delta\left(\Delta_{ij}^{*}-\sum_{\sigma} \hat{f}_{j\sigma}^{\dagger} \hat{f}_{i\sigma}\right) \tag{3034}$$

We introduce pseudo-fermion variables via the Dirac delta function

$$1=\int d\Delta_{ij}\, d\Delta_{ij}^{*}\, d\phi_{ij}\, d\phi_{ij}^{*} \exp\left\{\phi_{ij}^{*}\left(\Delta_{ij}-\sum_{\sigma} \hat{f}_{i\sigma}^{\dagger} \hat{f}_{j\sigma}\right)+\phi_{ij}\left(\Delta_{ij}^{*}-\sum_{\sigma} \hat{f}_{j\sigma}^{\dagger} \hat{f}_{i\sigma}\right)\right\} \tag{3035}$$

So, the partition function takes the form

$$Z=\int d\left[\hat{f}^{\dagger}\right] d\left[\hat{f}\right] d\left[\Delta^{*}\right] d[\Delta] d\left[\phi^{*}\right] d[\phi] d[\lambda] \exp\{-S\} \tag{3036}$$

where the effective action functional

$$S=\int_{0}^{\beta} d\tau L \tag{3037}$$

and the Lagrangian L:

$$L=\sum_{i\sigma} \hat{f}_{i\sigma}^{\dagger}\left(\partial_{\tau}-i\lambda_{i}\right) \hat{f}_{j\sigma}+\frac{J}{4} \sum_{ij\sigma} \eta_{ij} \hat{f}_{j\sigma}^{\dagger} \hat{f}_{i\sigma}+\frac{J}{4} \sum_{ij\sigma} \eta_{ij}^{*} \hat{f}_{i\sigma}^{\dagger} \hat{f}_{j\sigma}-\frac{J}{4} \sum_{ij}\left(\eta_{ij}^{*} \Delta_{ij}+\Delta_{ij}^{*} \eta_{ij}-2\left|\Delta_{ij}\right|^{2}\right)+i \sum_{i} \lambda_{i} \tag{3038}$$

where

$$\frac{J}{4} \eta_{ij}=\frac{J}{4} \Delta_{ij}+\phi_{ij} \quad , \quad \frac{J}{4} \eta_{ij}^{*}=\frac{J}{4} \Delta_{ij}^{*}+\phi_{ij}^{*} \tag{3039}$$

The Feynman path integration in (3036) now may be conveniently taken, yielding the effective action functional

$$S_{\text{eff}}=\operatorname{Tr} \ln\left[\mathbf{G}^{-1}\left(\eta^{*}, \eta, \lambda\right)\right]+\frac{J}{4} \sum_{ij}\left(\eta_{ij}^{*} \Delta_{ij}+\Delta_{ij}^{*} \eta_{ij}-2\left|\Delta_{ij}\right|^{2}\right)-i \sum_{i} \lambda_{i} \tag{3040}$$

Here, **G** is the Green's function of the pseudo-fermion.

14

Nonequilibrium Quantum Field Theory

Introduction

The recent development of experimental techniques permits us to observe details of time evolution of various electron systems far from equilibrium. These developments also allow many interesting measurements of various natural or artificially prepared structures. Mesoscopic (nanoscopic) systems with quantum processes, in particular, play a key role. The domain of nonequilibrium mesoscopic systems is flooded with open questions some of which are solved but still others have provisional or incomplete solutions. Generally, a complete description of nonequilibrium many-body systems, and in particular mesoscopic (nanoscopic) systems, has yet to be achieved due to their complex behavior. This complex behavior can be better understood and interpreted by improving methods of quantum field theory, many-body physics, statistical physics, and quantum transport theory via a nonequilibrium Green's function technique initially introduced by Schwinger [78, 105, 106]. This is an essential technique for investigating quantum dynamics of many-particle systems that are neither in their ground state nor in thermal equilibrium and for generalization of equilibrium many-body techniques to nonequilibrium systems. This technique finds its strength in several fields of physics including plasma, laser, chemical reactions, heavy ion collisions, and ultracold quantum gases, semiconductors, and superconductors.

Our focus is on the description of time evolution of expectation values of observables with the expectation values calculated from the time evolution of the density matrix. Generally, any density matrix can be expressed with the set of n-point Green's functions. Hence, knowledge of the dynamical evolution of all Green's functions amounts to a complete dynamical solution. Our main focus will be on the dynamic evolution of the lowest-order correlation functions because most of the experimentally relevant observables are determined by these functions. For a fermionic case, there is no macroscopic field, and the lowest correlation function of principal interest is the two-point function.

In quantum field theory, the path integral quantization allows us to use c-numbers instead of operators to efficiently tackle the time evolution. In this chapter, we examine the Schwinger-Keldysh technique [103, 107, 108], applying path integrals, and the functional method of 2PI effective actions—with the latter serving to construct powerful symmetry-preserving nonperturbative approximations for equilibrium as well as nonequilibrium descriptions of strongly interacting quantum many-body systems. This also provides a rigorous mathematical basis for exploring the quantum mechanical basis of thermalization and decoherence.

14.1 Keldysh-Schwinger Technique: Time Contour

A correlation function encodes the correlation between the field operators at different times and is an inference to all thermodynamic quantities. Also, knowledge of variations in correlation functions with respect to the external (source) fields provides answers to all questions about a given quantum system.

Because the pivot of every quantum theory is inherently probabilistic in nature, this is the reason why the quantum field theory statistical treatment of quantum fields focuses on determination of correlation functions and their properties. **The object encoding information of all the correlation functions is the density operator $\hat{\rho}(t)$, which is also known as the density matrix.** The time-dependent expectation value of the operator of any physical quantity $\hat{A}(t)$ in the interaction representation can be expressed via this density matrix:

$$\langle A(t)\rangle = \mathrm{Tr}\left\{\hat{A}(t)\hat{\rho}(t)\right\} \tag{3041}$$

This expectation value typically is a type of quantity of interest for a macroscopic system (i.e., a system consisting of a huge number of particles). The interpretation of $\hat{\rho}(t)$ as a probability distribution is imposed by the following normalization condition on the density operator at all times:

$$\mathrm{Tr}\left\{\hat{\rho}(t)\right\} = 1 \tag{3042}$$

We consider the time evolution of a quantum many-body system with density matrix $\hat{\rho}(t)$ starting at time $t = t_0 \equiv -\infty$, which is explicitly specified via the density matrix $\hat{\rho}(-\infty) = \hat{\rho}_0$ corresponding to the thermodynamic equilibrium for free particles in the absence of interaction and external fields.

Problems in nonequilibrium physics are treated according to whether

1. **They deal with open or closed systems.**
2. **They focus on the transient time evolution during and after some perturbation or on the stationary state that may develop when a system is driven by external fields.**

Open systems usually are coupled to a dissipative environment, whereas the dynamics of a closed system are completely described by the time-dependent Hamiltonian $\hat{H}(t)$. The closed system may include external fields but no coupling to a heat bath. Classical many-body theory kinetics usually are described by the Boltzmann's equation in phase space and, in the quantum field theory, by **the density matrix $\hat{\rho}$ governed by the von Neumann equation**:

$$i\frac{\partial \hat{\rho}}{\partial t} = \left[\hat{H}, \hat{\rho}\right] \tag{3043}$$

This is the quantum generalization of the classical-statistical Liouville equation and consists primarily of the commutator [13] between the density operator and the Hamiltonian describing the full-time dependent system:

$$\hat{H} = \hat{H}_0(t) + \hat{H}_{\mathrm{int}}(t) \tag{3044}$$

It is useful to note that the real-time Green's functions are generalizations of the density matrix. From the adiabatic switching of the interaction (external fields that are switched on realistically not adiabatically), we have the formal solution of equation (3043) in the interaction representation:

$$\hat{\rho}(t) = \hat{S}(t,-\infty)\hat{\rho}(-\infty)\hat{S}^{\dagger}(t,-\infty) = \hat{S}(t,-\infty)\hat{\rho}(-\infty)\hat{S}(-\infty,t) \tag{3045}$$

making use of the group property of the S-matrix defined as:

$$\hat{S}(t,t') = \hat{T}\exp\left\{-i\int_{t'}^{t}\hat{H}_{\mathrm{int}}(t_1)dt_1\right\} \quad , \quad \hat{S}^{\dagger}(t,t') = \breve{T}\exp\left\{-i\int_{t}^{t'}\hat{H}_{\mathrm{int}}(t_1)dt_1\right\} \tag{3046}$$

where $\hat{S}\left(\hat{S}^{\dagger}\right)$ is interpreted as the descriptor of quantum time evolution forward (backward) in time.

14.1.1 Basic Features of the S-Matrix (Operator)

$$\hat{S}(t_2,t_1)\hat{S}(t_1,t_0)=\hat{S}(t_2,t_0) \quad , \quad t_2>t_1>t_0 \tag{3047}$$

Equation (3047) expresses a composition rule for two subsequent time segments of S-matrix as well as establishing the relation between the interaction and the Heisenberg representations. Consider that the wave function is transformed according to the following equation:

$$\psi_I(t)=\hat{Q}(t)\psi_H \tag{3048}$$

where $\hat{Q}(t)$ is a unitary operator:

$$\hat{Q}^\dagger=\hat{Q}^{-1} \tag{3049}$$

This shows that the S-matrix is unitary. Hence, there is no restriction on the values of the times involved, and the intermediate time may, in fact, precede both terminal times and so on. Therefore,

$$\hat{Q}(t)=\hat{S}(t,t_0)\hat{Q}(t_0) \tag{3050}$$

or

$$\hat{Q}(t)=\hat{S}(t,\alpha)\hat{P} \tag{3051}$$

where the operator $\hat{P}$ is time independent and α is some time moment. We find the operator $\hat{P}$ by considering (3048), then

$$\exp\left\{i\hat{H}_0 t\right\}\psi=\hat{Q}(t)\psi_H=\hat{S}(t,\alpha)\hat{P}\exp\left\{i\hat{H}t\right\}\psi \quad , \quad \psi(t)=\exp\left\{-i\hat{H}t\right\}\psi_H \tag{3052}$$

Thus,

$$\exp\left\{i\hat{H}_0 t\right\}=\hat{S}(t,\alpha)\hat{P}\exp\left\{i\hat{H}t\right\} \tag{3053}$$

From

$$\hat{S}(\alpha,\alpha)=1 \tag{3054}$$

then

$$\hat{P}=\exp\left\{i\hat{H}_0\alpha\right\}\exp\left\{-i\hat{H}\alpha\right\} \tag{3055}$$

Suppose that at the time moment $t=-\infty$ the interaction is adiabatically "switched-on" at a finite time and adiabatically "switched-off" at $t=+\infty$. Though this assumption is purely formal, it permits us to obtain our results in the quickest manner. When $t=-\infty$, there is no interaction between particles; later, the interaction is switched on infinitely slowly and

$$\hat{P}=1 \quad , \quad \psi_I(t)=\hat{S}(t)\psi_H \tag{3056}$$

$$\hat{S}(t)=\hat{S}(t,-\infty) \tag{3057}$$

Considering equation (3047), we have

$$\hat{S}(t_2,t_1)\hat{S}(t_1,t_0)=\hat{S}(t_2,t_0) \quad , \quad t_2>t_1>t_0 \quad , \quad t_0=-\infty \quad , \quad \hat{S}(t_2,t_1)=\hat{S}(t_2)\hat{S}^{-1}(t_1) \tag{3058}$$

But

$$\hat{S}(t_2,t_1)\hat{S}(t_1,-\infty)=\hat{S}(t_2)\hat{S}^{-1}(t_1)=\hat{S}(t_2,-\infty)=\hat{S}(t_2) \tag{3059}$$

and

$$\hat{S}^{\dagger}(t)=\hat{S}^{\dagger}(\infty)\hat{S}(\infty,t) \tag{3060}$$

We see from (3056) that the operators in the interaction representation relate with those of the Heisenberg representation as follows:

$$\hat{\rho}(t)=\exp\left\{i\hat{H}t\right\}\hat{\rho}\exp\left\{-i\hat{H}t\right\}=\exp\left\{i\hat{H}_{\text{int}}t\right\}\hat{\rho}_I\exp\left\{-i\hat{H}_{\text{int}}t\right\} \tag{3061}$$

or

$$\hat{\rho}(t)=\hat{S}^{\dagger}(t)\hat{\rho}_I\hat{S}(t) \tag{3062}$$

14.1.2 Closed Time Path (CTP) Formalism

In this section, we develop a formalism based on path integrals that allow us to compute correlation functions via equation (3041) for a given density matrix $\hat{\rho}$. This is known as the closed time path (CTP) formalism, and it is developed in references [81, 107, 110]. The introduction of the CTP formalism is artificial and is only for technical reasons in order to provide a convenient way to obtain all the correlation functions needed to describe nonequilibrium physical situations.

The construction of the standard equilibrium many-body theory involves **switching on** the interactions adiabatically in the distant past and then switching them **off** in the distant future, with the essential assumption that by starting from the ground (or equilibrium) state of the system at $t=-\infty$, one reaches the same state at $t=+\infty$ up to some phase acquired along the way. This is not necessarily the case for a nonequilibrium situation. Generally, one starts with an arbitrary initial state, then switches on the interactions, and after a while turns them off. We have to see how to connect the time evolution of the S-matrix in the interaction picture with the relation between the **free** and **interacting states** of the Heisenberg picture. There is no guarantee that the system evolves into the state it was in prior to the switching on of the interactions. This implies the final state is dependent on the switching procedure. Lack of knowledge about the final state spoils the field theoretical formulation completely because the physical observables are described in terms of averages (or traces) of physical operators. To overcome this difficulty, we find a field theoretical formulation that avoids references to the final state at $t=+\infty$. Nevertheless, we still have to compute averages. Therefore, knowledge about the final state is still needed. J. Schwinger [93, 106, 111] and Kadanoff and Baym [112] suggested taking the final state as the initial one, which allows us to compute averages (or traces) of products of operators associated with the physical observables. This approach allows the quantum system to evolve forward in time and then to rewind its evolution back in time. Therefore, we construct a field theory with the time evolution along the two-branch contour C depicted in Figure 14.1.

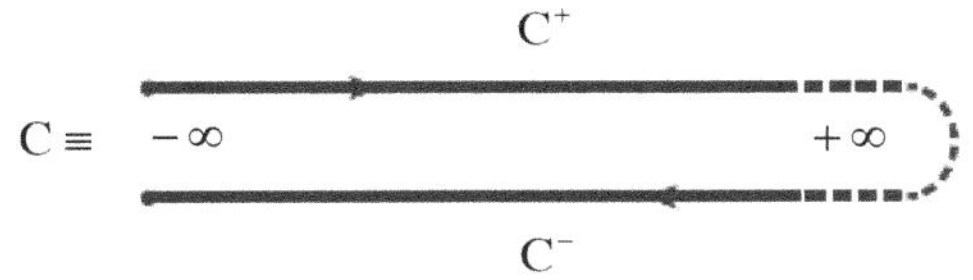

FIGURE 14.1 The Schwinger-Keldysh contour.

This allows the contour $C\left(=C^{+}\cup C^{-}\right)$ to start from $-\infty$ to ∞ (forward branch contour C^{+}) and back to $-\infty$ (backward branch contour C^{-}). This round-trip technique is known as **the Schwinger-Keldysh contour** (**Figure 14.1**).

Therefore, we consider that the interaction $\hat{H}_{int}(t)$ is adiabatically **switched on** between $t=-\infty$ and a finite time and then adiabatically **switched off** at $t=+\infty$. Considering that our system is also described by the time Schrödinger equation, then from its solution as $t=-\infty$, before the interaction begins, the wave function $\psi(t)$ coincides with the Heisenberg function and the wave function evolves as (**Gell-Mann-Low theorem [113]**):

$$\psi(t)=\hat{S}(t,-\infty)\psi_0 \tag{3063}$$

Our goal is to obtain the time-dependent expectation value of the operator $\hat{A}(t)$ of any physical quantity in the interaction representation:

$$\langle A(t)\rangle=\mathrm{Tr}\left\{\hat{A}(t)\hat{\rho}(t)\right\}=\mathrm{Tr}\left\{\hat{S}^{\dagger}(t,-\infty)\hat{A}(t)\hat{S}(t,-\infty)\hat{\rho}(-\infty)\right\} \tag{3064}$$

Formula (3064) implies the transition to the Heisenberg representation of the time-dependent operators averaged over a time-independent density matrix of noninteracting fields due to the adiabatic switching of the interaction. We consider the vacuum (ground) state by substituting $\hat{S}(\infty,t)$ in place of $\hat{S}^{\dagger}(t,-\infty)$. This is justified by the so-called **vacuum stability condition**:

Under the adiabatic transformation, the nondegenerate ground state can only be transformed into itself by multiplying that state by a phase factor.

Then

$$\hat{S}^{\dagger}(t,-\infty)=\hat{S}(-\infty,t)=\hat{S}(-\infty,t)\hat{S}(t,\infty)\hat{S}(\infty,t)=\hat{S}(-\infty,\infty)\hat{S}(\infty,t) \tag{3065}$$

Acting on the vacuum state, the first factor on the right-hand side of this relation is exactly the inverse of that phase factor— a c-number

$$\left\langle\hat{S}\right\rangle_0^{-1}=\left[\langle 0|\hat{S}(\infty,-\infty)|0\rangle\right]^{-1} \tag{3066}$$

So,

$$\langle A(t)\rangle=\left\langle\hat{S}\right\rangle_0^{-1}\langle 0|\hat{T}\left(\hat{S}(\infty,-\infty)\hat{A}(t)\right)|0\rangle \tag{3067}$$

For an arbitrary nonequilibrium state driven by an external (possibly time-dependent) field, the stability condition generally cannot be valid. Therefore, we use formula (3064), and the contour ordering occurs automatically. For the opposite time ordering in $\hat{S}$ and $\hat{S}^{\dagger}$, formula (3064) can be rewritten:

$$\langle A(t)\rangle=\mathrm{Tr}\left\{\hat{T}_C\left(\hat{S}_C\hat{A}(t)\right)\hat{\rho}_0\right\} \tag{3068}$$

Here, C is the contour that starts from $-\infty$ to time moment t and then back to $-\infty$; $\hat{T}_C$ is the time-ordering operator that orders the Heisenberg operators chronologically in the contour sense, with a factor of $(-1)^P$. Here, P is the number of permutations of the fermionic operators, and $\hat{S}_C$ is the S-matrix along the contour C. To extend all the integrals over the entire time axis, we insert the identity operator $\hat{S}(t,\infty)\hat{S}(\infty,t)$ into equation (3068).

Therefore, independent of the final state at $t = +\infty$, the system evolves backward to the known initial state at $t = -\infty$. In this case, there is no switching off of the interactions in the far future; instead, the interactions are switched on in the upper branch of the contour, which evolves forward in time, and off in the lower branch, which evolves backward. The contour $C\left(= C^+ \cup C^-\right)$ starts from $-\infty$ to ∞ (forward branch contour C^+) and goes back to $-\infty$ (backward branch contour C^-), which is known as **the Schwinger-Keldysh contour** (**Figure 14.1**). This method is shown to be better than the analytic continuation.

It is useful to note that this makes integration convenient when both branches of the contour propagate along the real-time axis. The complex time plane is eliminated from this consideration due to the usual nonanalytic time dependence of external fields (switching). Formula (3068) holds for averaged products of an arbitrary number of operators at different times. So, we have **real-time nonequilibrium Green's functions**. Positions of the times on different branches of the contour correspond to different time-ordering of operators because times on the reverse branch are oppositely ordered and always **later** than any time on the direct branch.

Therefore, we observe that for a nonequilibrium state one must use the Keldysh strategy. The beauty of the Keldysh time path is that the definition of the Green's function imitates the equilibrium case, and the price to pay will be several Green's functions to calculate. The Keldysh approach is free from any approximations and, hence, is transparent.

14.2 Contour Green's Functions

The study of single-particle Green's functions is a standard method for characterizing the spectrum and state in many-particle systems. The single-particle contour Green's functions for fermions are defined as contour-ordered expectation values in analogy to the equilibrium case. We introduce the Keldysh technique in the notation of Lifshitz and Pitaevskii [114]. We define a nonequilibrium Green's function as in the equilibrium case:

$$\mathbf{G}(1,1') = -i\,\mathrm{Tr}\left\{\hat{T}_C\left\{\psi_H(1)\psi_H^\dagger(1')\right\}\hat{\rho}_0\right\} \tag{3069}$$

Here, H denotes the Heisenberg representation, and the cumulative variables are denoted by numbers:

$$\alpha\tau \equiv 1 \quad , \quad \alpha'\tau' \equiv 1' \quad , \quad \psi_\alpha(\tau) \equiv \psi(1) \quad , \quad \psi_{\alpha'}^\dagger(\tau') \equiv \psi^\dagger(1') \tag{3070}$$

The beauty of the Keldysh time path is the definition of the Green's function, which plays an analogous role in the nonequilibrium formalism similar to the role the causal Green's function plays in the equilibrium theory. The price to pay is that there actually are several Green's functions to calculate. The difficulty now resides in the fact that time labels not only reside on a forward, but also on the backward, propagating time contour and consequently result in a doubling of the degrees of freedom. It is important to note that quantum transport occurs only in the frame of nonequilibrium Green's functions.

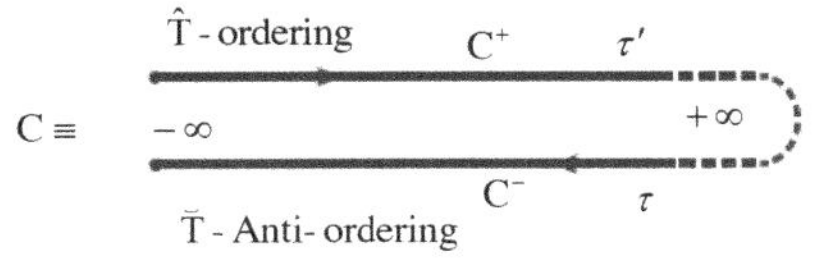

FIGURE 14.2 Keldysh contour [106].

A fermionic system in the normal state can be described by two independent Green's functions, $\mathbf{G}^{>}$ and $\mathbf{G}^{<}$, which are the greater and lesser functions, respectively:

$$\mathbf{G}^{>}(1,1') = -i\,\mathrm{Tr}\left\{\hat{\mathrm{T}}_{\mathrm{C}}\left\{\psi_{\mathrm{H}}(1)\psi_{\mathrm{H}}^{\dagger}(1')\right\}\hat{\rho}_0\right\} \quad , \quad \mathbf{G}^{<}(1,1') = i\,\mathrm{Tr}\left\{\breve{\mathrm{T}}_{\mathrm{C}}\left\{\psi_{\mathrm{H}}^{\dagger}(1')\psi_{\mathrm{H}}(1)\right\}\hat{\rho}_0\right\} \qquad (3071)$$

The extra minus sign in (3071) comes from the contour ordering. The lesser $\mathbf{G}^{<}$ and greater $\mathbf{G}^{>}$ Green's functions in (3071) are graphically represented in **Figure 14.3** as special cases of a more general Green's function defined on the round-trip contour going from τ_0 to τ_M **[107]**:

We use the lesser $\mathbf{G}^{<}$ and greater $\mathbf{G}^{>}$ Green's functions to define some auxiliary Green's functions:

$$\mathbf{G}^{+}(1,1') \equiv \mathbf{G}^{R}(1,1') = \theta(\tau-\tau')\left[\mathbf{G}^{>}(1,1') - \mathbf{G}^{<}(1,1')\right] \qquad (3072)$$

$$\mathbf{G}^{-}(1,1') \equiv \mathbf{G}^{A}(1,1') = -\theta(\tau-\tau')\left[\mathbf{G}^{>}(1,1') - \mathbf{G}^{<}(1,1')\right] \qquad (3073)$$

$$\mathbf{G}^{c}(1,1') = \theta(\tau-\tau')\mathbf{G}^{>}(1,1') + \theta(\tau'-\tau)\mathbf{G}^{<}(1,1') \qquad (3074)$$

$$\mathbf{G}^{a}(1,1') = \theta(\tau-\tau')\mathbf{G}^{<}(1,1') + \theta(\tau'-\tau)\mathbf{G}^{>}(1,1') \qquad (3075)$$

These auxiliary functions are referred to as retarded $(+)$, advanced $(-)$, chronological (c), and antichronological (a); $\theta(\tau)$ is the step or Heaviside function. It is useful to note that

$$\mathbf{G}^{A} = \left[\mathbf{G}^{R}\right]^{*} \qquad (3076)$$

We define the **Keldysh** or **kinetic** (K) **Green's functions**:

$$\mathbf{G}^{\mathrm{K}}(1,1') = \mathbf{G}^{<}(1,1') + \mathbf{G}^{>}(1,1') \qquad (3077)$$

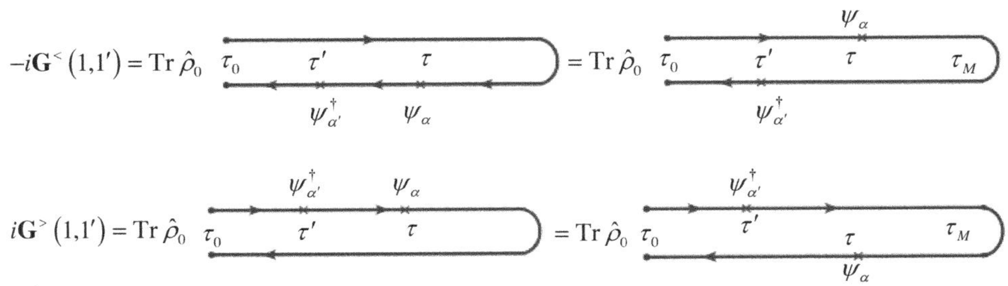

FIGURE 14.3 Graphical representation of the lesser $\mathbf{G}^{<}$ and greater $\mathbf{G}^{>}$ Green's functions where the lines correspond to $\hat{\mathbf{S}}$, cross signs, and field insertions; the arbitrary time $\tau_M > \max\{\tau,\tau'\}$.

Considering equation (3075), we have the following exact relations between the various Green's functions:

$$\mathbf{G}^{>}(1,1')-\mathbf{G}^{<}(1,1')=\mathbf{G}^{+}(1,1')-\mathbf{G}^{-}(1,1') \quad , \quad \mathbf{G}^{c}(1,1')+\mathbf{G}^{a}(1,1')=\mathbf{G}^{K}(1,1') \tag{3078}$$

$$\left[\mathbf{G}^{>}(1,1')\right]^{*}=-\mathbf{G}^{>}(1',1) \quad , \quad \left[\mathbf{G}^{<}(1,1')\right]^{*}=-\mathbf{G}^{<}(1',1) \tag{3079}$$

$$\left[\mathbf{G}^{c}(1,1')\right]^{*}=-\mathbf{G}^{a}(1',1) \quad , \quad \left[\mathbf{G}^{+}(1,1')\right]^{*}=\mathbf{G}^{-}(1',1) \tag{3080}$$

The purpose of developing the theory for the Keldysh Green's functions $\mathbf{G}^{K}$ first is to permit the uncorrelated initial condition in the distant past to be readily satisfied. In addition, it permits us to proceed with the basic structure of the equations of motion and the manner in which to solve them.

14.3 Real-Time Formalism

We observe that real-time formalism is appropriate for treating nonequilibrium situations; so, **the principal goal of nonequilibrium many-body theory is to calculate real-time correlation functions**. We observe that in the real-time formulation of the properties of nonequilibrium states, one encounters at least two types of Green's functions. For a physically transparent representation, the real-time matrix representation of the contour-ordered Green's functions is introduced to permit the quantum dynamics to do the doubling of the degrees of freedom necessary for describing nonequilibrium states. This will permit us to represent matrix Green's function perturbatively in a standard manner via Feynman diagrams. For that, we consider the $\pm$ notation that Keldysh refers to as the two branches of the time contour used as matrix indices in equation (3075). All time integrals extend only along the real time axis from $-\infty$ to ∞. The time integral corresponding to the matrix index "$-$" is taken with a "$-$" sign to account for the reverse direction of integration. The entire information about this system is now accounted for by the structure of Green's function matrices, with all real-time integrals that extend from $-\infty$ to ∞. Furthermore, matrices may be transformed by any canonical transformation that results in another equivalent description of the same system. Consequently, the matrix indices cease to be related to contour branches.

The matrix structure reflects the essence of the real-time formulation of nonequilibrium quantum statistical mechanics from Schwinger [88]—letting the quantum dynamics do the doubling of the degrees of freedom necessary for describing nonequilibrium states.

14.3.1 Real-Time Matrix Representation

So, instead of $\pm$ indices we use 1, 2, and the canonical transformation reduces the number of acting Green's functions because only two of them are linearly independent. For convenience, we introduce a matrix structure in the so-called Keldysh space. In particular, this simple transformation is obtained by transforming (3075) into a quadruplet that is conveniently arranged into a 2×2 matrix:

$$\hat{\mathbf{G}}(1,1')=\begin{bmatrix} \mathbf{G}^{c}(1,1') & \mathbf{G}^{<}(1,1') \\ \mathbf{G}^{>}(1,1') & \mathbf{G}^{a}(1,1') \end{bmatrix} \equiv \begin{bmatrix} \mathbf{G}^{++}(1,1') & \mathbf{G}^{+-}(1,1') \\ \mathbf{G}^{-+}(1,1') & \mathbf{G}^{--}(1,1') \end{bmatrix} \tag{3081}$$

We observe that the four explicit time components of the contour Green's function are not independent from each other. The off-diagonal elements $\mathbf{G}^{+-}(1,1')$ and $\mathbf{G}^{-+}(1,1')$ correspond to $\mathbf{G}^{<}(1,1')$ and $\mathbf{G}^{>}(1,1')$ of Kadanoff and Baym [115]. The diagonal elements can be expressed via the off-diagonal elements. From equations (3077) and (3078), we have the **Keldysh Green's function**

$$\mathbf{G}^{K}(1,1')=\mathbf{G}^{+-}(1,1')+\mathbf{G}^{-+}(1,1')=\mathbf{G}^{--}(1,1')+\mathbf{G}^{++}(1,1') \tag{3082}$$

and

$$\mathbf{G}^{-+}(1,1')-\mathbf{G}^{+-}(1,1')=\mathbf{G}^{+}(1,1')-\mathbf{G}^{-}(1,1') \quad , \quad \mathbf{G}^{++}(1,1')+\mathbf{G}^{--}(1,1')=\mathbf{G}^{K}(1,1') \tag{3083}$$

So, the **causality condition**

$$\mathbf{G}^{++}(1,1')+\mathbf{G}^{--}(1,1')=\mathbf{G}^{+-}(1,1')+\mathbf{G}^{-+}(1,1') \tag{3084}$$

Also from (3079) and (3080), we have

$$\left[\mathbf{G}^{-+}(1,1')\right]^{*}=-\mathbf{G}^{-+}(1',1) \quad , \quad \left[\mathbf{G}^{+-}(1,1')\right]^{*}=-\mathbf{G}^{+-}(1',1) \quad , \quad \left[\mathbf{G}^{++}(1,1')\right]^{*}=-\mathbf{G}^{--}(1',1) \tag{3085}$$

We may also do the denotation

$$\left[\mathbf{G}^{-+}(1,1')\right]^{*}=-\mathbf{G}^{-+}(1',1) \quad , \quad \left[\mathbf{G}^{+-}(1,1')\right]^{*}=-\mathbf{G}^{+-}(1',1) \tag{3086}$$

$$\left[\mathbf{G}^{++}(1,1')\right]^{*}=-\mathbf{G}^{--}(1',1) \quad , \quad \mathbf{G}^{++}(1,1')\equiv\mathbf{G}^{++}(1,1') \tag{3087}$$

We observe that these Green's functions are anti-Hermitian and from the previous relations, we have

$$\mathbf{G}^{R}=\mathbf{G}^{--}-\mathbf{G}^{-+}=\mathbf{G}^{+-}-\mathbf{G}^{++} \quad , \quad \mathbf{G}^{A}=\mathbf{G}^{--}-\mathbf{G}^{+-}=\mathbf{G}^{-+}-\mathbf{G}^{++} \quad , \quad \mathbf{G}^{-+}-\mathbf{G}^{+-}=\mathbf{G}^{A}-\mathbf{G}^{R} \tag{3088}$$

From here, we observe the matrix notation is broken down into two simple rules for the universal vertex structure in the dynamical indices, and this permits us to formulate the nonequilibrium aspects of the Feynman diagrams directly via various matrix Green's function components (retarded, advanced, and Keldysh), thereby establishing the real rules. In this respect, it shows how different features of the spectral and quantum statistical properties enter into the diagrammatic representation of nonequilibrium processes.

We show the equivalence of the imaginary time and the closed time path and the real-time formalisms. All are formally identical and are transformed into each other by analytical continuation.

14.4 Two-Point Correlation Function Decomposition

From the aforementioned equations, we introduce components of the two-point correlations functions with

$$\mathrm{F}(1,1')=i\left[\mathbf{G}^{<}(1,1')+\mathbf{G}^{>}(1,1')\right] \tag{3089}$$

being a **symmetric function that has statistical information about the system,** whereas the **so-called spectral function**

$$\rho(1,1') = \mathbf{G}^{<}(1,1') - \mathbf{G}^{>}(1,1') = \mathbf{G}^{A}(1,1') - \mathbf{G}^{R}(1,1') \tag{3090}$$

has spectral information. The functions $F(1,1')$ and $\rho(1,1')$ are explicitly **real**, as observed from the analyticity property of $\mathbf{G}^{<}(1,1')$ and $\mathbf{G}^{>}(1,1')$ in equation (3079), and they permit us to write the propagators $F(1,1')$ and $\rho(1,1')$ as follows:

$$F(1,1') = i\left[\mathbf{G}^{<}(1,1') + \mathbf{G}^{>}(1,1')\right] = i\left[\mathbf{G}^{>}(1,1') - \left[\mathbf{G}^{>}(1,1')\right]^{*}\right] = -\mathrm{Im}\,\mathbf{G}^{>}(1,1') \tag{3091}$$

$$\rho(1,1') = \mathbf{G}^{<}(1,1') + \left[\mathbf{G}^{>}(1,1')\right]^{*} = 2\mathrm{Re}\,\mathbf{G}^{>}(1,1') \tag{3092}$$

The statistical and spectral correlators also satisfy the symmetry properties:

$$F(1,1') = F(1',1) \quad , \quad \rho(1,1') = -\rho(1',1) \tag{3093}$$

14.5 Equilibrium Green's Function

In order to embark on the development of a general nonequilibrium perturbation theory and its diagrammatic representation starting from the canonical formalism, we consider a brief equilibrium theory and, in particular, the general property characterizing equilibrium. We derive a useful relation for the Green's function that clearly displays its physical content, which is known as the **Lehmann representation.** First, we consider the Green's function on the imaginary part of the contour. In this case, we set $t = -i\tau$, where τ runs from 0 to β. Without loss of generality, we set the time $t_0 = 0$. So, the Green's function:

$$\mathbf{G}(\alpha_1, -i\tau_1, \alpha_2, -i\tau_2) \equiv \mathbf{G}^{M}(\alpha_1\tau_1, \alpha_2\tau_2) \tag{3094}$$

or

$$\mathbf{G}^{M}(\alpha_1\tau_1, \alpha_2\tau_2) = \theta(\tau_1 - \tau_2)\mathbf{G}^{>}(\alpha_1, -i\tau_1, \alpha_2, -i\tau_2) + \theta(\tau_2 - \tau_1)\mathbf{G}^{<}(\alpha_1, -i\tau_1, \alpha_2, -i\tau_2) \tag{3095}$$

Consider the following greater Green's function $\mathbf{G}^{>}$:

$$\mathbf{G}^{>}(\alpha_1, -i\tau_1, \alpha_2, -i\tau_2) = -i\left\langle \hat{\psi}_{\alpha_1}(-i\tau_1)\hat{\psi}^{\dagger}_{\alpha_2}(-i\tau_2)\right\rangle \tag{3096}$$

Let $\{\epsilon_n\}$ be the eigenvalues of $\hat{H}_0$ with eigenvectors $\{|\epsilon_n\rangle\}$:

$$\begin{aligned}&\mathbf{G}^{>}(\alpha_1, -i\tau_1, \alpha_2, -i\tau_2)\\ &= -iZ^{-1}\sum_n \langle \epsilon_n | \exp\{-\beta\hat{H}_0\}\exp\{\tau_1\hat{H}_0\}\hat{\psi}_{\alpha_1}\exp\{-\tau_1\hat{H}_0\}\exp\{\tau_2\hat{H}_0\}\hat{\psi}^{\dagger}_{\alpha_2}\exp\{-\tau_2\hat{H}_0\}|\epsilon_n\rangle\end{aligned} \tag{3097}$$

Here, the partition function is defined as:

$$Z = \mathrm{Tr}\left[\hat{S}(t_0 - i\beta, t_0)\right] \tag{3098}$$

Inserting the resolution of identity in terms of the complete spectrum of $\hat{H}_0$ in equation (3097), we have

$$\begin{aligned} &G^{>}(\alpha_1,-i\tau_1,\alpha_2,-i\tau_2) \\ &= -iZ^{-1}\sum_{n,m}\exp\{-\beta\in_n\}\exp\{\in_n(\tau_1-\tau_2)\}\langle\in_n|\hat{\psi}_{\alpha_1}\exp\{-\tau_1\hat{H}_0\}|\in_m\rangle\langle\in_m|\exp\{\tau_2\hat{H}_0\}\hat{\psi}^{\dagger}_{\alpha_2}|\in_n\rangle \end{aligned} \tag{3099}$$

Further simplification yields

$$G^{>}(\alpha_1,-i\tau_1,\alpha_2,-i\tau_2) = -iZ^{-1}\sum_{n,m}\exp\{-\beta\in_n\}\exp\{(\in_n-\in_m)(\tau_1-\tau_2)\}\langle\in_n|\hat{\psi}_{\alpha_1}|\in_m\rangle\langle\in_m|\hat{\psi}^{\dagger}_{\alpha_2}|\in_n\rangle \tag{3100}$$

Similarly, we can evaluate for the lesser Green's function $G^{<}$:

$$G^{<}(\alpha_1,-i\tau_1,\alpha_2,-i\tau_2) = i\left\langle\hat{\psi}^{\dagger}_{\alpha_2}(-i\tau_2)\hat{\psi}_{\alpha_1}(-i\tau_1)\right\rangle \tag{3101}$$

or

$$\begin{aligned} &G^{<}(\alpha_1,-i\tau_1,\alpha_2,-i\tau_2) \\ &= iZ^{-1}\sum_{n}\langle\in_n|\exp\{-\beta\hat{H}_0\}\exp\{\tau_2\hat{H}_0\}\hat{\psi}^{\dagger}_{\alpha_2}\exp\{-\tau_2\hat{H}_0\}\exp\{\tau_1\hat{H}_0\}\hat{\psi}_{\alpha_1}\exp\{-\tau_1\hat{H}_0\}|\in_n\rangle \end{aligned} \tag{3102}$$

Inserting the resolution of identity in terms of the complete spectrum of $\hat{H}_0$ in the given equation, we have

$$\begin{aligned} &G^{<}(\alpha_1,-i\tau_1,\alpha_2,-i\tau_2) \\ &= iZ^{-1}\sum_{n,m}\exp\{-\beta\in_m\}\exp\{\in_n(\tau_2-\tau_1)\}\langle\in_n|\hat{\psi}^{\dagger}_{\alpha_2}\exp\{-\tau_2\hat{H}_0\}|\in_m\rangle\langle\in_m|\exp\{\tau_1\hat{H}_0\}\hat{\psi}_{\alpha_1}|\in_n\rangle \end{aligned} \tag{3103}$$

Further simplification yields

$$G^{<}(\alpha_1,-i\tau_1,\alpha_2,-i\tau_2) = iZ^{-1}\sum_{n,m}\exp\{-\beta\in_m\}\exp\{(\in_n-\in_m)(\tau_2-\tau_1)\}\langle\in_n|\hat{\psi}^{\dagger}_{\alpha_2}|\in_m\rangle\langle\in_m|\hat{\psi}_{\alpha_1}|\in_n\rangle \tag{3104}$$

We next check the antiperiodicity conditions. For $0 \le \tau_2 \le \beta$, we have:

$$G(\alpha_1,0,\alpha_2,-i\tau_2) = G^{<}(\alpha_1,0,\alpha_2,-i\tau_2) \tag{3105}$$

$$G(\alpha_1,0,\alpha_2,-i\tau_2) = -G^{>}(\alpha_1,-i\beta,\alpha_2,-i\tau_2) \tag{3106}$$

So,

$$\mathbf{G}^{<}(\alpha_1,0,\alpha_2,-i\tau_2) = -\mathbf{G}^{>}(\alpha_1,-i\beta,\alpha_2,-i\tau_2) \tag{3107}$$

From the explicit expressions of equation (3100), we observe that this relation is indeed satisfied. Considering finite electronic systems, we can often take the zero-temperature limit $\beta \to \infty$ and select the chemical potential such that $\in_0 < 0$ and $\in_n > 0$. If the ground state has N particles, we find the following expressions for $\mathbf{G}^{\lessgtr}$:

$$\mathbf{G}^{<}(\alpha_1,-i\tau_1,\alpha_2,-i\tau_2) = iZ^{-1}\sum_m \exp\{(\in_{N,0} - \in_{N-1,m})(\tau_2-\tau_1)\}\langle \in_0|\hat{\psi}^{\dagger}_{\alpha_2}|\in_{N-1,m}\rangle\langle \in_{N-1,m}|\hat{\psi}_{\alpha_1}|\in_0\rangle \tag{3108}$$

$$\mathbf{G}^{>}(\alpha_1,-i\tau_1,\alpha_2,-i\tau_2) = -iZ^{-1}\sum_m \exp\{(\in_{N,0} - \in_{N+1,m})(\tau_1-\tau_2)\}\langle \in_0|\hat{\psi}_{\alpha_1}|\in_{N+1,m}\rangle\langle \in_{N+1,m}|\hat{\psi}^{\dagger}_{\alpha_2}|\in_0\rangle \tag{3109}$$

Here $|\in_{N+1,m}\rangle$ denotes $N\pm1$-particle eigenstates of the system. The previous calculations could just as easily be carried out in real time on the real axis provided we do not switch on any time-dependent external fields. In that case, for simplicity in the zero-temperature limit, we have

$$\mathbf{G}^{<}(\alpha_1 t_1,\alpha_2 t_2) = iZ^{-1}\sum_m \exp\{i(\in_{N,0} - \in_{N-1,m})(t_2-t_1)\}\langle \in_0|\hat{\psi}^{\dagger}_{\alpha_2}|\in_{N-1,m}\rangle\langle \in_{N-1,m}|\hat{\psi}_{\alpha_1}|\in_0\rangle \tag{3110}$$

$$\mathbf{G}^{>}(\alpha_1 t_1,\alpha_2 t_2) = -iZ^{-1}\sum_m \exp\{i(\in_{N,0} - \in_{N+1,m})(t_1-t_2)\}\langle \in_0|\hat{\psi}_{\alpha_1}|\in_{N+1,m}\rangle\langle \in_{N+1,m}|\hat{\psi}^{\dagger}_{\alpha_2}|\in_0\rangle \tag{3111}$$

For the explicit time equilibrium Green's function properties we place t_1 , t_2 on the real-time branches at physical times prior to t_0, where the Hamiltonian is time independent at the given time regime seen previously.

Remark

In thermal equilibrium, correlation functions are dependent on the difference between the times, $t \equiv t_1 - t_2$ (i.e., they are invariant with respect to displacements in time).

We examine the diagonal Green's function for the case of fermions:

$$\mathbf{G}^{R}(\alpha_1 t_1,\alpha_2 t_2) \equiv \mathbf{G}^{R}(\alpha_1,\alpha_2,t_1,t_2) = \mathbf{G}^{R}(\alpha_1,\alpha_2,t_1-t_2) \tag{3112}$$

We use this to derive the so-called Lehmann (or spectral) representation for its Fourier transform $\mathbf{G}^{R}(\alpha_1,\alpha_2,\omega)$:

$$\mathbf{G}^{<}(\alpha_1 t_1,\alpha_2 t_2) = iZ^{-1}\sum_{n,m} \exp\{-\beta \in_m\}\exp\{(\in_n - \in_m)(t_2-t_1)\}\langle \in_n|\hat{\psi}^{\dagger}_{\alpha_2}|\in_m\rangle\langle \in_m|\hat{\psi}_{\alpha_1}|\in_n\rangle \tag{3113}$$

$$\mathbf{G}^{>}(\alpha_1 t_1,\alpha_2 t_2) = -iZ^{-1}\sum_{n,m} \exp\{-\beta \in_n\}\exp\{(\in_n - \in_m)(t_2-t_1)\}\langle \in_n|\hat{\psi}_{\alpha_1}|\in_m\rangle\langle \in_m|\hat{\psi}^{\dagger}_{\alpha_2}|\in_n\rangle \tag{3114}$$

or

$$\mathbf{G}^{<}(\alpha_1 t_1,\alpha_2 t_2)=iZ^{-1}\sum_{n,m}\exp\{-\beta\in_m\}\exp\{(\in_n-\in_m)(t_2-t_1)\}\langle\in_n|\hat{\psi}^{\dagger}_{\alpha_2}|\in_m\rangle\langle\in_n|\hat{\psi}^{\dagger}_{\alpha_1}|\in_m\rangle \tag{3115}$$

$$\mathbf{G}^{>}(\alpha_1 t_1,\alpha_2 t_2)=-iZ^{-1}\sum_{n,m}\exp\{-\beta\in_n\}\exp\{(\in_n-\in_m)(t_2-t_1)\}\langle\in_m|\hat{\psi}^{\dagger}_{\alpha_1}|\in_n\rangle\langle\in_m|\hat{\psi}^{\dagger}_{\alpha_2}|\in_n\rangle \tag{3116}$$

Letting

$$t\equiv t_1-t_2 \tag{3117}$$

and defining

$$\mathbf{f}^{\dagger}_{nm}(\alpha_2)=\langle\in_n|\hat{\psi}^{\dagger}_{\alpha_2}|\in_m\rangle \tag{3118}$$

we have the following

$$\mathbf{G}^{<}(\alpha_1 t_1,\alpha_2 t_2)=iZ^{-1}\sum_{n,m}\exp\{-\beta\in_m\}\exp\{(\in_n-\in_m)t\}\mathbf{f}^{\dagger}_{nm}(\alpha_2)\mathbf{f}^{\dagger}_{nm}(\alpha_1) \tag{3119}$$

$$\mathbf{G}^{>}(\alpha_1 t_1,\alpha_2 t_2)=-iZ^{-1}\sum_{n,m}\exp\{-\beta\in_n\}\exp\{(\in_n-\in_m)t\}\mathbf{f}^{\dagger}_{nm}(\alpha_1)\mathbf{f}^{\dagger}_{nm}(\alpha_2) \tag{3120}$$

We may therefore evaluate the retarded Green's function

$$\mathbf{G}^{R}(\alpha_1,\alpha_2,t)=\theta(t)\left[\mathbf{G}^{>}(\alpha_1,\alpha_2,t)-\mathbf{G}^{<}(\alpha_1,\alpha_2,t)\right] \tag{3121}$$

or

$$\mathbf{G}^{R}(\alpha_1,\alpha_2,t)=i\theta(t)Z^{-1}\sum_{m,n}\left[\exp\{-\beta\in_n\}+\exp\{-\beta\in_m\}\right]\exp\{i(\in_n-\in_m)t\}\mathbf{f}^{\dagger}_{nm}(\alpha_1)\mathbf{f}^{\dagger}_{nm}(\alpha_2) \tag{3122}$$

$$\mathbf{G}^{R}(\alpha_1,\alpha_2,t)=i\theta(t)Z^{-1}\sum_{m,n}\left[\exp\{-\beta\in_n\}+\exp\{-\beta\in_m\}\right]\exp\{i(\in_n-\in_m)t\}\mathbf{f}^{\dagger}_{nm}(\alpha_1)\mathbf{f}^{\dagger}_{nm}(\alpha_2) \tag{3123}$$

and its Fourier transform

$$\mathbf{G}^{R}(\alpha_1,\alpha_2,\omega)=\int_{-\infty}^{\infty}dt\exp\{i\omega t\}\mathbf{G}^{R}(\alpha_1,\alpha_2,t) \tag{3124}$$

or

$$\mathbf{G}^{R}(\alpha_1,\alpha_2,\omega)=-iZ^{-1}\sum_{m,n}\left[\exp\{-\beta\in_n\}+\exp\{-\beta\in_m\}\right]\mathbf{f}^{\dagger}_{nm}(\alpha_1)\mathbf{f}^{\dagger}_{nm}(\alpha_2)\int_{-\infty}^{\infty}dt\theta(t)\exp\{i(\omega+\in_n-\in_m)t\} \tag{3125}$$

or

$$\mathbf{G}^{R}(\alpha_1,\alpha_2,\omega)=-iZ^{-1}\sum_{m,n}\left[\exp\{-\beta\in_n\}+\exp\{-\beta\in_m\}\right]\mathbf{f}^{\dagger}_{nm}(\alpha_1)\mathbf{f}^{\dagger}_{nm}(\alpha_2)\int_{-\infty}^{\infty}dt\exp\{i(\omega+\in_n-\in_m)t\} \tag{3126}$$

or

$$\mathbf{G}^R(\alpha_1,\alpha_2,\omega)=Z^{-1}\sum_{m,n}\left[\exp\{-\beta\in_n\}+\exp\{-\beta\in_m\}\right]\frac{\mathbf{f}^{\dagger}_{nm}(\alpha_1)\mathbf{f}^{\dagger}_{nm}(\alpha_2)}{\omega+\in_n-\in_m+i\delta} \tag{3127}$$

This is the **Lehmann representation** of $\mathbf{G}^R(\alpha_1,\alpha_2,\omega)$. The singularities of $\mathbf{G}^R(\alpha_1,\alpha_2,\omega)$ are the poles located infinitesimally below the real axis at $\omega=\in_m-\in_n-i\delta$ provided $\mathbf{f}^{\dagger}_{nm}(\alpha_1)\mathbf{f}^{\dagger}_{nm}(\alpha_2)\neq 0$. We therefore see from the poles of $\mathbf{G}^R(\alpha_1,\alpha_2,\omega)$ that we can obtain information on the excitation energies $\in_n-\in_m$ associated with the eigenstates $|\in_m\rangle$ and $|\in_n\rangle$ that are connected via the creation operator $\hat{\psi}^{\dagger}_{\alpha_1}$ and $\hat{\psi}^{\dagger}_{\alpha_2}$. These are the eigenstates for which the state has a finite overlap with the state $\hat{\psi}^{\dagger}_{\alpha_1}|\in_n\rangle$ and $\psi^{\dagger}_{\alpha_2}|\in_n\rangle$. The eigenstate $|\in_m\rangle$ has a single particle more than the eigenstate $|\in_n\rangle$. Hence, $\mathbf{G}^R(\alpha_1,\alpha_2,\omega)$ provides information on the single-particle excitation spectrum.

14.5.1 Spectral Function

We calculate the single-particle spectral function that characterizes the excitation spectrum is given in terms of Fourier transforms:

$$A(\alpha_1,\alpha_2,\omega)=-\frac{1}{\pi}\mathrm{Im}\,\mathbf{G}^R(\alpha_1,\alpha_2,\omega)=\frac{1}{\pi}\mathrm{Im}\,\mathbf{G}^A(\alpha_1,\alpha_2,\omega) \tag{3128}$$

Considering

$$\frac{1}{s+i\delta}=\frac{P}{s}-i\pi\delta(s) \tag{3129}$$

Then, the equilibrium spectral function is

$$A(\alpha_1,\alpha_2,\omega)=Z^{-1}\sum_{m,n}\left[\exp\{-\beta\in_n\}+\exp\{-\beta\in_m\}\right]\mathbf{f}^{\dagger}_{nm}(\alpha_1)\mathbf{f}^{\dagger}_{nm}(\alpha_2)\delta(\omega+\in_n-\in_m) \tag{3130}$$

Let us express $\mathbf{G}^R(\alpha_1,\alpha_2,\omega)$ through the spectral function as follows:

$$\mathbf{G}^R(\alpha_1,\alpha_2,\omega)=\int d\omega'\frac{A_{\mathrm{eq}}(\alpha_1,\alpha_2,\omega')}{\omega-\omega'+i\delta} \tag{3131}$$

We perform the prior procedure for the advanced Green's function and find that

$$\mathbf{G}^A(\alpha_1,\alpha_2,\omega)=\left[\mathbf{G}^R(\alpha_1,\alpha_2,\omega)\right]^* \tag{3132}$$

Consider the number operator:

$$\hat{n}_{\alpha_2}=\hat{\psi}^{\dagger}_{\alpha_2}\hat{\psi}_{\alpha_2} \tag{3133}$$

then

$$\hat{n}_{\alpha_2}\hat{\psi}_{\alpha_1}=\hat{\psi}^{\dagger}_{\alpha_2}\hat{\psi}_{\alpha_2}\hat{\psi}_{\alpha_1} \tag{3134}$$

and from

$$\hat{\psi}^{\dagger}_{\alpha_2}\hat{\psi}_{\alpha_1}+\hat{\psi}_{\alpha_1}\hat{\psi}^{\dagger}_{\alpha_2}=\delta_{\alpha_1\alpha_2} \tag{3135}$$

Then

$$\hat{n}_{\alpha_2}\hat{\psi}_{\alpha_1} = -\hat{\psi}^{\dagger}_{\alpha_2}\hat{\psi}_{\alpha_1}\hat{\psi}_{\alpha_2} = \hat{\psi}_{\alpha_1}\hat{\psi}^{\dagger}_{\alpha_2}\hat{\psi}_{\alpha_2} - \hat{\psi}_{\alpha_1} = \hat{\psi}_{\alpha_1}\hat{n}_{\alpha_2} - \hat{\psi}_{\alpha_1} \tag{3136}$$

So

$$\begin{aligned} &\mathbf{G}^{>}(\alpha_1,\alpha_2,t) \\ &= -iZ^{-1}\sum_{m,n}\exp\{-\beta\in_n\}\exp\{i(\in_n - \in_m)t\}\delta_{\alpha_1\alpha_2}\langle\in_n\|\left[-\hat{\psi}_{\alpha_1}\hat{\psi}_{\alpha_1}\hat{\psi}^{\dagger}_{\alpha_2} + \hat{\psi}_{\alpha_1}\delta_{\alpha_1\alpha_2} - \hat{n}_{\alpha_2}\hat{\psi}_{\alpha_1}\right]\|\in_m\rangle\mathbf{f}^{\dagger}_{mn}(\alpha_2) \end{aligned} \tag{3137}$$

or

$$\mathbf{G}^{>}(\alpha_1,\alpha_2,t) = -iZ^{-1}\sum_{m,n}\exp\{-\beta\in_n\}\exp\{i(\in_n - \in_m)t\}\delta_{\alpha_1\alpha_2}\langle\in_n\|\left[\hat{\psi}_{\alpha_1}\delta_{\alpha_1\alpha_2} - \hat{n}_{\alpha_2}\hat{\psi}_{\alpha_1}\right]\|\in_m\rangle\mathbf{f}^{\dagger}_{mn}(\alpha_2) \tag{3138}$$

or

$$\mathbf{G}^{>}(\alpha_1,\alpha_2,t) = -iZ^{-1}\sum_{m,n}\exp\{-\beta\in_n\}\exp\{i(\in_n - \in_m)t\}\left(1-\left\langle\hat{n}_{\alpha_1}\right\rangle\right)\mathbf{f}^{\dagger}_{mn}(\alpha_1)\mathbf{f}^{\dagger}_{mn}(\alpha_2) \tag{3139}$$

14.5.1.1 Kubo-Martin-Schwinger (KMS) Condition

From the aforementioned we establish the following fluctuation dissipation relation or so-called Kubo–Martin–Schwinger boundary conditions:

$$\mathbf{G}^{>}(\alpha_1,\alpha_2,t) = i2\pi A_{eq}(\alpha_1,\alpha_2,\omega)\left(1-\left\langle\hat{n}_{\alpha_1}(\omega)\right\rangle\right) \tag{3140}$$

Similarly,

$$\mathbf{G}^{<}(\alpha_1,\alpha_2,t) = i2\pi A_{eq}(\alpha_1,\alpha_2,\omega)\left\langle\hat{n}_{\alpha_1}(\omega)\right\rangle \tag{3141}$$

These are fluctuation-dissipation theorems for the fermionic single-particle Green's functions with the Fermi-Dirac distribution function being

$$\left\langle\hat{n}_{\alpha_1}(\omega)\right\rangle \equiv n_F(\omega) = \frac{1}{\exp\{\beta\omega\}+1} \tag{3142}$$

From equations (3140) and (3141), we obtain **the so-called Kubo-Martin-Schwinger (KMS) condition [116], which is also referred to as the fluctuation-dissipation relation**:

$$\mathbf{G}^{>}(\alpha_1,\alpha_2,\omega) = -\exp\{\beta\omega\}\mathbf{G}^{<}(\alpha_1,\alpha_2,\omega) \tag{3143}$$

This should be a detailed balancing condition. Absence of the chemical potential in the exponential in (3143) shows the specification of the relationships in the grand canonical ensemble, where energies are measured relative to the chemical potential.

14.5.2 Sum Rule and Physical Interpretation

Let us show an example of the sum rule:

$$\int_{-\infty}^{\infty} d\omega A(\alpha,\omega) = 1 \tag{3144}$$

So

$$\int_{-\infty}^{\infty} d\omega A_{eq}(\alpha_1,\alpha_2,\omega) = Z^{-1}\sum_{m,n}\left[\exp\{-\beta \in_n\}+\exp\{-\beta \in_m\}\right]\mathbf{f}_{mn}^{\dagger}(\alpha_1)\mathbf{f}_{mn}^{\dagger}(\alpha_2)\int_{-\infty}^{\infty} d\omega\delta(\omega+\in_n-\in_m) \tag{3145}$$

or

$$\int_{-\infty}^{\infty} d\omega A_{eq}(\alpha_1,\alpha_2,\omega) = Z^{-1}\sum_{m,n}\left[\exp\{-\beta \in_n\}+\exp\{-\beta \in_m\}\right]\langle \in_m|\hat{\psi}_{\alpha_1}^{\dagger}|\in_n\rangle\langle \in_m|\hat{\psi}_{\alpha_2}^{\dagger}|\in_n\rangle \tag{3146}$$

We can rewrite this as

$$\int_{-\infty}^{\infty} d\omega A_{eq}(\alpha_1,\alpha_2,\omega)$$
$$= Z^{-1}\sum_{m,n}\left[\exp\{-\beta \in_n\}\langle \in_n|\hat{\psi}_{\alpha_1}|\in_m\rangle\langle \in_m|\hat{\psi}_{\alpha_2}^{\dagger}|\in_n\rangle+\exp\{-\beta \in_m\}\langle \in_m|\hat{\psi}_{\alpha_2}^{\dagger}|\in_n\rangle\langle \in_n|\hat{\psi}_{\alpha_1}|\in_m\rangle\right] \tag{3147}$$

or

$$\int_{-\infty}^{\infty} d\omega A_{eq}(\alpha_1,\alpha_2,\omega)$$
$$= Z^{-1}\sum_{m,n}\left[\exp\{-\beta \in_n\}\langle \in_n|\hat{\psi}_{\alpha_1}|\in_m\rangle\langle \in_n|\hat{\psi}_{\alpha_2}^{\dagger}|\in_m\rangle+\exp\{-\beta \in_m\}\langle \in_m|\hat{\psi}_{\alpha_2}^{\dagger}|\in_n\rangle\langle \in_m|\hat{\psi}_{\alpha_1}|\in_n\rangle\right] \tag{3148}$$

or

$$\int_{-\infty}^{\infty} d\omega A_{eq}(\alpha_1,\alpha_2,\omega) = Z^{-1}\sum_{m,n}\left[\exp\{-\beta \in_n\}\langle \in_n|\hat{\psi}_{\alpha_1}\delta_{mn}\hat{\psi}_{\alpha_2}^{\dagger}|\in_m\rangle+\exp\{-\beta \in_m\}\langle \in_m|\hat{\psi}_{\alpha_2}^{\dagger}\delta_{nm}\hat{\psi}_{\alpha_1}|\in_n\rangle\right] \tag{3149}$$

or

$$\int_{-\infty}^{\infty} d\omega A_{eq}(\alpha_1,\alpha_2,\omega) = Z^{-1}\sum_{n}\exp\{-\beta \in_n\}\langle \in_n|\left[\hat{\psi}_{\alpha_1}\hat{\psi}_{\alpha_2}^{\dagger}+\hat{\psi}_{\alpha_2}^{\dagger}\hat{\psi}_{\alpha_1}\right]|\in_n\rangle = Z^{-1}\sum_{n}\exp\{-\beta \in_n\}\delta_{\alpha_1\alpha_2} \tag{3150}$$

If $\alpha_1 = \alpha_2 = \alpha$, then

$$\int_{-\infty}^{\infty} d\omega A_{eq}(\alpha,\omega) = \int_{-\infty}^{\infty} d\omega A(\alpha,\omega) = Z^{-1}\sum_{n}\exp\{-\beta \in_n\} = 1 \tag{3151}$$

Some examples of the sum rule are not exact like the integral in (3151). If we examine the integral (3151), then $A(\alpha,\omega)$ can be interpreted as a probability density and $d\omega\, A(\alpha,\omega)$ is the probability that a fermion with momentum $\vec{\kappa}$ has an energy in an infinitesimal energy window $d\omega$ about ω.

Apart from the purely real-time Green's function, we introduce the mixed-time Green's functions. This is done by placing t_1 on either of the real-time branches and t_2 on $C_\beta = [t_0, t_0 - i\beta]$. Then, we have the following expression for the mixed Green's functions:

$$\mathbf{G}\left(\alpha_1, t_1, \alpha_2, \tilde{t}_0 - i\gamma\right) = i\int \frac{d\omega}{2\pi} A_{eq}\left(\alpha_1, \alpha_2, \omega\right) n_F(\omega) \exp\left\{-i\omega\left(t_1 - \tilde{t}_0 + i\gamma\right)\right\} \tag{3152}$$

$$\mathbf{G}\left(\alpha_1, \tilde{t}_0 - i\gamma, \alpha_2, t_2\right) = -i\int \frac{d\omega}{2\pi} A_{eq}\left(\alpha_1, \alpha_2, \omega\right)\left[1 - n_F(\omega)\right] \exp\left\{-i\omega\left(\tilde{t}_0 - i\gamma - t_1\right)\right\} \tag{3153}$$

Here, $\tilde{t}_0$ is the starting time of the real-time contours, $t_1, t_2 < t_0$, and $0 \le \gamma \le \beta$ is the imaginary time. If the spectral function is a continuous function of ω, then the Lebesgue-Riemann lemma bound to the continuity of the spectral function implies:

$$\lim_{\tilde{t}_0 \to \infty} \mathbf{G}\left(\alpha_1, \tilde{t}_0 - i\gamma, \alpha_2, t_2\right) = \lim_{\tilde{t}_0 \to \infty} \mathbf{G}\left(\alpha_1, t_1, \alpha_2, \tilde{t}_0 - i\gamma\right) = 0 \tag{3154}$$

This is due to the presence of the rapidly oscillatory factors $\exp\left\{\pm i\omega\tilde{t}_0\right\}$ in equations (3152) and (3153). So, the mixed Green's functions vanish for the limit $\tilde{t}_0 \to \infty$. In the contour time integrals, this implies ignoring the imaginary branch of C_{KB} and also ensuring:

$$\mathbf{G}\left(\alpha_1\tau_1, \alpha_2\tau_2\right)\Big|_{t_1, t_2 < t_0} = \mathbf{G}_{eq}\left(\alpha_1\tau_1, \alpha_2\tau_2\right) \tag{3155}$$

This is suitably applied by introducing the KMS conditions as boundary conditions. However, this may not be true for all cases. For example, say,

- **Finite systems have finite number of states and their spectral functions are isolated δ peaks.**
- **The noninteracting gas particles strictly obey the mass-shell condition, with the spectral function consisting of δ peaks. Here, our concern is on interacting systems with thermodynamic limits. This has continuous spectral functions based on physical grounds. So, we can adopt the time contour C to the Schwinger-Keldysh round-trip contour, though we set the limit $\tilde{t}_0 \to \infty$.**

14.6 Keldysh Rotation

A linear transformation on **G** to obtain another matrix

$$\hat{\mathbf{G}} = \begin{bmatrix} \mathbf{G}^{++} & \mathbf{G}^{+-} \\ \mathbf{G}^{-+} & \mathbf{G}^{--} \end{bmatrix} \tag{3156}$$

can be performed by the **so-called Larkin-Ovchinnikov (triangular) representation** [117] that is different in form from the original one by Keldysh [106]. Nevertheless, the physics is the same between the two transformations. In addition, the one by Larkin and Ovchinnikov is used more frequently in condensed matter physics, especially in the field of superconductivity [27]. Considering the components of the matrix in (3156), we observe that not all of these four propagators are linearly independent of one another because of

$$\mathbf{G}^{++} + \mathbf{G}^{--} = \mathbf{G}^{+-} + \mathbf{G}^{-+} \tag{3157}$$

Due to this fact, the **Schwinger–Keldysh rotation [106]** is often used to eliminate one of the four propagators, and the **retarded G^R, advanced G^A, and Schwinger–Keldysh propagator G^K** are introduced as a linear combination of the propagators in the matrix (3156). Moreover, the remaining three propagators are not completely independent of each other, while there is a relation between the retarded and advanced ones. So, we need only two independent functions to describe the nonequilibrium physics. In our case, these functions are the statistical propagator, F, and the spectral function, ρ. The equilibrium is considered as a special case of the nonequilibrium situation. For this special case, a universal relation between the statistical propagator, F, and the spectral function, ρ, is established. So, for the equilibrium situation, only one propagator is needed, whereas in the nonequilibrium situation, there are more degrees of freedom.

In condensed matter physics, a representation via trigonal matrices is often used after applying the **Schwinger–Keldysh rotation.** This is a unitary transformation by a unitary matrix **R**. The **so-called triangular representation** obtained by carrying out a $\frac{\pi}{4}$ rotation in Keldysh space of the matrix Green's function $\hat{G}$ in conventional **so-called triangular representation** amounts to

$$\hat{\hat{G}} = \mathbf{R}\hat{G}\mathbf{R}^T = \begin{bmatrix} G^K & G^R \\ G^A & 0 \end{bmatrix} = \begin{bmatrix} -iF & G^R \\ G^A & 0 \end{bmatrix} \tag{3158}$$

where, besides the usual retarded $\mathbf{G}^R$ and advanced $\mathbf{G}^A$ Green's functions, there is the Keldysh component

$$\mathbf{G}^K = \mathbf{G}^{++} + \mathbf{G}^{--} = \mathbf{G}^{+-} + \mathbf{G}^{-+} \tag{3159}$$

which is central to the nonequilibrium formulation. Note that $\mathbf{G}^K$, and only $\mathbf{G}^K$, does the bookkeeping of the initial distribution. In equation (3158), we introduce the unitary matrix **R** with

$$\mathbf{R}^{-1} = \mathbf{R}^T \tag{3160}$$

This matrix can be written:

$$\mathbf{R} = \frac{1}{\sqrt{2}}\begin{bmatrix} 1 & 1 \\ 1 & -1 \end{bmatrix}, \quad \mathbf{R}^T = \frac{1}{\sqrt{2}}\begin{bmatrix} 1 & 1 \\ -1 & 1 \end{bmatrix} \tag{3161}$$

For the **so-called triangular representation**, not only are they economical, but they also are appealing from a physical point of view because $\mathbf{G}^R$ and $\mathbf{G}^K$ appearing in their components contain distinct physically relevant information:

1. **The spectral function that relates G^R has information on the quantum states of the system and energy spectrum.**
2. **The Keldysh Green's function G^K has information on the occupation of these states for nonequilibrium situations.**
3. **It is minimal; the number of nonzero matrix elements cannot be reduced further by a canonical transformation.**
4. **It is symmetric (retarded as well as advanced functions) with respect to time—like the Feynman's G_{causal} in a vacuum.**
5. **It is symmetric in emission and absorption processes. For fermions, it is charge symmetric.**

Considering a miniature quantum system with only a single bosonic state of energy ω and Hamiltonian operator

$$\hat{H} = \omega\hat{\psi}^{\dagger}\hat{\psi} \tag{3162}$$

we apply the time contour

$$C:(t=0)\rightarrow(t=T)\rightarrow(t=0) \tag{3163}$$

The propagators in the real-time representation, considering $1\equiv\alpha t$, are:

$$\mathbf{G}^{++}(1,1')=-i\exp\{-i\omega(t-t')\}\big(\theta(t-t')+n(\omega)\big) \tag{3164}$$

$$\mathbf{G}^{--}(1,1')=-i\exp\{-i\omega(t-t')\}\big(\theta(t'-t)+n(\omega)\big) \tag{3165}$$

$$\mathbf{G}^{+-}(1,1')=-i\exp\{-i\omega(t-t')\}n(\omega) \tag{3166}$$

$$\mathbf{G}^{-+}(1,1')=-i\exp\{-i\omega(t-t')\}\big(1+n(\omega)\big) \tag{3167}$$

where $t=0$ and $t=T$ are the initial and final times, respectively, with $n(\omega)$ being the Bose distribution function. We apply the **so-called triangular representation** obtained by carrying out a $\frac{\pi}{4}$ rotation in Keldysh space of the matrix Green's function $\hat{\mathbf{G}}$:

$$\hat{\hat{\mathbf{G}}}=\mathbf{R}\hat{\mathbf{G}}\mathbf{R}^{\mathrm{T}}=-i\exp\{-i\omega(t-t')\}\begin{bmatrix}1+2n(\omega) & \theta(t-t')\\ -\theta(t'-t) & 0\end{bmatrix}\equiv\begin{bmatrix}\mathbf{G}^{\mathrm{K}} & \mathbf{G}^{R}\\ \mathbf{G}^{\mathrm{A}} & 0\end{bmatrix} \tag{3168}$$

14.7 Path Integral Representation

To construct the path integral representation of the Keldysh partition function, for convenience, we consider the aforementioned miniature quantum system with only a single bosonic state of energy ω. The partition function for coherent states can be represented as the Keldysh functional integral:

$$Z=\int d\left[\hat{\psi}^*\right]d\left[\hat{\psi}\right]\exp\left\{iS\left[\hat{\psi}^*,\hat{\psi}\right]\right\} \tag{3169}$$

where the action functional:

$$S\left[\hat{\psi}^*,\hat{\psi}\right]=\int_C\hat{\psi}^*(t)\hat{\mathbf{G}}^{-1}\hat{\psi}(t)dt \tag{3170}$$

Considering the following inverse Green's function:

$$\hat{\mathbf{G}}^{-1}=i\partial_t-\hat{\mathrm{H}}\quad,\quad\hat{\mathrm{H}}=\omega\hat{\psi}^*\hat{\psi} \tag{3171}$$

Then we consider the entire contour, and the action functional will be

$$S=\int_{-\infty}^{+\infty}dt\hat{\psi}^{*+}\left(i\frac{\partial}{\partial t}-\omega\right)\hat{\psi}^{+}+\int_{+\infty}^{-\infty}dt\hat{\psi}^{*-}\left(i\frac{\partial}{\partial t}-\omega\right)\hat{\psi}^{-} \tag{3172}$$

The Green's function

$$\mathbf{G}(t,t')=-i\int d\left[\hat{\psi}^*\right]d\left[\hat{\psi}\right]\exp\left\{iS\left[\hat{\psi}^*,\hat{\psi}\right]\right\}\hat{\psi}(t)\hat{\psi}^*(t')=-\left\langle\hat{\psi}\hat{\psi}^*\right\rangle \tag{3173}$$

and

$$\hat{\mathbf{G}}^{-1} = \begin{bmatrix} \hat{\psi}^{*+} & \hat{\psi}^{*-} \end{bmatrix} \begin{bmatrix} \hat{\psi}^{+} \\ \hat{\psi}^{-} \end{bmatrix} = \begin{bmatrix} \mathbf{G}^{++} & \mathbf{G}^{+-} \\ \mathbf{G}^{-+} & \mathbf{G}^{--} \end{bmatrix} \tag{3174}$$

We introduce a rotation that defines the **quantum and classical fields**, respectively:

$$\hat{\psi}^{\mathrm{qu}} \equiv \hat{\psi}_R = \frac{\hat{\psi}^+ - \hat{\psi}^-}{\sqrt{2}} \quad , \quad \hat{\psi}^{\mathrm{cl}} \equiv \hat{\psi}_{R^{\mathrm{T}}} = \frac{\hat{\psi}^+ + \hat{\psi}^-}{\sqrt{2}} \tag{3175}$$

then

$$\hat{\tilde{\psi}} = \frac{1}{\sqrt{2}} \begin{bmatrix} 1 & 1 \\ 1 & -1 \end{bmatrix} \begin{bmatrix} \hat{\psi}^{+} \\ \hat{\psi}^{-} \end{bmatrix} = \frac{1}{\sqrt{2}} \begin{bmatrix} \hat{\psi}^+ + \hat{\psi}^- \\ \hat{\psi}^+ - \hat{\psi}^- \end{bmatrix} = \begin{bmatrix} \hat{\psi}_{R^{\mathrm{T}}} \\ \hat{\psi}_R \end{bmatrix} = \mathbf{R}\hat{\psi} \tag{3176}$$

and

$$\hat{\tilde{\psi}}^{\mathrm{T}} = \frac{1}{\sqrt{2}} \begin{bmatrix} 1 & 1 \\ 1 & -1 \end{bmatrix} \begin{bmatrix} \hat{\psi}^{*+} & \hat{\psi}^{*-} \end{bmatrix} = \begin{bmatrix} \hat{\psi}^*_{\mathbf{R}^{\mathrm{T}}} \\ \hat{\psi}^*_{\mathbf{R}} \end{bmatrix} \tag{3177}$$

$$\hat{\tilde{\mathbf{G}}} = \mathbf{R}\hat{\psi}\hat{\psi}^{*\mathrm{T}}\mathbf{R}^{-1} = \mathbf{R}\hat{\mathbf{G}}\mathbf{R}^{-1} = \mathbf{R}\hat{\mathbf{G}}\mathbf{R}^{\mathrm{T}} = \begin{bmatrix} \mathbf{G}^{\mathrm{K}} & \mathbf{G}^{R} \\ \mathbf{G}^{\mathrm{A}} & 0 \end{bmatrix} \quad , \quad \mathbf{R}^{-1} = \mathbf{R}^{\mathrm{T}} \tag{3178}$$

$$\hat{\tilde{\mathbf{G}}}^{-1} = \begin{bmatrix} 0 & \left[\mathbf{G}^{A}\right]^{-1} \\ \left[\mathbf{G}^{R}\right]^{-1} & \left[\mathbf{G}^{\mathrm{K}}\right]^{-1} \end{bmatrix} \tag{3179}$$

So,

$$S\left[\hat{\psi}_{R^{\mathrm{T}}}, \hat{\psi}_R\right] = \int_{-\infty}^{\infty}\int_{-\infty}^{\infty} dt\,dt' \begin{bmatrix} \hat{\psi}^*_{R^{\mathrm{T}}} & \hat{\psi}^*_R \end{bmatrix} \begin{bmatrix} 0 & \left[\mathbf{G}^{A}\right]^{-1} \\ \left[\mathbf{G}^{R}\right]^{-1} & \left[\mathbf{G}^{\mathrm{K}}\right]^{-1} \end{bmatrix} \begin{bmatrix} \hat{\psi}_{R^{\mathrm{T}}} \\ \hat{\psi}_R \end{bmatrix} \tag{3180}$$

Considering the interaction Hamiltonian

$$\hat{\mathrm{H}}_{\mathrm{int}} = \lambda \sum_r \hat{\psi}^*_r \hat{\psi}^*_r \hat{\psi}_r \hat{\psi}_r \quad , \quad \lambda = \frac{4\pi a}{m} \tag{3181}$$

where a is the scattering length. The interaction action functional can be written

$$S_{\mathrm{int}}\left[\hat{\psi}^\dagger, \hat{\psi}\right] = -\lambda \int dr \int_C \left[\hat{\psi}^*(t)\hat{\psi}(t)\right]^2 dt = -\lambda \int dr \int_{-\infty}^{+\infty} dt \left[\left[\left(\hat{\psi}^+\right)^* \hat{\psi}^+\right]^2 - \left[\left(\hat{\psi}^-\right)^* \hat{\psi}^-\right]^2\right] \tag{3182}$$

Considering equation (3175), then

$$S_{\mathrm{int}}\left[\hat{\psi}_{R^{\mathrm{T}}}, \hat{\psi}_R\right] = -\lambda \int dr \int_{-\infty}^{+\infty} dt \left[\hat{\psi}^*_{R^{\mathrm{T}}} \hat{\psi}^*_R \left(\hat{\psi}^2_R + \hat{\psi}^2_{R^{\mathrm{T}}}\right) + \mathrm{c.c}\right] \tag{3183}$$

The corresponding diagrams are as follows (Figure 14.4).

FIGURE 14.4 Quantum and classical field interactions.

So, the total action functional

$$S\left[\hat{\psi}_{R^{\mathrm{T}}},\hat{\psi}_R\right]=S_0\left[\hat{\psi}_{R^{\mathrm{T}}},\hat{\psi}_R\right]+S_{\mathrm{int}}\left[\hat{\psi}_{R^{\mathrm{T}}},\hat{\psi}_R\right] \tag{3184}$$

14.7.1 Gross-Pitaevskii Equation

We find the Gross-Pitaevskii equation via the following saddle-point equation

$$\frac{\partial S\left[\hat{\psi}_{R^{\mathrm{T}}},\hat{\psi}_R\right]}{\partial\hat{\psi}^*_{R^{\mathrm{T}}}}=0 \tag{3185}$$

This follows $\hat{\psi}_R=0$, and for

$$\frac{\partial S\left[\hat{\psi}_{R^{\mathrm{T}}},\hat{\psi}_R\right]}{\partial\hat{\psi}^*_R}=0 \tag{3186}$$

which follows the celebrated **Gross-Pitaevskii equation**

$$\left(\left[\mathbf{G}^R\right]^{-1}-\lambda\left|\hat{\psi}_{R^{\mathrm{T}}}\right|^2\right)\hat{\psi}_{R^{\mathrm{T}}}=0 \tag{3187}$$

This is a nonlinear Schrödinger equation describing the classical physics of the evolution of a bosonic order parameter amplitude $\hat{\psi}_{R^{\mathrm{T}}}$ in self-interaction with its own density $\approx\left|\hat{\psi}_{R^{\mathrm{T}}}\right|^2$. This Gross–Pitaevskii equation imitates the time-dependent Ginzburg–Landau equation of the transition into a superfluid state. Alternatively, it imitates a nonlinear Schrödinger equation describing the wave function of a Bose–Einstein condensate with inhomogeneous solutions that describe the collective excitations of the condensate field.

For the Dyson equation, we consider

$$\mathbf{G}=-i\int d\left[\hat{\psi}_{R^{\mathrm{T}}}\right]d\left[\hat{\psi}_R\right]\exp\left\{iS\left[\hat{\psi}_{R^{\mathrm{T}}},\hat{\psi}_R\right]\right\}\hat{\psi}_{R^{\mathrm{T}}}\hat{\psi}_R \tag{3188}$$

and expand $\exp\left\{iS\left[\hat{\psi}_{R^{\mathrm{T}}},\hat{\psi}_R\right]\right\}$ around the classical saddle-point $\hat{\psi}_{R^{\mathrm{T}}}=0$.

14.8 Dyson Equation and Self-Energy

We can reformulate the Dyson equation in the form:

$$\hat{\mathbf{G}}=\hat{\mathbf{G}}_0+\hat{\mathbf{G}}_0\hat{\sigma}_3\hat{\Sigma}\hat{\sigma}_3\hat{\mathbf{G}} \tag{3189}$$

Hence, the Keldysh rotation makes Dyson equation more transparent. Equation (3189) imitates the product with $A \cdot B$ being an abbreviation for the convolution $\int_C d\tau A(\tau)B(\tau)$. We observe that the elements of exact Green's function $\hat{\mathbf{G}}$ may be expressed via the exact self-energy Σ:

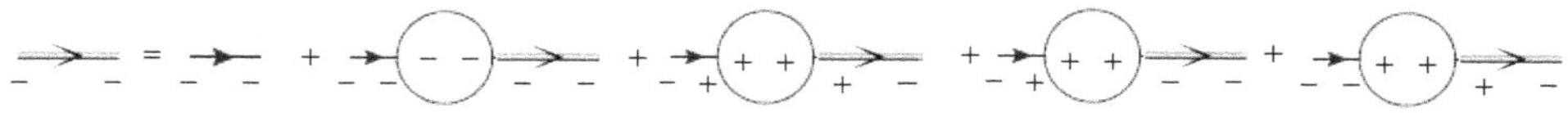
(3190)

See above the diagrammatic representation of the Dyson equation. This is similar to other elements of the exact matrix. The bare Green's functions are represented diagrammatically in Figure 14.5.

FIGURE 14.5 Diagrammatic representation of the bare Green's functions.

From the exact Green's functions and considering that each vertex in the diagram has a + or − sign, we have four exact self-energies. Then, the **exact self-energy** in a system of particles with pair interaction may be represented as follows:

$$\hat{\Sigma} = \begin{bmatrix} \Sigma^{++} & \Sigma^{+-} \\ \Sigma^{-+} & \Sigma^{--} \end{bmatrix} \tag{3191}$$

It is useful to note that the Keldysh technique allows the diagrams to be summed in **blocks**. By applying the $\mathbf{R}$ and $\mathbf{R}^{\mathrm{T}}$ operators to (3189), then

$$\mathbf{R}\hat{\mathbf{G}}\mathbf{R}^{\mathrm{T}} = \mathbf{R}\hat{\mathbf{G}}_0\mathbf{R}^{\mathrm{T}} + \mathbf{R}\hat{\mathbf{G}}_0\mathbf{R}^{\mathrm{T}}\mathbf{R}\hat{\sigma}_3\hat{\Sigma}\hat{\sigma}_3\mathbf{R}^{\mathrm{T}}\mathbf{R}\hat{\mathbf{G}}\mathbf{R}^{\mathrm{T}} \tag{3192}$$

This allows us to arrive at the following Dyson equation in triangular representation:

$$\hat{\hat{\mathbf{G}}} = \hat{\hat{\mathbf{G}}}_0 + \hat{\hat{\mathbf{G}}}_0\hat{\hat{\Sigma}}\hat{\hat{\mathbf{G}}} \tag{3193}$$

where,

$$\hat{\hat{\mathbf{G}}}_0 = \mathbf{R}\hat{\mathbf{G}}_0\mathbf{R}^{\mathrm{T}} \tag{3194}$$

From equation (3193), we observe that the matrix self-energy, $\hat{\hat{\Sigma}}$, from the perturbation theory, imitates the sum of diagrams that cannot be cut in two by cutting only one internal free propagator line. From this point of view, it is a functional of the bare as well as full matrix Green's functions $\hat{\hat{\Sigma}} = \hat{\hat{\Sigma}}\left[\hat{\hat{\mathbf{G}}}_0, \hat{\hat{\mathbf{G}}}\right]$ and also the sum of all skeleton self-energy diagrams. This implies that the diagrams cannot be cut in two by cutting only two full propagator lines. **The self-energy matrix $\hat{\hat{\Sigma}}$ has a triangular structure as the triangular form of the Green's function:**

$$\hat{\hat{\Sigma}} = \mathbf{R}\hat{\sigma}_3\hat{\Sigma}\hat{\sigma}_3\mathbf{R}^{\mathrm{T}} = \begin{bmatrix} \Sigma^{K} & \Sigma^{R} \\ \Sigma^{A} & 0 \end{bmatrix} = \begin{bmatrix} -i\Sigma^{F} & \Sigma^{R} \\ \Sigma^{A} & 0 \end{bmatrix} \tag{3195}$$

Here, we have used the definitions of retarded, advanced, and Keldysh self-energies:

$$\Sigma^R = \Sigma^{+-} - \Sigma^{++} = \Sigma^{--} - \Sigma^{-+} \quad , \quad \Sigma^A = \Sigma^{-+} - \Sigma^{++} = \Sigma^{--} - \Sigma^{+-} \quad , \quad \Sigma^K = \Sigma^{--} + \Sigma^{++} = \Sigma^{-+} + \Sigma^{+-} \tag{3196}$$

This implies that

$$\Sigma^{-+} - \Sigma^{+-} = \Sigma^A - \Sigma^R \tag{3197}$$

So, we write the Dyson equation:

$$\hat{\tilde{\mathbf{G}}} = \hat{\tilde{\mathbf{G}}}_0 + \hat{\tilde{\mathbf{G}}}_0 \hat{\tilde{\Sigma}} \hat{\tilde{\mathbf{G}}} = \begin{bmatrix} \mathbf{G}_0^K & \mathbf{G}_0^R \\ \mathbf{G}_0^A & 0 \end{bmatrix} + \begin{bmatrix} \mathbf{G}_0^K & \mathbf{G}_0^R \\ \mathbf{G}_0^A & 0 \end{bmatrix} \begin{bmatrix} \Sigma^K & \Sigma^R \\ \Sigma^A & 0 \end{bmatrix} \begin{bmatrix} \mathbf{G}^K & \mathbf{G}^R \\ \mathbf{G}^A & 0 \end{bmatrix} \tag{3198}$$

From here, we observe that the Keldysh Green's function may always be expressed via the retarded and advanced Green's functions and Keldysh self-energy:

$$\mathbf{G}^K = \mathbf{G}_0^K + \mathbf{G}_0^R \Sigma^K \mathbf{G}^A + \mathbf{G}_0^R \Sigma^R \mathbf{G}^K + \mathbf{G}_0^K \Sigma^A \mathbf{G}^A \tag{3199}$$

We also have the equation for the advanced Green's function

$$\mathbf{G}^A = \mathbf{G}_0^A + \mathbf{G}_0^A \Sigma^A \mathbf{G}^A \tag{3200}$$

We may obtain the corresponding equation for $\mathbf{G}^R$ from (3198) or by taking the Hermitian conjugate of (3200). Hence, the equation for the retarded Green's function is

$$\mathbf{G}^R = \mathbf{G}_0^R + \mathbf{G}_0^R \Sigma^R \mathbf{G}^R \tag{3201}$$

This is due to the Hermitian conjugate of (3200). The integral equation (3201) integrates the boundary condition via the bare propagator. It is instructive to note that the boundary conditions are not dependent on the initial condition specific to the selected uncorrelated initial state.

14.9 Nonequilibrium Generating Functional

Now, we are interested in deriving the Schwinger functional, $\mathbf{W}$, which is a generating functional for connected Green's functions in the presence of a classical source field η. To describe nonequilibrium physics with an initial state specified, we use the closed-time-path (CTP) formulation developed by Schwinger [106, 118] and relate it to n-point functions. In addition, we use a path integral formulation where time ordering is intrinsically built in and the n-point functions that are our focus are expectation values of time-ordered field operators.

So, our interest here is on the time-dependent correlation functions in nonequilibrium physics. Therefore, we use expectation values of Heisenberg field operators as in equation (3069). The nonequilibrium dynamics of the system are not only tailored by the model Lagrangian but also by its initial conditions encoded in an initial density matrix $\hat{\rho}_0(t_0)$ at time moment t_0. The expectation values of time-ordered Heisenberg field operators are given by the trace over them together with the initial density matrix $\hat{\rho}_0(t_0)$ as seen in equation (3069). Therefore, letting

$$\psi_\alpha(\tau) \equiv \psi(1) \quad , \quad \psi_{\alpha'}^\dagger(\tau') \equiv \psi^\dagger(1') \quad , \quad \psi_\alpha(t_0^\pm) = \psi_{0,\tilde{\alpha}} \quad , \quad \tilde{\alpha} = (\alpha, c) \quad , \quad c = \pm \tag{3202}$$

and

$$\psi^{\dagger}\left(t_0 \in C^{+}\right)=\psi_{0,+}^{\dagger} \quad , \quad \psi^{\dagger}\left(t_0 \in C^{-}\right)=\psi_{0,-}^{\dagger} \tag{3203}$$

then it is convenient to evaluate the trace on the basis of eigenstates of the Heisenberg field operators at the initial time, that is,

$$\psi_{H}(1)\left|\psi_{0,c}\right\rangle=\psi_{0,\bar{\alpha}}\left|\psi_{0,c}\right\rangle \tag{3204}$$

This is so that the matrix elements of $\hat{\rho}_0(t_0)$ can be evaluated considering condition (3204). It is worthy to note as previously seen that $\left|\psi_{0,c}\right\rangle$ denotes the eigenstates (so-called coherent states) and $\psi_{0,\bar{\alpha}}$ are the corresponding eigenvalues.

Equation (3069) can conveniently be formulated as a path-integral representation of the contour Green's function. This imitates the conventional Green's functions except that the time integrations are done on the Schwinger-Keldysh contour instead of the real (or imaginary) lines. As seen earlier, the path integral representation of fermionic fields is done via Grassmann fields and fermionic coherent states where the completeness relation is satisfied:

$$\int d[\psi]d\left[\psi^{\dagger}\right]|\psi\rangle\langle\psi|=\hat{1} \tag{3205}$$

with $\hat{1}$ being the identity operator in Fock space and $|\psi\rangle$ being a normalized fermionic coherent state with $\langle\psi|$ as its adjoint. We express the trace in equation (3069) as a Berezin integral over a fermionic coherent state $\left|\psi_{0,+}\right\rangle$ constructed from the field operators at t_0:

$$\mathbf{G}(1,1')=-i\int\prod_{c=\pm}d\left[\psi_{0,c}\right]d\left[\psi_{0,c}^{\dagger}\right]\exp\left\{iF_C[\psi]\right\}\left\langle\psi_{0,-}\right|\left\{\hat{T}_C\left\{\psi_H(1)\psi_H^{\dagger}(1')\right\}\right\}\left|\psi_{0,+}\right\rangle \tag{3206}$$

where we parametrize the density matrix via the following ansatz [26, 119]:

$$\exp\left\{iF_C[\psi]\right\}\equiv\left\langle\chi\psi_{0,+}\right|\hat{\rho}_0\left|\psi_{0,-}\right\rangle \tag{3207}$$

with F_C being a functional of fields that can be expanded in powers of fields. The quantity $\chi=-1$ for fermions and $\chi=+1$ for bosons in $\left\langle\chi\psi_{0,+}\right|$ considers the anticommutation of the Grassmann fields. We note the time-ordering operator is ill-defined if the field operators have the same time arguments. So, we circumvent the explicit appearance of it in our expressions. To get rid of the time-ordering operator, we rewrite the matrix elements as a path integral expression because path integrals are intrinsically time ordered. The path integration runs along the contour C instead of along the real line:

$$\left\langle\psi_{0,-}\right|\left\{\hat{T}_C\left\{\psi_H(1)\psi_H^{\dagger}(1')\right\}\right\}\left|\psi_{0,+}\right\rangle=\int_{\psi_{0,+}}^{\psi_{0,-}}d'[\psi]d'\left[\psi^{\dagger}\right]\psi_{\alpha}(\tau)\psi_{\alpha'}^{\dagger}(\tau')\exp\left\{iS_C[\psi]\right\} \tag{3208}$$

where

$$\int_{\psi_{0,+}}^{\psi_{0,-}}d'[\psi]d'\left[\psi^{\dagger}\right]\equiv\lim_{\Delta t\to 0}\prod_{k=1}^{n}\int\prod_{c=\pm}d\left[\psi(t_0+k\Delta t,c)\right]d\left[\psi^{\dagger}(t_0+k\Delta t,c)\right],\; S_C[\psi]=S_0[\psi]+S_{\text{int}}[\psi] \tag{3209}$$

and

$$S_0[\psi] = \int_C d\tau d\tau' \psi^\dagger(1) \mathbf{G}_0^{-1}(1,1') \psi(1') \tag{3210}$$

$$S_{int}[\psi] = -\int_C d\tau U_{\alpha_1\alpha_2\alpha_3\alpha_4} \psi^\dagger_{\alpha_1}(\tau) \psi^\dagger_{\alpha_3}(\tau) \psi_{\alpha_4}(\tau) \psi_{\alpha_2}(\tau) \tag{3211}$$

The bare inverse Green's function $\mathbf{G}_0^{-1}(1,1')$ is defined as:

$$\mathbf{G}_0^{-1}(1,1') \equiv -\hat{\tilde{\mathbf{M}}}_0 \delta_{\alpha\alpha} \delta_C(\tau,\tau') \tag{3212}$$

where

$$-\hat{\tilde{\mathbf{M}}}_0 = i\frac{\partial}{\partial\tau} + \frac{\nabla^2}{2m} - U(1) + \mu \tag{3213}$$

Substituting equation (3208) into (3206), we then have

$$\mathbf{G}(1,1') = -i\int \prod_{c=\pm} d[\psi_{0,c}] d[\psi^\dagger_{0,c}] \int_{\psi_{0,+}}^{\psi_{0,-}} d'[\psi] d'[\psi^\dagger] \psi_\alpha(\tau) \psi^\dagger_{\alpha'}(\tau') \exp\{i\Phi_C[\psi]\} \tag{3214}$$

where

$$\Phi_C[\psi] = F_C[\psi] + S_C[\psi] \tag{3215}$$

The fields $\psi_{0,c}$ are now not only integration fields appearing in the matrix elements of the initial density matrix but also the (upper and lower) limits of the dynamical functional integral with the measure $d[\psi]$ of this integral excluding these initial time fields.

Remarks on Nonequilibrium Two-Point Correlation Functions

1. **The functional integral (i.e., second functional integral) in equation (3214), with the action functional $S_C[\psi]$, embeds the quantum fluctuations (born out of the multitude of paths) of the quantum dynamics.**
2. **The weighted average with the initial time elements (i.e., first functional integral) in equation (3214) incorporates the statistical fluctuations.**
3. **Causality requires that the contributions from the time path vanish for all times exceeding the largest time argument of the n-point function. This can be achieved by setting the external two-point source field η to zero for these times such that the time evolution operators on the C^+ and C^- branch of the CTP cancel each other.**

Because the expectation value of a fermionic field operator vanishes due to the Pauli Exclusion Principle, we introduce a method to calculate $2n$-point correlation functions. This is done by taking the nth derivative of a generating functional (the so-called partition function):

$$Z[\eta] = \int \prod_{c=\pm} d[\psi_{0,c}] d[\psi^\dagger_{0,c}] \int_{\psi_{0,+}}^{\psi_{0,-}} d[\psi] d[\psi^\dagger] \exp\{i\Phi_C[\psi,\eta]\} \tag{3216}$$

where

$$\Phi_C[\psi,\eta] = F_C[\psi] + S_C[\psi] + \int_C d\tau d\tau' \left(\psi^\dagger(1)\eta(\gamma) + \eta^*(\gamma)\psi(1')\right) \tag{3217}$$

Apart from the action functional in the path integral (3216), an additional term is introduced that considers a nonlocal, two-point source term $\eta(\gamma) \equiv \eta_{\alpha\alpha'}(\tau,\tau')$, where $\gamma \equiv (1,1')$. The two-point source η is defined to obey

$$\eta_{\alpha'\alpha} = \chi\eta_{\alpha\alpha'} \tag{3218}$$

The path integral (3216) properly shows the main components that constitute the nonequilibrium quantum field theory. The first integral in (3216) represents the initial conditions of the system that describe the statistical fluctuations, while the second encodes information on the quantum fluctuations of the quantum dynamics via the action functional of the system $S_C[\psi]$. From the definition of (3216), we have

$$\mathbf{Z}[\eta=0] = \mathrm{Tr}\{\hat{\rho}(t_0)\} = 1 \tag{3219}$$

The full $2n$-point correlation function is generated via the following relation:

$$\mathrm{Tr}\left\{\hat{T}_C\left\{\hat{\psi}^\dagger(1_1)\hat{\psi}(1_1')\cdots\hat{\psi}^\dagger(1_n)\hat{\psi}(1_n')\right\}\hat{\rho}_0\right\} = \left.\frac{\partial^n \mathbf{Z}[\eta]}{\mathbf{Z} i\partial\eta(\gamma_1)\cdots i\partial\eta(\gamma_n)}\right|_{\eta=0} \tag{3220}$$

Through the partition function we can have the full correlation function, while the Schwinger functional gives only the connected Green's functions. Their generating functional, defined by the Schwinger functional

$$\mathbf{W}[\eta] = -i\ln\mathbf{Z}[\eta] \tag{3221}$$

as the two-point correlation function or Green's function, is connected in the case of fermions because there is no macroscopic fermionic field. So,

$$-\chi\frac{\partial\mathbf{W}[\eta]}{\partial\eta(\gamma_1)} = \mathbf{G}(\gamma_1) \tag{3222}$$

Hence, the Schwinger functional $\mathbf{W}$ is the generating functional for $\mathbf{G}$. It is important to note that the external two-point source η is introduced only for the technical reason of deriving the 2PI effective action. The physical situation corresponds to the absence of external two-point sources (i.e., to $\eta = 0$). However, systems exist with source terms that have physical meaning, such as open systems. For such cases, the source cannot be set to zero.

14.10 Gaussian Initial States

So far, we have derived a path integral formulation for describing nonequilibrium dynamics for arbitrary initial conditions. The nonequilibrium generating functional in equations (3216) and (3221) allows arbitrary initial conditions. To describe experimental relevant scenarios, it is enough to specify only a few of the lowest n-point functions. For **Gaussian initial states**, specifying up to two-point functions are usually sufficient. In this section, we show that we can combine the initial density matrix with the external two-point source η to write the generating functional, in equation (3216), in a much simpler form for Gaussian initial states.

Our focus will be on the initial conditions in the first integral in equation (3216), which allows us to find an alternative formulation for the initial density matrix. It is sufficient to specify the first lowest correlation functions that describe the experimental setups. For the case of fermions, there are no expectation values of the fields. This implies that there is no macroscopic fermionic field. Hence, we have to set the initial conditions for the two-point correlation functions. This choice of initial conditions is referred to as **Gaussian initial conditions**. As a consequence of this choice of initial conditions, we write the nonequilibrium partition function in a simpler or more compact manner.

We facilitate discussions by deriving the general and compact notation for the contour Green's function by using the parametrize density matrix (3207) [26, 119]:

$$F_C[\psi]=\Gamma_0+\sum_{n=0}^{\infty}\frac{1}{n!}\int_C \Gamma_n\left(\tilde{1}_1,\cdots,\tilde{1}_n\right)\prod_{m=1}^{n}d\tau_m\hat{\psi}\left(\tilde{1}_m\right) \quad , \quad \hat{\psi}_{\tilde{\alpha}_m}(\tau_m)\equiv\hat{\psi}\left(\tilde{1}_m\right) \quad , \quad \tilde{1}_m\equiv\tau_m\tilde{\alpha}_m \tag{3223}$$

where the compact notation is interpreted as follows:

1. The **charge index,** $\tilde{\alpha}=(\alpha,c)$, bundles the internal degrees of freedom α and defines the **charge implicit Grassmann new field**

$$\hat{\psi}\left(\tilde{1}\right)=\begin{cases}\hat{\psi}(1) & , \quad c=-\\ \hat{\psi}^{\dagger}(1) & , \quad c=+\end{cases} \tag{3224}$$

   The **charge explicit Grassmann fields** $\hat{\psi}$ and $\hat{\psi}^{\dagger}$ create a particle and a hole, respectively, and imply removing a particle from the system. This convenient notation allows us to treat $\hat{\psi}$ and $\hat{\psi}^{\dagger}$ on an equal footing, as seen earlier in this book.
2. The integral in (3223) is along the Schwinger–Keldysh contour. It is evaluated on the forward branch, C^+, and the backward branch, C^-, of the CTP with the boundary conditions:

$$\psi^{\dagger}\left(t_0\in C^+\right)=\psi^{\dagger}_{0,+} \quad , \quad \psi^{\dagger}\left(t_0\in C^-\right)=\psi^{\dagger}_{0,-} \tag{3225}$$

   The density matrix is dependent solely on the coherent states at the initial time t_0; so, the cumulants $\Gamma_n\left(\tilde{1}_1,\cdots,\tilde{1}_n\right)$ are solely nonzero at the initial times t_0 on both branches of the Schwinger–Keldysh contour, that is, they carry contour Dirac delta functions that make them nonzero solely at $t=t_{0,c}$.
3. The hermiticity of the density matrix $\rho_0=\rho_0^{\dagger}$ imposes further conditions on the cumulants. Because $\Gamma_n(1_1,\cdots,1_n)$ represents the initial correlation in the system, it is the initial correlation vertex where the zeroth cumulants Γ_0 sets the overall normalization of the density matrix considering the following physical requirement:

$$\mathrm{Tr}\left[\hat{\rho}_0\right]=1 \tag{3226}$$

   For the fermionic system, there are no field expectation values. Hence, the cumulant for $n=1$ can be set to zero. So,

$$\Gamma_1=0 \tag{3227}$$

   as

$$\langle\psi\rangle=\langle\psi^{+}\rangle=0 \tag{3228}$$

The Γ_2 describes the initial two-particle correlation that may be the number density, superconducting order parameter, and so on, whereas for the fermionic case, the initial correlation vertices with an odd number of external legs vanish. For this case of the vanishing cumulants, $\Gamma_n(1_1,\cdots,1_n)$, for all $n>2$ and at the initial times t_0 the so-called **Gaussian density matrix or Gaussian initial state** is achieved and the Wick theorem is applicable while making the Feynman-Dyson expansion of the propagator possible. The nonequilibrium generating functional for the Gaussian initial states is then given by

$$Z[\tilde{\eta}]=\int d[\psi]\exp\left\{i\Gamma_0+iS_C[\psi]+i\int_C d\tau d\tau'\psi^{\dagger}(1)\tilde{\eta}(\tilde{\gamma})\psi(1')\right\} \tag{3229}$$

where the new nonlocal two-point source, $\tilde{\eta}_{\alpha\alpha'}(\tau,\tau')$:

$$\tilde{\eta}(\gamma)=\eta(\tilde{\gamma})+\frac{1}{2}\Gamma_2(\tilde{\gamma}) \quad , \quad \tilde{\gamma}=(\tilde{1},\tilde{1}') \tag{3230}$$

The integration measure $d[\psi]$ also includes the fields $\psi_{0,c}$ at the initial time, t_0, in contrast to equation (3216). They occur as the limits of the path integral and no longer are basis states of the matrix representation of the density matrix—the application of the initial conditions is now completely determined by the cumulants.

We then absorb the cumulant Γ_0 into the integration measure because it does not affect the nonequilibrium dynamics of a system. This permits us to rewrite the partition function in a compact form:

$$Z[\tilde{\eta}]=\int d[\psi]\exp\left\{iS_C[\psi]+i\int_C d\tau d\tau'\psi^{\dagger}(1)\tilde{\eta}(\tilde{\gamma})\psi(1')\right\} \tag{3231}$$

Therefore, we are able to specify an arbitrary initial density matrix. However, higher correlation functions not specified at the beginning can be built up during the nonequilibrium time evolution. Generally, most physical setups are well approximated by Gaussian initial states, including experiments. If higher (say, up to n) cumulants need to be specified to describe the initial state, the most convenient way is the generalization of this approach to the nPI effective action. Alternatively, staying within the 2PI effective action approach could also specify an artificial past and describe the desired higher cumulants at time t_0 via the time integrals over the artificial past in the dynamic equations. Nevertheless, setting such a suitable specification is very difficult if not impossible.

The two main reasons to introduce the 2PI effective action in describing nonequilibrium dynamics are:

1. **The nonlocal source $\tilde{\eta}$ allows us to specify the initial conditions for the connected two-point Green's function G of a Gaussian initial state in a very elementary manner.**
2. **The Gaussian initial state simplifies the generating functional.**

14.11 Nonequilibrium 2PI Effective Action

In the previous section, we derived the nonequilibrium generating functional and then simplified it by considering Gaussian initial states in equation (3231). However, direct evaluation of the full quantum real-time path integral in the simplified generating functional is, in general, not feasible due to the oscillating complex measure, which is not positive definite and, so, represents a variant of the sign problem [120, 121, 22]. Therefore, analytical techniques are important in evaluating the dynamics in regimes where quantum fluctuations are relevant. This is specific for long-time evolutions where interactions are strong. However, if quantum fluctuations are small, the quantum part of the fluctuating fields can be integrated out and lead to a classical path integral. Our aim is to simplify the path integral in the generating functional in equation (3231). Instead of using the classical action $S_C[\psi]$ and a path integral over the fluctuating field ψ, we introduce an effective action $\Gamma(\mathbf{G})$ that incorporates the fluctuations.

Considering the system of interacting particles, perturbation theory is performed analogously to the equilibrium theory, and the diagrammatic expansion in the Keldysh formulation imitates the standard Feynman representation with the difference arising from the contour integration. The nonequilibrium generating functional seen in equation (3216) is a generalization of the equilibrium partition function. In thermodynamics, the Legendre transform of the logarithm of the partition function describes the same physics. So, as an analogy, in nonequilibrium quantum field theory, the Legendre transform of the Schwinger functional in equation (3221) presents another equivalent representation. The Legendre transform considering the one-point source term yields the 1PI effective action. The Legendre transform up to the two-point source term yields the 2PI effective action. We can generalize this procedure to an arbitrary order of source terms leading to the nPI effective action. All the generating functionals describe the physics equivalently, with the effective actions having an advantage in that:

1. **They are expressed in terms of correlation functions.**
2. **The initial values of correlation functions are easier to access than those of source terms.**
3. **The effective action obeys the variational principle that makes it easier to derive dynamic equations for the correlation functions.**

The derivation of the effective action conforming to the aforementioned advantages strongly imitates the least action principle in Lagrangian classical mechanics where the dynamical equations are obtained by requiring the stationarity of the classical action. So, we find a functional $\Gamma(\mathbf{G})$ such that it becomes stationary at the exact $\mathbf{G}$ [20]:

$$\frac{\partial\Gamma(\mathbf{G})}{\partial\mathbf{G}}=0 \tag{3232}$$

This gives rise to the Schwinger-Dyson equation for the nonequilibrium Green's function $\mathbf{G}$, as will be seen later in this section.

As seen in the previous section, because the variations of $\mathbf{W}(\tilde{\eta})$ over $\tilde{\eta}$ yield $\mathbf{G}$, we can trade $\tilde{\eta}$ for $\mathbf{G}$ via a Legendre transformation of the Schwinger functional given in equation (3221), that is, defining the 2PI effective action $\Gamma(\mathbf{G})$ [122–124]:

$$\Gamma(\mathbf{G})=\mathbf{W}(\tilde{\eta})-\chi\int_C d\tau d\tau'\tilde{\eta}(\tilde{\gamma})\mathbf{G}(\gamma)=\mathbf{W}(\tilde{\eta})+\mathrm{Tr}[\mathbf{G}\tilde{\eta}] \tag{3233}$$

and

$$\mathbf{G}(\tilde{\gamma})=-\frac{\partial\mathbf{W}(\tilde{\eta})}{\partial\tilde{\eta}(\tilde{\gamma})} \tag{3234}$$

The functional derivative of the 2PI effective action $\Gamma(\mathbf{G})$ with respect to $\mathbf{G}$ yields $\tilde{\eta}$:

$$\frac{\partial\Gamma(\mathbf{G})}{\partial\mathbf{G}(\tilde{\gamma})}=-\tilde{\eta}(\tilde{\gamma}) \tag{3235}$$

This is the stationary condition in equation (3232) where the physical Green's function is evaluated for vanishing nonlocal source fields $\tilde{\eta}$. The trace in equation (3233) denotes a summation over all field and spin indices as well as an integration over all spatial coordinates and over all times along the CTP.

To calculate the 2PI effective action we need an expression for the generating functional of the $2n$-point correlation function given in equation (3231). Generally, the action has a free and an interaction part:

$$S_C[\psi] = S_{C,\text{int}}[\psi] + \int_C d\tau d\tau' \psi^\dagger(1) i\mathbf{G}_0^{-1}(\tilde{\gamma})\psi(1') \tag{3236}$$

The bare inverse propagator $\mathbf{G}_0^{-1}(\gamma)$ appears in the free part of the action. Performing the Taylor expansion of the exponential containing the interaction action term $\exp\{iS_C[\psi]\}$ permits the generating functional in equation (3231) to be rewritten as

$$\mathbf{Z}[\tilde{\eta}] = \int d[\psi]\exp\left\{-\int_C d\tau d\tau' \psi^\dagger(1)\mathbf{G}^{-1}(\gamma)\psi(1')\right\} + \mathbf{Z}[\psi] \tag{3237}$$

Here,

$$\mathbf{Z}[\psi] = \int d[\psi]\exp\left\{-\int_C d\tau d\tau' \psi^\dagger(1)\mathbf{G}^{-1}(\gamma)\psi(1')\right\}\sum_{n=1}^{\infty}\frac{\left(iS_{C,\text{int}}[\psi]\right)^n}{n!} \tag{3238}$$

and

$$\mathbf{G}^{-1}(\tilde{\gamma}) = \mathbf{G}_0^{-1}(\gamma)\delta_{c_1,+}\delta_{c_2,-} - i\tilde{\eta}(\gamma) \tag{3239}$$

In equation (3237), the term corresponding to $n = 0$ (the one-loop-order term) is absent because it is a Gaussian path integral that is exactly solvable and yields a functional determinant expression. The second summand in equation (3237) includes the interaction of the underlying model and is at least of the order of two loops.

We focus on the one-loop part and derive the 2PI effective action at one-loop order. Performing the Gaussian path integral, we obtain for the partition functional at one-loop order:

$$\mathbf{Z}^{(1\text{loop})}[\tilde{\eta}] = \det\left[\mathbf{G}^{-1}(\gamma)\right] = \exp\left\{\text{Tr}\ln\mathbf{G}^{-1}(\gamma)\right\} \tag{3240}$$

The trace implies summation over all field indices and integration along the Schwinger–Keldysh CTP. From here, the Schwinger generating functional for connected correlation functions at one-loop order follows:

$$\mathbf{W}^{(1\text{loop})}[\tilde{\eta}] = -i\ln\mathbf{Z}^{(1\text{loop})}[\tilde{\eta}] = -i\,\text{Tr}\ln\mathbf{G}^{-1}(\gamma) \tag{3241}$$

So, the 2PI effective action at one-loop order:

$$\Gamma^{(1\text{loop})}[\mathbf{G}] = \mathbf{W}^{(1\text{loop})}[\tilde{\eta}] + \text{Tr}[\mathbf{G}\tilde{\eta}] = -i\,\text{Tr}\left[\ln\mathbf{G}^{-1} + \mathbf{G}\left[\mathbf{G}_0^{-1} - \mathbf{G}^{-1}\right]\right] \tag{3242}$$

14.11.1 Luttinger-Ward Functional

To go beyond the one-loop order, the 2PI effective action $\Gamma[\mathbf{G}]$ is conveniently written as the one-loop contribution plus a rest term $\Gamma_2 \equiv \Phi[\mathbf{G}]$ known as the **Luttinger-Ward functional [26], which is a generating functional of the self-energy Σ**. It considers all the scattering effects, as discussed earlier. The rest term $\Phi[\mathbf{G}]$ includes everything per se beyond one-loop:

$$\Gamma[\mathbf{G}] = -i\,\text{Tr}\left[\ln\left[\mathbf{G}^{-1}\right] + \mathbf{G}\left[\mathbf{G}_0^{-1} - \mathbf{G}^{-1}\right]\right] + \Phi[\mathbf{G}] \tag{3243}$$

In this expression, we will neglect the constant term $\mathrm{Tr}\left[\mathbf{G}\mathbf{G}^{-1}\right]$ because it is irrelevant in the system's dynamics:

$$\Gamma[\mathbf{G}] = -i\,\mathrm{Tr}\left[\ln\left[\mathbf{G}^{-1}\right] + \mathbf{G}\mathbf{G}_0^{-1}\right] + \Phi[\mathbf{G}] \tag{3244}$$

The $\Phi[\mathbf{G}]$ term in the 2PI effective action $\Gamma[\mathbf{G}]$ is observed to be a functional of the full propagator $\mathbf{G}$ and mathematically implies effective action mapping of the full propagator to a scalar. This is because it has no indices or time arguments. From the perspective of the Feynman diagrams, the effective action contains only closed diagrams. It is easy to show that the term $\Phi[\mathbf{G}]$ is the sum of all possible 2PI Feynman diagrams collected from bare vertices and full propagators, $\mathbf{G}$, and it implies that for the 2PI term for Feynman diagrams, we can break two lines and the Feynman diagram is still connected. **This term is the so-called 2PI part of the effective action** and is formally written as

$$\Phi[\mathbf{G}] = -i\left.\left\langle \sum_{n=1}^{\infty} \frac{(iS_{\text{int}})^n}{n!} \right\rangle\right|_{2\text{PI},\mathbf{G}} \tag{3245}$$

Here, $\langle\cdot\rangle|_{2\text{PI},\mathbf{G}}$ denotes the expectation value of n interaction terms S_{int}. This is on the condition that the corresponding Feynman diagrams have to be 2PI, with the lines being the full propagators $\mathbf{G}$. The interaction term of the action determines the exact appearance of the 2PI Feynman diagrams. Hence, it is determined by the underlying model under consideration.

Considering a fermionic system at low temperatures and energies, no relativistic effects such as the annihilation and creation of particles will occur. In such a case, no bound states exist; therefore, we assume a two-to-two body scattering. This implies that at each interaction vertex, four full propagator lines meet. For this interaction term, the Feynman diagram representation of the 2PI part of the effective action is explicitly shown in equation (3246).

$\Phi[\mathbf{G}] =$ 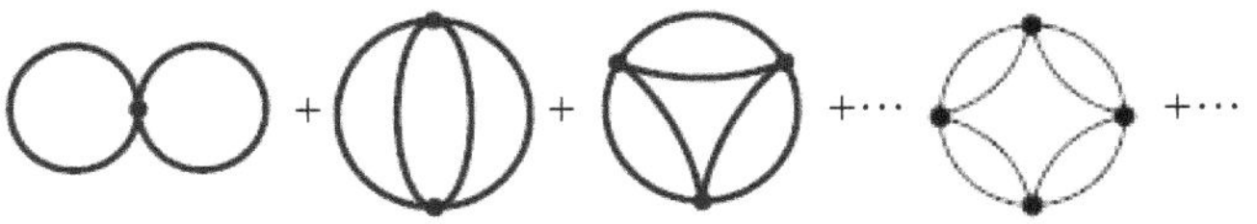 (3246)

FIGURE 14.6 Diagrammatic expansion of the 2PI part $\Phi[\mathbf{G}]$ of the effective action (3244) in terms of closed 2PI diagrams. The solid black lines represent the full propagator, $\mathbf{G}$, and the black dots are the bare vertices. We have omitted all factors determining the relative weights of the diagrams.

We observe that the perturbation series is organized graphically as an expansion in successive orders with respect to the number of closed propagator loops. The solid lines represent the full propagator, $\mathbf{G}$, and the dots represent the bare vertices. These Feynman diagrams do not distinguish between spin up and spin down because the focus is solely on the general structure of the diagrams.

14.12 Kinetic Equation and the 2PI Effective Action

We have studied the evolution of many-body systems and the form of operators representing the physical properties of a system, all of which are embodied by the quantum field. In this section, we study the quantum dynamics of many-body systems, which can also be embodied by the quantum fields. We consider the quantum dynamics of a system instead of its being described via the dynamics of the states, or the evolution operator (i.e., as previously done via the Schrödinger equation) can be carried instead by the quantum fields. So, quantum dynamics is expressed via the correlation or Green's functions of the quantum fields evaluated over some state of the system.

From the system of interacting particles, the perturbation theory in $\hat{H}_{int}$ is performed analogously to the equilibrium theory. The diagrammatic expansion in the Keldysh formulation is analogous to the standard Feynman representation. The difference is the contour integration that corresponds to a summation over the upper and lower branches at each internal vertex. The Dyson equation seen earlier for the self-energy forms a closed set of self-consistent equations for Green's function. By solving them, we can trace the nonequilibrium time evolution of the system. Those equations can be reduced to obtain quantum transport equations in the phase space via the standard prescriptions of the **Wigner transformation [125, 126]** and a subsequent gradient expansion. Therefore, some variables of Dyson equations can be removed that are irrelevant in many cases. The approximation numerically holds excellently over a wide range except for an initial time interval that is much shorter than the time scale for thermalization.

The 2PI effective functional technique provides a powerful tool for dealing with controlled nonequilibrium dynamics with nonsecularity and late time universality. The 2PI effective action relates statistical as well as quantum fluctuations. How does it relate the dynamic equation for the two-point Green's function? It gives the Kadanoff-Baym equation, which describes the dynamics for quantum fluctuations of fields. Because no approximations are made in the derivation of the 2PI effective action, the derived dynamic equation, via the stationary condition, should be exact. Gaussian initial conditions are not an approximation of the quantum dynamics of the system. Only at the beginning, the higher correlations are set to zero. Nevertheless, with time evolution, these higher correlations (that are very important for equilibration processes) can build up.

14.12.1 The Self-Consistent Schwinger–Dyson Equation

Earlier, we saw the diagrammatic expansion rules for **Luttinger-Ward functional $\Phi[\mathbf{G}]$**, as a functional of two-point Green's function $\mathbf{G}$, imitate the thermodynamic potential. Based on this functional approach, the analytic continuation of the two-point Green's function $\mathbf{G}$ in imaginary time formalism to real time analyses can be done via the Kadanoff-Baym formalism, which is reformulated in terms of a variational principle in a round-trip technique—**the Schwinger-Keldysh contour**. We do this by introducing the so-called $\Phi[\mathbf{G}]$-derivable approximation given by a truncated set of closed 2PI diagrams. The main advantage of this approximation is the charge, energy, and momentum conservation of the system in the resulting equations. The 2PI action technique is a finite temperature version of Wick theorem, which satisfies nonequilibrium systems and is applicable in areas of physics, such as cosmology, ultra-relativistic heavy ion collisions, or condensed matter physics for the Bose-Einstein condensate (BEC), because the 2PI approach is a candidate with properties of gapless excitation and conservation laws. The 2PI approach with the Kadanoff-Baym equation is also useful in understanding the thermalization processes toward quark-gluon plasma formation in high-energy heavy-ion collisions.

So, our focus will be on the 2PI approach where substituting the full 2PI effective action in equation (3245) into the stationary condition in equation (3235) yields the **well-known real time Schwinger–Dyson equation** for the one-particle Green's function in differential form [100]:

$$\mathbf{G}^{-1}(\gamma)=\mathbf{G}_0^{-1}(\gamma)-i\tilde{\eta}(\gamma)-\Sigma(\gamma;\mathbf{G}) \tag{3247}$$

where the 1PI self-energy:

$$\Sigma(\gamma;\mathbf{G})=-i\left.\frac{\partial\Phi(\mathbf{G})}{\partial\mathbf{G}(\gamma;\tilde{\eta})}\right|_{\mathbf{G}=\bar{\mathbf{G}}} \tag{3248}$$

and $\bar{\mathbf{G}}$ is the solution of the **stationarity condition** (3232). We observe from equation (3248) that the self-energy is obtained by taking a functional derivative of the Luttinger-Ward functional $\Phi(\mathbf{G})$ that has to be 2PI with respect to a Green's function. This implies that taking the functional derivative with respect to the full propagator **G** is equivalent to graphically breaking one full propagator line, **G**, in the corresponding Feynman vacuum diagram. So, the diagrams do not split into allowed subgraphs by simply breaking any single propagator line. This reduces the order of the particle irreducibility by one. Because the self-energy $\Sigma(\gamma;\mathbf{G})$ is 1PI, the diagrammatic expansion of $\Phi(\mathbf{G})$ only has 2PI diagrams. So, as seen earlier, the functional $\Phi(\mathbf{G})$ is identified as the sum of 2PI connected vacuum diagrams, with **G** in place of $\mathbf{G}_0$.

The Schwinger–Dyson equation, (3247), obtained from the stationary condition of the 2PI effective action, is not easily solvable numerically for given initial conditions. So, to derive a partial differential equation as a dynamical equation for the two-point Green's function **G** that is suitable for initial-value problems, we convolve equation (3247) with the full propagator, **G**. This allows us to obtain the equivalent Kadanoff-Baym equation of motion [127, 128]:

$$\int_C dz\,\mathbf{G}_0^{-1}(\alpha t,\gamma z)\mathbf{G}(\gamma z,\beta t';\tilde{\eta}) = i\delta_C(t-t')\delta_{\alpha\beta} + i\int_C dz\big(\Sigma(\alpha t,\gamma z;\mathbf{G}) + i\tilde{\eta}(\alpha t,\gamma z)\big)\mathbf{G}(\gamma z,\beta t';\tilde{\eta}) \quad (3249)$$

The two-point source $\tilde{\eta}$ brings information about the density matrix into the given equation for the initial time, t_0. For a closed system (i.e., a vanishing external two-point field $\eta = 0$), $\tilde{\eta}$ is only nonzero at the initial time, t_0, at both branches on the Schwinger–Keldysh contour, where it is determined by the initial-time density matrix. Therefore, the source term $\tilde{\eta}$ fixes the initial values for the two-point function **G** (i.e., the term does tailor the dynamics of the system). However, when a system is open and can interact with its environment, there is a change. Generally, the nonlocal two-point source term can be nonzero at all time moments and also can strongly tailor the dynamics of the system.

Though the dynamic equation (3249) is exactly solvable, it requires knowledge of the self-energy and, therefore, of the 2PI part $\Phi(\mathbf{G})$ of the effective action. Truncations of the series of 2PI diagrams are chosen for practical computations. For spin degenerate Fermi gases, some possible truncation schemes are treated for a Kondo lattice gas. It is essential that these approximations are made at the level of the effective action (i.e., on the level of a functional). Deriving the approximated self-energy from an approximated functional and via a variational procedure has the advantage that conservation laws associated with the symmetries of the original effective action are automatically fulfilled by the resultant approximated dynamic equation.

For a nonrelativistic system, the bare inverse propagator $\mathbf{G}_0^{-1}$ has a first-order time derivative. It is also diagonal in the time space. This permits us to perform the integral and then write the Kadanoff-Baym equations of motion as an integro-differential equation. This also permits us to isolate the local contribution of the self-energy from the nonlocal proper self-energy:

$$\Sigma(\alpha t,\beta t';\mathbf{G}) \to -i\Sigma_0(\alpha t,\beta t;\mathbf{G})\delta_C(t-t') + \Sigma(\alpha t,\beta t';\mathbf{G}) \quad (3250)$$

The local contribution to the self-energy is combined with the one-body Hamiltonian appearing in the bare inverse propagator, $\mathbf{G}_0^{-1}$:

$$\hat{\mathbf{M}}(\alpha t,\beta t';\mathbf{G}) = \delta_C(t-t')\big(\delta_{\alpha\beta}\mathrm{H}_\alpha^{1B}(t) + \Sigma_0(\alpha t,\beta t;\mathbf{G})\big) \quad (3251a)$$

Substituting this into equation (3249) and performing integration over the bare inverse propagator and the local contribution of the self-energy. Excluding the term $\int_C dz\tilde{\eta}(\alpha t,\gamma z)\mathbf{G}(\gamma z,\beta t';\tilde{\eta})$ because $\tilde{\eta}$ enters the dynamic equation in the same manner as the self-energy, the Kadanoff–Baym equation of motion now becomes

$$\left(i\frac{\partial}{\partial t} - \hat{\mathbf{M}}(\alpha t,\beta t';\mathbf{G})\right)\mathbf{G}(\alpha t,\beta t';\tilde{\eta}) = i\delta_C(t-t')\delta_{\alpha\beta} + i\int_C dz\Sigma(\alpha t,\gamma z;\mathbf{G})\mathbf{G}(\gamma z,\beta t';\tilde{\eta}) \quad (3251b)$$

14.13 Closed Time Path (CTP) and Extended Keldysh Contours

Next, we focus on systems initially in thermal equilibrium. So far, we have seen the Keldysh contour and the nonequilibrium Green's function that is a particular example of a function defined on a contour. We examine systems in the thermal equilibrium at $t = t_0$. The equilibrium is disturbed due to the presence of time-dependent terms in the Hamiltonian. We associate with any observable quantity A a Hermitian operator $\hat{A}$:

$$A = \mathrm{Tr}\left\{\hat{A}\hat{\rho}_0\right\} \tag{3252}$$

Due to the adiabatic switch-on process, $\hat{\rho}_0(t_0)$ is identified as the density matrix of the system in the remote past when the nonequilibrium perturbation was turned off. It then evolves in the usual way by the S-matrix $\mathbf{S}(0,t_0)$ to a nonequilibrium density matrix $\hat{\rho}$. Hence, we usually choose $\hat{\rho}_0(t_0)$ to be an equilibrium distribution and compute it via the grand-canonical ensemble:

$$\hat{\rho}(t_0) \equiv \hat{\rho}_0 = \frac{\exp\left\{-\beta\left(\hat{H}(t_0) - \mu\hat{N}\right)\right\}}{\mathrm{Tr}\left[\exp\left\{-\beta\left(\hat{H}(t_0) - \mu\hat{N}\right)\right\}\right]} \tag{3253}$$

Here, $\hat{N}$ is the number operator and μ is the chemical potential. The trace denotes a sum over a complete set of states in Hilbert space. If the system is isolated, then the Hamiltonian is time independent, and the expectation value of any observable quantity A is constant provided

$$\left[\hat{H}_0, \hat{\rho}_0\right] = 0 \tag{3254}$$

We have just constructed the perturbation expansion such that the Green's function is expressed as an average over the density matrix of the system in the remote past because the adiabatic switch-on process in equation (3253) is explicitly known. Indeed, the notion of the Keldysh technique is to avoid reference to the general nonequilibrium density matrix $\hat{\rho}$ that is not usually explicitly known but not mandatory. With knowledge of the density matrix $\hat{\rho}$ of the system at $t = 0$, the use of an equation for computation is sufficient:

$$\mathbf{G}(1,1') = -i\,\mathrm{Tr}\left\{\hat{T}_C\left\{\mathbf{S}_C(0,0)\hat{\psi}(1)\hat{\psi}^{\dagger}(1')\right\}\hat{\rho}\right\} \tag{3255}$$

Here, $\hat{\rho}$ is the full density matrix at $t = 0$. The contour over which the $\hat{T}_C$ operator and the S-matrix $\mathbf{S}_C$ are defined is the **so-called closed time path (CTP) contour** C going from 0 to the latest of t or t' and then back to 0. In addition, the time $t = 0$ is not all that special. The Heisenberg and interaction pictures could have been defined via some other initial time t_0. So, the average over the density matrix at time t_0 would be:

$$\mathbf{G}(1,1') = -i\,\mathrm{Tr}\left\{\hat{T}_C\left\{\mathbf{S}_C(t_0,t_0)\hat{\psi}(1)\hat{\psi}^{\dagger}(1')\right\}\hat{\rho}(t_0)\right\} \tag{3256}$$

The CTP contour now goes from t_0 to the latest of t or t' and then back to t_0 (**Figure 14.7**).

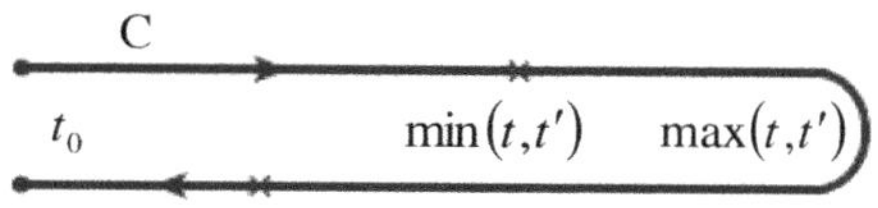

FIGURE 14.7 Closed time path contour C.

In order to recover the Schwinger-Keldysh contour, we take the limit $t_0 \to -\infty$ and then insert a factor $\mathbf{S}(t,\infty)\mathbf{S}(\infty,t)$ or $\mathbf{S}(t',\infty)\mathbf{S}(\infty,t')$ in the perturbation expansion of the Green's function in (3069). We then extend the contour to ∞, depending on whether t or t' is the latest. The density matrix is now $\hat{\rho}(-\infty)$. Therefore, we invoke the adiabatic switch-on process:

$$\hat{\rho}(-\infty) = \frac{\exp\left\{-\beta\left(\hat{\mathrm{H}}(-\infty) - \mu\hat{N}\right)\right\}}{\mathrm{Tr}\left[\exp\left\{-\beta\left(\hat{\mathrm{H}}(-\infty) - \mu\hat{N}\right)\right\}\right]} \tag{3257}$$

14.14 Kadanoff-Baym Contour

Consider equation (3256) is in a suitable form for a perturbation expansion. This is provided $\hat{\rho}(t_0)$ is a noninteracting density matrix and the field operators $\hat{\psi}$ and $\hat{\psi}^\dagger$ in the interaction picture evolve with a noninteracting Hamiltonian $\hat{\mathrm{H}}$. These are indeed the conditions of applicability of Wick theorem and are satisfied in the Keldysh theory if $\hat{\mathrm{H}}$ is noninteracting because $\hat{\rho}(-\infty)$ is a one-particle density matrix. We see more when $\hat{\mathrm{H}}$ contains interactions.

For the time t_0, we take the density matrix in equation (3253), with $\hat{\mathrm{H}}$ being a general interacting Hamiltonian. For $t_0 \to -\infty$, we recover the Keldysh theory at the end. Assuming the system is in the thermal equilibrium at t_0 as well as at all times prior to t_0, then

$$\hat{\mathrm{H}}(t < t_0) = \hat{\mathrm{H}}(t_0) = \hat{\mathrm{H}}_0 \equiv \hat{\mathrm{H}}_{\mathrm{eq}} \tag{3258}$$

The equilibrium is disturbed at times $t > t_0$ due to a possible time-dependent mechanical perturbation described by the time-dependent Hamiltonian $\hat{\mathrm{H}}(t)$ applied to the system:

$$\hat{\mathrm{H}}(t) = \hat{\mathrm{H}}_0 + \hat{\mathrm{H}}_{\mathrm{int}}(t) \tag{3259}$$

Here, $\hat{\mathrm{H}}_{\mathrm{int}}(t)$ is the interaction Hamiltonian. The expectation value of $\hat{\mathrm{A}}$ at $t > t_0$ is then given by the average on the initial density operator $\hat{\rho}_0$ of the operator $\hat{\mathrm{A}}$ in the Heisenberg representation:

$$\mathrm{A}(t) = \mathrm{Tr}\left\{\hat{\mathrm{A}}_{\mathrm{H}}(t)\hat{\rho}_0\right\} = \mathrm{Tr}\left\{\hat{\mathrm{S}}(t_0,t)\hat{\mathrm{A}}\hat{\mathrm{S}}(t,t_0)\hat{\rho}_0\right\} \tag{3260}$$

Here, the operator in the Heisenberg picture is time dependent:

$$\hat{\mathrm{A}}_{\mathrm{H}}(t) = \hat{\mathrm{S}}(t_0,t)\hat{\mathrm{A}}\hat{\mathrm{S}}(t,t_0) \tag{3261}$$

The S-matrix or evolution operator $\hat{\mathrm{S}}(t,t')$:

$$\hat{\mathrm{S}}(t,t') = \hat{\mathrm{T}}\exp\left\{-i\int_{t'}^{t} \hat{\mathrm{H}}(t_1)dt_1\right\} \quad , \quad \hat{\mathrm{S}}^\dagger(t,t') = \breve{\mathrm{T}}\exp\left\{-i\int_{t}^{t'} \hat{\mathrm{H}}(t_1)dt_1\right\} \tag{3262}$$

If the Hamiltonian is time independent in the interval between t and t', then the S-matrix operator:

$$\hat{\mathrm{S}}(t,t') = \exp\left\{-i\hat{\mathrm{H}}(t-t')\right\} \tag{3263}$$

If the system is initially in thermal equilibrium with an inverse temperature β and chemical potential μ, then we have the initial density matrix in (3253). We assume $\hat{H}_0$ and $\hat{N}$ to commute; hence, $\hat{\rho}_0$ can be rewritten via $\hat{S}$ with an imaginary time-argument $t = t_0 - i\beta$ reminiscent of the Matsubara formalism:

$$\hat{\rho}_0 = \frac{\exp\left\{-\beta\left(\hat{H}_0 - \mu\hat{N}\right)\right\}\hat{S}(t_0 - i\beta, t_0)}{\mathrm{Tr}\left[\exp\left\{-\beta\left(\hat{H}_0 - \mu\hat{N}\right)\right\}\hat{S}(t_0 - i\beta, t_0)\right]} \tag{3264}$$

Inserting (3260) in (3041), we have

$$A(t) = \frac{\mathrm{Tr}\left\{\exp\left\{-\beta\left(\hat{H}_0 - \mu\hat{N}\right)\right\}\hat{S}(t_0 - i\beta, t_0)\hat{S}(t_0, t)\hat{A}\hat{S}(t, t_0)\right\}}{\mathrm{Tr}\left[\exp\left\{-\beta\left(\hat{H}_0 - \mu\hat{N}\right)\right\}\hat{S}(t_0 - i\beta, t_0)\right]} \tag{3265}$$

Here, the S-matrix $\hat{S}(t_0 - i\beta, t_0)$ evolves the density matrix along a contour $[t_0, t_0 - i\beta]$ on the imaginary axis and describes interactions. From the time arguments of the S-matrix $\hat{S}(t_0 - i\beta, t_0)$ of equation (3265) from left to right, we might say that the system evolves from t_0 along the real time axis to t after which the operator $\hat{A}$ acts. Then the system evolves back along the real axis from time t to t_0 and finally parallel to the imaginary axis from t_0 to $t_0 - i\beta$.

14.14.1 Green's Function on the Extended Contour

With this observation, we rewrite the Green's function in equation (3256), which now will be in the form:

$$\mathbf{G}(1,1') = -i\frac{\mathrm{Tr}\left\{\hat{T}_C\left\{\hat{S}_C(t_0, t_0)\hat{\psi}_H(1)\hat{\psi}_H^{\dagger}(1')\right\}\exp\left\{-\beta\left(\hat{H}_0 - \mu\hat{N}\right)\right\}\hat{S}(t_0 - i\beta, t_0)\right\}}{\mathrm{Tr}\left[\exp\left\{-\beta\left(\hat{H}_0 - \mu\hat{N}\right)\right\}\hat{S}(t_0 - i\beta, t_0)\hat{S}_C(t_0, t_0)\right]} \tag{3266}$$

The S-matrix $\mathbf{S}_C(t_0, t_0)$ describes external fields, and the evaluation of the trace with given weight $\exp\left\{-\beta\left(\hat{H}_0 - \mu\hat{N}\right)\right\}$ allows Wick theorem. We have also shown that the field operators still evolve according to the full Hamiltonian $\hat{H}$ and can achieve the current interaction picture via $\hat{H}_0$. This can be achieved via the S-matrix, such as in equation (3261). So, equation (3266) now becomes

$$\mathbf{G}(1,1') = -i\frac{\mathrm{Tr}\left\{\hat{T}_C\left\{\hat{S}_C(t_0, t_0)\mathbf{S}(t_0, t)\hat{\psi}(1)\mathbf{S}(t, t_0)\mathbf{S}(t_0, t')\hat{\psi}^{\dagger}(1')\mathbf{S}(t', t_0)\right\}\exp\left\{-\beta\left(\hat{H}_0 - \mu\hat{N}\right)\right\}\hat{S}(t_0 - i\beta, t_0)\right\}}{\mathrm{Tr}\left[\exp\left\{-\beta\left(\hat{H}_0 - \mu\hat{N}\right)\right\}\hat{S}(t_0 - i\beta, t_0)\hat{S}_C(t_0, t_0)\right]} \tag{3267}$$

14.14.2 Kadanoff-Baym Contour

Considering the arguments in the numerator of equation (3267) from the right to the left, we can design a three-branch time contour $C_{KB} = C^{+} \cup C^{-} \cup C_{\beta}$ (Kadanoff-Baym contour) **[118]**, where $C_{\beta} = [t_0, t_0 - i\beta]$. This contour goes from t_0 to the latest of t and t', back to t_0, and then down to $t_0 - i\beta$, **Figure 14.8**.

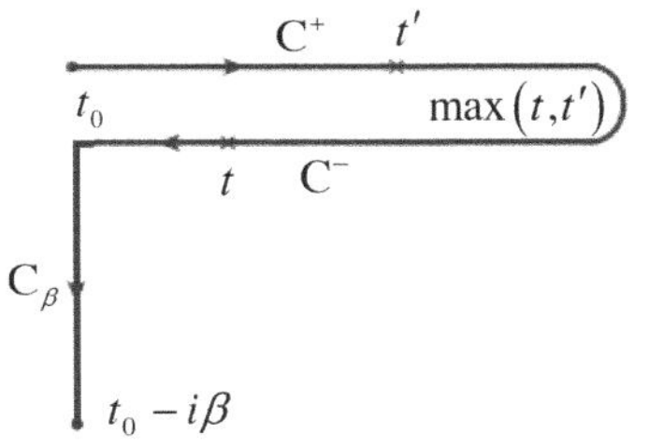

FIGURE 14.8 The Keldysh contour (in the complex time plane) starting at t_0 and ending at $t_0 - i\beta$, with t on the backward branch and t' on the forward branch. Any point lying on the directed vertical line is later than a point lying on the forward or backward branch.

We therefore define an ordering operator $\hat{T}_{CKB}$ along this contour so that we move the thermal S-matrix $\hat{S}(t_0 - i\beta, t_0)$ into the contour-ordered product. So,

$$\mathbf{G}(1,1') = -i\frac{\mathrm{Tr}\left\{\hat{T}_{CKB}\left\{\hat{S}_C(t_0,t_0)\hat{\psi}(1)\hat{\psi}^\dagger(1')\right\}\exp\left\{-\beta\left(\hat{H}_0 - \mu\hat{N}\right)\right\}\hat{S}_{CKB}(t_0 - i\beta, t_0)\right\}}{\mathrm{Tr}\left[\hat{T}_{CKB}\exp\left\{-\beta\left(\hat{H}_0 - \mu\hat{N}\right)\right\}\hat{S}_{CKB}(t_0 - i\beta, t_0)\hat{S}_C(t_0,t_0)\right]} \tag{3268}$$

where

$$\hat{S}_{CKB}(t_0 - i\beta, t_0) = \hat{T}_{CKB}\exp\left\{-i\int_{CKB}\hat{H}(t_1)dt_1\right\} \tag{3269}$$

The partition function is written as

$$Z \equiv \mathrm{Tr}\left[\exp\left\{-\beta\left(\hat{H}_0 - \mu\hat{N}\right)\right\}\hat{S}_C(t_0 - i\beta, t_0)\right] = \mathrm{Tr}\left[\hat{T}_{CKB}\exp\left\{-\beta\left(\hat{H}_0 - \mu\hat{N}\right)\right\}\hat{S}_{CKB}(t_0 - i\beta, t_0)\hat{S}_C(t_0,t_0)\right] \tag{3270}$$

This is due to the fact that the S-matrices are already time-ordered on their respective contours. Also, the times on the $[t_0, t_0 - i\beta]$ trip are always later than times on C. So,

$$S_C(t_0,t_0) = 1 \quad , \quad \hat{S}_{CKB}(t_0 - i\beta, t_0) = \hat{S}(t_0 - i\beta, t_0) \tag{3271}$$

even inside the $\hat{T}_{CKB}$-ordered product. We also observe that the time arguments in equation (3268) are defined on the three-branch Kadanoff-Baym contour C_{KB}. The perturbation theory can now be applied to equation (3268) because the averages are via a one-particle density matrix $\hat{\rho}_0$. The field operators evolve according to the noninteracting Hamiltonian $\hat{H}_0$; therefore, Wick theorem is applicable. So, the expression (3268) can now be used to study the behavior of the system out of equilibrium at times $t > t_0$, after an external perturbation has been switched on at time t_0 while taking into account the initial correlations at $t = t_0$.

14.15 Kubo-Martin-Schwinger (KMS) Boundary Conditions

The so-called **KMS boundary conditions [116]** are for the Green's function $\mathbf{G}$ in equation (3268). These conditions originally were derived for the equilibrium Green's functions and now have been extended to systems out of equilibrium. We derive the boundary condition that couples the values of the Green's function $\mathbf{G}$ at the initial time t_0 and at the final time moment $t_1 = t_0 - i\beta$ by using the cyclic property of the trace:

$$\mathrm{Tr}\left\{\hat{\psi}(1_1)\hat{S}(t_0 - i\beta, t_2)\hat{\psi}^\dagger(1_2)\hat{S}(t_2,t_0)\exp\left\{-\beta\left(\hat{H}_0 - \mu\hat{N}\right)\right\}\right\}$$

$$= \mathrm{Tr}\left\{\hat{S}(t_0 - i\beta, t_2)\hat{\psi}^\dagger(1_2)\hat{S}(t_2,t_0)\hat{\psi}(1_1)\exp\left\{-\beta\left(\hat{H}_0 - \mu\hat{N}\right)\right\}\right\}$$

and so

$$\mathbf{G}(\alpha_1,t_0-i\beta;\alpha_2t_2)=-\mathbf{G}(\alpha_1t_0,\alpha_2t_2) \tag{3272}$$

The Green's function defined in equation (3268) therefore obeys the boundary condition (3272). The property

$$\mathbf{G}(\alpha_1t_1,\alpha_2t_0)=-\mathbf{G}(\alpha_1t_1;\alpha_2,t_0-i\beta) \tag{3273}$$

for the other argument is equally easily verified. These boundary conditions are sometimes referred to as the **so-called Kubo-Martin-Schwinger (KMS) conditions [60, 110, 129]**.

14.15.1 Remark on KMS Boundary Conditions

The KMS conditions state that the Green's functions G are antiperiodic or periodic in imaginary time. This depends on the particles being fermions or bosons, with the interval of periodicity being set by the inverse temperature, which is a critical observation for the Euclidean or imaginary-time formulation of quantum statistical mechanics.

Similar boundary conditions are satisfied by the equilibrium temperature Green's function that is obtained for the special case where the time arguments are located on the contour along the imaginary axis t_0 to $t_0-i\beta$ with the Hamiltonian being time independent. Similarly, for the one-particle Green's function, we can further define the n-particle Green's function that satisfies the **Kubo-Martin-Schwinger boundary conditions**. We can derive a set of so-called hierarchy equations of motion that relate the n-particle Green's function to the $n\pm1$-particle Green's functions.

14.15.2 Generalization of an Average Value

This section aims to pave the way for a quantum kinetic equation for the single particle density matrix $\hat{\rho}(t)$ from the nonequilibrium Green's function (NGF) equations. We generalize equation (3265) and let z be a time-variable (real or complex) on the contour C_{KB} unless otherwise stated. Letting the variable $\bar{z}$ run along this same contour, equation (3265) can formally be:

$$A(z)=\frac{\mathrm{Tr}\left\{\exp\left\{-\beta\left(\hat{H}_0-\mu\hat{N}\right)\right\}\hat{T}_{C_{KB}}\exp\left\{-i\int_{C_{KB}}d\bar{z}\hat{H}(\bar{z})\right\}\hat{A}(z)\right\}}{\mathrm{Tr}\left[\exp\left\{-\beta\left(\hat{H}_0-\mu\hat{N}\right)\right\}\hat{T}_{C_{KB}}\exp\left\{-i\int_{C_{KB}}d\bar{z}\hat{H}(\bar{z})\right\}\right]} \tag{3274}$$

If z lies on the directed vertical line, then there is no need to extend the contour along the real axis. Instead, we have

$$A(z)=\frac{\mathrm{Tr}\left\{\exp\left\{-\beta\left(\hat{H}_0-\mu\hat{N}\right)\right\}\exp\left\{-i\int_{z}^{-i\beta}d\bar{z}\hat{H}_0\right\}\hat{A}\exp\left\{-i\int_{0}^{z}d\bar{z}\hat{H}_0\right\}\right\}}{\mathrm{Tr}\left[\exp\left\{-\beta\left(\hat{H}_0-\mu\hat{N}\right)\right\}\right]}=\frac{\mathrm{Tr}\left\{\exp\left\{-\beta\left(\hat{H}_0-\mu\hat{N}\right)\right\}\hat{A}\right\}}{\mathrm{Tr}\left[\exp\left\{-\beta\left(\hat{H}_0-\mu\hat{N}\right)\right\}\right]} \tag{3275}$$

Here, we have applied the cyclic property of the trace. The right-hand side (RHS) of (3275) is independent of z and matches with the thermal average in (3252). We can easily show that (3274) gives exactly the same result for $A(t)$. Here t is real if the Hamiltonian is time independent, that is, $\hat{H}(t)=\hat{H}_0$ as well as for $t>0$.

From the aforementioned we can summarize: For equation (3274), the variable z lies on the contour of Fig. 14.8. The RHS of (3274) gives the time-dependent statistical average of the observable A when z lies on the forward or backward branch. If z lies on the directed vertical line, the statistical average is taken before the system is disturbed.

The powerful Kadanoff-Baym formalism is adequate for investigating initial correlations (i.e., effect for times $t > t_0$ and an interacting density matrix at time t_0). The price to pay for this formalism is that the Green's function is defined on the C_{KB} contour. It has a complicated expression via simultaneous perturbation expansion of two S-matrices. Nevertheless, for many practical purposes, this is overkill. For steady-state problems, the effect of initial correlations is irrelevant, and for the majority of cases, we assume the correlations decay in time. So, there exists no signature of the correlations in the initial density matrix $\hat{\rho}(t_0)$ for the limit $t_0 \to -\infty$ at any finite time $t >> t_0$. The **Bogoliubov principle of weakening correlations** is a general principle in nonequilibrium statistical mechanics. Therefore, in some cases, initial correlations may persist for long times as a result of metastable states.

14.16 Neglect of Initial Correlations and Schwinger-Keldysh Limit

We can safely ignore the initial correlations for most practical purposes when taking the limit $t_0 \to -\infty$. Neglecting initial correlations results in neglecting the imaginary strip $[t_0, t_0 - i\beta]$ in the Kadanoff-Baym contour and, consequently, the Schwinger-Keldysh contour. This is due to extension of the closed time path contour. So, the S-matrices in the denominator of equation (3268) are trivial and the Green's function:

$$\mathbf{G}(1,1') = -i\,\mathrm{Tr}\left\{\hat{\mathrm{T}}_{\mathrm{C}}\left\{\mathbf{S}_{\mathrm{C}}(-\infty,-\infty)\mathbf{S}_{\mathrm{C}}(-\infty,-\infty)\hat{\psi}(1)\hat{\psi}^{\dagger}(1')\right\}\hat{\rho}_0\right\} \quad (3276)$$

Here, $\hat{\rho}_0$ is defined in (3253), and we obtain $\mathbf{S}_{\mathrm{C}}(-\infty,-\infty)$ from the Kadanoff-Baym contour-ordered S-matrix of equation (3269) by neglecting the third branch of the contour and setting $t_0 \to -\infty$:

$$\mathbf{S}_{\mathrm{C}}(-\infty,-\infty) = \hat{\mathrm{T}}_{\mathrm{C}}\exp\left\{-i\oint_{\mathrm{C}}\hat{\mathrm{H}}_{\mathrm{int}}(\tau_1)d\tau_1\right\} \quad (3277)$$

Here, all the ordering takes place along the Schwinger-Keldysh contour, C.

Considering equation (3276) for a system at zero and finite temperature, for an appropriate expectation value:

$$\mathbf{G}(1,1') = -i\frac{\mathrm{Tr}\left\{\hat{\mathrm{T}}_{\mathrm{C}}\left\{\mathbf{S}_{\mathrm{C}}(-\infty,-\infty)\mathbf{S}_{\mathrm{C}}(-\infty,-\infty)\hat{\psi}(1)\hat{\psi}^{\dagger}(1')\right\}\exp\left\{-\beta\left(\hat{\mathrm{H}}_0 - \mu\hat{N}\right)\right\}\right\}}{\mathrm{Tr}\left[\exp\left\{-\beta\left(\hat{\mathrm{H}}_0 - \mu\hat{N}\right)\right\}\right]} \quad (3278)$$

This is the important starting point for calculations in Keldysh theory, and the perturbation expansion is now applicable with both nonequilibrium and interaction terms via equation (3269). It is instructive to note that the strength of the Kadanoff-Baym and Keldysh approaches to nonequilibrium field theory is based on their structure, which imitates the usual equilibrium many-body theory, albeit with a time evolution and the corresponding perturbative expansion defined on a Kadanoff-Baym contour (more general contour). So, most of the tools of quantum field theory are applicable and—in particular, Feynman diagrams—integral equations for vertex functions such as the Dyson equation and so on. We apply the Schwinger-Keldysh technique to quantum spin systems.

14.16.1 Equation of Motion for the Nonequilibrium Green's Function

In terms of Bose quantum field operators, we represent the many body Hamiltonian

$$\hat{H}' = \hat{H} - \mu\hat{N} \tag{3279}$$

describing bosons interacting via a two-body potential $v(\vec{r}-\vec{r}')$:

$$\hat{H}' = \int d\vec{r}\hat{\psi}^\dagger(\vec{r})\left(-\frac{\nabla^2}{2m}-\mu\right)\hat{\psi}(\vec{r})+\frac{1}{2}\int d\vec{r}\,d\vec{r}'\hat{\psi}^\dagger(\vec{r})\hat{\psi}^\dagger(\vec{r}')v(\vec{r}-\vec{r}')\hat{\psi}(\vec{r})\hat{\psi}(\vec{r}') \tag{3280}$$

The equation of motion for the quantum field operator $\hat{\psi}(\vec{r},t)$ in the Heisenberg representation:

$$i\frac{\partial\hat{\psi}(\vec{r},t)}{\partial t} = \left[\hat{\psi}(\vec{r},t),\hat{H}'\right] \tag{3281}$$

with solution

$$\hat{\psi}(\vec{r},t) = \exp\left\{i\hat{H}'t\right\}\hat{\psi}(\vec{r})\exp\left\{-i\hat{H}'t\right\} \tag{3282}$$

Relation (3281) can be interpreted as the equation of motion for the field operators $\hat{\psi}$ for imaginary times τ within the region:

$$0 \le \tau \equiv it \le \beta \tag{3283}$$

Letting $i \equiv (\vec{r}_i, t_i)$, then from equations (3280) and (3281), we have the following equation of motion for the quantum field $\hat{\psi}(1)$:

$$-\hat{\mathbf{M}}_0\hat{\psi}(\vec{r},t) = \int d\vec{r}'\hat{\psi}^\dagger(\vec{r}',t)v(\vec{r}-\vec{r}')\hat{\psi}(\vec{r},t)\hat{\psi}(\vec{r}',t) \tag{3284}$$

where

$$i\frac{\partial}{\partial t}+\frac{\nabla^2}{2m}+\mu = -\frac{\partial}{\partial it}+\frac{\nabla^2}{2m}+\mu \equiv -\frac{\partial}{\partial\tau}+\frac{\nabla^2}{2m}+\mu \equiv -\hat{\mathbf{M}}_0 \tag{3285}$$

or

$$\hat{\mathbf{M}}_0 \equiv \frac{\partial}{\partial\tau}-\frac{\nabla^2}{2m}-\mu \tag{3286}$$

Multiplying equation (3284) by $\hat{\psi}^+(1')$, and applying the $\hat{T}$-operator, and taking the expectation value of the resulting equation, we then have

$$-\left\langle\hat{T}\hat{\mathbf{M}}_0\hat{\psi}(\vec{r},t)\hat{\psi}^\dagger(1')\right\rangle = \int d2\varpi(1-2)\left\langle\hat{T}\hat{\psi}(1)\hat{\psi}(2)\hat{\psi}^\dagger(2')\hat{\psi}^\dagger(1')\right\rangle\Big|_{t_2=t_1} \tag{3287}$$

Considering that

$$\left\langle\hat{T}\frac{\partial}{\partial\tau_1}\hat{\psi}(1)\hat{\psi}^\dagger(1')\right\rangle-\frac{\partial}{\partial\tau_1}\left\langle\hat{T}\hat{\psi}(1)\hat{\psi}^\dagger(1')\right\rangle = \delta(1-1') \tag{3288}$$

we then have the equation of motion for the Green's function $\mathbf{G}(1,1')$:

$$-\hat{\mathbf{M}}_0\mathbf{G}(1,1') = \delta(1-1') + i\int d2\varpi(1-2)\mathbf{G}_2(1,2,1',2')\big|_{t_2=t_1} \tag{3289}$$

Here, the two-particle nonequilibrium Green's function

$$\mathbf{G}_2(1,2,1',2') = \frac{1}{i^2}\left\langle \hat{\mathrm{T}}\hat{\psi}(1)\hat{\psi}(2)\hat{\psi}^{\dagger}(2')\hat{\psi}^{+}(1')\right\rangle \tag{3290}$$

and

$$\varpi(1-2) \equiv v(\vec{r}_1-\vec{r}_2)v(t_1-t_2) \tag{3291}$$

The notation for the interaction ϖ is changed so as to incorporate its time dependence (i.e., an instantaneous action). Similarly, and taking the adjoint equation of motion-containing derivatives with respect to the 1′ variables, we have:

$$\breve{\mathbf{M}}_0\mathbf{G}(1,1') = \delta(1-1') + i\int d2\varpi(2-1')\mathbf{G}_2(1,2',1',2)\big|_{t_2=t_{1'}} \tag{3292}$$

where

$$\breve{\mathbf{M}}_0 \equiv \frac{\partial}{\partial\tau'} + \frac{\nabla^2}{2m} + \mu \tag{3293}$$

Equations (3289) and (3292) satisfy the boundary condition in (3143) and have the desired form of an equation of motion. But they are not closed because they contain an unknown Green's function of a higher-order. The dynamics of the system, specified by the time dependence of the one-particle Green's function, is therefore described via higher-order correlation functions in the field operators. The equation of motion for the one-particle Green's function therefore yields an infinite hierarchy of equations for correlation functions containing ever-increasing numbers of field operators describing the correlations set up in the system by the interactions: The two-particle Green's function $\mathbf{G}_2$ describes propagation of two particles added to the gas. For this Green's function, we could obtain an analogous equation containing a three-particle Green's function, and so on. This permits us to develop an infinite chain of equations, the so-called Martin–Schwinger hierarchy.

Because there is no closed set of equations for reduced quantities such as Green's functions, in practice, approximate techniques are needed to obtain information about the given system. Occasionally, the system provides a small parameter that tailors the approximations. In less tailorable situations, we yield to the tendency of higher-order correlations to average out for a many-particle system with such average properties as densities and currents, so that the hierarchy of correlations can be broken off self-consistently at low order.

Hence, we resolve to apply methods of converting Equations (3289) and (3292) to closed equations for the single-particle Green's function. The single-particle Green's function has very useful dynamical and statistical mechanical information of the system. Examples are seen in equations (3140) and (3141), where $\mathbf{G}^{<}$ is proportional to the equilibrium expectation value of the density of particles $n(\vec{r},t)$. Higher-order Green's functions, defined similarly to $\mathbf{G}$ and $\mathbf{G}_2$, describe processes involving more than two particles.

14.16.1.1 Nonequilibrium Green's Function Equation of Motion: Auxiliary Fields and Functional Derivatives Technique

If we apply an external field to a system initially in thermodynamic equilibrium, then it will subsequently evolve in time. It is this time dependence on which we are focused. If our interest is in the effect of disturbances due to the external time-dependent fields, we must introduce the nonequilibrium Green's functions. The system in the presence of an external perturbative field $v(t)$ can be described by the following Hamiltonian determinant:

$$\hat{H}(t) = \hat{H}' + v(t) \tag{3294}$$

For the external perturbative field, we consider an auxiliary external time-dependent scalar field $U(\vec{r},t)$ (disturbance) that couples to the particles' local density $\hat{n}(\vec{r},t)$:

$$v(t) \equiv \int d\vec{r}\, U(\vec{r},t)\hat{n}(\vec{r},t) \tag{3295}$$

Here, the particles' local density $\hat{n}(\vec{r},t)$:

$$\hat{n}(\vec{r},t) = \hat{\psi}^{\dagger}(\vec{r},t)\hat{\psi}(\vec{r},t) \tag{3296}$$

We assume the given nonequilibrium perturbation vanishes for the time moment $t < t_0$. This implies that the system is in thermal equilibrium prior to the time moment t_0. For equilibrium, the time dependence of the operators in the Heisenberg representation is given by relation (3282); for nonequilibrium, we must include the time-dependent external potential $v(t)$. This situation is resolved by generalizing the well-established equilibrium formalism in a manner such that it acts as a basis for the nonequilibrium theory. Through the imaginary-time nonequilibrium Green's functions, we can obtain results correlated with the real-time physical response functions needed to describe nonequilibrium experimental probes. Though physical response functions are defined for real times, they are difficult to obtain directly because they do not satisfy a simple boundary condition, such as in (3143). So, instead of the real-time Green's functions, the Kadanoff-Baym formalism is appropriate for nonequilibrium Green's functions defined on the imaginary time domain in (3283) and satisfies the same boundary condition in (3143), at $\tau = 0$ and $\tau = \beta$, as equilibrium Green's functions. Therefore, applying an analytic continuation to the equations of motion for the imaginary-time Green's functions, we obtain the equations of motion for the real-time response functions of physical interest. This is the **so-called Martin–Schwinger approach.**

For the nonequilibrium situation, the Schrödinger equation for the evolution of a system under the perturbation $v(t)$ for $t > t_0$ is as follows:

$$i\frac{\partial|\phi(t)\rangle}{\partial t} = \left(\hat{H}' + v(t)\right)|\phi(t)\rangle \tag{3297}$$

We write the time dependence of the eigenvector $|\phi(t)\rangle$ in the interaction representation

$$|\phi(t)\rangle = \exp\left\{-i\hat{H}'(t-t_0)\right\}|\phi_I(t)\rangle \tag{3298}$$

and substitute it in (3297); then we have

$$i\frac{\partial|\phi_I(t)\rangle}{\partial t} = v_I(t)|\phi_I(t)\rangle \tag{3299}$$

Here, the subscript I labels the interaction representation and the time evolution of the external perturbation $v_{\mathrm{I}}(t)$:

$$v_{\mathrm{I}}(t)=\exp\left\{i\hat{\mathrm{H}}'(t-t_0)\right\}v(t)\exp\left\{-i\hat{\mathrm{H}}'(t-t_0)\right\} \tag{3300}$$

The general solution of (3299) can be written:

$$\left|\phi_{\mathrm{I}}(t)\right\rangle=\hat{\mathbf{s}}(t,t_0)\left|\phi_{\mathrm{I}}(t_0)\right\rangle \tag{3301}$$

where

$$\hat{\mathbf{s}}(t,t_0)=\hat{\mathrm{T}}\exp\left\{-i\int_{t_0}^{t}dt'v_{\mathrm{I}}(t')\right\} \tag{3302}$$

The time-dependent expectation value of the operator of any physical quantity $\hat{\mathrm{A}}(t)$ under the external perturbative field $v(t)$ (in the interaction representation):

$$\left\langle\phi(t)\right|\hat{\mathrm{A}}\left|\phi(t)\right\rangle=\left\langle\phi(t_0)\right|\hat{\mathrm{A}}_U(t)\left|\phi(t_0)\right\rangle \tag{3303}$$

Here,

$$\hat{\mathrm{A}}_U(t)=\hat{\mathbf{S}}^{\dagger}(t,t_0)\hat{\mathrm{A}}\hat{\mathbf{S}}(t,t_0) \tag{3304}$$

The subscript U on the operator $\hat{\mathrm{A}}$ denotes the external scalar potential. The quantity $\hat{\mathbf{S}}(t,t_0)$ has the form reminiscent of the S-matrix in the interaction representation defined:

$$\hat{\mathbf{S}}(t,t_0)=\hat{\mathrm{T}}\exp\left\{-i\int_{t_0}^{t}dt'\hat{\mathrm{H}}(t')\right\}=\exp\left\{-i\hat{\mathrm{H}}'(t-t_0)\right\}\hat{\mathbf{s}}(t,t_0) \tag{3305}$$

It is easy to show that $\hat{\mathrm{A}}_U(t)$ satisfies the following equation of motion:

$$i\frac{\partial\hat{\mathrm{A}}_U(t)}{\partial t}=\left[\hat{\mathrm{A}}_U(t),\hat{\mathrm{H}}'\right] \tag{3306}$$

The time evolution of the operator $\hat{\mathrm{A}}_U(t)$ can be expressed via the operator $\hat{\mathrm{A}}_{\mathrm{I}}(t)$ in the interaction representation:

$$\hat{\mathrm{A}}_U(t)=\hat{\mathbf{s}}^{\dagger}(t,t_0)\hat{\mathrm{A}}_{\mathrm{I}}(t)\hat{\mathbf{s}}(t,t_0) \tag{3307}$$

where

$$\hat{\mathrm{A}}_{\mathrm{I}}(t)=\exp\left\{i\hat{\mathrm{H}}'(t-t_0)\right\}\hat{\mathrm{A}}\exp\left\{-i\hat{\mathrm{H}}'(t-t_0)\right\} \tag{3308}$$

The S-matrix $\hat{\mathbf{s}}(t,t_0)$ satisfies the following conditions:

$$i\frac{\partial\hat{\mathbf{s}}(t_0,t_1)}{\partial t_1}=\hat{\mathrm{H}}'(t_1)\hat{\mathbf{s}}(t_0,t_1)\quad,\quad\hat{\mathbf{s}}(t_0,t_0)=1\quad,\quad\lim_{t_0\to-\infty}\hat{\mathbf{s}}(t_0,t_1)=\hat{\mathbf{s}}(t_1)\quad,\quad\lim_{t_0\to-\infty}\hat{\mathbf{s}}(t_0,t_0-i\beta)=1 \tag{3309}$$

It is instructive to note that the perturbation is due to the additional external field. So, for the transition to the Dirac picture, all field operators have the full Heisenberg time dependence, including the interactions. The thermal expectation value of an operator $\hat{A}$ evolving in the presence of external time-dependent perturbation $\hat{H}'(t)$ is given by

$$\left\langle \hat{A}_U(t) \right\rangle = Z^{-1} \sum_n \exp\left\{-\beta(\epsilon_n - \mu N_n)\right\} \langle \epsilon_n | \hat{A}_U(t) | \epsilon_n \rangle \tag{3310}$$

or

$$\left\langle \hat{A}_U(t) \right\rangle = Z^{-1} \sum_n \exp\left\{-\beta(\epsilon_n - \mu N_n)\right\} \langle \epsilon_n | \hat{s}^\dagger(t,t_0) \hat{A}_I(t) \hat{s}(t,t_0) | \epsilon_n \rangle = \left\langle \hat{A}(t) \right\rangle_U \tag{3311}$$

The grand canonical partition function:

$$Z = \sum_n \exp\left\{-\beta(\epsilon_n - \mu N_n)\right\} \tag{3312}$$

In equation (3311), the expectation value written without the subscript U denotes the equilibrium expectation value, and $|\epsilon_n\rangle$ is the initial state at t_0 when perturbation is switched on. The eigenstate with energy ϵ_n evolves in time under the effect of external perturbation $v(t)$ as follows:

$$|\phi_n(t)\rangle = \hat{\mathbf{S}}(t,t_0)|\epsilon_n\rangle \tag{3313}$$

So, all operators develop in time as they would in an equilibrium ensemble.

We introduce a new Green's function incorporating the field U in the imaginary-time interval $[t_0, t_0 - i\beta]$:

$$\mathbf{G}(1,1',U,t_0) = -i \frac{\mathrm{Tr}\left\{\hat{T}\hat{\psi}(1)\hat{\psi}^\dagger(1')\hat{\mathbf{S}}(t_0,t_0 - i\beta)\right\}}{\mathrm{Tr}\left[\hat{\mathbf{S}}(t_0,t_0 - i\beta)\right]} \tag{3314}$$

Here, the evolution operator $\hat{\mathbf{S}}(t_0,t_0 - i\beta)$ is defined in (3305), and the generalized response function $\mathbf{G}(1,1',U,t_0)$ satisfies the following boundary condition:

$$\mathbf{G}(1,1',U,t_0)\Big|_{t_1=t_0} = \exp\{\beta\omega\}\mathbf{G}(1,1',U,t_0)\Big|_{t_1=t_0-i\beta} \tag{3315}$$

This may be compared with the case of an equilibrium Green's function

$$\mathbf{G}(1,1')\Big|_{t_1=0} = \exp\{\beta\omega\}\mathbf{G}(1,1')\Big|_{t_1=-i\beta} \tag{3316}$$

Physical response functions defined for real times are described by the real-time Green's functions:

$$\mathbf{g}(1,1',U) = -i\,\mathrm{Tr}\left\{\hat{T}\hat{\psi}_U(1)\hat{\psi}_U^\dagger(1')\right\} \tag{3317}$$

With the basic assumptions underlying the Kadanoff-Baym nonequilibrium theory outlined previously, we now derive the equations of motion for the imaginary-time nonequilibrium Green's functions via

$$\left\{\hat{T}\hat{\psi}(1)\hat{s}(t_0,t_0 - i\beta)\right\} = \left\{\hat{T}\hat{\psi}(1)\exp\left\{-i\int_{t_0}^{t_0-i\beta} d2U(2)\hat{n}(2)\right\}\right\} \tag{3318}$$

or

$$\left\{\hat{T}\hat{\psi}(1)\hat{s}(t_0,t_0-i\beta)\right\}=\left\{\hat{T}\exp\left\{-i\int_{t_1}^{t_0-i\beta}d2U(2)\hat{n}(2)\right\}\right\}\hat{\psi}(1)\left\{\hat{T}\hat{\psi}(1)\exp\left\{-i\int_{t_0}^{t_1}d2U(2)\hat{n}(2)\right\}\right\} \quad (3319)$$

We take the time derivative of (3319):

$$i\frac{\partial}{\partial t_1}\left\{\hat{T}\hat{\psi}(1)\hat{s}(t_0,t_0-i\beta)\right\}=i\left\{\hat{T}\frac{\partial\hat{\psi}(1)}{\partial t_1}\hat{s}(t_0,t_0-i\beta)\right\}+\left\{\hat{T}\hat{\psi}(1)\hat{s}(t_0,t_0-i\beta)\right\}U(1) \quad (3320)$$

From here, we obtain the equation of motion

$$-\hat{\tilde{\mathbf{M}}}_0\mathbf{G}(1,1';U,t_0)=\delta(1-1')+\int d2i\varpi(1^+-2)\mathbf{G}_2(1,2,1',2^+;U,t_0)\Big|_{t_2=t_1} \quad (3321)$$

where

$$-\hat{\tilde{\mathbf{M}}}_0=i\frac{\partial}{\partial t_{1'}}+\frac{\nabla_{1'}^2}{2m}-U(1')+\mu\equiv-\frac{\partial}{\partial it_{1'}}+\frac{\nabla_{1'}^2}{2m}-U(1')+\mu=-\frac{\partial}{\partial\tau_{1'}}+\frac{\nabla_{1'}^2}{2m}-U(1')+\mu \quad (3322)$$

Similarly, taking the adjoint equation of motion containing derivatives with respect to the 1 variable, we have:

$$\breve{\tilde{\mathbf{M}}}\mathbf{G}(1,1';U,t_0)=\delta(1-1')-\int d2i\varpi(1^+-2)\mathbf{G}_2(1,2,1',2^+;U,t_0)\Big|_{t_2=t_1} \quad (3323)$$

where

$$\breve{\tilde{\mathbf{M}}}_0\equiv\frac{\partial}{\partial_{\tau_1}}+\frac{\nabla^2}{2m}-U(1)+\mu \quad (3324)$$

The equation of motion for $\mathbf{G}$ depends on the two-particle Green's function, $\mathbf{G}_2$, defined as:

$$\mathbf{G}_2(1,2,1',2';U,t_0)=\frac{1}{i^2}\frac{\mathrm{Tr}\left\{\hat{T}\hat{S}(t_0,t_0-i\beta)\hat{\psi}(1)\hat{\psi}(2)\hat{\psi}^{\dagger}(2')\hat{\psi}^{+}(1')\right\}}{\mathrm{Tr}\left[\hat{S}(t_0,t_0-i\beta)\right]} \quad (3325)$$

We define the functional derivative with respect to the functional variable U by a relation generalizing the notion of a full differential of a function of many variables expressed via partial derivatives. The functional $\mathrm{F}[U]$ changes by $\delta\mathrm{F}$ due to an infinitesimal change in the external potential:

$$U(2)\to U(2)+\delta U(2) \quad (3326)$$

So,

$$\delta\mathrm{F}[U]=\int d1\frac{\delta\mathrm{F}}{\delta U(1)}\delta U(1)+\text{higher order terms} \quad (3327)$$

The coefficients at $\delta U(1)$ define the functional derivative as a function of the variable 1. For example, if we write

$$U(1)=\int d1'\delta(1-1')U(1') \tag{3328}$$

from (3327), we then have the identity

$$\frac{\delta U(1)}{\delta U(1')}=\delta(1-1') \tag{3329}$$

Instead of propagating the equation of motion with some approximate form of $\mathbf{G}_2$, we have to introduce the electronic self-energy Σ. To derive this equation of motion for the nonequilibrium Green's functions for a system subjected to an external time-dependent potential, we find the change in $\mathbf{G}$ resulting from an infinitesimal change in the external potential (3326) and the S-operator changes as follows:

$$\delta\hat{\mathbf{s}}=\delta\left\{\exp\left\{-i\int_{t_0}^{t_0-i\beta}d2U(2)\hat{n}(2)\right\}\right\}=\hat{\mathbf{s}}\frac{1}{i}\int_{t_0}^{t_0-i\beta}d2\delta U(2)\hat{n}(2) \tag{3330}$$

So, from here, we find the change in $\mathbf{G}$ resulting from an infinitesimal change in the external potential U:

$$\delta\mathbf{G}(1,1';U,t_0)=\int_{t_0}^{t_0-i\beta}d2\left[\mathbf{G}_2\left(1,2,1',2^+;U,t_0\right)-\mathbf{G}(1,1';U,t_0)\mathbf{G}\left(2,2^+;U,t_0\right)\right]\delta U(2) \tag{3331}$$

Thus, the functional derivative of $\mathbf{G}(1,1';U,t_0)$ with respect to $U(2)$ is as follows:

$$\frac{\delta\mathbf{G}(1,1';U,t_0)}{\delta U(2)}=\left[\mathbf{G}_2\left(1,2,1',2^+;U,t_0\right)-\mathbf{G}(1,1';U,t_0)\mathbf{G}\left(2,2^+;U,t_0\right)\right] \tag{3332}$$

This is the basis for generating the density response function via a single-particle Green's function. From (3332), we rewrite the equation of motion in (3321) as a closed differential equation:

$$\left\{-\hat{\mathbf{M}}_0-\int d2i\varpi\left(1^+-2\right)\left[\mathbf{G}\left(2,2^+;U,t_0\right)+\frac{\delta}{\delta U(2)}\right]_{t_2=t_1}\right\}\mathbf{G}(1,1';U,t_0)=\delta(1-1') \tag{3333}$$

This equation is satisfactory in that:

1. **It is a nonperturbative equation that closes the Martin–Schwinger hierarchy at the one-particle level.**
2. **The mean field (Hartree) potential is separated out.**

Considering the physical case $U\to 0$, we may compare the starting equation (3289) with the final form

$$\left\{-\hat{\mathbf{M}}_0-\int d2i\varpi\left(1^+-2\right)\left[\mathbf{G}\left(2,2^+,t_0\right)+\frac{\delta}{\delta U(2)}\right]_{t_2=t_1}\right\}\mathbf{G}(1,1';U=0,t_0)=\delta(1-1') \tag{3334}$$

Equation (3289) is observed to be linear (first link in the chain of linear Martin–Schwinger equations), whereas the autonomous equation (3334) is nonlinear, as observed from the self-consistent nature of the mean field term. The term relating the functional derivative integrates all beyond the mean field that are all exchange and correlations. This inner many-particle dynamical structure of the Green's function is observed to be tailored by a response function probing the reaction of the system to the U field.

There are two serious technical drawbacks of the solution of equation (3334) that together symptomatically are the essential physical constituents of the problem:

1. **Unfortunately, the quantity $\frac{\delta \mathbf{G}}{\delta U}$, which is the pair correlation function given by equation (3332), has no direct treatment available.**
2. **The proper inclusion of the initial/boundary conditions fortunately can be resolved by employing the Keldysh initial conditions. With these initial conditions, the techniques for how to solve equation (3334) can be developed in close parallel analogy to procedures known for equilibrium systems.**

14.16.1.2 Keldysh Initial Condition

This initial condition is imposed on the nonequilibrium Green's function via the unperturbed Green's function. It is introduced by relations similar to equation (3314):

$$\mathbf{G}_0(1,1',U,t_0) = -i\frac{\mathrm{Tr}\left\{\hat{\mathrm{T}}\hat{\psi}(1)\hat{\psi}^{\dagger}(1')\hat{\mathbf{S}}(t_0,t_0-i\beta)\right\}}{\mathrm{Tr}\left[\hat{\mathbf{S}}(t_0,t_0-i\beta)\right]} \tag{3335}$$

Considering the bare Green's function, $\mathbf{G}_0$, the differential equation (3333) for $\mathbf{G}$ may be converted to an integral form:

$$\mathbf{G}(1,1';U,t_0) = \mathbf{G}_0(1,1';U,t_0) + \int d2d2' i\varpi(2'^{+}-2)\mathbf{G}_0(1,2';U,t_0)\left(\mathbf{G}(2,2^{+};U,t_0) + \frac{\delta}{\delta U(2)}\right)\mathbf{G}(2',1';U,t_0) \tag{3336}$$

This equation serves as a basis for the iterative procedure yielding $\mathbf{G}$ as a series in powers of $i\varpi$. The bare Green's function $\mathbf{G}_0$ satisfies the following equations

$$-\hat{\tilde{\mathbf{M}}}_0\mathbf{G}_0(1,1';U,t_0) = \delta(1-1') \tag{3337}$$

$$\breve{\mathbf{M}}_0\mathbf{G}_0(1,1';U,t_0) = \delta(1-1') \tag{3338}$$

where

$$-\hat{\tilde{\mathbf{M}}}_0 = i\frac{\partial}{\partial t_1} + \frac{\nabla_1^2}{2m} - U(1) + \mu \equiv -\frac{\partial}{\partial \tau_1} + \frac{\nabla_1^2}{2m} - U(1) + \mu \tag{3339}$$

and

$$\breve{\mathbf{M}}_0 \equiv \frac{\partial}{\partial \tau_{1'}} + \frac{\nabla_{1'}^2}{2m} - U(1) + \mu \tag{3340}$$

So, the Green's function in equation (3336) is readily verified to satisfy the full equation of motion in (3333). Furthermore, the integral form (3336) integrates the boundary condition set by the bare Green's function $\mathbf{G}_0$ and, in particular, the asymptotic initial condition common to the bare, as well as the full, Green's function and the external fields. Equation (3336) can only be easily solved by iteration starting from the zero-order solution:

$$\mathbf{G}^{(0)} = \mathbf{G}_0 \tag{3341}$$

The first iteration is given by the following equation

$$\mathbf{G}^{(1)}(1,1';U,t_0) = \mathbf{G}_0(1,1';U,t_0) + \int d2d2'i\varpi(2'^{+}-2)\mathbf{G}_0(1,2';U,t_0)\left(\mathbf{G}_0(2,2^{+};U,t_0) + \frac{\delta}{\delta U(2)}\right)\mathbf{G}_0(2',1';U,t_0) \tag{3342}$$

with the following functional derivative, which is an analogue of (3332), again playing the vital role:

$$\frac{\delta \mathbf{G}_0(1,1';U,t_0)}{\delta U(2)} = \left[\mathbf{G}_{02}(1,2,1',2^{+};U,t_0) - \mathbf{G}_0(1,1';U,t_0)\mathbf{G}_0(2,2^{+};U,t_0)\right] \tag{3343}$$

Here, $\mathbf{G}_{02}$ is defined as in equation (3325). In further iterations, $\mathbf{G}_{03}$ enters as well as others thereby progressively yielding the full Martin–Schwinger hierarchy of the unperturbed n-particle Green's functions.

Our focus will be on a particular class of initial states with no inner correlations (**uncorrelated initial states**) and defined as states, for which the bare, two-particle Green's function factorizes to an antisymmetric product of a pair of one-particle Green's function. So, the functional derivative of $\mathbf{G}_0$ is expressed via $\mathbf{G}_0$ itself. Hence, from (3343), we have $\mathbf{G}_{02}$ as being uncorrelated:

$$\frac{\delta \mathbf{G}_0(1,,1';U,t_0)}{\delta U(2)} = \mathbf{G}_0(1,2;U,t_0)\mathbf{G}_0(2,1';U,t_0) \tag{3344}$$

because

$$\mathbf{G}_{02}(1,2,1',2^{+};U,t_0) = \mathbf{G}_0(1,1';U,t_0)\mathbf{G}_0(2,2^{+};U,t_0) - \mathbf{G}_0(1,2;U,t_0)\mathbf{G}_0(2^{+},1';U,t_0) \tag{3345}$$

The expression (3344), for the functional derivative of $\mathbf{G}_0$, makes it feasible to obtain the conventional Feynman diagrams. So, for each iteration, equation (3336) yields closed expressions for $\mathbf{G}$ via $\mathbf{G}_0$ and ϖ. This makes the many-body perturbation expansion feasible.

14.16.1.3 Perturbation Expansion and Feynman Diagrams

We now develop the perturbation expansion for the Green's function that results in the Dyson equation for $\mathbf{G}$. We generate the Feynman diagrams using the alternative functional derivative technique by slightly rearranging equation (3336) and the functional derivative of $\mathbf{G}_0$ in equation (3344). Hence, expansion in powers of the interaction ϖ:

$$\mathbf{G} = \mathbf{G}^{(0)} + \mathbf{G}^{(1)} + \mathbf{G}^{(2)} + \cdots + \mathbf{G}^{(n)} + \mathbf{G}^{(n+1)} \tag{3346}$$

This is sequentially obtained via the recurrent relation

$$\mathbf{G}^{(n+1)}(1,1';U,t_0) = \int d2d2'i\varpi(2'^{+}-2)\mathbf{G}_0(1,2';U,t_0)\left[\sum_{k=0}^{n}\mathbf{G}^{(k)}(2,2^{+};U,t_0)\mathbf{G}^{(n-k)}(2',1';U,t_0) + \frac{\delta \mathbf{G}^{(n)}(2',1';U,t_0)}{\delta U(2)}\right] \tag{3347}$$

Because all perturbation corrections up to $\mathbf{G}^{(n)}$ are expressed via $\mathbf{G}_0$ and ϖ, it is also true for $\mathbf{G}^{(n+1)}$ considering (3344). So, the Green's function is a functional of $\mathbf{G}_0$ and ϖ to all orders of the perturbation expansion, or simply $\mathbf{G}[\mathbf{G}_0,\varpi]$. The perturbation expansion in equation (3346) can be represented by standard Feynman diagrams and, in particular, the following three elements of the diagrams:

$$\mathbf{G}_0(1,1';U,t_0) = \tag{3348}$$

$$i\varpi(1,1') = \tag{3349}$$

$$\int d2 = \tag{3350}$$

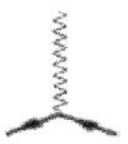

where, (3348), (3349) and (3350) are respectively the free propagator line, interaction line and the vertex.

Considering (3347) and (3344), we have the following first-order correction that will show the basic overall features of the diagrammatic expansion of the given Green's function:

$$\mathbf{G}^{(1)}(1,1';U,t_0) = \int d2d2'i\varpi(2'^+ - 2)\mathbf{G}_0(1,2';U,t_0)\mathbf{G}_0(2,2^+;U,t_0)\mathbf{G}_0(2',1';U,t_0)$$
$$+\int d2d2'i\varpi(2-2')\mathbf{G}_0(1,2;U,t_0)\mathbf{G}_0(2,2'^+;U,t_0)\mathbf{G}_0(2',1';U,t_0) \tag{3351}$$

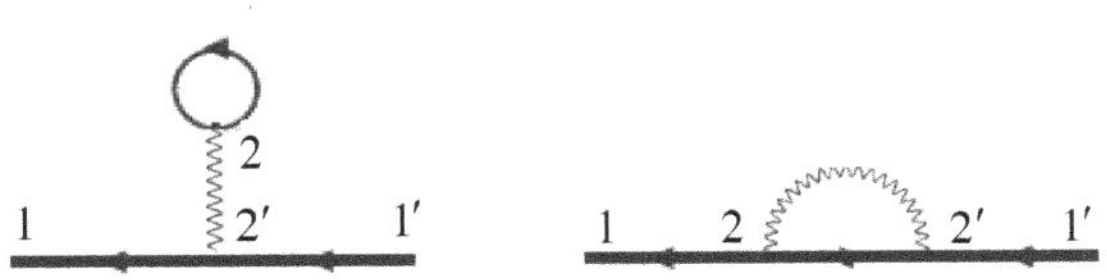

FIGURE 14.9 The diagrams correspond to two integrals in equation (3351).

Remark

So, from (3347) and (3344), then

- **Diagrams of any order are recurrently obtained following purely mechanical rules established from the analytical equations.**
- **A single representative is obtained for each set of topologically equivalent diagrams.**
- **All diagrams are connected in the diagrammatic expansion of the Green's function.**

Each such diagram consists of a chain of bare propagator lines linked by inserts of propagators as well as interaction lines and having exactly two end points. The entire diagram is **reducible** because it can be split into disconnected parts by cutting a single propagator line. The inserts cannot be split by such a single cut and so are the **irreducible parts.**

Letting the notation IR be the irreducible parts, the full Green's function is then rewritten as the sum:

$$\mathbf{G} = \mathbf{G}_0 + \mathbf{G}_0 \sum_{k=1}^{\infty} \sum_{\mathrm{IR}_1} \cdots \sum_{\mathrm{IR}_k} \mathrm{IR}_1 \mathbf{G}_0 \cdots \mathrm{IR}_k \mathbf{G}_0 \tag{3352}$$

This sum may be rearranged as

$$\mathbf{G} = \mathbf{G}_0 + \mathbf{G}_0 \sum_{\mathrm{IR}_1} \mathrm{IR}_1 \left\{ \mathbf{G}_0 + \mathbf{G}_0 \sum_{k=2}^{\infty} \sum_{\mathrm{IR}_2} \cdots \sum_{\mathrm{IR}_k} \mathrm{IR}_2 \mathbf{G}_0 \cdots \mathrm{IR}_k \mathbf{G}_0 \right\} \tag{3353}$$

Because all the sums are infinite, then for any IR_1 we have

$$\left\{ \mathbf{G}_0 + \mathbf{G}_0 \sum_{k=2}^{\infty} \sum_{\mathrm{IR}_2} \cdots \sum_{\mathrm{IR}_k} \mathrm{IR}_2 \mathbf{G}_0 \cdots \mathrm{IR}_k \mathbf{G}_0 \right\} = \mathbf{G} \tag{3354}$$

and so, from (3353), we have the following Dyson equation

$$\mathbf{G} = \mathbf{G}_0 + \mathbf{G}_0 \Sigma \mathbf{G} \tag{3355}$$

letting the sum of all irreducible two-end diagrams be the self-energy

$$\Sigma = \sum_{\mathrm{IR}} \mathrm{IR} \tag{3356}$$

with the lowest-order approximation

$$\Sigma^{(1)}(1,1';U,t_0) = i\int d2\varpi(1-2)\mathbf{G}_0(2,2^+;U,t_0)\delta(1-1') + i\varpi(1-1')\mathbf{G}(1,1'^+;U,t_0) \tag{3357}$$

or diagrammatically as

$\Sigma^{(1)}(1,1';U,t_0) =$ 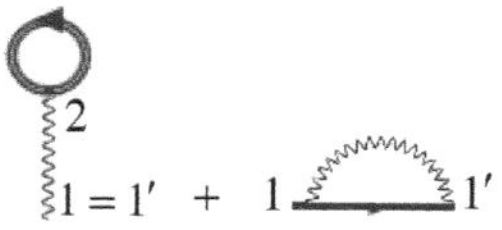

(3357)

FIGURE 14.10 Diagrammatic representation of the lowest-order approximation of the self-energy.

Letting the single-particle self-energy be Σ in the presence of the external potential U, we then have the **right-hand Dyson equation** in Keldysh space:

$$\mathbf{G}(1,1';U,t_0) = \mathbf{G}_0(1,1';U,t_0) + \int d1''2''\mathbf{G}_0(1,1'';U,t_0)\Sigma(1'',2'';U,t_0)\mathbf{G}(2'',1';U,t_0) \tag{3358}$$

Similarly, we have **the left-hand or conjugate Dyson equation**

$$\mathbf{G}(1,1';U,t_0) = \mathbf{G}_0(1,1';U,t_0) + \int d1''2''\mathbf{G}(1,1'';U,t_0)\Sigma(1'',2'';U,t_0)\mathbf{G}_0(2'',1';U,t_0) \tag{3359}$$

Remark

- **The pair interaction can be incorporated into the nonequilibrium Green's function by means of a perturbation series with individual terms classified by Feynman diagrams identical to those known from the equilibrium many-body theory.**

- **The technique of summation of infinite subsets of Feynman diagrams is applicable in the same manner as in the case of equilibrium theory.**

14.16.1.4 Right- and Left-Hand Dyson Equations

The self-energy also could have been introduced just as well via the equations of motion, (3321) and (3323), without applying the perturbation expansion. For convenience, we introduce the inverse bare Green's function:

$$\mathbf{G}_0^{-1}(1,1';U,t_0) = -\hat{\tilde{\mathbf{M}}}_0\delta(1-1') \tag{3360}$$

where

$$\delta(1-1') \equiv \delta(\vec{r}_1 - \vec{r}_1')\delta(t_1 - t_1') \tag{3361}$$

Here, the inverse matrix $\mathbf{G}_0^{-1}(1,1'';U,t_0)$ is defined via

$$\int_{t_0}^{t_0-i\beta} d1''\mathbf{G}_0^{-1}(1,1'';U,t_0)\mathbf{G}_0(1'',1';U,t_0) = \delta(1-1') \tag{3362}$$

$$\int_{t_0}^{t_0-i\beta} d1''\mathbf{G}_0(1,1'';U,t_0)\mathbf{G}_0^{-1}(1'',1';U,t_0) = \delta(1-1') \tag{3363}$$

Considering the equations of motion, (3321) and (3323), and letting the single-particle self-energy be Σ in the presence of the external potential U, we then have the **right-hand Dyson equation** in Keldysh space:

$$\int d1''\left[\mathbf{G}_0^{-1}(1,1'';U,t_0) - \Sigma(1,1'';U,t_0)\right]\mathbf{G}(1,1';U,t_0) = \delta(1-1') \tag{3364}$$

Similarly, we have **the left-hand or conjugate Dyson equation**

$$\int d1''\mathbf{G}(1,1';U,t_0)\left[\mathbf{G}_0^{-1}(1,1'';U,t_0) - \Sigma(1,1'';U,t_0)\right] = \delta(1-1') \tag{3365}$$

We also consider these equations as a definition of the inverse full Green's function:

$$\mathbf{G}^{-1}(1,1';U,t_0) = \mathbf{G}_0^{-1}(1,1';U,t_0) - \Sigma(1,1';U,t_0) \tag{3366}$$

$$\int d1''\mathbf{G}(1,1'';U,t_0)\mathbf{G}^{-1}(1'',1';U,t_0) = \int d1''\mathbf{G}^{-1}(1,1'';U,t_0)\mathbf{G}(1'',1';U,t_0) = \delta(1-1') \tag{3367}$$

14.16.1.5 Self-Energy Self-Consistent Equations

We examine the functional equation (3333) that will permit us to generate self-consistent equations for the self-energy:

$$\int d2\Sigma(1,2;U,t_0)\mathbf{G}(2,1';U,t_0) = -\int d2i\varpi(1^+-2)\left[\mathbf{G}(2,2^+;U,t_0) + \frac{\delta}{\delta U(2)}\right]\mathbf{G}(1,1';U,t_0) \tag{3368}$$

With the help of the equation

$$\delta \mathbf{G} = \delta \mathbf{G}\left(\mathbf{G}^{-1}\mathbf{G}\right) = \mathbf{G}\left(\delta \mathbf{G}\mathbf{G}^{-1}\right) = \mathbf{G}\left(-\mathbf{G}\delta \mathbf{G}^{-1}\right) \tag{3369}$$

that follows from equation (3367), we then write the following equation:

$$\frac{\delta \mathbf{G}(1,1';U,t_0)}{\delta U(2)} = -\int d2'd2''\mathbf{G}(1,2';U,t_0)\left[\delta(2-2')\delta(2'-2'') + \frac{\delta\Sigma(2',2'';U,t_0)}{\delta U(2)}\right]\mathbf{G}(2'',1';U,t_0) \tag{3370}$$

From equations (3368) and (3370), we let

$$\Sigma_{\mathrm{HF}}(1,1';U,t_0) = -\int d2 i\varpi\left(1^+-2\right)\mathbf{G}\left(2,2^+;U,t_0\right)\delta(1-1') + i\varpi(1-1')\mathbf{G}(1,1';U,t_0) \tag{3371}$$

$$\Sigma_{\mathrm{H}}(1,1';U,t_0) = -\int d2 i\varpi\left(1^+-2\right)\mathbf{G}\left(2,2^+;U,t_0\right)\delta(1-1') \tag{3372}$$

$$\Gamma(2,1';2';U,t_0) = \delta(2-1')\delta(2-2') + \frac{\delta\Sigma(2,1';U,t_0)}{\delta U(2')} \tag{3373}$$

$$\Sigma_{\mathrm{coll}\,1}(1,1';U,t_0) = \int d2d2' i\varpi\left(1^+-2'\right)\mathbf{G}(1,2;U,t_0)\frac{\delta\Sigma(2,1';U,t_0)}{\delta U(2')} \tag{3374}$$

and

$$\Sigma_{\mathrm{coll}\,2}(1,1';U,t_0) = \int d2d2' i\varpi(1-2')\mathbf{G}(1,2;U,t_0)\Gamma(2,1';2';U,t_0) \tag{3375}$$

We have the following variants of the self-energies:

$$\Sigma(1,1';U,t_0) = \Sigma_{\mathrm{HF}}(1,1';U,t_0) + \Sigma_{\mathrm{coll}\,1}(1,1';U,t_0) \tag{3376}$$

and

$$\Sigma(1,1';U,t_0) = \Sigma_{\mathrm{H}}(1,1';U,t_0) + \Sigma_{\mathrm{coll}\,2}(1,1';U,t_0) \tag{3377}$$

These results can be easily generalized to deal with a Bose-condensed system by the formal change of the Green's functions and self-energies into 2×2 matrices.

Remark

The self-energy

- **Is obtained from (3357) by renormalizing $\mathbf{G}_0$ to $\mathbf{G}$ as well as by renormalizing one of the vertices in the exchange term from a simple point-like bare vertex to a three-point full many-body vertex. The vertex correction $\frac{\delta\Sigma}{\delta U}$ tailors the correlation effects, that is, everything "beyond the Hartree–Fock (HF)."**

- **Has the first-order Hartree (Hartree–Fock) contribution $\Sigma_H(1,1';U,t_0)$ ($\Sigma_{HF}(1,1';U,t_0)$) that describes the mean-field effects of an interaction and $\Sigma_{coll}(1,1';U,t_0)$ denotes the collisional (or exchange-correlation) contributions that are second and higher order in the interatomic interaction ϖ. It is the collisional integral in a Boltzmann equation.**
- **Equation (3376) can generate higher-order, self-consistent approximations by a simple procedure of iteration of the first-order HF self-energies.**
- **Is expressed uniquely via G and ϖ, and its U dependence is thus mediated via G. This implies that the self-energy $\Sigma[G]$ is a functional of the one-particle Green's function G. This functional dependence supplements the right-(left-) hand Dyson equation in (3364) ([3365]) (that is simply an identity among G_0, Σ, and G) and constitutes a set of equations that could be solved self-consistently once the functional dependence of Σ on G is known.**

We compare Σ with the lowest-order iteration, $\Sigma^{(1)}$, of the perturbation series in equation (3378):

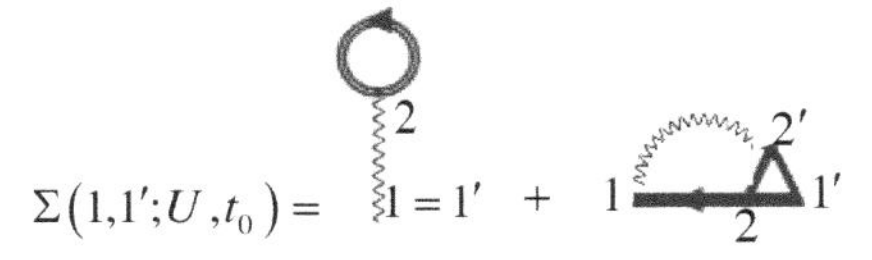

(3378)

FIGURE 14.11 Diagrammatic representation of the Hartree and Fock exchange terms.

The two diagrams in Figure 14.11 are both expressed via the unperturbed Green's function as well as the interaction. If the self-energy functional $\Sigma[G]$ is known, then the auxiliary U field is irrelevant and a system of self-consistent equations for the nonequilibrium Green's function can be achieved. Considering equations (3377) and (3370), we find a closed equation for the vertex Γ:

$$\Gamma(1,1';2;U,t_0) = \delta(1-1')\delta(1-2) + \frac{\delta\Sigma(1,1';U,t_0)}{\delta U(2)} \tag{3379}$$

or

$$\Gamma(1,1';2;U,t_0) = \delta(1-1')\delta(1-2) + \int d2'd2'' \frac{\delta\Sigma(1,1';U,t_0)}{\delta \mathbf{G}(2',2'';U,t_0)} \frac{\delta \mathbf{G}(2',2'';U,t_0)}{\delta U(2)} \tag{3380}$$

Letting $U \to 0$, the integral equation for Γ is achieved:

$$\begin{aligned}\Gamma(1,1';2;U,t_0) &= \delta(1-1')\delta(1-2) + \\ &\int d2'd2''d3'd3'' \frac{\delta\Sigma(1,1';U,t_0)}{\delta \mathbf{G}(2',2'';U,t_0)} \mathbf{G}(2',3';U,t_0)\mathbf{G}(3'',2'';U,t_0)\Gamma(3',3'';2;U,t_0)\end{aligned} \tag{3381}$$

So, from equations (3377) and (3364) and considering the presence of the slowly varying external field U, we achieve a self-consistent system of equations for an arbitrary nonequilibrium process starting from a Keldysh initial condition:

$$-\hat{\mathbf{M}}_0 \mathbf{G}(1,1';U,t_0) - \int_{t_0}^{t_0 - i\beta} d2\Sigma(1,2;U,t_0)\mathbf{G}(2,1';U,t_0) = \delta(1-1') \tag{3382}$$

and

$$\Sigma(1,1';U,t_0) = \Sigma_H(1,1';U,t_0) + \int d2d2' i\varpi(1-2')\mathbf{G}(1,2;U,t_0)\Gamma(2,1';2';U,t_0) \tag{3383}$$

also

$$\breve{\tilde{\mathbf{M}}}_0\mathbf{G}(1,1';U,t_0) - \int_{t_0}^{t_0-i\beta} d2\mathbf{G}(1,2;U,t_0)\Sigma(2,1';U,t_0) = \delta(1-1') \tag{3384}$$

For a convenient solution of the self-consistent system of equations, we chose a physical approximation for the four-point vertex $\frac{\delta\Sigma}{\delta\mathbf{G}}$.

From the equation of motion for the Green's functions (3382) and (3384), we can obtain, for the matrix Green's function, the equation of motion **from the left**:

$$-\hat{\tilde{\mathbf{M}}}_0\mathbf{G}(1,1';U,t_0) - \Sigma(1,1';U,t_0)\mathbf{G}(1,1';U,t_0) = \delta(1-1') \tag{3385}$$

and **from the right**

$$\breve{\tilde{\mathbf{M}}}_0\mathbf{G}(1,1';U,t_0) - \mathbf{G}(1,1';U,t_0)\Sigma(1,1';U,t_0) = \delta(1-1') \tag{3386}$$

Introducing the inverse bare Green's function

$$\mathbf{G}_0^{-1}(1,1';U,t_0) = -\hat{\tilde{\mathbf{M}}}_0\delta(1-1') \tag{3387}$$

equations (3385) and (3386) can be expressed via the inverse bare matrix Green's function of the **non-equilibrium Dyson equation from the left**

$$\left(\mathbf{G}_0^{-1} - \Sigma\right)\mathbf{G} = \delta(1-1') \tag{3388}$$

and **from the right**

$$\mathbf{G}\left(\mathbf{G}_0^{-1} - \Sigma\right) = \delta(1-1') \tag{3389}$$

The matrix nonequilibrium Dyson equations include the three coupled equations for $\mathbf{G}^{R,A,K}$:

$$\left(\mathbf{G}_0^{-1} - \Sigma^{R(A)}\right)\mathbf{G}^{R(A)} = \delta(1-1') \tag{3390}$$

and

$$\mathbf{G}_0^{-1}\mathbf{G}^K = \Sigma^R\mathbf{G}^K + \Sigma^K\mathbf{G}^A \tag{3391}$$

Similarly, from equation (3389), we have

$$\mathbf{G}^{R(A)}\left(\mathbf{G}_0^{-1} - \Sigma^{R(A)}\right) = \delta(1-1') \tag{3392}$$

and

$$\mathbf{G}^K\mathbf{G}_0^{-1} = \mathbf{G}^R\Sigma^K + \mathbf{G}^K\Sigma^A \tag{3393}$$

Remark

Comparing the given results with the perturbative technique, we note that for the perturbative technique, the self-energy is generated as a functional $\Sigma[G_0]$, while in the present case, it is dressed as $\Sigma[G]$, with the advantage of this transition being:

- **The bare Green's function G_0 with spurious meaning at finite times is eliminated.**
- **The relation $G \leftrightarrow \Sigma$ is nonperturbative and self-consistent.**
- **The many-body vertex structure is separated from single particle fields.**
- **In $G_0\varpi \to G\varpi$, the Green's function G is dressed (renormalized), while ϖ is bare.**

We now examine the case where interactions are renormalized. In particular, we focus on the Coulomb interaction, which accounts for the all-important screening effects. Considering a local spin-independent interaction as the Coulomb force, we have

$$-\int d2i\varpi\left(1^+-2\right)\mathbf{G}\left(2,2^+;U,t_0\right)=\int d2\varpi\left(\vec{r}_1-\vec{r}_2\right)\left\langle n\left(\vec{r}_2,t_1;U\right)\right\rangle=V_H\left(1;U\right) \tag{3394}$$

We introduce the screened field:

$$U_{\text{eff}}\left(1\right)=U\left(1\right)+\left[V_H\left(1;U\right)-V_H\left(1;U=0\right)\right] \tag{3395}$$

Then, from equation (3333), we have

$$\left\{-\tilde{\hat{\mathbf{M}}}_0-\int d2i\varpi\left(1^+-2\right)\frac{\delta}{\delta U(2)}\right\}\mathbf{G}\left(1,1';U,t_0\right)=\delta\left(1-1'\right) \tag{3396}$$

where

$$-\tilde{\hat{\mathbf{M}}}_0=i\frac{\partial}{\partial t_1}+\frac{\nabla_1^2}{2m}-V_H\left(1;U=0\right)-U_{\text{eff}}\left(1\right)+\mu\equiv-\frac{\partial}{\partial\tau_1}+\frac{\nabla_1^2}{2m}-V_H\left(1;U=0\right)-U_{\text{eff}}\left(1\right)+\mu \tag{3397}$$

From equations (3396) and (3368), we have

$$\int d2\left(\Sigma\left(1,2;U,t_0\right)-\Sigma_H\left(1,2;U,t_0\right)\right)\mathbf{G}\left(1,2;U,t_0\right)\equiv\int d2\Sigma_{XC}\left(1,2;U,t_0\right)\mathbf{G}\left(1,2;U,t_0\right) \tag{3398}$$

or

$$\int d2\Sigma_{XC}\left(1,2;U,t_0\right)\mathbf{G}\left(1,2;U,t_0\right)=-\int d2d2'i\varpi\left(1^+-2\right)\frac{\delta U_{\text{eff}}\left(2'\right)}{\delta U\left(2\right)}\frac{\delta\mathbf{G}\left(1,1';U,t_0\right)}{\delta U_{\text{eff}}\left(2'\right)} \tag{3399}$$

where U_{eff} is a new variational variable instead of U, and we introduce the renormalized interaction:

$$\varpi_s\left(1-1'\right)=\int d2\varpi\left(1^+-2\right)\frac{\delta U_{\text{eff}}\left(1'\right)}{\delta U\left(2\right)} \tag{3400}$$

as well as the renormalized vertex

$$\Gamma_s\left(2,1';2';U_{\text{eff}},t_0\right)=\delta\left(2-1'\right)\delta\left(2-2'\right)+\frac{\delta\Sigma_{XC}\left(2,1';U_{\text{eff}},t_0\right)}{\delta U_{\text{eff}}\left(2'\right)} \tag{3401}$$

then

$$\frac{\delta \mathbf{G}(1,1';U,t_0)}{\delta U(2)} = -\int d2'd2''\mathbf{G}(1,2';U,t_0)\Gamma_s(2',2'',2;U,t_0)\mathbf{G}(2'',1';U,t_0) \tag{3402}$$

and

$$\Sigma_{XC}(1,1';U,t_0) = -\int d2d2'i\varpi_s(1-2')\mathbf{G}(1,2;U,t_0)\Gamma_s(2,1';2';U_{eff},t_0) \tag{3403}$$

From equations (3403) and (3401), we achieve an expansion of Σ_{XC} via $\mathbf{G}$ and ϖ_s. Insertions between interaction lines are absent, and all diagrams correspond to vertex corrections of the increasing topological complexity in the diagrammatic series. The renormalized interaction

$$\varpi_s(1-1') = \varpi(1-1') + \int d2d2'\varpi(1-2)\Pi(2.2')\varpi_s(2'-1') \tag{3404}$$

imitates the Dyson equation, whereas

$$\Pi(2.2') = \int d2''d3\mathbf{G}(2,2'';U,t_0)\mathbf{G}(3,2;U,t_0)\Gamma_s(2'',3;2';U_{eff},t_0) \tag{3405}$$

is the **polarization operator** and the screened or renormalized vertex:

$$\Gamma_s(1,1';2;U_{eff},t_0) = \delta(1-1')\delta(1-2) + \int d2'd2''d3d3'\frac{\delta\Sigma_{XC}(1,1')}{\delta\mathbf{G}(2',3)}\mathbf{G}(2',3)\mathbf{G}(3',2'')\Gamma_s(3,3';2) \tag{3406}$$

So, we confirm that the self-energy is a functional of $\mathbf{G}$ with no explicit dependence on U. Considering (3396) and (3399), the Dyson equation can now have the form:

$$-\hat{\tilde{\mathbf{M}}}_0\mathbf{G}(1,1';U,t_0) - \int d2\Sigma_{XC}(1,2;U,t_0)\mathbf{G}(2,1';U,t_0) = \delta(1-1') \tag{3407}$$

14.17 Kadanoff-Baym (KB) Formalism for Bose Superfluids

An elaborate Green's function argument provides a means of describing transport phenomena in a self-contained manner, beginning from a dynamical approximation. The theory simultaneously provides a description of the occurring transport processes and determination of the quantities appearing in the transport equations. The treatment of the nonequilibrium dynamics of a Bose condensed gas may be undertaken via the 2PI effective action and the Schwinger–Keldysh closed-time path formalism; minimizing the 2PI effective action yields the equations of motion for the condensate wave function and noncondensate Green's functions. To do this, we first relate the imaginary-time Green's function $\mathbf{G}(1,1',U,t_0)$, defined in (3314) to the **real-time** Green's functions $\mathbf{g}(1,1',U)$ and defined in (3317) as:

$$\mathbf{G}(1,1',U,t_0) = \mathbf{G}^<(1,1',U,t_0) = -i\frac{\mathrm{Tr}\left\{\hat{\mathbf{T}}\hat{\mathbf{S}}(t_0,t_0-i\beta)\left[\hat{\mathbf{S}}^\dagger(t_0,t_{1'})\hat{\psi}^\dagger(1')\hat{\mathbf{S}}(t_0,t_{1'})\right]\hat{\mathbf{S}}^\dagger(t_0,t_1)\hat{\psi}(1)\hat{\mathbf{S}}(t_0,t_1)\right\}}{\mathrm{Tr}\left[\hat{\mathbf{S}}(t_0,t_0-i\beta)\right]} \tag{3408}$$

and

$$\mathbf{g}^<(1,1',U) = -i\,\mathrm{Tr}\left\{\hat{\mathbf{T}}\hat{\mathbf{s}}^\dagger(t_1)\hat{\psi}(1)\hat{\mathbf{s}}(t_1)\hat{\mathbf{s}}^\dagger(t_{1'})\hat{\psi}^+(1')\hat{\mathbf{s}}(t_{1'})\right\} \tag{3409}$$

assuming

$$i(t_1 - t_0) < i(t_{1'} - t_0) \tag{3410}$$

It is obvious that the real-time Green's function $\mathbf{g}(1,1',U)$ is identical to $\mathbf{G}(1,1',U,t_0)$ in the limit $t_0 \to -\infty$. This implies the fundamental formulae of the KB formalism, which relates the imaginary time to the real-time response function:

$$\lim_{t_0 \to -\infty} \mathbf{G}^{<}(1,1',U,t_0) = \mathbf{g}^{<}(1,1',U) \quad , \quad \lim_{t_0 \to -\infty} \mathbf{G}^{>}(1,1',U,t_0) = \mathbf{g}^{>}(1,1',U) \tag{3411}$$

So, in the KB formalism, the equations of motion for response functions defined by the real-time Green's functions $\mathbf{g}$ are obtained via the equations of motion for the related imaginary-time Green's functions $\mathbf{G}$. Consequently, the real-time Green's functions are obtained by analytic continuation. The imaginary-time Green's function is expressed as a Fourier series over the discrete frequencies. Beginning with the Fourier coefficients defined on the discrete set of imaginary frequencies, an analytical continuation applied on all frequencies yields the required functions on the real frequency axis and, consequently, the real-time Green's functions $\mathbf{g}(1,1',U)$.

We denote the collision self-energy Σ_{coll} that is the second and higher order contribution in $\mathbf{g}$:

$$\Sigma_{\text{coll}}(1,1',U,t_0) = \begin{cases} \Sigma^{>}_{\text{col}}(1,1',U,t_0) & , \quad i(t_1 - t_{1'}) > 0 \\ \Sigma^{<}_{\text{col}}(1,1',U,t_0) & , \quad i(t_1 - t_{1'}) < 0 \end{cases} \tag{3412}$$

Because physical response functions relate the correlation functions $\mathbf{G}^{<}$ and $\mathbf{G}^{>}$, considering equation (3410), we have the equation of motion for $\mathbf{G}^{<}$:

$$\begin{aligned} -\hat{\tilde{\mathbf{M}}}_0 \mathbf{G}^{<}(1,1';U,t_0) = {} & \int_{t_0}^{t_1} d1'' \Sigma^{>}_{\text{coll}}(1,1'';U,t_0) \mathbf{G}^{<}(1'',1';U,t_0) + \int_{t_1}^{t_{1'}} d1'' \Sigma^{<}_{\text{coll}}(1,1'';U,t_0) \mathbf{G}^{<}(1'',1';U,t_0) + \\ & + \int_{t_{1'}}^{t_0 - i\beta} d1'' \Sigma^{<}_{\text{coll}}(1,1'';U,t_0) \mathbf{G}^{>}(1'',1';U,t_0) \end{aligned} \tag{3413}$$

where

$$-\hat{\tilde{\mathbf{M}}}_0 = i\frac{\partial}{\partial t_1} + \frac{\nabla_1^2}{2m} - U_{\text{eff}}(1,t_0) \tag{3414}$$

The effective mean field $U_{\text{eff}}(1,t_0)$ is the sum of the external potential U, the Hartree–Fock part of the self-energy Σ_{HF} and the chemical potential μ. The real-time Green's function equation (3317) of motion can be achieved via the limit of equation (3413) for $t_0 \to -\infty$.

14.17.1 Kadanoff-Baym Equations

From the aforementioned we can confortably establish the Kadanoff-Baym equations by considering first the equation

$$-\hat{\tilde{\mathbf{M}}}_0 \mathbf{g}^{<}(1,1';U) = \int_{-\infty}^{t_1} d1'' \mathbf{a}(1'',1';U) \mathbf{g}^{<}(1'',1';U) - \int_{-\infty}^{t_{1'}} d1'' \Sigma^{<}_{\text{coll}}(1,1'';U) \Gamma(1'',1';U) \tag{3415}$$

Here,

$$\Sigma^{<}(1,1'';U) \equiv \Sigma^{<}(1'',1';U,-\infty) \tag{3416}$$

and

$$\mathbf{a}(1'',1';U) = \Sigma^{>}_{\text{coll}}(1,1'';U) - \Sigma^{<}_{\text{coll}}(1,1'';U) \quad , \quad \Gamma(1'',1';U) = \mathbf{g}^{>}(1'',1';U) - \mathbf{g}^{<}(1'',1';U) \tag{3417}$$

Similarly, considering equation (3415), we have the equation of motion for $\mathbf{g}^{<}$ with respect to the variable $1'$:

$$\breve{\mathbf{M}}_0 \mathbf{g}^{<}(1,1';U) = \int_{-\infty}^{t_1} d1'' \Sigma^{<}_{\text{coll}}(1,1'';U)\Gamma(1'',1';U) - \int_{-\infty}^{t_{1'}} d1'' \mathbf{a}(1'',1';U)\mathbf{g}^{<}(1'',1';U) \tag{3418}$$

where

$$\breve{\mathbf{M}}_0 = -i\frac{\partial}{\partial t_{1'}} + \frac{\nabla^2_{1'}}{2m} - U_{\text{eff}}(1') \quad , \quad U_{\text{eff}}(1') \equiv U_{\text{eff}}(1';-\infty) \tag{3419}$$

The previous procedure is done for $\mathbf{g}^{>}$, assuming

$$i(t_1 - t_0) > i(t_{1'} - t_0) \tag{3420}$$

and we have the following equations of motion:

$$-\hat{\mathbf{M}}_0 \mathbf{g}^{>}(1,1';U) = \int_{-\infty}^{t_1} d1'' \mathbf{a}(1'',1';U)\mathbf{g}^{>}(1'',1';U) - \int_{-\infty}^{t_{1'}} d1'' \Sigma^{>}_{\text{coll}}(1,1'';U)\Gamma(1'',1';U) \tag{3421}$$

and

$$\breve{\mathbf{M}}_0 \mathbf{g}^{>}(1,1';U) = \int_{-\infty}^{t_1} d1'' \Sigma^{<}_{\text{coll}}(1,1'';U)\Gamma(1'',1';U) - \int_{-\infty}^{t_{1'}} d1'' \mathbf{a}(1'',1';U)\mathbf{g}^{>}(1'',1';U) \tag{3422}$$

Relations (3413), (3418), (3421), and (3422) are the well-known Kadanoff-Baym equations [112]. We have a minimum set of two coupled integro-differential equations, with the Green's functions $\mathbf{g}^{>}$ and $\mathbf{g}^{<}$ as the two unknowns. These Kadanoff-Baym equations are written in a manner close to the transport equations. A drawback of the Kadanoff-Baym equations is that the spectral and statistical aspects are not distinct, and there is no direct way of separating both.

We generalize the aforementioned equations of motion to describe the 2×2 real-time Green's function matrix $\hat{\tilde{\mathbf{g}}}$ for the noncondensate atoms:

$$\int d1'' \left[\hat{\mathbf{g}}_0^{-1}(1,1'') - \hat{\Sigma}_{\text{HFB}}(1,1'')\right]\hat{\tilde{\mathbf{g}}}^{\lessgtr}(1'',1') = \int_{-\infty}^{t_1} d1'' \hat{\Gamma}(1,1'')\hat{\tilde{\mathbf{g}}}^{\lessgtr}(1'',1') - \int_{-\infty}^{t_{1'}} d1'' \hat{\Sigma}^{\lessgtr}_{\text{coll}}(1,1'')\hat{\mathbf{a}}(1'',1') \tag{3423}$$

and

$$\int d1'' \hat{\tilde{\mathbf{g}}}^{\lessgtr}(1'',1')\left[\hat{\mathbf{g}}_0^{-1}(1,1'') - \hat{\Sigma}_{\text{HFB}}(1,1'')\right] = \int_{-\infty}^{t_{1'}} d1'' \hat{\mathbf{a}}(1,1'')\hat{\Sigma}^{\lessgtr}_{\text{coll}}(1'',1') - \int_{-\infty}^{t_{1'}} d1'' \hat{\tilde{\mathbf{g}}}^{\lessgtr}(1,1'')\hat{\Gamma}(1'',1') \tag{3424}$$

where $\hat{\Sigma}_{HFB}$ is the Hartree–Fock-Bogoliubov self-energy matrix; the 2×2 matrices $\hat{\mathbf{a}}(1,1'')$ and $\hat{\Gamma}(1'',1')$ are defined by the matrix elements:

$$\mathbf{a}_{\alpha\beta}(1'',1';U)=\Sigma^{>}_{\alpha\beta}(1,1'';U)-\Sigma^{<}_{\alpha\beta}(1,1'';U) \quad , \quad \Gamma_{\alpha\beta}(1'',1';U)=\tilde{\mathbf{g}}^{>}_{\alpha\beta}(1'',1';U)-\tilde{\mathbf{g}}^{<}_{\alpha\beta}(1'',1';U) \tag{3425}$$

We split the single-particle self-energy in (3423) and (3424) into two parts:

$$\hat{\Sigma}(1,1'')=\hat{\Sigma}_{HF}(1,1'')+\hat{\Sigma}_{coll}(1,1'') \tag{3426}$$

Here, the Hartree–Fock self-energy $\hat{\Sigma}_{HF}$ is given by equation (3423) and (3424), and the self-energies $\Sigma^{>}_{coll}$ and $\Sigma^{<}_{coll}$ correspond to the second-order collision self-energy $\hat{\Sigma}_{coll}$.

So far, we have obtained the nonequilibrium form of the Dyson–Beliaev equations of motion for the noncondensate atoms that depend on the nonequilibrium condensate wave function.

14.17.1.1 Fluctuation-Dissipation Theorem

Quantum systems initially set far-from-equilibrium evolve towards equilibrium. This implies the systems are time- and space-translation invariant. Notwithstanding this does not imply the systems are thermalized to describe systems with a (grand) canonical density matrix as they can equilibrate without being thermalized. The fluctuation-dissipation relation gives the possibility to verify if the equilibrated state is as well thermalized. In thermal equilibrium, the Keldysh or kinetic Green's function and the retarded and advanced Green's functions, or rather the spectral weight function, are related for the case of fermions or bosons according to

$$\mathbf{G}^{K}=\mathrm{F}(\epsilon)\left(\mathbf{G}^{R}-\mathbf{G}^{A}\right) \tag{3427}$$

as well as

$$\left[\mathbf{G}^{-1}\right]^{K}=\mathrm{F}(\epsilon)\left(\left[\mathbf{G}^{-1}\right]^{R}-\left[\mathbf{G}^{-1}\right]^{A}\right) \tag{3428}$$

Basically, this is the **fluctuation-dissipation theorem**, and

$$\mathrm{F}(\epsilon)=\begin{cases}\coth\dfrac{\epsilon}{2T} & , \text{ Bose}\\[2ex] \tanh\dfrac{\epsilon}{2T} & , \text{ Fermi}\end{cases} \tag{3429}$$

From equation (3427), we also have

$$\Sigma^{K}=\mathrm{F}(\epsilon)\left(\Sigma^{R}-\Sigma^{A}\right) \tag{3430}$$

So, the equilibrium relation of the Fermi gas is valid in general, and all Green's functions can thus be specified once. For example, let us say that the retarded Green's function is known. The quantum statistics of the particles is then reflected in relations governed by the fluctuation–dissipation type relationship such as in equation (3430).

Hence, in the nonequilibrium situation, the fluctuation–dissipation relation is no longer valid. Because the Green's function is a traced quantity, a closed set of equations cannot be achieved, and we obtain complex equations for an infinite hierarchy of the correlation functions. If the hierarchy is broken at most at the two-particle correlation level, we obtain quantum kinetic equations. This implies equations imitating kinetic equations but with quantum features that are not included in the classical Boltzmann equation. Despite the complex structure of these equations, there is little progress in their analytic solution. Nevertheless, the diagrammatic technique is promising. We now embrace the operations leading to a form of quantum kinetic equation that imitates classical kinetic equations. This is done via **Wigner coordinates.**

14.17.1.2 Wigner or Mixed Representation

The Green's function **G** is observed to oscillate rapidly with the difference $\vec{r} = \vec{r}_1 - \vec{r}_2$ on the scale of the inverse Fermi wave vector κ_F. So, if our interest in variations is on much longer length scales, we then perform a transformation to center-of-mass coordinates, (R,T), and difference coordinates, $(\vec{r},t)$ [131]. To derive quantum kinetic equations imitating the form of classical kinetic equations, we introduce the **mixed or Wigner coordinates [131]:**

$$\vec{R} = \frac{\vec{r}_1 + \vec{r}_2}{2} \quad , \quad \vec{r} = \vec{r}_1 - \vec{r}_2 \tag{3431}$$

and time variables

$$T = \frac{t_1 + t_2}{2} \quad , \quad t = t_1 - t_2 \tag{3432}$$

in order to separate the variables, $(\vec{r},t)$, describing the microscopic properties driven by the characteristics of the system from the variables, (R,T), describing the macroscopic properties driven by the nonequilibrium features of the state under consideration, such as a result of the presence of an applied potential.

14.18 Green's Function Wigner Transformation

The Wigner representation, introduced in 1932, considers quantum corrections to classical statistical mechanics [132, 133] and permits us to extend the concept of **phase space** in classical statistical mechanics to quantum statistical mechanics. It also closely relates to the Weyl quantization and has a reasonable impact in clarifying the foundation of quantum mechanics, and also has a connection to classical mechanics [134]. In addition, it provides an indispensable tool for deriving quantum transport equations. The **Wigner transformation** corresponds to the Fourier transforms of all functions with respect to the relative coordinates t and $\vec{r}$ and, in particular, with respect to the Green's function.

It is observed that the lesser Green's function $\mathbf{G}^<$ is most closely related to the density matrix. We use this fact to derive the **Quantum Boltzmann equation** by subtracting the lesser component of the right-hand Dyson equation in equation (3364) and its conjugate in equation (3365) from one another:

$$\left[\left[\mathbf{G}_0^{-1}(1')\right]^* - \mathbf{G}_0^{-1}(1)\right]\mathbf{G}^<(1,1') = \mathrm{St}[\lambda,\Sigma] \tag{3433}$$

with **the collision integral** on the right-hand side of the Boltzmann equation being

$$\mathrm{St}[\lambda,\Sigma] \equiv \mathrm{I}_{\mathrm{coll}}[\lambda,\Sigma] = \int d2\left[\mathbf{G}^R(1,2)\Sigma^<(2,1') + \mathbf{G}^<(1,2)\Sigma^A(2,1') - \Sigma^R(1,2)\mathbf{G}^<(2,1') - \Sigma^<(1,2)\mathbf{G}^A(2,1')\right] \tag{3434}$$

and

$$\left[\mathbf{G}_0^{-1}(1')\right]^* - \mathbf{G}_0^{-1}(1) = -i\left(\frac{\partial}{\partial t_1} + \frac{\partial}{\partial t_{1'}}\right) - \frac{1}{2m}(\Delta_1 - \Delta_{1'}) = -i\left(\frac{\partial}{\partial T} - \frac{i}{m}\nabla_{\vec{R}}\cdot\nabla_{\vec{r}}\right) \tag{3435}$$

The collision term accounts for electron correlation and introduces memory effects and dissipation.

We now show how the lesser Green's function is most closely related to the density matrix. We write the $\mathbf{G}^<$ in terms of $X \equiv (T,\vec{R})$ and $x \equiv (t,\vec{r})$:

$$\mathbf{G}^<(1,1') = \mathbf{G}^<\left(T+\frac{t}{2},\vec{R}+\frac{\vec{r}}{2};T-\frac{t}{2},\vec{R}-\frac{\vec{r}}{2}\right) \equiv \mathbf{G}^<\left(X+\frac{x}{2},X-\frac{x}{2}\right) \equiv \mathbf{G}^<\left(T,\vec{R};t,\vec{r}\right) \tag{3436}$$

For the Fourier transform,

$$\mathbf{G}^<(X,p) = \int dx\mathbf{G}^<\left(X+\frac{x}{2},X-\frac{x}{2}\right)\exp\{-ipx\} \tag{3437}$$

where

$$p = (\omega,\vec{p}) \quad , \quad xp = -\omega t + \vec{p}\vec{r} \tag{3438}$$

We find the distribution function $n(\vec{R},T)$ as well as the charge density $\rho(\vec{R},T)$ via mixed variables

$$n(\vec{R},T) = -i\int\frac{d\omega}{2\pi}\mathbf{G}^<(\omega,\vec{p},\vec{R},T) \tag{3439}$$

and

$$\rho(\vec{R},T) = -2ie\int\frac{d\vec{p}}{(2\pi)^3}\int_{-\infty}^{\infty}d\omega\mathbf{G}^<(\omega,\vec{p},\vec{R},T) \tag{3440}$$

The integration over $\frac{d\omega}{2\pi}$ is equivalent to setting $t = 0$, and the factor of two in $\rho(\vec{R},T)$ is from the spin of the particles, for example, electrons. We also express the average electric current density in the presence of a vector potential $\vec{A}$:

$$\vec{j}(\vec{R},T) = -\frac{e}{m}\int\frac{d\vec{p}}{(2\pi)^3}\int_{-\infty}^{\infty}d\omega\left(\vec{p} - \frac{e}{c}\vec{A}(\vec{R},T)\right)\mathbf{G}^<(\omega,\vec{p},\vec{R},T) \tag{3441}$$

The distribution function (3439) considers contributions from all independent energies ω. So, a perturbation scheme constructed for $\mathbf{G}^<$ enables us to keep the energy as an independent variable until the selection on how to determine the energy of a particle from its position in phase space. With this argument, we will find a proper distribution function and avoid the problems with the high-momenta tails of Wigner's function. The presence of the independent energy permits us to discriminate two very different contributions in the transport equations for the correlation function $\mathbf{G}^<$:

- **The on-shell contributions, for which a dispersion relation between the energy and the position of the particle in phase space holds true**
- **The off-shell contributions, for which no such relation exists**

This discrimination leads to the formulation of the perturbation schemes that better suit the conditions of the quantum kinetic equation, which is derived as an asymptotic limit of the equation for $\mathbf{G}^<$. This asymptotic equation is not closed for the Wigner distribution—only for the on-shell part of $\mathbf{G}^<$ that can be interpreted as the quasiparticle distribution.

Considering equation (3433), we take the Fourier components (3437) on each side and set $t = 0$; from equation (3439), we have

$$\left(\frac{\partial}{\partial T} + \frac{\vec{p}}{m}\nabla_{\vec{r}}\right) n\left(T, \vec{R}, \vec{p}\right) = \mathrm{St}[n] \tag{3442}$$

This is a kinetic equation of the Boltzmann type because it is an equation for a distribution function $n\left(T, \vec{R}, \vec{p}\right)$ **that has no independent energy variable.** In this equation, $\mathrm{St}[n]$ is the collision integral that is some functional of the distribution function. Equation (3442) relates $\mathrm{St}[n]$ with $\mathrm{St}[\lambda, \Sigma]$. **Therefore, any derivation of an asymptotic kinetic equation from the Kadanoff-Baym equations implies finding an auxiliary functional** $\mathbf{G}^<[n]$ **for which the independent energy becomes fixed and relates to the phase-space variables.**

References

1. Dirac, P.A.M. and N.H.D. Bohr, "The Quantum Theory of the Emission and Absorption of Radiation." *Proceedings of the Royal Society of London. Series A*, 1927. **114**(767): 243–265.
2. Jordan, P. and E. Wigner, "Über das Paulische Äquivalenzverbot." *Zeitschrift für Physik*, 1928. **47**(9): 631–651.
3. Negele, J.W. and H. Orland, *Quantum Many-Particle Systems*. Addison-Wesley, Redwood City (1988).
4. Festa, R.R., "Wolfgang Pauli (1900–1958): A Brief Anecdotal Biography." *Journal of Chemical Education*, 1981. **58**(3): 273.
5. Kohn, W., "Image of the Fermi Surface in the Vibration Spectrum of a Metal." *Physical Review Letters*, 1959. **2**(9): 393–394.
6. Schulman, L.S., *Techniques and Applications of Path Integrals*. 1981, New York: Wiley.
7. Klauder, J.R. and B.-S. Skagerstam, *Coherent States: Applications in Physics and Mathematical Physics*. 1985, Singapore: World Scientific.
8. Martin, J.L., "The Feynman Principle for a Fermi System." *Proceedings of the Royal Society of London A: Mathematical, Physical and Engineering Sciences*, 1959. **251**(1267): 543–549.
9. Tempere, J. and J.P. Devreese (2012), *Path-Integral Description of Cooper Pairing*, in Superconductors - materials, properties and applications, Alexander Gabovich, IntechOpen, pp. 1–32, DOI: 10.5772/48458.
10. Berezin, F.A., *The Method of Second Quantization. Nauka, Moscow*, 1965. *Tranlation: Academic Press, New York, 1966. (Second edition, expanded: M. K. Polivanov, ed., Nauka, Moscow, 1986.).*
11. Nagaosa, N., *Quantum Field Theory in Condensed Matter Physics*. 1999, Springer-Verlag Berlin Heidelberg.
12. Feynman, R.P., "Space-Time Approach to Non-Relativistic Quantum Mechanics." *Reviews of Modern Physics*, 1948. **20**(2): 367–387.
13. Fai, L.C. and M. Wysin, *Statistical Thermodynamics: Understanding the Properties of Macroscopic Systems*. 2012, Boca Raton, FL: CRC Press, Taylor & Francis Group.
14. Feynman, R.P. and A.R. Hibbs, *Quantum Mechanics and Path Integrals*. 1965, New York: McGraw-Hill.
15. Zee, A., *Quantum Field Theory in a Nutshell*. Second ed. 2010, Princeton, NJ: Princeton University Press.
16. Ashcroft, N.W. and N.D. Mermin, *Solid State Physics*. First ed. 1976, Philadelphia: Saundrers College.
17. Lifshitz, E.M. and L.P. Pitaevskii, *Statistical Physics, Part 2. Course of Theoretical Physics. Vol. 9.* 1980, Oxford: Pergamon Press.
18. Dupuis, N., "Nonperturbative Renormalization-Group Approach to Fermion Systems in the Two-Particle-Irreducible Effective Action Formalism." *Physical Review B*, 2014. **89**(3): 035113.
19. Dupuis, N., "Renormalization Group Approach to Interacting Fermion Systems in the Two-Particle-Irreducible Formalism." *The European Physical Journal B: Condensed Matter and Complex Systems*, 2005. **48**(3): 319–338.
20. Schwinger, J., "On the Green's Functions of Quantized Fields, I." *Proceedings of the National Academy of Sciences*, 1951. **37**(7): 452–455.
21. Schwinger, J., "On the Green's Functions of Quantized Fields, II." *Proceedings of the National Academy of Sciences*, 1951. **37**(7): 455–459.
22. Dyson, F.J., "The S Matrix in Quantum Electrodynamics." *Physical Review*, 1949. **75**: 1736–1755.

23. Alkofer, R. and L. von Smekal, "The Infrared Behaviour of QCD Green's Functions: Confinement, Dynamical Symmetry Breaking, and Hadrons as Relativistic Bound States." *Physics Reports*, 2001. **353**(5–6): 281–465.
24. Abrikosov, A.A., L.P. Gorkov, and I.E. Dzyaloshinski, *Methods of Quantum Field Theory in Statistical Physics*. 1963, Englewood Cliffs, NJ: Prentice Hall.
25. Lifshitz, E.M. and L.P. Pitaevskii, *Statistical Physics, Part 2: Theory of the Condensed State*. 1980, Oxford: Pergamon Press.
26. Luttinger, J.M. and J.C. Ward, "Ground-State Energy of a Many-Fermion System, II." *Physical Review*, 1960. **118**(5): 1417–1427.
27. Cornwall, J.M., R. Jackiw, and E. Tomboulis, "Effective Action for Composite Operators." *Physical Review D*, 1974. **10**(8): 2428–2445.
28. Gell-Mann, M. and F. Low, "Bound States in Quantum Field Theory." *Physical Review*, 1951. **84**(2): 350–354.
29. Fröhlich, H., "Theory of the Superconducting State. I. The Ground State at the Absolute Zero of Temperature." *Physical Review*, 1950. **79**(5): 845–856.
30. Cooper, L.N., "Bound Electron Pairs in a Degenerate Fermi Gas." *Physical Review*, 1956. **104**(4): 1189–1190.
31. Hubbard, J., "Calculation of Partition Functions." *Physical Review Letters*, 1959. **3**(2): 77–78.
32. Stratonovich, R.L., "On a Method of Calculating Quantum Distribution Functions." *Soviet Physics Doklady*, 1957. **2**: 416.
33. Lindhard, J. (1954), "On the properties of a gas of charged particles," *Dan. Mat. Fys. Medd.* **28**, 1–57.
34. Fetter, A.L. and J.D. Walecka, *Quantum Theory of Many-Particle Systems*. 2003, New York: Dover.
35. Gell-Mann, M. and K.A. Brueckner, "Correlation Energy of an Electron Gas at High Density." *Physical Review*, 1957. **106**(2): 364–368.
36. Fai, L.C., C.S. Cabisov, and A. Mody, *Theoretical Physics: Volume 1, Relativistic Theory and Electrodynamics*. Second ed. 2010, Navi Mumbai, India: Shroff.
37. Wilson, K.G., "Renormalization Group and Strong Interactions." *Physical Review D*, 1971. **3**(8): 1818–1846.
38. Landau, L.D. and Lifshitz, E.M. (1971), *The Classical Theory of Fields* (Volume 2 of *A Course of Theoretical Physics*), Oxford: Pergamon Press.
39. Patashinskiĭ, A.Z. and V.L. Pokrovskiĭ, "The Renormalization-Group Method in the Theory of Phase Transitions." *Soviet Physics Uspekhi*, 1977. **20**(1): 31.
40. Pokrovskiĭ, V.L., "Similarity Hypothesis in the Theory of Phase Transitions." *Soviet Physics Uspekhi*, 1968. **11**(1): 66.
41. Garelick, H. and J.W. Essam, "Critical Behaviour of the Three-Dimensional Ising Model Specific Heat Below T_c." *Journal of Physics C: Solid State Physics*, 1968. **1**(6): 1588.
42. Udodov, V.N., "New Consequences of the Static Scaling Hypothesis at Low Temperatures." *Physics of the Solid State*, 2015. **57**(10): 2073–2077.
43. Essam, J.W. and M.E. Fisher, "Padé Approximant Studies of the Lattice Gas and Ising Ferromagnet below the Critical Point." *The Journal of Chemical Physics*, 1963. **38**(4): 802–812.
44. Gaunt, D.S., et al., "Critical Isotherm of a Ferromagnet and of a Fluid." *Physical Review Letters*, 1964. **13**(24): 713–715.
45. Widom, B., "Degree of the Critical Isotherm." *The Journal of Chemical Physics*, 1964. **41**(6): 1633–1634.
46. Widom, B., "Relation Between the Compressibility and the Coexistence Curve near the Critical Point." *The Journal of Chemical Physics*, 1962. **37**(11): 2703–2704.
47. Bogoliubov, N.N., "A New Method in the Theory of Superconductivity." *Journal of Experimental and Theoretical Physics*, 1958. **7**(41).
48. Bogoliubov, N.N. On the Theory of Superfluidity (in English); К теории сверхтекучести" (in Russian). *Journal of Physics* **11**, 1, 23–32; Известия АН СССР, физика, 11, 1,77, 1947.

49. Mahan, G.D., *Many Particle Physics (Physics of Solids and Liquids)*. 3rd ed. 2000, India: Kluwer Academic/Plenum Publishers-Springer.
50. Feynman, R.P., *Statistical Mechanics: A Set of Lectures*, ed. J. Shaham. 1972, Reading, MA: W.A. Benjamin.
51. Lee, T.D., K. Huang, and C.N. Yang, "Eigenvalues and Eigenfunctions of a Bose System of Hard Spheres and Its Low-Temperature Properties." *Physical Review*, 1957. **106**(6): 1135–1145.
52. Täuber, U.C. and D.R. Nelson, "Superfluid Bosons and Flux Liquids: Disorder, Thermal Fluctuations, and Finite-Size Effects." *Physics Reports*, 1997. **289**(3): 157–233.
53. Penrose, O. and L. Onsager, "Bose-Einstein Condensation and Liquid Helium." *Physical Review*, 1956. **104**(3): 576–584.
54. Bardeen, J., L.N. Cooper, and J.R. Schrieffer, "Microscopic Theory of Superconductivity." *Physical Review*, 1957. **106**(1): 162–164.
55. Bardeen, J., L.N. Cooper, and J.R. Schrieffer, "Theory of Superconductivity." *Physical Review* 1957. **108**: 1175–1204.
56. Onnes, H.K., "The Resistance of Pure Mercury at Helium Temperatures." *Communications from the Laboratory of Physics at the University of Leiden*, 1911. **12**: 120.
57. Abrikosov, A.A., et al., "Possibility of Formulation of a Theory of Strongly Interacting Fermions." *Physical Review*, 1958. **111**(1): 321–328.
58. Langreth, D.C. and J.W. Wilkins, "Theory of Spin Resonance in Dilute Magnetic Alloys." *Physical Review B*, 1972. **6**(9): 3189–3227.
59. Anderson, M.H., et al., "Observation of Bose-Einstein Condensation in a Dilute Atomic Vapor." *Science*, 1995. **269**(5221): 198–201.
60. Ketterle, W., et al., "Bose–Einstein Condensation of Ultracold Atomic Gases." *Physica Scripta*, 1996. **1996**(T66): 31.
61. Lifshitz, E.M., "Superfluidity," in *Perspectives in Theoretical Physics*, L.P. Pitaevski, Ed. 1992, Pergamon: Amsterdam. 413–424.
62. Alexandrov, A.S., *Theory of Superconductivity: From Weak to Strong Coupling*. 2003: Philadelphia, PA: Institute of Physics.
63. Jochim, S., et al., "Bose-Einstein Condensation of Molecules." *Science*, 2003. **302**(5653): 2101–2103.
64. Greiner, M., C.A. Regal, and D.S. Jin, "Emergence of a Molecular Bose-Einstein Condensate from a Fermi Gas. *Nature*, 2003. **426**(6966): 537–540.
65. Devreese, J.T. *Proceedings of the International School of Physics "Enrico Fermi" Course CXXXVI*. 1998. Amsterdam: IOS Press.
66. Alexandrov, A.S. and S.N. Mott, *Polarons and Bipolarons*. 1995, Singapore: World Scientific.
67. Fai, L.C., A. Fomethe, V.B. Mborong, S.C. Kenfack, J.T. Diffo, S. Domngang, and Ashok Mody, "Screening Effect on The Polaron by Plasmons in The Field of Self-Action Potential in a Planar Nanocrystal." *Superlattices and Microstructures*, 2010. **47**: 631–647.
68. Landau, L.D., "Research in the Electron Theory of Crystals." *Physikalische Zeitschrift der SOWJETUNION*, 1933. **3**: 664.
69. Fai, L.C., A. Fomethe, A. J. Fotue, V. B. Mborong, S. Domnganga, N. Issofa, and M. Tchoffo, "Bipolaron in a Quasi-0D Quantum Dot." *Superlattices and Microstructures* 2008. **43**(1): 44.
70. Mott, N.F., *High Temperature Superconductivity*, ed. D.P. Tunstall, W. Barford 1991: Bristol: Adam Hilger.
71. Alexandrov, A.S., A.M. Bratkovsky, and N.F. Mott, "Transport Properties of High-Tc Oxides in the Bipolaron Model." *Physica C: Superconductivity*, 1994. **235–240**, **Part 4**: 2345–2346.
72. Bartenstein, M., Altmeyer, A., S. Riedl, S. Jochim, C. Chin, J. Hecker Denschlag, and R. Grimm, "Crossover from a Molecular Bose-Einstein Condensate to a Degenerate Fermi Gas." *Physical Review Letters*, 2004. **92**: 120401.
73. Zwierlein, M. W., C. A. Stan, C. H. Schunck, S. M. F. Raupach, A. J. Kerman, and W. Ketterle, "Condensation of Pairs of Fermionic Atoms near a Feshbach Resonance." *Physical Review Letters*, 2004. **92**: 120403.

74. Sa de Melo, C.A.R., M. Randeria, and J.R. Engelbrecht, "Crossover from BCS to Bose." *Physical Review Letters*, 1993. **71**: 3202.
75. Tempere, J., S.N. Klimin, J.T. Devreese, and V.V. Moshchalkov, "Imbalanced d-wave superfluids in the BCS-BEC crossover regime at finite temperatures." *Physical Review B*, 2008. **77**(13): 134502.
76. Fulde, P. and R.A. Ferrell, "Superconductivity in a Strong Spin-Exchange Field." *Physical Review*, 1964. **135**(3A): A550–A563.
77. Inguscio, M., W. Ketterle, C. Salomon, *Ultra-cold Fermi gases* 2007, Amsterdam, The Netherlands: IOS Press.
78. Gubbels, K.B. and H.T.C. Stoof, "Imbalanced Fermi Gases at Unitarity." *Physics Reports*, 2013. **525**(4): 255–313.
79. Shankar, R., "Renormalization-Group Approach to Interacting Fermions." *Reviews of Modern Physics*, 1994. **66**(1): 129–192.
80. Bighin, G., et al., "Pair Condensation of Polarized Fermions in the BCS–BEC Crossover." *Journal of Physics B: Atomic, Molecular and Optical Physics*, 2014. **47**(19): 195302.
81. Huang, K. and C.N. Yang, "Quantum-Mechanical Many-Body Problem with Hard-Sphere Interaction." *Physical Review*, 1957. **105**(3): 767–775.
82. Frederick W. Byron, JR., Robert W. Fuller. *Mathematics of Classical and Quantum Physics*. 1992, Mineola, NY: Dover.
83. Palestini, F. and G.C. Strinati, "Temperature Dependence of the Pair Coherence and Healing Lengths for a Fermionic Superfluid Throughout the BCS-BEC Crossover." *Physical Review B*, 2014. **89**(22): 224508.
84. Abrikosov, A.A. and Gor'kov L.P., "On the Theory of Superconducting Alloys." *Journal of Experimental and Theoretical Physics*, 1959. **8**: 1090.
85. Anderson, P.W., "Theory of Dirty Superconductors." *Journal of Physics and Chemistry of Solids*, 1959. **11**(1): 26–30.
86. Weiss, P., "L'hypothèse du Champ Moléculaire et la Propriété Ferromagnétique." *Journal of Theoretical and Applied Physics*, 1907. **6**(1): 661–690.
87. Onsager, L., "Crystal Statistics. I. A Two-Dimensional Model with an Order-Disorder Transition." *Physical Review*, 1944. **65**(3–4): 117–149.
88. Néel, L., R. Pauthenet, and B. Dreyfus, *Chapter VII The Rare Earth Garnets, in Progress in Low Temperature Physics*, C.J. Gorter, Editor. 1964, North-Holland Publishing Company, Interscience Publishers (a division of John Wiley & Sons, Inc.): Amsterdam, The Netherlands. p. 344–383.
89. Fallot, M., "Paramagnétisme des Éléments Ferromagnétiques. *Journal de Physique et Le RADIUM*, 1944. **5**(8): 153–163.
90. Collet, P. and G. Foëx, "Propriétés Magnétiques du Platine. Différentes Variétés. Influence du Champ. Passage d'un État à un Autre." *Journal de Physique et Le RADIUM*, 1931. **2**(9): 290–308.
91. Callen, H.B. and E. Callen, "The Present Status of the Temperature Dependence of Magnetocrystalline Anisotropy, and the l(l+1)2 Power Law." *Journal of Physics and Chemistry of Solids*, 1966. **27**(8): 1271–1285.
92. Mermin, N.D. and H. Wagner, "Absence of Ferromagnetism or Antiferromagnetism in One- or Two-Dimensional Isotropic Heisenberg Models." *Physical Review Letters*, 1966. **17**(22): 1133–1136.
93. White, R.M., *Quantum Theory of Magnetism*. Springer Series in Solid-State Sciences. 2007, Berlin: Springer-Verlag.
94. Wilson, K.G., "Renormalization Group Methods." *Advances in Mathematics*, 1975. **16**(2): 170–186.
95. Wilson, K.G., "The Renormalization Group: Critical Phenomena and the Kondo Problem." *Reviews of Modern Physics*, 1975. **47**(4): 773–840.
96. L. D. Landau and E.M. Lifshitz, *The Classical Theory of Fields*. 3 ed. Course of theoretical physics Vol. **2**. 1971, Oxford: Pergamon Press.
97. He, Wen-Bin, Yang-Yang Chen, Shizhong Zhang, and Xi-Wen Guan, "Universal Properties of Fermi Gases in One Dimension." *Physical Review A*, 2016. **94**(3): 031604.

98. Bertaina, G., "Two-Dimensional Short-Range Interacting Attractive and Repulsive Fermi Gases at Zero Temperature." *The European Physical Journal Special Topics*, 2013. **217**(1): 153–162.
99. Eriksson, G., et al., "Vortices in Fermion Droplets with Repulsive Dipole-Dipole Interactions." *Physical Review A*, 2012. **86**(4): 043607.
100. Abrikosov, A.A., "Electron Scattering on Magnetic Impurities in Metals and Anomalous Resistivity Effects." *Physics Physique Fizika*, 1965. **2**(1): 5–20.
101. Kondo, J., "Resistance Minimum in Dilute Magnetic Alloys." *Progress of Theoretical Physics*, 1964. **32**(1): 37–49.
102. David Pines and P. Nozières, *Theory Of Quantum Liquids*. 1 ed. Vol. **1**. 1989, London: Taylor & Francis Group, CRC Press.
103. Keldysh, L.V., "Diagram Technique for Nonequilibrium Processes." *Journal of Experimental and Theoretical Physics*, 1965. **47**(4): 1515.
104. Kiselev, M., K. Kikoin, and R. Oppermann, "Ginzburg-Landau Functional for Nearly Antiferromagnetic Perfect and Disordered Kondo Lattices. PHYSICAL REVIEW B, 2002. **65**(18): 184410.
105. Onnes, H. Kamerlingh, *The coefficients of viscosity for fluids in corresponding states*. Communications from the Kamerlingh Onnes Laboratory Nr. 1–12 (1885–1894), University of Leiden.
106. Rammer, J. and H. Smith, "Quantum Field-Theoretical Methods in Transport Theory of Metals." *Reviews of Modern Physics*, 1986. **58**(2): 323–359.
107. Babadi, M., "Non-Equilibrium Dynamics of Artificial Quantum Matter." PhD dissertation, 2013, Harvard University.
108. Špička, V., B. Velický, and A. Kalvová, "Electron Systems out of Equilibrium: Nonequilibrium Green's Function Approach." *International Journal of Modern Physics B*, 2014. **28**(23): 1430013.
109. Schwinger, J., "Brownian Motion of a Quantum Oscillator." *Journal of Mathematical Physics*, 1961. **2**(3): 407–432.
110. Craig, R.A., "Perturbation Expansion for Real-Time Green's Functions." *Journal of Mathematical Physics*, 1968. **9**(4): 605–611.
111. Starostin, A.N., et al., "Quantum Corrections to the Particle Distribution Function and Reaction Rates in Dense Media." *Plasma Physics Reports*, 2005. **31**(2): 123–132.
112. Berges, J., "Introduction to Nonequilibrium Quantum Field Theory." *AIP Conference Proceedings*, 2004. **739**(1): 3–62.
113. Garny, M. and M.M. Müller, "Kadanoff-Baym Equations with Non-Gaussian Initial Conditions: The Equilibrium Limit." *Physical Review D*, 2009. **80**(8): 085011.
114. Williams, R.A., et al., "Raman-Induced Interactions in a Single-Component Fermi Gas Near an S-Wave Feshbach Resonance." *Physical Review Letters*, 2013. **111**(9): 095301.
115. Kadanoff, L.P. and G. Baym, *Quantum Statistical Mechanics: Green's Functions Methods in Equilibrium and Nonequalibrium Problems*. 1962, Menlo Park, CA: W.A. Benjamin.
116. Carrington, M.E., H. Defu, and J.C. Sowiak, "Kubo-Martin-Schwinger Conditions for 4-Point Green Functions at Finite Temperature." *Physical Review D*, 2000. **62**(6): 065003.
117. Kieu, T.D. and C.J. Griffin, "Monte Carlo Simulations with Indefinite and Complex-Valued Measures." *Physical Review E*, 1994. **49**(5): 3855–3859.
118. Baym, G., "Self-Consistent Approximations in Many-Body Systems." *Physical Review*, 1962. **127**(4): 1391–1401.
119. Wigner, E., "On the Quantum Correction for Thermodynamic Equilibrium." *Physical Review*, 1932. **40**(5): 749–759.
120. Martinez-Morales, J. L., "The Feynman Measure As a Limit of Complex Measures." *Journal of Mathematical Analysis and Applications*, 2013. **397**(2): 494–502.
121. Hillery, M., R.F. O'Connell, M.O. Scully, and E.P. Wigner, "Distribution Functions in Physics: Fundamentals." *Physics Reports*, 1984. **106**(3): 121–167.

122. Baym, G. and L.P. Kadanoff, "Conservation Laws and Correlation Functions." *Physical Review*, 1961. **124**(2): 287–299.
123. Kubo, R., "Statistical-Mechanical Theory of Irreversible Processes. I. General Theory and Simple Applications to Magnetic and Conduction Problems." *Journal of the Physical Society of Japan*, 1957. **12**(6): 570–586.
124. Martin, P.C. and J. Schwinger, "Theory of Many-Particle Systems. I." *Physical Review*, 1959. **115**(6): 1342–1373.
125. Semkat, D., D. Kremp, and M. Bonitz, "Kadanoff-Baym Equations with Initial Correlations." *Physical Review E*, 1999. **59**(2): 1557–1562.
126. Moyal, J.E., "Quantum Mechanics as a Statistical Theory." *Mathematical Proceedings of the Cambridge Philosophical Society*, 1949. **45**(1): 99–124.
127. Popov, V. N. and S. A. Fedotov, "The Functional-Integration Method and Diagram Technique for Spin Systems." *Journal of Experimental and Theoretical Physics*, 1988. **94**: 183–194.
128. Kiselev, M. *Semi-Fermionic Approach for Quantum Spin Systems*. 2003, Basel: Birkhauser Verlag.
129. Regal, C.A., M. Greiner, and D.S. Jin, "Observation of Resonance Condensation of Fermionic Atom Pairs." *Physical Review Letters*, 2004. **92**(4): 040403.
130. Leduc, M., J. Dugué, and J. Simonet, "Laser Cooling, Trapping, and Bose-Einstein Condensation of Atoms and Molecules." *AIP Conference Proceedings*, 2009. **1119**(1): 37–42.
131. Gehm, M.E., et al. "Stability of a Strongly-Attractive, Two-Component Fermi Gas," presented at *Lasers and Electro-Optics/Quantum Electronics and Laser Science Conference*. 2003, Baltimore, MD: Optical Society of America.
132. Zhu, Q.-Z. and B. Wu, "Superfluidity of Bose-Einstein Condensates in Ultracold Atomic Gases." *Chinese Physics B*, 2015. **24**(5): 050507.
133. Stringari, S., "Bose-Einstein Condensation in Ultracold Atomic Gases." *Physics Letters A*, 2005. **347**(1): 150–156.
134. Peng, Shi-Guo, Shi-Qun Li, P.D. Drummond, and Xia-Ji Liu, "High-Temperature Thermodynamics of Strongly Interacting S-Wave and P-Wave Fermi Gases in a Harmonic Trap." *Physical Review A*, 2011. **83**(6): 063618.

Index

R

S

T

U

V

W

Y

Z

Milton Keynes UK
Ingram Content Group UK Ltd.
UKHW051905071024
449327UK00025B/2096

9 780367 779597